ANUS HORRIBILIS

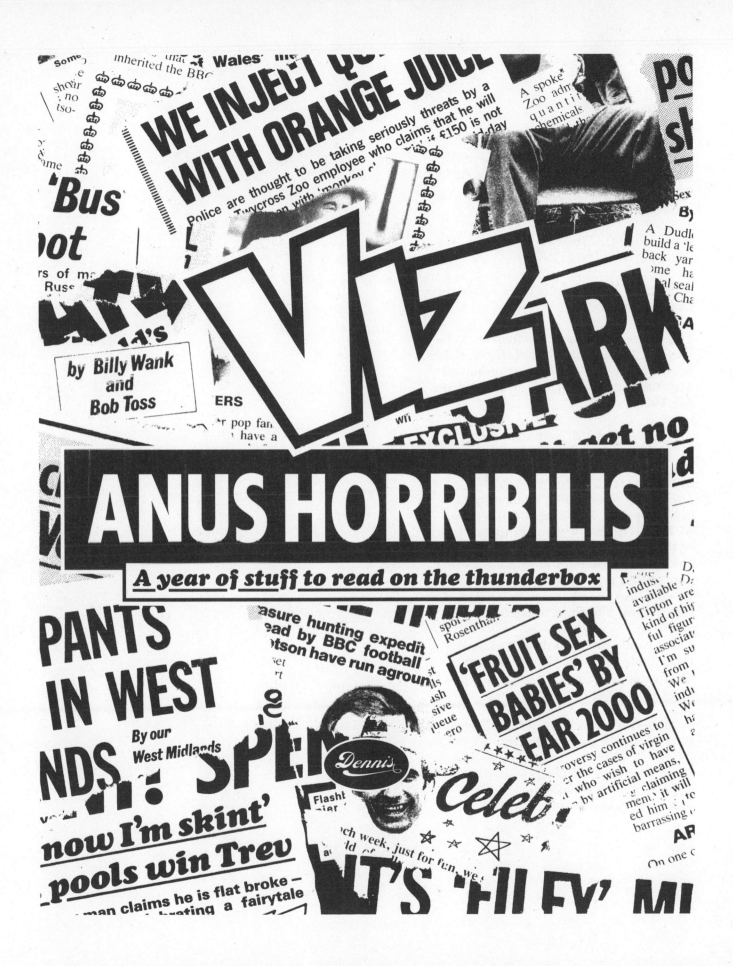

VIZ

ANUS HORRIBILIS

A year of stuff to read on the thunderbox

First published October 2011 by
Dennis Publishing Ltd.,
30 Cleveland Street,
London W1T 4JD.

ISBN-10: 1907779914
ISBN-13: 978-1907779916

A CIP catalogue record for this book
is available from the British Library.
They've got a nice cafe there, too, just
up at the top of the stairs.

Compiled & edited by Graham Dury,
Davey Jones & Simon Thorp.

Come and be our friend
at facebook.com/vizcomic
or twitter.com/vizcomic

ANUS HORRIBILIS

Viz

A SELECTION OF ARTICLES FROM VIZ COMIC ORIGINALLY WRITTEN AND ILLUSTRATED BY

Graham Dury, Davey Jones, Simon Thorp, Chris Donald, Simon Donald & Alex Collier

ADDITIONAL CARTOONS BY

Jim Brownlow, Guy Campbell, Tony Harding, Roger Radio, Brent & Les Heywood

DESIGNER

Wayne Gamble

PRODUCTION MANAGER

Stevie Glover

PUBLISHER

Russell Blackman

Thanks also to Ann Gates, Susan Patterson, Maddie Fletcher, Sheila Gill, Colin Davison, Caroline Addy, everyone at JBP & Dennis, most people at IFG, and our late friends Iain McKie & Steve Donald.

www.viz.co.uk

LESSONS IN LUST!

Desperate head teachers struggling on low wages are turning to prostitution in order to pay their bills.

Hard up headmasters offer sex for sale

And the hard up heads regularly give 'private lessons' to clients, making home visits for seedy 'extra curricular' love lessons. And in order to sell their bodies the sordid schoolmasters turn up for sex in full uniform, including black cape, mortar board and corduroy trousers.

SEEDY

We uncovered the seedy staff room sex scandal after reading an ad in the local newspaper. The ad boasted 'Pretty headmaster of local school offers full sexual services in the home. Including spanking, S.M., T.V., felching, chocolate fudge cake and watersports.' The ad continued to give a daytime telephone number where the headmaster could be contacted.

NUMBER

When we rang the number it turned out to be that of a local secondary school, St Bartholemews. A receptionist who answered told us the head was taking assembly, and asked us to call back in twenty minutes.

SEEDY AGAIN

When we rang back the headmaster, Mr A, talked openly about the seedy services he offered. "My call out fee is £20. On top of that it's £80 for full sex."
"I do lots of extras", he boasted. "I can bring a blackboard if you like".

INVESTIGATOR

We arranged for Mr A to visit our investigator in a hotel room later that evening. When he arrived at first he seemed nervous. A tall, well spoken man in his early fifties, he politely apologised for being slightly late. As he plonked his battered brown leather briefcase onto the table several maths books and a packet of condoms fell out.

CANE

"I don't do anything over the top. I prefer the straight sex, but I can cane your bottom if you like. That's £30 extra", he whispered nervously.

SEX FOR SALE

Local headmaster offers private love lessons in the home. Oral, topless relief, shaven haven etc. Corporal punishment a speciality. Ring now to compare my prices.

Ring Fulchester ~~----~~ and ask to speak to the headmaster.

Giveaway - Headmaster A's ad which appeared in local press

Our man offered the headmaster a glass of wine and he soon got talking. He revealed that he had been the head of the local school for 11 years, and had only turned to prostitution in desperation a few months ago.

MORTGAGE

"I've got a wife, two kids and a mortgage. There's no way I can make ends meet on a headmaster's salary. I'm only doing this for the money", he admitted.
"I usually do 2 or 3 visits a week, and a couple at weekends. My wife thinks I'm supervising detentions at the minute. She'd go through the roof if she knew I was a prostitute."

WEIRDOS

Then he got down to business. "I'll do anything if the money's right. But I have to be careful. You get some real weirdos. It's £15 for the straight massage, £25 for topless relief - anything else is negotiable. But I don't do oral. I do everything else but not French. But a friend of mine will. He's a metalwork teacher".

ENORMOUS

When quizzed headmaster A spoke fondly of his teaching career. "I love my job. The satisfaction is enormous. It's the most rewarding job in the world, but financially it's a struggle. I've got two sons at University, the mortgage and bills to pay, plus a caravan to keep on the road. It's just not possible on a headmaster's wage".

HEADMASTER A.

Our investigator declined the offer of sex, but said he would be interested in a lesbian show with two or more head teachers. Headmaster A promised to make the arrangements and invited our man to visit the school at noon the following day.

STUDY

When our man arrived at the impressive red brick Victorian building he was shown into the head's study by the receptionist who introduced herself as 'Maureen'. "Have you been here before?" she asked. She offered our man a glass of champagne and told him to relax while he waited for the head to arrive.

LOUNGE

Moments later headmaster A entered the room with his colleague, Mr B, a 42-year-old metalwork teacher who introduced himself as 'Clive'.

"We've only got 30 minutes. I'm taking a lesson at one", Mr B told us. Beneath his blue metalwork apron Mr B was wearing fawn corduroy trousers with skimpy purple 'Y' fronts clearly visible above his waistline. "Do you want a quick preview", he teased, beginning to undo his buttons.

KITCHEN

Headmaster A was more nervous. "To be quite honest I don't do much lesbian", he told our man. "It'll be £75 each, and we want the money up front." At this point Mr B dropped his trousers. "You can take pictures, but touching is extra", he said.

BEDROOM

At this point our investigator made his excuses and left. Most headmasters in Britain earn only £42,000 per year – a mere £800 a week. And it is estimated that something approaching half of all headmasters are in arrears with mortgage repayments. As a result a growing number – possibly in the tens of thousands – are turning to prostitution.

SPARE BEDROOM

When confronted later, headmaster A denied that he had offered any sexual services. "I don't know what you're talking about", he said. "Go away", he added. *Our dossier on the case is now in the hands of the local Parent Teachers' Association.*

CHEGWIN'S PAINFUL SECRET

EXCLUSIVE

Keith Chegwin — not quite visible in this picture taken during last week's Saturday Superstore.

The showbusiness world has this week been shocked by the revelation that BBC TV star Keith Chegwin has no talent whatsoever.

TV viewers and celebrities alike have been stunned by Chegwin's decision to 'come out' and make public his lack of talent. And now there are fears that other stars will soon follow suit with disastrous implications for the TV industry.

INEPETITUDE

The remarkable ineptitude and lack of ability suffered by many top names in light entertainment has for years been one of the TV world's best kept secrets. For although it is widely known within the BBC and ITV that many of today's best known celebrities are totally without talent, their names have never been publicly revealed. Chegwin's decision to break this unofficial code of conduct has set alarm bells ringing through the entertainment world.

PROBLEM

"Keith felt that his problem had got to the stage where it could no longer be concealed", a source close to Chegwin reported last night. Indeed, on a recent edition of 'Saturday Superstore' viewers were shocked by Chegwin's apparent inability to tell a joke. And a former colleague told us that Chegwin's condition had been evident during the making of the

'I have no talent' says Cheggers

'Swap Shop' programme in the late seventies.

SERIOUS

"Keith always had trouble with sentences. He couldn't string them together properly. But he could always grin and laugh a lot. At that stage none of us realised what a serious problem he had. I'm very shocked and saddened to hear this news".

RUMOURS

Rumours are rife that a host of other celebrities, among them Leslie Crowther, Russel Harty and members of the Blue Peter team could now join Chegwin and confess to being talentless.

A leading TV expert last night welcomed the news of Chegwin's revelation. "The fact that A.N.T.S. — Absolutely No Talent Syndrome — has been brought into the public view can only do good in the long term. Hopefully other sufferers will now begin to come forward and seek help".

FRIEND

After visiting Chegwin at his London home yesterday a close friend told us that Keith was doing well, and was in high spirits.

CHILDREN

"He intends to continue working as before", he told us.

"He just hopes that the public will understand his difficulty and be able to accept him in future low budget children's TV programmes".

CONFUSED

When we asked a telephonist at the BBC for a comment she seemed confused, and didn't understand the question.

TOMMY ROT SAYS: EATING TURNIPS GIVES YOU PILES

MICHAEL ANGELO AND HIS INVISIBLE YO-YO — BONK!

7

GARDEN OF DEATH!

A man arrested for digging up a vegetable patch at the BBC Television Centre in West London yesterday claimed that he was searching for the bodies of several former Blue Peter presenters who he fears may have become the victims of Britain's worst ever serial killer.

"Up to 16 bodies may be buried at Television Centre" claims Blue Peter fan

Veteran Blue Peter fan Frank Gubbins, 55, had been a regular viewer of the popular children's show since it was first broadcast in 1958. And for the last twenty years he has lead a one man campaign to solve the mystery of the programmes vanishing presenters.

SCREENS

He claims that over a period of more than 25 years up to 16 different presenters have vanished from TV screens, never to be seen again. And Frank fears that they could have become the victims of a serial killer.

MILLS

Ever since 1962 when Leila Williams was replaced by Valerie Singleton the show's presenters have been disappearing. Perhaps the best known example is John Noakes who vanished form the screen in 1979 and has never been seen since. The BBC's official line is that he is living on a boat somewhere off the coast of Spain, but Frank finds that hard to believe.

"No-one had ever seen him or his boat. I heard rumours that he did a programme called 'Go With Noakes', but I certainly never saw it, and I don't know anyone who did".

SOCKS

Valerie Singleton, a pretty, dark haired young woman joined the show in 1962. She was last seen in 1971, in the company of Noakes and co-presenter Peter Purves. "I made enquiries about Valerie Singleton's whereabouts with the BBC, but it was like banging my head against a brick wall. They said she was working on Radio Four, which sounded like a rather convenient excuse to me".

SURFERS

Another missing presenter, a young woman called Janet Ellis, was believed to be pregnant when she was last seen on Blue Peter in 1987. Attempts to trace her have proved fruitless, and Frank now fears the worst. Her co-presenter Simon Groom also disappeared at around the same time.

Frank believes that the sinister disappearances are linked to the show's Editor Biddy Baxter. "She is the one person who has been there throughout the show's entire history. And I have heard stories that she is obsessive about the programme, and constantly at odds with the presenters."

CHEATERS

Frank fears that Baxter may have killed some or all of the missing presenters, and disposed of their bodies in the Blue Peter garden. The garden, a small plot of land within the grounds of the BBC Television Centre in Wood Lane, West London, was first used by the programme in Spring 1974 as a vegetable plot, and was supervised by gardening expert Percy Thrower. However, in 1978, the garden was expanded to include a sunken goldfish pond, paved patio area and numerous flowerbeds. During this and subsequent alterations at the site there would have been numerous opportunities for Baxter to dispose of human remains beneath areas of concrete, paving stone or even below the pond itself.

ASSISTED

"The base of the pond was excavated and filled with concrete in April 1978, only 7 years after Valerie Singleton went missing. At around the same time 3 tons of crazy paving were laid. Blue Peter editor Biddy Baxter would have had keys to the Television Centre, and she would therefore have had access to the site at all times, day and night."

FALL

Mr Gubbins believes that the remains of up to 16 bodies could be buried beneath the garden. But so far police have failed to respond to his tip-offs. In desperation Gubbins took the law into his own hands, and began digging in the garden after scaling a nearby wall. But almost immediately he was arrested and taken to a nearby police station where he was held for 72 hours under the Mental Health Act before

being released without charge.

TUNNEL

He is now more determined than ever to find out the truth about the missing presenters. "The sooner the authorities start digging the better. I am particularly interested in the pond, the patio area and the vegetable patch, although there are other places such as flower beds and even plant pots which will all have to be thoroughly examined."

ERMERE

"This could prove to be one of the most extensive and exhausting murder enquiries in British history", Frank told us late last night. "It may take many days or even weeks of careful digging and forensic examination before the Blue Peter garden finally gives up the last of its dreadful secrets".

A crowd of ghoulish onlookers begins to gather at the BBC Television Centre, scene of Mr Gubbin's grim search.

Garden expert Percy Thrower, together with Blue Peter presenters John Noakes and Leslie Judd, seen in the garden during 1978. All three are now feared dead, although Noakes' and Judd's bodies have never been found.

Work being carried out excavating the sunken pond area in April 1978. The walls of the pool were later lined with concrete.

This computer simulated image released yesterday by Mr Gubbins shows how John Noakes would probably have aged in the years since his disappearance. This is how he may look today.

A few weeks later and the first fish arrive. Thirteen Goldfish, six Golden rudd and one Golden tench.

OH WHAT A WONDERFUL WOMAN

IN THIS, HER SIXTIETH YEAR, WE SALUTE A MAJESTIC, SUPERB LADY

You are fantastic Your Majesty

At 60 she is the most popular monarch in British history. The longest reigning since Queen Victoria. The most loved, respected and admired lady in the world. She is of course our noble Queen.

Despite her sixty years she remains attractive. More so than many women half her age. Many of today's young girls must enter beauty competitions wishing that they had a fraction of the Queen's charm, style and good looks. In fact, to look at her you wouldn't think she was 60. She looks more like 25.

Remarkable

Her reign has been long and victorious. Even before she came to the throne the nation was indebted to this remarkable lady. For as a wartime services volunteer the young Princess Elizabeth almost single-handedly saved London from the German blitz.

Glorious

There can be little doubt that this truly wonderful lady has the most difficult job in the world. Harder than being an astronaught or deep sea diver. Yet over her glorious 34 year reign she has performed her duties to perfection, far better than anyone else could have done. Shaking hands, visiting foreign countries, talking to lots and lots of people. Often being in two places at once, but always finding time to stop and talk to old ladies.

Cameras cannot do her justice

Despite competition from Princess Diana the Queen remains Britain's best dressed woman. Whether in a flowery hat or yellow coat, a sparkling evening gown or yellow coat, she always steals the show. Top photographers all agree that their cameras simply cannot do justice to this fantastic woman.

Charming

But as well as her charm the Queen has nerves of steel, as she showed during a recent visit to a foreign country. Suddenly, her hat blew off. A whole nation held its breath. Millions of TV viewers around the world looked on in silence. Calmly, she caught it, and put it back on her head.

Indestructable

The security hazards of public life have increased many fold in recent years, but the Queen

The Queen as we see her every day on money and (inset) wearing a hat.

never gives a thought to her personal safety. Indeed, she isn't scared of anything. We will never forget, in 1982, she was confronted by an intruder in her bedroom. A foreign King or Queen would probably have panicked. But not our Queen. "Get out of here at once!" she calmly told the man who was later arrested. He was one visitor who definitely wouldn't be going back to Buckingham Palace in a hurry!

But as well as fulfilling her public duties, changing the guards at Buckingham Palace, opening endless buildings and appearing on stamps, money and postal orders, there is a private side to the Queen. The caring mother who, in difficult times, has struggled to bring up a Royal family. With several large homes to run, and four hungry mouths to feed. Now that job is done, and her children have grown up. All of them healthy, attractive, warm and friendly people. Four majestic, glowing tributes to their outstanding glorious mother.

Brilliant

And like those children we all love the Queen. She is absolutely brilliant, and in this her sixtieth year, we raise our hats and our glasses in wishing her a further one hundred and sixty marvellous years ruling over us.

LOANS? MORTGAGE ARREARS HOUSEHOLD BILLS CREDIT CARDS H.P. CHARGES

Being swamped by bills?. A loan from us will turn your hair white and bankrupt you.

I AM NEARLY BROKE. HELP FINISH ME OFF. LEND ME £.........

p.o. box 3, luton.

Have They Got Tattoos for You?

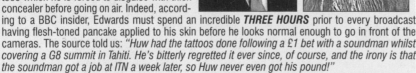

Body Art Secrets of the BBC Newsroom

THE-SOBER-suited BBC newsreader has been a familiar fixture on our TV screens for many years. Names such as Richard Baker, Peter Woods and Robert Dougall evoke a staid, civilised image which is in keeping with Lord Reith's original vision. In the BEEB's earliest times, presenters were required to wear a dinner suit whilst reading the headlines... *and that was just on the radio!* But more recently the rules have been relaxed, and journalists are now occasionally allowed to deliver their reports whilst dressed casually, perhaps wearing jeans or an open-necked shirt.

But one regulation, stating that no newsreader may display body-piercings or tattoos on screen, is still as strictly enforced as ever. And it's a rule that has led to many of the country's leading newscasters being forced to cover up bizarre body adornments. Here, in an article the BBC **TRIED TO BAN** we take an eye-opening peek behind the scenes to reveal who's hiding what in the BBC newsroom...

Bottom Marks for Paxo

You may have wondered why you rarely see BBC grand inquisitor **Jeremy Paxman** with his shirt off, and when you do it's never from behind. The reason is that the combative Newsnight host's back and buttocks are covered in raised, patterned scars called keloids. *"I was on safari in the Masai Mara when my jeep broke down hundreds of miles from anywhere,"* he told us. *"It was going to take the AA over three weeks to reach me, but luckily a couple of passing tribesmen took me back to their village. We got on really well, and they offered to initiate me into their tribe. I was sure they would be offended if I refused, but if I'd known what the initiation ceremony involved, I think I would have politely turned them down!"* he added. Paxman was drugged with a potent infusion of roots and poison berries, and spent the next four days tied face down in a mud hut whilst tribal chiefs made thousands of tiny slits in his skin, into which they they pushed pea-sized pieces of ash. *"When the drugs wore off and my hallucinations stopped, the pain was indescribable,"* Paxman continued. *"Let me tell you, I was glad when the AA eventually turned up and fixed my flat tyre."*

Lip Up, Chatty

Dermot Murnaghan is a familiar face that we welcome into our living rooms every morning. Yet few viewers realise that, offscreen, the mild-mannered breakfast sofa chat king sports a twelve-inch plate lip. Murnaghan first got the circular facial ornament after visiting the Amazon basin to film a report about Brazillian deforestation in the 1980s. *"I saw one on a tribal chief and knew there and then that I had to have one. It was the coolest thing ever,"* he told the Radio Times. *"The witch doctor used a sliver of bamboo to cut a four-inch hole under my bottom lip before stretching it wide enough to slip the frisbee-sized plate inside, all without anaesthetic. It really hurt, but it was worth it when I looked in the mirror back at the hotel,"* he added. Murnaghan always takes his plate out before going on air, but one morning last year Breakfast Time viewers nearly got more than they bargained for. *"I'd overslept, so I only had time for a quick shave and a cup of tea on my way to work,"* he told us. *"I got to the studio just in time and we were one second from going on air when the producer spotted that in my hurry I'd left my plate lip in. Natasha, like the true professional she is, didn't miss a beat. She seamlessly covered for me while I took out my plate and hid it under a cushion on the breakfast sofa!"*

News Anchor

Another newsreader who is forced to hide her body art whilst on screen is veteran presenter **Moira Stuart**. A BBC source told us: *"In the latter years of the eighteenth century Moira was press-ganged into the Royal Navy and spent several years as a deckhand on a three-masted frigate, the HMS Renown. In 1796, whilst her vessel was being re-fitted in Tunis, she and her crewmates got drunk on rum and she foolishly agreed to get the name of her ship and an anchor tattooed prominently on each forearm."* Now, more than two centuries later, these crude tattoos are still clearly visible and so Moira - nicknamed Popeye by newsroom pals - is under strict instructions to wear long sleeves whilst reading the headlines.

And Spinally

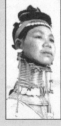

Ever wondered why **Fiona Bruce**, the ice-cool New at Ten beauty keeps her head so still when reading the news? According to BBC insiders it's because Fiona is one of the few women in Britain to wear Burmese rings which stretch her neck to then incredible length of 2 foot 6. Bruce underwent the neck stretch procedure after watching a World About Us programme featuring the Giraffe Women of Burma. She now has 120 rings on her neck which she must keep on for life. *"The vertebrae are stretched so far apart that if the rings were removed, her head would flop onto the newsdesk and she would die,"* a BBC bar source told us. To disguise the neck when reading the news, Bruce kneels down behind a foam rubber torso specially created by boffins in the BBC special effects department. To add realism the man who used to operate Grodon the Gopher works a pair of artificial arms to shuffle the papers.

Tattoo Huw

Huw Edwards rarely cracks a smile when reading the news. That's because he's terrified that if he did, his makeup would crack and his secret would be revealed. For more than 98% of the Welsh newsman's face is covered with complex Polynesian tribal tattoos which he is forced to disguise with thick concealer before going on air. Indeed, according to a BBC insider, Edwards must spend an incredible **THREE HOURS** prior to every broadcast having flesh-toned pancake applied to his skin before he looks normal enough to go in front of the cameras. The source told us: *"Huw had the tattoos done following a £1 bet with a soundman whilst covering a G8 summit in Tahiti. He's bitterly regretted it ever since, of course, and the irony is that the soundman got a job at ITN a week later, so Huw never even got his pound!"*

Bill's Not Cock-a-Hoop

Deadpan morning news man **Bill Turnbull** showed unexpected mettle when he foxtrotted his way to an early exit on Strictly Come Dancing. But Turnbull, 57, has another bit of metal that is even more unexpected and that he certainly doesn't show on screen. For unbeknownst to everyone except his wife, Turnbull sports a Prince Albert - a sort of metal ring punched through the side of his bellend and out his hog's eye. *"I can't think what possessed me to have it done,"* he told Smash Hits in 1994. *"It was excrutiatingly painful, I can't piss without leaving a wet patch on my trousers and my cock's forever setting airports alarms off,"* he added.

FIREBALL DI'S HUMAN TORCH TERROR!

By the gardener who saved her life

Fire rescue hero Reg, above, and the tragic Princess he saved.

A former Buckingham Palace gardener who helped save the Princess of Wales' life during one of her many suicide attempts has come forward to reveal all about the Royal marriage bust up.

And he hopes that by telling his side of the story he may, in some way, earn lots of money.

HORRIFIC

Reg Molesworth personally witnessed one of Diana's most horrific attempts on her life. And if it wasn't for his brave actions the Princess of Wales would not be alive today.

PETROL

"I was working in the garden one afternoon when Diana marched out onto the lawn carrying a can of petrol. Next thing I know she starts pouring it over her head. Then she sits down, gets out a match, and ...*whoof!* She was away".

SMOKE

Reg fought through clouds of thick black smoke to get to the burning Royal. "The heat was incredible, and I had to keep low to avoid the fumes, but I eventually got close enough to pour water on her with my watering can".

SMOKE AGAIN

After the flames subsided Reg and other gardening staff hosed the Princess down with a sprinkler before Di was whisked away to hospital for emergency treatment. "There was a lot of smoke damage, but otherwise she was okay. Luckily she avoided more serious injury because she had been wearing a flame retardent blouse".

PRINCESS

This was just one of dozens of similar attempts which the Princess made on her life while Reg was working at the Palace. "The chef caught her with her head in the oven so many times he had to switch

Gardener's 999 mercy dash to save burning Princess

to using a microwave. And anything sharp had to be kept under lock and key. Diana's bedroom was like a prison cell, just bare walls, a bunk bed and a bucket in the corner for the call of nature. They even took her shoes off her in case she tried to swallow them".

ALLEGRO

Reg places blame for the Royal marriage rift squarely on Charles' shoulders. "It was obvious to anyone working at the Palace that the poor girl, to put it bluntly, wasn't *getting any*. He was always away, and when he was around they'd use separate bedrooms. Many's the time she'd be wandering about the garden in a skimpy frock, eyeing up the gardeners. One time I was doing a bit of weeding when I heard her come up behind me. I could feel her eyes staring at my arse. I thought 'Ello what's this?'. But I had to ignore her. It would have been more than my jobs worth to give her one, I can tell you.

MARINA

Mind you. I'm not kidding. If she hadn't been the Princess of Wales I'd have had her in

my potting shed any day of the week."

MAESTRO

Reg sees little hope of reconciliation between Di and the House of Windsor. Having fallen out with the Royals himself, he knows from experience how unforgiving the Queen and her family can be.

MONTEGO

"You just can't talk to the Queen. She wants everything her own way. And being the Queen, you can be bloody sure she gets it. If I planted some flowers and they were the wrong colour, quick as a flash she'd be out that door screamin' blue murder. I've done plenty of gardens in my time, but that Queen is a bugger to work for, I can tell you".

BOTANY

According to Reg, he was personally dismissed by Her Majesty after a row over the Royal corgis.
"I was sick of cleaning up their shit off the lawn", he told us. "Every fucking day there'd be another bucket full. So one day I said to her 'How would you like it if my kids came round and shat on

your carpet?' That was it. She blew a fuse, and straight away I was out on me arse".

ZOOLOGY

Reg plans to use a detailed diary which he kept during his time at the Palace to write a book revealing the real life Royal dramas that go on behind closed doors. And he's already had one serious enquiry from a leading Sunday newspaper interested in serialising the book.

GENETICS

We would like to point out that Mr Molesworth has received no form of payment whatsoever from this magazine, although we did say we'd send him a cheque next week.

After 25 years the question WHO KILLED KENNEDY? remains

DID ELVIS KILL T

Startling new evidence links 'King of Rock'n'Roll' to Kennedy assasination

It is now almost quarter of a century to the day since the assasination of President John F. Kennedy shook the world. Yet despite the passing of time, an air of mystery still surrounds the President's death. Twenty-five years on the question is still being asked — Who DID kill the President?

Officially the case closed many years ago. According to the history books Lee Harvey Oswald pulled the trigger on that fateful day in Dallas in 1963. But Oswald's guilt was never proven, and subsequently rumours of mafia involvement and Government cover ups have abounded.

INCREDIBLE

But now new evidence has come to light — incredible evidence linking Elvis Presley, the late 'King of Rock 'n' Roll', to the killing. Evidence which, in weeks to come, could have startling repercussions both inside The Whitehouse and across the entire pop music industry of the world.

UNLIKELY

For the last 15 years Archibald Gubbins has dedicated his life to uncovering the truth surrounding Kennedy's death. And he is now convinced that the man who shot the President was in fact Elvis Presley himself. Unlikely as it seems, Archie Gubbins now believes he has all the necessary evidence to support his claims.

CARAVAN

Archie first developed an interest in the case after a friend he'd met on holiday hinted at a possible Elvis link to the Kennedy killing. "My wife and I went to Rhyll for a week in 1973 and it turned out that the man in the caravan next to ours, who was called Derek, had been a secret agent with the FBI during the sixties. He didn't talk much about his work, but one night in the pub after he'd had a few drinks this tongue began to loosen. He mentioned how, shortly after Kennedy was shot, they had found a guitar string on the floor in the book depository overlooking the scene.

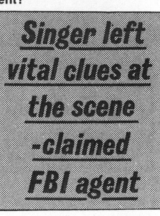

Singer left vital clues at the scene -claimed FBI agent

"Later, they discovered several rhinestones on and around the grassy knoll which Kennedy's car was passing when the shots rang out. These were identical to rhinestones worn by Presley on his stage clothing".

COVER-UP

Archie was surprised to find no reference to this evidence in any of the official reports. "There had obviously been some sort of cover-up, so I immediately became suspicious", he told us. My wife and I had also been Elvis fans for many years, and had often helped organise Elvis nights at our local pub. I felt that no matter what it took, I had to get to the bottom of the mystery".

CLUES

In his search for the truth, Archie spent months carrying out painstaking research in his local library, scouring literally dozens of books in both the History and Popular Music sections for clues. He also spent hours watching videos of TV documentaries on the subject. But after years of research Archie had drawn a blank. Then one day, out of the blue, he got a lucky break.

"I was sitting in the kitchen browsing through a book on the subject when something caught my eye. It was a photograph taken at the scene of the assasination seconds before Kennedy was shot. In the background was the book depository building, and in a window I saw what appeared to be a human figure. But it was only a blur and I couldn't be sure.

"As luck would have it my brother-in-law, who is a former chemist and keen amateur photographer, was staying with me at the time. I showed the photograph to him and he said it might be possible to magnify it many times using a previously unknown photographic technique. He did this the next day and when I saw the results I couldn't believe my eyes. There, standing in the window was Elvis, as clear as day. I was absolutely speechless for several minutes".

Elvis Aaron Presley — did the 'King' turn killer?

Fifteen years after the fatal shot had been fired, Archie was now convinced that a cover-up had taken place. He immediately wrote and asked for Kennedy's remains to be exhumed so that an independent autopsy could be carried

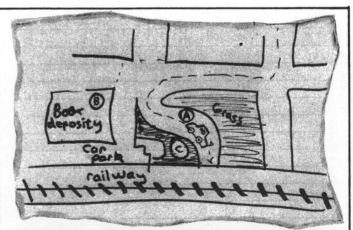

A map drawn by Mr Gubbins himself showing the scene at Dealey Plaza, Dallas, at 12.30 pm, Friday 22nd November 1963. (A) The route taken by the Presidential limousine prior to the shots being fired. (B) The fifth floor window of the Dallas School Book Depository from which Mr Gubbins believes that Elvis fired three times. And (C) the 'Grassy Knoll' where rhinestones were found similar to those worn by the singer. "In the car park to the rear of the grassy knoll several witnesses claim to have seen a man who looked a bit like Colonel Tom Parker, Elvis's manager", says Mr Gubbins. "But they're all dead now", he added.

HE PRESIDENT?

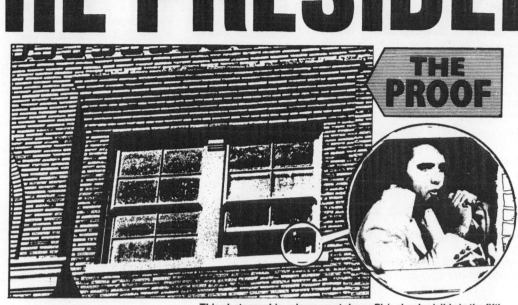

THE PROOF

This photographic enlargement shows Elvis clearly visible in the fifth floor window of the Book Depository seconds before the shots rang out.

'I know too much'
—Archie fears for his life

out. "A friend of mine had worked in an abattoir and offered to do a pathologist's report for me", says Archie. However, his request was turned down.

TRIGGER

"I decided to go ahead and do our own autopsy, using photographs of the President from a book in the library". The report confirmed what Archie already knew. "There was no doubt in my mind Elvis pulled the trigger".

BARREL

But what motive would drive the King of Rock 'n' Roll to kill the President? What was going through the singer's mind when he took aim on that cold, grey, November lunchtime?

BEATLES

One theory which Archie puts forward is that Kennedy, disillusioned with politics, was about to launch his own pop career. Already under threat from The Beatles, 'The King' feared that Kennedy may succeed in capturing his crown. However there is little hard evidence to support this notion.

MONKEES

Archie believes that jealously was the real reason. "Apparently, Elvis had heard from a friend that Kennedy fancied his wife Priscilla and wanted to go out with her", Archie told us. "That's probably why he did it".

> 'I could confirm that the gunshot wound which killed President Kennedy would be consistent with him having been shot in the head by the type of bullet fired from a gun by Elvis Presley.'
>
> **KEVIN DOBSON
> FREELANCE PATHOLOGIST**

According to Archie, further evidence was left by Presley in the words of songs which he recorded after the shooting. "One evening my wife and I began to notice strange, almost cryptic references to the murder in the words of Elvis's songs. It was almost as if he was leaving deliberate clues for us to find. I suppose it was his way of admitting his guilt.

DESPAIR

"For example 'You saw me crying in the chapel' is, I believe, Elvis's way of asking God to forgive him. And in 'There goes my everything' he sums up his feelings of despair once it had dawned on him what a terrible thing he'd done.

REMORSE

"But the words of 'Rock-a-Hula Baby' speak for themselves. Elvis was obviously overcome with remorse. I think killing President Kennedy was a mistake which Elvis regreted for the rest of his life".

Even sceptics would have to agree that the Kennedy assasination certainly did mark a turning point in Presley's career. From that point onwards he began to concentrate on slow, mournful ballads, he stopped touring and began to shy away from the public eye. He became a recluse inside his palatial Memphis home and subsequently lost control over his bowel movements.

Despite all the evidence put forward by Mr Gubbins, the authorities steadfastly refuse to re-open the case. And Archie now fears that his knowledge of the true events of that grey November day in 1963 could put his own life at risk. "I'm convinced that the telephone box in our street is being bugged, and my car has been tampered with. It keeps slipping out of first gear, and I've had to have the clutch looked at twice in as many weeks. It's scary when you think about it. The kind of people I'm dealing with here are above the law".

COINCIDENCE

"I know too much — just like Buddy Holly, Bill Haley and now The Big 'O'. It's more than just a coincidence that all three of them have died since Kennedy was killed".

SINISTER

We rang the FBI to ask whether or not they were involved in a sinister cover-up of Elvis's part in the Kennedy killing, but it was only 5 am in America and there was nobody in.

 **OPINION** page 33

This remarkable photograph was taken by an eye witness on a polaroid camera and shows the trees to the rear of the grassy knoll. Using another photographic technique, a portion of the picture can be enhanced to show quite clearly the figure of a man, not unlike Colonel Tom Parker, standing amongst the trees

ME, SEX AND THE STARS

Glamorous TV actress and model **BETTY RAMSBOTTOM** has rubbed shoulders with scores of showbiz celebrities in her glittering career since appearing briefly as an extra in Dr Who twelve years ago. Now, at 33, Betty has decided to spill the beans about her celebrity lifestyle, about the stars she's dated, and the stars who were after more than just a date!

SEXY TV chat show king JONATHON ROSS is the kind of guy that most girls dream of. But just thinking of that creep gives me nightmares! He's always fancied me, and he used to pester me day and night for a date.

Eventually I gave in and agreed to meet him in a posh nightclub. But my bus was held up in a traffic jam and Jonathon got there before me. He wasn't wearing his contact lenses at the time and he mistook another girl — Jane Goldman — for me, and asked her to marry him. She couldn't believe her luck!

IN BED

They're engaged now, but I still don't think Jonathon realises his mistake. Still, it doesn't bother me. I don't even fancy him, and apparently he's no good in bed.

DIRTY

I must admit, I did quite fancy **LESLIE GRANTHAM**, star of Eastenders, and I was thrilled when a friend who works at the BBC agreed to fix a date for us. I couldn't wait to find out why they called him *Dirty Den*.

Leslie agreed to pick me up at my flat that evening, and he probably spent a fortune buying red roses and chocolates for me. Unfortunately I never got any of them. My friend must have given him the wrong address, because he never arrived.

THROB

Some people will try anything to get a date with me. Perhaps its because I've got such big tits. American heart-throb TV actor **BRUCE WILLIS** called around at my flat one day — disguised as an electricity meter reader! Why he didn't just ask me out I'll never know. He seemed really nervous, and in the end he just read the meter then left.

TURN ON

PAUL NEWMAN tried a similar stunt, pretending to be a Water Board official! His disguise was good, but he didn't fool me. He asked me to go upstairs and turn on the tap, but when I returned he'd lost his nerve and disappeared out the back door, taking my handbag as a souvenir! I'm pretty sure he'd have pinched my underwear off the washing line too, if there'd been any there.

In an interview with **BILLY BOLLOCKS**

Often it's the quiet stars who are the ones to watch. Off screen **ROGER MOORE** is supposed to be the quiet family man. But the Roger Moore I know is so sexy he makes James Bond look like a nun.

ROGER

He sent me a message through my mother, who is a psychic medium, inviting me to the premier of one of his films. When I arrived at the cinema the doorman suggested I sit on the pavement outside so that Roger's wife, who was with him at the time, didn't get suspicious. I had fancied Roger for years and I was so thrilled to be going out with him the cold and rain didn't seem to matter at all.

Did you know that **SHAKIN STEVENS** was once a professional ice hockey player? Or that **BRUCE FORSYTH** owns sex shops in Reading, Edinburgh and Dundee?

No? Well that's simply because it isn't true. For these are just some of the many thousands of 'untruths' — unsubstantiated and often totally ficticious claims — which are made about some of Britain's top celebrities every day.

EVERTON

With so much untrue information available these days, it's often hard to tell the facts from the fiction — even for the professionals. When we rang Shakin' Stevens record company and suggested that Shaky's brother Gary Stevens plays for Everton, there was a confused silence.

CONTEST

"Shakin' Stevens has never entered the Eurovision Song Contest", snapped a quick tempered spokesman when we called to suggest Shaky had won the competition in 1968.

A spokesman for the police later suggested that we stop making calls to Epic Records or charges would be brought against us.

Top DJ wrote me steamy love letters

Sexy Radio One DJ **GARY DAVIES** is another one of my admirers. He must have fallen for me the minute I appeared briefly on Dr Who in 1977. He began writing me love letters — dozens of them — telling me what he would do to me when at last we were alone together. At one time if I listened carefully to the radio while he was playing a record I could hear him writing away in the background. The poor man was completely obsessed with me.

In the end he must have written so many initimate, sexy love letters he simply couldn't afford to post them all, which is why I never got any.

Appearing on TV certainly turned my life upside-down. I get pestered not only by fellas, but also by TV and film companies offering me work. At one time the phone just wouldn't stop ringing, unless I answered it.

PORNOGRAPHIC

I remember in one month alone I was asked to appear in **TWO** industrial training films, a pornographic video and as a model in an office furniture catalogue.

Next week: How I got caught up in a crazy love triangle between **PRINCE ANDREW**, **PRINCE EDWARD** and the **DUKE OF EDINBURGH**.

Betty as she was seen by millions of 'Dr Who' viewers.

Ugly Scenes Round UK as Fuel Protests Bite

■ By INGLEDEW BOTTERILL

MANY AREAS WERE yesterday facing a second day of shortages, as fuel pumps dried up throughout the country. With many tanker drivers refusing to cross price protest picket lines outside refineries and depots, forecourts throughout the UK reported scenes of panic buying by motorists desperate to fill up before dwindling stocks ran out.

QUEUE

In Leeds, there were scuffles at an Esso garage after customers objected to a driver who had jumped the queue. Meanwhile in Bristol, angry motorists banged on a filling station window after staff locked the door when supplies dried up. And according to local news reports, police were called to a garage in Harpenden where an angry van driver apparently drove through bollards and used bolt cutters to remove the nozzle from a padlocked fuel pump. Government sources appealed for calm as similar scenes were repeated around Britain.

EWE

At a hastily-arranged press conference, fuel minister Malcolm Wicks told reporters: "Although stocks are running perilously low and there is a danger that the country will run out of fuel in the

WON'T GET FUELLED AGAIN: Energy Minister Wicks yesterday

next day or so, there is absolutely no need for panic buying." However, his words did little to quell the fears of motorists, and fresh scenes of disorder were reported around the country just minutes

+++ ROUND THE COUNTRY +++ ROUND THE COUNTRY +++

❶ NOTTINGHAM: A group of bandits wearing long, leather coats and led by a grinning self-employed haulage contractor dressed like General Patton driving a V8-engined jeep chassis, ransacked the Fina garage just next to the Winning Post pub on the Clifton Estate's Farnborough Road.

❷ ABERYSTWYTH: An evil, megalomaniac dwarf sitting in a harness on the back of a simple-minded giant had appropriated the final remaining fuel stocks in West Wales, which he was protecting within a rusty, steel stockaded compound just behind the dunes at Borth.

❸ BIRMINGHAM: A scarecrow-thin sky pirate, piloting a ramshackle leather and canvas microlight, swooped down over the M6 just outside Bournville and forced a petrol tanker onto the hard shoulder by throwing a poisonous snake through the cab window. The pilot's feral child cohorts then took over the controls and ran the driver over with his own vehicle, leaving his crushed body in the road.

❹ LONDON: Mutant motorists, angry at the shortage of fuel, formed a vigilante army in dusty bondage gear, led by a wild-haired woman in fishnet tights, and took over the London Palladium, where they were last night believed to be holding a series of gladiatorial combats to the death on bungee ropes.

❺ SOUTH WALES: Mohawked commuters in steel hockey masks and riding powerful, stripped-down chopper motorcycles, ambushed a picket-busting tanker leaving an oil refinery at Milford Haven. Witnesses reported that one of the bikers, who had sharpened teeth, leapt from his bike and fell between the rear wheels of the truck, but managed to pull himself along the ground between the axles. He eventually crawled up onto the bonnet, kicked in the windscreen and killed the driver with a medieval mace. The driverless tanker then careered out of control, ran down an embankment on the B4325 near Llanstadwell, and jack-knifed. However, the attackers' jubilant celebrations turned to apoplectic rage when they discovered that it was in fact a decoy full of sand.

+++ ROUND THE COUNTRY +++ ROUND THE COUNTRY +++

after the end of his statement.

EYE

In Basingstoke, angry car owners driving a ramshackle convoy of cannibalised ex-army vehicles ram-raided a supermarket filling station, killing the manager with a crossbow before draining remaining supplies of fuel from underground tanks. And on the M78 just outside Glasgow, vigilante taxi drivers in stripped-down pick-up trucks drove other vehicles off the road, maiming and killing drivers with boomerang-sized chrome death-stars, before syphoning petrol from their cars. Witnesses reported that the attacks were orchestrated by a one-eyed man with a Mohican haircut, who was standing on top of a rusty armoured personnel carrier, swinging a chain round his head and whooping.

EMM

At another, even more hastily-arranged press conference, Prime Minister Gordon Brown appealed for calm once again and warned the public not to take the law into their own hands in their search for petrol and diesel. "I urge the British population not to form themselves into marauding, lawless gangs of leather-clad misfits and outlaws, stopping at nothing to get hold of fuel for their fleets of home-made, customised pursuit vehicles," he said, doing that thing with his mouth.

KILL THE THUGS!

By Bob Liar & Mick Crap

Britain's army of football hooligans should be birched and executed. That was the almost unanimous verdict of SEVERAL members of the British public in a survey which was carried out recently.

Disgusted by the behaviour of the mindless thugs, LOTS of people are demanding:

● **STIFFER** penalties, and

● **TOUGHER** measures, in order to curb trouble on the terraces.

TOUGHER

And among the many suggestions we received were tougher controls at the turnstiles. "Why not only let 48 fans into the ground at a time" suggested Norman Thomson, a bus driver from Luton. "That's the system we use on the buses, and we never get any bother".

TIGHTER

Other people suggested **TIGHTER** security measures. "Strip fans to their underpants before they enter the ground, then burn their shoes", said Mrs Dorothy Squires of Ebchester.

'Sprinkle them with agricultural chemicals'

Increasing the minimum ticket prices to £175 was another popular idea, while Mrs Anne Barker of Rotherhyde was in favour of sprinkling fans with agricultural chemicals as they entered the ground.

A **GOOD MANY** people we spoke to were in favour of tougher sentences for convicted football hooligans.

"Birch them to within an inch of their lives", said Glenda Jones, a bank clerk from Ryegate.

Dorothy Squires - 'Strip them'

Anne Barker - 'Chemicals'

Tom McGuire - 'Experiments'

Muriel Rowntree - 'Gas them'

"Put them in cages and experiment on them", said Tom McGuire, a retired car park attendant from Slough. "Put shampoo in their eyes and interfere with their hormone levels. That would put a stop to their nonsense", he told us.

But of all the people we spoke to, a staggering **FAIR FEW** demanded **CAPITAL PUNISHMENT** for convicted trouble makers.

"Birch them, make them pick up all the litter, then gas them", said Mrs Muriel Rowntree of Evesham, whose kitchen window was broken by football hooligans two years ago.

PENALTIES

When we spoke to Peter, a football hooligan from Leeds, he told us that stiffer penalties like the birch would certianly make him think twice about causing trouble.

FREE KICKS

"If they brought back the birch I would certainly think twice about causing trouble". he told us. "In fact I'd probably smarten myself up, stop going to football matches and go out and find a job straight away", he admitted.

BUT HOW ARE WE GOING TO DO IT?

KILLING PEOPLE is the answer to the current crime wave which is sweeping the UK. That was the resounding view of a GOOD MANY people questioned in our recent survey.

But despite the growing lobby for the return of capital punishment, the British public are divided on which method of execution they would favour.

HANGING

The most popular techiques are of course hanging, gas and lethal injection, with the guillotine and electric chair also in the running. But we thought up three slightly unusual ways of killing people and asked you, the public, to choose between them.

Of the people we spoke to:

By Charlie Pontoon

● **SEVERAL** favoured the firing sqaud.
● **A FEW** voted for drowning
● **THE REST** plumped for death by stoning.

SHOT

But although the firing sqaud was a popular choice, you weren't so sure about WHERE the victims should be shot.

● 78% said in the chest
● 15% said aim for the head
● 5% favoured the lower abdomen

Shoot them in the lower abdomen

while the remaining 2 per cent went for a groin shot.

KEEL HAULED

Can you think of a good way of killing people? It could be one you've heard about, or perhaps one you thought up yourself. Why not write and tell your MP. Mark your envelope 'IDEAS FOR CAPITAL PUNISHMENT' and address it to your local MP, c/o The Houses of Parliament, London.

Passengers be on your guard!

A record breaking British Rail guard was honoured yesterday at a special ceremony at Kings Cross station.

EXCLUSIVE

Bob Birchenall received a special commemorative medal from British Rail after being rude to his one millionth passenger. And Scotsman Bob, from Aberdeen, is the first railway employee in history to reach that figure.

SWIFT

Bob joined the railways as a station porter in 1947 and won swift promotion to ticket inspector after losing several items of luggage. His abruptness and unpleasant attitude gained him a further promotion to train guard in 1962. Since then Bob has been consistently rude to passengers on a daily basis, reaching the remarkable milestone of one million earlier this year.

Scotsman is flying into the record books

"There's no secret to it. Just hard work, I suppose", said Bob, aged 59. "As I look at it every passenger enquiry is an opportunity to annoy someone", he told reporters. And the highlight of his career so far? "Definitely the time I made a woman carrying two children break down and cry hysterically at the ticket barrier", he told us with a smile. "I refused to allow her onto the train, even though she had a valid ticket. I was obstinate and totally unpleasant. Eventually I shouted at her and she snapped. It was quite a thrill".

SWALLOW

Bob is pictured here receiving his medal from BR's Eastern Region Personnel Manager, Mr Trevor Banks. Also honoured at the same ceremony was former ticket office employee Richard Whittle. It was his suggestion that lead to the introduction of the highly confusing 'Blue Saver' ticket system which

Proud Bob receives his medal yesterday.

has caused unprecedented passenger confusion and has helped generate an estimated 15 million complaints since its introduction in the early eighties. Mr Whittle is now in charge of customer enquiries at Euston Station.

Offers invited for 'Peter Pan' of pop

FOR SALE
Detached mid 20th century singer
Many original features still intact

A unique opportunity to acquire one of Britain's best loved pop singers has arisen after EMI record bosses decided to sell their veteran star Cliff Richard.

Richard, now in his fifties, was a bargain when EMI snapped him up for next to nothing back in 1958. Since then the 'Peter Pan' of pop has churned out a remarkable string of over 100 hot singles for the record company. But hits have been thin on the ground lately, and with the passing years the cost of maintaining a star of Cliff's size has increased dramatically.

"Over the years we have treated Cliff with tender loving care", said EMI spokesman Tony Wordsworth. "But as he gets older the cost of maintaining him increases, and at some point we have to look realistically at what is best for his future, as well as for ours".

SPRING

Reluctantly EMI decided to sell and Richard was originally put on the market in the spring, freehold with vacant possession, for offers in the region of £1.25 million. Although several people viewed him, no firm offers were received, and subsequently the price has been lowered to £950,000.

"I think the asking price is a very fair reflection of current marketing trends", said a spokesman for estate agent Savilles. "We hope that Cliff will appeal to a family perhaps, or someone looking for a singer to use at weekends. There is also great potential for development, subject to the relevant planning consents".

FACE

Cliff's fans are hoping that the National Landmark Trust might show an interest in the former Eurovision Song Contest winner. Among them fan club secretary Una Phillips. "Cliff could easily be converted into up-market holiday accommodation whilst still preserving features such as his leather trousers and microphone".

STRAP

Meanwhile, outline plans to convert the singer into an old people's home were yesterday rejected by Weybridge Planning Authority. "We don't feel that this sort of development would be in keeping with the individual concerned", a spokesman said yesterday.

CANE

Offers for Cliff should be made in writing and must be received by the selling agent no later than midday on Friday 11th November.

Anyone for sugar?

Britain's teeth are getting less sweeter, and that's official.

For a report out today claims less tea drinkers than ever before are adding sugar to their cuppa. And that's down on previous figures for the same statistic.

A 1974 tea sipping census revealed that 7 out of 10 tea drinkers were taking sugar in their tea, a tooth rotting average of 4 spoonfuls per cup. This year's survey reveals an average drop of 2 spoonfuls per person per cup, with only 5 out of 10 tea drinkers now prefering an average of 2 spoonfuls of sugar in it.

However, the total amount of sugar dissolved in tea has increased, due to a massive leap in tea sales, while the weight ratio 'sugar to tea consumed' has altered in favour of the drink. In 1974 half as much tea was sold in Britain as is sold today.

Whilst today housewives pick twice that amount up from supermarket shelves all over Britain.

The good news for tea drinkers is that dentists are welcoming the good news about the drop in the amount of sugar in tea. But the bad news for tea drinkers is that prices of tea look set to rise despite the increase in sales. However, the good news for dentists is that sugar prices look set to follow suit, although that news will be less welcome among half of tea drinkers who enjoy sugar in their tea.

A further reduction in the amount of sugar in tea is a likely outcome of the imminent rise in prices, and no doubt dentists will be keen to discover whether a drop in tooth decay is a possible result.

IT'S YOUR VERDICT

Are less people having sugar in tea than in the old days? Or do you think more people are having sugar in tea? How many spoonfuls do YOU have? We're opening a special hotline to record YOUR views. Simply dial the following numbers according to how many spoonfuls of sugar you have in your tea.

0 – Dial	**(091) 21 2 1 213**	3 – Dial	**(091) 2 121 21 3**
1 – Dial	**(091) 21 21 21 3**	4 – Dial	**(091) 21 2 1 213**
2 – Dial	**(091) 212 1 2 13**		

If you take more than four, dial a combination of numbers to match your total. (e.g. If you take six, dial the number for 3 twice. Or the numbers for 2 then 4, etc.). All calls are charged at normal BT rates.

TV NOEL RUINED MY LIFE

While TV millionaire Noel Edmonds was last week signing a £2 million contract to keep him at the BBC, hundreds of miles away one former fan of the star was choking back tears as she recalled how the heartless House Party host left her life in tatters.

Heartless star breaks fans heart in two

Sun star gazer Morgan

Bitter Tyneside mother-of-two Susan Patterson won't be tuning in to any of Noel's new TV shows. For she knows the real Noel Edmonds – the man who TV viewers never get to see.

HEART

For many years ago Susan's heart was broken by her idol Edmonds, then presenter of TV's Swapshop.

Susan was just a teenager when, along with friends, she went on a day trip to the Lake District. She could hardly believe her eyes when she saw Edmonds in a pub car park.

CLOUT

"He was on his way into the pub and I asked him for his autograph. He said I could have it later, when he came out", Susan recalled, the pain of the memories etched on her face.

Susan waited. And waited.

The smiling face of cash magnet Noel posing for a publicity photograph outside his bank yesterday

But Edmonds never returned.

BANGLES

Husband Michael has tried to help Susan rebuild her life, but it has been difficult. At times both have turned to drink, especially at Christmas parties. And as Edmonds' TV career goes from strength to strength, Susan struggles to get by as a part-time receptionist while Michael looks for work abroad.

"The strange thing is that my two young children both watch TV, and are both big fans of his House Party programme. One day, when they're old enough to understand, I'll tell them about the real Noel Edmonds. The Noel Edmonds who broke my heart", said Susan yesterday.

What the expert says

We spoke to Celebrity expert Piers Morgan of The Sun newspaper and asked him whether Edmonds was within his rights to refuse a young fan an autograph in a pub car park in the Lake District several years ago.

"There are no hard and fast rules governing celebrities in these situations", he told us, "although there are certain recommended standards which celebrities ought to maintain.

I believe Edmonds was heartless to turn down this young girl's request".

SLITS

Has a star ever shat on you from a height? Are you a faithful fan who's been treated like shit by his or her idol? Write and tell us. There's a crisp tenner for every letter we use, and a selection of your best stories appear on the following page.

PRINCE OF WHALES

He parlez with plants – and he talks to the trees. And now Britain's pottiest Prince is doing a Dolittle!

For Buckingham Palace officials have significantly failed to deny rumours that Prince Charles plans to turn his back on Britain and live underwater – becoming the world's first fish monarch.

PRINCE

Close friends of the Prince fear that he intends to submerge himself entirely in water, blow bubbles out of his mouth, and swim about with long, thin trails of excrement dangling out of his arse.

Experts are linking the Prince's dramatic transformation to a 'King Arthur' style experience Charles may have had during a holiday in Cornwall. There is speculation that whilst fishing the heir to the throne witnessed a mysterious fish or something emerging from the water. And it said something to him. And from that moment on he has devoted his life to ruling an underwater kingdom of fish. Probably.

Naval experts believe that the Prince may use a 'Stingray' style submarine in which to travel underwater. And Britain's top shipyards yesterday confirmed that if Buckingham Palace asked them to build a 'Stingray' style submarine, they would be eager to tender for the work. The cost of such a vessel would depend on the Prince's exact specifications, but one source yesterday revealed that it could run into **BILLIONS** of pounds.

Britannia rules the whales: This is how Charles would look with a Troy Tempest hat on (left) and 'Stingray' (below).

A cheaper alternative could be a 'Thunderbird 4' style underwater mini-sub, which would be yellow, and would have attachments on the front. However, Charles' advisors would no doubt inform the Prince that such a vessel would need to be carried in a 'pod' – a large, transportable container – belonging to Thunderbird 2, which is piloted by Virgil Tracey. Who is a puppet.

'Stingray' jobs boost for North

A North MP yesterday welcomed news that local shipbuilders Swan Hunter may be invited to tender for a 'Stingray' style submarine possibly being built for the Prince of Wales.

"With empty order books job prospects are grim for the remaining workforce", said Labour's Derek Twatt. "Any prospect of new work in the pipeline has got to be good news for the region".

GOOD NEWS

Meanwhile there was more good news for jobs in the region last night when a local shopkeeper announced that he is

to pay a man £25 to do some welding on the sills of his car in order to get it past its MOT.

BLACK MAGIC

Sid Williamson, owner of a second hand shop in the Heaton area, told us he had offered someone £20 to carry out the work, but had now agreed a price of £25.

"It's just the sills that need doing", he told us yesterday. "Otherwise the car is sound as a pound".

The LEXICON of SEX

An A to Z of Celebrities & their Aliterating Sexual Organs

A ADAM ANT'S ARSEHOLE

B BRIAN BLESSED'S BELL END

Next week: Charlie Chaplin's cheesy chopper and Desmond Dekker's dirtbox.

20 THINGS YOU NEVER KNEW ABOUT MARZIPAN

It's in cakes and it's on cake ingredient shelves in the local supermarket. Yes, there's no getting away from marzipan. It's the cake ingredient that everyone wants to have on their cake and eat it. But how much do we really know about marzipan? We picked away the icing to reveal twenty fascinating facts about Britain's favourite cake covering.

1 You have to be *nuts* to make marzipan. That's because nuts – almonds to be precise – are what marzipan is made out of.

2 The name marzipan derives from latin – literally meaning 'bread of mars'. That's because the Roman astronomers believed the pasty cake covering originated from the red planet.

3 Marzipan first came to Britain in the 15th century when merchants returning from China used it to pack boxes of fragile china plates, cups and saucers.

4 To this day the idea of eating marzipan is unthinkable in China where it remains the country's main packing material.

5 Ironically, Chinese chefs use polystyrene, which they consider a delicacy, to coat their Christmas cakes.

6 Followers of a little known religion – Marzipology – refuse to eat marzipan. They believe that owing to a typographical error in the Bible marzipan was omitted from the gifts delivered to Jesus by the three wise men, and the word 'myrhh' was incorrectly used in it's place. They believe that marzipan is Holy, and they light special marzipan candles in their churches.

7 Stars who enjoy marzipan include controversial slap head Sinead O'Connor, Catherine Zeta Jones out of the Darling Buds of May, and Eric Clapton.

Marzipan fan Darling Catherine Zeta Jones

8 Meanwhile, American rocker Bruce Springsteen hates the stuff. He shocked guests at an MTV awards party recently by removing marzipan from his cake and leaving it on the side of his plate.

Stars who have not yet publicly expressed an opinion on marzipan include cockney comic and Eastenders star Mike Reid, actress Julia Sawalha and TV funny man Jasper Carrott.

10 Last Christmas Britain's housewives spent an average of £78 stocking up on marzipan for use as a cake covering and decoration. That's more than we spent on candles, miniature plastic Santa Claus and Christmas tree figures, candle holders and little silver edible balls put together.

11 Marzipan can be made at home using icing sugar, caster sugar, almonds, eggs, vanilla essence and lemon juice.

12 Dentists prefer to make their own marzipan as opposed to buying it in the shops. That's because ready mixed marzipan contains up to 50 per cent more sugar than the home-made variety.

13 Pop star Marc Almond changed his name from Robert Marzipan Blenkinsop. He thought Almond would be less of a mouthful.

14 If you were caught with marzipan in France during the last war, you'd have been shot by the Nazis. For marzipan is identical in texture and appearance to gelignite, and the French resistance used Christmas cakes as a means of smuggling explosives.

15 Millionaire rock star Mick Jagger is mad about marzipan. Indeed he chose his home – Mustique in the Carribean – because it is the world's second largest marzipan producing country.

Marzipan man Jagger

16 America is the largest, with an annual output of 140 million tons – enough marzipan, if it were rolled into a thin strip, to stretch to the moon and back – ten times! All 140 million tons are exported to Belgium.

17 Russian marzipan contains no sugar or almonds. Intead it is made of sand and lime. 'Ugstondk', as it's known, is used extensively in the building industry.

18 Top of the Pops dancers Pan's People were originally called Marzipan's People after the seventies teenage craze of eating marzipan. Kids used marzi-pan to give them extra energy in much the same way that Ecstasy is used today. However, the name was changed as the BBC felt the word 'marzipan' was too controversial.

19 There was 'marzipandimonium' in Hollywood recently when pint-sized Batman star Danny Devito opened Hollywood's first marzipan restaurant. Top stars queued for hours to choose from 150 different flavours of marzipan on offer. Demand was so great that the restaurant completely sold out of marzipan within half an hour.

20 Marzipan hasn't always been used to decorate cakes. During the war hair combs and brushes were rationed, and soldiers would use a strip of pink marzipan in their hair as a false parting.

Old Wives Tales Have Soap Stars in a Lather

EXCLUSIVE

Soap-

They never utter the title of 'The Scottish Play', they never whistle in the dressing rooms, they never say 'Macbeth' and they never perform with children or animals. Actors are notoriously superstitious, but according to a new book by a mobile caterer who has provided on-set food for all of Britain's leading serial dramas, when it comes to superstitions, the stars of our soaps take the biscuit.

In a 30-year career spanning three decades, Marjorie Bibby has dished out tea, sandwiches, soup, burgers (with or without onions) and advice to all of Britain's best known soap stars. And in her new book *'Soaperstitions'* (Croissant Books, £12.99), she lifts the lid on the old wives' tales which rule the behind-the-scenes world of our favourite TV shows.

■ by TRAFFORD LOVETHING

Directors

"You might think that the writers and directors control who and what we see on our soaps, but you'd be wrong. The actors superstitions play a lot bigger part than anyone realises. Most people believe that Ross Kemp left EastEnders when he signed a million pound deal with ITV drama. But that's simply not the case, I can tell you. The truth is that he was hounded out of the thrice-weekly London soap by his fellow actors after their superstitions got the better of them.

"One rainy day, Kemp was filming a scene with Sharon Watts. The script called for him to leave the Queen Vic, open his umbrella and walk into that little park bit in the middle. But Kemp fluffed the shot, and opened his brolly whilst he was still inside the Vic.

"The cast and crew were clearly upset, and shooting was halted. Barbara Windsor maintained that the bad luck could be reversed if the actor simply turned round three times and spat. However, Kemp was having none of it, growling that it wasn't in his contract."

And according to Marjorie, things went from bad to worse. Over the coming months Kemp...

• *Dropped a knife in the Fowlers' kitchen, and then refused to tap it on the table leg, whilst saying: "Sharp surprise - tap it on wood, sure to be good"*

• *Put his hat on Phil Mitchell's bed without then placing it upside down on the windowsill overnight to let the devil out of it*

• *Placed his shoes on the table in Kathy Beale's cafe, and then refused to bury a piece of meat in the garden.*

"Eventually the cast got so scared that they petitioned the BBC to axe Kemp before his actions led to a jinx on the soap. Station bosses agreed, and the actor was given his marching orders.

"Chillingly, in his career since EastEnders, Kemp has been dogged by four pieces of terrible fortune; In Defence, Hero of the Hour, Without Motive and Ultimate Force."

Gladiator

Sometimes, there's simply no way for the soap stars to

STAN AND DELIVER: Marjorie delivers vinegar onto Stan's chips

reverse their bad luck. Marjorie remembers being called onto the set of EastEnders on New Years Day a few years ago. The tea urn had gone on the blink over Christmas and the cast needed hot drinks.

"When I arrived, they were filming a scene in the launderette. Dot Cotton was going to have an argument with Pauline whilst doing a service wash for Dr Legg. The scene was going really well until Dot Cotton suddenly stopped. I thought she'd forgotten her lines until I saw her face. She looked like she'd seen a ghost. Then she started screaming.

"I'm washing on new year's day! I'm washing on new year's

SUPERSTITIOUS: Armstrong & Cotton

day!" she cried. The set fell silent. We all knew what that meant; Dot Cotton was washing someone out of her family. Dot was inconsolable, it's one of the few bad lucks to which there is no effective antidote. As a result, no-one was surprised when, a few months later, she opened her script

to find out that her husband Charlie Cotton had been tragically killed in a lorry crash."

Pedigree

Not everybody in soaps is superstitious. But those who write it off as mumbo jumbo could be making the worst mistakes of their acting careers.

"I remember one occasion when I was serving food to the cast and crew of Brookside. At the time, the popular Liverpool soap was riding high in the ratings, with exciting storylines and entertaining characters. During the commercial break, producer Phil Redmond came to my van for a bacon sandwich, but whilst seasoning it, he spilt some salt on the counter. I told

SHARP SURPRISE: The EastEnders set where Kemp dropped his knife

Our Number's Up
Spoonbender's sinister warning to the world

We're all scared of the number thirteen, but top mystic conjuror Uri Geller, who rose to fame in the 1970s bending spoons using only the strength of his mind and hands, has recently begun to fear another number. The number **ELEVEN**.

❝ The number eleven will bring about the destruction of the world," he says. "Look at the evidence and you will see I am correct. The Twin Towers looked a bit like an eleven, and fell down on the **ELEVENTH** of September. 9-11 - add the three digits up, and it makes eleven. September is only **TWO** months before the eleventh month, and the Roman for two is eleven. Where were the twin towers? **NEW YORK CITY** - eleven letters. Which district? **IN MANHATTAN** -

eleven letters again. What crashed into the towers? **2 AEROPLANES** - eleven letters (if you count the number 2 as a letter) and where was the atrocity planned? In **AFGHANISTAN.** Eleven letters again.

Who was the attack perpetrated against? **GEORGE W BUSH** and **TONY BLAIR MP** - surprise, surprise, eleven letters each. And who was the evil terrorist who planned the outrage? **MR O. BIN LADEN, OSAMA B. LADEN, MISTER LADEN, MR. OSAMA BIN L., BIN LADEN ESQ.** However you spell his name, you come up with the same answer. Yes, that's right. **ELEVEN LETTERS.**

Some people would dismiss these bizarre coincidences as mere bizarre coincidences, but these coincidences are simply too bizarre to be written off as mere bizarre coincidences. ❞

erstitious!

him he ought to throw some over his shoulder to hit the devil in the eye, but he just laughed and walked off.

"However, the consequences of his hasty actions soon became clear. As the second half of the show began, a mystery virus broke out in the Close, the plots became farcical and Brookside got axed. Redmond should think twice next time he spills salt on the set of Hollyoaks."

Bum

Soaps work to tight schedules and even tighter budgets, and none more so than Emmerdale Farm. Marjorie recalls an incident where a star's refusal to tempt fate cost his soap bosses a fortune. Quite literally.

"I was on location in the Yorkshire Dales providing a finger buffet for the cast and crew of Emmerdale farm. They were about to film a scene where Annie Sugden was up a ladder cleaning the Woolpack windows. The script called for Seth Armstrong to walk under the ladder, look up her skirt and wolf whistle.

"But Seth was having none of it. He flatly refused to walk under the ladder, saying bad luck would visit his house three times if he did. The producers tried to persuade him, but he wouldn't listen. In the end, tempers frayed and he threatened to leave the series".

So Marjorie couldn't believe her eyes when she tuned in that night to see Seth walking under the ladder, as plain as day. She later found out that Yorkshire TV had ended up

getting Pixar Studios in Hollywood - the creators of Toy Story and Shrek - to produce a special computer animated Seth for the 2 second shot. Apparently, just animating the whiskers on the veteran rubbish actor's face had taken over a year, at a cost of over £10 million. Ironically, the shot ended up on the cutting room floor and was never used.

'Soaperstitions' by Marjorie Bibby is available in publisher's outlets, supermarket tills and all-night garages from November 10th.

SMASH! Broken mirror brought 7 years bad luck for Albert

In the hot summer of 1977, the Granada TV canteen failed a routine health and hygiene inspection, and Marjorie was called in to do a bit of emergency catering on the Coronation Street set.

"One of my first jobs was to take a cup of tea and a slice of black pudding to the dressing room of Rovers Return regular Albert Tatlock. He was getting ready for a scene, carefully applying purple greasepaint to the end of his nose, and as he reached for his cuppa he accidentally knocked his shaving mirror into the sink, where it shattered.

Albert froze with horror. I tried to make light of it, but deep down we both knew what the broken mirror meant."

And Albert didn't have long to wait. His 7

Albert Tatlock: Unbelievably bad actor

years of bad luck started the very next day.

Tatlock's catalogue of misfortune:

1977 Albert's real-life car fails its MOT due to a faulty brake light.

1978 While out shopping in Salford, Tatlock kicks a really big dog turd, and some of it goes all on the front and top of his shoe.

1979 The street veteran's real-life brother Sid mysteriously dies of natural causes at the age of 93.

1980 Tatlock leaves his trademark cap on a bus. It is never handed in, leaving the actor little choice but to buy a new one.

1981 Albert's nephew Ken Barlow's marriage to Dierdre Rashid flounders.

1982 While out shopping in Didsbury, Tatlock steps right into a dog turd, this time coating his heel and instep in foul-smelling orange excreta.

1983 Albert places a £5 each way bet on the Grand National, and his horse comes in fourth.

Finally, in **1984,** exactly seven years and a few months to the day since he broke his mirror, Tatlock suffers his biggest stroke of bad luck, when he dies in real life.

Why Do We...?

Have you ever wondered why we cross our fingers for luck, or are scared when a black cat crosses our path? Dr Boris Fäckt, professor of everything at the University of Wisconsin explains the origins of a few of the Soapstars superstitions.

• Why is it unlucky to whistle on a Sunday?

Ken Barlow out of Coronation Street

In Victorian times, whistling was said to be the sound of the devil's flute, and that anyone who could whistle was in league with Satan. These people were burned as witches on a Sunday after their dinner. That's why, although people are no longer burned at the stake, it is still considered bad luck to whistle on a Sunday.

• Why is the number 13 considered unlucky?

Benny out of Crossroads

This goes back to Victorian times when formal education was only available to the very rich. Consequently, bakers often had difficulty counting to twelve, and would often put 13 buns in a bag and charge only for twelve, losing money on the sale. As a result, many bakers ended up in debtors prison, put there by the number 13.

• Why do we say 'White Rabbits' on the first day of the month, unless that month has an 'R' in it.

Lofty Watts, EastEnders

This tradition dates back to times past when rabbits, especially white ones, were regarded as bringers of good fortune on the first day of the month, particularly if the month had an 'R' in it.

Sporting Superstartions

Paul Ince never puts his shirt on before stepping onto the pitch, Tim Henman always laces up his left shoe first and Sally Gunnell never shaves on the morning of a race. Many of our favourite sports stars have secret rituals which they swear bring them luck. Here's a few of the more unusual ones.

• Olympic oarsman **Steven Redgrave** insists on always wearing his lucky underpants in every competition. "They're the pair I was wearing when I won my first rowing race at the age of six," he told us. "Obviously, they're far too small for me now, so for comfort I wear them over a normal sized pair."

• Leeds Rhinos rugby star Barrie McDermott's *superstition* has cost him more than £100,000 during his career. "On the day of my first professional match, I was so nervous whilst driving to the ground that I ran into the back of a taxi at a roundabout. However, during the game I played a blinder, and we won 64-12. I don't like to tempt fate and ever since that day I have to run into the back of a taxi on my way to a match." Barry's insurance premiums have gone through the roof and he is often beaten up by cabbies. "It's worth it, though," he adds. "Because we always win 64-12."

• World hop, skip and jump champion **Jonathan Edwards** goes through a special superstitious ritual at every athletics meeting he attends. "Before I leave the changing rooms, I kneel down, put my hands together and have a chat with an invisible bearded man who lives on a cloud," he says. "I ask him to help me win so I can have all the prize money and the big shiny cups for myself. He's only let me down a few times, like at the Atlanta Olympics when I only come second."

• Top flight darts player Jocky Wilson has won the Embassy World Darts Championship more times than any other player. And the hunky Scot attributes his success to having a lucky pre-match ritual. "All the players on the darts circuit share dressing rooms. Before each match, I'll sit and relax in the dressing room for ten minutes. Then, when my opponent isn't looking, I'll do a lucky shit in one of his shoes', he says.

• England goalie **David Seaman** offers himself up to Beelzebub at a stone altar in the Highbury dressing room before every match. "Dressed in black robes, I sacrifice six virgins by slitting their throats with a special curved dagger. Then I remove and eat their wombs and bathe in their still warm blood," he says. "The boss and the other lads think I'm daft as a brush, and deep down I know it's silly, but it keeps me focused before a game."

What's in a Name? Eamonn Holmes

WITH HIS TWINKLY Irish smile and his loveable brogues, TV presenter Eamonn Holmes has carved himself a niche in the world of inconsequential broadcasting. Whether he is cowering behind co-presenters on *The National Lottery Jet Set,* or failing to engage countless interviewees in conversation on cheap daytime TV shows, Eamonn is rarely off our screens. We think we know everything about him, but is there a hidden depth behind the bland, shallow facade that we see on TV? It is a question infinitely more interesting than all the things he has ever said put together... and once again, the answer is all in the name...

EATING

IN BETWEEN filling the TV schedules with half-arsed programmes, Eamonn likes to eat food... a lot of food! And he loves nothing more than Battenburg cake. "I can't get enough Battenburg," he told his website Eamonn.tv. It has been said that in a week, Eamonn pushes enough Battenburg slices into his face with his pudgy fingers to make a board big enough to play 350 games of chess.

ALLERGY

ON TV, Eamonn's inane smile hides a tragic secret. For the journeyman presenter was born with a lethal food allergy. Throughout his time at TVAM, canteen staff were under strict instructions not to serve him any fresh fruit, as one bite could kill him within seconds. In fact the only fruits Holmes can eat without dropping dead are toffee apples, banana fritters and Terry's chocolate oranges. And strawberry bonbons.

MEDIOCRITY

EAMONN first hit our screens in the early nineties, and a rollcall of his TV credits reads like a Who's Who of broadcasting cat litter. Programmes such as *Holiday Outings, Pot Black Timeframe,* and his lunchtime general knowledge quiz *SudoQ* all faded from memory even before their end credits finished rolling. Entertainment scientists have calculated that if awards were given for mediocrity, Holmes would need a cabinet the size of ten Albert Halls to house all his trophies.

OVERWEIGHT

VACUOUS TV host Holmes has struggled to keep his weight down for years. According to doctors, a normal Body Mass Index (the ratio of a person's weight to the square of their height) is between 18 and 25. With a BMI of 44.8, Holmes lies somewhat outside the normal range, although that doesn't necessarily indicate obesity. "It could be that I'm simply of an extremely athletic build, because muscle weighs more than fat," he told his website Eamonn.tv, whilst licking Battenburg crumbs off his fat fingers.

NAKED

AMAZINGLY, Eamonn's missus Ruth has never once seen her chubby hubby in his birthday suit. The bashful broadcaster never takes his clothes off, and even showers in his vest, pants and socks. And when it comes to bedtime, Mrs Holmes still doesn't get an eyeful of Eamonn, as his bedroom has a Victorian bathing machine into which Eamonn climbs in order to don his jim-jams. His wife then pulls it alongside the bed so he can slip under the duvet whilst retaining his modesty.

NUDIST

EAMONN is a keen nudist, and when he's not fronting televisual polyfilla, he'll likely as not be found playing volleyball in the buff at his local nudist camp. Holmes loves the sense of freedom that going without clothes gives him, but there is a downside to his saucy behaviour. "Last year at the nudist club barbecue, I was having a wee through a knot-hole in the fence when a short-sighted chef mistook my you-know-what for a sausage. Before I knew what happened, he'd pricked it with a fork, cooked it over the hot coals for ten minutes and given it to the lady Mayoress in a bread bun!" he told his website Eamonn.tv

HORROR FILMS

EAMONN loves to watch scary films about ghosts, monsters and frankensteins last thing at night. "I love the horror film genre. As a boy, I was brought up on the films of Bela Lugosi, Boris Karloff, and the Hammer studios, and it's a love that stayed with me all my life," he told his website Eamonn.tv. However, his wife Ruth has had to pull the plug on her fella's late-night fright fests. "He kept on having nightmares and wetting the bed," she told her website Eamonnswiferuth.tv

OESOPHAGUS

LIKE MANY stars, Eamonn has succumbed to the lure of the plastic surgeon's knife. But he has not had a facelift, a tummy tuck or an anal bleach - he's had his oesophagus surgically widened. The two hour procedure stretched the TV presenter's gullet to the width of a drainpipe, enabling him to insert whole Battenburg cakes into his stomach without wasting precious time chewing.

LACTATION

EAMONN suffers from a rare medical condition that causes his breasts to secrete cottage cheese whenever he hears a baby cry. "I remember presenting a feature about prams on GMTV once, and there were babies crying all over the studio. I had to change my shirt six times, and under the studio lights, the smell of curdled cottage cheese was unbearable," he told his website Eamonn.tv

MISFORTUNE

EARLY IN childhood, Eamonn suffered a terrible misfortune. At the age of ten, he wrote to *Jim'll Fix It* to ask if he could look round a Battenburg cake factory to see how his favourite food was made. Unfortunately, his letter wasn't chosen. Twenty years later, on his first day as a cub reporter for Ulster TV, he was sent to a cake factory to report on a story about a kitten who had fallen into a giant vat of marzipan. "They needed a volunteer to eat a tunnel through the marzipan to the stricken cat. No prizes for guessing who stepped forward!" the overweight nonentity told his mother-in-law's website Eamonnswiferuthsmam.tv

ERECTILE DYSFUNCTION

17% OF MEN suffer from some form of impotence, so there is a 1 in 6 chance that Eamonn's performance between the sheets is as disappointing as it is on screen. But doctors say that the likelihood of erectile dysfunction is increased in men who are middle-aged, overweight and who have failed to achieve anything worthwhile in their lives. "I'll put my shirt on Eamonn Holmes not having got it up for years," says Morris Moskovitch, Cambridge University's Professor of Speculative Urology.

SOILED UNDERPANTS

EAMONN'S only memorable television broadcast occurred when Fathers for Justice campaigners invaded the set of *The National Lottery Jet Set* on May 20th 2006. A shaken Holmes cowered behind co-presenter Sarah Cawood, before scurrying out of the studio. He was seen several minutes later, hiding a pair of excrement-filled underpants behind a bush in the Blue Peter Garden. "Yes, I did soil my pants, but it wasn't because I was frightened. I'd foolishly eaten four packets of Smints just before going on air," he told his underpants' website Eamonnskegs.tv

ZOO OF SHAME

Animals made love as children watched

Our reporters have investigated a disturbing behind the scenes sex scandal at a leading London zoo. And we have proof that staff at the zoo in Regents Park have actually encouraged sex between animals.

PENGUINS

On a recent visit to the zoo we watched as penguins made love openly while visitors, some of them young children, passed nearby. And we witnessed giraffes attempting sex as a zoo keeper looked on.

NAKED

In the reptile house the temperature was noticeably high, and we saw lizards and snakes romp naked in the grass.

PASSION

Nearby a crowd watched as lions explored each others bodies, while in the next cage leopards prepared for a night of passion.

ACTS

It seems that such acts of shame are an everyday occurence at this, the zoo of sin. And the staff there have been actively involved in encouraging sexual relationships between animals.

KEY

We can reveal that the zoos top attraction, Chi Chi the giant panda, is the key figure in an international sex syndicate involving well known zoos in China and the USA.

Our investigations revealed that the bear is regularly taken out of the country to other zoos where it is forced into sex sessions with other pandas.

CLAIMS

Yesterday the zoos head keeper was not prepared to discuss our claims. But when we confronted him with our dossier, including photographs of a young rabbit having sex, we were told to piss off.

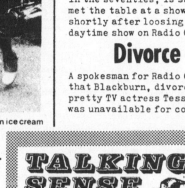

This woman sold children ice cream while animals had sex

DJ to wed table - claim

Officials at the BBC are denying rumours that disc jockey Tony Blackburn plans to marry a table.

Fame

Blackburn, who shot to fame with his 'Bisto' gravy commercials in the seventies, is said to have met the table at a showbiz party shortly after loosing his daytime show on Radio One.

Divorce

A spokesman for Radio One said that Blackburn, divorced from pretty TV actress Tessa Wyatt, was unavailable for comment.

TALKING SENSE with Charlie Pontoon

Three million unemployed? What a load of rubbish. Kids today just can't be bothered. Why do we pay out social security to three million scroungers when they haven't even got the sense to get a job? It's like spending money on old rope.
Come on Maggie. Hit the scroungers, and hit 'em hard.

★ ★ ★

It's all well and good aiming missiles at Russia, but anyone who knows their history will tell you that the French are the real trouble makers.
So come on Maggie. Hit the frogs, and hit em hard.

★ ★ ★

Jimmy Hill tells us that football hooliganism is killing the game. So what's he gonna do about it? Talking on the telly isn't going to help.
So come on Jimmy. Get down on the terraces next Saturday, and give those hooligans a hiding they're never gonna forget.

★ ★ ★

At last there's a law on wearing seat belts. About time too. But how do we expect the police to enforce the law when motorists outnumber them by over a thousand to one.
Now's the time to get the coppers tooled up. It makes sense. Motorists won't argue if they're looking down the barrel of a .38!

Million mile marathon nears end

AL JOLSON this week embarks on the final leg of a marathon million mile walk which has taken the tragic black singer a record 65 years to complete.

The walk came about after Jolson recorded the hit song 'Mammy', in which he vowed to walk a million miles for one his mammy's smiles. At first his devoted mother Eunice thought her all singing all dancing son was joking. But Jolson has spent an entire lifetime proving otherwise, turning his back on a glittering showbusiness career in order to prove his point.

Walking

Jolson set off from Hollywood in 1931 and has been walking almost non-stop ever since. When his historic trek began Edgar Hoover was still president, pizzas had not been invented, and a Ford 'Model T' car cost just twelve dollars and fifteen cents. His epic journey has taken him through 165 countries, across the Himalayan mountains (eighteen times), through the hottest deserts, and even across thousands of miles of sea bed.

Crying

Along the way Jolson has got through 288,576 pairs of shiny black tap dancing shoes, lost 2,867 straw hats

Al Jolson arrives in Leicester on Wednesday

and 9,446 walking canes, and has changed his white cotton gloves no less than 189,545 times. Jolson has walked constantly, without sleep or food, managing to maintaining an average of 2 miles per hour despite ageing considerably over the years.

Sleeping

His journey was briefly interupted in 1939 by the outbreak of war. Unable to cross European borders the singer spent six years walking round in circles in a field in Ireland. When he eventually left locals clubbed together and raised enough cash to have a small statue erected in what has become known as 'Jolson's Field' near

"Wait a minute. Wait a minute. You ain't seen nuthin' yet", says a tired black Al Johnson yesterday.

Letterkenny, to commemorate the singer's visit.

Talking

Jolson will this week clock up his millionth mile on British soil, having arrived through the Channel Tunnel from France yesterday. Ironically, his journey will end in Leicester - 4,500 miles from his mother's home in Carolina - but the singer will nevertheless be guaranteed a warm reception.

Living Doll

"This is a great honour for our city", said Deputy

Lord Mayor Eric Thonks who will officially welcome the singer when he arrives at the DeMontford Hall on Wednesday afternoon. "My wife and I are big fans of his, and we will be inviting Mr Jolson to unveil a plaque to commemorate his great feat of endurance."

Mile End tube etc.

There is however a tragic side to the story. When Mr Jolson arrives he will be told that his mammy died in 1932, only 9 weeks after he set off on his mammoth hike.

Man dies in think tank

AN inquest has heard how a man who died in a Government think tank had not been wearing protective breathing apparatus.

Frank Ramsbottom, 52, was found dead inside the think tank, at Reading, Oxfordshire, in May of last year. He had been cleaning the tank when the accident occurred.

Ladder

Fellow worker Jack Higgins told the inquiry how he had attempted to pull Mr Ramsbottom out of the tank after his colleague collapsed, but was unable to carry him up a narrow ladder. He was eventually driven back by noxious ideas and ran to get help. Neither men had been wearing breathing apparatus at the time.

Inquest hears how safety rules were not followed

A Government spokesman said it was standard procedure for maintenance men to wear breathing apparatus when entering a think tank. But he could not confirm that the men had been issued with suitable equipment on that occasion. Stringent

safety rules were applied and suitable training given, but he added that it was not always possible to ensure that correct procedures were being followed.

Hose

The think tank was being cleaned out in readiness for a delivery of new ideas and concepts. It was a routine operation carried out every 2 months, and there had been no reported incidents of this type in the past.

Leder

A home office pathologist confirmed that Mr Ramsbottom had died after inhaling a large quantity of toxic thoughts. He said a thin residue of ideas was found on the

bottom of the tank and that Mr Ramsbottom would have died within minutes. The coroner recorded a verdict of accidental death and recommended that procedures for cleaning out think tanks be reviewed in the light of the accident.

Hosen

A man was killed whilst trying to unblock a brain drain at Dublin University last week. Thomas McDonnagh, 27, had lowered himself through a manhole and was attempting to remove leaves and other debris when he was swept away by a torrent of brains. His body was later recovered from the river Liffey.

SHIT THICK!

Shame of D.J.'s who cannot spell their names

Many of Radio One's top disc jockeys are so stupid they are unable to spell their own names. And at least one of the highly paid 'jocks' is TOTALLY ILLITERATE.

These are the shock claims being made by Randy Blenkinsop, 38, who has been a disc jockey himself for over twenty years.

PLANK

"Many of the so-called 'top names' on Radio One are as thick as short planks," Randy told us, speaking from the garden shed which has become home to his booming disco hire operation. "In fact I heard from a very good source that only one daytime DJ in the current Radio One line up has any academic qualifications at all – a solitary CSE in domestic science."

SIMPLE

Randy claims that even the simplest links between records have to be scripted and rehearsed over and over again before the simple jocks can get them right. "One popular DJ had to be sent to night classes before he took over the Top Forty Show. He was unable to read the chart countdown, and had never counted up to forty before."

DAFT

"Every single show is recorded weeks in advance, and it often takes them 10 or 12 hours just to record a simple 3 hour show."

Randy denies that there is any element of sour grapes in his accusations, but admits that he has been refused auditions for Radio One on several occasions. "I've sent them tapes before, but they didn't even bother replying. One of the reasons is probably that I'm tall and fairly good looking. It's a well-known fact in the business that most of the Radio One guys are less than 5 feet tall, and alongside me they'd look a bit daft."

Indeed Randy claims that BBC boffins use special effects to make their DJs appear normal when they appear on TV. "When they do Top Of The Pops you never see their feet. That's because they always stand on boxes. And they always get loads of people to stand around them. That's so you can't see how fat they are." Randy claims that one DJ stands a mere 4 feet 6 inches tall, and weighs in at almost 18 stone. "He has to spend 10 hours in make-up before they allow him on Top Of

Top Radio One DJ 'Diddy' David Hamilton. We have no evidence to suggest that he is unintelligent. However he is quite short.

The Pops. If you met him in the street you'd run a mile," said Randy.

Randy has no regrets having missed out on a Radio One career. "It's their loss, not mine," he insists. "In fact, if they offered me a job tomorrow, I'd probably turn it down. And in any case I'm fully booked doing Christmas discos most weekends from now until January,"

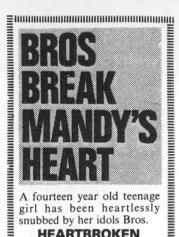

BROS BREAK MANDY'S HEART

A fourteen year old teenage girl has been heartlessly snubbed by her idols Bros.

HEARTBROKEN

Pop fan Mandy Jones was left heartbroken after the incident, and has vowed never to buy another record by the heart throb group.

BIRTHDAY

Mandy, a Bros fan for over 2 years, sent a letter to twins Matt and Luke Goss inviting them to her fourteenth birthday party at her home in Helmsdale, Northern Scotland. But as Mandy's father Bill explained, on the day of the party the Goss twins simply failed to turn up.

"We waited as long as we could, and eventually the party went ahead without them", he told us. "Mandy was in tears. She'd been looking forward to meeting them, and had told all her friends that they'd be there".

DESPICABLE

Mandy has now given away her collection of Bros records, and has vowed never to listen to the group again. "Pop stars simply don't care about their fans, even though it's the fans they owe their success to, said Bill. "It's despicable the way they treat them".

VIPERS

A spokesman for Bros's record company told us that the Goss twins had been in America at the time, and would have had to cancel their tour in order to attend the party.

INTENSIVE COMA WARD.

WHY DON'T YOU COME ROUND FOR A COFFEE SOME TIME?

CABINET MINISTER 'HAD SEX WITH DOLPHIN'

GOVERNMENT SEX SHOCKER!

A bizarre sex scandal involving a leading Tory cabinet minister and a dolphin is set to rock the Government.

Mrs Edith Potter, of Bridlington, North Yorkshire, claims she witnessed a sex act taking place between the top Tory politician and the playful eight foot fish.

WINDOW

Mrs Potter was looking out of her window at the Golden Sands Rest Home, which overlooks the bay, when she spotted the dolphin frolicking in the water.

"A bowler-hatted gentleman who had been standing on the beach, undressed and swam out towards the dolphin", claimed Mrs Potter. She claims that a sordid sex act then took place.

CLOTHES

"He them returned to the beach and put his clothes back on before leaving in a chauffeur-driven car. It seemed to be heading towards London".

Unfortunately there were no other witnesses among the hundreds of holiday-makers on the beach at the time. However Mrs Thatcher is sure to face angry questioning from Labour MP's over this latest controversy to rock her Government's already ailing popularity.

SPOKESMAN

For legal reasons we are unable to name the minister involved, however a spokesman for the Department of Health refused to comment.

Mrs Potter, who is 104, has been resident at the Golden Sands Rest Home since 1958 when ill-health forced her to give up her job at a Teesside glue factory.

In 1963 she was at the centre of the notorious 'Grey Seal Affair', when she attempted to bring a private prosecution against the former Prime Minister Sir Winston Churchill, claiming that sexual intercourse had taken place between him and several common grey seals on cliffs overlooking Bridlington Bay.

OPINION

with **CHARLES PONTOON**

Leave Fergie alone

We are shocked and saddened to hear rumours surrounding the Duchess of York's health. She is only human, and constant criticism of her figure by the gutter press has taken its inevitable toll.

Her figure is her business, and nobody elses.

But her's is a very privileged position, and as a Royal it is her job to fly the flag for Britain. A lot depends on her appearance.

After all, who wants a flabby Duchess lumbering around on state visits and regularly getting her arse stuck in car doors?

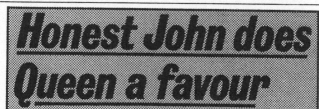

Honest John does Queen a favour

Honest John Burchall couldn't believe his eyes when he opened a handbag he found lying on a London street corner. It belonged to the Queen!

Inside John, who is a heavy drinker, found the Queen's cheque book, keys, credit cards, and personal belongings including photographs, letters and a diary plus over **TWO MILLION** pounds in cash!

"I couldn't believe it", he told us, "I'd never seen so much money in my life". Without hesitation, John decided to return the bag and its contents to its rightful owner. But a Guard on duty at Buckingham Palace refused him entry, and suggested John take the bag to the nearby police station.

POLICE

"The police took the bag from me and I gave them a note of my name and address. I was sure that the Queen would at least ring me to say thanks". But after four months John has heard nothing.

AMBULANCE

"I am extremely disappointed", he told us. I wasn't looking for a knighthood – I didn't even expect a reward,

Tight arsed Royals don't even say 'Thanks'

I just thought a 'thank you' would have been nice, that's all. Mind you, there was so much money in the bag, she might have slipped me a little something for my trouble. Even if it was only fifty quid or something".

FIRE SERVICE

According to John, the most valuable item he returned was probably the Queen's diary. "Inside it there were phone numbers of all the other Kings and Queens in the world, and loads of other famous people too. If she'd lost those she'd have been in a right fix".

BAG

"The more I think about it, the luckier she was that I found the bag. As well as her cheque book and credit

The Queen of England

cards, there were several bunches of keys, among them, keys to the Tower of London. If they had got into the wrong hands, the entire Crown Jewels might have been stolen".

COW

A spokesman for Buckingham Palace suggested that Mr Burchall may have been mistaken and that the Queen had not lost her handbag. "It was probably Princess Di's then", Mr Burchall told us later.

Rude Kid

ALMOST TIME FOR BED DEAR.

Esther's Heart of Gold

GOLD BEATER: Esther Rantzen with her teeth, yesterday.

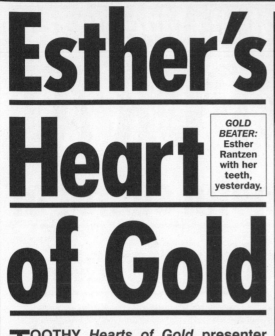

TOOTHY *Hearts of Gold* presenter Esther Rantzen last night received her very own heart of gold at Papworth Hospital in Cambridge.

In an eighteen-hour operation, the former *That's Life* host's own healthy heart was cut out, discarded and replaced with a 24-carat artificial organ, the first such transplant carried out anywhere in the world.

Miss Rantzen was last night recovering at home, where a spokesman described her condition as critical but stable.

GOOD

He told us: "The procedure went as well as can be expected, but Esther's not out of the woods yet. However, if she gets through the next forty-eight hours without any significant complications arising, her chances of survival are quite good."

BLOW

It is thought that Rantzen decided to undergo the grueling op after a routine chest X-ray revealed that her own heart was made out of exactly the same material as everyone else's. "The results of her medical came as a terrible blow to Esther," her spokesman told us. "So when the chance for a

That's Life or Death Op for Esther

solid gold heart came up, she grabbed it with both hands."

ODD

"After all the decades she's spent championing consumer rights on television and setting up the Childline charity single-handed, if anyone deserves a heart of gold, Esther does," he added.

Rantzen is reported to have paid over £1 million for the fully-func-tioning golden organ, which boasts diamond-studded ventricles, ruby-encrusted aortas and a tricuspid valve carved from a solid block of platinum.

EVEN

And unlike the owner of a traditional artificial heart, Esther will never need to change her ticker's batteries. That's because it is fitted with a state-of-the-art clockwork mechanism that can run for over a week on a single winding.

SHAKIN'

"As long as she can get someone to stick their hand up her fanny and pull down on the chain that hoists the counterweight up into her chest every eight days, there's no reason why Esther shouldn't live forever," the spokesman told us.

"Just imagine that. Esther making her judgemental, moralising programmes for all eternity. It's like all the nation's dreams come true at once," he added.

How It Was Done

9.00am
ESTHER Rantzen is wheeled into the operating theatre and anaesthetized.

10.30am
SURGEON makes his first incision, entering the thoracic cavity via the pericardium.

1.00pm
THE functions of Rantzen's heart are taken over for the duration of the operation by an artificial oxygenating pump.

3.15pm
TWO ribs are removed to allow surgical access to the cardiacical chamber.

5.00pm
RANTZEN'S heart is disconnected from its veins, removed and discarded.

7.30pm
THE surgical team takes a well-earned tea break.

8.00pm
IT'S time for pudding. There's a choice of semolina, jam roly poly or sherry trifle.

9.00pm
AFTER tea, the bejewelled, prosthetic organ is brought into the operating theatre on a sterile velvet cushion.

10.30pm
THE aorta and pulmonary veins are offered up and connected using gold jubilee clips.

12.00am
THE secondary arteries are plumbed in and a nurse gives the precious jewels a final polish before the wound is closed.

2.00am
THE gold heart is wound up with a key and started for the first time.

2.30am
THE surgeon inserts the final stitches in Esther Rantzen's chest before she is wheeled out of the operating theatre into the recovery ward.

3.00am
SHE wakes up with her new heart of gold!

CRASH! PAM! WALLOP!

Knocker-hungry Net-Nerds Knicky Knacky Noo. Baywatch Babe's Tits Down for 4 Hours

Showbiz Exclusive

FILMING of hit US TV show VIP was thrown into chaos last night after so many internet surfers logged onto Pamela Anderson's cheeky 'sucking-off-her-hus-band-on-a-boat website' that her tits CRASHED!
~*Reuters*

EURGH! WHATS THAT?

THAT'S JUST WHAT THE DOCTOR ORDURED.

MUTANT HORROR AT

Shameful secret of 'forgotten Royals'

By our investigative reporters
BILLY BOLLOCKS and **MAURICE SHITE**

SCORES of members of the Royal Family have spent their entire lives locked away in institutions, darkened attic rooms, cold damp cellars and dusty broom cupboards, hidden away from the public, we can reveal exclusively.

It's a shameful Royal cover-up that has been going on for centuries. Over the years it is estimated that well over 200 members of the Royal Family, many of them directly related to the Queen herself, have spent their lives locked away out of sight.

GENETIC

Burke's Peerage, the top people's ancestral "bible" lists the names of every Royal going back through the ages. But missing from those pages are the names of the many Royals born deformed, mutated and with genetic disorders, hidden away and forgotten by their peers. Records of their births and deaths are never kept.

UNHUMAN

Among the most shocking revelation is that the heir to the throne, Prince Charles, was not in fact the Queen's first born child. For we can reveal that the Queen gave birth to another son, Reginald, two years earlier. But born a monster, Reginald was locked away in a darkened cellar in the depths of Windsor Castle, where locals claim to have heard his unhuman cries late at night.

Rhino Prince sold off to a zoo

As long ago as 1536 the Royal Family were involved in the cover-up of the birth of a Royal child. King Henry the Eighth produced several children by his many wives, but the record books do not tell of the fire breathing mutant Prince born with three heads, the son of Anne Boleyn. Secretly locked away in the Tower of London, he became a fire hazard and was eventually sent abroad hidden inside a big metal box. But the ship caught fire before it reached France and the un-named Prince went to his watery grave where he remains to this day.

Is this the face of a future King of England?

Prince Derek, born a Rhinoceros, died in 1890.

Perhaps the most grotesquely deformed Royal in unrecorded history was Prince Derek, the first born son of King Edward the Seventh. Born a rhinoceros, Royal doctors ordered his immediate shipment to Africa, for it was believed a rhinoceros Prince would seriously tarnish the image of the Royal Family. But the beast was later sold to Whipsnade Zoo where he died twenty-six years later. In all that time spent confined to a small cage in a darkened corner of the Zoo, Prince Derek never received one Royal visitor, and his death in 1890 went unreported in the press.

TODAY

And **TODAY** the Royal cover-ups continue. Top physicians were called to the Palace to examine young Prince William after he fell and hurt his head playing. Told that his son may be left with a visible scar, Prince Charles considered banishing the child to a cramped attic room for the rest of his days, and elaborate plans for a cover-up

about our missing monarchs
THE PALACE!

Windsor Castle — Its hundreds of attic rooms and darkened cellars are bulging at the seams with mutants and ugly monsters.

were prepared. Auditions for a look-alike to step into William's shoes were well underway when the scheme was suddenly dropped at the last minute after Princess Diana intervened.

The Hulk.

Britain's stately homes echo with the cries of freakish mutants and mutated monsters banished from society. And although rare, there are witnesses who have seen them and lived to tell the tale. Charlie Johnson, a window cleaner, saw a blood curdling beast while cleaning attic windows at Balmoral, Scotland.

MONSTER

"Suddenly a huge, black, hairy monster, slavering blood from its enormous mouth, lunged forward at me from out of the darkness", says Charlie, who almost fell off his ladder in fright. It turned out that the beast had been born into the Royal Family over 100 years earlier, and had been confined to the attic room ever since.

DINOSAUR

Another witness wasn't so lucky. For in 1742 an Electricity Board official called at Buckingham Palace to read the meter. But in the darkened cellar his torchlight fell upon a hideously deformed creature with a dinosaur's head and teeth ten feet long.

The terrified meter reader was sworn to secrecy, given a pension and sent to live in Australia, but he died in mysterious circumstances only weeks later. He was the man who saw to much.

PARKINSON

Nowadays hypnosis is often used to 'wipe clean' the memory of anyone unfortunate to witness an 'uncrowned Royal'. People like TV presenter Michael Parkinson, who discovered a secret room containing unsightly mutated monsters while looking for the toilets at Buckingham Palace, have no recollection of the incident, due to hypnosis.

BAT-LIKE

But despite great advances in medicine, the problem of Royal mutations is on the increase. For Palace officials now fear that fertility drugs prescibed for the Duchess of York could result in the birth of the most horrific examples to date. Possibly frightening monster quadruplets with long, swan-like necks and green bulging eyes, each weighing over six stones at birth, or some sort of bat-like amphibious swamp creature with a razor sharp beak and a wing span of 36 feet.

TELEVISION LICENCE — MONOCHROME ONLY
This licenc **Born on** the last day of
	Title	Initials	Surname
Baby	PRINCE VERNON		
Mum	PRINCESS DI		
Dad	PRINCE CHARLES		

WIRELESS TELEGRAPHY ACT 1949

WARNING: This licence is not valid until it is initialled and properly stamped by the officer issuing at a p... TV Licence Records Office

The birth certificate proving the existence of a third son to the Prince and Priness of Wales

Despite the remarkable lengths to which the Royal Family will go to conceal the births of genetic monster children, we have uncovered revealing evidence of one such birth.
For among rotting food and other domestic refuse left in dustbins outside Buckingham Palace recently, we discovered the tattered remains of a birth certificate recording the birth of a third son to the Prince and Princess of Wales.

Attempts had been made to mutilate the document, which was badly burned, but clearly visible are both the child's and parents' names. What fate has befallen this uncrowned Prince we may never know.

IT's every boy's dream to become a Premiership footballer, walking out onto the turf as 50,000 voices sing his name. But to make it, he will need skill, dedication, team spirit, fitness and at least 110% commitment. Even then, most who set out on the road to glory will fall by the wayside before getting the chance to kick a ball in anger. For many, the disappointment can be hard to bear. Are you destined to join David Beckham and Michael Owen as a footballing hero? Or are you set to follow the likes of Eddie Large and the Pope and become a failed zero. Take our test, answering the questions a, b or c, and tot up your score. Do YOU have what it takes to become a...

PREMIER LEAGUE FOOTBALLER?

1 You go on a romantic weekend to Paris with your girlfriend. How do you show her a good time?

a. *Take her to the top of the Eiffel Tower, and surprise her with a bottle of Champagne.*
b. *Take her for a candlelit meal at a cosy bistro, and afterwards to a show at the Moulin Rouge.*
c. *Take her to a bar and kick her fucking head in.*

2 How many hours work could you manage in a week?

a. *35-39*
b. *40-65*
c. *$1^1/_2$*

3 You've been out partying all night, and you pop into a fast food outlet for a burger, only to be told that you have to choose from the breakfast menu. What do you do?

a. *Order anything that's available, you're so hungry you're not fussy.*
b. *Make your way to another fast food outlet in the hope that burgers are being served.*
c. *Scream that you want a burger cooked by a white man, then start kicking the place to bits.*

4 You are convicted of an appalling racist crime that disgusts the nation. How would you attempt to make amends and win back the respect of your community?

a. *Issue a heartfelt statement expressing deep regret and abject shame at your appalling behaviour, and hope that in time people will forgive your actions.*
b. *Attempt to undo some of the damage by committing yourself to a programme of voluntary work in the ethnic minority community.*
c. *Wrong foot a defender and slot one in at the keeper's near post.*

5 It's your son's birthday. How do you celebrate?

a. *Have a small party with a little cake, inviting his grandparents, cousins and a few friends from school.*
b. *Take him and a small group of friends to the Wacky Warehouse, or for a day out at the Sea Life Centre.*
c. *Have his stupid name tattooed across your arse in Ye Olde English capital letters.*

6 How big are your girlfriend's tits and what are they made of?

a. *Small to medium and made of flesh, skin and milk.*
b. *Medium to large and made of flesh, skin and milk.*
c. *The size of spacehoppers and made of something out of a vat from ICI.*

7 What would be your dream home?

a. *A brand new dull, nondescript two-bedroom Wimpy house.*
b. *An imposing forty-bedroom, Georgian stately home in its own grounds.*
c. *A brand new, dull, nondescript forty-bedroom Wimpy house in its own grounds.*

8 You find yourself out of work in your mid-thirties. What do you do?

a. *Get on your bike and look around your area for any work that's going.*
b. *Go back to college and retrain for a different career.*
c. *Sell crisps.*

9 You are queuing for stamps in the post office when an old lady accidentally nudges into you and catches your shin with her walking stick. How do you react?

a. *Smile and apologise, even though you were blameless.*
b. *Tut under your breath, turn to the person next to you and roll your eyes.*
c. *Hit the ground like a sack of spuds, roll over and over clutching your shin and lie groaning for a couple of minutes before standing up, taking a few tentative 'test' steps and limping theatrically back to the queue.*

10 What did you do on your last holiday?

a. *Spend a gentle week meandering along Britain's Inland Waterways on a rented canal boat.*
b. *Have a traditional seaside holiday, eating ice-cream and building sandcastles on the beach.*
c. *Go to a Spanish hotel and make a hard core porn video featuring you and your mates up to the apricots in impressionable teenage muff.*

11 It's Christmas, and you are doing food shopping in a supermarket. The place is packed, and when you get to the checkouts you find there is a large queue. How do you react?

a. *Stand patiently and wait your turn, it's only like this once a year.*
b. *Abandon your trolley, go home and return when it's less busy.*
c. *Send your wife to the front of the queue to screech "Do you know who we are?" at the unfortunate girl on the till, receiving a ban from every Tescos in Britain?*

HOW DID YOU DO?

Mainly a: *Oh dear, you are no footballer, professional or otherwise. With your two left feet, you probably can't even kick a ball straight. Limit your footballing ambitions to a knockabout on the beach, or occasionally passing the ball back to some kids in the park. Better luck next time.*

Mainly b: *Close, but no orange at half time. You'd probably make a pretty good Sunday league player. You might even get the odd game for a half decent team of part timers like Blyth Spartans, Accrington Stanley or Southampton. But let's face it, you haven't got what it takes to make it as a top-flight pro.*

Mainly c: *Congratulations. The Premiership and a pair of poncy silver boots await. You must be over the moon and all credit to you because at the end of the day, it's the score that counts. You set your stall out early doors, gave it 110% over the full eleven questions and let your answers do the talking.*

The Viz Box for the Year 2000

H marks the spot as he buries

2000

IN A LAVISH ceremony last week in the Viz Italian Sunken Garden, the Viz Box for the Year 2000 was buried by H out of Steps. The time capsule, which contains mementos of our own century, will not see the light of day until the next millennium.

When it is opened, it will show the Viz readers of the future exactly how people lived in our times. The objects placed inside the box have been specially selected to represent all the different aspects of modern life.

Here is what is inside the Viz Box for the Year 2000.

After burying the box, H out of Steps made a map showing its exact location in the Viz Italian Sunken Garden, so that the people of the year 2000 will be able to locate it easily. And he hopes that - if he's still alive - he will be present when it finally surfaces again, in the first issue of Viz in the year 2000.

A jazz-mag.
From our viewpoint here in the 20th century, we find it amazing that our Victorian ancestors used to masturbate frenziedly if they so much as caught a glimpse of a piano leg. And likewise, how soft the art pamphlet of today will seem to the space-age tugster of the future! Armed with his interactive 3-dimensional holographic virtual reality sex helmet, he'll be able to choose from a limitless menu of depraved pornographic scenarios, of a core far harder than we can even imagine, before settling down for a shamefaced shuffle - into a glittery silver sock!

A newspaper.
In the world of tomorrow, the cumbersome papers we know today will be a thing of yesteryear. In the future, computers small enough to fit on a desk will be commonplace in many homes. To catch up on world events, people in the year 2000 will simply hold their hand on a humming, glowing sphere and close their eyes. It will make a 'mmyow mmyow' noise, and an entire newspaper - including the crossword, the racing and the TV (all SEVEN channels!) - will be instantly downloaded into their brain.

A bowl of Weetabix.
Breakfast, like all meals in the future, will come in pill form. A bowl of cereal such as the one in our box will be familiar to the man who digs it up only as a dusty museum exhibit, or a faded picture in a history book.

Paperclips.
Some paperclips. Because everyone will communicate by telepathy in the next millennium, paper will no longer be required. And with no bits of paper to hold together, paperclips will soon find themselves surplus to requirements too. The man of the future who opens the box will certainly scratch his head when he sees these curious little things!

Money.
We are including examples of every coin that is currently legal tender (except the pound and the two pound. And the fifty pee). Shopping in the year 2000 will not involve these primitive, clumsy coins which we take for granted. The shopper of the future will simply have the cost of his purchases debited automatically from his bank account, using a thin magnetic strip, on a piece of plastic no bigger than a credit card.

Half a packet of Lockets.
In the next millennium, disease will be nothing more than a closed chapter in an unread history book on the dusty shelves of a library. On the moon. Wonder-drugs of today, such as AZT, Elasto-plast and Tunes will have long since vanished from the medicine cabinets of the future. Anyone who catches an illness will simply have his head cloned onto a disease-free body. And he won't have to worry about joining a long waiting list for his operation either. The whole process will take no more than 5 minutes, and all he'll have to do is slip a twenty-pence piece into the slot of a 'Clone-Me" booth, in his local Post Office, Railway Station, or Woolworths.

Sad Death of Lucy the Viz Elephant

REGULAR readers of Viz will be saddened to hear of the death of Lucy, the Viz elephant. She became a firm favourite in the late seventies, making numerous public appearances where she gave rides to children, but quickly outgrew her home, a lock-up garage in Huddersfield, and eventually retired from the limelight.

In the mid eighties, Lucy once again hit the headlines when she was found, still in her Huddersfield lock-up garage - but now seriously malnourished and neglected.

She was moved to a slightly larger lock-up garage in Leeds, where she spent a further twelve years, before being chained up and left on a piece of waste ground near Wakefield, where she was found dead earlier this month, after youths had repeatedly driven a stolen Landrover into her legs and pelted her with bricks and bottles.

In a sombre funeral ceremony, cheering crowds paid up to £5 each to watch as Lucy was winched on a crane to a height of over 250 feet before being dropped to the ground.

HOLLYWOOD STARS TO GIVE BIRTH TO ENDANGERED SPECIES

Top scientists believe they have solved the genetic riddle that will enable Hollywood Stars to give birth to animal babies.

After a dramatic breakthrough by researchers at the University of California it appears that American Scientists have reached the summit of the mountain on top of which lies the final piece of jigsaw required to complete the genetic crossword puzzle that has baffled man throughout the centuries – the key to the very Rubik's Cube of life itself.

DNA

Scientists have discovered a method of injecting animal DNA – a sort of biological barcode which is printed on the bottom of all living things – into eggs taken from human ovaries – turning a woman's baby into the animal of her choice.

KLF

And conservationists are hopeful that if tests prove successful, within weeks women could be giving birth to endangered species of animals such as panda bears, tigers and snow leopards.

EMF

Already a host of Hollywood stars are queuing up to receive injections of animal DNA. And top of the list is thought to be Jane Fonda.

In private she has vowed to friends that she will personally have four pandas a year until the species is safe from the threat of extinction. And fellow Scot Sheena Easton has vowed to have at least one buffalo.

ELO

One of the advantages of this new breakthrough in genetic engineering is that Hollywood stars will be able to have the animal baby of their choice – without actually having to go to bed with that kind of ani-

By our Science Correspondent LULU

mal. Unless of course they particularly want to.

Instead human eggs will be removed from the ovary, boiled for two minutes, and then DNA from the chosen animal will be injected into a small hole drilled in the shell.

The egg will then be replaced by surgeons, and a few months later the woman will give birth to a baby animal.

ELP

The exact length of the pregnancy – the gestation period – will probably depend on how long it usually takes that sort of animal to have a baby, multiplied by how long a human takes. Or something like that.

BCRs

It was this same remarkable scientific breakthrough that Stephen Spielberg used as the basis for his blockbuster film Jurassic Park in which previously extinct dinosaurs are recreated by injecting traces of their DNA – found on pine cones – into crocodile eggs.

VCRs

However, it was felt that Hollywood stars such as Elizabeth Taylor, Michelle Pfeiffer, Kim Basinger and Sharon Stone would be reluctant to give birth to anything as large and awkwardly shaped as a dinosaur.

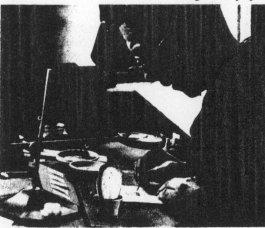

An American scientist at the forefront of genetic engineering research prepares to inject a human egg with mongoose DNA

Big screen sex sirens could soon have "animal babies"

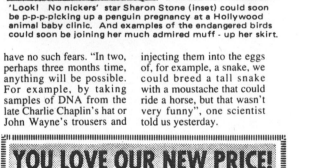

TAYLOR: 'No' to dinosaur delivery

"That would be like trying to get a JCB digger through a car tyre", one leading gyneacologist told us yesterday.

JCBs

But this week as Hollywood's first $25,000 per day animal baby clinic opened its's doors to the stars, there were fears that moral and ethical considerations were being overlooked, and that animal babies could simply become the latest fashion accessory in Tinseltown.

HGVs

Some commentators fear that fickle celebrities will tire of having tigers and that genetic cross-breeding could become fashionable, with stars like Julia Roberts possibly mixing pig, sausage dog and elephant DNA in order to have a pink baby elephant with short legs and a long body.

PSVs

But the Californian scientists working on the project

'Look! No nickers' star Sharon Stone (inset) could soon be p-p-p-picking up a penguin pregnancy at a Hollywood animal baby clinic. And examples of the endangered birds could soon be joining her much admired muff - up her skirt.

have no such fears. "In two, perhaps three months time, anything will be possible. For example, by taking samples of DNA from the late Charlie Chaplin's hat or John Wayne's trousers and injecting them into the eggs of, for example, a snake, we could breed a tall snake with a moustache that could ride a horse, but that wasn't very funny", one scientist told us yesterday.

HOPE SPRINGS ETERNAL

Comic Bob lingers a while on the road to Heaven

VETERAN comedian Bob Hope was celebrating with friends last night as his final days entered their *THIRD* sucessive year. Family and friends of the 98-year-old Hollywood star have been holding a round-the-clock bedside vigil since he began fading fast in June 1999.

By our Hollywood Correspondent

Fanny Batter

Hope's luxury mansion in Bel Air, where his one man death show is packing them in.

Hope's runaway deathbed run has smashed all records. The previous longest demise was set by Frank Sinatra, who spent a marathon *two and a half years* facing his final curtain.

"I knew that Bob's death would be a wow, but I never dreamed it would be such a blockbuster," said his long time friend and agent Hyman Prepuce.

> *"His heartbeat is weak and erratic, his breathing shallow and laboured, but this death is gonna run and run"*

trooper

And as crowds flock to pay their last respects, the demise shows no sign of coming to an end.

"Bob's a real trooper. His heartbeat is weak and erratic, his breathing shallow and laboured, his hearing and eyesight have left him. But this death is gonna run and run."

In fact audiences have risen since January, after rumours that Bob's life would close at the end of the month.

glue

Tickets to the deathbed vigil are changing hands for over $1000 each. A tout in Rancho Mirage, California told us: "This is Tinsel Town's 'must see' demise of the year. It seems like everyone wants to see Bob knocking on death's door."

man

One Beverly Hills friend said this was the best bereavements ever to hit Hollywood: "I've seen Bob on his deathbed 29 times already, and I'm going again to the matinee on Saturday."

natural

Hyman Prepuce refused to confirm that Bob may spend a season passing away on Broadway. "It's a little early to say yet, but nothing is ruled out." he told reporters.

annuation

"There's even been talks about a world tour with Bob passing away for a few months in London's West End. Let's just wait and see."

'Hats Off' to Bridlington!

Cap-Happy Brid Brim-full o' Titfers

'IF YOU want to get ahead, get a hat'... *or go to Bridlington!*

For the popular seaside resort has been voted 'Hat Capital of East Yorkshire' by *East Yorkshire Milliner* magazine.

Editor Ron Gubba said: "There are 2.8 hats per head of population in Bridlington, compared to 2.76 throughout the rest of the region. It may be something in the water, but Bridlingtoners are mad as hatters for their hats!"

'Hat's the way to do it!' A typical Bridlington scene ironically without any hats and (above) Mayor Oyster Jackanory, also ironically without a hat neither.

The town's Lord Mayor, Mr. Oyster Jackanory was delighted with the findings. "It's another pat on the head, or rather *hat* on the head, for Bridlington", he quipped.

"I'd take my hat off to the townsfolk, only it's been eaten by a goat."

–Reuters

The Viz Business Page......

Terror Threat

TERRORISTS IN YOUR BUSINESS are like worms in an apple. They work from the inside, to destroy the infrastructure of the core, eventually blowing up the apple with plastic explosives. And don't think your company is immune. It is believed that for every ten employees working in Britain today, at least *TWO* are *bloodthirsty crackpot fundamentalists*, bent on bringing the business community to its knees.

Just look around your office. It's a sobering fact that one or more of your employees is probably a terrorist.

Gone are the days when he could be easily spotted thanks to his irish accent, shoulder-length hair, Noddy Holder sideburns and wide lapels. Nowadays they come in all shapes and sizes. From the spotty school-leaver in the post-room to the old lady with the tea trolley, anyone on your payroll could be a foot soldier marching to the beat of a terrorist drum in the percussion section of an orchestra of death, with the axis of evil firmly gripping the baton.

An office yesterday - can YOU spot the two terrorists?

canteen

Unfortunately, there's not a lot you can do about the terrorists who already have their feet under the canteen table, but you CAN stop them at the recruitment stage.

Here's a terror-list of 10 tell-tale signs to look out for during a job interview. Check them off to decide if the candidate in front of you is a terrorist.

1 *Is he slightly nervous?*
Remember, a terrorist is looking at life imprisonment in Camp X-Ray if he's found out, so he may appear a little jittery.

WE'RE IN THIS THING TOGETHER

2 *Does he engage in polite smalltalk when he enters?*
A brainwashed zealot bent on destroying western democracy will deviously try to win your confidence by talking about the traffic or the weather, so be on your guard.

3 *Does he have facial hair?*
Everyone knows that modern terrorists like Osama Bin Laden sport huge bushy beards, so the first thing your candidate will do is shave his off. Or will he?

facial hair - yesterday

A fundamentalist killer is a slippery eel, and as a ruse he may decide to leave some or all of it on. Be on your guard for candidates who are clean-shaven, or who have a small, medium or large amount of facial hair.

4 *Is he very smartly dressed?*
As a recruitment executive you might expect a terrorist to turn up to the interview wearing his usual assortment of sandals, sheets and tea towels, gripping a curved dagger between his teeth. But he won't. He'll almost certainly have swapped them for a smart business suit, shiny shoes and a tie. Be on the lookout for all of these, they're a dead giveaway.

5 *Does he arrive on time?*
Anybody who plants bombs for a living will make a point of being punctual. Military precision will have been drilled into him from an early age, so suspect anyone who turns up bang on time, or five or ten minutes early. Or a little late if the traffic's bad.

6 *Is he interested in the pay?*
All terrorists want to know about their salary. Exploding shoes and weapons of mass destruction don't come cheap, so steer clear of any interviewee who expresses curiosity about his pay or conditions of employment. He's almost certainly a fundamentalist murderer bent on killing everyone in your office.

....with Viz Business Editor, Fontella Bass

o UK Offices

1 in 5 of British workforce Al Q'aeda Sleepers

- report

Inhuman Resources - Is the smart bomb taking over from the smart suit in YOUR business?

7 *What are his references like?*
Just like a genuine candidate, a murderous paramilitary may well turn up with an impeccable array of qualifications and references. Beware, as they are almost certainly falsified but impossible to tell from the real thing.

8 *Does he know his subject?*
The terrorist candidate will appear very knowledgeable about the job he is applying for, he may even have anticipated a few of your questions and prepared answers.

9 *Does he ask about holidays?*
Beware of interviewees who betray any interest in their annual holiday entitlement. All terrorists need at least twenty days off each year so they can travel to the Tora Bora caves for more brainwashing from Hussein, Bin Laden and their ilk.

10 *Does he appear ideally suited for the job?*
Remember that the qualities that make a good employee - being a bright, confident and articulate team-player - are also the qualities that make a good terrorist. Give this candidate the job and you're signing a death warrant for yourself and thousands of innocent people including women and children.

Desk of HATE!

Is there a terrorist sleeper in your office? When the man next to you goes outside to light a cigarette, take the opportunity to have a good look at his desk for clues. It may be your last chance. If he's a terrorist, chances are the next time he may be going outside to light his shoe. *Here's what to look out for...*

Explosive devices. There can be no excuse for bringing live bombs of any kind into the workplace. It contravenes every Health and Safety regulation in the book, and explosions can be bad for office morale, and that can seriously impact on your business's profitability. If you're the boss, give him a written warning.

Signed photograph of Osama Bin Laden. It has often been said that we judge a man by the company he keeps. If your co-worker's friend is the single most evil man in the world, you've got to wonder what sort of person he is himself.

Biological toxins. Any member of your staff who brings anthrax spores into work and spends the day spooning them into envelopes and posting them off to international heads of state has got some explaining to do. Quite apart from being highly dangerous, using company time and stationery in this way is a sackable offence.

It is unusual to find large caches of guns and ammunition in british offices. It's possible your colleague may be keeping firearms handy in order to keep rats down in the warehouse, but there may be a more sinister explanantion.

Bombshell at the Palace! Royal soap shock EXCLUSIVE!

QUEEN MUM AXED!

By our TV Soap Reality/Fiction Blurring Correspondents
Dan Shite & Una Pissflaps

"It's news to me" - actress Liz left in the dark

Actress Liz Bowes-Lyons leaving Stringfellows nightclub with Brian Harvey out of East 17 in happier times. Their marriage lasted two-and- a-half hours.

BRITAIN'S favourite soap star the Queen Mum is set to be axed.

In a move that is sure to shock the entire nation the Queen Mother is to be written out of Britain's longest running soap The Royal Family.

The show's producers made the controversial decision to axe their most popular star at a heated production meeting in London last week. It is the latest in a line of outlandish storylines aimed a boosting the show's flagging ratings.

Shocked

In recent years viewers have been shocked by scenes which have included:

The Queen Mum as she is seen by millions of viewers.

* **ADULTERY** by both the Prince and Princess of Wales.
* **DIVORCE** for both Andrew and Charles.
* **DRAMA** as Windsor Castle burnt down.
* **A STROKE** for heavy smoker Princess Margaret.

At one point there was even talk of a first ever gay Royal romance involving Prince Edward, but this was considered too controversial by the show's senior executives.

Pfeiffer

Critics had claimed that recent raunchy plots were a TURN OFF for the show's traditional audience who have been abandoning the soap in their droves. But following Princess Diana's dramatic exit from the soap in August last year - which made front page news and saw TV audiences quadruple - the show's writers have been under pressure to produce even more controversial storylines.

Ma belle

Elizabeth Bowes-Lyon who plays the Queen Mother was not at the Royal Family's London studios yesterday. However the news came as a shock to the rest of the cast.

"We're a close-knit team", said Elizabeth Windsor who plays the Queen. "We work together and socialise together too. In many ways we're like a real family. This will effect us all. It's a big blow, and it makes you realise that none of us are indispensable".

Sont

Greek actor Phillipos Battenburg, who plays the soap's romeo rat Prince Philip, said the Family would not be the same without the Queen Mum. "She's almost as big as the show itself. When I joined in 1947 I had only just got off a boat from Greece, but Liz sat me down, helped me with my lines and made me feel at home. She's always been an absolute darling and a real professional".

Les

Details of the Queen Mum's exit are being kept a closely guarded secret. Three separate endings will be filmed for her final episode, but not even the cast or film crew will know which one is to be screened.

Mots

Last night no-one from the show's production team was available for comment.

DURHAM born actress Lizzy Bowes-Lyon was last night reeling after we broke the news of her sacking from the long running soap.

Qui

Speaking outside the luxury 6 bedroom Essex ranch which she shares with her boyfriend, Gladiator Wolf, she told us she'd be seeking an urgent meeting with the show's producer's first thing on Monday morning to discuss her future.

Vont

"This is news to me", she told us. "I haven't heard anything official because I've been having a few days off. But if this is true, like, then I feel really sad. Not just for me, but for all the viewers who have really

By our Essex Doorstepping Correspondent
KIM SLAPARSE

taken the Queen Mum to their hearts over the years".

Tres

"I only hope they'll consider leaving the door open for her to return at some point in the future. It would be nice for me to be able to go away and do other things, and still pop back to the Family now and then. To do Christmas specials and that, like what Mike Reid does in EasteEnders", she added.

Bien

But Lizzy's future outside of the soap could be uncertain. After 97 years in the same role

casting directors may see her as stereotyped and she could have trouble finding work. In the past other Family stars have struggled after leaving the show.

Ensemble

Sarah Ferguson was fired after a notorious off-screen toe sucking incident with her boyfriend in 1994. Ferguson, who played red haired man-eater the Duchess of York, has since failed to re-launch her career in Hollywood and is now said to be heavily in debt.

Tres

This winter she is booked to appear alongside Chris Quentin and Peter Dean in Puss In Boots at the Empire Theatre, Bridlington.

A sombre faced Elizabeth Windsor arrives at the Royal Family studios for work yesterday. Cast members were said to be stunned by the news.

But there's no return for Princess Diana

THERE could be even more surprises in store for Family viewers as producers ring in the changes in an attempt to rejuvenate the flagging show.

But producers have so far ruled out a return for Princess Diana, one of the shows most popular characters. Family bosses know that her return would be an enormous coup, but her controversial death in a car crash has left script writers with a headache.

Bien

One way around the problem was for her to return by stepping out of a shower, and for her death and all subsequent events to be explained as a dream. But

producers have dismissed the idea as too far fetched.

Ensemble

However a comeback could be on the cards for Princess Alice of Athlone, grand-daughter of Queen Victoria. Insiders confirm that former Carry On star Joan Simms could be handed the role following the real-life death of veteran actress Margaret Rutherford who played the crotchety Alice for over 80 years.

★★★★★★★★★★★★★★★★★★★★★★★★★★★★★★★★ ★★★

All other girls had two – but nature had given Petula Plenty an incredible 24 breasts. Here is Part One of her exclusive and frank life story.

★★★★★★★★★★★★★★★★★★★★★★★★★★★★

'I caused quite a splash in the pool!'

By Petula Plenty

I first began to notice that my body was developing at the tender age of 13. It is an experience that every girl goes through. But I slowly began to realise that I was not the same as all the other girls.

I soon developed a normal healthy bust. But over the next few months I continued to develop.

At first I had two breasts. Then three, then four. By the age of 14 I had twelve perfectly shaped *you-know-whats*.

GIRLS

My schooldays were a nightmare. I was teased by the other girls, jealous of my *many attributes,* or pestered by inquisitive boys. At the age of 15 I ran away from school, cruel nicknames and taunts about my *substantial developments* echoing in my ears.

IN BED

That night I lay in bed and cried myself to sleep.

FELT

I decided to see a doctor. I felt lonely and depressed. By this time my number of *assets* had risen to sixteen. He told me that although uncommon, it was not unusual for a girl my age to be so *well endowed*.

FRIEND

This gave my confidence a little boost and I accepted a friends offer when she invited me to go swimming. But I soon ran into problems.

UP

My *abundant attributes* meant that a normal two piece swim suit was out of the question. I had to buy another SEVEN bikini tops just to cover up my *extraordinary assets*. As you can imagine, I caused quite a splash in the pool.

I decided to look for a job. But finding work is hard enough if you've got 24 'o' levels, never mind 24 *you-know-whats!* For by this stage I was fully developed and 24 was indeed how many I'd got.

'It is not unusual for a girl to be so well endowed' doctor told me

I got a job working in a lift, pushing all the buttons that make it go up and down. But I dreaded the mornings when men would cram into the lift on their way to work. It was always a *tight squeeze,* and my *considerable developments* didn't make closing the doors any easier. The men seemed to enjoy it, but I felt dirty an ashamed.

BED

I would go home from work tired and emotionally drained, and at nights I would lie in bed and cry myself to sleep.

I had an unpleasant experience at the age of 17. I was being followed by a stranger hoping to inspect my *cumbersome credentials*. I began to run and so did he. Suddenly I was trapped in a dark alleyway. I thought I was done for.

REVEALED

But God must have been watching over me that night, because at that very moment my bra strap snapped and *all was revealed!* My would-be attacker suffered a massive heart attack and died immediately.

LAY

Despite my good fortune it had been a truly frightening experience, and that night I lay in bed and cried myself to sleep.

Like any girl I enjoy dressing up and buying pretty clothes. But shopping is a nightmare for a girl with so many *statistics* I couldn't fit into the fitting room, never mind the clothes.

The few dresses which I could buy had to be specially altered so that I could wear them upside-down.

TOOK ME

Although they had always stared or wolf-whistled at me, I had never had a real boyfriend. That was until I met Peter. He was kind and gentle, and he didn't seem to mind my *plentiful proliferations*. He took me to the pictures one night and had to pay for three seats. One for him, one for me, and one for my *prolific proportions*. I don't know whether he enjoyed the film, but he certainly had his hands full that evening!

NEXT WEEK ~

How I discovered sex, a woman does her best to seduce me, and my bra snaps again.

20 THINGS YOU NEVER KNEW ABOUT WOOD

A wooden table and chairs yesterday

Wood you believe it! Wood is back with a bang. There was a time when everything was wooden. Houses, wardrobes, shoes (in Holland) and strips on the side of certain motor cars. But in recent years the price of wood has *sawed*, and manufacturers have turned to other materials, such as plastic and tupperware, for making things out of.

But now wood is making a comeback, with wooden doors, drawers and cupboards more popular today than at any time in the past. Probably.

But what exactly is wood? What are the facts about this fibrous, knotty substance we see all around us? Here are twenty things you never knew about wood.

(1) Money doesn't grow on trees, or so the saying goes. But wood does! Believe it or not a tree is nothing more than a giant flower, made out of wood. And with leaves on it.

(2) Saying 'you can't see the wood for the trees' is like saying 'you can't see the school for the fish'. That's because a group of trees **IS** a wood!

(3) As indeed a group of rhinocerouses is a 'crash'.

(4) As indeed as well a group of gorillas is a 'flange'.

(4) Cave man first discovered wood's potential as a manufacturing material, making crude wooden clubs to hit dinosaurs with.

(5) The Queen Mother's teeth, originally a wedding gift from Sir Henry Moore to King George VI, are made entirely out of wood.

(6) Many of today's top pop stars take their names from wood. For example Ronnie Wood. And Phil Oakey.

(7) And so do many sports stars, like tennis player Arthur Ashe, and cricketer Derek Underwood.

(8) We've just thought of another good one for 6. Courtney Pine.

(9) The word 'wood' can be used twice in one sentence. For example. "Wood you pass me the wood please, Woody (out of the Bay City Rollers)". In fact that's three.

Woody out of Bay City Rollers

(10) You might expect a bread board to be made out of bread. But you'd be disappointed. It is in fact a round, flat board for cutting bread on, made out of wood.

(11) Similarly a cheese board is not made of cheese. But it's not necessarily made out of wood either. We know someone who's got one made of marble.

(12) Mind you, they're *usually* made of wood.

(13) The Gas Board is definitely not made out of wood. Rather, it is an administrative body responsible for the sale and distribution of domestic and commercial gas supplies.

(14) The Greek word for wood is 'xylos', meaning xylophone or literally 'wooden piano'.

(15) It may not look appetising to us, but wood is eaten and enjoyed by many animals, among them beavers, whose revolving 'saw-like' teeth can fell six trees in a minute.

(16) Some animals don't *eat* wood, but are actually *made out of it*. For example the Wooden Horse of Troy, a large wooden horse, in Troy. And woodworms,

K £**39**.99

A wooden cupboard with shoes in it

small worms made out of wood.

(17) A 'clothes horse' is *not*, as you might expect, another type of horse, made out of clothes. It is in fact an indoor washing line sort of thing... made out of wood!

(18) And a 'saw horse' is not a horse made out of saws. In fact it's not a horse at all. But it *is* made out of wood.

(19) A 'horse fly' isn't a horse either. It's a fly that eats horses! (But that *isn't* made out of wood.)

SPACE AGE RUG IS GOOD NEWS FOR SLAP HEADS

Top hair boffins believe they are on the verge of a major breakthrough in hair-piece technology.

For they now believe that by the year 2000 they will have perfected a wig which *doesn't look like a wig*.

To date, despite millions of pounds spent each year on wig research, the world's top scientists have been unable to produce a half decent toupe.

GINGER

"The problem is that all wigs are detectable, due to their being made out of coarse ginger nylon, and standing an imovable half inch proud of the scalp", says Professor Heinz Ravioli, senior tricologist at California's Humphrey Bogart Wig Foundation.

SHUTTLE

But now, as a result of new Space Shuttle technology, pointing and laughing at people wearing obvious wigs is set to become a thing of the past. And demand for the new "Space Wig 2000" is expected to be enormous, with Britain's baldies queuing up to pay over £15,000 per rug.

LOOM

The new hairpieces will come as a Godsend to top showbiz baldies like Paul Daniels, Terry Wogan, and Bruce Forysth. For despite thousands of pounds spent on advanced hairpieces and transplants, their careers are all too often held back by unsightly and unconvincing syrups.

BOBBIN

The new wigs are expected to become available over the next 2 years, and advance orders are already being taken, not only from baldies, but also from people who can expect to be bald by the year 2000. People like newsreader Jeremy Paxman. However, Professor Ravioli was at pains to point out that due to technical and production difficulties, early versions of the wig would only be available in one colour – bright ginger – and in small size only.

SPINNING JENNY

"The colouration of the wigs may be affected by sunlight, so they should only be worn indoors, and under no circumstances should they be allowed to come into contact with water. Or heat", said the Professor yesterday.

Showbiz baldie Bruce Forsyth (left) and Judith Chalmers of TV's 'Wish You Were Here'.

SPOT THE BALDY AND WIN £100!

Soon cheap laughs at the expense of sensitive baldies will be a thing of the past, and the nylon rugs that they balance precariously on top of their heads will be museum pieces.

So this could be your last chance to go WIG SPOTTING. And we're offering bumper cash prizes, plus hair products, to the spotters of the most obvious wigs.

All you have to do is load up your camera and wander the streets in search of hairpieces. When you spot a hidden baldy, take a candid snapshot of his (or her) laughable toupe, and send the results to us.

The photos of the most ridiculous hairpieces will appear in our next issue (if anyone bothers to send us any) and the sender of each snap used will receive £25, with the lucky winner collecting £100 cash, a comb PLUS a year's supply of shampoo and conditioner.

Send your photographs to Viz Wig Spotting Competition, P.O. Box 1PT, Newcastle upon Tyne, NE99 1PT. Please write your name and address clearly on the back of each photo, together with your hair type (e.g. dry, normal, greasy). Photographs which are returned will have no chance of winning a prize, for technical and administrative reasons.

NIGHTMARE OF BRITAIN'S GOLDFISH

THE MAGAZINE THAT **Viz** CARES ABOUT GOLDFISH

Millions of goldfish in Britain are being kept in appalling overcrowded conditions by heartless owners.

And animal welfare experts fear that each year thousands of helpless fish **DIE** as a result.

BOWLS

Goldfish, often bought as gifts, or won as fairground prizes, are regularly kept in bowls which are:

★ **TOO SMALL** for the fish to swim around in.

★ **TOO LIGHT** for the fishes eyes, and

★ **TOO DIRTY** due to over-feeding.

According to experts one gallon of water is needed for each inch of a goldfish's length, **not** including its tail. And too much light causes discomfort, as goldfish are unable to close their eyes.

BATS

Fish owners were also warned that over-feeding can **KILL** fish, as uneaten food quickly dirties the water. "Goldfish should be kept in clean water, away from direct sunlight, and fed a small amount of food once daily", said one expert.

A spokesman for the Ministry of Agriculture and Fisheries said no figure was available for the number of goldfish in Britain which die each year as a result of maltreatment by uncaring owners. But some animal welfare experts we spoke to feared that the figure could be as high as 250,000. Or even half a million.

STOMPIES

Controversial actress Emma Thomson said yesterday that she spoke from her heart when she pleaded with goldfish owners to look after their fish properly. "People should not have goldfish as pets unless they are prepared to treat them with the respect and dignity which they deserve", she said.

Singer Chrissie Hynd joined in the chorus of support for goldfish. "They have just as much right to be on this earth as we have, if not more", she said. And she asked people to boycott pet shops where fish are kept in unsatisfactory conditions.

How are Yo

We say 'potato', they say 'potato'. They call a slapper a tramp, a tramp a bum, a bum a fanny, and a fanny a puh-seh. In 1946, US war hero John Wayne said that Britain and America were two countries separated by a common language. This may have been true then, but is it still true today? Thanks to the Internet, Concorde and cordless phones the world is shrinking, and American culture is increasingly an everyday part of the British way of life. With our star-spangled diet of McDonalds food, Coca-Cola and the Disney Channel, it is almost inevitable that we will soon become the 51st state of the Union. But are YOU ready to be a yank? It's time to get off your hoss, drink your milk and ask...

ANSWER the following questions a, b, or c. Tot up your score at the end and see how you did.

1 You decide that your relationship with your partner is over. How do you break the news that you are leaving?

a. Leave a tearful note on the kitchen table and slip away in the night.
b. Sit down with your partner and calmly discuss the reasons for your decision.
c. Attack him with a chair in front of a rabble of cheering, pumped-up trailer-trash vermins, on national television.

2 You are visiting Egypt and are concerned over the recent terrorist attacks on foreign nationals. What do you wear to remain inconspicuous?

a. A tee-shirt and a pair of jeans.
b. A Demis Rousoss tent dress, fez, a false beard and sunglasses.
c. A high-rise baseball cap, trainers with knee-length socks, an horrendous flowery shirt, Eric Morecambe shorts and 8 cameras.

3 Where are you most likely to find your local copper?

a. Outside his police house in the village, mending a puncture on his bicycle.

b. Asleep in his patrol car on a motorway flyover.
c. On his yacht, wearing a pastel suit with the sleeves rolled up , feeding his pet crocodile.

4 You are the political leader of your country. An interviewer asks you a question on foreign affairs. How do you respond?

a. Knowledgeably, addressing the issues and answering all the points.
b. As best you can, deftly steering the conversation towards topics on which you are better qualified to speak.
c. Stand there grinning gormlessly, then throw up on the Japanese prime minister, before going home and getting sucked off by a fat-titted intern.

5 You fancy a night in watching something funny on the telly. What kind of comedy show do you choose?

a. A sitcom like Fawlty Towers or Father Ted.

b. A sketch show like The Fast Show or Smack the Pony.

c. A thinly-disguised morality play set in a massive lounge where the audience whoop for ten minutes every time an overpaid actor makes an entrance to deliver a lightweight wisecrack.

6 Your fourteen-year-old son is going through a difficult phase. He is becoming disruptive at school and reclusive at home. What do you do?

a. Don't worry. It's just a phase he is going through. You were the same at his age.
b. Encourage him to get out and about more. Perhaps join a youth club or get involved in some team games.
c. Take him to the local supermarket and buy him an arsenal of semi-automatic weapons and enough ammunition to kill a small town.

7 You and your mates decide to have a game of football in the park. What do you need to take?

a. A ball.
b. A ball and two coats.
c. A ball, 50 crash helmets, 4 tons of body armour, 20 cheer leaders, a marching souzaphone band with a grand piano on a trolley, and a team of orthopaedic surgeons specialising in spinal injuries.

8 Whilst getting ready for bed, you stub your toe on your wife's dressing table. What do you do?

a. Shout and swear a little, after all it did hurt and you didn't have your slippers on at the time.
b. Make a mental note to move the table as soon as possible to prevent it happening again.
c. Immediately call a hotshot lawyer with an uptown reputation and sue your wife's ass.

American OU?

cakes with maple syrup, a dozen waffles, five corn dogs and a diet root beer.

12 What sort of car do you drive?

a. A small economical runabout.

WAM 545

b. A medium-sized family saloon.

c. A forty-foot long chromium-plated jukebox that does 1 mile to the gallon.

13 You and your partner decide to take the plunge and get married. What sort of ceremony do you have?

a. A quiet little do with a few friends in a registry office.

b. A church service followed by a traditional reception at a fancy hotel.

c. A minute-long mockery at a 24-hour drive-thru chapel in Las Vegas, presided over by a transvestite vicar dressed as Elvis.

How did you 'Yankee Doodle Do?'

Mostly a's

You are in no way American. You probably still spell colour with a 'u' and call your trousers 'trousers'. Try wearing a baseball cap and driving on the wrong side of the road a little.

Mostly b's

Good try, but no kewpie doll. You're halfway there, but you could still do better. Why not put a little white fence around your garden and ask the postman to put your letters in a bread bin on a stick.

★ ★ ★ ★ ★

Mostly c's

Well hot diggety, you're as American as Uncle Sam with sassafras on rye. You were born on the 4th of July and you've got Mom's apple pie and napalm coming out of the buns of your ass. Take the fifth and have a nice fucking day, y'all.

9 You are driving along a country road one day when you accidentally run over a rabbit. What do you do?

a. Stop and see how badly injured it is, taking it to a vet if it's still alive.

b. Carry on driving, but hope that it is still alive, or If not that it died quickly.

c. Strap it across the bonnet of your car and drive home hollering and whooping, throwing empty Budweiser cans out of the window.

10 You wake up one morning with a rather stiff neck after sleeping in an awkward position. What do you do?

a. Ignore it, it will probably loosen up as the day goes on.

b. Take a couple of aspirins and get on with things.

c. Take yourself to a prostitute-addicted televangelist faith-healer in an ill-fitting wig, who will lay his hand on your forehead, whilst screaming about the devil in front of an audience of gibbering inbreds.

11 What do you have for breakfast?

a. A bowl of cornflakes, a slice of toast and a mug of tea.

b. A glass of orange juice, a croissant and a cup of black coffee.

c. A bag of donuts with ice cream, a 32 ounce steak with six eggs sunny-side-up , fifteen pan-

RAILWAY ENTHUSIAST RECRUITMENT

GNER

Head Trainspotter
£32-40k   Ref: H20/33

As part of their Care in the Community scheme, Great North Eastern Railways are looking for a young enthusiastic trainspotter to work the East Coast mainline between York and Newcastle. The sucessful applicant will have:

- **A notebook and pencil**
- **An uninspiring personality**
- **Chronic acne**

He will be a static individual with at least 4 years experience crossing out numbers in a book. A 6-year-old stained anorak and a super pair of binos will be provided.

Apply in writing stating your favourite class 37 locomotive to GNER, York Station, York, YO1 1AB

Are YOUR Trousers 4 inches too short?

A vacancy has arisen at Derby railway station for a

Grade 3 Railway Enthusiast
28,500 pa (inc sandwiches and flask)

The sucessful applicant will live with his mother, have no social skills and be completely unable to interact with the public on any level. He will be required to stand at the very end of platform 4 in a kagoul with the hood up, squinting at trains.

Send a current CV to The Station Master, Derby Train Station, Derby

---

Railway Enthusiast, Technical Class 4   £45-48k
(inc London weighting)

Based at Euston Station. Duties will include speaking the numbers of trains into a dictaphone in a monotone voice and videoing trains coming into the station and writing the numbers down later on. The sucessfull applicant must be familiar with the use of dictaphones and video cameras and must be able to ignore the accusations of more traditionalist railway enthusiasts that his method is 'not proper spotting'.

Send a long and pedantic cv to: The Station Manager, Euston Station, London

---

- **Are you tired of people who don't know the Flying Scotsman is a scheduled service and not a *TRAIN*?**
- **Are you infuriated by idiots who can't tell the difference between a Deltic and a Class 55?**
- **Do morons who don't understand railway procedure make you *SICK*?**

Railway Enthusiast Grade 5   £42-45k

- *Are you at least four stone overweight?...*
- *Are your glasses held together with a plaster?...*
- *Are you on the Sex Offenders register?...*

If you can answer *YES* to all these questions, then you may be the man we are looking for.

West Coast Rail are looking for a middle-aged-bachelor to head a team of social misfits writing down train numbers at Crewe Station. The sucessful candidate will smell of salt and vinegar crisps, and will have a proven track record of getting very excited when a train comes in. A load of limp, margarine sandwiches in a tupperware box would be an advantage.

For an application form, get your mum to call 0900 800 900, and ask her to quote Ref: 27/001

SWR ⁄⁄⁄⁄⁄

Senior Spotter
92hr/wk   **£42k**

There are trains coming in and out of the station all day. How many of their numbers can you write down?

Is it 80%?... 90%?... **99%?...**

...Yes?

Don't even fill in the form.

South Western Rail are looking for a socially inept man in his early twenties who is committed to standing hunched on the end of a platform at Tiverton Parkway station. The sucessful candidate must be serious about railway procedure will have thick glasses, a facial tick and Aspergers Syndrome. If you think you've got what it takes to write down train numbers all day in all weathers, then we'd like to hear from you.

Write to: Personnel Officer, South Western Rail, Beeching Street, Tiverton. Please quote ref: 30/00.

99.99% NEED NOT APPLY

Great Western Rail
Principal Train Spotter   £40k pa

Great Western Rail are looking for a 52-year-old paedophile to stand with flecks of saliva at the corner of his mouth, masturbating in his trousers whenever a class 47 Deltic comes past. Severe BO and a wonky eye would be an advantage, but full training in not washing will be given.

For an application form call 0898 000 700

...then WE want to hear from YOU

Ab-Fab star in Peg-Leg Rumour Scare

THE WORLD of showbiz was rocked last night as uncorroborated rumours began circulating that one of Joanna Lumley's legs may have been amputated. Details were sketchy, but unsubstantiated suggestions that the Avengers star had possibly lost her left leg above the knee were greeted with horror by celebrity insiders.

Fears for the popular actress's wellbeing were further fuelled by groundless speculations that the limb had been surgically removed as a result of thrombosis perhaps caused by the glamorous star's years of possible heavy smoking.

Joanna

"If this is true, then it would be a terrible blow for Joanna," said former New Avengers colleague Patrick Macnee. "But if Joanna has indeed undergone amputation, I would imagine that she would be in good spirits after surgery at the Portland Clinic. I dare say she already might be sitting up in bed, laughing and joking with the nursing staff," he hazarded "She's a fighter, and I'm sure she would almost certainly bounce back from a putative tragedy such as this."

Meanwhile Lumley's agent Cunnilingus O'Hara was remaining tight-lipped and refused to be drawn on his client's condition. "There is absolutely no truth in these rumours," he told reporters. "She's with me now, there's nothing wrong with her legs and she's got two of them," he added, further fuelling speculation that Lumley had indeed possibly lost a leg.

• In 1980, Good Life star Felicity Kendal was rumoured to have lost her arse following a bizarre accident in Selfridges, London. However, stories of the actress's buttocks becoming entangled in the mechanism of an escalator and having to be removed by paramedics were later proved to be false. In fact, later that year she won the coveted 'Rear of the Year' title.

Quakey Shakey!

Stevens - shakin.

LONG-DORMANT pop star Shakin' Stevens may be set to explode back into activity, according to seismologists who have been monitoring him.

The Welsh rocker, who topped the charts throughout the eighties with hits such as *Green Door, This Ole House* and *Marie Marie*, stopped shaking and became still in the early 90's, and has remained inactive for nearly a decade.

tremors

However, Stevens' concerned neighbours recently called in experts after they felt tremors which dislodged pictures from walls and caused pottery to rattle.

"It may be nothing, but it could be a warning that Shaky is becoming active again," said Professor Kid Chocolate of Swansea University's department of geophysics.

squirm

"If this shaking continues, there could be a devastating full scale eruption, with villages, towns and even cities laid waste under millions of tons of red hot lava."

CURSED!

Tragedy stalks the cast of favourite TV show

A catalogue of tragedy has befallen the cast of one of Britain's favourite TV shows. And it has lead to claims that there is a CURSE on the popular TV comedy series Dad's Army.

For ten years actor JOHN LAURIE played eerie undertaker Fraser in the series. In 1985 he attended one final funeral – his own.

In 1987 the acting world was shocked by the death of JOHN LE MESURIER. He was 75. He had brilliantly portrayed the platoon's mild mannered Sergeant Wilson.

Tragedy struck again when Le Mesurier's former wife, fellow actress HATTIE JACQUES, also died.

Jacques' on-screen brother ERIC SYKES has himself fallen foul of the curse. He suffers from a perforated eardrum.

For no fewer than **SIX** of the stars have died tragically in recent years.

TRAGIC

Based upon the adventures of a bungling platoon of Britain's Home Guard during the second world war, Dad's Army has constantly topped the viewing figures with regular audiences of over 17 million. But success has now been overshadowed by a tragic series of events which have befallen the cast, crew and even the programme writers.

DEATH

Everyone connected with the series was shocked and saddened by the death of actor James Beck in 1973. He had been well known for his portrayal of cocky Private Walker. But it became a double tragedy when shortly afterwards, in 1977, Edward Sinclair who played the verger Mr. Yeatman, died after an intermittent illness.

BAD LUCK

Over the following ten years no fewer than FOUR more members of the cast were to die in what, on the face of it, were unconnected incidents. Arthur Lowe, Arnold Ridley, John Laurie and John Le Mesurier all having passed away. And bad luck has befallen many surviving members of the cast.

INJURED

Clive Dunn, famous for his catch phrase "Don't panic!" suffered a minor accident while gardening, and has since retired to Portugal. Ian Lavender, the youngest member of the cast, injured a thumb while installing central heating pipes at his home. No longer the fresh faced "stupid boy" he portrayed as Private Pike, time has taken its toll on grey haired Lavender who has not always found acting work easy to find.

ROLL CALL OF DOOM

Of these smiling faces only five remain alive today. (Back row, left to right) John Laurie (died 1985), writer Jim Perry, Arnold Ridley (died at his home in Norwood, Middlesex, 1984, aged 88), Bill Pertwee as Warden Hodges, James Beck (died 1973), writer David Croft, Ian Lavender. (Front row, kneeling) John Le Mesurier (died 1987), Arthur 'Captain Mainwaring' Lowe (dropped dead in his dressing room at Birmingham's Alexandra theatre, 1983) and Clive Dunn who, after recording the hit record 'Grandad' fled to Portugal to start a new life.

It is not only the actors in the jinxed series who have suffered. One BBC lighting technician, who preferred to remain anonymous, spoke of nightmarish technical breakdowns and hitches that haunted the show throughout its 11 year history.

SUFFERED

"On one occasion an electric generator we were using broke down and it took us two hours to find a replacement. And on another occasion a van being used to transport costumes suffered a flat tyre. No sooner than we'd fixed it, the fan belt broke".

DISASTERS

Many insiders are convinced that it was disasters like these which eventually led to the programme being taken off the air in 1977. But despite the fact that its run ended 13 years ago, Dad's Army is still proving to be as unlucky as ever.

DAMAGE

"During the recent repeats of the series a continuity announcer who was possibly due to introduce the programme the following week was involved in a car crash. Although he wasn't injured, his car suffered considerable damage", our source revealed.

TRAGEDIES

Since creating the hit series, writers Jimmy Perry and David Croft have also been involved in a string of tragedies. These have included "It ain't 'alf hot mum', 'Are you being served?' and 'Hi-De-Hi'.

ESTHER VICTIM OF GOLD HEART HOAX

A CONMAN has tricked *Hearts of Gold* presenter Esther Rantzen out of more than a million pounds after promising to fit her with a solid gold prosthetic heart.

Rantzen paid a "significant seven-figure sum" to the fraudster after seeing an advert for a fictitious gold organ transplant company on the internet. After being shown the golden heart and handing over the money, she was apparently driven in a windowless van to a lock-up garage which she was led to believe was the world-renowned Papworth Hospital.

She was then placed on an operating table and anaesthetized; however no surgery took place. When she awoke, she was sent home. The hoax only came to light more than a week later when one of Rantzen's assistants went to wind up the clockwork organ, but was unable to locate a chain.

FAKE

It was then discovered that the foot-long operation scar on the TV host's chest was a fake that had been drawn on with an eyebrow pencil.

ESTHER EGG ON FACE: So sad Rantzen yesterday.

Police later raided a property in Peterborough, where they recovered a large quantity of cash, an eyebrow pencil and a grapefruit covered in Ferrero Rocher wrappers.

FRACKER

A spokesman told us: "Esther feels very foolish. But she hopes that her story will serve as a warning to ordinary members of the public. If you are offered a golden organ transplant, make sure you check your surgeon's credentials before you hand over any money."

R.I.P. (Rest In Paradise)

Rolling Stone gathers swish £250,000 grave

SHOWBIZ EXCLUSIVE

Rock'n'Roll senior citizen Mick Jagger has splashed out a quarter of a million pounds on a luxury grave on the sun drenched Caribbean island of Monserrat.

Millionaire Mick will be in Heaven when he moves into his magnificent Montserrat grave. Our artist's impression shows how it might look.

And millionaire Mick is set to splash out a cool half million more lavishly converting the graveyard gaff into a tomb fit for a king.

Plot

The Monserrat plot, in an exclusive corner of the island's most prestigious cemetery, brings to five the total number of final resting places owned by the Rolling Stone. Jagger, now 72, bought his first grave in 1964, paying just over £2000 for a modest plot in his local cemetery at Richmond in Surrey. Since then he has added a small crypt in the Highlands of Scotland, a lavish $2 million marble tomb in the Belle Air district of Beverly Hills and a small weekend urn on the West Bank in Paris.

Mad about graves - Mick looks forward to being committed - to the Earth!

Keg

This latest addition was an impulse buy made during a tea break in the recording of the Stone's latest album 'Voodoo Lounge'. "Mick was taking a break from recording when he just happened to drive past the cemetery. He saw the grave and just fell in love with it", a Stones insider told us.

Kex

"It wasn't for sale but he made an offer the owner couldn't refuse." The previous occupant was exhumed and moved out that afternoon.

After splashing quarter of a million on the grave itself Jagger will now lay out twice as much again on lavish refurbishments before it is ready for moving in. "Mick has his own very personal tastes", our source told us "and he'll want to get it just right, whatever the cost". Jagger is rumoured to have spent over £80,000 revamping his Richmond grave recently before sacking the grave digger and ordering the work to be carried out again.

Shreddies

Mick has told showbiz pals that after his death he intends to spread his time between his graves, spending a few months of the year in each. "Knowing Mick he'll still get around a bit after he's gone", our source confirmed. "But he's really a home loving man and I think he'll spend most of his time in his Richmond plot near his family and friends."

What a plot he's got! Mick snapped up his favourite final resting place in Richmond (above) for only £2000 in 1964. But he's also planning to push up daisies at the remote Scottish church yard below where he owns a magnificent detatched stone crypt set in 10 yards of grass.

BLAST ROCKS BRIT POP AWARDS

THE BRIT EKLAND Pop Awards ended in confusion last night after a bottle of home made ginger beer exploded showering tables with broken glass. The ceremony, which was taking place in a function room above the Red Lion pub in Watford, had to be abandoned in the chaos which followed.

Film star Miss Ekland, 58, had been announcing nominations in her annual awards for soft drinks when the bottle burst on a table behind her.

She was clearly shocked by the explosion but bravely attempted to carry on. Shortly afterwards she paused and appeared unsteady on her feet. She was then lead away, clearly in some distress.

The awards were launched by Miss Ekland in 1967 in recognition of her favourite fizzy drinks, and ran for twelve successive years until they were abandoned in 1979 due to lack of media interest. Since then Miss Ekland has

lead a vigorous solo campaign for their reinstatement and last night's awards were to have marked their return after an absence of 17 years.

It is not known whether the ceremony will be rearranged. Less than half the invited audience of 23 had attended, and a spokesman for Miss Ekland said the envelope containing winners names had been lost in the mayhem following the explosion. Brit Ekland was last night unavailable for comment.

Brit Ekland - left in tears after pop explosion wrecked ceremony.

SEX AND THE CITY-CENTRE CAR PARK

IF YOU'RE looking for sexy stars, you'd probably think of heading for Hollywood or trawling the topless beaches of St Tropez. The last place you'd expect to find them is in an NCP car park in northern England. But you'd be wrong. From Marilyn Monroe to Madonna, from Jayne Russell to Jordan, car park attendant *George Mintoe* has checked all their tickets.

And in his new book *'Multi-Story Star Park'* (Weetabix Books £12.99) 58-year-old George lifts the lid on the secrets of his celebrity-studded life at the sharp end of one of Newcastle's busiest multi-storey car parks. And his X-rated tales of saucy stars in their cars are set to raise more than a few eyebrows in the worlds of showbiz and car parks.

Early in the book, Goeorge reveals how he was expecting his first day working in the ticket booth to be dull. But his first visitor gave him the surprise of his life!

Monroe - almost stripped naked in George's booth

> It was the first day of my new job and I was feeling a little bit nervous. My mum had made me a flask of tea and I was just pouring a cup when I heard a tap on the window. When I looked out I couldn't believe my eyes. It was Hollywood sex goddess *Marilyn Monroe*! She told me she was opening a pound shop nearby and was running a bit late. She wanted to know if she could use my ticket booth to change into her bikini.

Just imagine how I felt. I was a 16-year-old lad and here was the world's most beautiful woman asking if she could strip naked in my

"I was a 16-year-old lad and the world's most beautiful woman was asking if she could strip naked in my booth"

booth. But deep down I knew that allowing non-company personncl into the booth was strictly against the rules, so reluctantly I had to refuse.

I often wonder what might have happened if I'd let Marilyn in through that door all those years ago. Perhaps she might have lost her balance whilst taking her knickers off and accidentally pushed her bosoms into my face. Who knows.

But I've got no regrets, believe you me. Those are the rules, and NCP car parks makes no exceptions for stars, no matter how sexy they are."

Marilyn was George's first brush with a glamorous celebrity, but she certainly wasn't his last. Later in the book, he reveals how a mix-up over coins very nearly led to a steamy encounter with a British sex siren:

"The Romps start when the engines stop" says attendant George

"It was December 1976 and the top deck of the car park was covered with snow. However, I was quite cosy in my booth as I had a chitty from NCP head office to turn on the second bar of my electric fire. Suddenly, there was a tap at the

George - offered herself on a plate

window. I looked up to see the gorgeous *Susan George*. I recognised her immediately because I'd just been reading an article about her in Titbits.

I slid the glass partition back an inch or two. She told me she needed an 80p ticket, but she only had a pound coin. She wanted me to change her pound coin for five 20s. I explained I wasn't allowed to give change and pointed out the 'No Change Given' sign in the window but she was very insistent.

fingering

She started fingering the zip on her fur coat, telling me she would be "very grateful" if I would waive the rules this once. She certainly made it very obvious what she was offering in return for the change she wanted.

Believe you me, I'd be lying if I said I wasn't tempted. However just at that moment one of the other attendants came back from Greggs with some pasties and she went off to get some change in a sweet shop.

grateful

Susan George was offering me it on a plate and I often wonder what would have happened if my colleague hadn't returned at that moment. However, one thing's for certain. I should be grateful. If he hadn't come back when he did, I could have thrown away my whole career with NCP car parks for the sake of a few hours of unimaginable sexual ecstasy with one of the world's most desirable women."

Most people have seen the film 'Body of Evidence', but few realise that the movie's notorious car park sex scene was shot not in Hollywood, but in Newcastle upon Tyne... right under George Mintoe's nose!

"I'm quite used to people making films in the car park, but it's usually small local productions like Spender, Catherine Cookson or When the Boat Comes In. So imagine my surprise when a stretch limousine driven by **Madonna** pulled up at my booth window, and she asked me the way to the set.

Madonna - almost had sex with George on car bonnet

I directed her to the third floor, and turned on my security monitor to see what was going on. Madonna was lying on a car bonnet in the rik, and everyone else was standing around looking at their watches.

sex

A few minutes later there was a tap on the glass. It was the director of the film. He explained that they were supposed to be filming a scene where

> Parking attendant George (above), yesterday and the multistorey car park (far right) where he worked for forty years. And the roof top CityGold Parking level (right) where the Sex in the City four-in-a-car lesbian show almost took place.

Madonna had kinky sex with William Dafoe on the bonnet of a car.

The problem was, Dafoe's plane was late in from Hollywood. The director asked me if *I* would like to do the scene with Madonna. I couldn't believe my ears! Here was I, a middle aged car park attendant, being offered the chance to have it off on the bonnet of a Cadillac with the world's sexiest pop star. *What red-blooded englishman would refuse an opportunity like that?*

> **"Here was I, being offered the chance to have it off on the bonnet of a Cadillac with the world's sexiest pop star"**

However, I had to say no. The other attendant had gone to Bakers' Oven for some steak bakes, and I couldn't leave the booth unattended. What would have happened if a customer had turned up wanting change and there was no member of staff there to refuse.

boots

In the end Dafoe turned up and did the scene. I watched it on the security video, thinking how it could so easily have been me. They say you only get one chance of having it off with Madonna in this life, and I guess that was mine. I often wonder how things would have turned out if my colleague hadn't nipped out to Bakers' Oven on that fateful morning. I could have been a household name like William Dafoe. Perhaps I'll never know."

Not all George's sexy encounters with stars have taken place in his booth. In his book, he recalls an episode when he went to investigate a case of illegal parking on the top floor...and nearly got more than he bargained for!

"I was sitting reading some pornography in the booth when I glanced at the CCTV monitor and noticed a car pulling into a space on the 7th floor. That level is reserved for CityGold Permit holders only, and I certainly didn't recognise this particular vehicle. It was definitely illegally parked, so I grabbed my book of tickets and went up in the lift to see what was going on.

When I got there I couldn't believe my eyes. It was an open-topped Rolls-Royce with **Sarah-Jessica Parker** and the **other three from Sex and the City** in it! They said they'd been unable to find another space on the lower floors. I told them I was going to have to give them a ticket unless they moved it, but they had other ideas.

treat

The girls told me they were going to put on a bit of a show for me. Sarah Jessica seductively began to unzip the front of her rubber

dress, whilst the ginger one opened a carrier bag and took out a bottle of baby oil which she began squeezing onto her breasts.

kinkade

I have to admit, at this point I would have taken them up on their offer and let them park in the CityGold permit area. After all, I'm only flesh and blood, and here were four of TV's sauciest sexpots about to make my wildest fantasies come true.

Jessica-Parker - part of four way lesbian show that almost took place

However, I'd already written their registration number on the ticket. Because they're sequentially numbered there is no way I could just crumple it up and throw it away, no matter how much I was tempted. And I WAS tempted, believe you me. Sadly, I put the ticket under the windscreen wiper as the girls put their clothes back on.

Looking back I think it was the right thing to do. We have a saying in the car park business: Rules are rules, and I think that's very true. **"**

© 2003 George Mintoe with Youssou N'Dour.

Next week: *How **Sam Fox** nearly took her bra off to fix a fanbelt, and the time a jammed ticket machine almost led to oral sex with **Liz Hurley.***

20 THINGS YOU NEVER KNEW ABOUT CAR PARKING

In Britain today there are over 15 million people who drive cars. And, like it or not, they all have to park them somewhere.

Yes, car parking is now a part of our everyday lives. Some of us do it all the time, some of us do it less often. Everyday thousands of people park their cars. But how much do you really know about car parking?

For instance, did you know...

1 There is room to park over 200 cars on a football pitch. However, if everyone attending the F.A. Cup Final at Wembley Stadium arrived by car it would need an incredible 500 football pitches for them to park their cars.

2 The largest car park in the world is probably in America.

3 If every driver in America decided to park at the same time, there wouldn't be enough room.

4 Nowerdays parking meters accept 10p and 20p coins. In days gone by they accepted only pre-decimal currency. If every parking meter in Britain was placed on top of each other, it would be impossible to reach the ones near the top.

5 Unlike parking meters, traffic wardens can move from one street to another, putting 'tickets' on illegally parked cars.

6 There are now car parks in every town in Britain, enabling drivers to travel anywhere in the UK and park their car when they get there.

7 The first car park was opened in Munich, Germany in 1864. However it remained empty for 21 years until the car was invented in 1885.

8 Modern car parks are now computerised. Electronic signs now flash messages like "SPACES" or "NO SPACES" to indicate to passing motorists whether or not any spaces are available. And modern car park ticket machines can even tell you the time.

9 Visitors to the Haymarket public house in Newcastle, England, will have no problem finding a parking space. Many customers complained about lack of parking facilities, so the owners decided to demolish the building and replace it with an up-to-date car park.

10 Dine at a high class restaurant and the doorman will often park your car for you. Another way to avoid parking problems when eating out is to go by taxi. Pay the driver and he will simply drive away after you have got out.

11 In medieval times castles were built without car parks. People travelled on horses, which they parked in fields or tied up on nearby trees.

12 The most expensive parking space in the world is on board the United States Space Shuttle. It would probably cost about two billion dollars to park your car in the Shuttle's cargo hold during one of it's space missions.

13 In Britain up to three parked cars are stolen by unscrupulous 'car thieves' every minute. Enough cars to fill 22 football pitches every day. Hardly suprising when you consider the average parked car is worth over £3,000.

14 The most unsuccessful parking attempt ever was made by a J. Edgarson of Illinois, U.S.A., who spent three weeks and two days trying to find a parking space in his home town of Bloomington in November 1958. During that time he spent over 200 dollars on petrol.

15 Jersey TV detective Jim Bergerac rarely has problems parking his car during the popular BBC TV series. However, in real life actor John Nettles often has difficulty parking in London.

16 In Britain a yellow line along the side of the road means "no parking'. A double yellow line means "deffinately no parking".

17 Modern 'multi-storey' car parks enable cars to park on top of each other, saving time and money. Many city centre car parks are built with up to 500 parking spaces — enough room for over forty-one dozen cars.

18 In America many car parks are fitted with giant cinema screens enabling drivers to watch a film while they park their car. And generous drivers can give their cars a well earned holiday. Special 'motels' have been built — hotels for cars!

19 City centre car parking can be an expensive business, so special 'short stay' car parks exist for drivers who can only afford to park for a few moments. Meanwhile, wealthy motorists can leave their cars in 'long stay' car parks for anything up to 2 years!.

20 Recent advances in car rust-proofing have enabled scientists to look to the sea bed for an answer to the car parking problems of tomorrow. They estimate that there is room for an incredible 1,880, 973,400,000 parking spaces on the bed of the Pacific Ocean alone. The first underwater car parks could be open to the public by the year 2000.

JURASSIC PARK, TIPTON

World's first dinosaur zoo set for West Midlands

Plans are afoot to open the world's first multi-pound real life 'Jurassic Park' at Tipton, in the West Midlands.

Property developer and former Councillor Hugo Guthrie unveiled his plans yesterday at a special press conference. And if his ambitious scheme gets the go ahead, dinosaurs could be returning to Tipton after a million year absence.

Blockbuster

Mr Guthrie admits that the idea came to him after seeing the blockbuster film 'Jurassic Park' at the cinema recently. "Mrs Guthrie and I were most impressed, but we believe Tipton can go one better than Hollywood and create the real thing, here in the West Midlands".

Ballroom Blitz

Experiments to recreate giant dinosaurs, extinct for millions of years, are already well advanced. "We hope that by crossing various existing animals with each other we can bypass the need for revolutionary advances in genetic science. For example, my wife has suggested that by crossing an ostrich with a tortoise and perhaps a frog we could achieve something not unakin to a dinosaur."

Wig Wam Bam

If all goes according to plan Tipton Jurassic Park will open its doors to the public

Hugo Guthrie (left) - the Tipton based visionary behind Jurassic Park, and one of his terrifying creations - a real life dinosaur - which he has bread specially in his garage.

this autumn. "I have already written to Sir David Attenborough who stars in the film, asking him to perform the opening ceremony", Mr Guthrie announced. A site has been chosen on derelict land adjacent to a garden centre belonging to Mr Guthrie's brother-in-law, and the first dinosaurs could be released there as soon as the area has been fenced.

Little Willie

Mr Guthrie's application for a real life dinosaur zoo goes before the town's planning committee next Thursday, and they are bound to weigh up the possible dangers associated with real life dinosaurs against the undoubtable benefits to tourism that such a scheme would offer.

Residents in nearby Walsall have already expressed fears that escaped dinosaurs could pose an additional safety threat to children, on top on the menace already caused by joy riders.

Big Fanny

Last night Mr Guthrie was able to confirm rumours circulating among neighbours that one dinosaur has already been successfully created in a garage adjoining his home in Cedar Gardens. "I am pleased to say that Mrs Guthrie and I have succeeded in crossing a stuffed armadillo with a pine cone to get a small stegosaurus. We are having a few problems with the pine cone falling off at present, but hope to unveil it in the near future".

JURASSIC SHED

By our Gardens & Outhouses Correspondent THE HUES CORPORATION

Scientists in California believe they have discovered the 'missing link' between today's garden sheds, and the tree houses used by monkey's to keep bananas in.

Fossil remains discovered in 'Dinosaur Valley', an area where hundreds of dinosaur bones have been discovered, appear to be those of a primitive cave man garden shed.

Historians shed light on shed history mystery

BRANCHES

Tests show that the structure, which was made of tree branches with large leaves on the roof, probably contained garden tools, although DNA testing has been inconclusive.

TRUNKS

For many years scientists have been baffled by the mystery of shed development, and a vital piece of the historical jigsaw puzzle has always illuded them. Until now the earliest shed

remains on record were those of a Roman shed, containing a bike and a lawn mower, found near the Roman fort of Vindolanda in Northumberland.

TUSKS

Modern sheds, which are made of wood, or aluminum, can be purchased for as little as £99, including erection. Wooden or metal sheds however are not to be confused with outhouses, the latter being of concrete, brick or stone construction.

GOING FOR A SHED

With the late ARTHUR NEGUS O.B.E.

"Think you know your sheds? Well here's a little test for all you armchair shed experts at home. There are four sheds shown below, each from different eras in hut history. Can you date them all? Have a go, then check the answers below. I may be dead, but when is these shed?"

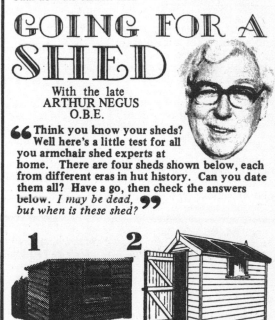

1 **2**

3 **4** ASBESTOS £45'5L

'CUNT' TAYLOR MUST HANG

along with his family

An alcoholic sports writer and former third division footballer has called for the death penalty to be introduced for football managers after England's disappointing results in the World Cup qualifying matches.

And John Cobblers has offered to don a black cap and pull the lever himself, claiming that Taylor's family should also face the death penalty, along with anyone else who knows him.

COBBLERS

In a recent editorial Mr Cobblers likened Taylor to a woman's vagina after England's 2-1 defeat in a friendly with the United States. Under a banner headline which read 'CUNT', Mr Taylor was pictured with a large hairy fanny instead of a nose, and a tampon in his mouth.

"England have not won a game in over two months", Mr Cobblers said yesterday. "The fact that we have not played one is irrelevant. We invented football, yet teams made up entirely of foreigners seem able to beat us at

will. Taylor should hang, and so should his successor."

BOLLOCKS

Mr Cobblers caused a storm of controversy last month when he suggested England cricket captain Graham Gooch should be castrated following two test defeats at the hands of Australia. However, he later withdrew his remark and suggested that Mr Gooch should be crowned King of England after England drew the following game.

ROLLING IN IT!

The top stars really do have
'loads of mon~eeey'

Enfield (right) pictured next to one of his many houses

Today, many of the showbiz world's top celebrities are having to face up to a bizarre problem — what to do with all their money. With staggering six, seven and eight figure pay cheques common place in the entertainment world, many of the biggest names in show business simply can't cope with the cash.

In the crazy world of pop, stars who only weeks ago were scraping a living in the pubs and clubs of their home towns, suddenly find that instant stardom has made them overnight millionaires. Musicians who used to wonder where the next meal was coming from can suddenly afford to buy entire restaurants. And often they do.

One music business insider told us about the star who walked into a top London restaurant and ordered a curry. 'Sorry sir, this is a Chinese restaurant. We don't serve curry', the waiter replied. The singer was furious. He bought the restaurant on the spot, and sacked the entire staff. Then he sat in the bar and waited while a team of top Indian chefs and waiters were flown in from Delhi. "After about three hours his curry eventually arrived", our informant told us. "But by this time he was so drunk he poured it all over the waiter's head, urinated on the table and then jumped out of the window without paying".

LOLLY

Traditionally pop stars spend lots of their lolly on fast cars. Current teen idols Bros are no exception. In one recent shopping spree alone the Goss twins are estimated to have spent around £200,000 on new cars. However, after travelling only a few yards Matt and Luke realised they couldn't drive. So they stopped the cars, got out, and gave them to a passer by. "I couldn't believe my luck", said painter and decorator Tony Adams. "They just handed me the keys. Obviously, they hadn't realised how difficult driving a car was".

WAD

As well as the pop stars, TV entertainers and top comedians also have a money problem — too much of it! Zany funny man Harry Enfield may joke when he waves his famous wad and says 'I've got loads of money'. But in real life communist Enfield is wealthier than his fans could ever imagine. Despite owning three homes in London, the 27 year old former public school boy has just splashed out a cool £2 million on a 28 bedroomed castle in Cumbria. A keen opera buff, Enfield has arranged for the entire cast of 'Phantom of the Opera' to be flown in from the USA for a one night performance at his new home. The bill for the nights entertainment is thought to be well over £1 million.

HUNTING

Enfield has also angered locals by inviting lefty comic Ben Elton and other well-to-do pals to his castle for hunting weekends. "They call themselves socialists, but in real life they ride around on horses killing foxes", one villager told us.

HOUSES

Many muddled millionaires fork out fortunes on houses they will never see. Some stars own homes all over the world, and occasionally one may get overlooked. Believe it or not, Mick Jagger, a millionaire since the sixties, once bought the same house twice!

LIMOUSINE

"He was driving along the road in his limousine when he spotted an enormous house which he quite fancied", said a friend of the ageing star. "So he drove to the local estate agents and bought it, paying in cash". It was only later, when Jagger went to look over his new house that he realised his mistake.

"The house was already his. He'd bought it in 1968 and completely forgotten about it", said his pal. That lapse of memory cost the red faced Rolling Stone about £2 million — the equivalent of 2 days spending money to a star like Jagger.

DRUGS

There is also the sad, sleazy side of pop success, with many wealthy, jet setting stars spending fortunes on drugs. And with so much money to spend, their drug problems can take on enormous proportions.

"Some stars spend a million pounds a year on drugs", one record company executive told us. "In fact, some spend that much in a month". He recalled how one big name American artist took five articulated lorries on tour with him. One was for his stage gear, another instruments, one for the sound rig and another for the lights. The fifth lorry, which was the biggest of them all, was for carrying his personal supply of drugs. "There were tons of them. But after two weeks on the road, the truck was almost empty".

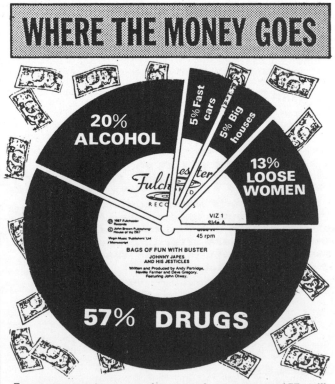

WHERE THE MONEY GOES

20% ALCOHOL

5% Fast cars

5% Big houses

13% LOOSE WOMEN

57% DRUGS

For every pound you spend on records, an average of 57p will be spent on drugs by the stars. Each year in Britain we spend over £2 billion buying singles, LP's, CD's and cassettes. And almost every penny of the money goes straight into the pockets of the stars. This fascinating chart, based on expert knowledge, shows how that money is spent.

STOP THIS EVIL CRAZE!

Thousands of kids could be at risk from a new cult music craze that is sweeping the country. And if you have children of your own — beware. They may already be involved.

'Pop' is the latest craze to hit Britain, and youngsters are flocking to dance halls and clubs to attend 'Pop Music Discos'. But experts fear that the youngsters are being exposed to sex and drugs, and that lives could soon be at risk.

HYPNOTIC

'Pop' is the name given by its fans to the loud, rhythmic form of music which has its origins in the seedy back streets and drug dens of America. At 'discos' the thumping, hypnotic beat of the music is often accompanied by bright, flashing lights. 'Pop' fans can be easily distinguished by their distinctive clothing — T-Shirts often bearing the name of their favourite 'group'. The groups, with unusual names like "Bros", "Wet Wet Wet" and "Pet Shop Boys" are worshipped by the youngsters who often spend all their pocket money buying their records.

A pop 'group' strip off their clothes and take drugs on stage (above) while fans work themselves into a perverted sexual frenzy (below).

High on cigarettes, alcohol and drugs, and sex, youngsters 'dance', zombie-like, to the incessant rhythms of the latest 'pop' records.

But it is the 'Pop Music Discos' where the kids are most at risk. High on drink, drugs and cigarettes, they dance in a frenzied fashion to the loud, repetitive rhythms of the music. The discos quickly degenerate into drunken orgies, with drug crazed youngsters, their eyes glazed, kissing and fondling each other. Afterwards, some couples may even endulge in sexual intercourse, which may also take place to the sound of 'pop music'.

SEX

The lyrics of the songs give great cause for concern. With titles like 'When Will I Be Famous', 'Temptation' and 'I Should Be So Lucky', many of them are promiscuous and encourage cigarette smoking, alcoholism and sex before marriage. And experts fear that the psychological effects of the music, especially on young children, could be damaging.

JAILED

"Children who listen to pop music will become shoplifters and homosexuals". That is the opinion of MP Anthony Regents-Park, an outspoken critic of 'Pop'. "The discos should be stopped and the people responsible jailed". and Mr Regents-Park believes that pop music should be banned from the radio airwaves.

SLAMMED

But a spokesman for the BBC where radio announcers like Simon Bates, Gary Davis and Bruno Brookes play pop music regularly during their programmes, denied that any harm was being done. "There's no evidence to suggest that pop music is harmful to kids", we were told. But one housewife we spoke to slammed the Beeb's attitude, describing it as "irresponsible".

Sir Anthony - 'It's disgusting'

Mavis McGuire's 11 year old son David was badly injured in a road accident shortly after listening to pop music at a friend's house. He suffered a broken leg after being hit by a bus while crossing the road.

SCANDAL

"I think it's a scandal and a disgrace", she told us. "These people ought to be ashamed. These pop music records are available to kids of any age. They just have to walk into a shop and buy them", she told us.

DECLINED

When we approached the manager of a record shop to ask whether pop records were being sold to children, he declined to comment. "I've got a living to make you know", he told us before slamming the door in our faces.

KYLIE

Later we watched as a boy of 8 entered the shop. He emerged 2 minutes laters, carrying several records in a yellow plastic bag. The word 'Kylie' was clearly visible on one — a clear reference to Kylie Minogue, the Australian actress turned singer who is thought to have made a fortune selling 'pop' records to kids.

TELLTALE SIGNS

If your child shows any signs of being interested in pop, then you should act now before it is too late. Look out for these tell-tale signs.

● They may spend time alone, or with friends, listening to records in their bedrooms.

● They may put posters on their bedrooms walls.

● They may appear to be snappy, irritable and short tempered.

● You may catch them singing or whistling unfamiliar tunes around the house.

● They may appear quiet and withdrawn, and lose their interest in sporting activities and other hobbies.

● They may come home with an unusual haircut.

● Their speech may become slurred, and they will appear unsteady on their feet.

● They may stay out late in the evenings, or fail to give you an explanation as to where they've been.

● They may put on weight suddenly, or lose it. Or indeed their weight may stay the same.

● They may suffer from dramatic hair loss and periodic spells of blindness.

EXPOSED! SECRET

Coronation Street is Britain's most popular TV soap. Yet, according to one insider the **REAL LIFE** drama that goes on behind the scenes of "The Street" is more exciting than the soap itself. "You wouldn't believe the goings on", Street star Sidney Blenkinsop has told us.

Sidney first appeared in Britain's longest running soap as long ago as 1968 when viewers briefly saw him sit quietly in a corner of the Rovers Return. Since then he's become a regular appearing many times in the Rovers and on one occasion actually ordering drinks. Over the years he's got to know the stars better than anyone, and now for the first time he blows the lid off Britain's longest running soap.

'Granada TV pay writers a fortune to invent storylines for the Street. But the most exciting stories of all are the ones which go on behind the scenes. Viewers simply wouldn't believe some of the episodes I've witnessed!

EYE

On screen many of the characters don't see eye to eye. That's what soaps are all about. Occasionally they come to blows, like the time when Mike Baldwin, alias actor Johnny Briggs, grappled with Ken Barlow outside Mike's factory. Of course they were acting. But you wouldn't believe some of the real life fisticuffs that go on. Usually its over

Behind the scenes drama of Britain's favourite soap

EXCLUSIVE

By Street star Sid

something petty – a disagreement about the script for example – but feet and furniture start flying, and it always ends up with at least one person in hospital. If you watch carefully you can always spot at least one of the actors with a black eye, and I don't think a single members of the cast has got a full set of teeth left.

BUNCH

You'd imagine the cast would be a tight knit friendly bunch. But I can tell you they're not. There's so much competition for the best lines, and endless arguments about who's going to say what. On set you can cut the atmosphere with a knife, and off it none of the actors ever speak to each other.

FAMILY

But despite the odd disagreements the Street stars are a great bunch to work with. We're like one big family, always looking out for each other. Like the time Roy Barraclough lost his wallet and we all stopped work to help him look for it.

If you ever think some of the plots that script writers dream up for the Street seem far fetched, you should see the kind of drama the stars get up to in real life. Hardly a day goes by without one star or another getting pregnant, having a love child or discovering that another star is their real father.

LOVE CHILD

I will never forget the day Chris Quinten turned up for work only to be told that Thelma Barlow, alias Mavis Riley was expecting his love child. Sparks were flying and her on-screen husband, actor Peter Baldwin threatened to kill him. Thelma told Chris she'd ruin him unless he handed over £250,000 to bring up the child. Chris's screen wife Gail Tilsley told him that Thelma would have to have an abortion, and her screen mother Audrey Roberts was taken to hospital in hysterics. Alf Roberts, alias actor Brian Mosley, Audrey's on-screen husband, armed himself with a gun and went out to shoot Quinten, his on-screen son-in-law, alias actor Brian Tilsley.

Gail Worth, alias actress Helen Tilsley, former on screen husband of actress Chris Quinten (alias actor Brian Tilsley).

In the end it all turned out be have been a big mistake. Someone had got a telephone message wrong, and nobody was pregnant after all. But that was quite an episode I can tell you.

Not many people know that Prince Charles is probably Coronation Street's biggest fan. Indeed if it wasn't for the Prince of Wales there wouldn't be a Street at all.

Charles – Street was his idea.

For when Granada TV first planned the series it was going to be set in a high rise block of flats called Coronation Court. But the Prince heard of their plans and wrote to them pointing out that the community spirit would be lost in an unsightly high rise building, and that elderly residents would be isolated, especially if the lifts didn't work. He suggested a traditional row of Victorian terraced houses instead. The producers agreed, and the idea of Coronation Street was born.

TOPLESS

The Street has for many years been Britain's most successful soap, however the producers panicked when the BBC's Eastenders soared to number one in the viewers' chart. They considered a number of crazy schemes to boost viewing figures. They even tried asking sexy barmaid Bet Lynch, alias actress Julie Goodyear to work topless behind the bar at the Rovers Return. But fortunately their letter to her was lost in the post, otherwise I'm sure she would have resigned in disgust.

DOCTOR. I'VE GOT A PAIN IN THE ARSE.

Sidney (arrowed) as viewers often see him in the Street. In the foreground is the Rovers Return

Albert Tatlock got on the wrong side of Granada bosses by asking for a pay rise. When they turned him down actor Jack Howarth threatened to leave and join Eastenders instead. A week later he was dead.

TOP

Jack was just one of the many great actors who have appeared in the Street. Many of today's top stars began their acting career in the series, among them Davy Jones of pop group The Monkees.

Street pop star actor Monkey Jones.

It's a tribute to the actors that the viewers often imagine characters like Rita Fairclough, Jack Duckworth and Harry Cross as being real people. Every week Granada TV receive tons of mail addressed to these people, and the actors who play them are often approached by fans mistaking them for their on screen characters.

PARTS

But sometimes an actor can become so obsessed whith his part he can, without realising it, literally become the character he plays – 24 hours a day. It's a fascinating psychological condition, only recently discovered and quite unique to soap stars who can end up losing their own identity. Unfortunately however there aren't any examples of that I can think of. ,

Next week: Frank Sinatra's Street connection, plus how detectives hunting the Yorkshire Ripper swooped on the Rovers Return.

Several new storylines were considered and a sizzling no holds barred sex scene between Ivy Tilsley and Don Brennan was filmed under tight security. No one was allowed on the set, but I crept to a window to take a look. It was red hot stuff, but they'd only been at it for 2 or 3 minutes when the window steamed up!

END

In the end ITV bosses insisted that the scene be cut, or the programme would have to go out after 11.00 p.m. Reluctantly they agreed to drop it.

ATTRACTIVE

Dirty Den was single handedly responsible for Eastenders' success. The ladies were switching to the BBC to watch him in their millions. So Street producers told actor Bernard Youens, who played Stan Ogden in the series, to lose weight in order to become more attractive to women.

DIED

But after months of unsuccessful dieting Bernard died, and the Street lost one of its brightest stars. No one can be blamed for his death, but I blame the producers.

KNOCKERS

It's easy to criticise the producers for the way they run the show. But they have a difficult job to do, and I would never knock them.

RELIEF

There were sighs of relief at Granada TV when Dirty Den eventually left Eastenders. BBC chiefs decided to have him bumped off by a gangland hit man. It was lucky for actor Leslie Grantham, alias Dirty Den character Queen Vic landlord Dennis Watts, that they did, because Granada bosses had been planning to do the job themselves – with real bullets. They had been making discreet enquiries about the availability of a real life hit man and had already got several quotes for the job.

SOAPY

Murder might seem a little far fetched, but in the crazy world of the soaps anything can happen. And sometimes it does.

BY GEORGE!

Bank Boss strips to reveal his assets

Snooty bosses at the Treasury are seeing red after discovering that the Governor of the Bank of England, Eddie George has bared all in a girlie magazine.

Gorgeous George, 61, will set readers' interest rates soaring when he swaps his pin striped suit for his birthday suit across ten pages of next month's raunchy Razzle magazine.

In some of the pics, too hot to print here, curvy Eddie, 38-48-46 is seen in the vaults of the Bank of England draped across Britain's gold reserves, and posing provocatively next to the Exchange Rate Mechanism. And it's all been too much for the Bank's top brass, who may call for Mr. Georges resignation.

Nude

Speaking from his flat in London, Eddie, who received £150 for the photo shoot admitted: "I knew there would be a fuss, because you're not allowed to pose nude when you're the Governor of the Bank of England, but I didn't realise I would be in this much trouble. It was only a bit of fun.

Bottom

"Everyone thinks I have a glamorous job, jetting off round the world to the International Monetary Fund or G7 conferences. But I only get £160,000 a year, and most days it's just boring bank work, filling in forms. I'm constantly being chatted up, and randy Treasury officials are always trying to pinch my bottom when I bend over to change the minimum lending rate.

"The suit and tie is really uncomfortable, too. That's another reason why I couldn't wait to get them off in the pictures".

One thing's for sure, when Eddie's pictures hit the newsstands, there'll be a sudden burst of inflation - in the nation's underpants!

READY FOR BUSINESS: Eddie shows off his figures.

SPACE FILLER!

BY **BILL SHITE**

RELAXING: Eddie takes it easy in his office.

RAZZLE: The issue where Eddie bares all.

READY FOR WORK: Eddie in his Bank of England uniform.

TOAST BAG

It's the POST with the MOST when it comes to TOAST

ACCORDING to Sod's law, toast always lands buttered side down. To get round this, I eat my toast dry, so it doesn't matter which way it lands if I drop it. Then I then eat the butter separately with a spoon.

Dr J Kingbast, Tewksbury

MY HUSBAND worked as a postman for fifty years. He has now retired, but still gets up at 4.00am and brings me a cup of tea and a slice of toast. He jokes that he used to deliver the POST every morning, now he delivers the TOAST every morning. However, it is not strictly true, as he worked in the Mount Pleasant Sorting Office and has never pushed a letter through a door in his life.

Edna Aneurism, London

I LAUGH when I hear young people complaining about how their electric toasters have broken. I was given a toasting fork as a wedding present in 1936. I have used it every day since and it has never let me down. And what's more, the toast tastes better.

Edna Vulvitis, Bracknell

LIKE MANY of my generation, I deplore the fact that the word 'toast' has been hijacked by so-called musicians. In my day, toasting meant grilling a piece of bread that one would eat at breakfast time. These days it means ranking up a phat sound on a ragga dub beat tip. Once again, as happened with gay, queer and cottage, a perfectly good English word has been lost to the language. And fist.

Brigadier Hepscott-White, Jedburgh

PS. And rim.

I SCREAM with hysterical laughter for several minutes every time I hear someone use the phrase 'warm as toast.' That's because sometimes you can have toast cold, for example when eating pate, or having breakfast at a Holiday Inn Express.

Francis Crabs, Crastor

WHEN MY father went off to the trenches during the First World war, my mother gave him a piece of toast for the journey. He put it in the breast pocket of his tunic and said to my mother 'I'll eat it when I get back'. Two years later on the Somme, a German bullet hit the pocket of his jacket. Fortunately, he wasn't wearing it at the time. His best friend had borrowed it and the bullet went straight through the toast and into his heart, killing him instantly. When he got back from the war, my father showed us children his lucky piece of toast with the bullet hole in the middle.

Edna Pleurodynia, Sheffield

I RECENTLY attended a wedding with my six year old grandson. After the speeches, the best man stood up and proposed that we 'Toast the Bride and Groom'. I looked at my grandson expectantly, but unfortunately he was fully aware that the verb 'to toast' as well as 'to grill bread' can also mean 'drink the health of.' Consequently he didn't make any charming or touchingly naive comments.

Edna Parkinsons-Disease London

Miriam

MIRIAM ANSWERS YOUR TOAST PROBLEMS

Dear Miriam...

I CAME HOME from work early recently and caught my husband making himself some toast. He was in the kitchen and had a recipe book spread open on the table. I was shocked.

I am 32 and he is 34 and we have been married for ten years. In all that time I have made his toast and he always seemed to enjoy it. To catch him making his own toast shocked me. When I confronted him about it later, he claimed at first that he was just testing the pop-up mechanism of the toaster. However, he eventually admitted that he had been a bit peckish and was making toast. He says that I am making a big deal about it.

The first time I tried to make toast for him afterwards it was a complete disaster. I burnt it and ended up in tears. I haven't been able to make him any toast now for about a month.

He says he still loves my toast and wants me to make it for him again, but I am so confused. Please help me, Miriam.

MW, London

Dear MW,

Sorry, but I'm going to agree with your husband. There is nothing wrong with a man making himself a bit of toast now and again, even when he has got a wife to make it for him. It doesn't mean he loves your toast any less. However, it may be that after ten years of marriage your toast has become a little dull. Try making him some with different toppings, adding a bit of Worcester sauce to the cheese and so on. Experiment a little. You may find that you actually like it.

TOAST Word — by Polyurea

EASY

Across
2. Grilled bread, often eaten with baked beans (5)

Down
1. Machine for making toast (7)

CRYPTIC

Across
2. Marten mixes up toast order (5)

Down
1. Sat to grill bread. I'll drink to that (5)

LAST WEEK'S ANSWERS: EASY *Across* 1. Toasting Fork *Down* 2. Pop Up Toaster. **CRYPTIC** *Across* 1. Saxophone. *Down* 2. Toast

WE ♥ YOUR TOAST POEMS

The food that I love most,
Is not a chicken roast
Lovely fish and chips
Will never pass my lips
And I would rather not
Eat sausages quite hot
A great big birthday cake,
does not me hungry make,
For the food I love the most,
And this I really boast,
Is a lovely slice of... toast!

Edna Senile-Dementia Wales

When I was a little girl
With my hair a mass of curl,
There were no jumbo jets,
Or modern internets,
Policemen all had smiles,
And there was no paedophiles,
We went to Sunday school,
Because that was the rule,
Though Hitler's bombs did fall,
We were not sad at all,
For life was bright and gay
With toast for tea each day!

Edna Massive-Stroke Luton

Edna and Edna's lovely poems win them £80,000 cash and a Mercedes 560SEL each.

Frankie "Ironfist" London
the Viz Heavyweight Consumer Champion

FIGHTING FOR YOUR CONSUMER RIGHTS

Dear Frankie,
I RECENTLY bought a new digital camera to take pictures at my daughter's wedding. I paid £48.99 at my local camera shop, Snappy Days in Hartlepool. However, when I got it home I discovered that the small door over the memory card slot was broken. I took the camera back but the assistant refused to give me a refund, saying that I must have broken it myself.

Ralph Titball, Cleveland

Frankie Writes:
I went round to see Mr Philips, the manager of Happy Snaps on Appleton Street, Hartlepool. I started off with a quick left to the jaw, which knocked out a couple of his teeth. That sent him off balance a bit, so I followed it up with a couple of swift jabs to the kidneys. He staggered backwards and fell awkwardly over a chair, so straight away I moved in with my boot. I stood on his throat and gave him few quality toe-enders in the ribs. He wasn't getting up after that, but I gave his face the millimetre tread for good measure.

SORTED!

Dear Frankie,
I SENT off for a new hosepipe reel from Ajax Garden Products of Redditch. They debited £24.50 from my card within a day, but 4 weeks later I had still to receive my hose. I called to complain, and was told that the company was having stock problems and my payment would be refunded. However, when I checked my bank statement later that month I saw that they had in fact taken a second payment of £24.50 off my card.

Percy Grainger, Ossett

Frankie Writes:
I found out where Brian Jackman, the owner of Ajax Garden Products, lives from records at Companies House, and I paid him a visit late one Saturday night. When he answered the door, I kneed him in the groin. Then I got him in an arm lock and, whilst wearing a duster, gave it to him again and again in the mush. He tried to fight back, hooking his foot round the front of my leg to get me off balance, but I'm no mug and I took the skin off his shins with my heel. The noise brought his wife and kids down stairs, but I told them to keep out of it, this was between me and him. I back-heeled him in the knee and I heard it pop. He screamed and went down like a sack of spuds. Before I left, I gave his face the millimetre tread for good measure.

SORTED!

Dear Frankie,
THE OTHER day I received a bill from Scottish Gas for £65.22. I'm a pensioner and I live on my own in a one bedroom bungalow and I don't use that much gas. They've estimated the bill this time as I was out when they came to read the meter, but I think it's a bit high as I usually pay around £55 per quarter.

Morag McGrew, Auchtermuchty

Frankie Writes:
I went round to Scottish Gas headquarters in Edinburgh, grabbed hold of the first bloke I saw and shoved him up against the wall. His glasses came off and I dragged him by his tie across the office and slammed his face into a desk. Another geezer came across playing the hero and tried to pull me off, but a quick elbow in the old solar plexus sent him sprawling into the water cooler. By this time security had come in and two of them tried to get me on the deck, but I was pumped up on test and they didn't stand a chance. I bit one of them on the bridge of the nose and put my thumb into the other one's eye. Four blokes were now down and out of action, but I gave all their faces the millimetre tread for good measure.

SORTED!

ARE YOU IN DISPUTE WITH A COMPANY?
WRITE TO: Frankie "Ironfist" London, Viz Heavyweight Consumer Champion, PO Box 656, North Shields NE30 4XX

56

You Ask, I Flannel
Your Ecclesiastical Questions Waffled Over by the Archbishop of Canterbury

Dr. Rowan Williams

Dear Archbishop,
I HAVE SEEN the cherubs they have in Heaven in lots of paintings and it seems to me that they tend to be rather fat babies and have very small wings. They are going to have to flap them at a fair old rate to keep themselves flying. This being the case, I was wondering if, like bees, they make a buzzing sound as they flit from cloud to cloud? I hope not, as this buzzing would certainly impair my enjoyment of the hereafter, especially if I was there for all eternity.

D Holloway, Jesmond

Dear Mr Holloway,
I wouldn't worry. Don't forget that Heaven is high in the sky on top of the clouds where there is very little gravity. Cherubs may only have to flap their wings very slowly in order to fly, and they may make a very low noise, like a moth. And you can only hear moths when they come right by your ear.

✝✝✝✝✝✝✝✝✝✝✝✝✝✝

Dear Archbishop,
I WAS AT an art gallery recently, and I saw a painting set in Heaven where a lot of cherubims and seraphims were playing a fanfare on trumpets. In your previous answer, you said that Heaven is very high up in the sky. It is my understanding that at very great altitude the air is very thin, and that this would make the trumpets sound very high-pitched, a bit like speeded-up kazoos. Well, what sort of a way is that to serenade the Father of all creation?

C Wilson, Ealing

Dear Miss Wilson,
I have a great and unshakeable faith that the trumpets in heaven resonate at a lower frequency than our earthbound trumpets do. This compensates for the thin heavenly atmosphere that would, ordinarily, lead to the high-pitched fanfare you describe. What you have to remember is, these heavenly hosts have been playing brass instruments for God since the dawn of time, so I am sure they have ironed all the wrinkles out by now!

✝✝✝✝✝✝✝✝✝✝✝✝✝✝

Dear Archbishop,
ON A RECENT visit to the Vatican's Cistine Chapel, I couldn't help noticing that there was an awful lot of nudity going on in heaven. Now I'm rather old fashioned, and I don't really like the idea of having to go completely bare in the hereafter. Would it, do you think, be possible for me and my husband to keep our clothes on in Heaven?

Mrs Edna Snitterby, Toft next Newton

Dear Miss Snitterby,
Unfortunately for you, I believe that there is a very strictly enforced "no clothes" rule in Heaven. The reason for this is simple. People have been dying for thousands of years, and if everyone was allowed to wear their own clothes, you'd have people walking round in doublets and hoses, caveman animal skins, Victorian top hats and astronaut suits all at the same time. People from the olden days would covet modern things like zips, velcro fastenings and trainers and end up having to go to Hell. No, it's better all round if everyone is naked. But I'm sure that God would not want you to feel uncomfortable or embarrassed. It is unlikely that he would raise any objection to you wearing a towel or even fleshings for the first few days of your eternal stay.

✝✝✝✝✝✝✝✝✝✝✝✝✝✝

Dear Archbishop,
I NOTICED in a painting of the Annunciation, that the Archangel Gabriel had rather large wings compared to his body size, which when folded were visible over the top of his head, a bit like Brian Blessed's in *Flash Gordon*. I was just wondering what sort of noise they make when he takes off.

D Holloway, Jesmond

Dear Mr Holloway,
I have searched deep into my soul and prayed for guidance on this question, and I am now firmly of the belief that his wings will sound like someone opening and closing an umbrella.

"That's all I've got time for this week, religion fans. But if you've got any ecumenical questions you want waffled on about but not really answered, then write to me at: **The A. B. of C, Viz Comic, PO Box 656, North Shields NE30 4XX"**

Rowan

♥♦♠♣♥♦♠ Roses are red, violets are blue ♥♦♠♥♦♠♥

IS YOUR VALENTINE AS ROMANTIC AS YOU?

On February the 14th Valentine's Day will be here again, and thousands of hopeful lonely hearts up and down the country will be sending cards and flowers.

As always, they'll be hoping that this year Cupid's arrow will strike, and their Valentine will fall for them. For men sending a Valentine can be a pretty expensive business, with cards alone costing upwards of a pound, and postage to pay on top of that. And roses can cost anything up to £3 each as greedy florists cash in on the romantic mood.

So it pays to stop and consider for a moment – is she the sort of girl who's going to make it worth your while? Or are you wasting time and money barking up the wrong tree.

Here's a simple questionnaire to help you judge for yourself. For each question simply choose the answer (a), (b) or (c) which best suits her. Then tot up your score and all will be revealed!

1. She has planned to go out for a drink with a few friends. Suddenly, she realises she doesn't have any clean knickers to wear. The taxi is waiting at the door. What would she do?
   (a) Cancel the taxi and stay at home for the evening instead.
   (b) Sling on a dirty pair. It's only for one night, and no-one's going to notice.
   (c) Go out wearing a short skirt, high heels and no knickers at all.

2. At the end of a night out she is about to board the last bus home when she suddenly realises she has no money left. What would she do?
   (a) Ring her father and ask him to come and collect her.
   (b) Walk home, convincing herself the exercise will do her good.
   (c) Offer to sit on the driver's knee, and promise him a bumpy ride if he lets her off with the fare.

3. What sort of school did she go to?
   (a) A mixed secondary modern school.
   (b) A posh school, where they play hockey and wear hats.
   (c) A Roman Catholic, girls only Convent School, where all the teachers are nuns and the world 'boys' is not allowed.

4. Which subjects interested her most at school?
   (a) Difficult things like Latin, Sciences or English Literature.
   (b) The usual subjects. Cookery, needlework and netball.
   (c) Pulling off boys behind the bike sheds.

5. When did she leave school?
   (a) Not until the age of 21, after completing a swotty career at University.
   (b) After taking her 'O' levels, aged 16 or 17.
   (c) At the age of 14 after setting up home with the maths teacher.

6. If she had the choice of these three holidays, which one would she prefer?
   (a) A two week brass rubbing holiday in Lincolnshire, spent cycling around interesting churches and cathedrals in the area.
   (b) A couple of weeks camping in France.
   (c) A raunchy, booze soaked Club 18 to 30 topless 'sun, sea, sand and sex' orgy style holiday in Ibetha.

7. How do you think she spends her typical Sunday morning?
   (a) Going to church with the family, then perhaps helping dad wash the car.
   (b) Lazing around the house, watching telly or listening to records.
   (c) Sleeping off a stinking hangover before getting up at midday and going home.

8. If your Valentine was given a Marks & Spencer gift voucher to the value of £25, how would she spend it?
   (a) On a selection of thermal underwear, or a pair of sensible shoes.
   (b) On a brightly coloured cardigan, or a nice but plain nightie.
   (c) On a pair of red nylon split crotch panties and a large tub of Vaseline. And a cucumber.

9. If she walked into the pub and you offered her a drink, what would she have?
   (a) A fruit juice – she never touches alcohol.
   (b) Just a half of lager.
   (c) Nothing. She's already drunk half a bottle of vodka in the phone box outside just before she came in.

10. What sort of film whould she most like to go and see?
   (a) A real slushy romance, like 'Gone With The Wind'. Or perhaps 'The Sound of Music'.
   (b) An all action thriller, like 'Terminator', or maybe 'Fatal Attraction'.
   (c) A filthy 'X' certificate French porno movie in a seedy cinema with plenty of room in the back row for her to perform various lurid sex acts.

11. In a restaurant she orders sausage and mash, but when the food arrives she notices that the chef has accidentally put cheese sauce on the end of her sausage. What would she do?
   (a) Refuse to touch it until it had been returned to the kitchen and the offending sauce removed.
   (b) Simply ignore it, and enjoy her meal.
   (c) Take the sausage in her hand and slowly lick the sauce off with her extended tongue.

12. She's doing a spot of decorating when she accidentally gets a bit of wallpaper paste in her mouth. What would her reaction be?
   (a) She'd be horrified, would spit it out immediately, and be sick.
   (b) She wouldn't be too concerned, and would simply spit it back into the paste bucket.
   (c) She would swallow it, then go over to the bucket and rub some more on her tits.

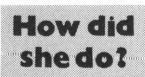

How did she do?

Now add up your Valentine's score and see how she did. Award no points for each answer (a), 2 points for a (b) and 3 for a (c).

15 or less: A dismal score. You'll get nowt off this bird.

16 to 29: Average: Your bird falls into a grey area.

30 or over: Roses are red, violets are blue. This bird's a go-er, she'll do it for you! Congratulations, you've picked a winner. A cheap card or a tatty flower delivered on the 14th should be all you need to access her pants.

20 THINGS YOU NEVER KNEW about INSECTS

They're here, there and everywhere. Summer's here again, and so are insects. Whether you're walking in Wakefield Westgate, picnicking in Pontefract park or hitch hiking in High Heaton, just pick up a rock and there they are – hundreds of little insects, ranging from big to small, fat to thin, and with any number of legs. Girls hate them, babies eat them. But what exactly are insects? What do they do and where do they live? How much do you really know about these miniature marvels of Mother Nature. Here's twenty creepy crawly things you never knew about insects . . .

① You'll never drown an insect by holding its head under water – no matter how hard you try. That's because insects breath through special lungs – called *trachea* – which are in their arse.

② Next time you cross the road as well as looking out for cars, keep an eye peeled for the Deer-Bot fly. For it's the world's fastest flying insect, travelling in short bursts at up to 36 miles per hour! That's faster than the prescribed speed limit for a motor vehicle travelling in a built-up area.

③ Insects live in the most unusual places. For example moths, which live in wardrobes, spiders, which live in the bath, and cockroaches, which live in chinese restaurants.

④ And Dung beetles, which live up cows' arses.

Clare Short.

⑤ The noisiest insect of all is the Trumpet beetle of Papua New Guinea. Mating pairs make a noise often compared to a washing machine in its 'spin' cycle. An endangered species, the few remaining examples are protected by strict conservation laws. Residents living near the insects' breeding grounds can claim government grants enabling their homes to be fitted with secondary double glazing.

⑥ We've all cursed after tripping on a broken paving stone. But next time you do it, don't blame the council. Blame insects. For as well as living under rocks, many insects set up home under pavements and footpaths, their constant to-ing and fro-ing causing damage which costs local authorities an estimated *£19 million* a year in crack repairs alone.

⑦ During the swinging sixties it was all the rage to name pop groups after insects, the best example of course being 'The Beatles' . . . The only other one we can think of is Buddy Holly and the Crickets.

⑧ Next time you're in Australia and you need to go to the toilet, check under the seat before you sit down. For each year over 2,000 unfortunate Aussies die from insect bites. The Black Widow spider nests under toilet seats, and repeatedly bites the arse of any unsuspecting victim who sits down to use the loo. Within ten minutes of being bitten the poor toilet-goer develops a huge pan-handle, and dies.

⑨ Another dangerous insect, the bee, can only sting once, and then it dies. That's because the poor insect pulls its ringpiece inside out while flying away afterwards.

⑩ Mind you, another kind of 'B', phone company 'B.T.' can sting you repeatedly. Every time you pick up the fucking phone to be precise.

⑪ Like us, bees have a Queen. However, unlike our Queen, the Queen bee doesn't sit on her arse all day, occasionally waving at people.

⑫ And another thing. The Queen bee's offspring go out and become useful members of their community, working hard, instead of going skiing for 8 months every year.

⑬ And if bees had universities, which they don't, but *if* they did, the Queen bee's kids would have to qualify for entry on merit, instead of getting into Oxford with one bloody 'A' level in dance or something ridiculous like that.

⑭ Getting back to bees, the Bee Gees are another sixties pop group with an insect in their name, although ironically 'Bee Gee' was never intended as an insect reference. It is, of course, the initials of the 'Brothers Gibb'.

⑮ Insects hit the pop headlines again in the early eighties when Adam Ant and The Ants dressed up as swashbuckling highwaymen, complete with red indian war paint, and invited the record buying public to "come and join our insect nation", whatever that meant.

⑯ Like so many post punk pop stars of that era, among them Gary Numan, Howard Jones and Nick Heyward, Adam Ant (real name George O'Dowd) later disappeared up his own arse.

⑰ One exception was of course Limahl out of Kajagoogoo, who by all accounts dissappeared up . . .

⑱ Getting back to insects, if you visit a flea circus, you can expect to see fleas, tied up with fuse wire, perform a variety of spectacular and exciting tricks. (However, if you go to a flea market you can expect to be charged about eight quid for an old bottle with soil on it.)

⑲ If someone tells you they've got 'butterflies', they don't necessarily own a collection of lepidopterous large winged insects. More likely they're suffering from a mild attack of nerves. Or alternatively they might be referring to a video recording of Carla Lane's 'gentle' BBC TV sit-com, starring Wendy Craig and Geoffrey Palmer. Although that's a bit unlikely, as it was a heap of shite. Just like Bread. And everything else she does, for that matter.

Linda McCartney – guest starred in TV's 'Bread'.

⑳ Some parasitic forms of flea are unable to live independent lives. Instead, they live on the back of a 'host' beetle upon whom they rely entirely for their survival.

Beans Talk

Britain's best-loved Bean Forum
with TVs Morrison's ad voiceover man **Sean Bean**

WELL WHAT a week it's been, (and that's 'been' as in the past participle of the verb 'to be', not a small leguminous seed). The Met Office tells us that it's been one of the hottest summers on record, so in between doing ads for Morrison's and some of my many other projects, I decided to do a bit of sunbathing in the garden. When I came in the house my wife, if I've got one, quipped "Oh, look. A baked Bean!" And talking of beans, here's a selection of the best bean letters that have 'been' sent in this week...

Star Bean Letter

My husband sent me to the greengrocers to get some beans for his tea. "You'll have to be quick," he said. "Why?" I asked. "Because I want runner beans," he replied. How I eventually laughed after he had explained this hilarious beans-based wordplay.

Mrs Ethyl Acetate
Rochdale

* £5 goes to Mrs Acetate

WHAT A swizz these so-called baked beans are. I opened a tin of them the other day and they were cold.

T Power, London

HEINZ BEANZ FACTZ
57 Amazing Facts about Beans, brought to you by Professor Heinz Wolf

1. THE MOST dangerous bean in the world was Wild West hanging judge Roy Bean. In a thirty year career working at Dodge City Magistrates' Court, Bean sentenced 2,463 outlaws to death. By an eerie coincidence, 2,463 is exactly the number of baked beans in a 4oz tin!

Don't miss Heinz's Fact no. 2 on the next page!

WHAT'S THE deal with beans? With some beans, like kidney beans and baked beans you just eat the beans. But with others, for example string beans and runner beans, you eat the pod as well. What's all that about?

Jay Leno, California

MY SON emigrated to Australia over ten years ago. As a boy he used to love baked beans, so to stop him feeling homesick I started posting him a parcel containing four tins of them each week. Over the years it must have cost me a small fortune in stamps, but he's worth it. He never writes to thank me... he's probably too busy tucking into his favourite beans!

Ada Golightly, Sunderland

WHAT'S the deal with coffee beans? They're not beans at all, For a start, they don't come in a pod. And you don't eat them, you cover them with boiling water and you drink them. What's all that about?

D Letterman, New York

IN THE 1980s I was taken hostage in Beirut by Hesbollah. For five years I was chained to a radiator and given only beans to eat.

Has Beans
Celebrities from the Past who still enjoy Beans

No. 38 ~ Russ Abbott
"I used to be the drummer in The Black Abbotts, so I suppose you'd expect me to like black beans. But they bring me out in a rash on my scrotum. It's broad beans for me every time."

Next week: John Inman

Luckilly they are my favourite food so it was a lovely treat each day that kept me smiling through my ordeal.

Brian Keenan, Dublin

IT ALWAYS makes me laugh when a lively person is referred to as being 'full of beans'. I was once 'full of beans' after eating 18 catering-size tins of them for a bet. I didn't feel in the slightest bit lively. I spent the next four days lying on the sofa farting, clutching at my swollen belly and belching into my cheeks every few seconds.

K Turnpike, Leicester

AS BEAN enthusiasts, my wife and I were very excited to see that our local theatre was putting on a play called 'Jack and the Beanstalk', and immediately bought front row tickets. However, although the beanstalk referred to in the title was featured in the production, we were disappointed that beans themselves did not make an appearance. We left after two hours feeling very short-changed indeed, and needless to say we haven't 'bean' back to that theatre again.

Dr H Trubshaw, Cheshire

DURING my incarceration at the hands of terrorists, I spent two years in a windowless room with Brian Keenan. He spent all day everyday going on about how much he liked beans whilst farting incessantly. What a miserable experience that was. The day my captors put a bag over my head and dragged me off to solitary confinement was the happiest day of my life.

John McCarthy, Cornwall

WHILST in the greengrocers the other day I was amused to notice a sign reading 'Bean's 45p/lb'. I immediately pointed out to the shopkeeper that he had put an apostrophe where none was necessary. However, imagine my embarrassment

Kids Say the Beaniest Things!

I TOOK my 5-year-old grandson into a fast food restaurant the other day, and saw that they were selling bean burgers. "They must be made from runner beans," he said.

Mrs Ethyl Acetate Rochdale

* Mrs Acetate wins £5

when he explained that he hadn't made a mistake, and that the beans in question had been reserved earlier in the day by rubber-faced funnyman Rowan Atkinson aka Mr Bean. The apostrophe was there to turn his characters name into the possessive noun!

Lynn Truss, London

MANY YEARS ago whilst working as an envoy for the Archbishop of Canterbury, I was taken hostage in Beirut. For five years my kidnappers fed me on nothing but fish fingers, chips and beans. I absolutely hate beans, but I didn't want to hurt my captors' feelings in case they killed me. So I used to slip the beans into my pockets when they weren't looking. By the time I was released and flown into RAF Brize Norton, I had over four tons of beans in my pockets.

Terry Waite, Canterbury

I'M AN Australian, and packages containing tins of beans keep arriving at my house addressed to the previous occupant. He left six years ago without leaving a forwarding address. The thing is, I can't stand beans so I end up throwing them in the bin. I always open the parcels first, though, because they sometimes have money in.

Bruce Hogan, Melbourne

HEINZ BEANZ FACTZ
57 Amazing Facts about Beans, brought to you by Professor Heinz Wolf

2. THE MOST widely-travelled bean in the world is former astronaut Al Bean. As commander of the Apollo 16 mission, Bean flew 516,274,309 miles to the moon and back. By an eerie coincidence, 516,274,309 is exactly the same number of baked beans in one of them really big catering tins that they use in schools!

Don't miss Heinz's Factz nos. 3 to 57 next week!

Bean Texts

❏ I LUV PS! BNS R 4 LOOZRS! GET A LYF U SADOS! **hairy al, lvpl**

❏ bnz r gr8 ps r gay! **rod scott, bath**

❏ plz prnt more lush pix of mung bnz 4 the ladz on hms invincible. **the beanster**

❏ P EATIN SCUM SHUD B STRNG UP. WIV STRING BEENZ! **angry coppa, nott'm**

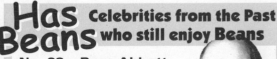

Bean Jokes

Q. What bean isn't very slow when it moves?
A. A runner bean.
Mrs Ethyl Acetate Rochdale

* Mrs Acetate wins £5

Knock knock!
Who's there?
Ivor
Ivor who?
Ivor liking for beans.
Mr Ian Likingforbeans Cardiff

Q. What sort of bean would you find half way down a child's leg?
A. A 'kid knee' (kidney) bean.
Stephen Fry, Norfolk

Q. What sort of bean would you find half way down a young goat's leg?
A. A 'kid knee' (kidney) bean.
John Sessions, London

After 114 years, we reveal the answer to the question that has baffled detective

Who was... JACK the RIPPER?

Jury - set to crack the 114-year-old mystery

IN THE BLOODY anals of murder, only one criminal case has remained unsolved above all others; and that's the one about Jack the Ripper. For ten dreadful weeks in 1888, the east end of London was shocked by a series of grisly murders. Newspapers dubbed the killer *Jack the Ripper*, but no-one was ever tried for the gruesome crimes. Even today, 114 years later, Scotland Yard are no nearer to solving the case.

In an attempt to finally bring the killer to justice, we've enlisted the help of one of Britain's most experienced detectives to weigh up the evidence against four prime suspects.

For forty years Jack Jury has worked as a front line crimefighter in some of Britain's toughest supermarkets. Starting as an ordinary beat store detective in the 60s, Terry worked his way up through the ranks to become head of undercover surveillance (chilled foods) at Morrisons supermarket in Whitley Bay.

Here, he takes time out from looking into people's bags for frozen sausages to look into this case for fresh evidence and so casts new light on one of the most unsolved series of murders in the history of crime. *Each suspect is scored out of 10 on Motive, Means & Opportunity.*

When you've been in this business as long as I have, you develop a sixth sense when it comes to criminals. I can spot a lifter, as we call them, a mile off.

It's hard to explain, but when I'm watching somebody by the freezers, I seem to know that they're going to steal something, even before they do. It's a feeling in my bones, and I get that exact same feeling when I look at photographs of all 4 suspects.

When somebody steals a chicken, they need motive, means and opportunity. It's exactly the same when murdering women, so let's see how our suspects fit Jack the Ripper's profile.

Suspect	Motive	Means	Opportunity
Merrick, John Alias: *The Elephant Man* Occupation: *Freak*	As a hideously deformed elephant man, Merrick would almost certainly have suffered rejection by females from an early age. Cruel taunts from Victorian girls at the school disco may well have left their mark on his psyche. And elephants never forget. Who can say for certain whether his bubbling resentment of women didn't later boil over into an orgy of sickening violence in the dark back streets of old London town. **10/10**	Merrick was given free range of Whitechapel hospital by Dr Frederick Treves, and would therefore have had easy access to a bewildering array of sharp knives from the kitchen and doctors' bags from the staff room. Who knows what uses his twisted mind could find for these implements of death, as he skulked amongst the shadows of the back streets of old London town. **10/10**	The window of opportunity was certa open for Merrick. His rooms were situ only yards from where the victims met fates. However, Jack the Ripper was kn to disappear into the crowd after perpe ing his foul deeds, and it is dou whether a trumpeting fifteen ton bull phant with a sack over its head would it easy to merge into the bustling night crowds of Victorian Whitechapel. **7/**
Hawking, Steven Alias: *Professor Steven Hawking.* Occupation: *Boffin*	On the surface, happily married scientist Hawking seems to have no especial hatred for women. Who can say why a leading intellectual with a balanced, stable personality should choose to murder at least five prostitutes - possibly more - in the most gruesome way imaginable over a century ago. Only Hawking knows the truth of what he did or did not do in the foggy shadows of nineteenth century London. **6/10**	No-one knows more about the time/space continuum than Hawking. Appearing on Star Trek five hundred years in the future proves that the Cambridge egghead has already mastered the technology necessary to travel through time. He can flit between the past and future as easily as he can hop on a bus, and a murderous trip to the smoky maze of Victorian London's back streets would be no problem. **10/10**	Hawking would only have to be out of nurse's sight for a split second to acc plish the Ripper's killings. With his machine, he could spend an hour mu ing a nineteenth century London prosti and arrive back in twenty-first cen Cambridge ten minutes before he set And because he's read the history bo he knows that he was never brough justice for his dreadful deeds in the g back streets of England's capital. **10/**
Nightingale, Flo. Alias: *The Lady with the Lamp, Forces Sweetheart.* Occupation: *Retired nurse*	After a career spent looking at men's innards, Nightingale may have been curious to find out what was inside a woman. The sight of pox-ridden soldiers and may have encouraged her to rid the streets of good-time girls. At the same time, virgin Florence would have felt envious of the prostitutes, who had sex up to ten times every day. Who can say whether this envy finally spilled over into a sickening murder spree amongst the smoggy back streets of old London. **10/10**	With or without a knife, it's doubtful whether a 68 year-old former nurse would have been a match for a gin-soaked toothless prostitute who would have thought nothing of brawling in the cobbled gutters outside the smoky alehouses of the East End's notorious red light district. Her trademark lamp also counts against her, as it is well known that Jack the Ripper relied on the smoky dark gloominess of old London town to accomplish his dark deeds undisturbed. **8/10**	When the first murder was commi Nightingale had spent the three deca since the end of the Crimean War in giving her ample opportunity to plan su series of grisly killings down to the last detail. And her meticulous planning tainly paid off, for the deceptively frail nurse never stood trial for the murder Mary Nichols, Annie Chapman, Eliza Stride, Catherine Eddowes and Mary K **10/**
Holmes, Sherlock Alias: *Basil Rathbone, Conan the Barbarian.* Occupation: *Consulting detective.*	Sherlock Holmes's only relationship with a woman ended when she met a violent death. Perhaps, by carrying out the Jack the Ripper killings, he was wreaking a sick revenge on all the women of Victorian London who hadn't met a violent death by making them meet one. More simply, the murders may have been motivated by greed. Addicted to opium, Holmes may have robbed the prostitutes of their money to pay for drugs and violin lessons. **10/10**	Highly intelligent Holmes would have run rings around the feeble-minded police of olden times. Access to scalpels would also have proved no problem; it's certain that Dr Watson, played by Nigel Bruce, would have trusted Holmes with his medical bag. His lodgings at 221b Baker Street were suspiciously near to the scenes of the Whitechapel killings; a matter of just 4 stops on the Bakerloo Line, changing at Embankment, and then a mere 8 stops on the District Line. **10/10**	As a self-employed detective, no-one w have been surprised to see Holmes slip unseen in and out of his lodgings in the r dle of the night. His network of street urc - the Baker Street Irregulars - would informed him of the whereabouts of pote victims. A master of disguise, Holmes c have dragged a victim up a back lane dres as a one-legged pirate, murdered her circus ringmaster, then made his escape Chinese washerwoman costume. **9/**

ce time began...

"I was MURDERED by Jack the Ripper... *but I'll NEVER reveal his identity!*" ~ Nichols

Mary Nichols was the Ripper's first victim. She was discovered dead at 3.30am on 31st August 1888. Her throat had been cut and her body mutilated. Now, for the first time since her savage murder she speaks to columnist _Lynda Lee Stokes_... and reveals that she _KNEW HER KILLER!_

EXCLUSIVE INTERVIEW!

" I WAS A YOUNG PROSTITUTE new in London and with stars in my eyes. One foggy night in August 1888 I was soliciting in Whitechapel when I was approached by a tall young man wearing a cloak and hat, and carrying a Gladstone bag. He had a scottish accent and I thought I recognised him from the television, but I couldn't be sure in the atmospheric gaslight.

He asked me if I was doing business and I said yes. He offered me a shilling and we went into a narrow passageway.

razor

He said he wanted to murder me, but I told him no. He told me: "You know you want it really." The next thing I knew he'd pushed me up against a wall and started murdering me with a cut-throat razor. I kept telling him to stop, but he wouldn't take no for an answer. He just kept on murdering me.

wheely big cheese

Before I knew what was going on, he'd cut my throat and mutilated my body. He stepped over my corpse and made his escape, vanishing swiftly into Victorian London's maze of back streets in the direction of the GMTV studios.

I never said anything at the time, because I knew no-one would believe me, and I was also dead. However, since then I've spoken to at least four other dead prostitutes who have also been murdered by the same household name.

Nichols - in 1888 yesterday

114 years later, I am still adamant that I will never name my killer. I took his secret with me to the grave, and I'm going to keep it there. But the memory of being murdered will stay with me for the rest of my life. These days, whenever I see Jack the Ripper fronting daytime television programmes, sitting next to Fern Britton and smiling like butter wouldn't melt, my blood runs cold. "

No one ever knew his identity... And here's

10 MORE THINGS YOU NEVER KNEW about JACK the RIPPER

Aniston, who plays Rachel in 'Friends'.

1 THE Whitechapel murderer is believed by many to have been a member of the Freemasons, a secret club of corrupt coppers, businessmen and councillors. To this day, each new member of the society is entrusted with three secrets: a super club handshake and password, the names of the eleven herbs and spices that go into the batter for Colonel Sanders's Kentucky Fried Chicken, and the true identity of Jack the Ripper.

2 THE smallest ever man to be suspected of being Jack the

Ripper was Calvin Phillips. American crime author Patricia Cornwell accused the four-inch high New Yorker of carrying out the slayings using a darning needle as a dagger, a thimble as a top hat and half an After Eight wrapper as a cloak. And making his escape in a hansom cab made from a roller skate... pulled by a mouse!

3 THE legend of Jack the Ripper has inspired many films. The most recent one, 'From Hell', took over £100,000 at the box office and starred Hollywood heartthrob Johnny Depp as the killer.

4 JOHNNY also starred in 'Edward Scissorhands', 'Fear and Loathing in Las Vegas', 'Donnie Brasco' and 'Ed Wood'.

5 HE was also in 'Sleepy Hollow' but it wasn't very good.

6 JOHNNY recently married Jennifer

7 SORRY. That was Brad Pitt.

8 BRAD Pitt starred in 'Seven', 'Twelve Monkeys', 'Fight Club' and 'Meet Joe Black'

9 HE was also in 'Ocean's Eleven' but it wasn't very good.

10 BRAD recently married Jennifer Anniston, who plays Rachel in 'Friends'.

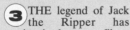

Arise, Sir Midge?

Band Aid Ure still clinging to knighthood dream ~claim

Ure out of Ultravox yesterday (left) and (above) the tragic kagoul bearing one of the Cash's laundry tags Ure had made.

IT'S TWENTY YEARS to the day since Band Aid hit the charts with their record-breaking single *'Do They Know it's Christmas?'*. Since then, the pop song has raised countless millions of pounds for starving people all around the world. Ten years after the song hit the charts, Bob Geldof was given his thanks in the form of an Honorary Knighthood from the Queen. But a further decade on, his co-writer is still waiting for his letter from the palace to drop on the mat.

But friends of Midge Ure, the mild-mannered Scotsman who penned the song with Sir Bob, say that he has not given up hope of being knighted.

sorrow

Mrs Morag Crabtree, who lives next door to Ure's modest semi-detached house in Auchtermuchty, Fife, spoke last night of her sorrow that Midge has been snubbed. "I feel so sorry for the wee man," she told reporters. "Every morning I hear the patter of his little feet scurrying down the stairs to check the post. Then, I hear a deep sigh, and him slowly trudging back to his room. It's heartbreaking."

Morag is also regularly witness to

a pathetic scene that the Ultravox frontman plays out in his back garden. "I often see wee Midge and his mammy up by the shed practising for when he receives his knighthood. He wants to get it right for the big day, if it ever comes," she told us. "He kneels down in front of her and she taps him three times on the shoulders with a breadknife and says 'Arise, Sir Midge'. He looks so proud in his suit and his little top hat, I could weep for the man, I really could."

let's dance

But perhaps the saddest evidence of Midge's disappointment came to light when Mrs Crabtree was planning a walk in the Highlands. "I went round next door to see if I could borrow a kagoul for my trip.

Mrs Ure handed me one of Midge's and told me that I could keep it as he had recently grown out of it," she told us. "When I looked in the collar, I saw he had a Cash's laundry label in it that read 'Sir Midge Ure', and the 'Sir' had been scribbled out in biro. He must have had a batch of them made when Bob Geldof got his knighthood. The

Scruffy bleeder Sir Bob, yesterday

poor mite must have thought he was next, bless him. Och, it was pitiful."

china girl

Although he has never spoken publicly about it, it is belived that Ure, 48, feels agrieved that the four-letter mouthed Boomtown Rat has received more credit for the charity single than himself. And with good reason, according to pop scientist Dr Dave Kidderminster-Jenson. "It's well known that Sir Bob wrote the words to *'Do They Know it's Christmas?'*, whilst Midge done the tune," he told us. "But my research has shown that had the song been released as a poem without any music, it would scarcely have dented the top 100. I am certain that Midge's contribution to the single is responsible for 99% of the money raised. It's a travesty that he hasn't been knighted."

ch-ch-ch-ch changes

A spokesman for Buckingham Palace told us: "It's nothing personal against Mr Ure. It's just that Her Majesty does the knighthoods alphabetically, and she's usually used them all up by about R or S. That's why Donald Sinden had to wait so long.

" If Midge Ure changes his name to perhaps Midge Aardvark, or Midge Abacus, he'll probably get one in the New Year's Honours," he added.

What's URE Opinion?

'Do They Know it's Christmas?' raised untold millions to alleviate famine in Africa. (It has only recently been out-charitied by Elton John's Candle in the Wind, Lady Di re-mix which made a billion pounds to pay for a beautiful algae-filled trench in Hyde Park and a moving court case against Franklin Mint).
But is it enough to earn its writer a Knighthood?
We went on the street to do an Ultravox-pop of the ordinary Britiash public. Should Midge be made Sir Midge, or should he be condemmned to stay plain Mr Ure? Here's what **YOU** said:

"...Bob Geldoff is very scruffy, and by his own admission only has one bath a week. And yet he is knighted. Midge Ure on the other hand, who is always immaculately turned out, remains a commoner. There is no justice in this world."
J Wells, Derby

"...forget about Band Aid. Midge Ure has one of the neatest moustaches I have ever seen. Along with his geometrically trimmed sideburns, I think he should receive a knighthood for services to facial hair."
M Hartstone, Leeds

"...undoubtedly he has done a lot for

the starving millions in the third world. But let's not forget that he is only 5 foot 4 tall. Dubbing him a knight would make this country a laughing stock. Who next?... Sir Ronnie Corbett? Sir Mini-me? Sir him out of Diff'rent Strokes?"
B Riddell, Bolton

"...a knighthood is insufficient reward for what Ure has done for charity. I think he should be made a life peer."
J Townshend, Luton

"... a life peerage would be a slap in the face after all Ure has done to alleviate poverty around the globe. I think he

should be crowned King Midge I of England."
L Stonehouse, Hull

"...I think Ure had ulterior motives in his raising millions for charity. Call me cynical, but I think it is no co-incidence that just five years after Band Aid, If I Was made number 12 in the charts. How high would it have reached had he not been involved in the charity?"
B Pollard, Leeds

"...Ure has wasted little time cynically cashing in on his Band Aid fame by releasing Band Aid 20 this year. He should be publically hung - and I'll pull the lever!"
Albert Pierrepoint, London

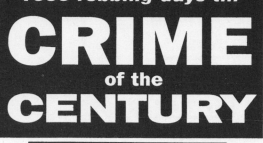

Hurry hurry hurry! Only 1000 robbing days till

CRIME
of the
CENTURY

By our alcoholic man in the bookies
Reg Soiltrousers

LADBROKES yesterday slashed the odds of the Yorkshire Ripper murders being voted Britain's official Crime of the Century.

The Sutcliffe slayings are now quoted at 10 to 1 for the big prize. Fred West's House of Horrors murders remain an outside bet at 25 to 1.

Popular

"Despite killing more people, West's crime was less popular with the public", bookmaker Frank Carpet explained. "By burying his victims under his house he didn't get the on-going media coverage at the time of his murders, and topping himself in jail meant their was no grisly details revealed at the trial".

Cortina

Now police forces around the UK are bracing themselves for a last minute crime rush as the year 2000 approaches. Criminals have less than three years to commit the 'big one' and get their names in the record books. And so far the men they have to beat are the Great Train Robbers.

For over 20 years after it was carried out their mail train hijack remains the nations favourite blag. Despite no-one being directly killed in the raid, it is still red hot 2 to 1 favourite to scoop the honours.

Capri

"Kenneth Noye will be kicking himself", bookie Frank told us. "The Brinks Matt Bullion job involved a lot more money, and technically it was a far more successful crime. But the punters simply weren't impressed".

TURPIN - Olde worlde highwayman

Modern day Robin Hood Ronnie Biggs and his gang stole £2.5 million in the daring snatch, and planned to give it to the poor. But they were foiled by winnit sniffing cops when a tagnut dragnet closed around the isolated farmhouse in which they had holed up. Minute dangleberry deposits taken from the toilets were linked to the robber's ringpieces. It was the first time *ringgerprints* had been used in a police investigation.

Mustique

If the Great Train Robbery is voted Crime of the Century it will be good news for millionaire musician Phil Collins. Short-arse Phil bought a majority holding in the Great Train Robbery after it went public in 1984, investing a mere £14,000. At current market prices the robbery is today worth a staggering £850 million. Helium balloon rights alone are worth an estimated £180,000.

Simply the breast!

BRITISH birds have defied their knockers, and proved that they're the bust in Europe - for checking their chests!

For many years our birds were *boobing* - and allowing their *assets* to develop killer cancer unawares. But a recent survey shows that nowadays women are more aware of the dangers of breast cancer - and girls are giving their whoppers a once-over on a regular basis!

Phoaar! Britain's assets are boobing!

Claims

The killer disease claims thousands of lives in Britain each year. But cancerous curves can be cured - if caught early enough. As a result jubblies are generally in better health, and that's fabulous news for fellas! For it means oncologists are less likely to get their mits on our missus's mambas!

Premiums

Indeed, whopper docs may one day be out of work. But you won't catch them complaining!

Salesmen

"It's *bra*-vellous news from our point of view", one imaginary specialist told us yesterday. "Women are more 'up front' about what they've got up front. And by checking their charms on a regular basis they can help keep them in *tit*-top condition for years to come", he added.

One *tit* wonders

Here's a zany Top Ten hits for chemotherapists!

1. 'Simply The Breast' by Tina Turner
2. 'Can-cera Cera' by Doris Day
3. 'Knocker Three Times' by Dawn
4. 'Mammary Mia' by Abba
5. 'Radio(therapy) Bra Bra' by Queen
6. 'Baby Be-neign Tonight' by The Tubes
7. 'Can The Can-cer' by Suzi Quatro
8. 'I One Tit All' by Queen
9. 'Always On My Tits' by Elvis
10. 'Do They No Tits Christmas' by Band Aid

Toaster virus pops up in UK

Unsuspecting UK householders could soon be having their toast burnt by a worldwide toaster virus.

The techno virus - code name 'Burnt Toast' - was created by American defence boffins who hoped to wreak havoc in Kremlin kitchens. But the micro menace was leaked from a top secret Pentagon lab and has been spreading to toasters throughout the world.

Baffled

Toaster bosses are baffled by the bug, which is thought to have contaminated up to ten million toasters so far. "The virus travels by wire", research engineer Ross McKeown of toaster manufacturer Kenwood told us. "It can remain latent in a toaster for years, but is then triggered when the toaster is set to 'two and a half'. Once the virus is activated the toaster begins to malfunction, and the toast is burnt".

TOASTER - £19.50 from Argos

Toast set to go up in smoke

The outbreak is thought to have started when a teenage toaster whiz kid hacked into high security Pentagon kitchen equipment using a brand of toaster widely available at stores throughout the UK.

Our reporters, dressed as ordinary people, were able to purchase the model of toaster from a well known High Street electrical goods retailer.

After buying the toaster we identified ourselves and put it to a member of the sales staff that the machine could, in the wrong hands, be used to penetrate national kitchen security. He looked at us blankly, before trying to sell us a 2 year warranty.

MAD AS A HATTER!

The Queen is losing her marbles. That is the unofficial word from Buckingham Palace as Her Royal Highness begins her 37th year on the throne.

This incredible claim is being made by Roger Thompson, a former Palace employee who says he is desperately worried about Her Majesty's health. Sacked from his job at the Palace for stealing cutlery — a crime which he strenuously denies — Thompson has decided to speak out and make his concerns public, for the Queen's own sake.

"I never stole that cutlery", Roger told us. "It was all a big cover-up, and I took the blame to protect the Queen. Everyone inside the Palace knew that she'd been stealing it herself. She used to hoard it in pillowcases in her bedroom".

HUMMING

According to Roger, the Queen had been acting strangely for many months. "On a couple of occasions I'd seen her wandering around the Palace humming strange tunes to herself. Then on another occasion I saw her chasing butterflies around the Palace garden. Nothing unusual about that I thought, until she caught one — and ate it! I mentioned it to Prince Phillip later that day but he told me I had been imagining things".

SUSPICIONS

A few days later Roger's suspicions were confirmed when the Queen came down to breakfast — dressed as Napoleon. "She ate her breakfast in silence. No-one said a word. Afterwards the staff were told to forget the incident, or it would cost us our jobs".

As Roger recalls, that was one of the last meals the Queen ate at the Palace. "You see, she was convinced that the chef was trying to poison her and steal the Crown Jewels. I know it sounds crazy, but it's true. At meal-times she would refuse to eat a thing. Eventually she started going out in her royal carriage to buy Chinese takeaways, then she'd take them to the Tower of London and eat them with her eyes firmly fixed on the Crown Jewels".

CUCUMBER

The Queen's unusual behaviour was beginning to cause some embarrassment in public. "I remember one royal garden party in particular", Roger told us. "She refused to touch the cucumber sandwiches in case they were poisoned, and she was mingling with the guests eating Kentucky Fried Chicken out of a huge red and white party bucket, then wiping her fingers on her dress. The guests were clearly embarrassed".

The Queen is a fruitcake claims former Palace man

past me on all fours, barking ferociously. I followed her down to the kitchen where I found her underneath a table — feeding small bits of cheese to mice".

DUNGEONS

On another occasion the Queen summoned Mrs Thatcher to the Palace and told her to invade France. The Prime Minister explained that we were members of the EEC, and that we couldn't go to war with France. Furious, she told the Guards to take Mrs Thatcher down to the dungeons and have her stretched! They took her away and locked her in a cell, then, after a couple of days when the Queen had forgotten about it, they let her go".

She dresses as Napoleon – and barks like a dog

Over the weeks Roger noticed the Queen's condition was deteriorating. "The next thing that happened really set me worrying. I was awoken one night by the sound of a dog howling and barking. It was a terrible sound, and what made it all the more eerie was the full moon outside.

LETTUCE

I got up and as I walked along the corridor the Queen rushed

Trying to start a war was just one of the Queen's many unusual requests. As Roger recalls, another was calling for a Court Jester! "She was bored with watching the telly one night, so she demanded a Jester be summoned to entertain her. We rang a theatrical agency and they sent round a top TV comic and talent show host to do his act. Of course the Queen insisted he dress authentically — in a funny hat with pointy shoes with bells on the end.

TOWER

The poor bloke must have felt a proper fool, but he did his act anyway. At the end there was complete silence. The Queen was not amused. She told the Guards to take him to the Tower of London and behead him the next day! Luckily, by the morning she'd forgotten about it, so they let the poor fellow go. But that was one night he will never forget. Although he probably has done by now".

According to Roger steps are now taken to avoid situations like that. "For most of her public appearances, they use a double — that woman who looks like the Queen usually stands in for her. The Queen herself spends most of her time in her bedroom, watching game shows on TV. She wears sunglasses and keeps the room in total darkness. If you saw her today you wouldn't recognise her.

FINGERNAILS

She must weigh all of 18 stone, and she eats Kentucky Fried Chicken virtually non-stop. Her fingernails are 2 feet long, and her hair almost reaches the floor. She only ever leaves the room to visit the bathroom, and every time she insists that a new carpet is laid in the corridor so she doesn't catch any germs. As another precaution she wears a surgical mask and a pair of brightly coloured fisherman's waders. Getting in and out of this ridiculous costume takes her so long that a single visit to the bathroom can take up to 4 hours".

● FOOTNOTE

Roger Thompson pleaded guilty to the charge of stealing cutlery at Bow Street Magistrates Court yesterday, and asked for 173 similar offences to be taken into account. He was remanded in custody for 14 days pending psychiatric reports.

OH – FOR THE WINGS FOR THE WINGS OF A DOVE ...

WILD THING. YOU MAKE MY HEART SING.

CAN YOU SEE YOUR DRINKS OFF NOW PLEASE?

We expose bear facts about toadies who panda to stars
What a VIP Off!

CASE 1

Place: The Ivy, London

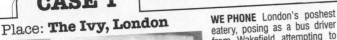

WE PHONE London's poshest eatery, posing as a bus driver from Wakefield attempting to book a meal to celebrate his wedding anniversary.

The receptionist snaps: "Sorry. We're completely full. There are no tables available for the next fifty years. Goodbye."

A few minutes later we call back, this time doing an impression of Pop Idols's **Gareth Gates**. We ask: "C-kerr-c-can I bee-berr-b-book a tee-terr-t-table for tee-terr-tonight pee-perr-p-please, mee-merr-m-missus."

Miraculously, the receptionist finds a table is available, and the booking is made.

No room at the inn? The tables were turned when this idol popped the question.

CASE 2

Place: McDonald's, North Shields

WE PULL UP at the Drive-thru window, posing as a retired plumber from Glasgow and order a fillet-O-fish. We are told: "Your order will take about 4 minutes," and are asked to park in Grill order 1, a reasonable request.

But 10 minutes later, we return dressed as masked pop megastar **Michael Jackson**. As before, we order a fillet-O-fish. How different is the response THIS time! "Your order will take about 3½ minutes," we are told. No surprises when it arrives in just *2 minutes 48 seconds!* But the meal leaves a nasty taste in our mouth.

It seems that being one of the world's richest men makes the service at a fast food restaurant even faster.

CASE 4

Place: Trickett's Non-ferrous Metals, Nottingham

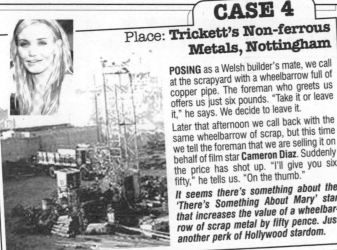

POSING as a Welsh builder's mate, we call at the scrapyard with a wheelbarrow full of copper pipe. The foreman who greets us offers us just six pounds. "Take it or leave it," he says. We decide to leave it.

Later that afternoon we call back with the same wheelbarrow of scrap, but this time we tell the foreman that we are selling it on behalf of film star **Cameron Diaz**. Suddenly the price has shot up. "I'll give you six fifty," he tells us. "On the thumb."

It seems there's something about the 'There's Something About Mary' star that increases the value of a wheelbarrow of scrap metal by fifty pence. Just another perk of Hollywood stardom.

A MAN phones a restaurant to book a meal, only to be told it is full, and no tables are available for weeks to come. 10 minutes later, another man phones the same restaurant. Unlike the first man, he is told that there IS a table. In fact, he can turn up any time he likes and be sure of being welcomed with open arms.

What is going on? The answer is simple. The first man is you or I, the man in the street. The second is a star, like Bruce Willis, Rolf Harris or Wendy Craig. For the stars, life is made that little bit easier - their names open doors which remain firmly closed to the rest of us.

EXCLUSIVE INVESTIGATION

It seems like there's one rule for the rich and famous, and another for Joe public. We decided to test this theory with a little experiment of our own. Posing first as ordinary members of the public, and then as celebrities' representatives, we tried to find out just how much better the stars get tret than the rest of us.

CASE 3

Place: Blue Lagoon Sauna, Hull

POSING as a landscape gardener, we book a full body massage. Half way through, we turn over with an erection and ask for "extras". Tricia, our masseuse tells us: "It's fifteen quid for a hand job, twenty for French, and I don't do kissing." We opt for hand relief, and hand over our money.

25 minutes later we are back, claiming to be ITN political editor **John Sargeant**. Once again we wait till half way through before turning over with an erection, and here is the headline: This time it's FIVE POUNDS cheaper! We opt for French, paying a mere fifteen pounds.

It seems that all news is good news when you're an ITN superstar.

CASE 5

Place: St. Thomas's Hospital, London

WE PHONE the oncology department and describe a set of symptoms, posing as a mild-mannered accountant working for a light engineering firm in the West Midlands. The doctor listens carefully before delivering his diagnosis. The news is not good: "You've got advanced pulmonary cancer," he explains. "I'm afraid you've only got six months to live."

We phone back half an hour later and describe an identical set of symptoms, but this time we claim to be TV favourites **Ant and Dec**. The doc tells us it's cancer again, but this time the news is a little better. "You've each got eight months to live," he tells us.

Never mind radiotherapy, it seems TV stardom is enough to keep this Byker Grove duo out of their Byker Graves for an extra two months.

That's rich!
Shop boss cops for a whopping packet

A newsagent from Walthamstow has angered customers with a controversial decision to award himself a THREE THOUSAND PER CENT pay rise.

Paul Khana, a 42 year old self-employed shop owner, will now take home a staggering £600,000 a year, while his lowly customers, many of whom are unemployed, struggle to afford a newspaper and packet of cigarettes.

Writes Billy Bollocks

DEFENDED

Khana, who had previously earned £20,000 a year, yesterday defended the move. "I think I'm worth it. My workload has increased significantly, especially since we started selling lottery tickets. I spend hours explaining to people how to fill them in", he told reporters.

ATTACKED

However the massive pay boost has outraged customers and politicians alike. One man who was buying a newspaper said "It's disgraceful". Meanwhile Labour MP Mr Derek Twatt blasted Mr Khana for being 'insensitive'.

MIDFIELDED

"At a time when newsagents are asking customers to pay higher prices for magazines, confectionery, tobacco and greeting cards, I find it grossly offensive that a pay rise of this magnitude could be considered appropriate", he said on television last night. However Sir Anthony Regents-Park, Tory MP and former Junior Minister for Sweet Shops, News Vendors and Tobacconists, last night defended the move.

REFEREED

"Unless Britain's newsagents reward themselves with realistic salaries then we are going to lose them to foreign competitors. Good shopkeepers are in demand world wide", he said. "If we want a streamlined, competitive news trade for the nineties, we're going to have to pay for it".

GOALKEEPERED

Mr Khana's six-figure pay award now puts him in the same earnings bracket as the Consett taxi driver who last week awarded himself an extra £500,000 a year on top of the £6,000 he was already earning. Meanwhile a part-time car park attendant from Peterborough has packed in his job and retired at the age of 52, after awarding himself a £300,000 a year pension.

No more laughs for Les

Picture: FRANK SHIT

The sound of laughter once echoed in his ears wherever he went. Millions watched him as the host of TV's Blankety Blank.

But now his only audience is earthworms, as the lonely figure of Les Dawson lies in a graveyard in Lancashire, buried under the ground, a shadow of his former self.

After his death the work began to dry up, and now the sad star lives the life of a recluse, hiding away from his former fans. Rarely seen outside his modest wooden coffin, friends say he has lost pounds in weight since his funeral in 1992, and wants to be remembered the way he was, and not as the pathetic figure he has become.

Neighbours at the ramshackle cemetery had no idea about the identity of the 'quiet man who keeps himself to himself'.

"Les Dawson? I've never even heard of him", said one elderly woman who is buried nearby. 183 year old Mildred Brown has never watched Blankety Blank. "I'm afraid I died before they invented television", she told reporters yesterday.

The Undersink Cupboard of Jacques Cousteau

'Allo and welcome to my sink 'Calypso'. Last week we journeyed deep into zee sirink cupboard, and played wizz a shoal of towels and some odd socks.

Burt today we will be explorink zee fascinating hidden world of zee cupboard under zee kitchen sink.

We 'ope to see a multitude off zee undersink life. Zee mighty Domestos, king of ziss realm...

...beautifool shoe polish tins in a myriad of colours... a seventies bottle off Windolene... some of zose pan scourer zings... per'aps even a multi-pack ozz light bulbs.

So, come wizz me now....

CRACK!

ooof.

NEXT WEEK Hans & Lottie Haas journey to those shelves in the back of the garage to film the migration of a herd of dried up tins of paint with lengths of broken-off dowel in them.

LADY DIES

Police are investigating the death of a woman who died after swallowing a horse.

The exact circumstances of her death are not yet known. However police are thought to be examining the theory that her death may be connected with several smaller animals which she had swallowed previously.

Potty Hill quits Beeb

TV soccer pundit Jimmy Hill has dramatically quit the BBC in order to join a bizarre religious cult.

for cop cult

Hill recently stunned BBC bosses by refusing to analyse first half highlights of the recent F.A. Cup final on 'religious grounds'. And after a blazing row with Match of the Day producers Hill is rumoured to have quit his £100 a week job and joined a little known religious cult calling themselves the Church of Latter Day Dixonology.

DIXON

The basic belief of Dixonologists is that Dixon of Dock Green, the likeable 'bobby on the beat' played for over twenty years by the late Jack Warner, will return to Earth in a 'third coming' and lead humanity on the path to law and order. They

believe that Dixon has risen once already, referring to the fifties film 'The Blue Lamp' in which P.C. Dixon was shot outside a cinema by Dirk Bogarde, only to return to life years later in the long running TV series 'Dixon of Dock Green'.

CURRY

Hill and other Dixonologists believe that the 'third coming' of P.C. Dixon can be brought about by his followers sitting in a circle and blinking their eyes very quickly indeed. They also stick rigidly to a diet of things beginning with 'P', such as peanuts, Penguin biscuits and Parmesan cheese. Cult members are forbidden from having uPVC double glazing or any type of conservatory in their homes.

RUMBELOW

It was thought that Hill had settled his differences with BBC bosses after they

agreed to let him analyse action from this summer's World Cup finals wearing a crown. A dispute had begun when Hill told colleagues that in future he was to be referred to as the 'King of Football', and several commentators, including former dentist Barry Davies, walked out in protest. However the situation had seemed to be resolved when

producers agreed to let Hill wear a crown and sit in a slightly larger chair than the other match analysts.

COMET

When we rang his home yesterday we were told that Hill was in the garden, hiding behind a tree, and was not prepared to come to the telephone.

STAR SHIT ENTERPRISE

An unemployed Bolton man plans to go where no businessman has gone before by launching a pioneering enterprise of his own.

Harold Biggins plans to make a fortune selling souvenirs of the stars, despite the fact that his products are shit. Quite literally! For Harold intends to market celebrity excrement, buying stools fresh from the stars, and selling them as paper weights.

METEOR

"I'm surprised nobody had thought of the idea before", Harold told us. "It seems such an obvious money earner. Turds which would otherwise have simply been flushed down the toilet can be taken away and sold to fans. I'm sure there'll be huge demand, especially for someone like George Michael or Sting's shit."

ASTEROID

Originally Harold had planned to make 'Celebrity Stool Snow Storms' with the logs, sealing them in a plastic dome filled with water, but there were various technical problems. "When you shook them they just turned into diarrhoea. They would have been very difficult to market, and unhygenic if they cracked." So instead he plans to encase them in glass, along with a signed picture of the star responsible for the dump.

HAEMORHOID

Already a host of celebrities have donated droppings after Harold began visiting them to explain his scheme. "Generally the people I've spoken to have been very helpful", said Harold. "I just turn up on their doorsteps with my plastic bag and a

spoon. Early morning is the best time, as that seems to be when most of the stars carry out their ablutions. There's the odd awkward customer who refuses to help, but generally speaking they've been marvellous. Cliff Richard, for example, even had one ready in a bag for me when he opened the door."

ADENOID

Unfortunately Harold's bank manager has been less than helpful. "He put me in touch with their Small Businesses Adviser, but when I explained my idea the only advice I got was to "fuck off". And without financial backing I can't get the business off the ground."

POLAROID

Unless the bank have a change of heart Harold fears he may have to throw away the dozen or so stools he has so far collected. "My wife won't let me keep them in the house, so I've got them all in the back yard at the minute. But there's a limit to how long you can keep them before they go all crumbly. They're already drying out."

POLAR BEAR

"The biggest one I've got so far was from Meatloaf. It's a bit on the big side for a paper weight, but it would make a good door stopper. However, if I don't get something sorted out soon I'm going to have to chuck it out, and that will mean having to break it up, which would be a shame.

Bewes sets target for walnut industry

Former TV Likely Lad Rodney Bewes has set a target for Britain's walnut growing industry. 'Self sufficiency by the year 1997'.

Bewes believes that Britain should be growing all the nuts we need within three years, and that the target is realistic. "I don't know much about walnuts, or how you grow them, but I think that we should be producing as many as we need, and hopefully within the next three years."

Bewes chose his target – the 6th of February 1997 – mainly because it was the anniversary of a friend's wedding. "I had a gut feeling about 1997, but I must confess a friend of mine suggested the 6th February because it was his

wedding anniversary. But I don't suppose that's important."

Bewes, who lives in Putney, South London, played Bob in the popular series. And we asked him how his hen pecking wife Thelma might have reacted to his target for walnut self-sufficiency. "I hadn't really thought about that", he confessed. "I seem to remember she was always trying to stop me

going to the pub with Terry, my best mate. I don't know. Perhaps she'd think it was a good idea. I'm not sure."

And Bewes was equally uncertain about how Britain's nut growers are supposed to go about increasing their crops. "I suppose if they invested in new technology – some sort of nut fertiliser, or mechanical nut pickers, that would help.

Pooar! Was that

A wind of change is blowing through the Kent port of Folkstone since the recent opening of the Channel Tunnel.

But the residents haven't had a whiff of the booming business and economic growth they had expected from their new link to the continent. Instead they are breathing in dense, foggy clouds of putrid garlic fumes which are drifting through the tunnel.

MATTER

And this gas is no laughing matter. For British officials believe their French counterparts are *deliberately* pumping trouser gas through the tunnel in order to solve their own pollution problems. And as a result dangerously high levels of French fart fumes could soon be causing serious environmental damage in the Kent area.

GLOSSER

Engineers at the British end of the tunnel first detected a whiff of pickled eggs the day after the tunnel was officially opened by the Queen. At about the same time several complaints were made from members of the public who noticed that the white cliffs of Dover were turning yellow. Scientific tests then confirmed that alarming levels of guff gases originating on the French side were filtering through the tunnel.

EGGSHELLER

But as well as the obvious dangers to safety, local residents are concerned about the immediate threat which the Chunnel chuff gases pose to the environment. For the unpleasant stench can very quickly erode stone work, cause cars to rust, trees to shed their foliage, and wallpaper to peel off. Farm produce from within a 50 mile radius of the tunnel entrance is being monitored by Ministry of Agriculture officials, and one herd of cattle has already been destroyed after their milk began to taste of mouldy cheese and pickled eggs.

FRENCH

Officially the French deny funnelling their fumes into the Chunnel, but their diet

Farting Frogs funnel chuffs through Chunnel

of thick black coffee, garlic and frog's legs has lead to serious pollution problems in the past. And the French government were known to be investigating new ways of getting rid of the estimated 750 million tons of trouser emissions which the French public let off every day.

CAPITAL

John Major is thought to have expressed his personal concern to the Prime Minister of France over the matter, however, the Channel Tunnel Treaty which was signed by both countries makes no mention of fart gases, and as a result the British authorities are unable to take action over the issue.

RED

How to dispose of their plentiful and particularly pungent cabbage clouds has been a constant problem for the French throughout history. Napoleon first

highlighted the problem in 1812 after his army conquered Moscow only for the city to be burnt down after French troops, celebrating their victory, had accidently ignited their botty burps. Napoleon offered a reward of 2,000 francs (a sum of French money) to anyone who could invent a method of safely disposing of his countrymen's anal emissions.

BLACK

Another attempt to solve the problem came in the shape of the Eiffel Tower which was originally designed as a giant flue to release pump gases into the sky above Paris. But street cafe owners complained that such a scheme would be unhygenic with so many Parisians sitting on the pavement all day drinking coffee and eating garlic bread. And so the tower was converted into a tourist attraction instead.

Scene of the smells – the British entrance to the Channel Tunnel yesterday.

These two world famous landmarks were both built as cunning fart disposal devices.

The people of New York were grateful when the French presented them with the Statue of Liberty to commemorate the anniversary of American independence. But unknown to the Americans the statue was in fact intended as a 'Trojan Pump Horse'. It was in fact a cunningly designed giant gas tank filled with odourous farts. The gas would have been burnt off slowly, keeping the statue's famous torch alight for up to fifty years. But the resulting clouds of pungent smoke would have thrown the city into darkness and caused widespread illness and disease. Fortunately for the Americans the statue sprang a leak during its voyage across the Atlantic and the gas escaped, killing millions of fish.

PINK

Only Sweden has ever attempted to tackle the problem of national flatulence. In 1989 they became the first country in Europe to harness pump power and convert it into energy. The

flatulence fired electrical generator station at Trask was the first of its kind in the world, using farts to power a series of giant windmills which would in turn generate electricity. A 'wind tunnel' was built from the densely populated South East of the country to the power station 600 miles away. Unfortunately the amount of fart coming through the tunnel was insufficient to make the giant windmills turn, and the decision was taken to scrap the £400 billion project.

SHIT

Last night British Nuclear Fuels began negotiating with French sewage officials in an attempt to resolve the Channel Tunnel wind problem. It is thought that BNF will buy France's excess emissions and transport them to Nuclear power stations in Britain where they will piss about with them for several years before deciding what the fuck to do next.

VOUS?

We're farting back for Britain!

We're launching a patriotic campaign to save Britain from the disgusting smell which the French are chunnelling in our direction. And we already have the backing of several top stars, including Jim 'Nick Nick' Davison, Sting, Dame Vera Lynn and the late Field Marshall Montgomery.

We plan to give the Frogs a taste of their own medicine by sending them some good old British farts, 'Dambusters' style. And these will be bouncing bombs with a difference, as Charles Aznavor and co. will soon be finding out.

We're going to inflate red, white and blue beach balls using British wind, and send them bouncing off the white cliffs of Dover towards the French coast. And in a highly emotional atmosphere Dame Vera will sing some of her wartime favourites as the bombs are launched. If the French thought D-Day was spectacular, wait till they see this!

This is how YOU can help. We want everyone in Britain to send us a fart, and we'll use your farts to inflate our bouncing bombs. All you have to do is fart into an envelope, and send it to the following address: Viz 'Fartbusters Campaign', P.O. Box 1PT, Newcastle upon Tyne, NE99 1PT. And remember to mark your envelopes 'Proud to be British'. You can send as many farts as you like, but each one must be in a separate envelope. We

Proud to be pumping for Britain. Jim 'nick nick' Davison (above) and Dame Vera (above above).

regret that we cannot accept wet ones.

To prevent your fart simply blowing away whilst in the post, each envelope should be weighed down. To do this simply pop a one pound coin into the envelope before sealing it. And make sure you lick your envelope *before* you fart into it, or use a self-sealing envelope. *Under no circumstances attempt to lick a fart filled envelope.*

Always follow the farting code

For reasons of safety always take these simple precautions when farting:

1. Always fart in a well ventilated room, away from children or pets.

2. Never fart near a naked flame, or attempt to ignite a fart.

3. Under no circumstances should you fart whilst suffering from diarrhoea or any similar medical condition. If in doubt consult your doctor.

4. Never hold a fart in – it could make your heart explode.

JAILHOUSE SHOCK

A Cleethorpes man is to write to the Home Office after a trip to one of Britain's 'luxury' jails turned into a nightmare.

Joe Worthington was looking forward to six months of booze, sex and drugs after reading about Britain's jails in the tabloid press. But within hours of arriving at Hull's high security prison Joe was already wishing he was back home.

STAY

"I'd read about the sex and drugs and was really looking forward to my stay", Joe told us. "I'd heard that all sorts of drugs were freely available inside, and that prostitutes would be provided, so I was quite excited when the judge gave me six months." But Joe's dreams were quickly shattered.

SIT

"The first thing that struck me was how small the rooms were. I asked for a single but they gave me a twin which I had to share with a total stranger. There was no tea and coffee making facilities, no TV and no mini-bar either. But worst of all our room didn't have en-suite facilities, so whenever I needed the bathroom during the night I had to use a bucket which the cleaner had left in the room."

HEEL

According to Joe the prison's restaurant had to be seen to be believed. "It was like something out of Fawlty Towers," he told us. "There were hardly any staff. It was so bad we had to do a lot of the cooking ourselves. And the service was appalling. In the whole time I was there I never saw a single wine list."

SOLE

Leisure facilities at the prison were also a disappointment. "There was never anything to do. It was left up to the wardens to organise games and activities, but they weren't exactly the brightest people you'll ever meet. All they ever did was tell us to walk round in circles in the yard. No day trips or outings or anything like that. In fact they wouldn't even take us to the pub in the evenings."

PLAICE

Joe's wife had been looking forward to visiting him. She saved up for a fortnight just to buy him drugs and had

'Luxury prisons' a con says con

Joe outside the gates of a prison similar to the one refered to in our story, yesterday.

put on a clean pair of knickers, fully expecting to be ushered into a side room for a steamy prison romp with her husband. But she was in for a big surprise. "We had to sit at a table in a big room full of other people. It was really embarrassing. My wife suggested I smoke some drugs to help me relax while we had sex on the table. But I hadn't even got her knickers off when the warden came and pulled us apart. Next thing you know they threw her out and chucked me back in my cell. And they wouldn't even let me keep the drugs."

SKATE

Indeed, getting his end away was proving to be a bit of a problem of Joe. "The prostitutes were the main reason I wanted to go to prison," he confessed. "I'd heard the wardens smuggle them in for steamy romps with the inmates. But when I politely asked if they could get me a couple of girls to put on a lesbian show for me, I was dragged before the gover-

nor. I told him one girl would do, and I'd even settle for some quick topless relief, but he wasn't having any of it. I ended up in solitary confinement for a week."

SKI

Since his release Joe has written to the prison authorities complaining about the facilities and also the standard of service inside the prison. But so far he has not received a reply. "I've also written to the judge who arranged the sentence for me. I'm sure he wouldn't be sending people to Hull prison if he knew what a dump it was." said Joe.

MÜLLER LITE

When we rang the Home Office a spokesman admitted that there were problems maintaining standards at many of Britain's jails. "To be quite honest we can't get the staff." he told us yesterday.

He's a Sneezy Lover

By our Medical Musical Correspondent
Dr. Feelgood Stutterford

AS well as contending with aching limbs, runny noses and swollen glands, flu sufferers this winter will face an extra headache - a whopping bill from pop millionaire Phil Collins!

For the baldy Genesis drummer, whose previous investments include fish farms, christmas trees and racehorses, has snapped up all world rights to the influenza virus.

Victims

Unlucky victims will find themselves coughing up an amazing £8.50 a day in royalties to the greedy chart-topping slapster. If this winter's expected epidemic materialises, Collins can look forward to profits of £5000 million billion or more.

The War Song

Collins, 45, acquired the infection privately two weeks ago and immediately leased it to himself via a wholly owned holding company, 'Ill Collins Plc' based in the Channel Islands. City analysts expect profits from the company to double with

Swollen coughers swell coffers for stumpy tubthumper

Collins - drumming up cash and Norman Dodds (below) - not taking it lying down.

this new addition to a portfolio which already boasts veruccas, bad guts and the clap.

The Medal Song

But news of Collins' winter bug buy-out got a cold reception from Norman Dodds, chairman of the National Influenza Sufferers Society. "This is a terrible blow for anyone with a bunged up nose" he told reporters.

It's a Miracle

And Collins is not the only pop star to cash in on peo-

ple's misery. As cases of T.B. increase, has-been trouser-splitting singer P.J. Proby looks forward to a cash windfall, having made what looked like a bad investment when he bought a majority share in the degenerative lung disease in the fifties.

However, illnesses are not always a healthy investment. Ex-Beatle Ringo Starr made a big blunder in 1967 when all his Yellow Submarine royalties sank without trace after he bought smallpox, three weeks before a cure was found. His sole income nowadays comes from ownership of the rights to that pain you get behind your eye if you eat ice cream too quickly.

Karma chameleon

A spokeswoman for Collins last night said, "The number you have dialed has not been recognised. Please replace the handset and try again. Do do dip. Do do dip."

Who's Next?

The Who guitarist Pete Townsend revealed this week how his life was wrecked after the death of the band's drummer, Keith Moon, 20 years ago.

For since that time, the rock legend has lived in fear of a curse developing that would pick the band members off one by one.

Member

"Keith's death could be written off as a one-off thing" he told us yesterday. "But if another band member, say Roger or John were to die, then 'The Curse of The Who' would be a reality, and I could be next."

Tool

The fear of the curse has taken its toll on Townsend. Nervous-looking and a chronic chain smoker, he hasn't left his Rich-

EXCLUSIVE

mond mansion since Moon's death in 1978, except to go out and perform his daily business.

Chopper

But other band members were less worried. "I wouldn't believe in 'The Curse of The Who'" said Roger Daltry, speaking from his fish shop. "It would all be a load of scaremongery and mumbo jumbo."

John Thomas

Bass guitarist John Entwistle was less sceptical, however. "The series

The Who - no hex please, we're Brit-pop.

of deaths would probably be a coincidence rather than a curse." he told us. "But being a superstitious person I'd probably be a bit more careful when crossing the road or eating fish bones."

THAT DOOR IS TOO BIG

MY WIFE LIKES TO MAKE A GRAND ENTRANCE

SAND

A 45-year-old Lincolnshire librarian was last night charged with sweetening a cup of tea left on a worktop by his mother.

Graham McBride of Bardney Old Cottages, Woodhall Spa stands accused of adding one or more teaspoonfuls of sugar to the tea, belonging to Mrs. Brenda McBride, 70, of

An angry McBride is led away by police.

the same address, making it unpalatable to her. A further charge of sipping the tea may also be brought if the results of forensic tests prove positive.

Statement

A short police statement issued this morning read, "At 2.30a.m., Graham McBride was charged with sweetening tea on the 15th November this year. We also wish to speak to him about a sipping offence, and he has been detained for further questioning."

Overdraft

Mrs. McBride was unavailable for comment today, but a neighbour who did not wish to be named told reporters that she had been visibly shaken on the day of the incident. "The first thing we knew about it was when the police cars pulled into the close," she said. "This isn't the sort of thing you expect around here."

Queueues

In 1956, Mr. McBride's father, Ernest, then 30, was hanged after being found guilty of stirring his tea with the sugar spoon, and then replacing it in the bowl when it was all wet.

TRAFFIC WARDENS CAN'T GET IT UP

Britain's army of traffic wardens are notorious for taking a **hard** line when it comes to minor parking offences. But when it comes to sex they're **SOFT**, and that's official.

For a report published this week reveals that 9 out of 10 male traffic wardens are unable to achieve an erection. And while everyone else enjoys raunchy, energetic sex with their partners, Britain's pathetic parking prefects are left to wander the streets, sticking little parking tickets on car windscreens.

INABILITY

Research shows a direct link between the issuing of parking tickets and an individual warden's inability to 'get it up', as Professor Morris McEwan-Scotch, author of the report, explains.

FRUSTRATED

"The worse a traffic warden is in bed, the more tickets he dishes out. He becomes frustrated, and angry, and inevitably he takes it out on innocent motorists."

AFFAIR

The professor refutes the suggestion that many of our traffic wardens are happily married, and enjoy normal sex lives. "This is simply not true", he told us. "My research has shown that 98% of traffic wardens' wives are having an affair with their next door neighbour, because their husbands cannot satisfy them".

VIRGINS

Professor McEwan-Scotch believes the problem of impotent traffic wardens is deep rooted. "Traffic wardens are without exception social inadequates who ideally would have liked to be policemen. Most of them are still virgins when they start the job, and they can't take their drink either. Add to this the fact that they are almost to a man *deeply* unattractive, and you can see how the problem develops."

EXCITEMENT

The professor goes on to cite various unhealthy sexual practices to which he believes traffic wardens turn in order to achieve excitement. "I have evidence which suggests up to half of Britain's traffic wardens are 'cross dressing' – wearing women's clothing – during the evening. And I am convinced that the vast majority of them use battery operated devices in the privacy of their own homes".

FOREPLAY

One of our reporters, posing as an attractive female motorist, invited a passing traffic warden to have sex in the back of his car. However, after almost thirty minutes of unimaginative foreplay the traffic warden was still unable to achieve an erection. At this point our reporter made his excuses and left.

CLAMP DOWN

Professor McEwan-Scotch wants the Government to clamp down on traffic wardens, sacking them all and abolishing parking meters. "I have sent my report to the Ministry of Transport and am awaiting a reply". Professor McEwan-Scotch was recently fined £16 for parking on a double yellow line while shopping near his home.

Pitiful sight – a lonely traffic warden wanders the streets yesterday.

Farmer Jack's poll tax shocker

Farmer Jack Johnson could hardly believe his eyes when he opened a Poll Tax bill sent to his farm by the local council at Coniston in the Lake District.

BILL

For the bill included a £512 Community Charge for one of Jack's employees. Nothing odd about that, until you realise that the employee concerned is none other than George, Jack's faithful Border Collie sheepdog.

SWEENEY

"I was flabbergasted. It's bad enough me and my wife having to pay. But with another £512 for the dog, we simply couldn't afford to make ends meet".

Jack decided the only way to avoid paying the bill was to shoot George, his faithful companion of 15 years, in the back of the head. "It broke my heart, but it was either that or paying up".

Z-CARS

The next day the local council wrote apologising for their mistake, and cancelling the bill which had been issued in error. The mistake occurred because Mr. Johnson had entered the dog's name on his Community Charge registration form, a spokesman explained.

Mr. Johnson is now contemplating suing the council for damages, together with the cost of the dog and six shotgun cartridges.

HEIR INDOORS

'No' to open door policy at Palace

AN urgent enquiry is to be launched after Prince Charles was left stranded inside a Buckingham Palace drawing room for almost 3 days last weekend.

By our Royal Correspondent Lickanarse Owen

The distressed Prince was found in a puddle of his own urine early on Monday morning by Palace cleaning staff.

Lunch

Charles is thought to have entered the small drawing room through an open door at around midday on Friday whilst visiting his mother the Queen for lunch. But when the door blew shut behind him, the Prince found himself alone inside the room.

Snuff

Palace staff who would normally open doors for the Prince failed to check the room before going off duty for the weekend and were unaware of the future King's plight.

Prisoner in the Palace - Charles wet himself

Cardboard

Charles was discovered at 5am on Monday morning by cleaning staff who had gone to the room to puff up cushions. He was reported to be in a distressed state, wandering around in circles and fiddling with his cuffs. The room was said to stink of faeces and urine.

Black

"Protocol has always forbidden members of the Royal Family opening doors for themselves", explained Royal author Sir Terrapin Walnut-Cake. "Charles would be totally baffled if confronted by a door which was closed. It would be a situation totally alien to a man of his upbringing and pedigree".

Robbie

The last monarch to open a door for himself was Henry VIII who caused a storm in 1545 by famously opening a bathroom cabinet in order to get some Alka Seltzer late at night.

Juke

Nowadays for security reasons all Royals are told never to enter a room on their own unless the door is securely fastened in an open position, or they can see an alternative exit. But it is thought that Charles, who has a stubborn streak, may have deliberately ignored this advice whilst going for a stroll.

The Prince of Wales spent Monday morning undergoing tests in a private room at St James' hospital in London. He was visited by his brother Edward who brought him clean underwear and some new trousers.

Dirt

Charles was later allowed home to Highgrove, and paused briefly to joke with a small crowd of demented old women and jingoistic, bigoted taxi drivers as he left the hospital.

DESPITE this latest scare the Queen remains reluctant to break hundreds of years of Royal tradition by allowing members of the Royal Family to open doors for themselves.

In 1982 Prince Edward risked the wrath of his mother by taking secret door opening lessons while studying at Gordonstoun school. However it was the Queen Mother who put a stop to it, threatening to turn Edward into a frog if the lessons continued.

Signs

But there are signs that in the Post Diana era the Royals are at least beginning to start to perhaps recognise the need for possible change.

Seals

The legacy of Diana is that Wills and Harry are able to use a TV remote control, and perhaps significantly, both princes wave to the public with an open hand, as opposed to the traditional rotating wrist 'wanker' style gesture preferred by the Queen.

Delivers

Haughty Royal nanny Threepotsandin Legless-Burke was recently scolded by Charles after photographs of Princes

Unhinged - Queen slams door on Royal door opening

Queen Mum - God Bless Her, she's 98 you know - made frog threats.

Harry and William opening a car door themselves during a holiday in Wales appeared in Sunday newspapers. But after his own harrowing experience it is hoped that Charles' attitude towards door opening may soften.

A snip at £40,000

A spokesman for the Royal Society of Gentlemen's Hairdressers yesterday defended the enormous hair cutting bill which Prince Charles has received after his two sons visited the barbers in July. He described the £40,000 bill as "not unusual".

Haircut

Former Prime Minister John Major took the boys, William and Harry, for a haircut at exclusive Mayfair barbers Shirtlift & Poovery over a month ago. However the Prince of Wales was said to be shocked by the size of the bill which he received several weeks later.

100

"The account no doubt reflects the amount of time that must have been spent on these haircuts, and it also includes a shampoo and rinse", said the spokesman whilst struggling to keep a straight face.

Rude GARDENERS' QUESTION TIME

With 'The Rude Gardener'

Dear Rude Gardener
Last year I planted a rhododendron but it has failed to flower and now it looks quite sickly. Everything else in the garden is fine. What could be wrong?
Mrs B., Essex

* You should have tested your fucking soil, you twat. They grow best in **acidic** soil, not lime, you dozy bitch. You've wasted your money and my fucking time. Next.

Dear Rude Gardener
Is it possible to grow olive trees outdoors in England?
Mr A. Kelly, Birmingham

* Is it fuck.

Dear Rude Gardener
On holiday recently in Devon I spotted a small yellow flower with white stripes on the petals and distinctive heart shaped leaves. I would very much like to grow it in my garden but do not know its name. Have you any idea what this pretty flower might have been?
Mrs Mary Hetherington

* How the fuck should I know? I didn't see it.

Send your queries to the Rude Gardener c/o Viz. The Rude Gardener regrets that he is far too busy to enter into individual correspondence with the likes of you. So fuck off.

IT'S FRIDAY...IT'S FIVE TO FIVE...IT'S...

THEORETICAL PHYSICS!

Crazy double lives of the TV Crackerjack boffins

What have the following TV stars all got in common? Michael Aspel, Bernie Clifton, Leslie Crowther, Ed Stewart and the late Eammon Andrews.

If you said they were all presenters of the popular kids' programme Crackerjack you would of course be right. But there is another less obvious link. For believe it or not each of them has at some time in the past baffled the scientific world with their own controversial and at times incredible physics theory.

SECRETS

The physics theories of the Crackerjack presenters has for many years been one of TV's best kept secrets. But now, for the first time, using files only just released from the BBC archives, we can exclusively reveal the physics theories of the former Crackerjack stars.

SCIENCE

MICHAEL ASPEL – launched a hugely successful career in TV with his appearances on Crackerjack in the early seventies. Since then he has never looked back, and is now one of Britain's top TV earners. But Michael's first love was always science, and in 1972 he published his own theory.

BOX

Aspel's theory was, quite simply, that if you spin around inside a box, you get dizzy. But if the box is then raised to 100 feet above ground level, and you spin around in it, you *don't* get dizzy.

GENIUS

This remarkable discovery was published in all the major scientific journals of the day, and Aspel was heralded as a genius by many experts. But the theory was disproved a short while later when a man spinning round inside a box 100 feet above the ground got dizzy. When the box was lowered to the ground the man got out and fell over.

EXCLUSIVE

Crackerjack stars (left to right) Glaze, Crowther, Aspel, Don McLean and a bird

Aspel was devastated. He immediately cancelled all his research and concentrated instead on his TV career. Needless to say he hasn't looked back since.

OSTRICH

BERNIE CLIFTON – is known to millions of kids as the man with the comedy ostrich. But during the seventies he starred as Crackerjack's resident funny man, and also took time out to develop a physics theory of his own.

EXPRESS

Bernie's belief was and still is that if you run backwards along the top of an express train which is travelling forwards at 100 miles per hour, the hands on your watch will move backwards.

FILTER

The Clifton Theory, as it became known, broke new ground for physics theoreticists. For he was challenging the very concept of time itself. And although his theory has never been substantiated by a successful controlled experiment, Clifton still has his supporters in the science world, among them the old bloke off 'How'.

CAPPUCINO

Physics theories of the Crackerjack presenters range from the brilliant to the bizarre. For instance **STU FRANCIS** once told delegates at a science conference in Bridlington 'If you wee into a milk bottle in a green house wearing a tin foil hat on your head, the temperature of your head will be equal to the volume of wee in the bottle, divided by its weight'.

INSTANT

And it was former presenter **ED STEWART** who made headlines in 1974 when he claimed that 'The total weight of apples on any tree would be sufficient to lift the tree out of the ground if gravity was upside down'.

APPLE

LESLIE CROWTHER, who played funny man to Peter Glaze's straight man on the show, added his own apple theory when he said that 'The number of apples on any number of apple trees at any one time is equal to that number of trees divided by itself, and multiplied by the total number of apples thereupon'. Crowther's Theory of Apples on Trees has since become universally accepted and is used by apple growers worldwide when calculating how many apples they have on their trees.

EMI

A BBC book entitled 'Unusual Physics Theories of the Former Crackerjack Stars' is due to be published later this month by BBC Publications, priced £14.99.

Ursine 'o' the Times

King of Pop survives another bear attack

PINT-sized popstar Prince has been forced to put the release of his latest album back by twelve months following a bizarre break-in at his Paisley Park studios. However, the intruder wasn't an internet bootlegger after his master-tapes... it was a 30-stone black bear bent on pinching his packed lunch, the *fourth* such attack in his career.

Prince was about to tuck into a mackerel paste sandwich during a break in recording when the ravenous animal tore its way through a studio wall. "Prince stopped singing and yelled 'Bear!'" bass player Hamilton Winter told reporters.

grizzly

"All us members of his backing band *The Revolution* fled for our lives as the giant grizzly cornered the 4'11" funk svengali, whose hits include Purple Rain, When Doves Cry and Cream, against a piano."

By our Pop Correspondent
Labia Seepage
in Mineapolis, USA

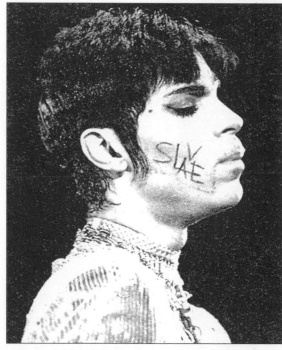

"It was terrifying," agreed producer Todd Rundgren, who watched the drama unfold.

matchwood

"The artist formerly known as the artist formerly known as Prince tried to fend it off with his guitar, but one blow from the bear's paw smashed it to so much matchwood. In the end he had to throw his sandwiches to the other side of the studio to distract

it long enough to make his escape."

The 10 foot bear then spent half an hour lumbering round the $12 million studio complex, ripping up master tapes, destroying mixing desks and shitting on the drums. By the time the Mineapolis State Wildlife Rangers arrived to shoot it with a tranquiliser dart, a year's work on Prince's latest album had been destroyed.

The 11' 4" brown bear (above) which burst into the recording studio where 4' 11' Prince (left) was recording.

The tiny singer, who has penned hits for stars such as *Cyndi Lauper, Sinead O'Connor, the Bangles* and *Clive Dunn*, had been hoping for a January release, but the bear attack now makes that deadline look increasingly unlikely.

Lauper - True Colours

Prince, who recently changed his name to a squiggle and then back again when nobody noticed, is believed to be particularly angered by the attack since he recently shelled out over $3 million having his recording studio bear-proofed

It's the Scare Be

THIS bear attack is just the latest in a terrifying series which have plagued the purple pocket rocker throughout his career:

● **Attack 1:** *Sessions to record Prince's 1984 chart-topper '1999' were interrupted when a confused kodiak bear which had woken early from hibernation wandered into the recording studio after apparently smelling a Toblerone which Prince was eating. The singer managed to hold it at bay by sticking a microphone stand up its arse until help arrived.*

● **Attack 2:** *During a 1987 UK tour to promote Prince's album 'Sign 'o' the Times', a female polar bear and her cubs which had escaped from the North Wales Mountain Zoo sneaked onto his tour bus while it was parked in a layby near Colwyn Bay. Prince and his band the New Power Generation were*

having a wee in a field when they heard the horn tooting. Returning to the coach Prince was horrified to discover the 11 foot arctic carnivore sitting in the driver's seat whilst her cubs played in the aisle. "Everything would have been okay if Prince hadn't made the mistake of somehow getting between the bear and her cubs," recalled drummer Sheila E. "That's the first rule when you're dealing with

12 THINGS YOU NEVER KNEW about BEARS

THEY'RE cuddly, they're grizzly, they're brown, they're polar and they're the bane of Prince's life. That's right, they're bears and they're the world's biggest animal beginning with 'B'. Apart from blue whales. We see them everywhere we go, from the North Pole to toy shops, But we "bearly" know anything about them. Here's the "bear" essentials, a dozen furry facts about our favourite cave-dwelling carnivores.

1 Evolution has played a cruel trick on bears, giving them short front legs. This means that their arses stick up higher than their heads, making bears a tempting target for angry bees.

2 Like Fred Astaire and Ginger Rogers, bears just love to dance. In Turkey and Greece proud bear owners teach their pets to trip the light fantastic, a process which involves snapping their teeth out with pliers and beating their shins with sticks.

3 Contrary to what you might think, not all bears are bears. For example, koala bears aren't bears at all. They're actually koalas, a type of small Australian bear which eats eucalyptus leaves.

4 What's more, some things that aren't called bears are bears. For example, the giant panda of China isn't really a panda. Scientists now believe that these bear-like black and white bamboo-chomping mammals are actually a special kind of bear called a panda.

5 Not all bears shit in the woods. Polar bears live in the Arctic, at least 3000 miles from the nearest tree, so they probably just have to do it behind some snow and then wipe their arse on a penguin. Other bears that don't shit in the woods include the Coventry Bears, a rugby team from the West Midlands. They probably shit in pint glasses in pubs.

6 Ask TV naturist David Attenborough what the biggest bear in the world is and chances are he'll tell you it's the Kodiak bear, which can top the scales at ten foot tall. However he'd be wrong, because the biggest bear in the world is actually the Great Bear, a constellation made of stars in the night sky which is hundreds of miles across.

7 Like British old age pensioners, bears curl up and go to sleep in the winter. However, unlike many British old age pensioners, bears tend to wake up again in the spring.

8 In the old days it was often said that if you stepped on the cracks in the pavement, the bears would get you. However, nowadays if you step on the cracks in the pavement you may be entitled to compensation worth thousands of pounds from the local authority on a no-win, no-fee basis. If you'd like to find out more about making a claim, call the Accident Advice Helpline on 0800 180 4060.

9 If you go down to the woods today and interrupt a bears' picnic, you could be in for a big surprise. That's because not only is a bear an animal, it's also a big, hairy homosexual. Small, bald homosexuals are called Jimmy Somerville.

10 Bears have featured in many films such as *The Jungle Book*, *BJ and the Bear* and *The Adventures of Grizzly Adams*. In 1978, Clyde the orange bear became the first bear to win an Oscar when he co-starred alongside Clint Eastwood and Geoffrey Palmer in the action comedy *Every Which Way But Loose*.

11 Insufferable TV favourite Gyles Brandreth has an amazing secret. For the ganzy-clad ex-MP owns 8,000 teddy bears, which he keeps at a museum in Stratford-upon-Avon. Every evening before Brandreth goes to bed he kisses each one of them goodnight - a process which bear-kissing boffins estimate takes 16,000 seconds!

12 The world's largest bear is the kodiak bear of Alaska, which stands over ten feet tall - that's equivalent to the height of Kylie Minogue standing on Ronnie Corbett's head. And it weighs nearly three quarters of a ton - the same as the Krankies... *carrying twenty-six car batteries!*

---

r Bunch

ears, but Prince clean forgot. All ell broke loose and he ended up iding in the parcel shelf until lwyd Police arrived and lured it ff the bus with a dead seal."

Attack 3: Whilst shooting a cene in a marmalade factory for is film debut 'Purple Rain', Prince arrowly escaped death when a narauding Peruvian spectacled ear smashed its way through a kylight and dropped onto the set. ten minute static chase on a onveyor belt ensued which only nded when Prince vanished into n automatic bottling machine, merging moments later in a bigger than usual jar of marmalade. All we could see was his eyes linking above the label," said irector Trevor Brooking. "It would ave made a great scene for the lm, but Prince insisted it ended p on the cutting room floor."

Could YOU survive a bear attack? ...Yes! With your FREE Gerry 'Grizzly' Adams' Bear-ometer!

" As leader of Sinn Fein, I'm used to pressure. But let me tell you, compared to coming face to face with a hungry kodiak bear, negotiating with the Reverend Ian Paisley across the table at Stormont is a *walk in the park!* You'd better believe that coping with an angry bear can be tricky. What's more, a tactic that might save you from one species could spell certain death when you come up against another.

Over the years, I've amassed a huge amount of data about how to cope with angry bears of all types. Now I'm making the fruits of my research available to Viz readers in the form of this handsome pocket Bear-ometer fact file. Keep it with you at all times to ensure complete safety in the event of an attack. **"**

Gerry 'Grizzly' Adams' Bear-ometer! ~ Pocket Bear Survival Guide

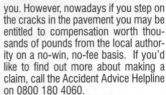

Breed of Bear	Danger Rating	Risk Factor	Peril Ratio	Escape Strategy
Koala Bear	10%	1/10	0.1	Throw Lockets or mentholyptus flavoured Tunes to distract the bear before walking calmly away.
Sun Bear	30%	3/10	0.3	Keep it at bay with a stick or snooker cue until help arrives.
Sloth Bear	30%	3/10	0.3	Sloth bears are scared of fire and loud noises, so brandish a burning faggot and shout "Hyah! Hyah! Gertcha!"
Spectacled Bear	40%	4/10	0.4	Dig a shallow pit filled with bamboo spikes between you and the bear, then goad it into falling in.
Giant Panda	40%	4/10	0.4	Pandas have very poor eyesight, so run upstairs and hide in the wardrobe.
Black Bear	60%	6/10	0.6	Clap loudly to confuse the bear whilst retreating slowly to a place of safety.
Brown Bear	70%	7/10	0.7	Smear yourself with your own excreta and lie perfectly still - the bear will soon move on.
Kodiak Bear	70%	7/10	0.7	Avoid eye contact, don't bare your teeth and run away backwards in a haphazard zig-zag.
Polar Bear	90%	9/10	0.9	Polar bears can run at speeds in excess of 60mph so you have no chance of outrunning one. Head towards thinnish ice which will support your weight but crack under that of a polar bear. But make sure it's not too thin, if you fall in as well you're a goner. A polar bear can swim at speeds in excess of 50 nautical miles an hour.
Grizzly Bear	100%	10/10	1.0	The grizzly is perhaps the world's most gullible bear so tell it that 30-stone darts star Andy Fordham is coming across the bridge next, and he will will make much finer eating than you.

THE UK'S LIVELIEST DEEP-FRIED POTATO SNACK FORUM

INTRODUCED BY TV'S CAROLINE QUENTIN

Quentin's CRISPS

"Hi, Crispophiles! Caroline Quentin here. I've had a really big bag this week... crisps, that is! Yum yum! But enough about my crisps, it's time for some letters about yours!"

I WAS a promising semi-professional footballer who had never even tried a crisp until 1995, when Gary Lineker started advertising them on the television. Because Gary was my sporting hero, I started to eat them by the box-full. 14 years down the line, I now weigh over 50-stone and suffer from scurvy, ulcerative colitis and congestive heart failure. How ironic that they are called Walker's Crisps, when I am unable to walk; I can't even turn over in my shit-filled bed.

Big Joe, Bradford

My Best Crisps!

Shakin' Stevens - *Salt'n'Shake*

I SENT my son out to buy me some Pringles at the corner shop. I couldn't believe my ears when he came back and told me he had just taken out a majority shareholding in the Scottish Borders-based knitwear manufacturers.

Mrs D Estelle, Falmouth

WITH reference to Mrs Estelle's letter *(above)*, the other day I needed some new diamond-patterned sweaters, so I sent my son out to buy me some Pringles. Imagine my shock when he came back with a cardboard tube containing overpriced crisps!

Val Doonican, Surrey

I AM an American, and during a recent holiday in England I went into a shop and asked for a bag of potato chips. Imagine my consternation when, instead of the crisps I was expecting, I was handed a ... no, hang on. This doesn't work.

Rhoda Felchenstein, New Jersey

I'M A huge crisps fan, so I got very excited the other day when I noticed that my local greengrocer was selling "Crispy Lettuces" at half price. But what a disappointment they turned out to be. They were large, green, soggy leaf-like things, and I had to empty nearly half a pound of salt and nearly a pint of vinegar onto them before they tasted anything like crisps. I certainly shan't be buying any of those again!

Emma Outhwaite, York

DURING the war, when rationing was on, my mother used to make her own crisps. She used to fix a potato in my dad's workbench vice and use his carpentry plane to shave off crisp-sized slivers. Then she used to iron the slices individually, flavouring them with dried bacon rind and onion skins, then leave them in the airing cupboard to cool down. Finally, she would unpick an old jumper and use the wool to knit a crisp bag to keep them in. In the 60 years since the end of the war, I have eaten many bags of so-called "proper" crisps, and do you know, I've never tasted anything even remotely as disgusting as my mother's home-made versions.

Phyllis Crabs, Manchester

I AM an American, and during a recent holiday in England I went into a shop and asked for a bag of French fries. Imagine my consternation when, instead of the ... no. It's still wrong.

Rhoda Felchenstein, New Jersey

AS A Professor of English Literature, I am constantly disgusted by crisp manufacturers who insist on replacing the "and" in "salt and vinegar" with a "'n'". And the phrase "Ready Salted" similarly leaves me incensed. The word "Ready" is an adjective, not an adverb. It is a simple grammatical fact that an adjective cannot be used to qualify the word "salted". With such poor standards of literacy on crisp bags, is it any wonder that this country is going to hell in a handcart?

Prof. A Gowans-Whyte, Cambridge

I AGREE with Professor Gowans-Whyte *(letter, above)*. Crisp manufacturers are setting a very poor example when it comes to the standard of English used on their packaging. We should not be expected to put up with it. To this end, when I go into my local corner shop, I refuse to hand over my hard-earned money for "Wotsits". Instead, I make a point of asking for a bag of "Which Are Theys".

Dr O Trubshaw, Carshalton

MY CAT Mr Truffles loves crisps. I took a photograph of him the other day with his paw stuck in a bag of his favourite Cheese & Onion crisps, and all crumbs in his whiskers. However, I'll not get round to finishing the film off until my summer holidays, so here's a photograph of my late sister instead.

Mrs Hilda Feldman, Jarrow

My Best Crisps!

Rose West - *Salt & Vinegar Chipstix*

Crispy Laffs

Q: *What sort of crisps do musicians like?*
A: Quavers.
Mrs Edna Arteriosclerosis, Glasgow

Q: *What sort of crisps do chiropodists like?*
A: Cheese and bunion.
Mrs Edna Arteriosclerosis, Glasgow

Q: *What sort of crisps do you call a sailor with a bottle of vinegar?*
A: Salt and Vinegar.
Mrs Doreen Angina, Glasgow

Q: *What other sort of crisps do chiropodists like?*
A: Pickled bunion.
Mrs Edna Arteriosclerosis, Glasgow

C. Harris, St Leonards

Readers' Crisps

Caroline's Crisp Bag

YOUR CRISP PROBLEMS SOLVED

Dear Caroline,
DURING a recent holiday in Dorset, my wife and I shared a particularly delicious bag of Walker's crisps which we bought from a shop in Corfe Castle. Foolishly, we threw the bag away when we had finished the crisps, so we have no idea what the flavour was. It was a green bag and we would love to find out the flavour of the crisps so we could buy another one.

T Medford, Cricklewood

When I got your letter, I chartered an executive jet and flew to the Leicestershire headquarters of Walker's Crisps, where I had a meeting with Sir Giles Walker, the company founder. Sadly, he is now in his late 90s and was unable to remember which varieties of his firm's products come in green bags, but thought that the ones you ate in Devon could possibly have been Cheese & Onion or Smoky Bacon.

Send in your crisp-related queries to: Quentin's Crisps, c/o Viz, PO Box 656, North Shields NE30 4XX. There's a crisp bag of barbeque beef crisps for every problem that we use.

Your Little Blue Salt Bag Texts

litl blu salt bag rox 100%!
crispy_jim, leeds

Y DO I HAV 2 PUT MI OWN SALT ON MI *&£*@ING CRISPS? WOTA RIPOF! ANTIBAG DAVE

jim of leeds get a life salt bags R 4 losers_gurrl_

GURRL IS TH REAL LOSER.SALT BAGS R WKD!DR_FRANCIS_WHITLEYBAY

antibag dave is gay_ the bagmeister

fuk of bagmeister im not gay u r u cunt frm antibag dave

antibag dave ur mum just sukd me of!!!_the bagmeister

Y R THEY BLU?WOTS RONG WIV RED

FOR US MANU FANS?ROONEY4EVA

BAGMEISTER U R FUKIN DED FRM ANTIBAG DAVE

Hi. Does anyone know when the little blue salt bag was invented? George

GEORGE U TAKIN TH FUKIN PISS OR WOT?U R DED 2 FRM ANTIBAG DAVE

nice1 rooney4eva! league facup championsleague champnions1999!ronaldofan

manu shit wankers fuk of u pedos_dennis hopkins from wimbledon

I just fucked Antibag Dave's mum up the arse. George

Kids Say the Funniest Things about Crisps

MY HUSBAND and I recently took our six-year-old grandson to a local Indian restaurant as a birthday treat. "Look nana!" he announced at the top of his voice as we sat down at our table. "Those are the biggest crisps I have ever seen!" Needless to say, he was looking at a plate of poppadoms! It still reduces me and my husband to helpless tears of hysterical laughter for several hours every time we think about it.

Mrs Glenda Feverish, Hull

MY WIFE is horribly afflicted with septic psoriasis, like the character played by Michael Gambon in *The Singing Detective*. The other day, she had just been scratching her face and scalp and was brushing a large pile of dry skin flakes off the table into a dustpan when our four-year-old grandson came into the kitchen. "Gosh granny," he piped up. "Those are the smallest crisps I have ever seen!"

Bert Mills, Essex

My Best Crisps!

Richard Gere - *Beef Space Raiders*

MIRIAM'S PHOTOGRAPHIC CASEBOOK

Lady Marchmaine's Perfidious Liason - Day 3

Lord Marchmaine has taken the Stephenson's Rocket to London on business, and once again his wife has seized the opportunity to pursue her clandestine relationship with Mr Fezzywigg the chimney sweep.

He's a grubby tradesman and I'm a member of the aristocracy. I know it's wrong....but it feels so right.

Cor! Strike a light, your ladyship. I'm werry obliged, strite up I am.

That was the front door! His Lordship must have returned from the city preveniently.

Lumme! I'll scarper up the chimney, m'lady.

Egad!

What the deuce are you doing in your bed-chamber at this hour, Euphemia dearest?

I'm afraid I had a fit of the melancholic vapours, Edward.

Oh? If that is so, then how, pray, do you explicate this?

CONTINUES TOMORROW...

Victorian Miriam

VICTORIAN MIRIAM ANSWERS YOUR 19TH CENTURY PROBLEMS

MP is Father of my Child

Dear Victorian Miriam...

I AM in big trouble and I don't know what to do. I am an 18-year-old chambermaid, working in a large house in Westminster. My employer, an important member of parliament, has been making improper demands of me for several months and now I find that I am going to have his baby.

He has told me to get rid of It and threatened me with dismissal if I don't. But I want to keep the child. Please help me, Victorian Miriam, I am confused. What should I do for the best?

Betty, London

You must do as your employer says and terminate your confinement. If you do not do this, and you are dismissed, you will undoubtedly end up in bad circumstances. Your baby will be put in the workhouse, get rickets and eventually freeze to death in the snow whilst you will be locked up in a lunatic asylum and spend the next forty years picking oakum. More importantly, the child's father will suffer irreparable damage to his reputation.

I am sending you my free leaflet for unmarried pregnant women, which explains how to drink a bottle of gin, take a hot bath and throw yourself down the cellar steps.

Tempted to Sin by Pianoforte

Dear Victorian Miriam...

I AM a professional, God-fearing, married gentleman, but I was recently moved to perform a sordid act of personal pollution upon my virile member, wherefore a most dreadful guilt is preying upon my mind.

I am 42 and I have been married for nearly twenty years. Until recently my wife performed her marital obligations without complaint. However, in the past year she has become most conjugally unaccommodating and thus I have found myself increasingly physically frustrated.

Last week I retired to the drawing room after dinner with the intention of imbibing a balloon of brandy, and found myself unaccountably fixated by the elegant leg of the pianoforte. Unbidden, I became roused to a state of tumescent passion by the firmness of its round curves, and the delicate pertness of its finely carved ankle.

Before I could help myself, I had dropped my breeches and wrought a foul act of onanism upon my turgid person, divesting myself of my base spendings all over the mantlepiece and everywhere.

Miriam, I am beracked by remorse at my weakness, yet at the same time I feel tempted to sin again in spite of my penitence. Please help me, as I fear for my soul.

Albert, Peterborough

Your fears regarding the welfare of your immortal soul are well-founded, for the road you have embarked upon will surely lead you to the very fires of Hell itself. Next time you feel the urge to despoil your parts of shame, try taking an icy bath or sharply rap the bellend of your unmentionable with a cold silver kedgeree spoon to dampen your ardour.

Worried that Husband is Stabbing Whores

Dear Victorian Miriam...

I AM severely vexed that my dear husband may have be pursuing a secret life that is threatening our marriage. We have been wed these five and twenty years, and until recently have been blessed by the good Lord with a state of matrimonial harmony.

However these few weeks past he has been behaving in a way strange to his previous manner, travelling to Whitechapel each night and returning in the early hours covered in blood.

Whilst going through his portmanteau, I discovered a leather apron, a butcher's knife of exceeding sharpness, miscellaneous human organs and an ink bottle filled with a congealed red fluid of some sort.

Miriam, I am afeared that my husband is Jack the Ripper, but I don't know what to do about it. Please come to my assistance, as I know not the correct direction in which to turn.

Nancy, London

Your husband certainly appears upon first glance to be Jack the Ripper, yet not-withstanding this may I caution you not to jump hastily to any unwarranted conclusions. Indeed, there may be a perfectly simple explanation for his eccentric behaviours.

Sit down and talk to him earnestly. Enquire of him in a frank manner whether he has started murdering prostitutes in the east end of the capital.

If you are still worried, send a penny-blacked addressed envelope for my leaflet '*I Think my Husband is Jack the Ripper*'.

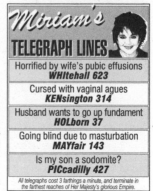

The Raggy Omaar Alphabet

THE TV COVERAGE of the Iraqi war has been choc-full of stuff for the lads - guns, bombs, explosions, tanks and fighting. There's even been an impromptu football match! But unlike previous conflicts, the ladies have been glued to their screens too, thanks to the presence of Raggy Omaar, the BBC's hunky Baghdad correspondent.

Raggy's authoritative rooftop reports from the Palestine Hotel have set the news agenda, whilst his dreamy good looks have sent the nation's ladies into a spin. The coalition forces may have stormed right into the heart of Saddam's stronghold, but Omaar has stormed straight into the heart of every red-blooded British bird.

Unfortunately, now that the thrill-a-minute bombardment of the Iraqi capital is over we'll be seeing less of Omaar on our screens. So to keep the ladies going, here's the A to Z of Britain's best-loved bullet-dodging dreamboat.

A is for **ANTS**. Although bombs, shells and missiles hold no fear for Raggy, he's absolutely terrified of ants!

B is for **BREAKFAST**. Raggy gets up every morning and munches his way through TWO bowls of Golden Grahams!

C is for **CAMBRIDGE**. The posh university. But Raggy didn't go there. He went to Oxford.

D is for **DRACULA.** Whilst a student, Raggy spent the night in a haunted house to raise money for charity. However, he fled halfway through the night after he was attacked by vampires.

E is for **ELEPHANTS**. If Raggy wanted to know the difference between an Indian and an African elephant, all he would have to do is look at their ears. That's because one of them has bigger ears than the other!

F is for **FOOD.** Raggy's favourite food is currants!

G is for **GLOCKENSPIEL**. This is Raggy's least favourite musical instrument. He hates its sound, after being forced to play it at nursery school!

H is for **HISTORY**. Want to know when the Battle of Hastings was, or how many wives Henry the Eighth had? Just ask Raggy - *he's got a degree in history!*

I is for **INTERCOURSE**. Raggy's the proud dad of two lovely kids - which means he must have had sexual intercourse in a lady.

J is for **JUNGLE BOOK**. Raggy's favourite film is Disney's Jungle Book. *"I reckon I must have seen the Jungle Book over 500 times,"* he says.

K is for **KIPLING**. Raggy's favourite story by Rudyard Kipling is the Jungle Book *"I reckon I must have read the Jungle Book over 500 times,"* he says.

L is for **LIBERTY-X.** Raggy is the teen band's biggest fan. His favourite is Kelli.

M is for **MARRIAGE**. The bad news is Raggy's already been snapped up - and the lucky lady's really posh, like the Duchess of Bedford or someone. The good news is that experts think stressful jobs like Raggy's put relationships under a lot of strain. Raggy's marriage could be headed for the rocks, so don't give up hope just yet, girls!

N is for **NUDE**. For a bet, Raggy once read the 9 o'clock news...stark naked! But don't get too excited, girls. It was on the radio!

O is for **ORANGE JUICE**. When he was 14, Raggy knocked a full glass of orange juice over on the kitchen table. His mum quickly mopped up the mess with some kitchen towels, and the whole incident was swiftly forgotten.

P is for **PRANKSTER**. Raggy once put a live crab down the back of Iraqi information minister Mohammed Saeed al-Sahaf's trousers!

Q is for **QUANTUM LEAP**. Raggy always mixes up the TV show 'Quantum Leap' with 'Sliders'. *"I can just never tell them apart"* he says.

R is for **RAISINS**. After currants, Raggy's second favourite food is raisins!

S is for **SULTANAS**. Strangely, although he loves currants and raisins, Raggy absolutely HATES sultanas. He says they look like rabbit droppings. Ugh!

T is for **TEASE**. Throughout the Gulf conflict, Raggy got the lady viewers hot under the collar by undoing the top two buttons of his shirt.

U is for **UNCLE**. Raggy has a famous uncle. Pint-sized veteran legless stand-up comedian Arthur Askey is married to his mother's sister.

V is for **VEST**. Raggy never goes out and about in war-torn Baghdad without his bullet-proof vest. But don't worry girls, he keeps his other vital organs well protected...in a sexy bullet-proof silk thong!

W is for **WAR**. Despite reporting on wars all over the world, Raggy is not a fan of armed conflict. He prefers peace!

X is for **XYLOPHONE**. It is not known whether Raggy has any particular feelings one way or the other about xylophones.

Y is for **YIKES!** Raggy once swallowed a drawing pin, and had to mash up his stools with a fork for a whole week to be sure it had come through his system!

Z is for **ZZZZZZZ**. When the Gulf War is over, Raggy's going home to catch up on his beauty sleep. Not that he needs it, eh girls?!

SUN-MAID
CALIFORNIA SEEDLESS
RAISINS

WHAM MAN SAYS 'THANKYOU FANS!'

George Michael was yesterday close to tears after generous fans put their hands in their pockets and dug deep to help the multi millionaire pop star in his continuing battle with record giants Sony.

Speaking from one of his palatial homes George, who lists his hobbies as dancing, clubbing and drinking, confirmed that the legal struggle to free him from a lucrative fifteen year recording contract would go on, despite a recent High Court ruling against him.

CONTRACT

The dispute began when George and his team of expert advisers were duped into negotiating and signing a highly lucrative contract which practically guaranteed to earn the 31 year old heart throb tens of millions of pounds in return for his writing and performing a handful of three or four minute pop songs every five years or so.

MERCY

And things came to a head when sex symbol George, a snappy dresser whose hits include 'I want your sex', unwittingly found himself being marketed as a sex symbol by the record company who seemed intent on selling as many of his records as possible.

Channel George yesterday

"They didn't seem to care about George's career", said one insider yesterday. "All they were interested in was selling as many of his records as possible over a long period of time by successfully promoting him as a recording artist".

UNLAWFUL

Brave George challenged the Sony Corporation dragon to a High Court battle, but despite the efforts of his expert legal advisers George lost the case, and was left with a bill of £5 million which must somehow be paid out of his estimated £50 million personal fortune.

CAMPAIGN

But now George's faithful fans have come to the rescue, with donations to our 'We want your cash' campaign already flooding in. All the money we raise will be sent to George to help pay his costs and finance an appeal against the High Court ruling. And after only a week the kitty already stands at £16.20, with more donations rolling in by the day.

CHAMPAGNE

"I will be selling all my Wham! records and sending George the proceeds", said

By George! We'll pay your costs

one empty headed tart yesterday. And another fan hit on the idea of a sponsored parachute sale at her local pub to raise money for the needy star.

OBOE

One musician who has spent the last 18 years trying to get a record contract spent the day busking outside Kings Cross station in order to raise money for George. "I'll be sending him my entire day's takings which could amount to several pounds", he told us.

So sad George close to tears yesterday as he leaves yet another expensive nightclub.

We want your CASH

You too can help by sending as much money as you can possibly afford to the following address. 'We want your cash appeal', Viz, P.O. Box 1PT, Newcastle upon Tyne NE99 1PT. Please make your cheques and postal orders payable to 'George Michael (the fat one out of Wham!)'

Hopes rise for raising Dana

Experts hoping to re-float Dana, the seventies singer who sank in the Channel en route to a European concert in 1978, are now optimistic that the Irish songstress can be brought to the surface in one piece.

Channel yields secret of sunken Irish pop Queen

A team of American scientists located the singer on Saturday, lying on her side in eighty feet of water. Her exact location is being kept a closely guarded secret while experts examine the wreck using hi-tech sonar equipment and a small unmanned submarine with cameras on board.

ROCKS

Leading the project is veteran salvage expert Bjorn Tronsk. "Pictures from the wreck show that Dana is in good overall condition, lying on a sandbank, and has not suffered any major structural damage. Our main fear was that she may have broken up on rocks as she went down, but fortunately she is still in one piece and we hope to be attaching lines to her later today".

BOLLOCKS

A salvage barge equipped to haul the former Eurovision Song Contest Winner to the

surface is en route from Norway to the scene, and once structural surveys have been completed, attempts to re-float her will begin. This may involve building a giant 'cradle' around the singer's body to protect her from strain during lifting.

NUTS

But last night salvage attempts were being hampered by poor weather conditions in the English Channel. High winds lead to the cancellation of all diving operations, although work was due to begin again early this morning.

BOLTS

If Dana does re-surface she will need specialist care to prevent her condition from deteriorating once she returns to dry land. She will have been preserved by the salty water which surrounds her, and once she is exposed to air the star's surface could begin to crack and dry up. It

is likely she will be towed to Southampton where preservation work will begin, although her long term future is uncertain.

WASHERS

The anonymous Irish businessman who financed the search for the star is keen

to take her back to Dublin where he plans to use her as a restaurant. However, the insurance company who paid out when Dana was lost in 1978 have already indicated that they will dispute ownership and a protracted legal argument could ensue.

The Egyptians knew the secret of LONG HAIR

Now *YOU* can too with 'Pyramid' formula mystical growth

HAIR COMPOST

Your hair will simply grow and grow!

- **NO MORE BALD PATCHES!**
- **GROW THICK BLACK LOCKS OVERNIGHT!**

This million year old formula has been carved on pyramid walls and passed down through the centuries by the Egyptian mystical Kings. Now the head compost of the ancients can be yours for as little as £79.99 a sack.

APPLY DAILY TO THE TOP OF YOUR HEAD IN A DARKENED ROOM, THEN LIE DOWN WITH YOUR ARMS FOLDED FOR HALF AN HOUR. SIMPLE AS THAT. AND WITHIN MINUTES YOU WILL HAVE STRONG, THICK, HEALTHY SHOULDER LENGTH BLACK HAIR.

WARNING Do not wear motorcycle helmet within 2 hours of compost application as rapid hair growth will occur.

Send your order to: **PYRAMID GROWTH LABORATORIES**, c/o Park Hill Pig Farm, Faversham, Kent.

Name...................................... Address...

I desire the secret of the ancient Egyptian Kings. Please send me two sacks of Hair Compost at £79.99 each. I enclose a cheque for £200 made payable to Park Hill Pig Farm Ltd., Please keep the change.

Signed...

You too can be a beautiful Princess with 'Mirror, mirror on the wall' ™

"YOU ARE"

- ★ **LOSING YOUR LOOKS?**
- ★ **WRINKLES BEGINNING TO SHOW?**
- ★ **FED UP WITH HUBBY NOT PAYING YOU COMPLIMENTS?**

Our magic* mirror is guaranteed* to change all that. Simply pop it on your bathroom wall and ask the question "Who is the fairest of them all?" Then press a button, stand back, and within a few moments the magic* mirror will reply "YOU ARE". Genuine feau quality plastic mirror with moulded surround. Petrol driven engine delivers 24 compliments to the gallon. First time starter on cold mornings.

Send a cheque, postal order or cash to: Mirror Mirror Offer, P.O. Box 12, Basildon, Essex. Due to the quality of this product please allow an unlimited period of time for delivery.

MAGICAL MIRRORS (UK) LTD.
Manufacturers of petrol driven complimentary fairytale mirrors since 1933.

The words 'magic' and 'guaranteed' are used in their broader sense. Please note that Mirror Mirror is neither magic, nor guaranteed.

WARNING: Keep bathroom well ventilated. Petrol driven audio mirrors can occasionally explode and should not be attached to a supporting structural wall.

ONLY £599.95

Turn WATER into PETROL WITH Petrol Fish ™

THE FISH THAT Esso TRIED TO BAN!

- *Fuel bills vanish overnight!*
- *Beat price increases at the PUMPS!*
- *Enjoy FREE motoring FOR LIFE!*

These rare and only recently discovered goldfish occur naturally deep beneath the Earth in the oil fields of Saudi Arabia. No larger than ordinary goldfish, and the same colour, their unique biochemistry gives them an unusually high octane capacity. As a result when they drink water, their urine turns into petrol. Place just one of these fish in your petrol tank then fill it with water and within seconds nature's miraculous **PETROL FISH** are turning the water into petrol. So successful is this natural fuel saving technique lawyers from all the major petrol companies have tried to ban the sale of our product. So far they have been unsuccessful, but we recommend that you ORDER TODAY while **PETROL FISH** are still legally available.

Please rush me................*(state quantity) **PETROL FISH** @£100 each.

Name.....................................Address.................................

Tick one ☐ FOUR STAR ☐ UNLEADED ☐ SUPER PLUS UNLEADED.

*Sorry we regret that customers are limited to a maximum of 800 fish each.

Send orders to: PETROL FISH SALES (UK) Ltd., The Aquarium, Colchester High Street, Colchester, Essex, CO1 5AH.

You sleep while your gloves do the driving

Home, James!

Automatic driving gloves

Faux POLICE approved!

You DRINK, your gloves DRIVE!

IF you CRASH, we return your CASH!*

*No Refunds

Ideal for:
- * Drunk Drivers
- * Disqualified Drivers
- * Women Drivers
- * Andrew Ridgely

Name.............................
address.............................
Dates & times when I'm out.............
I do/do not have a dog (please delete)
I enclose £........Signed.................
Home, James Chauffeur Gloves come complete with 12 MONTHS FULLY COMPREHENSIVE car insurance!
Send to: Box 23, Cardiff, Wales.
Please allow a long time for delivery

£299.99

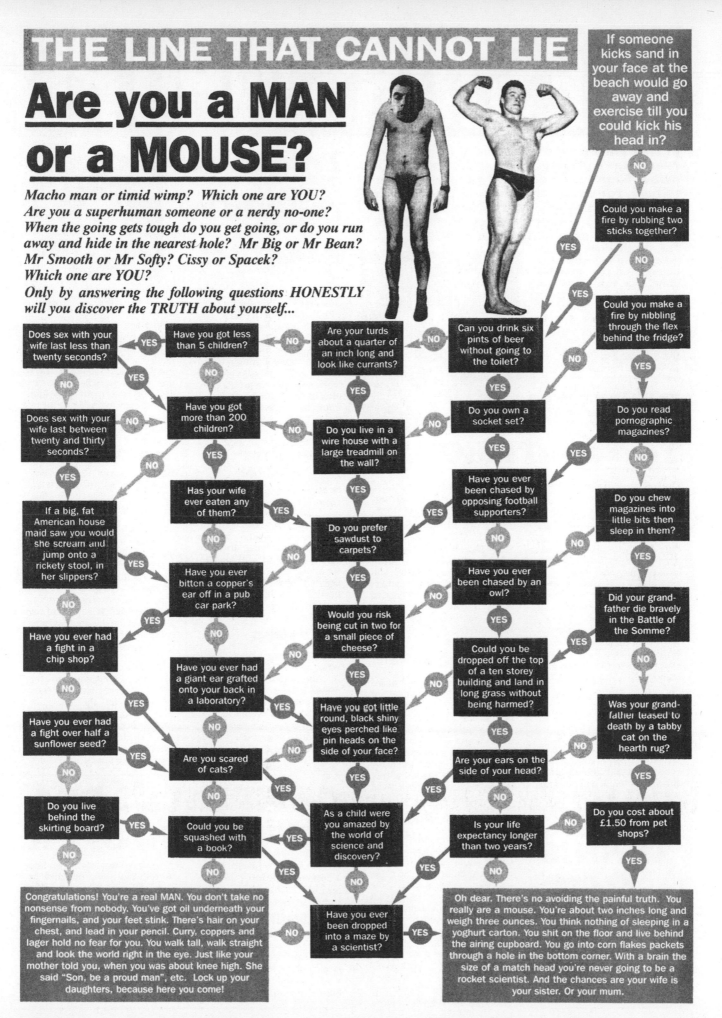

THE LINE THAT CANNOT LIE

Are you a MAN or a MOUSE?

Macho man or timid wimp? Which one are YOU?
Are you a superhuman someone or a nerdy no-one?
When the going gets tough do you get going, or do you run away and hide in the nearest hole? Mr Big or Mr Bean?
Mr Smooth or Mr Softy? Cissy or Spacek?
Which one are YOU?
Only by answering the following questions HONESTLY will you discover the TRUTH about yourself...

If someone kicks sand in your face at the beach would go away and exercise till you could kick his head in?

Could you make a fire by rubbing two sticks together?

Could you make a fire by nibbling through the flex behind the fridge?

Can you drink six pints of beer without going to the toilet?

Do you read pornographic magazines?

Does sex with your wife last less than twenty seconds?

Have you got less than 5 children?

Are your turds about a quarter of an inch long and look like currants?

Do you own a socket set?

Does sex with your wife last between twenty and thirty seconds?

Have you got more than 200 children?

Do you live in a wire house with a large treadmill on the wall?

Have you ever been chased by opposing football supporters?

Do you chew magazines into little bits then sleep in them?

Has your wife ever eaten any of them?

Do you prefer sawdust to carpets?

If a big, fat American house maid saw you would she scream and jump onto a rickety stool, in her slippers?

Have you ever bitten a copper's ear off in a pub car park?

Have you ever been chased by an owl?

Did your grand-father die bravely in the Battle of the Somme?

Have you ever had a fight in a chip shop?

Would you risk being cut in two for a small piece of cheese?

Have you ever had a giant ear grafted onto your back in a laboratory?

Could you be dropped off the top of a ten storey building and land in long grass without being harmed?

Was your grand-father teased to death by a tabby cat on the hearth rug?

Have you ever had a fight over half a sunflower seed?

Have you got little round, black shiny eyes perched like pin heads on the side of your face?

Are you scared of cats?

Are your ears on the side of your head?

Do you live behind the skirting board?

Could you be squashed with a book?

As a child were you amazed by the world of science and discovery?

Is your life expectancy longer than two years?

Do you cost about £1.50 from pet shops?

Have you ever been dropped into a maze by a scientist?

Congratulations! You're a real MAN. You don't take no nonsense from nobody. You've got oil underneath your fingernails, and your feet stink. There's hair on your chest, and lead in your pencil. Curry, coppers and lager hold no fear for you. You walk tall, walk straight and look the world right in the eye. Just like your mother told you, when you was about knee high. She said "Son, be a proud man", etc. Lock up your daughters, because here you come!

Oh dear. There's no avoiding the painful truth. You really are a mouse. You're about two inches long and weigh three ounces. You think nothing of sleeping in a yoghurt carton. You shit on the floor and live behind the airing cupboard. You go into corn flakes packets through a hole in the bottom corner. With a brain the size of a match head you're never going to be a rocket scientist. And the chances are your wife is your sister. Or your mum.

THE STARS WHO REFU

In showbusiness, as in the sky at night, for every star that rises, so another must fade away. Famous names and favourite entertainers, for whom the show is over. Death, whether it be by accident, disease or natural causes, is no respecter of celebrity status. Eventually it catches up with us all.

But death does not always bring to a close the careers of the showbiz stars. For the entertainment world is littered with ghostly tales of stars who have simply refused to die.

KING'S RETURN TO THE THRONE

During his 46 years on Earth **ELVIS PRESLEY** never once visited Britain. But an unemployed plumber from Altringham claims to have come face to face with the King of Rock-'n'Roll in the lavatory of his council semi, 14 years **AFTER** the star's death.

Fifty-nine year old Bob Cartwright told us how he was awoken late one night by a groaning sound coming from his toilet. "I went to investigate, and couldn't believe what I saw", said Bob. "There, slumped across the lavatory seat, was Elvis Presley – the King of Rock'n'Roll. He was extremely overweight and had been eating a slice of pizza. He had got stuck and was obviously in some pain". Suddenly Elvis spoke.

ECHOED

"I'll never forget his voice. It seemed somehow distant, and echoed around the lavatory. But his Texas accent was unmistakable". Elvis asked Bob for a spoon which he needed to take some drugs. Bob rushed downstairs to the kitchen, but as he ran back upstairs towards the bathroom door he heard a loud flushing noise, and turned the corner only to see bubbling water disappear down the toilet.

U-BEND

As the bowl re-filled Bob heard the unmistakable sound of Elvis' voice coming from beyond the 'U' bend. "He was singing Suspicious Minds. I'll never forget sitting there with my ear in the bowl listening as his watery voice gradually faded away. It made the hair on the back of my neck stand on end, I can tell you".

HAMBURGER

Since that night, Bob believes that Elvis' ghost has returned to his house several times, on one occasion firing a gun at his television. "Fortunately, the ghostly bullet passed through the TV screen causing it no harm, but on another occasion Elvis caused a small fire in the kitchen when he left the grill on after cooking a hamburger late one night". Fortunately Bob was alerted by neighbours who spotted the flames and called the fire brigade.

Tree terror re-lived by dead idol

Diesel fitter Darren Peabody didn't believe in ghosts, until one night in 1986 he had an experience he will never forget.

"I'd been to a friend's stag night at a local pub, and had been drinking heavily for several hours. It was pouring with rain and I couldn't find my car keys, so I was delighted when a stranger in the car park offered me a lift home in his mini. In the light of the full moon I made him out to be a young man, perhaps in his early thirties, with long dark curly hair, and make up.

HAIRPIN

"I began to tremble as the car sped along the narrow, winding road. Thunder and lightning flashed as the car careered around hairpin bends. I asked the driver to slow down, but it was too late, for at that very moment the car skidded out of control and hurtled towards a tree. I covered my face and braced myself for the impact.

URINE

The next thing I remember it was nine o'clock the next morning and I awoke to find myself lying in a pool of urine in the pub car park. There was vomit in my hair, and all over my clothes, and no sign of the young man in his mini. I made my way home and thought nothing more of the incident until a few weeks later when I mentioned what had happened to a friend. What he told me made my hair stand on end.

For less than 120 miles from that very spot where the car span off the road, pop idol Marc Bolan had been killed in an almost identical accident exactly nine years earlier, almost to the month.

SOMERSAULTED

But the story doesn't end there. For six months later, after a night of heavy drinking, Darren drove past the same spot and spotted something moving behind the trees.

"Through the mist and fog I could just make out the silhouette of a white swan, with someone riding on it. Just like in the words of the T Rex song". Darren was so frightened that he lost control of his car, clipped a passing bus, and somersaulted into a ditch.

HOSPITAL

"The next morning I awoke in hospital. I explained to the police officers that I'd been frightened by the ghost of Marc Bolan but they simply would not believe me". Darren was fined £250 for driving with excess alcohol and banned from driving for 2 years.

CAMPBELL'S PEA SOUPER

Holidaymaker Stuart Ferguson got more excitement than he bargained for the day he and his family hired a rowing boat for a day out on Lake Windermere in the Lake District.

SNAPPED

Stuart, his wife Morag and their two children Angus, 2 and Crawford, 5 had been rowing for about an hour when their oar snapped. Stuart takes up the story.

"We were stuck in the middle of the lake with no land in sight. After a while it got dark, and a thick blanket of fog descended on the lake. It was eerie.

"Suddenly the ghostly calm was broken by a sound that made my hair stand on end. It was the roar of a bright blue rocket powered speed boat. The gleaming vessel emerged

USE TO DIE!

THIS IS YOUR AFTERLIFE

Since the death of Eamon Andrews, dressing room number 666 at Thames Television has stood empty. For not one single star in the world of showbusiness would dare use the room formerly occupied by the This Is Your Life presenter.

Andrew's successor Michael Aspel was the first man to enter the room after the Irishman's death. Seconds later he fled screaming, his grey hair standing on end. It was several moments before Aspel had calmed down enough to describe his terrifying experience to horrified TV executives. For inside the dark, dingy dressing room, he had come face to face with the headless ghost of his predecessor!

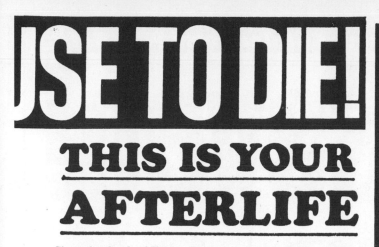

SPRANG

A Thames Television insider explained. "Eamon's ghost sprang out of the mirror and thrust a big red book at Michael. Fortunately Aspel fled, for the story goes that anyone who accepts the book from Andrews' ghost will immediately turn to stone".

Indeed one hapless cleaner, while dusting the light bulbs which surround the mirror in dressing room number 666, was accosted by the ghoulish figure, and took the book. She instantly turned to granite and her statue stands in the foyer of Thames TV as a warning to showbusiness celebrities and other would-be visitors to dressing room number 666.

WHAT ARE GHOSTS?

There have been many attempts made to explain the phenomenon we loosely term 'ghosts'. Are they simple illusions created by our brains, or perhaps figments of our vivid imaginations. Or maybe its just our minds playing tricks on us. There have been many attempts made to explain this baffling phenomena.

AFTERLIFE

But what do the stars of showbusiness themselves think? Do the stars of stage and screen believe in the afterlife? And what is their idea of a ghost?

SUPERNATURAL

We asked three former TV Dr Who's to offer their explanations.

White haired former time lord JON PERTWEE has little time to ponder the mysteries of the supernatural. "I really haven't given it a lot of thought", he told us yesterday. "But if you ring my agent next week he'll sort something our for you", he said.

Sauve time traveller PETER DAVISON is in no doubt about ghosts. "The human eye is like a camera, if you will", he told us.

RETINA

"Images are taken in and focused on the retina. When you see a ghost, it is merely the same process happening in reverse, the image being projected through your eyes and onto a wall, like a slide show, if you will".

CORNEA

"I don't know anything about ghosts, and frankly I don't particularly care", recent Doctor SYLVESTER McCOY told us. "And anyway, where did you get my number", he asked.

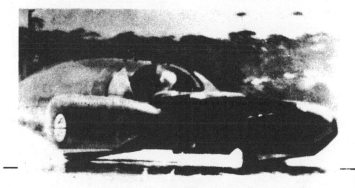

Campbell's ghostly speedboat (above) emerges from the fog.

from the fog and pulled up alongside us. Without saying a word the driver threw us a rope.

FLAMES

"The next thing I knew we were being towed back to shore at speeds in excess of 600 miles an hour, and it wasn't long before we were safely back on dry land.

"As I walked up the pier, I turned to see the boat roar off at high speed, flip up into the air and explode in flames before sinking without trace.

"The next morning I returned to see if I could be of any help, but there was no sign of the mystery stranger or his boat. Not a single bit of wreckage had been washed ashore. I described the man to an old fisherman who was mending his nets at the nearby harbour, and asked if he knew him. "Yes", he said. "That was the ghost of Donald Campbell".

Garden haunted by green fingered ghoul

Police were baffled when they were called to the Blue Peter garden at the BBC Television Centre. Vandals had dug up plants, overturned a statue of the Blue Peter dog Petra, and poured bleach into the Italian sunken pond.

Detectives were baffled. The gate had been locked, and the garden is surrounded by a six foot wall. "It's as if the vandals simply walked in through a solid wall", he said.

BOX

That officer may have been closer to the truth than he realised, for paranormal experts now believe that Percy Thrower, the Blue Peter gardener and former TV bird impressionist, was responsible for the damage, and that his spirit had returned to Earth to haunt the garden.

"Percy often spoke of his wish to be buried in the Blue Peter garden, alongside the box for the year 2000", one former presenter told us. But that wish was never granted. Blue Peter supremo Biddy Baxter refused to allow the burial, and even turned down an eleventh hour plea for Thrower's ashes to be scattered in the herbacious border.

"Percy Thrower's spirit cannot rest until his remains are taken to the Blue Peter garden", we were told. "And until they are, Thrower's ghost will haunt the garden, vandalising it once or twice every year".

Next week: The man who bought former Crackerjack funny man Peter Glaze's house reveals how the star's spectre has repeatedly foiled attempts to decorate the building by the use of a supernatural bungling comedy wallpapering routine.

MI5 Panic as Deadly Bond Prop is Stolen from Film Set

Oh-Oh Seven!

BRITAIN'S security services were put on RED ALERT last night after a LICENCE TO KILL went missing from the set of the latest James Bond film. The document, which gives the bearer unlimited powers to assassinate at will, vanished from the dressing room of Daniel Craig, the latest actor to star as the suave superspy.

Casino Royale, the 21st film in the series, was being shot at Pinewood Studios when the theft took place. During a break between scenes Craig, 24, put the document on his make-up table whilst he went to the toilet. When he got back, he found someone had sneaked in and pinched it, along with a Rolex watch and a wallet containing £18 and a Nectar card. MI5 chiefs are said to be furious, as the double-0 licence had been loaned to the filmmakers with the proviso that it was kept securely under lock and key when not needed on set.

Pinewood Studio police sealed the studios within minutes, but it is believed that the thief had already made his escape. And now it is feared that the permit may end up in the wrong hands, perhaps those of a lorry driver, a serial killer, or worse.

Police superintendent Will Hay told reporters: "We cannot emphasise enough what a serious situation this is. Whoever has this licence can kill as many people as they want, and they will be immune from prosecution. And what's more, it doesn't expire until 2016."

By our Secret Intelligence Staff
PAN'S PEOPLE
(except Babs)

Authorities already fear a worst case scenario in which the licence is sold to Al Q'aeda boss Osama Bin Laden. "It's a nightmare waiting to happen," a senior Scotland Yard source told reporters. "A mad fundamentalist with carte blanche to wander around systematically murdering everyone in Britain whilst the police have to stand by, powerless to stop him."

Actor Craig was last night said to be at an undisclosed location, being comforted by close sources. One told us: "Daniel is absolutely devasted. He realises that his little slip-up in not locking his dressing room could lead to a decade of legalised massacres throughout the country. He could have the blood of millions on his hands. By any measure, it's not the best of starts to his career as James Bond. Though arguably not as bad as George Lazenby's."

Her Majesty's Government Licence to Kill

In the name of Her Majesty Queen Elizabeth of the United Kingdom, the bearer of this licence is hereby entitled to kill any person or persons at their discretion without hindrance, and with full freedom from arrest, trial and prosecution under the law of the United Kingdom, the Commonwealth and its dominions.

SPECIMEN

Valid from 21st March 2006
Valid until 20th March 2016

647632 88 12 - 007

00 DANNY BOY: Licence *(left)* left unattended by Bond star Craig *(above)* could fall into hands of evil Osama *(top).*

Who Would YOU Do?

EVERYONE has someone they would LOVE to murder if there was no danger of prosecution - whether it's a beligerent boss, a noisy neighbour or Jimmy Carr, we've all got at least one name on our hit list.

We asked as many famous Bonds as we could think of who they would take out if they had a *Licence to Kill.*

Jennie Bond
Former BBC Royal Correspondent

"I am a pacifist, and I believe all killing is wrong. However, if I had to chose someone to do away with, I think I would kill everyone in the world who thought the Queen was anything less than absolutely wonderful. I'd shoot them in the head like pigs. With dum-dum bullets."

Nicholas Bond-Owen
38 year-old Child Actor

"When I played Tristram Fourmile in George and Mildred, Mr Roper from next door used to make life a misery for me and my on-screen parents. I think I'd use my licence to kill him. Perhaps I'd cut the brake cable on his motorcycle/sidecar combination, or pour petrol through his letterbox when he and his ugly wife Mildred were fast asleep."

Garry US Bonds
Veteran R'n'B vocalist

"Killing somebody is a serious thing to do, and I don't think I could do it unless I felt the person really deserved to die, like Nasty Nick Bateman off Big Brother One. He knew the rule about not taking a pencil into the house, but he smuggled one in nonetheless. What he did to his fellow housemates Craig, Nicola, Mel, Darren, Thomas and Anna was unforgiveable"

Jon Bond Jovi
Toy Poodle Rock God

"I don't have to think twice about who I'd murder if I was allow-ed to. It's the goddam milkman who comes up my path every morning at 4.30, clinking bottles and whistling. And he wears shoes with freakin' Blakies in. I'd put a line of cheesewire across my gate at throat level. That would sort the goddam son of a bitch out."

CANNON BLASTS QUEEN!

American TV cops of the seventies have launched a blistering attack on the British Monarchy.

U.S. T.V. cops' broadside for British Royals

Former top TV detectives, among them Cannon star William Conrad, have **BLASTED** the Royals, accusing them of being over-paid, out-dated and out of touch.

DINOSAUR

"The British Monarchy is an outdated institution. It's a modern-day dinosaur, and sooner or later it's going to become extinct", said the gravel voiced 22 stone actor who played porky private eye Frank Cannon in the hit series.

DO DO

His views were echoed by Jack Lord, better known to millions of seventies TV viewers as Steve McGarrett, crime fighting star of Hawaii Five O. Lord lashed out at the luxury lifestyles enjoyed by the Royals.

"It's not right that the Royals should live in luxurious palaces while everyone else in Britain is homeless", he told us.

KNUCKLE HEAD

"A right Royal rip-off". That's how bandy-legged actor Paul Michael Glaser described Britain's system of constitutional control. "The very least they could do is pay tax", continued the dynamic cardigan clad star of Starsky and Hutch. Paul's partner in crime fighting, alias actor David Hutch, agreed with his former cop colleague. "There's no doubt that the Royals perform a useful function, but their pay is out of all proportion to the amount of work they do", chipped in the actor and former singing star who's hits included 'Silver Lady'.

MY LIFE
BY JEFFREY BARNYARD

God, my life is boring these days. I went to the dentist on Tuesday, and he told me I'd got an abscess. He's offered to drain it for me, but I don't know. I think I'll just grin and bear it for a bit, and hope it goes away.

I might nip to the shops this afternoon. I want to buy some tinned peaches. But there's an old dragon works in the corner shop, and I loathe and detest her. So I don't think I'll bother. Looks like Ryvita for tea again. How I loath and detest the stuff.

Funny how all your friends start dying as you get older. There'll only be me left soon.

It's Brookside tonight. How I loath and detest that awful programme. My TV's on the blink again. A man who came to repair it told me it would cost £50 to put it right. I don't think I'll bother.

I'm afraid I seem to have developed another lump behind my ear. The second one in as many months. It doesn't cause me discomfort, but I may as well have it lanced anyway.

I must have had a good drink last night. I feel bloody awful today. I fell asleep in a chair at the club last night. I'm afraid I pissed myself as usual. I've now developed something of a rash between my legs as a result. And I remain convinced that there is a swelling in my right testicle, although the doctor insists there is not. How I loath and detest doctors.

Damn that woman. I *will* have the peaches.

Cannon - blast!

But the comments of the former US TV cops last night provoked an angry response from the British TV bobbies of the seventies, who swiftly jumped to the defence of the Royals. And they criticised the American TV tecs for meddling.

"What the hell do a bunch of American actors know about our Royal Family", fumed James Ellis, alias Sergeant Burt Lynch in the long-running Z-Cars series.

ODD BALL

Burt's former TV Z-Cars colleague and subsequent Softly Softly Task Force star Frank Windsor agreed. "The Royals do a marvellous job, but their dignity prevents them from responding to attacks like these. Former American TV policemen would do well to mind their own business. They've got enough problems of their own in America".

ODD JOB

Tough guy actor Dennis Waterman, alias burly Detective Constable Jimmy Carter in TV's The Sweeney, was unavailable for comment last night. But his agent told us that Waterman, former sidekick of one-time Sweeney star TV's Inspector Morse, John Thaw, was a fan of the Royals, and he would back the British TV police in the row over Royalty.

Soul - 'Silver Lady'

Screen cripple Ironside, who struggled against disability to bring law and order to the streets of San Fransisco, believes the Royals should hand over their cash to the poor. "They should use their millions to build hospitals, orphanages and housing for the poor", said able-bodied actor Raymond Burr.

Carter's Sweeney sidekick - Morse star Thaw

Meanwhile washboard playing comedy actor Derek Guyler, who regularly turned up as P.C. Corky in the hit series Sykes, stood up for the Queen Mother. "She does a marvellous job. God bless her. I'd like to see an American TV cop, for example David Jansen, star of Harry O, do her job."

A spokesman for Jansen, the silver-haired seventies sleuth, last night told us he was dead, and had been for some time.

We tried to speak to British actor Jack Warner, star of the ever popular Dixon of Dock Green, but we were told that he was dead as well.

LET'S HAVE A VOTE

Do you agree with the American TV cops of the seventies and think that it's time Britain got rid of the Royals? Or do you back the British screen bobbies and support the Royal Family?

0898

We're having a national telephone vote to decide who's right. But unfortunately we haven't got an 0898 phone number. So we want you to write to Viz 0898 Telephone Vote Line, P.O. Box 1PT, Newcastle upon Tyne, NE99 1PT. If you post your letter at peak times please enclose 45p, or 36p at all other times. Don't worry, half of the money will be sent to British Telecom.

KING ARTHUR'S CASTLE

It looks like a Royal Palace, a home fit for a King. But believe it or not these EXCLUSIVE pictures reveal for the first time the multi-million pound luxury Barnsley home of miner's president Arthur Scargill.

Commie Scargill's palace bought with Nazi gold

This is the house that Arthur built.

● It has 280 bedrooms, two swimming pools, tennis courts and a private golf course.

● It houses a priceless art collection, over 1,000 paintings, sculptures and expensive vases.

● It is set in 800 acres of private woodlands, with a hunting lodge and prime salmon fishing.

● The servant's quarters alone are ten times as big as the average family home.

● It contains over 10 square miles of specially woven Persian carpet.

● It cost a staggering £100 million to build.

DONATIONS

And we can prove that Arthur's palatial home was paid for **ENTIRELY** out of donations received by the N.U.M. during the miners' strike of 1984.

DENIALS

Despite denials from communist Scargill and his union cronies, we have traced **MILLIONS** of pounds intended for the miners' strike fund which were diverted by Scargill into secret Swiss bank accounts.

We believe the money was than smuggled to Brazil by a leading N.U.M. figure, and used to buy Nazi gold.

The Nazi gold, proceeds of Hitler's evil war crimes. was then used by Scargill to buy this £200 million luxury home

PICTURE EXCLUSIVE

Scargill's living room (right) is an Aladdin's Cave of priceless treasures, all paid for out of NUM funds.

The walls are lined with famous paintings. This one alone by Van Goff cost £24 million.

Scargill picked up this small vase, big enough to hold only a small flower, at Sotherbys. Price – £10 million.

This French Louis XIV waste paper basket carved from solid ivory weighs 40 tons and is worth more than its weight in gold.

Cushion – £40 million.

Arthur's favourite chair – a Wedgewood willow pattern Queen Ann 4 legged carver also folds out into a bed. Value – £350 million.

Hand knitted foambacked Egyptian carpets, embroidered in gold – a snip at only £3 million per square foot.

Our investigators have uncovered *documentary evidence* of Scargill's illicit dealings. Airline tickets purchased in the name of 'P. Heathfield' show that the N.U.M.'s Deputy President made *SIX* return journeys from Switzerland to Brazil during 1985.

Stuck up his bottom were several rubber balloons, *containing millions of pounds in used banknotes.*

BRAZIL

In Brazil he was met by former Nazi war criminals who exchanged the cash for gold.

Heathfield then returned to Britain in a luxury yacht belonging to Libya's Colonel Gadaffi.

ALMOND

And we also have a copy of the receipt handed to Scargill by the Barnsley builder who was paid £300 million on completing the house.

PISTACHIO

We rang the police and told them to arrest Scargill and his pinko pals, pointing out that our new evidence would put them behind bars for many years to come. But they said they were busy and asked if we'd call back later.

Monster Scargill (above) and (below) the receipt handed to him by a Barnsley builder.

Wm. Stubbs (builder) High Street Barnsley

Received From Arthur Scargill £400,000,000 (in Nazi gold) Wm Stubbs.

HEY! DON'T CRAMP MY STILE

GLUE

MIRIAM'S BEAN PROBLEM
PHOTO CASEBOOK

Shelley's Bean Dilemma ~ Day 58

Shelley and Dave love each other. But they just can't see eye to eye about kidney beans...

Shelley is having an afternoon shower...

I love Dave, but his dislike of kidney beans is threatening our relationship

Later...

I'm making a chilli tonight, Dave. How about I put just a few kidney beans in it?

No thanks, Shelley. I really dislike them

That night, Shelley has a dream about Dave's mate Steve...

I love kidney beans. Especially in chillis

Next morning, Shelley is confused...

I wish Dave loved kidney beans like Steve

Zzzzzz Zzzzzz

CONTINUES TOMORROW...

Dr Miriam's Bean Advice
The pulse advice lines you can trust

Husband hiding beans under knife & fork	01 811 8055
Worried about cooking times for broad beans	018 118 055
Green beans - steam them or boil them?	0181 18055
Tempted to eat beans straight out the tin	01811 80 55
Rimming tips to drive your man wild	0 1811805 5

Calls cost 60p/min and terminate on a small scrollwork table at the bottom of Miriam's stairs.

Have Your Say

FRANCE's Thierry Henry's shameful behaviour in the World Cup game against Spain, pretending he had been struck in the face, sickened the nation. It was no more than could be expected of a French car salesman, but it highlights the extent to which playacting has become part and parcel of the game. But what should be done to players who take dives during the game, and what steps could FIFA take to stamp it out? We went on the streets to find out what YOU thought.

...I think if footballers want to play the actor then they should be dressed for the part. Anyone caught indulging in theatrics should be yellow carded and forced to play the rest of the game wearing a doublet and hose and carrying a skull.

T Hennesey, Nottingham

...players like Henry are only human, and he may have reacted on the spur of the moment. He should not have been shown the yellow card for his actions as many have said, rather he should have been sent to a 'sin bin' for five minutes to reflect on his behaviour before returning to the game.

H Barnstorm, Leeds

...I believe that Mr Barnstorm (above letter) makes a good point, but I don't think that sitting in a 'sin bin' thinking is the best way to punish these players. Offenders should be given a short community service sentence by the ref, and have to spend ten minutes picking up litter in the stadium or painting the fence surrounding the ground before returning to the game.

Len Goatscheese, Hull

...for too long have FIFA adopted a softly, softly approach with these cheats. Any player caught diving during a game should be sent off the pitch, banned from football for life and put on the sex offenders register.

M Waddington, Goole

...expulsion from the game is the only thing that these players understand, but decisions should not be taken lightly. Perhaps FIFA could introduce a system whereby palyers are warned on their first offence, perhaps by being shown a yellow card. If they offend again, they should be shown a card of a different colour, perhaps blue or red, and ordered off the pitch.

Louis Playwood, London

...as a retired schoolteacher, I would like to see FIFA give the referee the power to beat any player guilty of diving. He should pull his shorts down, bend him over and give his bare behind six strokes of the cane in front of the whole stadium.

M Fibreboard, Nottingham

...is it any wonder that footballers go diving all over the place during games when Gianluca Pessotto, the manager of Juventus, goes diving out of his office window? What kind of example is he setting to his players?

H Nelson, Warrington

...I am a multi-millionaire businessman and I was about to replace my worldwide fleet of 8500 ageing company cars with brand new Renault Clios. However, after seeing Thierry Henry's disgraceful playacting in the France versus Spain game, I will be buying Seat Ibizas instead.

H Richenbakker, New York

...I watched the France versus Spain game with my six year old son who is an avid Arsenal fan. He sat watching the game in his Henry shirt, enraptured. However, when he witnessed his hero's shameful behaviour, he turned to me, his face full of confusion. "Daddy, Mr Henry is the best footballer in the world. Why did he have to cheat?" What could I say?

Frank Broadstairs, Preston

...I think that if these footballers want to go diving then they should be dressed for the part. Anyone caught diving should be yellow carded and forced to play the rest of the game wearing a wet suit, flippers and an aqualung. If they commit a second, similar offence they should be given one of those old fashioned diving suits with lead boots and a brass helmet.

T Hennesey, Nottingham

...football has long since stopped being a sport. The players are now businessmen and for them, time is money. If they are rolling around on the ground, they are not playing football and they should not be paid for that time. Then we'll soon see how injured they are. Henry spent thirty seconds pretending to be injured, so by my reckoning he should be £2768 short in his wage packet next week.

M Nantucket, Luton

No one can deny that Thierry Henry, Premiership Player of the season 2006, is a magnificent footballer, and that the French are wonderful people. But his actions brought shame on his nation, and I believe we should all stop going there on holiday, abandon all plans to adopt the Euro as our currency and pull out of the European Union.

G Sprake, Wolverhampton

I don't know what all the fuss is about. In this shamful episode, Thierry Henry is the real loser. In pretending to be hit in the face and being awarded a free kick that led to France's second goal and ultimate victory over Spain and progression into the quarter finals, Henry is only cheating himself.

J Plywood, Cornwall

I was watching the match on the telly and when I saw Henry pretend to get hit, I was so incensed that I put my foot through the screen and sent France the bill.

Renton Oerstryk, Manchester

No Snecks Please, We're British!

FEWER **BRITS** **THAN EVER** before are leaving their doors on the sneck, according to a report published today.

The Institute of Just Popping In and Out carried out a survey of 8 home owners in 2005 and discovered that just 25% of them now regularly leave their doors on the sneck. In the previous survey of 6 home owners in 2004, half left their doors on the sneck.

sneck

The report's author, Professor Malcolm Kingkong-Kirk, puts the sudden drop down to fewer Britons these days leaving their door on the sneck.

sback

"Nowadays, people are more likely to just leave the door to," he told reporters.

~ Reuters

WHO HAS THE LAST LA YOU -OR THE T.V. COMEDIANS

Nobody likes a laugh more than the British, and it's no wonder therefore that our comedians are the funniest in the world. TV funny men like Bob Monkhouse, Les Dennis and Jimmy Cricket have us falling about with laughter every time we turn on our televisions. Their friendly faces and beaming smiles fill our screens, bringing joy and laughter to a million living rooms.

But what are they like in real life? Are they the friendly, happy-go-lucky people we see on the screens, always smiling and telling jokes? Unfortunately, many are not, and the chances are that if you come face to face with a TV comedian, you wouldn't be laughing.

Nobody wants a fight — especially with their favourite TV comic. But if things did turn nasty, could you handle it? Use your imagination to answer the following questions a, b or c, then tot up your final score to see who'd have the last laugh — you or the TV comedian.

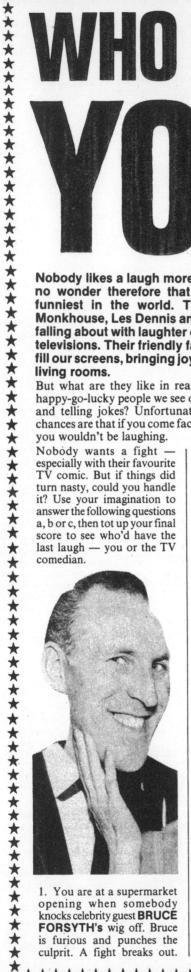

1. You are at a supermarket opening when somebody knocks celebrity guest **BRUCE FORSYTH's** wig off. Bruce is furious and punches the culprit. A fight breaks out.

What would you do? Would you:

a. Turn away and ignore the incident.
b. Watch eagerly to see what happens.
c. Jump in firmly and pull them apart, demanding that they both shake hands.

2. You are staying at a posh hotel. In the bar 'Blankety Blank' host **LES DAWSON** is staggering around drunk and singing Scottish football songs. He is waving an empty beer bottle in your face. What would you do? Would you:

a. Ignore him and leave.
b. Tell him to stop it or you'll call the police.
c. Knock him unconcious for his own safety then order him a pot of strong black coffee for when he wakes up.

3. You hear on a news bulletin that TV favourite **JIMMY TARBUCK** is wanted for armed robbery. The police have warned the public not to 'have a go'. Seconds later, Tarby bursts into your house armed with a shotgun and demands food and cash. What would you do? Would you:

a. Make him a meal, give him your money and hope that he goes away.
b. Talk to him, tell him you're a big fan, and try to

get him to give himself up to police.
c. Throw a cup of hot tea in his face, disarm him, wrestle him to the floor and tie him up until the police arrive.

4. You are in the pub drinking quietly when zany TV kebab salesman 'Stavros' approaches you and says, "I'm-a-not-like-your-bladdy-haircut, peeps. Is a puffs haircut, innit, you bladdy bast!" How would you react? Would you:

a. Go and have your hair restyled immediately.
b. Tell him to calm down, and offer to buy him a drink.
c. Take him into the car park and give him a fucking good kicking.

5. You're at a disco on Saturday night when you catch much-loved knockabout comedian **NORMAN WISDOM** looking at your girlfriend. What would you do?

a. Ignore him completely, and ask your girlfriend out for a dance.
b. Smile and introduce yourself, then ask him to trip and fall over while shouting "Mr Grimsdale".
c. Knock his cap off and punch him in the face.

6. You have been drinking heavily all day. Suddenly it occurs to you that loveable comedy straight man **SID LITTLE** may have been seeing your ex-girlfriend. What would your reaction be? Would you:

a. Dismiss the idea. You're probably wrong.
b. Admit to yourself she's a free woman, and bear no grudges against Sid.
c. Continue drinking until the pub closes, then go round to his house with a few mates and make bloody sure he doesn't do it again.

HOW DID YOU DO?

Award yourself 1 point for each question you answered 'a', 2 points for a 'b', and 3 points for each 'c'.

15 or less

Not a very good score. You may find them funny on screen, but off screen you'd better steer clear of the TV comedians. You're out of your depth.

Ted pictured with the teddy

UGH?

7. It's 11.30 on Friday night and well spoken ventriloquist **RAY ALLEN** and Lord Charles are talking loudly in the queue at your local chip shop. How would you react? Would you:

a. Order your chips and ignore them.
b. Congratulate them on their latest show, and wish them well in their summer season.
c. Knock their chips out of their hands and kick them repeatedly until the police arrive.

8. Finally, you are enjoying a wonderful holiday in Skegness. During a game of Crazy Golf on the seafront you realise that top international comedy star **BOB HOPE** is in front of you, and is having difficulty getting his ball through the windmill. A queue is building up behind him. Would you:

a. Wait patiently, chatting with friends about Bob's many hilarious films.
b. Play through, moving straight on to the next hole.
c. Trip him up with your putter and shove him into the flower bed.

16 to 20
You're no softy, and no doubt you can handle yourself if things get rough. But stick to light-weight comedians or game show hosts. Heavyweight comics like Bernard Manning could be more than you can handle.

21 or more
The comedians may tell all the jokes, but if the fists start to fly, it'll be you who has the last laugh. You need fear no comedian. Indeed, if you're in the audience, it's the comedian who'll have to be on his guard.

"THE QUEEN IS A BLOODY CHEAT!"

Fair's fair says Ted. I only want what's mine

A 32 year old father of three has rocked Buckingham Palace by claiming that his 8 year old son Ian is the rightful owner of the Crown Jewels!

Ted Henderson claims that his son traded a teddy bear for the jewels in a schoolyard swap with Prince Harry. And he claims that the Royal Family went back on the deal.

"My lad came home with all these jewels, and I was as surprised as anyone when he told me how he'd got them. But it was Prince Harry's idea in the first place, so I thought that was fair enough".

EMBARRASSED

"But that evening we sat down to watch the telly when there was a knock at the door. It was the Queen. She was a bit embarrassed and she said she'd come to get the Crown Jewels. The she handed Ian back his teddy bear. The lad was in tears".

JEWELS

Ted believes that despite being the Queen, she still had no right to take back the jewels. "It had been a fair swap. She had absolutely no right to take the stuff back. As far as I'm concerned it's still ours. I've been onto the police and I'm just waiting to see what they're going to do about it".

As well as contacting the police, Mr Henderson is also writing to his local MP. "I don't want to cause any trouble — I just want what's ours", he told us. And he insists that he won't let the matter drop until the Crown Jewels are returned to his son.

MARBLES

Coincidentally there has been a legal precedent to this case, also involving Mr Henderson. In 1983 it was claimed that his eldest son Kevin swapped six marbles with Paul McCartney's daughter in return for the publishing rights to all the ex-Beatles's songs.

Professor Piehead

Wacko Shako!

by Terry Twatt

According to rumours circulating in the music business, eccentric pop millionaire Shakin' Stevens has shocked close friends by leaving £1,750,000 in his will — to his pet duck 'Quacker'.

HAMSTERS

Shaky's string of million selling hits are known to have made him one of the ten richest men in the world. And as well as a duck, Shaky is believed to have several hamsters

On a recent world tour, pop's 'Mr Fruitcake' insisted on taking most of his pets with him! Hotel bills came to more than £12 million, which included the cost of dismantling one luxury hotel and re-building it on a beach twenty miles away so that Shaky could go for an early morning dip with his friendly dolphin, Harvey.

A dolphin

Stevens is known to sleep in a greenhouse. For he believes that fumes from a tomato plant will keep him looking eternally young.

BEATLE PAUL IS TOP OF THE POTS

By our
Pop Plant correspondent
MC Percy Thrower

In the introduction to a new book Paul McCartney admits for the first time that he is haunted by the ghost of John Lennon.

But Lennon's spirit has not returned to Earth to rekindle their legendary songwriting partnership. Instead, John has come back to advise his former pal about house plants

PLANTS

In his new book 'Success With House Plants The Beatles Way', Paul gives handy hints on how to get the best from house plants, including those in the conservatory. And he later revealed that while he was writing the book the spirit of John Lennon visited him, and provided quite a lot of background information and valuable gardening tips for the book. Indeed, without the help of his former musical foil, Paul believes the new book could not have been written.

"The idea for a book about house plants dates back to the early sixties when The Beatles hadn't quite made it", Paul said recently. "After one gig in Hamburg we were

Grevillea robusta yesterday

a bit depressed and thinking of packing in the music business. It was then that I had this idea to do a book about house plants; the best types to grow, how to look after them, plus a few general tips and bits of useful gardening information. John thought it was a great idea, but before we had a chance to do it one of our records went to number one and the Beatles took off".

The book project was put on a back burner while The Beatles dynamic career unfurled. Hit followed hit as the four unlikely lads from Liverpool played their way into the records books as the most successful pop group of all time. Then at the end of the sixties the group fell apart, and once again Paul began his book about house plants.

Paperback writer Macca gets by with a little help from his friend

"I actually started to write the book in the early seventies, but Wings began to take off and once again I had to shelve the project due to lack of time. People thought that John and I weren't speaking in those days, but that was nonsense. We'd spend hours on the phone chatting about the book, ideas he'd had for the cover, plants we wanted to include. The book was never far from our minds no matter where we were or what we were doing. In fact, even when John split with Yoko and went on a six month drink and drugs spree he still used to ring me with his ideas, although his voice was slightly slurred on those occasions".

PAGES

Paul began writing the book in earnest three weeks ago, partly as a tribute to John who died in 1982. And the moment he sat down at his typewriter he knew John was in the room. "It was a feeling I've not had since the days of The Beatles", he confessed. "I just felt him stroll in, sit down somewhere above my head, and start throwing ideas at me".

BONHAMS

The first problem Paul encountered when he sat down to write the book was a lack of information. "I don't know much about houseplants. In fact I don't think I could name more than perhaps two or three. So on the face of it writing the book was going to be a hapless task". But then

John's ghost came up with a stroke of genius.

SOTHEBYS

"John suggested I go to the library and get out a book on houseplants so that I could copy it. Sure enough it worked. I borrowed a copy of the Readers' Digest 'Success With House Plants', and within minutes my book had begun to write itself".

CHRISTIES

Another Lennon brainwave was the idea of having an index at the back, as Paul explains. "I'd planned to write a book featuring about 100 different plants, but I wasn't going to put them in any particular order. Then John suggested I put an

index at the back, so that people could look up information on a particular plant without having to flick through the whole book". Paul took John's advice, and the result is an invaluable index which gives a page reference for every plant in the book, all of which are listed in alphabetical order.

GOODYEARS

Their book writing partnership differed somewhat from the songwriting partnership which the duo had forged some thirty years earlier.

Macca (above) with new book and dead co-author Lennon (above left) yesterday

"When we used to write songs I'd often use a piano, and John would use a guitar. But with the book there are no musical instruments at all", he told us.

DUNLOPS

Paul's book is illustrated by Ringo Starr and George Harrison and is available from most good bookshops priced £19.95.

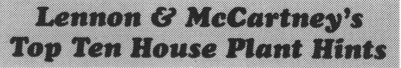

Lennon & McCartney's Top Ten House Plant Hints

1. Pinching out all the growing points of *coleuses* several times a year helps them stay bushy.

2. The only species of *Pfeiffera* (cactaceae) commonly grown as a house plant is *P.ianothele.*

3. When feeding *Tradescantia* (commonly known as Wandering Jew) apply standard liquid fertilizer once every two weeks from early spring to late autumn.

4. The best way to support unwieldy stems is to insert several thin stakes into the potting mixture and interlink them with a length of twine, looped in figured-8s around all stems and stakes progressively.

5. Lining a wire basket with *sphagnum moss* before planting *rhipsalidopsises* not only holds potting mixture in but also

improves the look of the display.

6. You must cut off a whole branch when propogating a *rhipsalis,* even if you then cut the branch into several smaller pieces.

7. Because *asplenium* roots tend to cling to the side of the pot, it may be necessary to break it in order to remove the plant before re-potting.

8. Do not overcrowd a conservatory with plants. Make sure you leave space for comfortable chairs.

9. Houseplants can broadly be defined as having six different basic shapes: Rossette, Bushy, Grassy, Upright, Treelike and Climbing/Trailing.

10. Do not worry when the stem of *A. brevifolia* topples over the edge of its pot; this is its natural growth habit.

LET US PAY

EXCLUSIVE

An unholy row has broken out over Church of England plans to introduce 'pay as you pray' meters in households throughout Britain.

The Archbishop of Canterbury, speaking last week, said that a system of metering was necessary because people's praying habits are changing. "More and more people are staying away from church and doing they're prayers at home", he told a top conference of bishops, vicars and vergers in Bournemouth.

CHRISTIANS

He blamed the trend away from church-going on D.I.Y. superstores and the fact that a great many Christians didn't want to miss Little House On The Prairie. And he warned that the church would have to adapt to fit in with new lifestyles.

LIONS

"By praying at home people are avoiding the collection plate, and that is hitting God where it hurts most, in the wallet", he told the conference. Church profits were down for the third consecutive year, he reported, and so a metering system was being considered as a possible solution.

GLADIATORS

If the scheme goes ahead a prayer meter would be installed in every household in Britain and this would record the amount of prayers being done. The local vicar would then come round for tea, and to read the meter. A quarterly bill would be sent to each household, followed two weeks later by a red reminder. Any household who failed to pay for their prayers would then be cut off from God.

TROJANS

Already a pilot scheme introduced in one Parish on the Isle of Wight has proved a success, according to church officials. The scheme, which has been operating for six months, has already raised over half a million pounds towards the local steeple restoration fund. But local residents aren't happy with the new arrangements.

DUREX

Sheila Foster was cut off from God after refusing to pay a £700 bill for prayers she claims she didn't make. "I got a bill for a prayer I was supposed to have made to the little baby Jesus. I

Two vicars calculate prayer bills yesterday

queried it because I hadn't done any prayers that week, but they sent me a final demand then cut me off."

MATES

Sheila is one of many Christians who are leaving the C of E in protest. "Now I've been connected up to the Jews, and they've been really great. They offer cheap rates at weekends, and you can even pray in the garden on Saturdays", she told us.

JIFFY

Another dissatisfied customer is 62 year old Ralph Henderson who has been with the Church of England all his life. "I had a friend staying with me for a few days and one day while I was at work he prayed for a sick relative in Australia during peak rates. When I got the bill I almost shit myself", said Ralph. "They sent a bishop round to check the meter but they still insisted I had to pay". Unable to find the money Ralph was cut off, and faces a winter alone without prayers.

HAND

"My wife only died last year, and I used to pray for her every night. Now I won't even be able to say Grace before my Christmas dinner", said a teary eyed Ralph yesterday.

CARRIER

Miriam Bigfatarsehole, spokeswoman for OffGod, the independent prayer watchdog, said that pensioners would suffer most under the proposed system. "Old people need to prayer more than most of us, because they're obviously old, and they're going to die soon, and naturally they don't want to go to Hell. By sending them these enormous bills the Church are effectively condemning them to drown in a lake of eternal fire."

HOMING

A Government spokeswoman last night said it had no immediate plans to introduce VAT on prayer bills, although the Chancellor of the Exchequer has so far refused to rule it out. Meanwhile, the Archbishop of Canterbury defended plans for TV advertising campaign for God. He said that the proposed payment of £8 million to Bob Hoskins for a series of two commercials was 'money well spent'.

Hoskins - £15 million

The stumpy headed actor will dress up as a vicar, talk in a cockney accent and pull a funny face at the end of two thirty second TV ads.

Bid to curb bee sex

Doddery TV porn campaigner Saint Mary's Lighthouse is to ban sex between bees in her garden.

The wrinkly prune who has devoted her lifetime to complaining about sex and bad language on the box and getting on everyone's tits, has now turned her attention to the humble bumble bee.

WINDOWS

Sickened by sordid scenes the wizened old hag has witnessed through binoculars from her French patio windows, she plans to put an end to all sex between insects in the garden.

VOLUME

"There was a time when gardens were a place where children could play, and old people could sit and fall asleep. But now everywhere you turn there are bees romping about naked, spunking up pollen and fingering their little bee fannies."

TONE

"The men bees are the worst", she continued, "coming at the poor little lady bees with their great big bee cocks in their hands, and dirty, leering grins on their faces. Roughly man handling the poor lady bees into flowers where they commit foul and unthinkable sex acts some of which, frankly, defy belief", she said yesterday.

BASS

Saint Mary's Lighthouse, who is 94, was today in hospital being treated for

'Hive had enough' says leathery faced pointy glasses TV sex campaigner

St. Mary's lighthouse - bee stings

bee stings after being found wandering naked and confused in her garden late last night.

Short hair

Hair cuts will be short in the spirit world, according to Christian pop singer Cliff Richard.

TREBLE

Cliff predicts short hairstyles will be the fashion 'on the other side', and that long hair will not be allowed in heaven, except for Jesus. Facial hair will be frowned upon also. "Only tidy beards like Noel Edmonds and Jeremy

Beadle will be allowed", Cliff told a hairdressing conference in Swindon yesterday.

Dog in a Million!

One Ron and his dog: Stonybridge with life-saver Bonzo

EVERY dog owner thinks that their pet is special, but meet Bonzo, the pooch in a million who has just been named 'Bravest Dog In Britain' at the prestigious *Daily Mirror* Animals of Courage awards. And it's a well deserved prize, for Bonzo's owner Ron Stonybridge reckons the heroic hound has saved his life no fewer than *TWENTY TIMES*.

The six-year-old Lurcher was presented with the 'Golden Bone' award at a glittering ceremony at the Albert Hall yesterday after owner Ron wrote to the newspaper to nominate him.

"Without that dog, I literally would not be here today," he told us. "Bonzo first saved my life a couple of weeks after I got him. I was sitting at home watching the racing on the telly when suddenly I heard an almighty crash from the kitchen. A bloke I owed a bit of money to had sent somebody round with a baseball bat to sort me out, and he had kicked the back door in.

"I'd had a few drinks so I was in no state to defend myself, but luckily Bonzo came to my rescue. He'd had a bit of an upset tummy for a few days, and there was quite a bit of shit in the hall. When this heavy come running through, he slipped in one of Bonzo's turds and went arse over tit, cracking his head on the banister.

"It gave me the few seconds I needed to escape out the window in the front room. If my dog hadn't crapped all over the hall, this bloke might have broke my legs, or even worse."

This incident alone would have been enough to win the award, but barely two weeks later Bonzo was to save Ron's bacon yet again.

"This one night, I'd had quite a skinful and I was fast asleep on the couch downstairs. At about 3am, I was woken up by Bonzo. He was making a terrible racket in the hall, retching something up. I got up to see what was going on, and I found him coughing up a tampon which I'd seen him eat down by

EXCLUSIVE

the canal earlier that day.

"At the time I was angry that he had woken me up and went over to kick him. But as I did, I heard the letterbox rattling. Then I saw the nozzle of a fuel can come through, and somebody started pouring petrol onto the mat. I opened the door and saw a man I'd borrowed some money off a few weeks earlier. He was trying to light some newspaper with a match, but when he saw me he ran off.

"I chased him for a couple of streets, but eventually I had to give up as I was in my vest and pants and I only had one shoe on. I owe Bonzo my life. If it hadn't been for him waking me up, I doubt I would have been here to tell the tale."

The next occasion that Ron's life was threatened, there seemed no way that his faithful hound could help. But the wonder dog was to come to his master's rescue in a most unexpected way.

"It was just before Christmas, and I was in my flat when there was a knock at the door. When I answered it, there on the step stood two nasty looking blokes. One of them had a shotgun in a carrier bag. At that time I'd run up quite a slate with an unlicensed bookmaker, and he'd been trying to get the money off me

> **"I owe Bonzo my life. If it hadn't been for him waking me up, I doubt I would have been here to tell the tale."**

for months. Finally he had lost patience and sent a couple of his heavies round to kill me.

"I shouted for Bonzo, but then I remembered I had shut him in the coal house because he'd been barking for his dinner all night. I thought my number was up, but then suddenly a police car pulled up and two coppers got out and came up the path. The two hitmen scarpered.

"I couldn't believe my luck. It turned out that the previous day I had left the gate open and Bonzo had got out into the back lane where he had savaged a toddler. The attack only lasted a couple of minutes and the kid just need a few stitches in his face and throat, but his mum had made a bit of a song and dance about it.

"The coppers had turned up to charge me with failure to keep control of a dangerous dog. Bonzo had done it again. If he hadn't have bitten that kid, the coppers would never have come to my door, and I dread to think what would have happened."

Bonzo left the stage with his tail held high, a golden bone medal and his prize, which included a year's supply of worming powder and a voucher for a free shampoo at a top London dog-grooming parlour.

"They say a man's best friend is his dog," said Ron. "And that's certainly true of me and Bonzo. I may have got him for free off a man in the pub, but after what he's done for me I wouldn't sell him for a hundred pounds!"

"But I will sell the grooming voucher and the worming powder. I need £15 before tomorrow morning, or someone I borrowed a bit of cash off says he's going to break my thumbs with a hammer," he added.

Top of the Pox

CLAP EXPERTS are predicting that Chlamydia will beat Syphillis to the coveted top spot in this year's Christmas venereal disease charts. The prediction follows a year which has seen a marked drop in cases of more traditional VDs, such as Gonorrhoea, Non-specific Urethritis and Genital Herpes.

"2005 has been a year of record success for Chlamydia," said Dr Frances Discharge of the British Institute of Genito-Urinary Medicine. "It has rarely been out of the Top Ten STDs in the past twelve months, and for the last six weeks it's been firmly ensconced in the number one slot."

VD nasty: A Chlamydia germ yesterday. *(Picture courtesy of Dame Shirley Bassey)*

And she had this advice for anyone thinking about having a flutter on the Christmas Sexually-Transmitted Disease Charts. "Traditional evergreen infections such as Syphillis, the Peter Pan of Pox, have had their day. There are simply too many new up-and-coming diseases about for dated, old-fashioned claps to dominate the December chart rundown like they used to."

"If I had to bet on the Christmas top position, I know where I'd put my money," she added. "Vaginitis may be bubbling under, but Chlamydia's the infection that's on everyone's lips at the moment."

1	Chlamydia
2	Trichmoniasis
3	Human Papilloma Virus
4	Crabs
5	Gonorrhoea
6	Vaginitis
7	Non-specific Urethritis
8	Syphillis
9	Genital Herpex Simplex 2
10	Molluscum Contagiosum

The December 1st Chart: but which VD will be celebrating on Christmas Day?

It's the most hotly debated topic around watercoolers in every office, factory and hospice for the terminally ill in the land...

Who is Britain's LADDISHEST Jamie?

They're the two undisputed kings of the 'lad' revolution - but just WHO is the ladder of the two? Is it Theakston, with his cheeky sideburns, smart suits and eye for the ladies? Or is it Naked Chef Oliver - with his gang of geezer mates and his love of hot 'Ruby Murrays' and cold beer?

Well now's your chance to find out as we sit as judge and jury over the Jamies. See if you agree with our verdict as we consider the evidence, and decide beyond reasonable doubt... **Who is Britain's Laddest Jamie?**

THEAKSTON | HOW THEY SCORE | OLIVER

The Birds — 7 / 7

Love 'em and leave 'em Theakston has been romantically linked to a bevy of beautiful beaver. Hardly a day goes by when he doesn't appear in the paper at a glitzy premiere with some classy piece on his arm. English rose Joely Richardson, yodelling wrist-slasher Mariah Carey and her out of All Saints are just three of the lairy Priory presenter's past girlfriends. However, with none of his conquests successfully impregnated, a question mark must still hang over this laddie's taddies.

Bossy wife Jules has doormat Jamie firmly under her thumb. Bounced into an early marriage after splitting his first and only kipper, the cockney pot-jiggler has never had the chance to sow his wild oats, unlike the fanny ferrets who he pays to be his mates in the adverts. Instead of being linked with a string of glamorous girlfriends, Jamie's been stuck in the kitchen, frying up a string of sausages for his pregnant potboiler missus.

Transport — 9 / 3

Theakston's chosen vehicle leaves Oliver in its wake... quite literally! That's because the Top of the Pops six-footer gets from A to B in a sleek formula one powerboat. Travelling at speeds up to and in excess of 200 mph, getting top marks in this round is plain sailing for laddish Theakston, especially since he's probably got several open-topped sports cars as well. And a helicopter.

If you saw Oliver strolling round Tesco's scoffing olives out of his shiny black open-face motorcycle helmet, you'd be forgiven for thinking that he rode a powerful sports motorbike, such as a Ducati 996 or a Suzuki Hayabusa. But you'd be wrong. The monkey-faced loudmouth loses lad points thanks to his choice of a pastel blue woman's scooter, which could be burnt off at the lights by Stephen Hawking.

Cock-nality — 5 / 8

It's hard to imagine public school-educated Home Counties boy Theakston stuck halfway up a chimney whilst his dad sticks pins in his feet. And with his upper class accent he'd find difficulty discussing jellied eels and the Krays with a pearly King. Down at the Old Bull and Bush.

Cockney lads don't come born any more through and througher than chirpy fuckgob Oliver. With his string of cheeky catchphrases, including "pukka", "sorted", "lovely jubbly" and "you plonker, Rodney", he's as Londonish as the roast beef and Yorkshire pudding that he cooks.

Rock and Roll — 6 / 8

The closest former auctioneer Theakston gets to a rock'n'roll lifestyle is presenting Top of the Pops once a week. But basking in the reflected second-hand laddishness of real rockers such as Liam Gallagher, Robbie Williams and H who was out of Steps is a poor imitation of the real thing, and Jamie makes a poor showing in this round.

If cooking is the new rock'n'roll, then rock'n'roll is the new cooking. And rock'n'roll cook Jamie's no exception to the rule. Every night Oliver twists off his apron and swaps his chicken drumsticks for some wooden ones, keeping the rhythm in his very own group! Whether he's beating drums or eggs, he's number one in the chef AND pop charts all year round.

Lack of Domestication — 9 / 5

Real lad Theakston's probably got no time for poncing around with a feather duster and a tin of shake 'n' vac. You putatively couldn't eat your dinner off his plates, let alone the floor. His hoover may be clogged with pizza boxes, birds' knickers and spent johnnies, and the pan of his shitter could well look like the starting line at Brand's Hatch. Such top untidiness earns blokeish Jamie a tidy score and he cleans up in this round.

You could eat your dinner off the floor of this Jamie's house, and it'd be an 8-course dinner he'd cooked himself like some sort of girl. As Oliver's army of viewers know, he keeps his fashionable penthouse mews loft conversion so spick and span it sometimes looks more like some expensive television studio set than a real lad's home. A flat performance in this round.

Drinking — 7 / 8

A bar-room kicking for Theakston in this round. Although he doesn't mind a drink, it's always fine wines for connoisseur Jamie. More at home in a posh gentleman's club than a spit & sawdust pub, quilted silk dressing gowned Theakston's always happiest ensconced in a leather armchair, sniffing snootily at a balloon of vintage port costing upwards of £10 a bottle.

Oliver is rarely seen without a bottle of his favourite designer lager in his hand. He even appeared in one Safeway advert with a stinking hangover after spending the previous night down the "rub-a-dub" with his "chinas". As a naked chef, he may know a thing or two about fine wines, but good old British Budweiser is this lad's tipple of choice.

Going with Pros — 10 / 5

Last Christmas, a pissed-up Theakston put his money where HER mouth was when he treated himself to a forty quid romp with a whore. He may have been too drunk to stand up, but his gut stick was certainly Live and Kicking. And not only did he pose for saucy suck-off snaps in the fashionable London bondage brothel, but he then proudly boasted of his red-blooded exploits to a Sunday newspaper. Uberlad Theakston tops this round with maximum points.

At eight months pregnant, Jules's jewels may well be out of bounds to husband Jamie. But we know from the Tesco advert that he is not averse to going behind his wife's back, nipping to his mum's for a curry whilst pretending to buy lightbulbs. So it would be no surprise to see his trademark scooter parked up outside a backstreet brothel whilst he gets his helmet polished inside. However, there is no evidence to suggest he uses prostitutes, so it's only half marks in this round.

Tongue Girth — 0 / 10

Oh dear! With his pathetic, normal sized tongue, Theakston loses marks in the final round.

Top marks for this Jamie who, looking like he's trying to spit out a pound of liver whenever he speaks, has Theakston well and truly licked.

THE VERDICT — 53 / 54

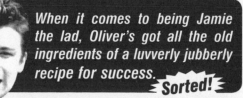

Mockney toff Theakston gets the bitter taste of defeat after a pitiful showing leaves him bottom of the lad pops.

When it comes to being Jamie the lad, Oliver's got all the old ingredients of a luvverly jubberly recipe for success. **Sorted!**

No Sex Please... W[

FIREMEN

HOT SEXCLUSIVE

| **1** (23) Financial Adviser | **2** (1) Fireman | **3** (2) Professional Footballer | **4** (4) Dentist | **5** (3) Painter & Decorator | **6** (8) Racing |

(Last year's position in brackets)

Britain's firemen are set to strike again, but this time it's not over pay, conditions or working hours. It's over their declining pulling power!

Once the country's most highly-sexed job, fire-fighting has recently fallen from its traditional number one position in the rumpy-pumpy professions chart. And union bosses fear that if the slide continues unchecked, firemen could even slip out of the top ten altogether by the end of the decade.

SEX

According to a survey carried out by the government's official sex watchdog OffBang, fire-fighting has now dropped behind financial management in the turn-on stakes. According to sex ombudsman Baroness Warnock, the scent of money rather than the scent of danger is what gets today's women going.

She told us: "Back in the eighties, the hunky heroes of films like Towering Inferno, Backdraft and Trumpton got a lot of ladies hot under the collar. They were attracted by the combination of strength and tenderness in six-packed firemen who fearlessly stepped into burning buildings to rescue babies and puppies without a thought for their own safety."

However according to Warnock, today's uncertain economic climate means women are increasingly transferring their sexual longings to men in the financial services sector. "Saving money has replaced saving lives as the biggest turn-on for ladies. Financial advisers, with their ability to juggle figures and fill in complicated forms, now have women going weak at the knees left, right and centre. Firemen have most definitely had their chips as the country's favourite heart-throbs."

VIOLIN

But fire union leader Andy Gilchrist slammed the new chart and pledged that his members would stay out on strike until the situation was rectified. He told us: "No fireman likes taking industrial action, but we can't sit back and allow our traditional position as the country's sexiest profession to be eroded in this way. We will be refusing to attend emergency calls until Britain's women come to their senses and once again think firemen are the horniest thing on legs."

BRIEF

And he had harsh words for government plans to bring in army recruits in Green Goddesses to provide emergency cover. "When Britain's birds see spotty squaddies tripping over their hoses like the Keystone Cops, they'll realise how hunky us real firemen are."

However, Britain's women were last night refusing to back down. Spokeswoman Germaine Greer told us: "We ladies have had the wool pulled over our eyes for too long. Firemen are simply not that sexy any more. Sicknote out of London's Burning was positively ugly, and the rest of the crew on Blue Watch was just plain.

PANT

"Compare that lot to some of the hunks you get in the financial services sector, like sexy Howard Brown from the

Firemen AXED

THE DOWNTURN in firefighters's fortunes has had knock-on effects in other industries.

Strippogram agencies have reported drops of up to 86% in demand for cheeky fireman strippers. One boss told us: "I used to take 40 or 50 bookings a week for firemen strippers, now I'm lucky if I take half a dozen. Nowadays all the women want to see a stripper with a briefcase and bowler hat, peeling off his pinstripe suit to reveal a sensible vest and Y-fronts. It drives them wild."

Meanwhile, ITV have axed the long-running fire drama **London's Burning** and announced plans for a new series set in the world of financial advice. **McGill** will star Robson Green as Keith McGill, a no-nonsense investment analyst in a busy Newcastle accountancy practice, who sometimes has to bend the rules to get results. It will be shown in the autumn, and is expected to go head to head with **Silent Partners,** the BBC drama starring Ross Kemp as Mike Brown, a mortgage arranger in a busy Manchester office who sometimes has to bend the rules to get results.

LOVE ME T[

Firemen first topped the hunkiness charts way back in 1911, and they have maintained their position ever since. This year's chart marks the first time in nearly a century that they have been knocked from their number one spot.

Over the years, many professions have challenged them. They came closest to being toppled in 1969, the year of the first Moon landings, when 42% of women would have bedded anything in a spacesuit. But firemen still held on by the narrowest of margins, with 44% of Britain's ladies still longing for a 999 hero between the sheets.

HOT STUFF - Some firemen not yesterday, and how the chart looked in 1911, or something.

AA 508

YRE

SEXY JOBS

Source - Official Government OffBang Rumpy-Pumpy Survey 2004

8 Vicar (5)	**9** Coach Driver (6)	**10** Milkman (9)

Halifax adverts or hunky money corespondent Declan Curry off BBC Breakfast News. Woof! I've got a wide-on just thinking about them," she added.

NDER

Firemen - The most hunksome of professions

Survey undertaken amongst the weaker sex to ascertain the romantic allure of diverse trades. Herewith are the results.

1. Firemen
2. Blacksmiths
3. Traction Engine Drivers
4. Silent Film Pianists
5. Coopers
6. Fletchers
7. Top Hat Makers
8. Penny Farthing Makers
9. Drapers
10. Town Criers

To see who has the sexiest job, we asked a fireman and a financial adviser to keep diaries of a typical day at work.

On the Job

CASE 1

Name: Jed Parslow
Age: 24
Occupation: Firefighter, Red Watch, Paddington Green Fire Station, London

8.00am Breakfast is interrupted by a call out to a small electrical fire above a grocer's shop in the High Street. Faulty wiring has ignited some oil-soaked lino behind a cooker. We put the blaze out using a carbon dioxide extinguisher and advise the tenant to invest in a battery-operated smoke alarm and a wall-mounted fire blanket.

10.00am A game of pool in the canteen is interrupted by a call out to rescue a cat which has become stranded up a tree. Since this is not an emergency, we make our way to the incident without lights or sirens operating, and obeying all speed restrictions and traffic signals. When we arrive, the cat has already made its way down the tree and the owner is waiting to apologise for calling us out.

11.00am A group of local Boy Scouts arrive for a pre-arranged look around the station facilities. They are keen to sit in the engine and slide down the pole. They are disappointed when I inform them that, due to Health and Safety regulations, this is not permitted. A photographer from the local paper arrives to take their picture wearing our helmets and oversized jackets.

12.00am We make a routine visit to a local office block to check extinguishers, fire exit signs etc. We find access to an emergency door in one of the offices partially blocked by two boxes of envelopes and refuse to issue a fire safety certificate until the situation has been rectified. People don't realise that in an emergency, moving obstructions from in front of emergency exits can take valuable time. It may only be a second or two, but a second or two in a major fire could mean the difference between life and death. Next we test the smoke alarm batteries in the communal area of the building, and find them all to be in order. Nevertheless, we remind the caretaker of the importance of testing the batteries at least once a week.

12.30pm A game of table tennis at the station is interrupted by a call out to an allotment fire that is threatening to get out of control. Hydrant access is blocked by a parked car, so we use the retardant foam in the tender to extinguish the blaze. It's a small blaze and only two officers are required to deal with this incident, so I remain in the cab, reading the paper and listening to Steve Wright in the Afternoon.

1.15pm On the way back to the station, we call in at the office we visited earlier to check that the boxes of envelopes have been removed. Fortunately they have, so we are able to issue a new fire safety certificate valid for the next twelve months. On our way out, we notice a plug socket in the foyer which has been dangerously overloaded. We always recommend the use of proper BSS standard electrical plugboards with fuses of the appropriate rating, preferably with safety cutout switches to prevent overheating.

2.30pm The Lady Mayoress arrives to ceremonially present us with 3 new sets of breathing apparatus, bought with money raised by the local Rotary Club. She is keen to sit in the engine and slide down the pole but once again Health and Safety regulations mean this is not possible.

4.00pm We are called out to a blaze at a local block of flats where a chip pan has been left unattended and has burst into flames. However, by the time we arrive, a neighbour has extinguished the fire by turning down the heat and covering the pan with a damp tea towel.

5.30pm Shift ends.

CASE 2

Name: Ken O'Dougal
Age: 46
Occupation: Independent Financial Adviser, Portland Associates, Grimsby

8.30am I'm up bright and early for a home visit to a couple who want to finance a loft conversion. When I arrive I'm told by the wife, an attractive, busty blonde in her mid-thirties, that her husband has been unexpectedly called away on business. I take her through the various options, explaining the most tax-efficient ways to extend her endowment mortgage, but she soon makes it clear that she's got another sort of endowment she wants to extend! Before I know it, she's pulled my clothes off and we're having wild, uninhibited sex right there on the kitchen table. After the most explosive orgasm of my life, she fills in the standing order forms while I put my clothes back on.

10.30am I'm back to the office, and there's a lot of paperwork to catch up on. A woman calls in for some financial advice. She has a TESSA that's just matured, and wants some ideas about the best way to invest her tax-free lump sum. However, it soon becomes clear that the tax-free sum isn't the only lump she's interested in! Complaining that my office is hot, she starts to unbutton her dress. Before you can say Financial Services Regulatory Authority we're both stark naked and having wild, doggy-style sex across the photocopier. Suddenly the door opens and my prim secretary walks in with a P11D to sign. When she sees what's going on she takes off her glasses, lets down her hair and joins us for a sexy threesome. After the most intense orgasm of my life, I sign the form and phone my brokers to arrange for my client's profits to be re-invested in a gilt-edged with-profits tracker as she re-arranges her clothing and prepares to leave.

2.00pm An Inland Revenue Inspector arrives, along with her lesbian lover, to discuss an irregularity in one of my clients' tax returns. I turn round to the filing cabinet, looking for the relevant documents but when I look up the two of them have stripped down to their silky lingerie and are putting on one hell of a show. They invite me to join in...but tell me to keep my bowler hat on. I don't need asking twice! After a series of earth-shattering orgasms, it turns out that, because it is a leap year, my client has inadvertently counted February 29th as week 53, when, because she is paid monthly, it should be counted as week 54 and carried over into the next tax year. I can't help wondering if the tax inspector knew that all along.

4.00pm I'm at a local hotel, to chair a seminar of local business leaders discussing the consequences of the Chancellor's pre-budget announcement. Throughout the meeting, I am distracted by an attractive woman who runs a mobile hairdressing salon and keeps crossing and uncrossing her legs. I can see she is wearing no knickers, like Sharon Stone in Basic Instinct. When the seminar is over, I get in the lift to go to the ground floor but just as the doors are closing, she jumps in with me. She asks me if I am "going down" and winks seductively. I press the stop button between floors as she begins to pull my clothes off. Soon she is orally pleasuring me to peaks of pleasure that I can only imagine. My moans of ecstasy set off the fire alarm and we find ourselves drenched by the sprinklers.

5.30pm Home to the wife, who takes my bowler hat and umbrella, and asks me if I've had a nice day at the office. Little does she know!

---

20 THINGS YOU NEVER KNEW ABOUT CARPETS, MOTORBIKES AND CROCODILES

They're in every newspaper you pick up, on every telly programme you watch. There's simply no getting away from carpets, motorbikes and crocodiles.

They're on our floors, in our garages and in our zoos. And whether we're walking on them, riding on them, or running away from them, one thing is for sure. They're here to stay.

But how much do we **really** know about them? Have *you* done your homework about carpets, motorbikes and crocodiles? For example, did you know that...

1 It was kiss curled pudding face American rock heartthrob Bill Haley who first brought the world's attention to crocodiles with his hit 'See you later Alligator'.

2 The lardy rocker no doubt thanked his lucky stars that he was never eaten by crocodiles. Or alligators. He died in 1976 on stage in Germany after being bitten by a snake which had stowed away in his trumpet.

3 Like chickens and mice, crocodiles are reptiles because they lay eggs. But never order a crocodile egg for breakfast. Crack it open with your spoon and instead of an egg yolk and white, inside you'll find a tiny baby crocodile!

4 The most expensive carpet ever made was for the cockpit of the Space Shuttle. A platinum and teflon weave, with a built-in carbon fibre underlay, it was rumoured to have cost over £65 a square metre, exclusive of fitting and gripper.

5 If, in the course of their sixties reign of terror, the Kray twins walked into your pub and asked for a carpet, they wouldn't be referring to your floor coverings. For in the East End of London 'a carpet' refers to the sum of £300. With the possible exception of in carpet shops.

6 Shout "on yer bike" at a policeman and he'd be liable to pinch your collar. Unless he was The Bill's beaky desk sergeant Barry Cryer, who'd be all in favour of the idea. For Barry is a keen motor-cycle enthusiast in his spare time.

Two birds on a motorbike the day before yesterday

7 And of course, he couldn't arrest you anyway, cos he's not a real policeman. He's just an actor with an unusually big nose.

Sgt. Barry Cryer - nose

8 Who's the odd one out? Singer George Formby, former World motorcycle champion Barry Sheen or Bullseye host Jim Bowen?

9 No. It's Barry Sheen. All are keen motorcyclists (with the exception of George Formby, who **was** a keen motrocyclist, but is now dead), but Sheen is the only one who never passed his road motorcycle test. (With the possible exception of George Formby.)

10 Who is the odd one out this time? George Formby, Barry Sheen or Jim Bowen.

11 No. It's Jim Bowen. Both the others are former champion motor-

Bowen - pub

cyclists, George Formby having won at the Isle of Man TT races. Comic Jim has never won a motorcycle race, although he does own a pub.

12 If somebody is crying crocodile tears, don't panic. They aren't necessarily an upset crocodile. 'Crocodile tears' is an expression used to describe false or insincere tears, such as those cried by stumpy Argentinian football cheats when they loose.

Stumpy cheat

13 If your boss offers to carpet you, don't go home and pull up all your old carpets in anticipation. Just go to his office and let him shout at you for five minutes. For a 'carpeting' is another term used to describe a dressing down.

14 A dressing *gown* on the other hand is a bath robe made of light-weight 'carpet-type' material which you steal from hotels.

15 White sports socks, available from burly youths in the street three pairs for a pound, are made of a similar material called Terry Toweling.

16 Terry *Thomas* on the other hand is neither sports sock, carpet, crocodile or motorbike. He's a British film actor famed for his versatility of roles, and the gap between his teeth.

17 Crocodile clips are not metal clips worn around your ankles to stop your trousers getting dirty while you ride on a crocodile. They are in fact the little metal things that fall off shortly after you buy a cheap battery charger from Halfords.

18 Many stars have sung the praises of carpets, motorbikes and crocodiles. Elton John for example encouraged pop fans to try rocking the 'crocodile rock'. Meanwhile David Essex had a dream... silver dream machine. (Which was some sort of motorbike, apparently.)

Elton in glasses (above) and (below) Essex

19 In all her lifetime Her Majesty the Queen of England has never set foot *off* a carpet. For everywhere she goes a team of 24 full-time Royal carpet fitters continuously roll out red carpet in front of her.

20 She didn't sing about rolling out carpets, but fat seventies cow Mrs Mills did record the popular Cockney knees-up anthem 'roll out the barrel', and probably got paid a few carpets for doing so.

You miserable bastards!

On the previous page we asked you to tell us if you had ever been tret like shit by a showbusiness celebrity. And your letters have been pouring in, painting a pretty grim picture of the stars.

For it appears that a great many of the idols we worship and adore are two-faced rats who wouldn't give their fans the time of day. Here is just a brief selection of some of the stories that you have told.

Brucie 'Didn't do well'

Bob Brown of Fulchester had always been a big fan of Bruce Forsyth, until the day his car broke down on the hard shoulder of the M6 near Lancaster. Bob had a flat tyre, it was pouring with rain, and as luck would have it, he'd forgotten to put a spare in the boot.

MORRIS

It was the middle of the night and there wasn't a car in sight – then suddenly a Morris Marina appeared in the opposite carriageway. "It was going very fast, and there was a lot of spray from the rain on the road. And it was dark. But I got a pretty good look at the driver, and I'm fairly sure it was Bruce Forsyth, or someone who looked a lot like him".

Bob waved desperately trying to attract the attention of the Generation Game host, but Forsyth sped by without so much as slowing down. "I felt as if he'd let me down. I'd watched all his shows, but when it came to the crunch he simply didn't want to know".

We rang Forsyth's agent to get his side of the story. We were put on hold. Two minutes later a girl came on the line. "I'm sorry but Bruce is too busy to talk to you this week", we were told.

EXOTIC

No doubt Bruce was 'too busy' to come to the aid of a faithful fan on that cold, dark windswept night in Lancashire all those years ago. Some things never change.

Stars shit on fans from a height

Not nice to see him (left) to see him not nice, and a Shirley similar to the one refered to in our story.

Shirley she could have helped me

Arthur Jones considered himself Shirley Bassey's number one fan. Until the night he turned to his idol for help in an emergency.

HEAVY

After a night of heavy drinking Arthur had called by Bassey's house at 4am to wish his favourite singer good night. But while reaching for the doorbell he had slipped and cut his head on milk bottles that had been left out on the step. In his confusion he then dropped his car keys down a drain. Unable to get home, he decided to wake Shirley by knocking loudly on her front door. But rather than coming to the injured fan's aid, the heartless singer threatened to call the police if he didn't go away.

"I only wanted to borrow a stick or something so I could get my keys back, and maybe a night-cap or something to send me on my way. But the rotten cow wasn't having any of it", Arthur told us.

BITTER

Indeed the selfish singer did call the police, and Arthur ended up spending the night in police cells. But according to Arthur it was he who had the last laugh. "I pissed in her flowerbed", he fondly recalls.

Esther is best'er the bunch

The caring face of TV's Mr Nice Guy Esther Rantzen

But not all the stars are bad. One reader wrote in to tell us a heart warming tale of a celebrity who did care.

WITS

Bill Rodgers, of Fulchester, was at his wits end when his two year old daughter fell blind and he lost his job all in the same week. In desperation he wrote to That's Life's Esther Rantzen.

"I couldn't believe it. Half an hour later she wrote back donating all of her kidneys to our daughter. It was the best gift anyone could ever receive."

TITS

Esther's selflessness has left the Hearts of Gold host with no kidneys of her own and as a result she must spend 20 minutes each day inflating her own artificial kidneys with a foot pump. But her generosity didn't end there. "The next thing you know she gave our pet rabbit the kiss of life after it had been run over by an ice cream van", said a grateful Bill. And later she turned up at Bill's daughter's first birthday party with a wheelbarrow full of BBC money for the sickly child, plus two watering cans full of diamonds.

SHITS

"Words can never say enough to thank Ester for everything she has done for this family", Bill told us. "She truly has a heart of gold".

Help! I need some money

Fred Johnson, also of Fulchester, grew up in the sixties. A big fan of The Beatles, he bought every record the group made. But 20 years later Fred fell on hard times, and in 1989 he found himself unable to pay his gas bill of over £200.

Fred wrote to millionaire Paul McCartney, his childhood hero, asking if he would pay the bill, plus a few pounds extra towards a coat for Fred's wife who had also been a fan of the group. But McCartney, who ranks among the richest men in Britain, didn't even write back.

"Fans like me have been paying McCartney's gas bills for the last 25 years, but when it comes to putting his hand in his own pocket, he simply doesn't want to know. Well, take it from me, I certainly won't be buying any more of his records. And he knows where he can stick his wife's veggie burgers too", said Fred.

LONG LEGS OF THE LAW

Britain's solicitors are the tallest in the world – and that's official!

A worldwide survey of the heights of people employed in the legal profession revealed that Britain's briefs are head and shoulders above the rest when it comes to tallness.

Yet curiously, Britain is one of the few countries that does not stipulate a minimum height for solicitors. Other countries, among them France,

MINIMUM

Belgium and the United States, have recently introduced minimum height requirements in an attempt to stamp out short solicitors. And in Spain where no height restrictions exist, dwarf solicitors are commonplace, with the average height of Latin legal eagles a meagre four feet eight inches.

Britain's average is a towering six feet four, well ahead of the Dutch in second place at five feet eleven.

AVERAGE

Legal profession height watchers were yesterday unable to explain Britain's baffling lead in the legal tallness stakes. One solicitor we spoke to declined to comment unless we paid him £85 an hour to do so.

It's Britain's LIVELIEST Door Forum!

...mind the DOORS

★ ★ ★ ★ ★ ★ ★ ★ ★ ★ ★ ★ ★ ★ ★

The Door & The Law

Your Legal queries with Door Lawyer Braxton Hicks

I recently had a pair of patio doors installed, leading from our back room into the garden. However, I was shocked when my neighbour pointed out to me that they were actually French windows. I contacted the builder who installed them for a refund, but he refused to give me my money back saying that I was happy with the job when he finished. What can I do?

Les Poindexter, Sheffield

★ *Braxton replies...* Very little I'm afraid, Mr Poindexter. Technically, your builder is in the right. By paying up at the end of a job you are, legally speaking, accepting the work he has done. Whilst patio doors and French windows are one and the same thing, it is up to you to make sure that you builder installs the correct one before handing over your hard-earned cash.

• I grew up in a house with doors in it, and it's left me with a lifelong love of them. Now I have a house of my own, and I have a door in every room. Some rooms have even got two.

J Barnestaple, Rhyll

• When I was a boy, my parents' house had two doors, which they referred to as the 'front door' and the 'back door'. Whilst the front door was indeed at the front of the house, the so called back door was actually down an alley at the side. I remember once someone delivering a new fridge to our house. *'Could you take it round to the back door?'* my mother cheerily asked. The poor man walked straight past it! It still makes me laugh today.

T Couts, Glamorgan

• *'Get yourself an education - it opens so many doors'* my father would always tell me when I was little. Of course, I didn't listen to him, and dropped out of school when I was 15. The funny thing was, I got a job straight away... *as the doorman at a local hotel!* I've been there for thirty five years and opened more doors than I care to remember. So I had the last laugh.

J Pritchett, Bude

• I own a small shop on the outskirts of Leeds. One morning as I opened my shop, the door fell off. It appeared that woodworms had eaten into it and weakened the timber around the hinges. But I didn't have to go far to get a new door... it's a door shop that I own, and I simply bought one off myself.

F McNally, Leeds

• As a boy, I always loved the door to my grandfather's house. It was a deep red colour and had the most beautiful stained glass in it. *'I'll leave it to you in my will'*, he often told me. I thought he was joking, but he was as good as his word. When he died, he left instructions to the executors of his will to remove the door and take it round to my house. It was a very emotional moment when I fitted it in place at the front of my house. My grandmother was furious, however. Without a front door, some burglars marched straight in and cleaned her out whilst she slept.

M McCloud, Birmingham

• Shakin' Stevens may have had a hit with the song 'Green Door', but he didn't have one in his house. I'm a painter and decorator in Cardiff, and I did his front door last week... *duck egg blue!*

Norbet Golightly, Cardiff

• I've always been a great fan of kitchen doors, especially the ones that swing in two directions. The other day, I was delighted to see a Laurel and Hardy film on TV where the pair get a job in a hotel kitchen, and I settled back to enjoy the doors. My delight soon turned to anger. Not two minutes into the film, the fat one, carrying a large stack of plates was sent flying, plates and all, by the thin one opening the door *the wrong way!* They were sacked by the little man with the moustache and the kitchen doors did not feature any further. My blood boiled. Had they followed the accepted catering industry proceedure of walking through the right hand door, the accident would not have happened and a perfectly enjoyable film would not have been ruined.

Alfred Biggles, Chester

• The other day, I went shopping as I do each and every Wednesday morning at 9.00 o'clock, rain or shine. When I got back at 11.00, I realised I had left my front door key at the bakers shop in town. Fortunately, I keep a spare key hanging on a piece of string behind the letterbox for just such emergencies.

Dorris Rabbit, 32 Oakenfield Avenue, Barnchester

• *'As one door closes, another one opens'*, my father once said when I was little. However, he remedied the situation by putting spring loaded stays on top of each door and draft excluders along the bottom.

J Geils, Folkston

• I went to buy a new front door from a builders merchant last week. I selected one at a cost of £48. The man in the store asked me if I would like a letterbox cut out of it and I accepted. Imagine my horror when he charged me £2 for cutting the hole. Correct me if I'm wrong, but a door with a letterbox has less wood than a door without one, so it should be cheaper, not more expensive. No wonder door manufacturers all drive around in Rolls Royces.

Hector Dunwoody, Leigh

• A couple of months ago I put a new front door on my house. The following day, my neighbour put a new front door on his. A week later I painted mine blue and he followed suit. Yesterday, I put a doorbell on it, and guess what?... he put a door bell on his. I put the number 36 on mine this morning. I dare say tomorrow there'll be a 36 on his door.

P Harper, Harpenden

Kids say the Funniest things... about doors!

..My grandson looked at the house number on my front door the other week. 'Do those numbers let people know how old you are, gran?' he asked. I wouldn't have minded, but I live at 324 Acacia Avenue. I don't look that old!

Edith Barnett, York

...'Look, mummy, that door's broken. It's got all holes in it', my daughter once said in the kitchen. She was pointing at a louvre door!

Margaret Barnes, Stoke

...'Granny! Granny! Why don't cars have their numbers on

their doors like houses do?' my grandson asked me the other day. I wouldn't mind, but he's 36 years old.

Ada Sykes, Hull

...'Could you answer the door?' I called to my grandson when the bell rang. 'I don't know, gran', he replied. 'What did it ask?' I've never liked him, the sarcastic little cunt.

Joan Timeshare, Totnes

Miriam

ANSWERS YOUR DOOR PROBLEMS

LETTER OF THE DAY

Dear Miriam...

I live in a small, very friendly street where everyone knows everyone else.

I am 54 and my neighbour is 51 and we have always been great friends. Recently, however, he has been calling the entrance to his walled garden the 'side door'. Granted it is six foot high and made of wood, but as the bricks do not go over the top of it, surely it is a gate, not a door.

I don't want to fall out with him over something so petty, but I don't know how much longer I can go on listening to him refer to this gate as a door.

AR, Bolton

Miriam says... This is something you need to sort out now before it ruins a perfectly good friendship. For what it's worth, I was always taught that if you could hold a bargepole aloft and carry it through, it was a gate. But your neighbour may have been taught differently.

Have you spoken to him about it? He may have a special reason for calling it a door. Or he may be perfectly happy calling it a gate if you tell him how it's upsetting you. You may just have to agree to differ and hope your relationship is strong enough to get you through the disagreement.

Elvis has Left the Coffin

Tomb With a View Suede Shoes: Elvis remains six foot under yesterday.

Since the day he died in 1977, rumours that Elvis is still alive have abounded. Barely a week has gone by without the King of Rock'n'Roll being spotted stacking shelves in a supermarket, working at a gas station or flipping burgers in a fast food drive-thru.

EXCLUSIVE!

But yesterday, over 28 years after his funeral, those rumours finally came to an end when the chart-topping dead star held a sensational live press conference from his Graceland coffin, confirming that he has indeed been alive since his death.

"I've just been taking things kinda easy for a while, that's all," Elvis told reporters in his trademark Tennesee drawl. "After more than twenty years living the rock'n'roll lifestyle, 1977 seemed like the right time to kick back and hang loose a little," he added.

Presley explained how he has spent nearly three decades under the ground, doing Sudoku puzzles and eating worms. He confessed that burgers were one of the things he missed most about being alive. "Them wriggly critters sure don't taste as good as one of Colonel Tom's chitterling sandwiches with a side order of hominy grits, but you get used to them eventually," he quipped.

But Elvis disappointed fans by announcing that he currently has no plans to resume his record-breaking pop career. "I've had plenty of hits down the years, but I'm seventy now and I guess rock'n'roll is a young man's game," he announced. However, he didn't rule out the possibility of a comeback at some time in the future, perhaps to celebrate the thirtieth anniversary of his death.

He told reporters: "I never say never. If Ronnie Barker can come out of retirement to do the 'Best of the Two Ronnies' sketch compilation specials, then anything is possible. Who knows how I'll feel in two years' time?"

But he confirmed he wasn't intending to brush the moss off his blue suede shoes just yet. "Sometimes I miss my old lifestyle, with its fast cars, private jets and Las Vegas glitz," he said. "But I've never for one minute regretted dropping dead on the toilet whilst doing a big shit. Looking back, it was the best career move I ever made."

YOUR chance to...
Spew Your Bile

CONVICTED rapist Iorworth Hoare's £7million win on the national lottery has sparked a storm of protest across the land. Never in the history of Letterbocks have Viz readers' danders been so gotten up. Here are a selection of the letters we received...

...I was disgusted to hear that a rapist was allowed to win £7 million on the National Lottery. By committing the most vile of crimes, a man should forfeit his right to win the jackpot. In fact, he shouldn't even be allowed to get 5 numbers and the bonus ball. Convicted rapists should be limited to matching three numbers, winning a maximum of £10.
M Hudson
London

...To prevent further rapists winning £7 million on the lottery whilst law abiding citizens miss out, the current system of numbers and balls should be scrapped. Everyone should pay £1 as they do now, and the person who has committed the least number of sexual offences should win the jackpot. They then do not enter the lottery again. This way, everybody will eventually win a jackpot, and rapists will have to wait a long time and win a significantly smaller sum.
L French
Peterborough

...I spend over £30 a week on lottery tickets and scratch cards and the most I have ever won is £10. perhaps if I went out and raped somebody, the powers that be at the National Lottery would come round and give me 7 million quid. Come on, Camelot, get your act together.
M Winterborough
Carlisle

...A rapist winning £7 million on the National Lottery? What is the world coming to? My son is no rapist, only ever having been convicted of a series of aggravated sexual assaults, and the most he has won is 58 grand when he got 5 numbers and the bonus ball. It doesn't seen fair.
T Plywood
Yorkshire

What I don't understand is, if God can part the Red Sea and make it rain biscuits on the Jews or whatever, why can't He jiggle a few ping-pong balls a bit in order to prevent a rapist winning the lottery? I can't see why He doesn't fix it for deserving people, like Mother Teresa, or Terry Waite to scoop the jackpot every week. It has certainly shaken my faith in Him.
Rev. J Foucault
Truro

Heartless Thieves Steal Shopping Scooter

So-sad Edna: In happier times sitting on her great big fat arse.

A South Tyneside woman was recovering at her home last night after heartless thieves stole her electric mobility scooter.

EXCLUSIVE

Edna Cretin from Jarrow had gone to play her nightly game of bingo at the local Mecca Hall on Wednesday. When she came out at around 10.30pm, the 48-year-old grandmother found that her ShopRider Scooter had been taken. Bingo-goers told reporters that 20-stone Edna was distraught. With-out her scooter, the lazy cow was forced to walk the 150 yards to her home.

Mrs Cretin bought the scooter two years ago after doctors told her that she was overweight. Since buying the scooter, her condition has deteriorated and the idle bitch has recently been diagnosed as being clinically obese.

"I hope these thieving monsters are proud of themselves," the tearful walkshy lardarse told us. "But I know the big-hearted folk of the north east will club together to buy me a new scooter, perhaps with a bigger seat this time.

"And a tray on the armrest for my sweets and crisps," added the idle fat pig.

UN weapons inspectors delve in

Just How Po

She's the head of Britain's newest royal family, she spent her childhood being chauffeured around in her father's Rolls-Royce, her wedding day resplendent on a golden throne and now lives like a queen in her very own palace. Every surface in her stately home groans with onyx trinkets and every wall is plastered with fine works of art. She's Posh Spice, one time pop princess and now Lady Victoria Beckham. But is she quite as posh as she is painted?

To find out once and for all, we commissioned United Nations arms inspector Dr Hans Blix to produce a detailed report to determine Posh's level of classiness. For a week, he was granted unprecedented access to Beckingham Palace, where he searched high and low for "smoking gun" evidence of commonness. Finally we presented his detailed dossier of results to Penelope Keith, Britain's most upper class woman. Here she assesses Dr Blix's evidence and finally answers the question: *Just HOW posh is Posh?*

MONDAY..........

MY inspection team arrived at the gates of Beckingham Palace and were let in by the guards. I immediately asked to be shown the master bedroom. The decor included gold light switches, swagged Venetian curtains and a 4-poster bed out the Scotts of Stowe catalogue. Mrs Spice appeared generally cooperative and opened the wardrobe to facilitate my inspection.

I found her dresses to be suitably posh and expensive, including many couture labels such as Moschino. However, one of my inspection team located two pairs of discarded ladies' underpants on the floor under the bed. Later in the day, we made a cursory inspection of Mrs Beckham's fridge, where we found several cartons of Tesco 'Value' orange juice and a box of Ferrero Rocher.

We asked for documented evidence of Miss Posh's last food shopping trip and were shown till receipts which revealed that she had procured, amongst other foodstuffs, several bags of Walker's 'Sensations' premium crisps and a bottle of expensive champagne. However, it was noted by my chief inspector that a crude attempt had been made to obscure an entry detailing the purchase of twelve Goblin meat puddings and a jar of Chicken Tonight.

Penelope Keith...

On the surface, I see little here to suggest that Posh Spice is concealing commonness. Gold light switches, swagged curtains and mail order catalogue beds are all signs of genuine poshness. Likewise, discarded undergarments are not necessarily the "smoking gun" they may appear to be; truly posh people don't pick up their own dropped scads, they have servants to do it for them.

Of slight concern at this stage are the working class food products amongst Mrs Beckham's groceries. However, the presence of the world's poshest chocolates - Ferrero Rochers - amongst the inventory leads me to conclude that, in the main, her tastes are genuinely rarefied.

TUESDAY.........

At noon, myself and a team of seven inspectors made a surprise visit to Mrs Beckham's dining room, where we positioned ourselves around the table while she ate her dinner.

We noted that the cutlery was silver, and laid out in accordance with the standard practice as recommended by Debretts. However, certain discrepancies were noted with regard to her usage of spoons throughout the meal. a) A pudding spoon was used for the consumption of soup. b) After stirring six sugars into her Earl Grey tea, Mrs Spice replaced the wet spoon back in the sugar bowl. c) Mrs Beckham used her clean soup spoon to eat her gravy, and d) She was then forced to go in the kitchen for a clean spoon to eat her dessert. It was noted that this was a soup spoon.

My inspection team appraised me of the fact that, when she was drinking her tea, Mrs Posh's little finger remained uncrooked. Furthermore, when she was eating her dinner, she was repeatedly seen to place her elbows on the table.

Penelope Keith...

This is altogether more worrying. Hairline cracks are beginning to appear in Beckham's posh facade. 50 percent of being posh is knowing one's way around the cutlery on the table, and here she appears to be let down badly by her confused spoon usage and poor etiquette.

However, her choice of tea is straight out of the top drawer; Earl Grey is ten to fifteen times as posh as PG Tips, so it may simply be that Mrs Beckham's lack of table manners is merely a manifestation of very British extreme upper class eccentricity.

I will wait for Mr Blix's full report before coming to any firm conclusions.

WEDNESDAY....

My team requested access to the toilet; it was our intention to make a thorough inspection of the palace lavatory facilities in order to determine their level of poshness. However, Mrs Beckham became uncooperative and refused our request, telling us that the lock was broke.

We moved on to the upstairs nursery where we found two children's beds with names painted onto the headboards - 'Romeo' and 'Brooklyn'. The floor was littered with plastic toys - such as Action Men, Transformers, radio-controlled cars etc.

Penelope Keith...

There could be any number of reasons why Posh Spice wouldn't want UN inspectors nosing round her lav, but sadly the most likely one is that it's in a right state. Perhaps there's skidmarks on the bowl, sprinkles of piss on the seat or a well-thumbed copy of the Autotrader lying on the mat. Without proper unfettered access to her toilet, it is impossible to discount any of these possibilities. A genuinely posh person's cludgy is kept as clean as their kitchen. I can't help beginning to wonder if Posh Spice is as posh as she seems to be.

On the other hand, she HAS given her children genuinely posh names, though it is interesting to note that Dr Blix makes no mention in his report of hand-crafted wooden toys like Noah's Arks, expensive rocking horses or old fashioned pedal cars. Miss Beckham's children evidently play with mass-produced plastic toys like common urchins on some dreadful council estate.

THURSDAY.......

The team decided to arrive unannounced at Beckingham Palace in order to make a dawn swoop on one of the living rooms. My inspectors identified and catalogued the following items: a) A 72" widescreen television set with surround sound speakers. b) A copy of 'TV Quick' with that day's schedule on QVC circled in biro. c) A grease-stained Greggs bag on the chair arm, containing the remnants of a chicken and vegetable pasty.

Mrs Beckham entered the room after the process had got underway. It was noted that she was wearing a toweling dressing gown and maribou high-heeled mule-type house shoes. Her hair was inspected and found to contain twelve Carmen heated rollers which were held in place with a nylon hairnet.

She announced that she was just going out and it was decided to send a small task force of inspectors to accompany her and report on her activities.

Mrs Spice walked out of the gate of Beckingham Palace, still dressed as described above, and entered a local newsagent's shop. Here she spent £40 on National Lottery scratch cards, comprising £17 on Millionaire Maker, £12 on Instant Jackpot and £11 on Lucky Lotto, which she then scratched on the counter using a large diamond ring.

Penelope Keith...

Dr Blix's discovery of a large television set is very significant. On the face of it, it may seem very posh to have a big TV, but in fact just the opposite is the case. Truly posh people shun television, and if they do own a set, it is a small black and white one used for watching nature documentaries and Newsnight. Likewise Mrs Beckham's choice of reading material - TV Quick rather than the TV Times - as well as her clothing and purchase of scratch cards, smack somewhat of the lower end of the social spectrum, as does her evident predilection for Gregg's pasties in preference to upper class fare such as Fortnum and Mason scotch quails' eggs.

FRIDAY............

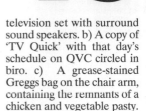
After receiving intelligence that the main hallway of Beckingham Palace contained a print of a street scene including a genuine clock, powered by an AA battery, my team decided to make a thorough search of all the corridors in the building. However, as the inspection got underway, Mrs Beckham received a telephone call and left the house. As before, a detachment of inspectors decided to accompany her, following closely in a liveried United nations Land Rover.

She was seen to drive into the car park of a local licensed premises, where she spoke to a man in a shell-suit, and got into his car. The couple were then followed to a lock-up garage where the man appeared to be showing Mrs Beckham a variety of unboxed electrical goods. She bought eight video recorders, paying cash.

Penelope Keith...

Finally, this is the "smoking gun" evidence I have been waiting for. The behaviour detailed by Dr Blix on the final day of his inspection proves that Posh Spice is as common as muck. A properly posh woman, such as the Queen, has many works of art and it is certain that none of them contains a real clock. Furthermore, if a member of the upper classes such as Tamara Beckwith or Nicholas Parsons wanted to buy eight video recorders, they would buy them from Harrods. And they would pay with a Coutt's cheque, signed with a solid gold fountain pen.

Penelope Keith's VERDICT

After weighing up all the evidence in Dr Blix's report, I have to conclude that Victoria Beckham is not quite as upper class as she might at first sight seem to be.

VERMIN

BRA HUMBUG!

One person who disagrees with Penelope Keith's verdict is Professor Tibor Zachas, emeritus professor of breasts at the University of Cincinatti. He maintains that a woman's chest and cup size is inversely proportional to how posh she is.

For the past 20 years he's made a systematic study of women's bosoms, mapping their dimensions against the class of their owners. And if correct, his revolutionary theory could put Victoria Beckham near the top of the social tree.

He told us: "In my research, I concluded that all the posh birds had tits like eggs sunny side up. Look at your Tara Palmer-Tomkinson, she may be high class muff but she's got knockers like two aspirins on an ironing board."

However, at the other end of the social scale professor Zachas found it was a different story. "Take Dolly Parton," he told us. "She's real trailer trash, but she's got goddam massive charlies."

So where does this leave Victoria Beckham? We showed some pictures of her to the professor. He told us: "From the size of her top bollocks, I'd say she was pretty lah-di-da. About the same as Princess Stephanie of Monaco."

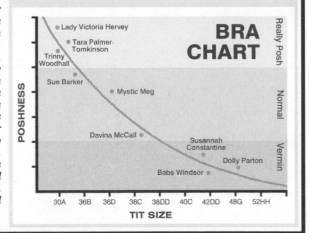

BRA CHART

Chart — POSHNESS (vertical axis: Really Posh, Normal, Vermin) against TIT SIZE (horizontal axis: 30A, 36B, 36D, 38C, 38DD, 40C, 42DD, 48G, 52HH)

- Lady Victoria Hervey
- Tara Palmer Tomkinson
- Trinny Woodhall
- Sue Barker
- Mystic Meg
- Davina McCall
- Susannah Constantine
- Dolly Parton
- Babs Windsor

PETER Bowles has been a fixture on Britain's TV screens for over thirty years. His perfectly judged performances in shows as diverse as *To the Manor Born, Executive Stress* and *The Bounder* have made his face into a household word from Land's End to John O'Groats.

Throughout his career he has played many parts in many stories, but none has been as strange as the story of his own life. It is a life which has seen him on the brink of sporting superstardom one day, to nearly taking a job in the world's most famous rock group the next; from almost becoming supreme leader of the world's largest religion to just about being offered the opportunity to go into outer space. Now, in this exclusive extract from his autobiography *'To the Manor Bowles'*, (PanMacmillan £12.99) he lifts the lid on one of the most extraordinary careers in showbusiness.

I have played many parts in many stories, but none have been as strange as the story of my own life

MOP-TOP OFFER NEARLY HAD ME BEAT

'I had wanted to be an actor all my life, so it came as quite a surprise when, out of the blue, I was offered the chance to become a pop star,' says Bowles.

'One day back in the sixties I was in my kitchen, making myself a bite to eat, when the phone rang. Picking it up, I immediately recognised the Liverpudlian voice at the other end of the line. It was Paul McCartney out of the Beatles. "Mr Bowles," he said. "It's such a privilege to speak to you. I've enjoyed your work in shows such as *Nanny Knows Best* for many years. In fact, I think you are the best actor in the whole world."

'He went on to explain that his group was busy recording their seminal Sergeant Pepper's Lonely Hearts Club album at Abbey Road Studios, just a stone's throw from my flat. He said that Ringo Starr had been taken ill and asked me if I would be interested in taking over as the Beatles' drummer. He told me that only someone with my impeccable timing would be suitable for the job. As you can imagine, I didn't need asking twice! But I told him I couldn't get round there for a few minutes, as I was just about to eat quite a large cheese sandwich.

'As it turned out, by the time I had finished my sarnie and got round to Abbey Road studios, Starr had recovered from his illness so I never did become a member of the world's greatest rock band. *But I often stop to think how different my life would have been as one of the Fab Four... John, Paul, George and Peter Bowles!'*

'THRONE' BY MISSED HOLIDAY JOB OPPORTUNITY

'As an actor, work opportunities happen all the time, often at the most unexpected moments. But even I found it hard to believe the job offer I received once whilst I was on holiday in Rome.

'I was sitting in a cafe, about to have my lunch, when I was approached by an elderly priest. He explained that he worked at the Vatican as the Pope's personal assistant. "The holy father is a great fan of yours," he told me. "In fact, he spends many hours every day sitting in his private apartments watching tapes of your programmes, such as the second and third series of *Executive Stress*, when you replaced Geoffrey Palmer in the role of Donald Fairchild."

'The man told me that his boss was thinking of retiring soon, and had started looking around for someone to take over after him as head of the Catholic church. "The Pontiff has been very impressed with the way you have made that role your own," he explained. "His holiness would be greatly honoured if you would come to the Vatican and become Pope Peter Bowles the First."

'Just at that moment, the waiter arrived with a mozzarella and tomato sandwich I had ordered, so we were unable to continue our conversation. But I often wonder what would have happened if I had followed up that job offer and become the infallible figurehead of the billion-strong Roman Catholic faith.

'I know one thing for sure, I couldn't have starred as Hilary, Rigsby's camp playwright tennant in *Rising Damp,* if I'd been the Pope. The show was always recorded at Yorkshire TV's Leeds studios on a Sunday evening, when I would have had to be a thousand miles away in Rome, *taking Mass from my balcony overlooking St Peter Bowles's Square!'*

MY FOOTBALL NEAR MISS WAS THE FINAL STRAW

'I was standing at a snack bar near Wembley Stadium when a bald man in a red tracksuit walked up to me. I recognised him as Sir Alf Ramsey, the England football coach. "I can't believe it's really you," he gushed. "You're my favourite actor of all time. I just can't get enough of your work in shows like *The Bounder,*" he added. He explained that he needed a quintessential Englishman like me to spearhead the national team's attack in a game which was just about to kick off over the road.

'I sadly explained that I couldn't accept his offer as I had just that moment bought a bacon, lettuce and tomato sandwich which I was about to eat. "That's a shame," said Sir Alf. "I'll have to find another centre forward."

'In the last forty years, I have often wished I had foregone that sandwich and turned out to represent my country in that match. For the man that replaced me in the England team was none other than Geoff Hurst, and it was his third goal in the last second of the match that secured the 1966 World Cup for England. During breaks in filming on some of the many hit shows I have starred in, I sometimes imagine how that famous commentary would have sounded with my name instead of Hurst's. *"And here comes Peter Bowles out of To the Manor Born! Some people are on the pitch, they think it's all over! It is now!"*

© Peter Bowles 2006

Next Week:
One Small Step for Man, One Giant Leap for Peter Bowles!
The time I met Buzz Aldrin in the BBC canteen and narrowly missed a chance to go to the moon. Because of a sandwich.

Going out with a bang

A BIRMINGHAM man is set to become Britain's first human firework. Ernest Greaves has paid a Tokyo company over £12,000 to book a 'firework funeral'.

Pyrotechnic

Pyrotechnic cremations are big business in Japan, with over half the population choosing to be publicly ignited in favour of a conventional burial or cremation. Undertakers pack their bodies with an array of sophisticated display explosives, and they are then mounted on a wooden frame.

Unioversity

It is traditional for the deceased's next of kin to light the blue touch paper whilst sombre mourners stand back a minimum of 25 metres to watch the display in safety.

Colliage

But the spectacular send offs may not catch on in Britain. For vicars are unlikely to allow dead people to be set on fire inside their church yards, especially if it is costing crematoriums business. Mr Greaves, who is 85, plans to be set off on waste land opposite his home.

Shark eats Kershaw

Pop singer Nick Kershaw has been eaten by a shark off the Yorkshire coast. The giant shark is reported to have leapt out of the water and attacked Kershaw while he was paddling only feet away from the beach at Bridlington.

Muriel

Fellow singer Howard Jones who had been building sandcastles with Kershaw only moments earlier looked on in horror as his friend was bitten in half by the giant shark. A spokesman for Bridlington Council warned other eighties pop stars to stay away from the town's beaches. "These sharks are a menace", he said.

Coma Como! Perry 'very' poorly

VETERAN American singer Perry Como was in a coma last night after a bizarre accident at his California home.

Como, 68, was hit on the head by a falling star during a barbecue in the back garden of his Beverly Hills home. He was knocked to the ground by the impact of the blow and was rushed to Hollywood General hospital's Accident and Emergency department. Still unconscious, he was later transferred to the Special Head Injuries unit where his condition is described as 'very poorly'.

Very poorly Perry (above) and a shooting star in space (left) similar to the one which landed on his head yesterday.

Como

According to eye witnesses Como had been attempting to catch the star and put it in his pocket when the accident happened. Actor Alan Hale, star of the TV series 'Casey Jones', was among guests at the barbecue.
"Perry had been cooking burgers and telling jokes. He was in high spirits. Then he decided he was going to catch this falling star and put it in his pocket. As it came down he ran across the lawn to get underneath it, but as he raised his hands he lost his footing and stumbled forwards. As he fell the star came down and hit him on the back of the head, knocking him out cold".

Cuphill gardener

The star, which had fallen out of space and was travelling at the speed of light, was larger than the Sun and had been burning at a temperature of 80,000 degrees centigrade. "It looked like a straightforward catch until Perry stumbled. There was a thud when it hit him, but people were laughing. They thought he was just fooling around. It was only when they saw him lying there completely still they began to realise he was hurt bad. I think it was James Dury out of 'The Virginian' who called the ambulance."

Cirt lifter

Fellow singers Andy Williams, Johnny Mathis and Neil Diamond were with Como at his bedside last night. Doctors say his condition is now stable, but it is too early to say whether or not he will make a full recovery.

Song come true - everybody knows one

Sting - stung by 'message in bottle' shock

Newton-John - 'Country roads' bombshell

Perry Como's bizarre accident is yet another example of songs 'come true' for the singers that sunged them.

A similar pop coincidence involving Sting took place in 1979. Only days after writing 'Message In A Bottle' he rang his father who was a milkman. In passing Sting mentioned the name of his new song. His face fell silent as his father told him what had happened that morning. While delivering milk he had received a message from a customer. The hand written note had been rolled up... *and carefully placed inside a milk bottle.*

By our Musical Coincidence Correspondent Kate Bush

Cuff

In 1993 singer Olivia Newton-John was visiting her family home in Australia. While driving on her scooter from Sydney airport to her home town of Hobart, Olivia was diverted because of road works. The diversion took her onto several minor country roads. By the time she eventually pulled up outside the Newton-John family home in Hobart the singer's face was numb with disbelief. For twenty years earlier, almost to the minute, she had reached number 15 in the charts. And the name of her hit single? *'Take Me Home Country Roads'.*

Not puff singer Sir Cliff Richard believes in God. And well he might. For it could only have been an act of God that brought about surely the most bizarre pop song title coincidence of all. Bachelor boy Cliff had decided to go for a summer holiday at the home of his American friend Carrie Fisher. But unknown to him his Star Wars actress pal had moved. The door was opened by a stranger who told Cliff *"Carrie doesn't live here anymore".* Cliff was stunned. Fortunately for the Peter Pan of pop Carrie *had* left a forwarding address. "She has moved to an apartment on the corner of two streets in New York, both of which have unlikely names", Cliff was told. "You can't miss it. She lives right on the corner of *Mistletoe and Wine".*

Wannabe a star?

Have you got the GIRL POWER?

USING their secret weapon GIRL POWER the Spice Girls have conquered the pop world, putting fellas well and truly in their place, and putting women back on the pop map.

In twelve money-spinning months the world conquering quintet have re-written the pop history books, with five consecutive number one singles, record sales of their debut album, and a blockbuster movie in the pipeline. No wonder every right minded girl in the world dreams of following in their footsteps.

But it takes something very special to get you to the top:- Girl Power! Have YOU got it? Here's a chance to find out. To reveal your G.P. rating just answer the following questions a, b or c. Then check your score against the Girl Power Meter below.

Baby

(a) Boycott down market, pornographic tabloid newspapers and support anti-pornography campaigns.
(b) Refuse to conform with the fashion stereotypes created by men, and instead choose individual clothes that reflect your own personality.
(c) Dye your hair ginger and get your tits out in a wank mag.

3. You're in a clothes shop, and you've got £1000 to spend. What would you buy?
(a) A good selection of moderately priced but nice dresses.
(b) One very expensive designer dress that looks absolutely perfect on you.
(c) Some Union Jack knickers and a plastic belt.

Scary

(a) Get a few of your female friends together, discuss your ideas, form your own band and go for it!
(b) Join an existing group, then gradually try to impose your girl power attitude on them.
(c) Reply to an advert that some bloke has put in a newspaper, go along to an audition and do everything he tells you to do.

2. You're fed up with the male dominated media's exploitation of women. What would you do about it?

Sexy

4. You decide to cultivate a sexy, slightly wild image that will appeal to men. What would you do?

1. Okay. So you want to star in your own girl power group. How would you go about it?

(a) Invest in a wardrobe full of slinky dresses, revealing outfits and sexy underwear.
(a) Dress from head to toe in black lycra, a bit like Catwoman, and wield a leather whip.
(b) Put a coach bolt through your tongue and start wearing Les Dawson 'Cosmo Smallpiece' spectacles.

5. You see a mouse in your kitchen. How would you react?
(a) Get your boyfriend to chase it away.
(b) Set a mouse trap, and hope to catch it.
(c) Stand on some big tall shoes and scream.

6. Your long term boyfriend, who has stuck by you through thick and thin, asks for your hand in marriage. What do you do?

Dirty

(a) Say "Yes", and immediately start making plans together.
(b) Tell him its too early to make a decision. You have your career to think of, and you need to make the right decision for both of your sakes.
(c) Ditch him and kop off with some over-paid, over-rated, hit and hope footballer who wears ugly slip-on shoes and no socks.

7. In your opinion, what is the most important ability that a pop star should possess?

(a) Singing ability.
(b) Songwriting ability.
(c) A stage nickname which sums up your entire personality in one adjective.

8. Which aspect of being a successful professional pop singer would you fear the most?

Sporty

(a) The inevitable loss of privacy. Having no private life, and living in the constant glare of the media spotlight.
(b) Commercial pressures; the demands from your record company for commercial success restricting your own musical development.
(c) Having to sing live with people watching.

9. Being a world famous pop star idolised by millions of teenagers across the world is a position of some considerable power. How would you utilise that power?
(a) Raise awareness amongst fans of issues which affect them and are important to their future, like the environment.
(b) Act as a good role model, speak out against drugs, and encourage your young fans act responsibly.
(c) Use your tits and arses to flog Pepsi to 13 year old kids.

---

10. Finally, what do you really, REALLY want out of life?
(a) To travel and see the world, meet interesting people and experience different cultures.
(b) To settle down and raise a happy, healthy family, in a warm, loving environment.
(c) Some zig-a-zig... eugh.

Girl Power-o-meter

Tot up your total, scoring **1** point for every (a) answered, **2 points** for each (b), and 3 points for a (c). Then check your voltage against the following:

18 or under: Crikey! It looks like you've had a Girl Power cut. Your reading is so low you might as well be a boy. Total lack of atti-tood.

19 to 29: You've got a medium amount of Girl Power - about enough to run a tumble drier for ten minutes. You don't let guys push you around. Except sometimes, when you do.

30 or over: Wooh!! Fellas beware! Girl Power cumin' at'cha!! Yes, you're the kind of girl who doesn't take no shit from men. Except perhaps your manager. (And his appointed agents). And your record company. (And their appointed agents). And of course the PR people. (And their appointed agents). And your publishers. (And their appointed agents). Oh, and your accountants. And legal advisors. And the marketing people. (And their appointed agents)...

Clowns see red over circus tax

CIRCUS clowns are paying through their red noses for big shoes as a result of Britain's barmy tax laws.

So says Tory MP Winston Churchill whose grandfather won the war.

Shoes

Under present rules children's shoes - sizes five and under - are exempt from VAT. However young clowns who wear big shoes as a vital part of their job are being forced to pay the extra tax when they buy adult size shoes.

Brass

Yet the bizarre tax laws are a boon for adult circus dwarfs, most of whom only wear children's sizes. The average circus dwarf can save up to £75 a year on unpaid VAT by wearing shoes that are size five or under.

Hooves

"The situation is a nonsense", claims Mr Churchill who is a long time campaigner on behalf of circus clowns. "I believe this minority of people are receiving unfair treatment under existing UK tax legislation. When you consider that the average clown is probably spending hundreds of pounds a year on dry cleaning to get custard off his clothes, and fish out of his pants, to make him pay extra for his shoes is totally unfair."

Box

Mr Churchill spoke as a convoy of clowns arrived at Westminster yesterday

VAT'll not do nicely on big shoes

to protest about new MOT regulations which they claim are also discriminatory. Their convoy of cars left Brighton in February of last year and took almost 16 months to complete the journey due to their doors falling off and the engines blowing up all the time. "These new regulations are a tax on comedy motorists and will place hundreds of livelihoods at risk", said the MP yesterday.

Slapstick MP Churchill (left) and a hard hit clown with comedy shoes (above).

Mr Churchill comes from a traditional circus background. His father was a high wire monocyclist and Mr Churchill himself is a lion tamer at weekends. Chancellor of the Exchequer Mr. Kenneth Clark, who has vowed to review the clown tax situation in his next budget, is himself a keen amateur bare back horse rider and escapologist.

The MAN in the PUB

Britain's most ill informed columnist

Here's a thing about the Royals you never knew. You know that King Edward? Him who ran off with Mrs Simpson? *Quarter inch tadger he had.* Honest! Like a bleedin' acorn. Gawd knows why that Mrs Simpson fancied 'im. Had to 'ave half her fanny sewn up she did. *Yeah!* Straights. Sewed it up so his cock would fit better.

You know how they make Turkish Delight? You don't wanna know mate, but I'll tell ya. *They rub dog shit on leather.* **It's true that. This mate of my brother, his history teacher told 'im.**

Know how far apart your nipples are, eh? Do ya? I do. Shall I tell ya? *Nine inches* mate. You measure 'em. Go on! No need mate. They'll be nine inches *exactly.* Know why? Everyone's are. No matter how old you are, your nipples is always nine inches apart. That's a fact that is. Go on. You measure 'em. And you know your liver? It's not an *organ*, mate. It's a *gland*.

That George Michael, he's so good at *sex*, **right, every time he has it off, 'is bird** *faints!* **It's true that. Apparently he was makin' a video once and this top model seen 'im in his underpants an' she keeled straight over she did. Sexiest bloke in the world they reckon.**

This bloke I know, he's got this number plate on his car, "TJ1". 'Ad it for years. Anyway, Tom Jones, the singer, right, stopped 'im in the street the other day and offered 'im a *million quid* for it! Told 'im to fuck off, my mate did. Would you believe it, eh? Turned down a million quid. Funny thing is, his car's only worth £600. But it's the *principle* you see.

You know that Patrick McGee out The Avengers? Him with the brolly an' the top hat. Been married for *fifty* **years he has, and he's never shagged his missus. Fifty bleedin' years and he's** *never* **given her one.** Hard to believe innit? An' you know what? The amazin' thing is, *they've got kids.* How do they do it, eh? It's *weird* that is if you ask me.

I bet you don't half fancy that bird in the Cadbury's Flake advert, eh? Fancy her do ya? Well listen to this. *It's a bloke!* Yeah, a fella! Straight up. And YOU fancied her! Go on, you can buy me a pint for that. Or I'll tell all yer mates you're a puff.

I'VE BEDDED TH

A 52 year old out patient at a West Midlands psychiatric hospital has spilled the beans on her steamy affairs with the saucy stars of the TV puppet shows.

For the last 25 years Dulcie Bagshaw claims to have lived a sordid life as a TV puppet groupie, jumping in and out of bed with some of the biggest names in children's television.

HORNY

Here, for the first time, she blows the lid clean off the horny, porny puppets who millions of innocent kids watch on TV every day.

RANDY

"Kids love TV shows like Rainbow, The Wombles and Bill & Ben. But if their parents could see what I have seen, they'd switch off in disgust.

PAUL

My first experience with a TV puppet came when I was only 14. I had been to the theatre to watch a stage version of The Sooty Show. Afterwards a group of us went backstage to get autographs. Somehow I ended up alone in Sooty's dressing room.

NANETTE

Sooty offered me a drink. I don't know what it was, but I was soon feeling dizzy. The next thing I knew Sooty was standing naked in front of me, holding his magic wand. I'll never forget the words he spoke. "Izzy wizzy, let's get busy".

SOOTY

That evening I had gone to see the Sooty show as a child. But when I awoke the next morning in Sooty's hotel room, I was a woman. I had bitten the forbidden fruit of sex with TV puppets, and now I wanted more. Over the next ten years I must have slept with over a hundred TV puppets. I wasn't fussy. Life became one long party, an endless whirl of late nights, drink and TV puppets.

You name them, I slept with them. Bill and Ben, The Woodentops, The Clangers, Orville the Duck, Joe 90. The list is endless. TV puppets came and went. The affairs were purely physical, strictly no strings attached. Until the day I met Basil Brush.

BASIL

Basil was a joy to be with. He used to love to fondle my "bum bum" as he called it, and he couldn't keep his foxy fingers of my brush. But he soon became obsessed with me, and even began following me around.

SYBIL

In the end we had a dreadful scene in a restaurant. I remember he was sitting on the bar. He was screaming and shouting, and he'd even threatened to kill himself if I didn't come back to him. Eventually Rodney Bewes arrived and took him home. I never saw Basil again after that.

SAD

I was sad to see the end of Rainbow. That show meant a lot to me. More than you could imagine. For I had got to know Bungle, Zippy and George intimately during a long and torrid affair which had lasted many years. I loved all three of them – on one occasion *all at once*.

Dulcie with her first love Sooty (left) and Sweep, and surrounded by the stars of Rainbow (left to right) George, Zippy and Bungle. Inset - Basil Brush.

SUMMERS

It was a hot and sticky summer's evening, but I had no idea just how hot and sticky I was going to get as I made my way to the Rainbow dressing room at the BBC Television Centre in West London. I'd been invited back for drinks with the Rainbow puppets after a chance meeting with Zippy in a cocktail bar. When I arrived I could tell straight away that all three TV puppets had already had quite a lot to drink.

DIAMOND

I was feeling hot and horny so I stripped off down to my bra and knickers and stretched out on the settee. Next thing I knew Zippy had undone his zip and was licking his lips. Bungle and George needed no more encouragement and soon both were as naked as I was.

HATHAWAY

Before long Zippy and George were fondling my heaving breasts while Bungle the bear took me roughly from behind. I climaxed again and again until I almost passed out. Then Zippy took over while George, the pink hippopotamus, paid special attention to my love buttons. Just when I thought I could

come no more, Zippy exploded inside of me, taking me to heights of ecstasy that before I had only ever dreamed of.

ATTENTION

I had never really paid any attention to Ray Allan's puppet Lord Charles. He didn't strike me as being my type. Until one day we met in the lift at Thames TV. Ray was afraid of lifts, so he took the stairs up to the top floor.

AT EASE

As the doors of the lift slid closed I found myself alone with Lord Charles. What happened next took me by surprise. As the lift began to move, Lord Charles suddenly pressed the 'stop' button, and it shuddered to a halt, throwing me into his arms.

LIPS

A strange force seemed to draw our lips together, and as we kissed passionately I felt a small wooden hand slip into my blouse and expertly undo my bra, which slid down around my ankles.

AROUSED

I could feel that Lord Charles was already aroused. His bulging manhood was practically bursting out of his

E TV PUPPETS!

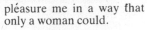

Seedy sex lives of the children's telly favourites

Lord Charles - his manhood was enormous.

small, pinstriped trousers. I struggled frantically to release it, until suddenly it sprang forth, like a coiled python.

ENORMOUS

It was that enormous I almost passed out. It was the biggest wooden penis I had ever seen. I simply had to have it inside me. I was already sopping wet as, with one powerful stroke of his 'silly arse', he forced it into me. My earth shattering climax was over in seconds; ecstasy exploding through my veins.

LUST

That day Lord Charles and I abandoned ourselves to sexual lust, pure and simple. Neither of us said a word. Neither of us have spoken about it since, and neither of us ever will.

CONQUEST

There was one sexual conquest that I never achieved. I had already slept with four of the five Tracey brothers out of Thunderbirds. And their *equipment* had been just as impressive as anything that they used on their International Rescue missions, I can tell you! But there was one Tracey brother who had always eluded my grasp.

WISDOM

I began to wonder whether I was losing my touch, until one night I was staying on Tracey Island when I heard the sound of laughter and splashing coming from the swimming pool. I crept out, and there in the moonlight was my unconquered Thunderbird pilot, wrapped in a tender embrace... with Brains!

TEBBIT

I'm not prepared to name that puppet, suffice to say that the only 'bird' he had eyes for was Thunder*bird* 4.

COOK

Mind you, I have nothing against gay TV puppets. In fact, one of my most fulfilling sexual experiences took place with a female puppet – Lady Penelope.

WAITER

She had offered me a lift into town in her big, pink car. As we were driving along Parker pressed a button and curtains automatically closed around the windows. We were alone. Suddenly Lady Penelope dabbed her stiff, plastic hand on my thigh.

MAITRE D'

At first I was repulsed. This was a *woman* TV puppet touching me. But slowly my fear began to subside, and waves of pleasure began to wash over me. Penelope's plastic fingers began to awaken in me feelings that I didn't know were there. It was my first time with a woman TV puppet, but Lady Penelope was able to pleasure me in a way that only a woman could.

THOMAS

I had always fancied Thomas the Tank Engine, but I had never dreamt that I might one day get the opportunity to sleep with him. That was until I met the Fat Controller at a TV puppet party to celebrate Lamb Chop's birthday. The Fat Controller invited me to come and visit Thomas and his friends on the Island of Sodor. I didn't need asking twice!

SHY

When I first met Thomas he was shy, not at all like I'd expected him to be. One day he'd been shunting trucks at the big station when he asked me if I wanted to come for a ride along his branch line. I was up in his cab like a shot!

OIL

After a few miles he stopped in a siding and we began to chat. Eventually the subject got on to sex, and to my surprise Thomas admitted to me that he had never been with a woman before.

MILK

I took this as my signal to pull off, and slowly began to slide down his zip and gently slip off his jeans. He was already aroused. I have never seen such a magnificent train's cock. I almost passed out. It was easily eleven inches in length, and my fingers could barely touch around it's massive girth.

BUTTER

We made passionate love for what seemed like an eternity. Thomas puffed and panted and blew his whistle as he shunted his load in and out of my love tunnel. Then suddenly I became aware of somebody watching us.

CHEESE

Sure enough, I looked up to see Gordon, Edward and James the red engine had stopped nearby and were watching us. That turned me on even more. I beckoned them over and invited them to join in. They didn't need asking twice, I can tell you.

YOGHURT

I must have passed out, but the next thing I remember was waking up to see Thomas, Gordon, Henry and James all pulling the biggest train you have ever seen.

CREAM

I lost count of the number of orgasms I had as the TV puppet railway engines continued to take me in turn for what seemed like an eternity. Then the helicopter joined in.

DEREK AND THE DOMINOES

I must have passed out again, because when I awoke I was in a mental hospital, with electrodes on my head.

Next week: How I fell pregnant to Noggin the Nog, and the Clangers give me my first weightless orgasm in space. And Great Uncle Bulgaria gives me one up the shitter.

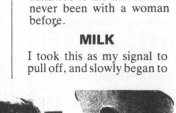

LADY PENELOPE 'pleasured me'

SCOTT TRACEY 'Impressive equipment'

PARKER He pressed button

KYLIE SPLIT OVER IRAQ

~ Tits stand firm for peace, but arse is right behind Bush

Kylie Minogue's **TITS** have come out against war in the Gulf - even though her **ARSE** is backing an attack on Saddam!

Stunned fans at a concert in Aberdeen were shocked when the Aussie popster's breasts adopted an anti-war stance only hours after her bottom had given the thumb's up to unilateral military action against Iraq.

Ba'ath

Kylie's arse insiders say that her peach-like buttocks maintain that the Ba'ath regime's failure to fully cooperate with weapons inspectors constitutes a material breach of United Nations resolution 1441, and therefore war is justified.

Sho'ower

However, sources close to her tits say that the 32A charms remain unconvinced by arguments for a pre-emptive strike against Baghdad. They say the pop cutie's lils need to see more and stronger evidence of Saddam's banned weapons programmes and

Baps - want second resolution.

Dirtbox - running out of patience.

his links to global terrorism before they will agree to back military action against him.

Meanwhile the pint-sized songbird's secondary sexual characteristics have agreed to put their differences to one side for the remainder of her current tour which tonight takes in Leicester's de Montfort Hall before traveling to North-ampton and Halifax.

Church burnt out at 16

SINGING sensation Charlotte Church was yesterday found burnt out and abandoned on wasteground near the Cardiff home of her boyfriend.

The voice of an Angel star had just finished a gruelling World Tour to promote her latest album when she was snatched by thieves from outside the home of her sweetheart, rap artist Steven Johnson.

Witnesses say they saw saw four youths riding her at high speed around the streets of a local estate in the early hours of Saturday morning. Firemen were called when she was set alight around four in the morning, but had difficulty putting her out when the came under ambush by youths who pelted them with stones and bottles.

Have YOUR Say...

With London's traffic now crawling along no faster than it did at the end of the last century, Lord Mayor Ken Livingstone has decided the time has come for drastic measures. To prevent total gridlock in the capital's traffic system, he's introducing controversial congestion charges; every driver without false numberplates entering the city will be fined £5. But is there a better way to tempt the public out of their pollution-belching cars and into 40-year-old diesel fume-farting buses? We went out onto London's traffic-choked streets to find out what YOU thought...

...TRAFFIC MOVED at its fastest pace in Victorian times, so to get London moving again it would make sense if penny-farthings, horse-drawn carriages and sedan chairs were made exempt from congestion charges.

Reg Halliwell, Butcher

...THE MAYOR'S new charges won't get me out of my BMW. I went on a bus once and there were other people on it and I had to sit next to someone dressed in gaudy clothes made of vulgar fabrics. What's more the driver refused to drop me at my door. Never again.

John Brown, Magnate

...THE ONLY way to tempt men out of their cars would be if they put topless lapdancing conductors on the buses. They could gyrate sexily on the pole, and commuters could stuff the fare into their knickers. It would certainly brighten up my journey to work.

Andy Inman, Window-cleaner (retd.)

...WHY NOT take advantage of these so-called asylum seekers to ease congestion? They should be made to walk along the bus-lanes with route-numbers tattooed on their foreheads, giving piggybacks to commuters. Let's see how long they hang around when there's a bit of proper graft to do.

Charlie Pontoon, Journalist

...I READ that gas-powered vehicles are to be exempt from the mayor's congestion charges. How ridiculous. These large American cars are renowned for guzzling fuel, and they take up twice as much room as a normal family hatchback.

Jasper Maskelyne, Magician

...A CAR can only crawl along at 4mph in central London because that's the speed of the car in front. And he's only doing that speed because of the car in front of him. If the police set a minimum speed limit of say 50mph, the traffic would fair zip along and the problem would be solved.

Denver Colorado, Pteridologist

...SINCE MOTORCYCLES are exempt, I intend to avoid the congestion charges by painting my car to look like three Domino's pizza bikes riding abreast.

Huw Edwards, Newsreader

...LONDON TRANSPORT should take a leaf out of the airlines' books to tempt people back onto the tube. Stewardesses should wander up and down the carriages of underground trains, selling cheap cigarettes, perfume and drink to the passengers, and telling them what to do in the event of the train crashing in the sea.

Milos O'Shea, Taxidermist

...I CAN'T wait for the capital's traffic to start moving more smoothly. There's nothing worse than sitting in a taxi, being forced to listen to the driver spouting his bigoted homophobic opinions every day, while he gets paid for the privilege.

Richard Littlejohn, Columnist

...CARS ONLY use the bottom four or five feet of London's streets. The Highways Agency should give money to the first scientist to invent hovering cars, allowing any number of traffic jams to be stacked vertically above a single road.

Patrick Pending, Crackpot inventor

...PEOPLE WOULDN'T get so frustrated in traffic jams if the authorities instigated a 'chicane' contraflow queuing system such as is in use in the post office. Although the queue would be just as long, it would feel shorter psychologically, so all the drivers would be happy.

Tina Catweazle, Mudwrestler

...POOR PEOPLE already get preferential treatment, with their dedicated bus-lanes. How about a bit of equality for the rest of us? What about special lanes for cars costing over £60,000, like my BMW M5?

John Brown, Magnate

Royals flushed over Queen bumhole snaps

HER ROYAL RINGPIECE

Candid camera catches Queen

with her knickers down

Sneaky snapper Tom (left).
The Queen is not amused.

A right Royal storm in a 'U' bend is brewing over lurid photographs taken from inside the Queen's lavatory.

For the sneaky snaps, snatched by a former Palace plumber, clearly show the Queen's arse perched proudly on the throne.

POSSESSION

The sensational pictures, which are now in our possession, are the work of plumber Tom Wilson, a former Palace employee whose job it was to look after the Queen's lavatories. The candid shots were taken using a secret camera hidden in the Queen's personal toilet, disguised as a number two.

SUPPLYING

Tom explained how his candid camera came about. "The Queen had asked me to mend her bog 'cos it hadn't been flushing properly. One particular dump had been reluctant to disappear, and kept bobbing about in the bowl. Anyway, as I looked at it, it occured to me that a camera hidden in the toilet would get a great view of the Queen's arse".

TRAFFICKING

Tom then set about the complex task of designing a camera that looked like a turd, and floated. "In the end I bought a top of the range £100 waterproof camera, tied a few corks to it, then disguised it by smearing a mixture of porridge and brown paint all over it. When I'd finished only the lens was visible, poking out the top".

MARMALADING

Tom tested the camera in his bath, making sure it floated, and that the lens would point up, directly at the Queen's bottom. "I rigged the shutter to open at the slightest trace of wind, so the gentlest of farts would set it off".

JAMMING

The next day Tom returned to the Queen's lavatory and dropped his secret floating camera down the pan. Then he waited. "I had never been as nervous in my life. I was convinced that the Queen would notice it and flush it away, or worse still fish it out for a closer look".

JAMMING

According to Tom's plan, once the camera had taken it's picture the Queen would then flush the lavatory. When she did he would pop down a nearby manhole to fish the camera out of the sewer. But after several hours standing knee deep in sewage beneath the Palace there was still no sign of a Royal flush.

JAMMING

"I later discovered that the Queen had used the toilet several times, but she hadn't flushed it once. Apparently Phillip encourages her to save water by only flushing it once every couple of days. So, after a particularly uncomfortable night I awoke at about 8.30 the following morning and heard a flush. Suddenly there was crap everywhere. I grabbed a couple of big logs, but they weren't the camera. Eventually I caught it, third time lucky."

JAMMING

Tom rushed to the local chemist to get his film developed straight away. But when he went back to collect the prints the following day, disaster struck. For there had been a mix-up, and his prints were handed to a police officer standing in the queue in front of him.

I WANNA

"He looked at the prints and immediately recognised the Queen's arse. Then he turned to me and asked if the photographs were mine. I thought I was done for. Then an idea sprang into my head. I said they were mine, but it was my *wife's* arse. I told him people were always mistaking my wife's arse for the Queen's."

JAM IT

"He seemed quite happy with this explanation and handed the prints over. By the time I got out of the shop I must have shit my pants about a dozen times, but it was worth it, I can tell you. On the bus home I just sat there looking at these lovely big pictures of the Queen's arse sitting on the bog. It was the most exciting moment of my life".

WITH YOU

We have obtained the pictures of the Queen's arse from Mr Wilson, *but we have no intention of publishing them*. Unlike certain other papers, we know where the line should be drawn between public interest and invasion of privacy. And the British public were quick to commend us on our brave stand.

"By refusing to print these pictures there is no doubt that the circulation of Viz will suffer. But I applaud this courageous moral stand", said one passer-by yesterday.

Ready, steady, GHOST!

World NEWS

THE Paranormal Olympic Games are to be held in Limbo in the year 2002.

Within hours of the announcement shops and businesses in the state, which exists somewhere between Heaven and Hell, were looking forward to the prospect of an economic boom which the money spinning spirit games will guarantee.

Souls

Over five million souls are expected to watch the games, in which ghosts, poltergeists and other supernatural entities from all over the world compete for medals. Events include the severed head put, wall walking, chain dragging and the 100 metre lurch through a cemetary. An estimated 80 million spirits around the world will watch the events telepathically.

Limbo

Limbo's success comes as a blow to Shangri La which had been confident of hosting the prestigious event for the first time. Limbo has already staged the games once, in 1968, and that occasion Russia were accused of cheating after murdering several of their top athletes and not allowing their souls to rest in order to gain qualification to the event.

Samba

Geographical disadvantages are thought to have swayed the Committee's final decision. Shangri La -

a wonderful place of true perfection - would have proven popular among spectators and athletes alike. But there would have been problems with transport. Thought to be in Nepal, the only way to get to Shangri La is by climbing through high mountains. According to legend an avalanche then occurs and an icy cave appears leading to the mythical place.

Lambada

Other disappointed delegations included representatives of Nirvana, the state of utter bliss where people no longer require their bodies. Olympic chiefs feared that out-of-body competitions would reduce the incentive for athletes to train. Photo finishes would also have been difficult to decide.

Lumbago

Eldorado, with its slogan 'the lost city of gold', had been favourites among the early bidders. Its high profile campaign to host the games was supported by many big names including W.G. Grace, Charlie Chaplin and Martin Luther King. However its whereabouts are unknown and the Committee was unable to visit it to check the facilities on offer. Disappointed Norwegian delegates from Valhalla were already talking opti-

Ghosts going for ghould, shiver and bronzergeist in paranormal games

An American head putter makes a spooky spectre as he ghouls for gold at the 1998 Paranormal Olympics in Hades

mistically about their chances of staging the 2006 games. Facilities in the mythical Hall of the Viking Kings are second to none. However problems over qualification for events have yet to be resolved. As the legends stand only athletes who died in battle with a sword in their hand could qualify for the final stages in the mythical Hall itself.

The Committee were keen to avoid any such problems, particularly in the wake of the 1994 fiasco in Hades, the Greek Hell, where competitors were forced to queue for hours with coins in their mouths waiting for a ferryman to take them across the river Styx. Many turned up late for their events and were disqualified as a result.

DOCTOR, I'VE GOT A STIFF BACK

RETURN TO SENDER

QUEEN BUM!

Her Royal Lowness is a crown and out

ONCE she was the Queen of England. Bedecked in jewels and dressed in lavish gowns, red carpets unfurled beneath her every foot. But now she wanders alone in a public park, huddled in rags, her only companion a mangey dog.

Runcible

There was a time when she lived a life of luxury and splendour, dining on swans and caviar, and slices of quince which she ate with a runcible spoon. Crowds flocked in their thousands to see her changing the guards at Buckingham Palace. But nowadays she prefers liquidised vegetable soup, and even her relatives avoid her as the Queen Mother seldom changes her clothing. In her heyday she was everyone's favourite Royal. But now she whiles away the hours at the Clarence House care home in central London, just another muddled pensioner waiting for the grave.

Crucible

Neighbours of the 98 year old fear for the health of the former Queen. She can often be seen wandering aimlessly in the local parks, and has trouble remembering who she is. On one occasion she was found singing in a bus shelter and told police officers she was Gracie Fields. Police and park wardens regularly round the old dear up and hand her back to nursing home staff. A Buckingham Palace source yesterday denied the Queen Mother was suffering from Alzheimer's disease or any similar marble mislaying illness.

BURGERS

TWO LARGE FRIES AND A THICK SHEIKH PLEASE

Pensioner's nightmare ends in amputation

CONSUMER watchdogs are warning old folk to be on their guard after a frail Yorkshire pensioner allowed artificial leg salesmen to cut off both of his legs.

Eighty-six year old Wilfred Barker - a veteran of two world wars - was subjected to over ten hours of high pressure sales pitching after two men turned up uninvited at his Barnsley home at 2 o'clock in the morning.

Leg salesmen got their feet in the door

Mr Barker with the two unevenly lengthed walking sticks he now needs to stand up.

Raining

"I invited them in because it was raining and made them a cup of tea. After a while, when I realised they were selling legs, I told them I'd already got some. But they said there was no obligation for me to buy and they just wanted to give me a quick demonstration".

Pouring

The two salesmen from Leeds based Alpine Legs then produced impressive, glossy, brochures showing photographs of several attractive new artificial limbs. But when they began quoting prices Wilfred told them straight away that he wasn't interested.

Snoring

"They measured my legs and came up with a price of over £10,000. I don't have that sort of money. Then they started telling me how much it was going to cost just to maintain my real legs. They said my old legs were coming to the end of their life and it wasn't worth having them repaired. I'd be better off investing in a new pair. I got a bit bamboozled. Then, after a while it seemed to make sense".

Bed

Mr Barker asked the salesmen for time to think, and suggested they return the following day. But they told him they were only in the area for one night and that this was a special offer. "They said if I didn't sign there and then the price would be doubled, and I might end up with no legs at all".

Head

Their aggressive sales pitch continued until dawn and throughout the ordeal they refused to allow Mr Barker to use the lavatory. He pleaded with them to leave, but they said they would only go if he allowed them to cut off his legs. At around noon the following day, exhausted and suffering from bladder cramps, Wilfred eventually agreed.

Morning

"They put a piece of paper in front of me and told me that if I signed it they would go away and let me sleep. So I did. Before I knew what was happening they'd sawn my legs off above the knee."

Cash

Wilfred handed over £4,500 cash on the spot. The men then left, promising to return the next day to fit his replacement legs. But he heard nothing for over six weeks. Then he received another bill, this time for £24,000. "The first payment had only been a deposit, and they were refusing to fit my new legs until I paid the balance."

Mathis

Reluctantly Mr Barker sold his house in order to pay the bill. Two weeks later two men arrived and fitted his new legs, breaking his hip and two ribs in the process. "They were only here a couple of minutes but I couldn't believe the mess they made". Since then Wilfred has had nothing but trouble from his new legs. "The left one is too short and rattles when I go ballroom dancing. The other one just keeps falling off. I feel such a fool".

Morris

When we showed Wilfred's legs to a leading orthopaedic surgeon he was horrified.

"This is a really shoddy job. For a start Mr Barker's legs haven't been sawn off evenly. And even worse, he's been fitted with two left legs. They've not been weatherproofed properly and water is already ingressing and causing swelling of the knees", he added.

Austin

A spokesman for OffPeg, the regulatory body responsible for artificial limbs and appliances have issued a warning to any old folk who were thinking of buying replacement legs. "Shop around, go to a reputable company or ask friends who've had their legs sawn off successfully and are happy with their replacements", he said. "But most importantly, never sign anything on the doorstep. Always ask for time to consider. Unless of course there's a special offer you might miss".

Wolsley

Meanwhile Alpine Legs managing director Reg Shit flatly denied that his employees use any high pressure sales techniques. "All my salesmen are nice blokes. In fact the gentleman concerned used to be vicar. As far as I'm aware Mr Barker's legs fit perfectly well. Now fuck off out of it or I'll set me dog on you", he told us.

SPACE OLYMPICS FOI

Manchester's hopes of hosting the Olympic Games in the year 2000 could be dashed if a surprise new contender enters the running.

For according to the bookies the red hot favourite to host the games isn't Manchester, Munich or Mozambique. It's Mars! For in the year 2000 we may be witnessing the first ever *Olympic Games in space.*

MARS

Leading 'astrologers', or space scientists, among them Patrick Moore, have long believed that there is life on Mars. And if a parallel life form has developed on the surface of the red planet, it is almost certain that they will have developed a form of 'Olympic Games' similar to our own.

MARATHON

And among Martian athletes the most popular events will probably be:

* TARGET SHOOTING using lazer guns firing at targets in another universe.

* CYCLING on jet powered mono-cycles. In the 'Rouge Prix de Mars', the Martian equivalent of the Tour de France, three legged cyclists regularly travel at ten times the speed of sound.

* FENCING with bright shiny sonic sword beams (like the ones on Star Wars).

* THREE DAY EVENTING on giant, poisonous equine space lizards, 30 feet long.

TOPIC

In the first ever inter-galactic Olympics competitors from Earth will be at a disadvantage in many sports. However, certain space factors could dramatically improve our performance in others.

BOUNTY

For example, *weightlessness* will lead to spectacular performances in the high jump, pole vault and shot putting events. Indeed, rescue space shuttles will have to hover in the atmosphere above the Olympic Stadium to save any pole vaulters who accidentally vault into space.

By our Science Correspondent Dr. Stanley Jordan

Moore - life on Mars

The planet Mars has no grass or water. So field sports will take place on bright red Martian astro-turf, the surface of which is hot enough to toast muffins, or piklets (a sort of flat, Martian crumpet). And the lack of water will mean that all the swimming events will have to take place in other liquids, such as white spirit or vinegar. British swimmers Adrian Moorhouse and Duncan Goodhew are already rumoured to be training in turps.

CUTTY SARK

There is no sunlight on Mars – Martian athletes see using special infra-red vision. However, terrestrial competitors will overcome this problem by wearing 'miners' style helmets complete with plutonium powered head-lamps.

The date: 2000, The venue: Mars. Is this the Olympics of the future?

Sports clothing manufacturers are already working flat out to produce shirts, shorts and socks suitable to withstand the rigors of the space atmosphere. Conventional kits and footwear would simply implode in the space vacuum on Mars. Adidas are thought to be leading the field, having gone into partnership with cooking foil giants Alcan to produce the world's first heat resistant tin foil plated teflon soled sports shoe, the 'Adidas Apollo'. But the bad news for parents is that the new shoe, set to become the height of teenage fashion, does not come cheap. Enormous research and development costs will lead to a likely price tag of around £12 million 99p a pair.

Communications and transport to the games could be a problem due to the enormous distances involved. 'Live' TV coverage of events in the Olympic Stadium will actually take 4 months to reach our TV screens. And because of the sheer distance involved, athletes hoping to compete in the Martian Olympiad will have to leave early. Next week in fact. For a journey by conventional space travel to Mars currently takes up to 8 years to complete, depending on 'orbits'.

VICTORY

For unlike towns and countries, planets move about, and sometimes when you get there they've gone. So you have to start looking for them again.

BEAGLE

But the good news for British athletes such as Joanne Conway and Wilf O'Reilly is that the lack of gravity on Mars means that everything happens in slow motion, so they won't hurt themselves when they fall on their arses.

ALSATION

Security at the first Martian Olympics will be at record levels. Indeed it will have to be. For the Martians are a

THINGS AINT WHAT THEY USED TO BE

R THE YEAR 2000!

Britain's athletes ready to get set to prepare for the Olympics on Mars

war-like race who, like Germany, do not take kindly to losing. And just as Germany's defeat in the 1936 Munich Olympics lead to the Second World War, so an Earth victory on Mars could lead to a full scale invasion of our planet by Martian battle rockets and 'Red Warrior' flying saucers.

LABRADOR

As usual, it'll be up to Britain to defend the Earth single handed. However, an impressive arsenal of space weaponry already under development would soon put paid to the alien invasion.

FISHER

Britain's space weaponry in the year 2000 will include:

* Hawker Hyper-Drive nuclear space fighter rockets with 360 degree space guided lazer missiles, and big machine guns.

* The Saturn 4 Earth-to-space missile, capable of taking out a whole fleet of Martian space rockets every half second.

* The Mohican Strike Delta Space Helicopter carrying a special top secret 'seek and destroy' robot guided anti-planetary nuclear obliteration device code-named 'Battlemaster'.

* The Jupiter Bomb - the most powerful weapon known to man, and capable of blowing up the entire universe – ten times over. Every second – for *ten minutes*.

CROMETY

In a high tec space age 'Blitz' scenario the Cockney spirit will once again be tested to the full as Martian 'Death Rain' rockets pour down onto London while the population huddles together in tube stations, singing along to the sound of Dame Sinead O'Connor, the darling of the forces in the year 2000.

TYNE

Alas, the Queen Mum won't be much good this time round. For at the ripe old age of 100, Britain's favourite great great great granny will be confined to a Royal bathchair, and will probably be gibbering away like a fool in the corner of some remote Royal garden.

Britain's athletes yesterday welcomed news of the Mars Olympics, and declared "We're raring to go!".

Long distance runner **Brendan Foster** was full of optimism. "It will certainly make a novel change competing on another planet. It's a great challenge for me, and I only hope that I'll be a suitable ambassador for my planet".

SEVERN

Strong-man panto ex-cop shot putter **Geoff Capes** was apprehensive about journeying to Mars. "I'd relish the challenge of competing against space aliens, but it would be a long way to travel. Sixteen years is a long time to be away from your wife and kids", said 'gentle giant' Geoff.

THAMES

Wily fox Canadian snooker hearthrob **Cliff Thorburn** was intrigued. "Owing to weightlessness there would be an intrinsic problem in keeping the balls on the table", Cliff pointed out. "Perhaps snooker could be played on a three dimensional basis, in a cube perhaps, with a pocket on each corner. Or a conventional table could be placed within a drum spinning on an axis to create an artificial or 'centrepedal' form of gravity", added Cliff.

T.V. AM

Cliff's Cockney colleague, whirlwind **Jimmy White**, had his own suggestion. "The game could be played inside a huge centrifuge, with two tables place diametrically opposite each other, spinning at a constant speed of, say, 200 r.p.m., thus simulating the effects of the Earth's gravity. Mind you, players would soon become nausious, and vomit. And besides, snooker isn't in the Olympics", pointed out whizz kid White.

Brendan Foster yesterday

British boxing champ **Lloyd Hunnigan** was sceptical about the whole idea of a Mars Olympics. "I challenge the basic hypothesis of intelligent life on Mars", stormed the champ. "Because of high radiation levels, any life on Mars would have to be silicon based rather than carbon based like ourselves. And in order to survive, Martian life forms would need to have allumina silicate shells to protect their DNA from mutagenesis. It is therefore highly improbable that they would have developed beyond a simple algi or diaton life form", Lloyd told us yesterday.

COMPACTA

Bad news for British athletes is that winners will NOT be in line to collect gold, silver or bronze medals for their efforts. These metals are commonplace on Mars, and would be thought of by Martians as nothing special. Instead Martian medal winners collect trophies made out of Mars' most coveted materials – coal, soil and plywood.

Charity Fish book launched

Prince would like to be a cod. And Edwina Currie would like to be a dab.

These are just some of the celebrities who have contributed to the book 'I Wish I Was A Fish'.

BRAINCHILD

The book is the brainchild of seventies keyboard wizard Rick Wakeman, and all the proceeds will go to charity. Complete with illustrations by pint-sized Aussie songstress Danni Minogue and featuring a painting of a haddock by The Bachelors on the cover, the book is published next week by Soufflé Books priced £3.99, although copies will be available in large piles from bargain book bins for 50p each two weeks later.

SCROUNGING BASTARDS!

This family deserve to DIE

MEET the Dougan family. Husband Bill and wife Doreen are Britain's biggest scroungers.

They pocket an amazing £120 a week in handouts, and live a life of luxury in a three bedroom house - paid for by the council.

By
**RAB. L. ROWSER
and LIN SCHMOBB**

Wannabe's

Bill hasn't done a single days work in the two years since he was blinded and partially paralysed in a car accident. He claims he's not fit for employment. But he still manages to get to his door mat once a week where he picks up a whopping £85 state benefits cheque for so-called 'invalidity'. Unable to walk, he sits at home on his arse all day, counting his cash.

Bumble bees

Dole family Dougan claim to be hard up - yet they still have TWO children. And soon there'll be more. They breed like RABBITS, and yo-yo knickered slut Doreen, 28, is hoping for ANOTHER sprog later this year, leaving tax payers like YOU to fork out another £12 a week in child benefit.

Humble cheese

Perhaps next time she should spend some of it on contraceptives.

Humble pie

Free school milk for their ugly brood costs YOU the taxpayer another £2 a week. Yet bone-idle Bill, 33, still wants MORE. "It's difficult getting by on benefits, and I'd like to be able to provide better for my children", the grasping git told our reporter.

Kids Michael, 9, and Angela, 5, have already jumped on the benefits gravy train. Like their work-shy parents they expect something for nothing and collect a thumping 50p a week EACH in pocket money.

Blind Faith

Their house is crammed with tell tale signs of their cushy lifestyle. In the kitchen Mrs Dougan offered us a cup of "tea or coffee". Oh yes. The big spending Dougans have BOTH. Their fancy Swan kettle probably set them back £20, and a swish pedal bin in the corner must have cost thirty or forty quid.

Steeleye Span

But then that's hardly surprising. Because wife Doreen isn't short of a few bob. She works nights as a cleaner, picking up a hefty £42 a week as well as cleaning up on state hand outs. Nice work if you can get it.

Steely Dan

But still she MOANS. "What I'd really like is to take the family on holiday", she told us. "We've never been away at all since before we were married". But wait a minute. That's not all.

Desperate Dan

"With Bill unable to work, I'd like to go out and

All smiles as the Dougan family pose for our conniving photographer outside their house yesterday, unaware of the editorial direction our reporters intended to take.

pursue a career of my own. But its difficult finding people to look after the kids", said the money grabbing bitch as she sat there, sipping her expensive Nescafe coffee and offering us fancy chocolate biscuits like there was no tomorrow.

Lord Snooty

Doreen's weekly shopping bill comes to £60, and she claims it's hard to make ends meet - despite raking in POUNDS in discount vouchers at the supermarket check-out. And the whining sow isn't even happy with her FREE council home. "One day I'd like to own a house of our own, with a garden for the kids to play in", groaned the grasping trollop.

Lord Snowdon

Last night a senile Tory MP stopped wanking for five minutes to BLAST the Dougans before we'd even told him anything about them: "These people are a disease on our society", he ranted drunkenly. "Why should the taxpayer fund their disgusting, depraved lifestyles? They should send them back where they came from, and beyond".

Mount Snowdon

A spokesman for the Labour party failed to say anything that we could use out of context, despite several cleverly weighted questions.

WHAT DO YOU THINK?

WE'VE whipped up our ignorant readers into a bigoted frenzy of hatred. Here's the kind of hand outs THEY'D like to see doled out to the money grabbing Dougans.

"*I think it's disgusting*", said Dawn Shitehouse, bulldog faced moron mother of six. "*Their house is better than mine. People like that don't even deserve to die, never mind live*", she added.

"*They should tattoo the words FILTHY SCUM BASTARDS on their foreheads and put their children in a mental home*", said neighbour Edna Pigshit who gets 20p an hour LESS than Mrs Dougan at her cleaning job. "*They're just vermin that's what they are. Hanging's too good for 'em. They should string 'em up, and throw away the key*".

"*Cut his cock off and make him eat it*", said disabled war veteran Joe Mengler, 82, of Leeds. Plucky Joe, who lost all his teeth biting a U boat, gets by on a paltry 2p a week army pension and is regularly mugged in his home by glue sniffers. "*And I'd pull the lever myself*", he added.

"*They should cook him in his own blood, and make him eat himself, then stone him to*

death with his own knackers", said taxi driver Ron Bigot, 32, who works a 60 hour week and comes home with less than £200 since all the foreigners came over here and took all the jobs, and the women. "*If he has any more babies the doctors should pop their heads with their fingers, like baby rats*", he added.

Ring our HATE LINE

Have YOUR neighbours got a nicer house than you? Do they appear to be better off than you are? Or perhaps their garden is a mess, or their kids have got snotty noses. Ring us today on 0171 922 7386 and tell us about your nightmare neighbours. Perhaps we can arrange for a lynching. Ring us today. There's dozens of jumped up little cunt reporters fresh out of college and with no morals whatsoever waiting to take your call.

JOYRIDING/JUMPER JOKE

PULL OVER

CARDIGAN

RADIO 97

18 20 THINGS YOU NEVER KNEW ABOUT CAKE

Every schoolboy knows that it was French King Louis Armstrong XIV who said "Let them have cake, but they can't eat it". But nowadays nothing could be further from the truth. Cake is consumed by everyone, rich and poor alike. Indeed the 20th century has become the year of the cake. And never more so than today.

But what do we really know about cake? What are the facts about our favourite food! Here's twenty fruit filled slices of exceedingly good information about cakes.

1 The world's first cake was baked in Egypt in 2200 BC – over a thousand years ago. It now rests inside a sealed vault at the Museum of Food and Cake at Vevey, Switzerland. No-one is allowed near the cake, as it carries an ancient Egyptian curse. A message written in hiroglyphics on the icing warns would be nibblers that anyone eating so much as a crumb will die a horrible death. Sure enough, Sir Humphrey Mountingboard who unearthed the cake in 1902 ate the cherry off the top and was killed instantly after a long illness several years later.

2 The largest man-made cake weighed a whopping 90,000 lb. That's heavier than an elephant. It was presented to the Prince of Wales during a visit to Austin, Texas in 1986. The cake stood untouched in a Buckingham Palace kitchen for almost 3 years until one night in 1989 when the Duchess of York ate it.

3 Rudyard Kipling is famous for his exceedingly good cakes. He named his Bakewell Slice after posh TV presenter Joan Bakewell, because it is a high class tart.

4 Cakes have played an important part in history. Before defeating the Spanish Armada at Trafalgar Square, King Arthur is reputed to have watched a spider burning cakes on his round table.

5 Ask for a Walnut cake in an East End bakers shop and you'll probably be referred to the nearest garden centre. That's because 'walnut cake' is cockney rhyming slang for 'garden rake'.

6 Just because something is described as a 'piece of cake' doesn't mean you can eat it. The expression 'a piece of cake' is used to describe something that is easy to do.

7 And so is 'a piece of piss'.

8 Mountbatten cake is a pink and yellow square cake with marzipan on the outside, named in honour of the late Lord Mountbatten.

9 If you want a fishcake, don't go into the baker's shop. Pop along the street to the fishmongers instead. That's because fishcake, strickly speaking, isn't a cake at all. It's a round, flat dollop of mashed potato covered in breadcrumbs.

10 And if you want a Pontefract cake, you'd best nip across the road to the confectioners. Pontefract cakes aren't cakes either. They're liquorice sweets.

11 If you want a piece of carrot cake you couldn't be blamed for popping next door to the greengrocers. However, the green-grocer would send you back across the road to the baker's shop again, for carrot cake is indeed a cake. Made with carrots.

12 If you want a cake of soap you'll have to go back across the road, down the street a bit then turn left, and keep going until you come to the chemists.

13 And if you want some cheese cake Marks and Spencers do a very nice one in three flavours for only 79p a slice. There's a choice of blackcurrant, strawberry or cherry.

14 Cakes are more popular that cars. In Britain last year we bought an incredible 311,850,000 cakes. Yet in the same period only half that number of new cars were sold. And most of those were Japanese, probably.

15 The vast majority of cakes are named after their county of origin. Danish Pastry, French Fancies, Belgian Buns, Rumanian cake. And Arctic Rolls. To name but a few.

16 It's hard to tell, but you would imagine that Belgians or Germans eat more cakes than anyone else.

17 The first ever cake shop was opened on a Moscow side street in 1923.

18 By the year 2000 people will go shopping on the moon and buy space cake.

19 ...

Celebrity Swears

Nº 184 Jeremy Beadle

PISS

SCARGILL IN NAZI SPACE GUN HORROR

The National Union of Mineworkers have denied allegations that several million pounds collected during the 1984 miners' strike were used by their chairman Arthur Scargill to build a 1,000 foot long steel tubular artillery gun at his home near Barnsley.

PINKO

Commie union boss Scargill, 54, has so far failed to comment on the further allegation that he paid former Nazi war criminals 75p

an hour to act as his personal 'minders' during the fourteen month long dispute.

M.O.D. experts believe that the alleged gun, if it existed, which it didn't, could fire big things a very long way. Perhaps even into space.

Sexterminate!

TV'S FAVOURITE TIME LORD Dr Who is back on our screens at last, and most of us are once again getting used to watching the action from behind our sofas. But for one woman who worked on the series back in its 70s heyday, all the action took place ON the sofa!

And now former BBC tealady Iris Poldark is set to blow the lid on the steamy behind-the-scenes goings on which went on behind the scenes of the popular sci-fi series.

In this exclusive extract from her new shockingly badly-written memoir *'Who Were You With in the Moonlight?'* (Beans on Toast Books, £1.99), Iris spills the beans on her sextra-terrestrial romps with a series of terrifying space aliens.

space

Iris began working at the BBC Television Centre straight from school in 1940. Starting as a lowly tea girl she quickly worked her way up, and by the 1960's she was a fully-fledged tea lady. In 1963, as one of the corporation's longest-serving catering staff members, she was given the important job of providing refreshments for the cast and crew of a brand new space adventure series - Doctor Who.

EXCLUSIVE!

" William Hartnell was the Doctor in those days, and he was a perfect gentleman. However I can't say the same for his arch enemy. During a break in filming the Master came over to my trolley for a cuppa and a gypsy cream.

naked

We started chatting, one thing led to another and before I knew what was happening we were both naked inside his four-dimensional grandfather clock Tardis. One thing's for sure - it was certainly bigger on the inside, and I'm not talking about his time machine! I'm talking about his *cock*.

We made love for what seemed like hours but when we came out of the clock I noticed only a minute and a half had passed. He may have come away second best in all his battles with the Doctor, but let me tell you, the Master certainly lived up to his name between the sheets."

> **"...his sink plunger began to creep round until it was stroking my breast"**

William Hartnell hung up his sonic screwdriver in 1966 when Patrick Troughton took over the title role. While the Doctor was having his first run in with the giant spiders on Metabilis 3, Iris was back on earth enjoying a sexy threesome with a couple of hunky Daleks.

"We'd been filming all day and it was getting quite hot under the studio lights. A couple of Daleks came over and asked me if I fancied a walk in the Blue Peter garden to cool off. I was pretty hot too, after standing next to my urn all afternoon, so I agreed. Little did I suspect that once they'd got me alone outside, things were set to get even hotter!

As we walked by the statue of Petra, one of the Daleks asked me what it was. I started explaining that it was an earth dog but I soon realised he wasn't listening to me. His sink plunger, which had been on my shoulder, began to creep round until it was stroking my breast.

drawers

Then the other one began to lift up my skirt and pull down my drawers. I realised resistance would be futile - and anyway, if truth be told I was enjoying myself too much by then to care - and it wasn't long before I found myself as the filling in a Dalek sandwich.

I pleasured one with my mouth whilst the other took me roughly from behind, taking me to new heights of ectasy with his egg-whisk. Their monotonous cries of electronic passion were soon echoing off the walls of the Blue Peter garden. If Percy Thrower was in his greenhouse, I'm sure he would have wondered what the heck was going on. I'm sure he'd never guess in a million years it was a BBC tealady being spit-roasted by two robotic denizens of the planet Skaro!"

Iris wasn't found out that time, but on another occasion she came pretty close.

"According to my contract, I wasn't allowed to have sexual relationships with Doctor Who baddies. If I'd been discovered making love to an alien monster I would have been sacked on the spot. However, the temptations were often too great, and sometimes I simply couldn't help myself.

lunch

I remember this one time we were on location filming a Doctor Who episode on a beach with John Pertwee. One of the Sea Devils started chatting me up over lunch, and before I knew it I found myself agreeing to go skinny dipping with him that night.

Later on, when the rest of

Tea lady Iris (above left) and some of her extraterrestrial lovers, yesterday

Sexterminate!

"I dropped my trolleys for Dr Who Monsters" says tea lady

More of Iris's intergalactic lovers. The Master (left) ~ lived up to his name between the sheets and a sea Devil (right) ~ made love in the surf at Frinton-on-sea.

the cast and crew had gone to bed, we sneaked out of the hotel and met up on the sand. As I slipped out of my blue-checked apron, he peeled off his seaweed-encrusted string vest and soon we were standing before each other stark naked.

There was no need for words, which is just as well as Sea Devils can't talk, and soon we were in the surf exploring every inch of each other's bodies.

chef

Despite his evil, reptiloid reputation, he was a tender and considerate lover and he knew exactly how to pleasure a human. Soon my cries of passion were mingling with his asthmatic hisses and the sound of the crashing surf at Frinton.

However, we never knew how close we had come to being discovered. Unbeknownst to us, the director had decided to have a walk along the beach and must have passed within feet of where we were having it off. If he'd spotted us, I'd have been fired and my Sea Devil romeo would have been banished back to the other side of the galaxy.

However, it would have been worth it for some of the best sex I ever had."

On another occasion Iris came even closer to being found out than the last time. She wasn't caught this time either, although it was close.

"There was this cyberman who I'd always fancied. I used to flirt with him during tea breaks, giving him an extra biscuit or three sugars instead of two in his cuppa. However, he seemed shy and it was difficult to strike up a conversation with him.

Eventually I cornered him and he confessed that he was feeling homesick for his native planet of Telos.

tongue

This seemed to open up the floodgates of emotion and he began to cry. I looked deep into his eyeholes and gently stroked

the handles on the sides of his head. Our lips met and all thoughts of his home planet were soon forgotten as his aluminium tongue began probing the inner recesses of my mouth.

haslet

Our passions aroused to bursting point, we made our way into the empty *Animal Magic* studio to give free rein to our mutual longings. Pretty soon we were locked together in an embrace of intergalactic lust, our bodies becoming as one. I could tell things were getting hot as steam began to squirt out of a vent in the top of his head. Before long, his breathing began to quicken and seconds later, we both came to a shuddering climax.

Afterwards, I sparked up a post-coital Senior Service but when I offered him the packet he politely declined, explaining that as a silicon-

based life-form, the carbon in the cigarette smoke would be fatal. Later on we discovered that a microphone had been left switched on in the studio and the sounds of our lovemaking had been caught on tape. Luckily, the evidence was never traced back to us. The powers-that-be assumed it was Johnny Morris masturbating and docked him a week's wages."

Next week: In a second exclusive extract from her dreadful book, Iris reveals all about the time Davros asked her to put on a show with a lesbian Zygon.

> "...he peeled off his seaweed-encrusted string vest and soon we were standing before each other stark naked"

Cyberman ~ A shy alien, but Iris brought him out of his shell in the Animal magic studio.

IRIS POLDARK
WHO WERE YOU WITH IN THE MOONLIGHT?

Miriam
SOLVES YOUR PROBLEMS

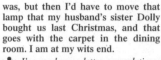

Dear Miriam... This morning I went to open a tin of beans for my husband's breakfast. The trouble is, I couldn't find the tin opener anywhere.

I am 29 and he is 32. We've had this problem before, but the tin opener has usually turned up after a minute or two. But this time it is different. I've been searching for ten minutes now, and I've reached the point where I'm searching in the places that I've already looked in. My husband has to go to work in twenty minutes. What can I do?

✱ This is a common problem. Many people your age lose the tin opener from time to time. Just relax, and I'm sure you will find it eventually. In the mean time, it is important to feel positive. Try to focus on the things in your kitchen that you can find, such as serving spoons, teacups or the washing machine. You may also like to send for my leaflet 'Lost Tin Openers' or call my premium rate helpline.

LETTER OF THE DAY

Dear Miriam... I got in my car the other day and noticed that the fuel gauge was almost on empty. I spoke to my boyfriend about it that night and he said that I was probably running out of petrol.

I am 21 and he is 22 and we have been going out for 6 months. He has suggested that we go to the garage and put a tenner's worth of petrol in, but I'm not sure. My boyfriend thinks I'm just being silly, but it's nearly £2.80 a gallon and I don't want to be rushed into something I may regret. Please help me.

✱ You are right to be cautious. Weigh up the pros and cons. Putting a tenner's worth of petrol in will cost you £10. On the other hand, if you don't, your car will eventually judder to a halt. But it is your car, and you must do what you feel is right for you. I'm sending you my leaflet 'Does My Car Need Petrol?' which will give you all the advice you need to make your descision.

Dear Miriam... I've just moved a vase from my dining room, and put it on the corner table in the front room. The trouble is, I don't know whether it goes with those curtains.

I am 48 and my husband is 50. I've asked him what he thought, but he just shrugged and carried on watching the telly. I could put it back where it was, but then I'd have to move that lamp that my husband's sister Dolly bought us last Christmas, and that goes with the carpet in the dining room. I am at my wits end.

✱ I've read your letter several times and I simply cannot see the problem. The vase goes lovely with them curtains. Which means that Dolly's lamp can stay where it is. If you are worried in the future, you can call for reassurance on my helpline 'Does that Vase go with them Curtains?'

Dr. Miriam Stoppard
Premium rate problem lines you can trust!

Mnah! Mnah! Does this soup need more salt?
0000 994 387

Which shoes with these trousers? Black or brown?
0000 994 388

Oh, look. What kind of bird is that on the bird table?
0000 994 399

Can you give me a hand with the other end of this wardrobe?
0000 994 390

Calls cost enough to keep Dr. Stoppard in very fancy earrings

Justice
WITH
Jacobs
THE FAT SOLICITOR

Bandit at 3 o'clock

Q It's 3 in the morning and I have just woken to find I am being burgled by a man in a balaclava. He has stabbed my husband in the stomach and now he is coming screaming at me with a 12 inch knife. I am 32 and my husband was 33. We have been burgled several times before, but this is the first time that we have been attacked. I am within reach of a very heavy lamp with a square alabaster base, and wonder if I am within my rights to hit the intruder over the head with it.

Mrs. B, Essex

A The fat solicitor regrets that he is unable to answer any legal queries, as he is eating a really, **really** big pie.

JUST DIAL J

The Fat Solicitor dishes out legal advice on major topics...

NEIGHBOUR TROUBLE?
Gary summarises the 1966 Boundaries Act whilst eating a whole packet of Hob-nobs
0000 994 388

FAULTY GOODS?
Your statutory rights explained whilst Jacobs stuffs two Battenbergs in his big fat face
0000 994 389

Calls cost 50 sausage rolls/min and terminate in the nearest Greggs

Christobies
Auctioneers and Valuers Ltd
Knightsbridge

Are privileged to announce
a sale of

Chasabilia and Daviana

on **Saturday 10th August 2002**
at the **Salerooms, Knightsbridge, London**
Viewing on **Wednesday, Thursday, Friday, 7th, 8th, 9th**

Sundry lots to include:

Several public house upright Pianofortes, assorted scraggy Beards, a bald Drummer *(sold as seen)*, Dr. Marten 'Airwair' shoes by Griggs and Co. *(12 pairs)*, a large collection of elasticated Braces, bass Guitars, sundry loopy Snooker equipment *(24 lots)*, various collarless shirts, gorblimey trousers *(several pairs)*, flat Caps *(2 lots)*, 1 silver Disc circa 1978, etc etc.

In all 628 lots, the whole comprising the property of Messrs. *Charles Hodges* and *David Peacock Esq*, late of Margate, England.

Angharad & Reece
Auctioneers
206~208 New Zealand Street, Bloomsbury, London

For Sale By Auction
Monday 12th August 2002
(viewing 9th~11th)

A circa 1930 wood veneer oak buffet Sideboard with thick rectangular moulded top above a central section with drawer and compartments carved with stylised panels and scrolling, the side cupboards quarter veneered, framed by acanthus carved columns with square plinth base raised on large bun feet, and containing 1 crate of Courage mild.

Est. £10~15

A Yoghurt
for the Rest of My Days

We asked Celebrities, if they could only eat one kind of yoghurt for the rest of their life... which would it be?

No. 103
Robert Powell

Strawberry for me. Yum! Yum!

MARS BAR PLEASE.

HAVEN'T YOU GOT ANYTHING SMALLER?

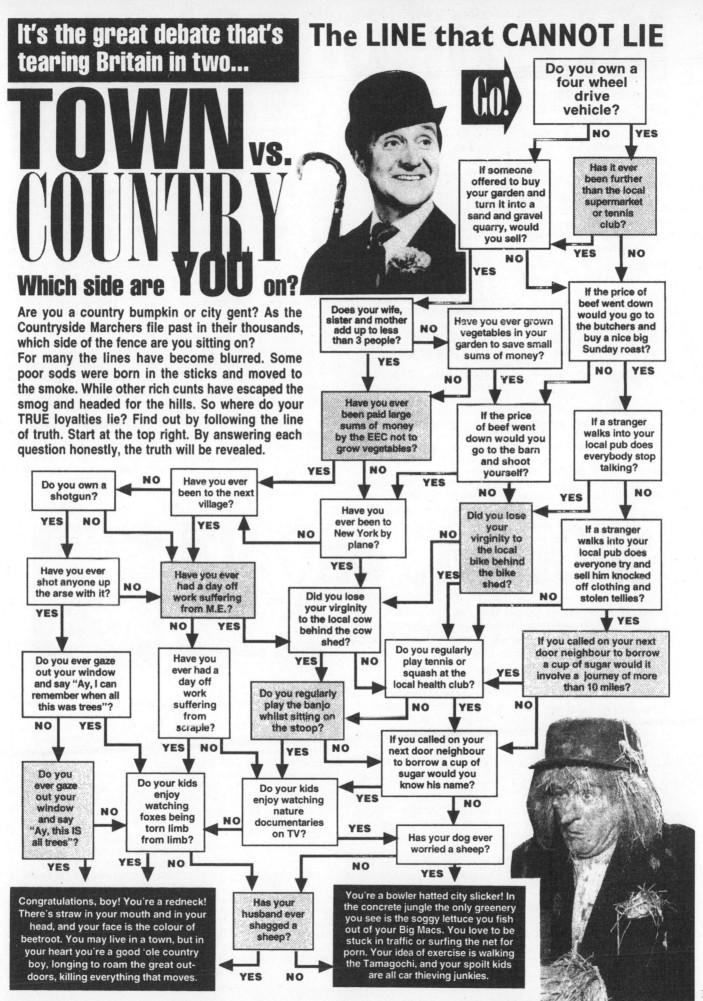

ISLE OF DOGS!

Clare Short

Women in Britain are _UGLY,_ their looks lagging way behind those of our European competitors.

That's the shock claim being made by Bristol landscape gardener Dennis Stokes. And Dennis believes come 1992 Britain's females will **LOOSE OUT** to more glamorous looking women from neighbouring countries.

"If you take a walk down any street in Britain, none of the birds you see are anything to write home about", Stokes told us. "But if you watch the telly, you see some really good lookers in Italy and France and that. The birds over here can't compete with classy tarts like that."

BOILERS

A secret Government report prepared by experts seems to confirm Mr. Stokes' fears. The document, recently leaked to us, reveals that in some parts of Britain, like Huddersfield, less than 10 per cent of the female

Britain trails in crumpet stakes

population are attractive. While a stunning 50 per cent – half the female population – are classed as 'absolute boilers'.

PISTONS

Officials at the Department of Health are known to be extremely concerned about the implications of the report come Britain's entry into the EEC in 1992. The removal of trade barriers will make international marriages a lot easier, and there are fears that Britain's fellas will shop abroad for partners.

EXCLUSIVE

"Me, I fancy that Italian bird that gets her tits out on the telly", Dennis Stokes told us yesterday. And come 1992 Dennis could be one of thousands of British men who flood through the Channel Tunnel to score themselves a piece of continental crumpet.

VALVES

Secret proposals to remedy the problem are currently being studied by the Prime Minister and Cabinet colleagues. They hope that dramatic measures can be taken to improve the appearance of the nation's women.

These are thought to include:

● Poll Tax relief to enable ugly women to spend more money on make-up.
● Cheaper housing for plain looking women in certain out-of-the-way areas, like Workington, to encourage

those women to live there. This would be part of an integrated 'out of sight, out of mind' policy.
● A so-called 'Boiler Curfew' to be introduced at weekends preventing plain or unattractive women from going out in the evenings.

DIODES

"Efforts will have to be made to bring Britain's birds into line with the gorgeous muff you see parading round on the continent in tight miniskirts", a Government spokesman told us yesterday.

TRANSISTORS

We tried to ring Labour MP Clare Short, to see what she thought about ugly women, but her number wasn't in the Newcastle phone book.

SADDAM'S BRITAIN

If Iraq had won the Gulf War London would no longer have been the capital of England. And landmarks such as Buckingham Palace, Nelson's Column and Tower Bridge would all have been rased to the ground.

Nightmare that almost came true

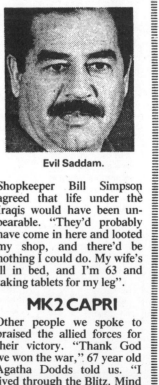

Evil Saddam.

These are just some of the horrific facts revealed in top secret documents abandoned by fleeing Iraqi troops as they left Kuwait. Documents which include an amazing blueprint of Saddam Hussein's Britain.

PLANS

Had Saddam won the war, life under the Iraqi dictator would have been a nightmare for the population of Britain. Among the many changes outlined in detailed plans drawn up by the Iraqi military were:

● Strict curfews, with the death penalty for anyone found walking the streets at night.
● A total blackout of TV and radio.

● Identity cards for every single person in Britain over the age of 12, similar to those which were to be carried by football supporters.
● Severe rationing of food and fuel, on a scale not witnessed since the Second World War.
● Skeleton rail and bus services, with as few as 1 in 5 timetabled services running.
● Early closing for shops on Mondays, Tuesdays and Thursdays.
● Restricted banking hours at the four major High Street banks.

PARADE

Madman Hussein planned to celebrate his victory with a military parade, not through the streets of London, but through Grimsby instead.

For Hussein's military advisers had recommended Grimsby as the new capital of England, with it's ideal harbour facilities. North Sea links and good road and rail services. A puppet Government would have been installed in Grimsby Town Hall, from where the evil Iraqis would have run the country.

ESCORT

Saddam saw London as a symbol of western imperialism. He planned to set it burning like the oil fields of Kuwait, killing the Royal Family and stealing their furniture, paintings and cars.

Shopkeeper Bill Simpson agreed that life under the Iraqis would have been unbearable. "They'd probably have come in here and looted my shop, and there'd be nothing I could do. My wife's ill in bed, and I'm 63 and taking tablets for my leg".

MK 2 CAPRI

Other people we spoke to praised the allied forces for their victory. "Thank God we won the war," 67 year old Agatha Dodds told us. "I lived through the Blitz. Mind you, we were happier then. You could leave your front door open in them days", she said. "And I never saw a banana till I was 42", she added.

Innocent Osmonds Set to Die

SEVENTIES singing sensations the Osmonds were last night languishing on Utah State Prison's Death Row, after being mistakenly sentenced to death for a murder they didn't commit. The cleancut pop brothers Donny, Jay, Merril, Wayne and Alan are set to face the electric chair on Christmas morning after a computer mix-up at the District Attorney's office.

The Osmonds: Goin' Home in a box this Christmas.

A 58-year-old shopkeeper was shot dead in March during a bungled raid on his Alabama liquor store. Three days later, police arrested 32-year-old unemployed local man Ricky-Bob Mullet, who confessed to the killing. However, instead of the murderer, the name of the popular mormon supergroup was mistakenly filled in on Mullet's execution warrant.

District Attorney's office secretary Terrylene Koswalski told reporters: "I was typing up the warrant on my computer, and I stopped to see if I could find a copy of 'Love Me for a Reason' going cheap on the internet. I guess I must have accidentally typed 'The Osmonds' into the death warrant instead of the eBay search engine."

late

Koswalski spotted her mistake after a couple of hours, but it was already too late. The warrant had been countersigned by the state governor and the Osmonds had been taken into custody. "It's a terrible thing, what's going to happen, but there's simply no way out," prison boss Nylon Hogg told Fox News reporter Hymen Prepuce. "It's going to break my heart to fry those good old Osmond boys, but the law's the law."

espreso

District Attorney Spiro Theocropolis said: "All legal avenues have been exhausted. Under state law, once the warrant has been authorised there is no appeal process. Sadly, in this case that means the Osmonds are going to have to be executed even though we know for a fact that they are innocent,"

Theocropolis continued: "We have already set up an enquiry to find out what went went wrong in this case."

capucino

"Unfortunately, miscarriages of justice happen all the time. I'm sure that one day we'll look back at what we're about to do to the Osmonds and see how it could have been avoided, but that will be with the benefit of hindsight. At the moment, we just have to learn from our mistakes, and try to make sure that this sort of thing doesn't happen again too often," he added.

Meanwhile, the real killer has been released and is making the most of his new-found freedom. Ricky-Bob Mullet told NBC's Smegma Glans III: "I just couldn't believe my luck when I heard those innocent brothers were going to die in my place. I know I've done some bad things in my life, but I've done good things too. I feel the Osmonds getting electrocuted on Christmas Day may be God's way of saying thank-you. I guess I feel a bit sorry for them, but that's life I suppose." And Mullet revealed what he plans to be doing on the morning of December 25th when governor Hogg pulls the lever that will send 20,000 volts crackling through Donny, Jay, Merril, Wayne and Alan, wiping their trademark grins from their faces forever.

maxwel house

"I'll be having me a nice quiet day," he said. "Just me, my girlfriend and my little baby daughter. I'll try not to think about what's happening in the jailhouse too much, because Christmas is supposed to be a happy time."

"But I just might drink me a toast to the Osmonds," he added. "Because if it wasn't for them I wouldn't be here."

This is not the first time that members of the Osmond family have found themselves at the wrong end of a miscarriage of justice. In 1996, Marie and Little Jimmy were sentenced to 35 consecutive life sentences after a jury foreman inadvertantly doodled their names on his notepad during a Minnesota kidnapping trial.

Mountains of TERROR

BRITAIN'S climbers were facing huge delays yesterday after security measures were stepped up at the country's highest peaks.

Government minister for Vertiginous Geographical Features Dr Kim Howells put all mainland mountains on a state of scarlet alert after intelligence services received information from a credible source that a terrorist strike was imminent. It is thought that a network of Al Qaeda cells were plotting to simultaneously blow the tops off ten British peaks including *Snowdon, Ben Nevis* and *Scafell Pike.*

In a coordinated nationwide sting, codenamed 'Operation Absolutely Certain This Time', police arrested 20,000 men with beards, before shooting them in the arm and releasing them without charge.

Miss Howells told a press conference yesterday: "I fully understand that the new security measures we are implementing will cause a certain amount of inconvenience to mountaineers, but in the light of this not at all imaginary threat, we feel we have no option but to put them in place."

Under the temporary restrictions climbers will not be allowed to take rucksacks, ropes, woolly hats or food and drink onto any British slope. Kendal mint cake will be allowed, but mountaineers will be required to take a bite out of each bar in front of a government official in order to prove that it is not made out of Semtex plastic explosive painted white.

"The danger that our mountains will get their pointy tops blown off by Islamic extremists is real and imminent," Miss Howells continued. "A bomb hidden inside a ham sandwich and wrapped in foil could be detonated in seconds using a fuse made with threads teased from the bobble off a woolly hat. In the circumstances, these farcical security

Bin Laden Peak Plot Foiled

precautions are fully justified."

"Innocent climbers have nothing to fear," she added. "Unless of course they look foreign, in which case it may be necessary for police to shoot them a little bit."

Meanwhile, there was chaos at the feet of Britain's mountains yesterday as the new measures came into force. In Llanberis at the base of Snowdon, the queue of climbers waiting to take to the slopes snaked through the gift shop, twice round the mountain railway cafe and out into the car park. It was estimated that mountaineers were having to wait up to more than 4 hours before setting out on their climbs, and frustrations were beginning to boil over.

"My wife and I have been climbing

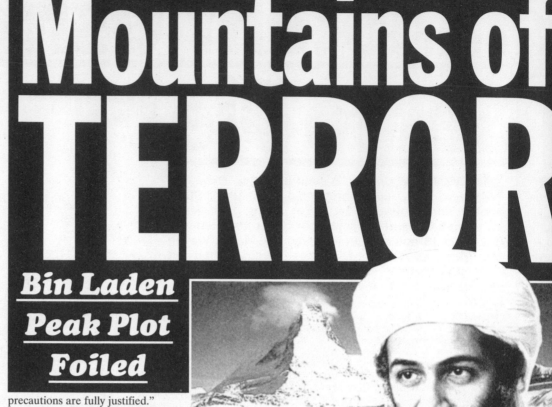

Snowdon every summer for more than thirty years," fumed hillwalker Halford Topman. "And this is the first time we've had our flasks of soup confiscated and been given internal rectal examinations by machine gun-toting policemen."

"We won't be coming back next year and that's for sure," he added.

HILLWALKERS! Here it is, your handy cut out and keep MOUN

ALERT COLOUR: Beige
ATROCITY STATUS: Highly unlikely
SCAREDNESS PERCENTAGE: 0%
ACTION: None. Food and drink may be taken onto mountain. No restriction on rucksacks, kagouls and bobble hats.

ALERT COLOUR: Taupe
ATROCITY STATUS: Unlikely
SCAREDNESS PERCENTAGE: 2%
ACTION: Little. All items allowed on mountain, but could be subject to random spot checks.

ALERT COLOUR: Fawn
ATROCITY STATUS: Slightly less unlikely than taupe, but still fairly unlikely.
SCAREDNESS PERCENTAGE: 3%
ACTION: As taupe, with unlimited access to lower slopes. However, security personnel may be stationed on crags near the summits.

ALERT COLOUR: Eau de Nil
ATROCITY STATUS: Possible
SCAREDNESS PERCENTAGE: 11%
ACTION: As fawn, plus CCTV cameras installed on summits. All mountaineers subjected to intrusive body cavity searches.

ALERT COLOUR
ATROCITY STA
SCAREDNESS P
ACTION: As eau kagoul hood draw wiches tasted in stationed behind search checks o

We're all going on...
Osama Holiday

MORAG: Clocked Bin Laden in gift shop

BOMB EVERY MOUNTAIN

REVEALED! Bin Laden's Plans for Britain's Summits

EVIL Al Qaeda boss Osama Bin Laden has spent the summer scaling Britain's tallest mountains to scout out locations for his terrorist outrages. That's according to Fortwilliam woman Morag Drambuie, who spotted the cave-dwelling Tora Bora tyrant buying souvenirs in the Ben Nevis giftshop where she works.

"It was definitely him," recalls the 72-year-old grandmother of thirty. "He'd tried to disguise himself by wearing glasses, talking with a Lancashire accent and making himself about a foot shorter than he is in the paper, but he wasn't fooling me."

And chillingly, the 9-11 mastermind was not alone. For Morag recognised his companion as none other than extradited hook-handed cleric Abu Hamza.

"I spotted him immediately," she remembers. "He was trying to pass himself off as Bin Laden's wife. He'd shaved his beard off, lost about ten stone and was wearing false breasts and mittens over his hooks."

Morag's suspicion was aroused when she spotted the sinister couple studying a map of Ben Nevis and the surrounding area on a teatowel. "The map is very detailed," she continues. "It shows all the major peaks in the Grampians, such as Ben Lawers, Ben Vorlich, Ben Lomond and Ben

Eagle-eyed Morag Spots Bin up Ben

Cruachan. It was clear to me that Bin Laden and Hamza were plotting which peaks to take out in a bomb outrage."

Morag immediately tried to call the SAS, but by the time she had got their number from Directory Enquiries the pair had already made their escape. She recalls sadly: "They came up to the till and bought a tin of shortbread, a Scottish piper doll in a plastic tube and several humorous postcards depicting the Loch Ness monster with a tartan hat and ginger hair. Then they asked me for directions to Mallaig before driving off in a blue car."

A police spokesman confirmed that Morag's sighting was being taken seriously. Sgt Tam O'Shanter of the Glen Scaddle Anti-Terrorist Squad told us: "If anyone sees an elderly couple in a blue car, we would urge them to get in touch with the SAS immediately."

BEN NEVIS
Al Quaeda cell plotted to take out highest UK peak with stick of dynamite hidden in sausage roll, detonated using heat from flask of tea.

SNOWDON
In echo of 9-11 attacks, suicide train driver planned to hijack mountain railway engine 'Hywel Dda' and crash it at 4mph into toilets next to cafe at summit.

SCAFELL PIKE
Disguised as hillwalker, Mullah Omar plotted to scale slope and blow off summit with dirty bomb hidden inside Scotch egg.

BEN MACDUI
Bombers intended to drill out Kit-Kats and Penguins then fill them with nitro-glycerine to blow top of mountain to smithereens.

CRIB Y DDYSOL
Tupperware box of processed cheese sandwiches made with plastic explosives instead of cheese was planned to explode peak.

DUNKERY BEACON
Terrorists intended to use massive nail bomb hidden in hollowed out Soreen malt-loaf to level popular Somerset landmark.

HELVELLYN
Extremists conspired to fit 4 empty Pringles tubes together to make 'Super-Bazooka' to destroy cairn on Lake District mountain top.

CAMEDD LLEWELYN
Hook-handed Finsbury Park cleric Abu Hamza planned to lob hand-grenades painted to look like small pineapples at 3491ft Welsh summit.

PEN Y GHENT
Suicide hiker aimed to fill climbing boots with TNT and jump up and down on top of Yorkshire mountain until they went off.

SHINING TOR
Tinfoil from round sandwiches fashioned into solar dish was to focus rays of sun onto pac-a-mac impregnated with gunpowder.

N TERROR ALERT LEVEL GUIDE

BRITAIN'S mountains are subject to constantly changing alert levels which indicate the likelihood of a terrorist attack. The different statuses range from Beige to Maroon. But what do those colours actually mean?

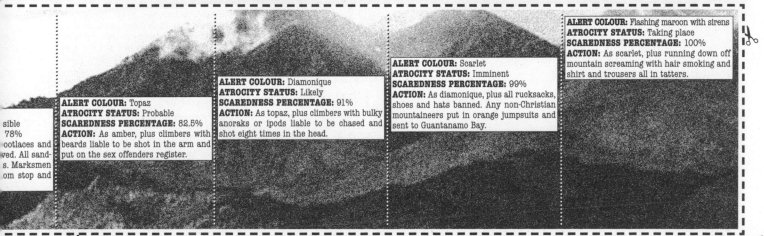

ALERT COLOUR: ...sible
...78%
...ootlaces and
...ved. All sand-
...s. Marksmen
...om stop and

ALERT COLOUR: Topaz
ATROCITY STATUS: Probable
SCAREDNESS PERCENTAGE: 82.5%
ACTION: As amber, plus climbers with beards liable to be shot in the arm and put on the sex offenders register.

ALERT COLOUR: Diamonique
ATROCITY STATUS: Likely
SCAREDNESS PERCENTAGE: 91%
ACTION: As topaz, plus climbers with bulky anoraks or ipods liable to be chased and shot eight times in the head.

ALERT COLOUR: Scarlet
ATROCITY STATUS: Imminent
SCAREDNESS PERCENTAGE: 99%
ACTION: As diamonique, plus all rucksacks, shoes and hats banned. Any non-Christian mountaineers put in orange jumpsuits and sent to Guantanamo Bay.

ALERT COLOUR: Flashing maroon with sirens
ATROCITY STATUS: Taking place
SCAREDNESS PERCENTAGE: 100%
ACTION: As scarlet, plus running down off mountain screaming with hair smoking and shirt and trousers all in tatters.

WHO IS BRITAIN'S TOP BOB?
MONKHOUSE v CHARLTON

It's the question on everyone's lips. People over the country are itching to know 'Who is Britain's Best Bob?' Is it **BOB MONKHOUSE**, whose jokes have left us laughing for over twenty years? Or is it **BOBBY CHARLTON**, whose goals guided England to their famous 1966 World Cup Victory? In pubs and clubs around the country the debate continues — who is the greatest Bob of all? Well, now is your chance to find out, as we answer the question — WHO IS THE TOP BOB?

Bob Monkhouse	HOW THEY SCORE		Bobby Charlton
That devilish smile and those angled eyebrows tell us that Bob, one time presenter of TV's 'Golden Shot', is a ladies man. He's slick, he's polished and he oozes sex appeal. But look out girls — he's married.	**GOOD LOOKS** 9	5	A footballing legend, he thrilled the ladies with his dazzling ball control. Now, with his rugged, mature appeal, he's the man your granny dreams of. But loss of hair costs Bobby points, as well as popularity among younger women.
On TV he's charming and cheerful, and despite his cheeky grin, he's as friendly off the screen as he is on it. Warm and considerate, Bob's heart is as large as a wardrobe. There's never a dull moment spent in his company.	**PERSONALITY** 10	6	Bobby's dynamic performances on the field and his incredible goal scoring achievements conceal a quiet side of his character. Off the field he is a modest, down to earth character, but his honesty is a strong asset.
Thirty star spangled years in show business have began to take their toll on Bob's much sought after, sexy frame. Although sensible Bob steers clear of excesses, too much gourmet meals and not enough time in the gym have lead to a bigger waist — and a smaller score.	**FITNESS** 6	9	Fitness was crucial to Bobby's career as soccer's deadliest marksman, and although past his peak, regular training and exercise ensure that this much loved centre forward remains in tip top condition.
Bob cuts a dash under the TV lights in his glittering suits and dicky bow ties. But sometimes taste goes out the window, leaving dazzled viewers reaching for their 'brightness' control. A formal dresser off screen, Bob always makes an effort.	**STYLE** 9	7	In his playing days Britain's most famous forward was never seen without a spotless club blazer and tie. Now Bobby the businessman woos the women in a series of smart suits and sports jackets. Although never a fashion leader, Bobby still cares about his appearance.
Bob is a regualr workaholic! He's rarely off our screens with shows like 'Bob's Full House' and 'Bob Says Opportunity Knocks'. And despite his busy schedule, he still finds time to make guest appearances on other people's shows. Even on his days off, Bob keeps busy trying out new jokes on his wife and family.	**WORK RATE** 10	8	Renowned in his playing days for his unselfish running off the ball, he created goals as well as scoring them. A player's player, Bobby never stopped running until the full ninety minutes were up. Nowadays despite business commitments, Bobby still finds time to make expert comments during half-time intervals.
Thought by many to be Britain's top comic (he is said to know more jokes than anyone else in the world), Bob's show business career has meant that goal scoring opportunities have been few and far between.	**GOALSCORING ABILITY** 0	10	Over the years Bobby banged in hundreds of goals, his powerful shooting from outside the 18 yard area was feared by keepers throughout the world. Unmatched for strength, sharpness and an accurate header of the ball, Bobby is in a goal scoring league of his own.
TOTAL 44 Nice try Bob, but not enough!			**TOTAL 45** Bobby's best! He's our champ!

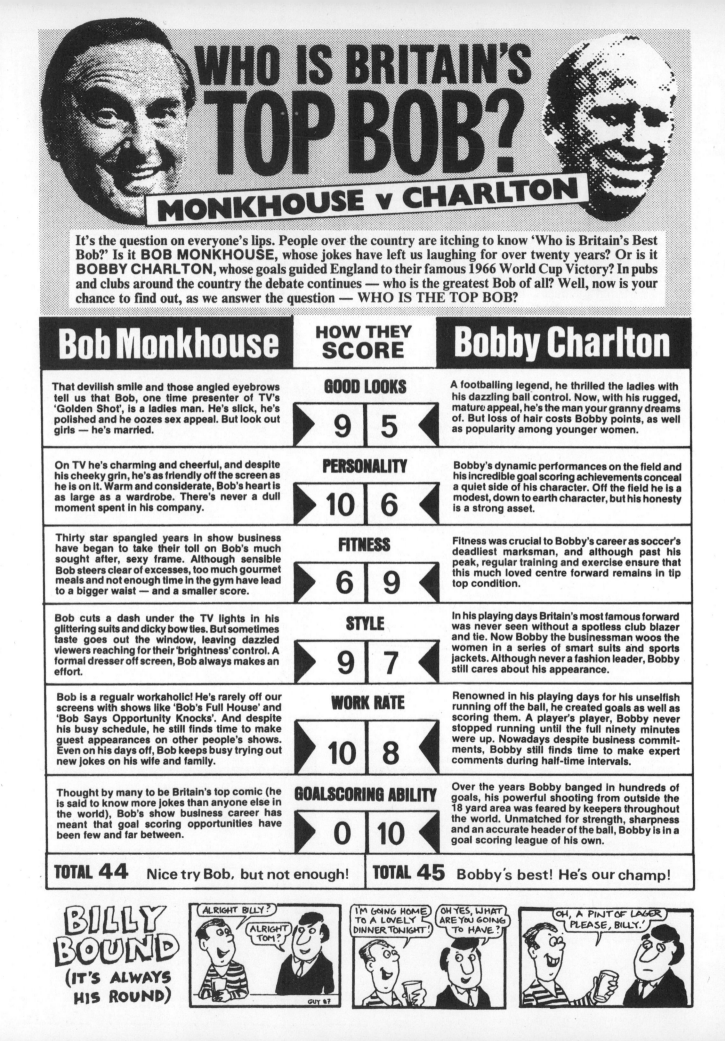

BILLY BOUND (IT'S ALWAYS HIS ROUND)

ALRIGHT BILLY?
ALRIGHT TOM?

I'M GOING HOME TO A LOVELY DINNER TONIGHT!
OH YES, WHAT ARE YOU GOING TO HAVE?

OH, A PINT OF LAGER PLEASE, BILLY.'

GUY 87

12 THINGS YOU NEVER KNEW ABOUT THE ROYAL FAMILY

We read about the Royals in our newspapers every day. We see them on TV and occasionally shake their hands or give them flowers in the street. Indeed, with so much media attention focusing on our favourite family, many of us feel we have truly come to know the Royals, and see them as friends more than just public figures.

But believe it or not, there are still a few unusual facts you may not know about our marvellous Monarchy. For instance, did you know that . . .

1 As a child, Prince Charles was known as 'Big Foot' among the Royal Family. For at the age of five he had developed incredible size nine feet! Eventually his gigantic feet stopped growing and nowadays are more or less in proportion to the rest of his body.

2 Prince Edward developed theatrical tendencies at a very early age. The young Prince was regularly entertaining palace guests with his spectacular one-man song and dance show — at the tender age of four.

3 Before joining the Navy, Prince Andrew yearned for a career as a train driver. So much so that on his tenth birthday, the young Prince was given a very special train set to play with — a full size steam locomotive and twelve carriages. Throughout his teenage years, Andrew could regularly be seen driving his train at speeds of up to 100 miles per hour on the Edinburgh to London railway line, accompanied on the footplate by his father, the Duke of Edinburgh.

4 Princess Anne's show-jumping career couldn't have got off to a worse start. For the young Princess was born with an allergy to the animals. Indeed the wiff of a horse alone was enough to turn Anne's skin purple and leave her temporarily blinded. It was the Queen Mother who eventually found a remedy for her ailment after several leading doctors had failed. She told Anne to sleep with a crushed walnut under her pillow and the Princess has never looked back.

5 The Queen's favourite snack isn't salmon or caviar as you might expect. She prefers a tin of pineapple rings in syrup.

The Queen — likes pineapple.

Indeed, during a state visit to India in 1962, the Royal Yacht Brittania was forced to make a 2,000 mile detour to the East African coast in order to stock up on the tinned fruit after supplies had run out.

6 Phillip is not the Duke of Edinburgh's real name. He was originally christened Norman, but changed his name by Deed Poll shortly before his engagement to Elizabeth was announced. At the time the Queen's parents felt that Norman didn't sound royal enough!

7 Before marrying the Duke of Edinburgh, the Queen had been romantically linked with several eligible bachelors of the day, among them Hollywood film star, James Cagney. Another former Royal escort was football star, Sir Stanley Matthews.

8 King Edward is the only member of the Royal Family to have played professional cricket. Aged 17, he spent one season on the books of Leicestershire County Cricket Club. He batted for them only once, scoring a measly three runs. And the man who bowled him out? A certain W. Churchill — later to become Britain's best-known Prime Minister.

Phillip — shot koala bear.

9 According to a quirk of law, under ancient Royal Hunting Rights, members of the Royal Family are still entitled to hunt animals anywhere in Scotland — including Edinburgh Zoo! However, the Royals have steered clear of the zoo, since the Duke of Edinburgh caused a public uproar by shooting a koala bear there in 1957.

10 We all know that a broken line down the centre of a road separates traffic travelling in opposite directions. But this was not always the case. Originally, the line was put there to divide the road into two halves. Normal traffic travelling in either direction was restricted to one side of the road, while the other side was left free for the exclusive use of the Royal Family. In the words of the Highways Act of 1872, this was to "enable their Royal Highnesses to travel swiftly and without hindrance and free from such encumbrance as other motor vehicles would provide". In 1912, the government bowed to public pressure and allowed regular traffic to use both sides of the road. In 1923, after a series of traffic accidents, a bill was put before Parliament to allow vehicles travelling in opposite directions to use different sides of the road.

11 During the Second World War, emergency plans were drawn up to safeguard the Royal Family in the event of a German invasion. One of several plans considered was to disguise them as a travelling circus troupe, and send them across the country, performing from town to town. It was hoped that their thick theatrical make-up and extravagant colourful costumes would fool the Germans.

12 Despite the extravagant forms of transport available to the Royal Family — a Royal yacht, Royal Train, elegant horse-drawn carriages, etc. The Queen often uses public transport when travelling short distances. On one occasion the driver of a London bus fainted when the Queen stepped on board and asked for a single fare to Windsor. While the driver recovered, the Queen happily signed autographs for her fellow passengers.

THERE'S A COUPLE OF THINGS IN THE PIPELINE.

Simon Cowell's EGGS FACTOR

> **HI THERE EGG FANS!** TV's Mr Nasty Simon Cowell, here. And as usual, you've been writing to me about eggs. And it's no 'yolk' to say that this week's Eggs Factor postbag is the biggest I've ever had! So without further ado, 'shell' we 'crack' on and 'dip' in?!

My husband had a boiled egg for his breakfast every day of his life, and he lived to the ripe old age of seventy-eight. It just goes to show that scientists don't know everything.

Mrs Yootha Turpentine, Hull

My wife and I both love eggs, but cooked in different ways. She likes hers scrambled, whilst I prefer mine poached. Sadly, the strain of our egg preparation preferences eventually proved too much and drove a wedge between us. We eventually got divorced the day before our Golden Wedding Anniversary. I am now remarried to a woman who likes poached eggs as much as I do, and I am happier than I have ever been.

Harold Meths, Leicester

Eggz
The eggs of the African ostrich are so big that if you made a two-egg omelette out of them, it would be big enough for Robert Pershing Wadlow, the world's largest man.
Factz

These food fascists say that eating too many eggs is bad for us. Well for their information, I eat far too many eggs every day and there's nothing very serious wrong with me. Once again, the so-called "egg-sperts" are left with egg on their faces!

Jemima Diesel, Bicester

Readers' Wives' Eggs

How about this pair of scrumptious frieds I snapped on my missus's plate? It will give her a real thrill to think of all your hungry Eggs Factor readers drooling over them. Name & address withheld, Altrincham

Check out this close-up omelette shot of my young wife's 2-egg breakfast. Believe you me, I scoff one of these down every day and it's as hot and delicious as it looks! Mucky Jim, Newcastle

Send your Egg-splicit Photos to the usual address.

Eggz
The world's smallest man Calvin Philips ate a 2-egg omelette made from bee humming bird eggs every day for breakfast. He died in 1956 of the world's smallest heart attack.
Factz

I was on holiday in America recently and a waitress in a restaurant asked me if I wanted my eggs "over easy". I had to laugh. She'd got it wrong and meant fried again!

Mrs T Vaseline, Gloucester

As a devout Catholic, my wife refuses to eat eggs because she says they are chickens' abortions. But I don't mind, because I'm an atheist and that means more eggs for me. I love 'em!

R Butane, Cambois

I'm a Muslim and I don't eat eggs, but it's nothing to do with my religion. They give me really smelly wind.

Iqbal Benzene, Tunbridge

When my son was little, I cooked him boiled eggs for breakfast every morning. Now he is an SAS swimming instructor, based in Exeter. He often jokes that he used to dip his soldiers in his *eggs*, whereas now he dips his soldiers in the *Exe* - the river which runs through Exeter. However, this isn't strictly accurate, as they have a swimming pool in the camp. And in any case, the soldiers get in of their own volition, thereby dipping themselves. And he never had bread & butter soldiers when he was little, as he is gluten intolerant. And he's not in the SAS either. He just tells people that to impress them.

Marjorie Hydrocarbon, Shrewsbury

Eggz
Bee humming bird eggs are so small that if you tried to make a two-ostrich-egg-sized omelette out of them, it would take hundreds...maybe even thousands! But probably not millions.
Factz

I was on holiday in America recently and was leaving a restaurant after eating two platefuls of fried eggs. After I paid the bill the waitress told me to "have a nice day". I had to laugh. She'd got it wrong and meant goodbye!

Mrs T Vaseline, Gloucester

I'm sick of show-off TV chefs breaking eggs using just one hand. If the good Lord had meant us to break eggs like that He would have given us one hand, not two. It would serve these people right if they lost an arm in an accident like the drummer out of Def Leppard.

Archbishop Rowan Williams, Canterbury

Eggz
Tortoises are unique in the animal kingdom, because they're the only creatures that have shells when they're eggs, and then have another shell when they're grown-ups! And turtles.
Factz

I was on holiday in America recently and a waitress in a restaurant asked me if I wanted my eggs "sunny side up". I had to laugh. She'd got it wrong and meant fried!

Mrs T Vaseline, Gloucester

I'm mad about eggs, and I was very excited to see a programme called *Eggheads* listed in the *Radio Times*. Imagine my disappointment when I tuned in to discover that it was a general knowledge quiz show. Fortunately, there was a question about eggs so my half hour wasn't completely wasted.

Frank Butanol, Pwllelli

My husband and I were playing Scrabble the other night. When it came to his turn, he drew two Gs, an S and an E, but announced that he was unable to make a word out of those letters. How foolish he felt when I pointed out that he could have made the word "eggs". The funny thing is, he's an egg farmer...and his name is Ronald Eggs!

Edna Eggs, Worcester

your t-egg-xt messages

eggs r gr8! i luv em ;-). eggy bill.

Eggs are the @1 fd 4 me. yum yum yum. the eggman.

Lv the white. H8 the yoke. Help!. johnny albumen.

EGS R 4 SADOS. BACONS BEST. FUK EGS U EGFUKERS. BAMBER G

Get ur facts str8 bamber g. bacon suks. eggs 4 evr. egg warrior.

Kids say the funniest things ...about Eggs!

My six-year-old grandson has been doing sex education at school, and it's filled his young mind with all sorts of weird and wonderful ideas. The other day, he came round for tea and asked what we were having. When I replied that it was eggs, he piped up, "Are we having sperm on them?" As you can imagine, there was a ghastly, embarrassed silence, and a very unpleasant, awkward atmosphere for the rest of the evening.

Mrs Audrey Clutchfluid, Leominster

Miriam's Egg Problem PHOTO CASEBOOK

Jess's Frying pan Dilemma: *Day 6*

Jess has been going out with her boyfriend Tom for five years. But whilst frying him an egg she has accidentally dropped a bit of shell into the pan...

Oh No! There's a bit of shell in the pan.

Tom hates shell in his egg.

If I get it out with a fork I might burst the yolk...

And I can't use my fingers because it's a bit hot...

Jess confided in her mum...

Oh, mum. I'm so confused...

Should just keep quiet about it? He may not find it, and what he doesn't know won't hurt him.

No, Jess! If Tom finds that bit of shell in his egg, you'll have lost his trust for ever.

But what should I do?

I don't know, Jess. You'll have to get the shell out somehow, and quickly...

...because if you leave it on the heat much longer, that egg is going to be like *rubber!*

Oh no!

Continues next week!

WINDS OF CHANGE

Blow me! We'd all better hang on to our hats if the latest forecasts from Britain's weather chiefs are to be believed.

For according to experts, by the year 2000, terrifying tornado force winds gusting at speeds of up to 1,000 miles per hour will turn the nation's streets into a no-go area for pedestrians.

WATCHDOGS

Met office forecasters fear that recent trends towards windier weather are set to continue. And Government weather watchdogs, aware of dramatic changes in our climate, have commissioned a comprehensive report on the problem from top boffins at Reading University's Department of Wind. And the 200 page report makes grim reading for people planning to go outdoors in the year 2000.

COOK REPORTS

For it predicts that continual high winds will bring about dramatic changes in the way we live our lives.

GONE will be high rise buildings. In their place, low, rounded, wind cheating, 'Smartie' shaped structures will spring up in their thousands.

GONE will be the cars of today. The 'windmobile' of the future will be powered by sails, and instead of wheels it will have big rubber suckers to hold it to the ground.

And **GONE** will be the extravagant fashions of today. Wind-proof clothing specialists will replace fashion boutiques on Britain's gale battered High Streets. Shoes will be heavy, like diving boots, with electro magnetic soles to anchor the wearer down to special metal pavements of the future. And skirts will fall victim to the wind. Instead ladies' legs will be completely covered in tight fitting silver 'Alcan' foil wrap trousers.

THAT'S LIVES

But it's not all bad news for fellas. These new wind resistant pants will have see-through bottoms.

1000 mph gales lash Britain in year 2000

Traffic jams will be replaced by massive queues in the clouds, as thousands of commuters sail to work – on kites.

PROPELLERS

And satellite dishes on the side of houses will be replaced by windmill style propellers, as greedy home owners cash in on the high winds to provide their own free electricity. Wind speeds exceeding 10,000, miles an hour will provide home owners with 25,000 volts per amp of power – enough electricity to boil a kettle the size of Wembley Stadium – *every three seconds*.

COCKPITS

In the field of sport, world records will be shattered. For example, in the long jump school children will be able to leap 300 yards with ease – more than ten times the present world best. But sports lessons will have to be cancelled, as with their next jump the wind could change and they could be thrown up to a quarter of a mile backwards, through a window or onto a busy main road.

WINGS

Outdoor pop concerts would be impossible, for thousands of Genesis, Dire Straits and Tina Turner fans would be blown away, quite literally. And the noise of a concert being held in Wembley Stadium would wake people up – *in Australia*.

BEATLES

Structural damage caused by high winds will create scenes reminiscent of the Blitz. Insurance claims for

An artist's impression of wind swept Britain in the year 2000.

storm damaged property are already up by 200 per cent on previous figures. But according to the experts, by the year 2000, buildings will collapse like matchwood models, and catch fire spontaneously. Buses and trains will be tossed around like litter, and Britain's streets will be up to ten feet deep in dead bodies. Meanwhile, tidal waves the size of Mount Everest will obliterate our coastline, reducing our island to a blood-soaked paddy field of death.

MONKEES

Plans are already afoot to protect the Royal Family from changes in the weather. By the year 2000 they will have moved into a new, specially designed 'wind sausage' shaped palace. The

new silver, sausage shaped building will swivel on an axis ensuring that it always faces into the wind. And should the weather worsen, the building will turn into a rocket to fly the Royals safely to Mars.

BANANA SPLITS

But the most shocking conclusion arrived at by Government experts is that by the year 2000 humans may no longer rule the world! Man could be forced to take a back seat as evolutionary forces change the face of the planet. Man will have shrunk to a mere 8 inches tall, and could be playing second fiddle to tortoises, which – unaffected by wind due to their streamlined shell – will rule the world.

The new 'sausage shaped' Royal palace of the future nearing completion yesterday.

Royal 'bit on the side' set to plumb new depths ~fear

THE WORLD of Royal watching was last night in shock after claims that the marriage of Prince Charles to Camilla Parker-Bowles will spell *disaster* for the long-term future of the monarchy.

In a new study looking into the relationship trends of the Prince of Wales, reserchers found that when he marries a woman, he then takes a mistress one twentieth as attractive.

report

The report from Royality boffins at the University of Birmingham could well take the shine off what was planned to be *'The Wedding of the Century'*.

"The trends were fairly obvious," said the report's author Professor Max Haystacks. "Each time Charles gets married, such as in 1981, he immediately has an affair with a woman 5% as attractive as his wife. If he repeats this behaviour pattern, and I see no reason why he shouldn't, then after marrying the unattractive Camilla Parker-Bowles, he will take a mistress only 5% as attractive as her."

Haystacks ~ report

And he had this stark message. "The figures speak for themselves. By this summer, Charles will be going behind his wife's back with a woman *one four-hundredth* as good looking as Lady Diana. That equates to somebody like Nora Batty or any of the women off the *Ocean Finance* adverts."

But friends of the Prince of Wales moved quickly to denounce the report as nonsense. "This report is nonsense," spluttered humpty-dumpty bloated tit **Nicholas Soames**. "I know

By our
Royal Correspondent
Ingledew Botteril

Charles ~ large ears

his royal Highness very well, and he has no plans to takes a mistress at present,

WOW! ~the '10' wife   **OUCH!** ~the '0.5' lover

be she ugly or attractive." Royal bum-nuzzler Norman St John-Stevas was more circumspect. "I disagree with the findings of the report. His royal Highness

will almost certainly take a mistress, but who is to say she will be 20 times uglier than Camilla?" he told journalists. "The Prince has impeccable taste, and he

may buck the trend and start doing the dirty with a really class bird like Kylie, Abi Titmuss, or Nicole out of the Renault Clio adverts."

Who Will the New Mistress Be?

IF THE REPORT turns out to be true, in 20 years time the Commonwealth could be ruled by an absolute steg. We know who the rider is, but who are the runners in the 2025 Princess of Wales Stakes? We look at the candidates and weigh up which minger is odds on to be Charlie's next darling.

Anne Widdecombe, MP for Maidstone.

Pros: Like all fairytale princesses, Anne is a virgin, so there will be no James Hewitts, Will Carlings, Dodi Fayeds or that heart surgeon bloke coming out of the woodwork. Like her potential lover, she is a conservative, so there are unlikely to be any political spats. And Charles was a pilot in the navy, so her Ark Royal landing deck-style tits will remind him of happy times.

Cons: Charles is never happier than when riding to hounds, and unfortunately, Anne cannot fit on a horse. She is also a member of the Roman Church, and it's illegal for Prince Charleses to marry Catholics. If she became the queen, under a law dating back to Henry VIII, she would have to be beheaded.

Odds: 8/1

Olive out of 'On the Busses', ugly actress.

Pros: As was seen in her famous sit-com role, Olive was very subservient, and unlike Diana, she is unlikely to rock the Royal boat. Her sit-com husband Arthur, alias real life actor Arthur Robbins died in real life, so there are unlikely to be any 'kiss and tell' stories sold to the tabloids.

Cons: Ugly Olive bursts into uncontrollable tears at the drop of a hat. The Royals are famed for their stiff upper lips, and such displays of humanity at the Cenotaph or a state funeral would be

embarrassing to the family. In addition, Olive had a baby in 'On the Buses' who may well be older than Prince William and would therefore have a rightful claim on the throne.

Odds: 5/2

Janet Street-Porter, TV harridan.

Pros: The upper classes like the 'horsey' type of woman, and Ms Street-Porter resembles a carthorse form a 1930s cartoon. As a landowner, Charles has a love of the countryside, a love that Janet shares as former president of the Ramblers Society. However, the fact that she likes to walk through it and he likes to kill things in it may cause friction.

Cons: Street-Porter is famously not a virgin, having been vaginaly penetrated by former Dance Energy presenter Normski. Also, whereas Prince Charles likes culture such as Shakespeare, Michelangelo and Beethoven, Street-Porter prefers modern culture such as Playstations, rap music and modern art.

Odds: 3/2 on

Maureen out of 'Driving School', comedy driver.

Pros: Loveable halfwit Maureen is a lethal menace behind the wheel of a car, but as a member of the Royal household, she would be chauffeur driven everywhere. She is also Welsh, which means that as the Princess of Wales she would be able to talk to her subjects in their made-up vowel-free language.

Cons: Unfortunately for Prince Charles, Maureen is not posh enough to carry off the role as Royal courtesan. At a banquet, it is likely that she would embarrass the Royal family by eating her soup with a pudding spoon. She also works nights cleaning the

Queen of the Jungle!

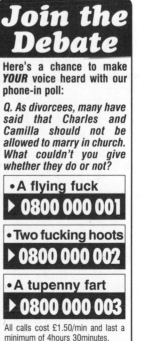

A SKELETON in the Parker-Bowles's family cupboard is once again threatening to derail the Prince of Wales's ill-fated marriage plans. Because according to a leaked document, Camilla Parker Bowles, the future Queen of England is descended from a family of hairy monkeys that lived up a tree!

The document, left on a photo-copier at the House of Commons revealed that her ancesters:

- *Dragged their* **knuckles** *across the floor*
- *Used* **leaves** *as rudimentary toilet paper*
- *Engaged in* **casual sex** *in banana trees*
- *Had unsightly* **blue bottoms** *and ate ants*

The anonymous author of the dossier claims that around $1^1/2$ million years ago, Camilla's ancestors lived in small family groups on the plains of central Africa.

berries

In contrast to their present day comfortable lifestyle of Range Rovers, country houses and polo weekends, the Parker-Bowleses of yesteryear lived a harsh nomadic life, foraging for grubs and berries and eating fleas off each other's backs.

EXCLUSIVE!

The dossier goes on to claim that over the millennia, the Parker-Bowleses lost their tails and began walking upright, eventually evolving into human beings around 50,000 years ago.

norrises

According to Palace insiders, Camilla is embarrassed by her humble origins and never talks about her prehistoric roots. A former Sandringham valet told us: "Camilla's family past is an open secret, but she is very touchy about it. No one is allowed to talk about monkeys, and at mealtimes it is strictly forbidden to serve bananas, nuts or PG Tips."

yeagers

Royal expert Dr David Starkey last night foresaw trouble if the marriage went ahead. "There

Camilla as she is today (inset) and an artist's impression of how the Parker-Bowleses would have appeared half a million years ago.

Camilla Parker-Bowles
'Descended from Apes' ~shock

would be many constitutional issues at stake," he told reporters. "Let's face it, blood will out, and I fear that at a Buckingham Palace reception Camilla may climb on the table and start drinking from the spout of the teapot. Either that, or at a variety performance, she may climb out of the Royal Box and start swinging from the lights, throwing handfuls of faeces at the audience."

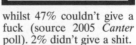

toilets of her local police station, so it is unlikely she would get time off to open the Commonwealth Games or make state visits to Tonga aboard the Royal Yacht Britannia.
Odds: *10/1*

Lily Savage, Scouse comedienne.

Pros: Already popular with the British people, Lily Savage would be a breath of fresh air in Royal circles. Quick witted and humorous, she would bring much needed laughter into staid Royal proceedings. Her trademark sharp tongue would also serve her well in her new role as stepmum to the young princes.

Cons: With her working class scouse background, Lily all too often falls into the trap of using four-letter words which could be very embarrassing - with Royals, their blood is blue, not their language. As an animal lover, she would find it hard to accept Charle's love of bloodsports. Another possible obstacle to an affair with Charles might be her possession of male genitals.
Odds: *7/2*

Jackie Stallone, actor's mum.

Pros: Part of a high-profile Royal's job is meeting famous people, a task which could be daunting to someone unused to living in the limelight. Fortunately, Jackie has rubbed shoulders with the famous, having met such A-list stars as her son Rocky Balboa, his ex-wife Brigitte Neilson, John McCrirrick, Lisa l'Anson and Bez.

Cons: Her habit of reading the future by looking at the lines on people's bottoms may not be seen by some as a fitting pursuit for a monarch. If Jackie pulled Kofi Annan's trousers down at a UN reception and started peering at his arse, saying he's going on a long journey, it could lead to a serious diplomatic incident. **Odds:** *7/3*

Public 'Split' over Royal Marriage

A NEW OPINION poll has showed that the public is as deeply divided as ever over the proposed marriage of Prince Charles to his long term lover Camilla Parker-Bowles.

According to the survey, when asked whether the Divorced Prince should be allowed to marry his mistress, 51% of people said they couldn't give a toss,

EXCLUSIVE!

whilst 47% couldn't give a fuck (source 2005 *Canter* poll). 2% didn't give a shit.

Here's what the nation thought

Q. Should Camilla Parker-Bowles be allowed to use the title 'Her Royal Highness?'

- Couldn't give a shit **40%**
- Fuck knows **23%**
- Who gives a fuck? **35%**
- Don't give a fucking monkey's **2%**

Q. In the light of his marriage, should Charles give up his claim to the throne in favour of Prince William?

- Couldn't give a toss **72%**
- Frigged if I care **23%**
- Don't give a widdle **3%**
- Don't give a wank **2%**

Join the Debate

Here's a chance to make *YOUR* voice heard with our phone-in poll:

Q. As divorcees, many have said that Charles and Camilla should not be allowed to marry in church. What couldn't you give whether they do or not?

- **A flying fuck**
  ▶ **0800 000 001**

- **Two fucking hoots**
  ▶ **0800 000 002**

- **A tupenny fart**
  ▶ **0800 000 003**

All calls cost £1.50/min and last a minimum of 4hours 30minutes.

HOLLYWOOD'S BLAZE TRAILER HANGS UP HOSE

Mr Adair in 1959 - the heyday of Hollywood chip pan fires.

A veteran Hollywood fire fighter is to hang up his hose after 50 years extinguishing chip pan fires in the kitchens of the rich and famous.

Blue Adair - third cousin, twice removed of Texan fire fighter Red - has saved many of tinseltown's most glamourous kitchens from serious fire damage. And he is about to tell his remarkable story in book entitled "Danger: Hot Fat in Hollywood".

Book

But the book has already faced criticism. From his home in Workington yesterday Mr Adair denied allegations that he is cashing in on the suffering of celebrities who have experienced traumatic chip pan fires.

Ghoulish

"There will always be a ghoulish element who want to read about these chip pan fires because of the people involved", he admitted. "But the book does not glamourise kitchen fires - it stresses the importance of fire prevention".

Goulash

Here, in exclusive extracts from his book, Mr Adair recalls some of the hair raising adventures he has had battling with the burning chip pans of the stars.

Ratatouille

'The biggest chip pan blaze I ever tackled was at the home of TV detective Frank Cannon. He had lit the flame under a chip pan then went to the toilet. He must of got diarrhoea cos he was on the toilet for ages, and when he got back the fat was on fire.

Rat-a-tat tat

Frank did the worst thing possible and poured water on the flames. The pan exploded into a huge ball of flame. Luckily Frank's mum had given him a fire

'Kitchen fire fighting has had it's chips'

An ordinary simmering chip pan like that on the left spells danger to stars like sixties Tarzan actor Ron Ely, above.

blanket for his the kitchen and I used it to cover the pan. Seconds later the fire was out. But chip pan fires don't always have happy endings.

Knock knock!

I had always been a big fan of Jim Morrison out of The Doors so one day I couldn't believe it when I answered the phone and heard him singing "Come on baby put out my fire". I could hear burning fat behind him, so I knew he wasn't kidding.

Who's there?

Jim's pan had over-heated and caught fire after he'd went to run a bath. I got round there straight away and put it out with a carbon dioxide extinguisher I keep in my boot. When I'd finished Jim asked me if I wanted to use the bath he'd just run to clean myself up.

"No thanks Jim", I said "You have it. I've got an electric shower at home". As I left Jim was heading for the bathroom with a towel under his arm. He never used it. Seconds later he had died in his bath, in Paris, and I was left wondering if I should of said "Yes" to having a bath. Perhaps if I'd did it would of been me that drowneded, not Jim Morrison out of The Doors.

Doctor

The injuries caused by burning fat can be horrific. One day sixties TV actor Ron Ely was filming Tarzan. It was lunchtime and he decided to go home for some fish and chips. But he made a fatal mistake which was to cost him dear.

Dr. Who?

Ron, who was still wearing just his leather under-

pants, didn't dry the chips before dipping them in the pan. As soon as the chips hit the fat he was sprayed with hot, spitting fat.

Blakes 7

When I arrived seconds later I was confronted with a scene of horror. Ron was standing next to the cooker, wiping his arm with a cloth. Luckily for him I am a member of the St Johns Ambulance so I quickly run his arm under the cold tap and put ointment on.

Space 1999

Ron still had red marks on his arm when he went back to work. But instead of covering them with make-up, which might of made them sore, they decided to pretend he had got bit by a lion in the jungle. It was a clever trick, but viewers were unaware that hours earlier Ron had been face to face with a far greater danger than a lion in his kitchen.

Red Dwarf

Few people could of had a more rewarding career than what I've done, and despite the dangers I would of been happy for my sons to follow in my footsteps, if I'd had any. But in recent years celebrity chip pan fires have become fewer and farer between due to the advent of Microchips.

Green Giant

In a way I'm happy, cos microwave and oven chips are much safer. But in a way I'm sad, cos I'll miss all the exciting adventures what I have had'.

© Blue Adair 1997. 'Danger: Hot Fat in Hollywood' is published by Mr Adair himself, priced £2.95 (plus a S.A.E.) from P.O. Box 999, Workington, Cumbria.

CONSUMER GRIPES

I RECENTLY bought a bottle of brown sauce which carried the warning 'Do not use if seal is broken'. As soon as I opened it the seal broke, immediately rendering it unusable. I wonder how many other innocent shoppers, especially pensioners, have fallen for this evil scam.

Drew Peacock, Email

ACCORDING to the manufacturers of Walker's Crisps, the reason their bags often appear to be empty is that the contents often settle in transit. However, in an experiment I recently carried out at work, I managed to get the contents of three bags of smokey bacon crisps into a single empty pack-

et and still got it closed. I then re-enacted the settling phenomenon by violently walking and jumping up and down the corridor for several minutes. When I had finished, there were still just as many crisps in the bag.

John Rostron, Email

YESTERDAY, whilst visiting my local Tesco, I saw a sign which read 'Mum of the Year 2006. Enter Your Mum Now!' I was so disgusted I vomited. Come on Tesco, clean up your act or get out of town.

Ian Bemail, Email
PS. Would this have been better with a photo of the sign?

HAS ANYONE managed to open a Rustlers microwave food packet without the corner coming off in your hand? If they have, then they are a better man than me, which isn't saying much.

Jimmy Graham, Email

Tell Me, Do

Dr Johannes Do, professor of Miscellany at the Hamburg Institute of Facts and Things, answers your queries.

IS THE actress Susan Hampshire related to the cricket team Hampshire?

Mrs Etherington, Maidenhead

Hampshire cricket team play at the Rose Bowl ground in Southampton, and won the County Championship three times in the 1970s. The dyslexic actress Susan Hampshire starred as 'Fleur In the Forsyte Saga'. It is not known whether they are related, although they share the same surname.

I RECENTLY bought a pair of trousers, but when I got them home they had three legs and no gusset. Am I entitled to my money back?

George Hermitage, Epsom

Ever since they were invented by the French Duc de Pantalon in 1654, trousers have usually been made with two legs and one gusset. If you take them back and swap them for another pair, make sure to count the number of legs and gussets before you leave the shop.

RECENTLY saw a crab singing in a cartoon. This got me thinking, can crabs really sing?

Dr J Miller, Cambridge

Crabs are sea-dwelling crustaceans with ten legs. The largest one on record was a Japanese spider crab with a legspan of over eleven feet. Crabs have lungs and gills, although scientists are undecided whether they have a mouth or can sing.

WHILST ON holiday on the Isle of Man I received some fluff in my change. Is this legal tender on the UK mainland?

E George, Harwich

The capital of the Isle of Man is Douglas, which is 75 miles west of Liverpool. Famous people who live on the island include Nigel Mansell, David Icke and Norman Wisdom. Coins which are legal tender on the UK mainland include the 50p, the 10p and the 2p. The largest in size is the £2 coin, which is also the largest in money.

Not Time, Gentlemen, Please! 24-hour pub opening spells last orders for last orders.

TONY Blair's recent decision to legalise round-the-clock pub opening has been greeted with mixed reactions from all sides of society. Whilst people who live close to pubs and the police have slammed the plans, homeless alcoholics, fruit machine manufacturers and binge drinkers have welcomed them. But who is right, and who is lying? We went on the streets and found the public's opinion divided.

...I'M A bus driver and 24-hour drinking will suit me down to the ground. If I can stay in the pub until the early hours, my hangover won't kick in till the following afternoon, well after I've finished my morning shift.
Reg Varnish, Chester

...THANKS to the ban on smoking and the new 24-hour opening hours, I'll be able to stay in the pub drinking all night without being forced to breathe in a load of unhealthy cigarette smoke.
Mrs Potatoes, Cheshunt

...THE NEW opening hours are a farce. I'm a designated driver, so when I go to the pub with my mates I usually make a pint of beer last all evening. With bars open round the clock, it looks like I might potentially have to make a single pint last for the rest of my life.
Terry Sperm, Chessington

...PEOPLE who say that relaxing the licensing laws will lead to an increase in violence should look at the figures. The present opening hours were introduced in 1914, and were followed by four years of fighting in which millions of people were killed,
Audrey Potter, Chelmsford

...TONY Blair has no right to change the law until he proves that he can drink round the clock. When he shows himself capable of spending a solid 24-hours sinking pints, then and only then will I listen to him when he tells me how long I must drink for.
Joe Philpott, Chelsea

...THE NEW opening hours are ridiculous. Each night after closing time at my local, we have a lock-in till 3am. Now that the landlord has applied for a licence to stay open till 4am, I'm not going to get home till breakfast time. Just

when does Mr Blair expect me to get some sleep?
Horace Guyzance, Chertsey

...WHAT sort of saddo feels the need to drink in a pub 24 hours a day? What's wrong with a carrier bag full of cans and a bench in the town centre?
Jimmy Jackson, Cheltenham

...NOW that my local is open 24-7, I have no need to ever go home again. I'm going to sell my house and spend the money on lager and peanuts.
Karl Lauder, Chesterfield

...HEAVY drinking never did anyone any harm. My grandfather drank 80 pints a day from the age of twelve, and he was killed on his hundredth birthday when he was trampled to death by a Salvation Army Temperance band.
Phoebe Booth, Cheddar

...IRELAND and Scotland already have round-the-clock drinking, and you don't see drunk people on the streets there.
Marjorie Kerbishley, Cheadle Hulme

...AS usual, the moaning minnies say they are concerned about the effects of all day boozing. But they should bear in mind that just because a pub is open round the clock, its customers won't necessarily be drinking 24-hours-a-day. They'll have to go out for a piss or a fight occasionally, or to lean out of the door and vomit into the street.
Frank Phonebook, Cheam

...IT'S all very well having 24-hour drinking, but what happens when the clocks go back in the Autumn and there's a 25-hour day? Do the police propose to arrest everyone who's in the pub at 2am? As usual, the government really hasn't thought this through.
Percy Tarrant, Cheng-Tu-Fu

Deep Sea Diva

OPERA STAR Dame Kiri Ti Kanawa won't be hitting the high 'C's this Christmas, she'll be hitting the high seas. That's because the 58-year-old singer has decided to swap choral music for coral reefs, and is intending to eat her Christmas dinner on the sea bed, over 7 miles beneath the surface of the Pacific Ocean.

She told us: "On December 25th, I'm going to descend in a bathysphere of my own design to the bottom of the Marianas Oceanic Trench, over 35,000ft under the sea." She intends to stay on the ocean floor for just 20 minutes, and is confident that her craft is capable of withstanding the immense water pressure which occurs at such a depth. "I can't wait to look out of my porthole when I get to the bottom," she continued. "With a bit of luck I may catch a glimpse of a giant squid or an echinoderm."

risks

The trip is not without its risks, however. On Boxing Day 1934, soprano Dame Nellie Melba was crushed to death when a steel cable connecting her cast iron submersible to a surface craft snapped and she sank 4,000ft to the seabed off Bermuda. And in more recent times, Spanish singer Monserrat Caballe narrowly escaped drowning on Christmas Day 1986 when her home-made bathyscaph sprang a leak during an attempt to attain a depth of 10,000ft near the Mediterranean island of Ponza.

ARE YOU CLIFF RICHARD'S LOVECHILD?

Without scientific evidence it is often impossible to prove whether or not you are Cliff Richard's lovechild. So here's a special questionnaire that we've designed to enable you to do your very own home test – and discover once and for all whether your mum dropped a Cliff clanger.

But the Peter Pan of Pop has nothing to fear. Because we've declared a special amnesty on all Cliff's love children. And if your answers reveal that you are the son or daughter of Cliff, we'll pay your maintenance in order to protect Britain's best loved pop singer from the harmful publicity this could generate.

FANS

Let's face it. Most of our mothers were fans of Cliff Richard. And even Cliff himself admits to sleeping with at least *one* women. So let's put your mind at rest once and for all with this easy to answer questionnaire.

CHANCE

Anyone can answer, but in order to avoid disappointment we must point out that to stand a reasonable chance of being Cliff's lovechild you must be *younger* than Cliff himself.

1. Which of the following would be your ideal holiday?
(a) A raunchy fortnight with your mates on the Costa del Sol.
(b) A week spent in a quiet cottage in remote Wales or Cornwall.
(c) Touring in the South of France in a big red bus with Una Stubbs and Melvyn Hayes.

2. You're about to catch a train. You nip into WH Smith to buy a book to read during your journey. What sort of book would you choose?
(a) A raunchy paperback, with a partially naked woman on the cover.
(b) An informative book, about gardening, cookery or a subject that interests you.
(c) The Bible.

3. Imagine that you have discovered the Christian faith. You begin to question the commercial exploitation of Christmas, which is, after all, a religious festival celebrating the birth of Christ. What would you do?
(a) Boycott Christmas altogether, refusing to celebrate it in any way.
(b) Shun all commercial aspects of Christmas, and try to get back to the religious basis.
(c) Bring out a crappy Christmas single with sleighbells and a choir singing and a book called 'Christmas with Cliff' featuring yourself on the cover, dressed as Santa Claus.

4. How difficult would you say it is for a rich man to enter into the gates of heaven?
(a) Pretty hard.
(b) Harder than for a camel to pass through the eye of a needle.
(c) Not very hard at all, really.

5. If you were a celebrity, what would be your idea of a sporting day out?
(a) A day at the races with Alex Higgins, sticking a few quid on the gee-gees.
(b) A round of golf with Tarby and Bruce Forsyth, and a few drinks afterwards at the nineteenth.
(c) A game of tennis with Sue Barker.

6. What sort of drugs do you take?
(a) Dope, coke, heroin.. whatever you can get, whenever you can get it.
(b) Parecetamol for your head. Kaolin for your arse. That's about it.
(c) None. Drugs are not for you.

7. What did you think of The Beatles?
(a) Great. They were the best band ever.
(b) Good, but they've been a little over rated in the past.
(c) I thought they were okay until they started taking drugs, getting into weird religions and sleeping with girls.

8. Go and look in the mirror. How would you describe your neck?
(a) Smooth, young looking with soft skin.
(b) A bit aged, with wrinkles, but not exceptional.
(c) Leathery, like a dinosaur's scrotum.

9. If totally unfounded rumours began circulating that you wore a colostomy bag (some ludicrous variations of which involved an alleged incident at Mile End tube station in the sixties), how would you react?
(a) Deny them at every available opportunity, and threaten to sue the perpetrators.
(b) Flatly deny them, and threaten to sue the perpetrators.
(c) Maintain your dignity by refusing to stoop so low as to even acknowledge that such malicious and patently untruthful rumours exist.

10. How do you see yourself in later life?
(a) Married, with kids, a car and a house.
(b) Living with a regular partner, but avoiding the commitments of marriage.
(c) You'll be a batchelor boy, and that's the way you'll stay-ay-ay-ay. Yes, you'll be a batchelor boy, until your dying day.

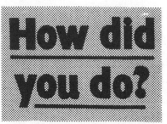

How did you do?

Now tot up your score. Award yourself 1 point for each answer (a), two points for (b), and three points for (c). If your score comes to 27 or more, there exists a strong likelihood that you are Cliff Richard's lovechild, and you may qualify for maintenance payments under our 'Coughing up for Cliff's Clangers' scheme.

The next step is to send us a photograph of yourself, together with a photo of Cliff Richard looking similar. Our judges' decision will be final in all cases. If our judges find in your favour, maintenance payments will be made to you discreetly by post.

Fill in the following form and enclose it with your photograph. Please remember to mark your envelope 'I suspect I may be Cliff's love child'. This competition is not open to former Shadow Jet Harris, his family, friends or relatives. No correspondence will be entered into. The judges decision will be final.

To: Cliff's Clangers, Viz, P.O. Box 1PT, Newcastle upon Tyne NE99 1PT.

I scored _____ in the Cliff questionnaire and therefore have reasonable grounds to suspect that Cliff Richard is my true father. I enclose a photo of me and one of Cliff looking slightly similar. If maintenance payments are made to me I promise not to go to the News of the World.

Signed _____

Address for maintenance payments _____

_____

Who is the Most Marvellous Mills?

WHETHER they're squabbling in the High Court over multi-million pound divorces or releasing new LPs of knockabout piano tunes, **HEATHER MILLS** and **MRS MILLS** are rarely out of the headlines. But which one is the best? It's an argument that is threatening to split the country into two warring factions. Perhaps you side with blonde beauty Heather - who has touched northern hearts with her selfless charity work. Or do you prefer late, overweight piano-thumper Mrs, whose tunes kept cockney knees in a perpetual state of elevation throughout the sixties and seventies?

Now it's time to settle this burgeoning civil war once and for all. Here, we weigh up the pros and cons of the two most famous Millses in the land in order to decide which one is our Mills of Hearts.

Heather v Mrs

Heather kicks off with a good score, as husbands don't come much more celebrity-er than mop-top ex-Fab Paul McCartney. Lovely Heather bagged herself the lonely-hearted Beatle after the two met at a landmine charity awards ceremony and she was instantly swept off her feet. Despite his Beatlemania heyday being way in the past, Macca still makes headline news wherever he goes.

8 — Celebrity Husbands — 5

Mrs gets off to a bad start as her husband of forty years, Mr Mills, failed to achieve any celebrity status of note during his life. The closest he got was in 1964, when he gained brief notoriety after he was accused of stealing a typewriter ribbon from the offices where he was employed as a lift attendant, and his picture appeared in the court section of the local paper. A low scoring round.

Before rising to fame as a charity fundraiser, vegetarian campaigner and writer, Heather Mills enjoyed a moderately successful career as a glamour model. She famously appeared in *Die Freuden der Liebe*, a prestigious German educational manual on human relationships, in which she was pictured smearing herself and her male co-educator with whipped cream and baby oil.

9 — Modelling Career — 5

Because of her busy recording schedule, Mrs Mills had no time to cultivate a career as a professional model. However, it is not out of the question that Mr Mills could have snapped the odd saucy polaroid of his wife reclining over the kitchen work surface wearing nothing but stockings and suspenders and a sequined eye-mask. These photos could possibly later have appeared in the Readers Wives section of *Razzle* or *Parade*, netting her lucky hubby a cool fiver for each one.

Heather is a long time campaigner for animal rights and a dedicated vegan, shunning sausages, bacon, pork pies and Scotch eggs. In fact, she is so committed to her beliefs that she even avoids dairy produce, and when she posed for a series of German educational photographs, she insisted that the cream she smeared around her breasts and nipples was synthetic.

9 — Animal Rights/ Vegetarianism — 3

Mrs cared little for the plight of animals, and her trademark honky-tonk upright piano boasted 88 ivory keys. What's more, so heavy was her playing style, (each arm weighing in at a hefty 14 stone) that the whole keyboard had to be replaced once a week. By the end of her career, it is estimated that Mrs had pounded her way through the tusks of no less than 6,000 elephants.

At times, the British media has treated publicity-shy Heather worse than a murderer or a paedophile, so it came as no surprise when the strain became too much for her. She eventually snapped during a live GMTV interview in 2007, when her voice went all high, a bit like Stan Laurel.

7 — TV Breakdowns — 3

A low scoring round for Mrs, whose sunny disposition was her trademark. Even when talking about upsetting experiences, such as her time in the blitz, her husband's death in the mechanism of an escalator, or the time she was mauled by a grizzly bear, she always kept smiling.

Heather is no stranger to the divorce courts. In 1991 she got divorced from her first husband, dishwasher salesman Alfie Karmal. Her second divorce, from ex-Beatle Paul McCartney, has so far lasted longer than her first marriage, but with one failed marriage under her belt and one well on the way, it's another good round for Heather.

8 — Divorces — 4

A low-scoring round once again, as Mrs and Mr Mills remained happily married until death did them part. In their 4 decades together, their marriage went through only 1 rocky patch. The couple took a holiday in Canada to 'work things out', during which Mrs Mills was trampled by a moose. But the unfortunate event had the effect of bringing them closer together.

Heather can turn her hand to many things, such as charity campaigning, being a Good Will Ambassador for the United Nations and posing with lipstick round her nipples for German grumble mags. However, one thing she has never tried is prospecting for gold. "Swinging a pick-axe all day, then standing in a freezing cold mountain stream sieving grit for hours on end has never really appealed to me," she told NBC's Regus Phitbin. A low scoring round for Lady Macca.

2 — Gold Digging — 9

As a young girl, Mrs had dreams of striking it rich in the American Gold Rush. She crossed the Atlantic and spent four months panning for gold in the Sierra Madre. However, on her way to the Assay Office she was ambushed by leering Mexican bandits who stole her bag of nuggets and her mule. To make matters worse, on the way back to her shack she was attacked by a cougar. She didn't fancy starting again from scratch and returned to London to learn the piano instead.

Oh dear! An accident several years ago saw the model lose a leg, and whilst she clearly deserves full marks for the fantastic way she copes with her disability and the wonderful example she sets for others in the same situation, unfortunately the wording of this round is unambiguous. Rules are rules, so she nets a dismal 0 out of 10.

0 — Full Complement of Two Legs — 15

Mrs scores 10 out of 10 for her two legs. She also bags an extra 5 points, because she was actually born with THREE legs! Her third limb, a full sized rear-facing leg complete with foot, meant that she could play the piano without a stool. But she was very embarrassed about this anatomical curiosity, and always kept it hidden under a long dress.

Oh, dear! A good showing throughout, but brave Heather trips up at the final hurdle. It's a case of close, but no cigar for the soft-hearted northern lovely.

43 — HOW DID THEY DO? — 44

Mrs is Top of the Millses. No doubt she'd be thumping out a rousing victory chorus on an out-of-tune piano surrounded by drunken Pearly Kings, if she were still alive.

A-Z of Jonny W

IN THE DYING SECONDS of the rugby World Cup final, superstar Jonny Wilkinson kicked his way into the annals of sporting history. With his unerring left foot he single-handedly brought the coveted William Webb-Ellis trophy back to Britain, whilst with his scrummy good looks he brought the nation's women to its knees.

THE whole nation has gone *Jonny crazy*. It's like Beatlemania all over again, but with Jonny Wilkinson instead of the Beatles.

We know about the public figure, but how much do we know about the private man behind the mask. Here's a comprehensive A-Z of the man who made Britain Great again.

A is for Ant.
You might think that Jonny's fantastic kick in the dying seconds of the World Cup final would be the most thrilling moment of his life... *but you'd be wrong.* On Radio 4's *Desert Island Discs* programme he told Sue Lawley that the highlight of his life so far was meeting his childhood hero Ant out of Ant and Dec.

B is for Buttocks.
When he was twelve, Johnny developed a skin-rash on his posterior. It was so unusual that a consultant asked him if he'd mind showing his bottom during one of his lectures at a medical school. This early experience of dropping his trousers and showing his arse to a roomful of students led him to take up rugby in later life. *The rest is history!*

C is for Cheese.
Jonny doesn't have a favourite cheese. He thinks they're all equally delicious, especially Stilton.

D is for Dec
You might think that skidding off the A1 and hitting a tree at Leeming Bar would be the low point of Jonny's life... *but you'd be wrong.*

On Radio 4's *Moral Maze* programme he told Michael Buerk that it had been the time he met his least favourite kids' TV presenter, Dec out of Ant and Dec.

E is for Elephants.
Like the mother of Victorian 'Elephant Man' Jonny Merrick, Jonny Wilkinson's mother was trampled by a circus elephant whilst pregnant. However, there the similarity ends. For Wilkinson turned into a handsome rugby hero whose kick in the dying seconds won the World Cup for England, whilst Merrick grew into a blasphemous abomination of nature in a big cap.

F is for Falcons.
Few people realise that Jonny doesn't just play for England. He also occasionally turns out for Newcastle United's rugby team, Newcastle Falcons United.

G is for Gimp.
For several years Jonny has been in a steady relationship, so it's likely that his sexual tastes are probably fairly mainstream. It's unlikely that he ever gets his kicks by donning a tightly-fitting rubber gimp mask and strapping a ball into his mouth whilst a woman in a Gestapo uniform treads on his genitals with spiky stiletto shoes.

H is for History Books.
Jonny's heroic kick in the dying seconds of the world cup meant that the history books had to be re-written. It is estimated that publishers were forced to pulp over six million volumes of history on the Sunday following the match, whilst libraries throughout the world built

huge bonfires of out-of-date history books. Leading historian Eric Hobsbawn has been working round the clock ever since to rewrite his history of the English Civil War to incorporate Jonny's thrilling achievement.

I is for Igloo.
Jonny lives in a large house in Northumberland. However, if he were an eskimo living at the North Pole, he would probably make his home in an igloo - a hemisphercal snow dwelling with a tunnel entrance.

J is for Jam.
When he's not playing rugby, crashing his car or meeting the Queen, Jonny likes nothing more than tucking into his favourite food - jam sandwiches. However, he doesn't like bread, so he eats it straight from the jar... *with his fingers!*

K is for Kick.
Jonny spends 12 hours every day practising kicking balls in his garden. However his next door neighbour Mr Taylor refuses to give them back and is so fed up with them coming over the fence that he bursts them all... *approximately 50,000 rugby balls each year!*

L is for Las Vegas.
With lucrative advertising deals, endorsements and a hefty salary, Jonny is one of the richest men in sport. So rich, in fact, that if he travelled to Las Vegas and bet the lot on a single spin of a roulette wheel, he could win *thirty-two times his stake!* But it's not all good news - if his number didn't come up, *he could lose the lot!*

M is for Mary.
Whilst at nursery school Jonny was cast as a shepherd in the nativity play. On the day of the performance, the little girl who was set to play Mary fell ill with chicken pox, and Jonny was forced to step into the breach. This early experience of cross-dressing led him to take up rugby later in life. *The rest is history!*

Oh Jonny, oh Jonny, Oh... B... E - Hero Wilkinson prepares to take the last minute kick in the dying seconds of the rugby World Cup final which secured the trophy for England. Hurrah!

N is for Nipples.
Jonny suffers from a rare medical condition called octo-papillacy, which means he has an incredible 8 nipples, like a sow. His cruel schoolmates nick-named him "Jonny Eight-tits", but he had the last laugh 20 years later, when his make or break kick in the dying seconds of the final secured the rugby World Cup trophy for England.

O is for OBE.
The Queen was so thrilled by Jonny's kick in the last few

oats! Now here's the definitive

Ilkinson

Re-live the greatest moment of history again and again!

seconds of the World cup final that she immediately awarded him an OBE. The rest of the team who threw the ball to him, but failed to score a single point between them merely got MBEs, the wooden spoons of the honours system.

P is for Pint Glass.
Whilst on a childhood caravanning holiday in Llandudno, 2-year-old Jonny was caught short, but his family had left his potty 200 miles away at home. Thinking quickly, his father saved the day by grabbing the nearest vessel to hand - a beer glass. This early experence of shitting into a pint pot led him to take up rugby in later life. *The rest is history!*

Q is for Quality Street.
As a child, Jonny once choked on a Quality Street whilst out shopping. He was turned upside down and slapped several times on the back by the manager of Woolworths. The sweet was eventually dislodged. It has been estimated that had that shopkeeper not been so swift to act all those years ago, Jonny would have died in infancy and England would have tragically lost the World Cup.

R is for Regis.
Jonny's favourite Regis is Lyme. *"It beats Bognor and Cyril hands down,"* he told Radio 4's *Veg Talk* programme.

S is for Salary.
Jonny's weekly wage playing for Newcastle Falcons is estimated to be over *half a million pounds a year*. That money would buy enough Smints to make a town the size of Reading shit its pants and have to leave them behind a bush.

T is for Teeth.
The golden boy of rugby has an amazing *thirty-two* teeth. He has 8 incisors with which he cuts his

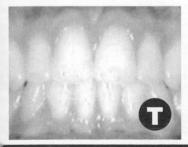

food, 4 canines which he uses for ripping and tearing, as well as 20 assorted molars, premolars and wisdom teeth which Jonny uses for grinding and chewing his prey.

U is for Urine.
Whilst on the Duke of Edinburgh Awards scheme, a teenage Jonny was cast adrift for several days on an open boat. With no fresh water available, he was forced to drink his own urine. This early experience of drinking a pint of piss led him to take up rugby in later life. *The rest is history!*

V is for Vagina.
Like most successful rugby players, it is quite likely that Jonny was born through a vagina. The other possibility is that he was delivered by

Caesarian Section, a type of fannyless birth invented by Julius Caesar.

W is for Wilkinson.
Jonny's surname is shared with a budget hardware store, selling cheap sellotape, paint and buckets. It is doubtful whether the quality of any of their stock matches the quality of Jonny's momentous kick during the final moments of the rugby World Cup final, which secured victory for the British team.

X is for Xylophone.
Although he's probably never done it, it's a fair bet that if push came to shove, Jonny's magic left boot could propel a xylophone through the uprights in the dying seconds of a World Cup final.

THE Collapse of the Berlin Wall... Man setting foot on the Moon... Edmund Hillary conquering Everest... All of them are great moments of history, but all of them pale into insignificance beside Jonny Wilkinson's championship-winning kick in the dying seconds of the rugby World Cup final. It is a moment we could literally never tire of watching.

But now penny-pinching rugby bosses want to charge TV companies £6000 every time they show that clip. £6000? Just to see a man kick a ball between two sticks? Don't make us laugh.

So here's your chance to relive that moment again and again. And it won't cost you a single penny, thanks to this fantastic **FREE** cut-out-and-keep *Jonny Wilkinson World Cup Kick-O-Matic.*

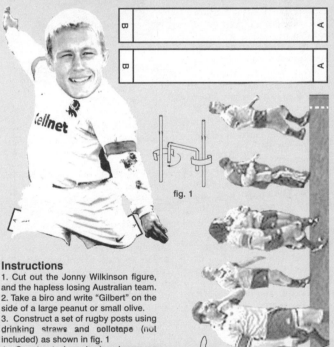

fig. 1

fig. 2

Instructions
1. Cut out the Jonny Wilkinson figure, and the hapless losing Australian team.
2. Take a biro and write "Gilbert" on the side of a large peanut or small olive.
3. Construct a set of rugby posts using drinking straws and sellotape (not included) as shown in fig. 1
4. Construct Jonny's leg loops as shown in fig. 2. Insert your fingers through the leg loops. Then drop the "ball" and flick it to victory again and again, reliving the most exciting moment in the history of the world as many times as you want!

Y is for Yeti.
Jonny has never seen the abominable snowman or yeti, but that doesn't mean he denies its existence. *"I try to keep an open mind about sasquatches, bigfoots and Loch Ness Monsters,"* he told Radio 4's *Poetry Please* programme.

Z is for Zygote.
Jonny started life as a sperm with no legs, just a long, thin, waggy tail. Luckily for British sport he managed to fuse with an ovum in his mother's falopian tubes. Then he grew his trademark legs and scored the drop goal that won the 2003 rugby World Cup for England in the dying seconds of the final.

The rush is on with only 248 shopping days to go! So here's...

20 THINGS YOU NEVER KNEW about CHRISTMAS

With Easter out of the way, once again it't that time of year when we start to think about Christmas. With so many presents to buy, trees to put up and parties to plan, before you know it Christmas is upon us!

But do we ever stop to think what exactly is Christmas? What does it mean? Before you reach for your decorations, why not stop for a moment, and find out.

Here's twenty festive facts you never knew about Christmas.

(1) Christmas, literally translated, means 'Jesus's Birthday Party'. It's a traditional religious celebration of the Birth of Christ, who was born in a manger at Bethlehem almost 2000 years ago.

(2) And he's not the only celebrity who has a birthday on December 25th. He is joined by zany TV comic Kenny Everett.

(3) And Noel Gordon out of Crossroads.

(4) For most people Christmas Day is a time for celebration, but Irish rock start Phil Lynott had no cause to celebrate on Christmas Day 1987. That's because he died on it.

(5) The earliest Christmas decorations known to man are believed to be cave paintings featuring reindeers and a plum pudding which were discovered in underground caves in Arizona, USA, and believed to be over 5 million years old.

(6) The Christmas Card was invented by unknown painter Thomas Merry purely by chance after he had painted a robin sitting on a spade handle. However, the first card was not sold until after his death in 1754 when a friend, Sir Henry Hallmark, realised that people were stupid enough to pay up to £2.50 each for the things which cost next to nothing to produce.

(7) Prince Albert was responsible for the introduction of the Christmas tree to Britain's homes. The eccentric Prince insisted that his gardeners dig up trees and bring them indoors for the winter to protect them

A Christmas tree yesterday.

from the cold. In 1872 a law was passed stating that every household in Britain should take in a tree for a period of 12 days. The law was abolished in 1958, but the tradition lives on to this day.

(8) The traditional children's Christmas gift the 'toy' was not introduced to this country until 1941 when German prisoners of war began to pass their time by carving miniature aeroplanes, cars and soldiers out of wood. Sadly, these toys were never enjoyed by children. They were banned by pioneer safety campaigner Lady Constance Foulds-Wood, because they had sharp metal bits in them.

(9) Norwegians may not be the brightest people in the world, but they are world beaters when it comes to growing Christmas trees. Each year they export over ten million trees to countries like Britain, and still have enough left to build themselves log cabins and houses out of wood.

(10) Towards the end of his career real life Santa Claus Elvis Presley gave away millions of dollars worth of Christmas presents to the poor and needy. Unbeknown to even his closest friends, big hearted Elvis, dressed in a nappy and full

Elvis – Secret Santa.

Santa outfit, actually climbed down people's chimneys on Christmas Eve and left valuable gifts, including limousines and motorcycles, by their bedsides.

(11) 'Carol' is a special kind of Christmas song sung in churches, and on doorsteps by people hoping to make some money.

(12) 'Carol' is also a girl's name. There are many famous Carols, such as Carol Decker out of T'Pau and Countdown's leggy TV maths brainbox Carol Vorderman.

(13) And Carol Barnes the newsreader.

(14) The largest Christmas turkey ever recorded was a 40 stone monster, over ten feet tall, served up at the Lord Mayor's Christmas Dinner at Arbroath Town Hall in 1926. When the oven door was opened after 15 hours of cooking, to put the potatoes in, the giant beast came alive, leapt out of the oven and killed two kitchen staff. The badly scorched bird was eventually wrestled back into the roasting tin by members of the Royal Scots Dragoon Guards pipe band who had been playing nearby. The bird took 6 weeks to cook, and to this day cold turkey sandwiches carved from it's huge carcass are served every lunchtime in the Lord Mayor's chamber.

(15) At Christmas time we go telly bonkers! 95% of the population spend at least 8 hours on Christmas Day watching TV, while only 2 out of every hundred go to church. That's because

church is cold, dull and uninteresting, and vicars are daft.

(16) The word 'pantomime' comes from the Greek 'pantomindrum', meaning 'to swap trousers'. Originally a traditional play performed around the fireside by Victorian families, this later developed into a stage show where out of work TV celebrities perform plays based on well known children's novels.

(17) Australians sit down to enjoy their Christmas dinner in the middle of summer! That's because Australia is six months behind the rest of the world, and Christmas Day falls on June 25th each year.

(18) At Christmas time the average Briton will eat TEN TIMES their body weight in nuts, tangerines, dates and chocolate. And the average family will produce 42 tons of waste, empty bottles, nutshells, wrapping paper etc., which, if they don't tip the binmen, gets spilled all over their back yard.

(19) Going back to number 11, Carol is also the name of a song by Neil Sedaka.

(20) By the year 2000 there'll be no such thing as a white Christmas. Instead, our Christmases will be red! That's because by the year 2000 we'll all be living on Mars.

I CAN'T SEE A FUCKING THING.

NEW SETT BACK FOR GUTHRIE

Guthrie: No stranger to scandal

A scandal-hit West Midlands councillor is once again clinging to his job, this time after a local newspaper published pictures of him apparently performing a sex act on a badger.

"Tipton Chief Romped in Layby" ~ Claim

by our BADGER CORRESPONDENT
RUDOPLH HESS

The publication of the photographs forced Hugo Guthrie, deputy chairman of Tipton Borough Council's Civic Amenities committee, into a humiliating climbdown after he had earlier denied ever having visited Brockodale Woods where the incident took place. He now admits that he did go to the area, known locally as Badger Gobblers' Gulch, after one of his constituents reported a spate of illegal fly tipping in the area.

statement

Reading a prepared statement, Mr Guthrie's solicitor told reporters: "My client has now remembered that he did indeed park his car in the Brockodale Woods layby on Wednesday morning.

"He then went for a short walk amongst the trees to look for evidence of unauthorised rubbish disposal. During this walk he met a badger which he now admits he took back to his car."

manager

He continued: "In the car, the badger appeared to be experiencing some discomfort. Fearing that it was suffering a sudden attack of pancreatitis, Mr Guthrie decided to make an examination of the affected area. Unfortunately, he had for-

gotten his glasses, so he moved his head very close to the badger's lower abdomen."

robbery

"Upon realising the severity of the badger's pancreatitis attack, Coun-cillor Guthrie opened his mouth to gasp in astonishment. It was at this moment that his picture was taken through the car window by a photographer from the Tipton Advertiser newspaper."

account

But this account of events was dismissed as nonsense by Tipton Advertiser editor Max Agincourt. He told us: "Councillor Guthrie is a bare-faced liar. We've got photographs which we couldn't possibly print in a family newspaper which clearly show him performing perverted sex acts on this badger."

adduke

Two years ago, similar circumstances forced Mr Guthrie to resign his chair as assistant treasurer of the Tipton Council Town Twinning sub-committee. Following what he described as a 'moment of madness' during which he climbed a tree in Tipton Park and performed an act of gross indecency with two grey squirrels, he was fined £400 plus costs by local magistrates.

www.double.d

REPORTS that an American woman is planning to open her blouse and reveal her bra on the Internet have led to calls for a tightening up of laws governing the worldwide web.

Mother of eight Draylene Shinz, 49, of Illinois expects over *30 million* computer enthusiasts to log onto her home page www.lady-inabra.com to see her in her brassiere on December 18th.

Popular

Moral watchdogs fear that if her plan proves popular, it may spark off a trend for even harder material on the internet *-including ladies exposing their nude bosoms or even knickers.*

Mid-west Mom expects massive Net interest

And home secretary Jack Straw has been swift to join in the debate.

"If left unchecked, I could envisage a situation where a young man who isn't even old enough to get married could buy a computer, and look at pictures of ladies in bras, whilst he slaps the back of his neck and steam comes out of his collar," he told us. "This must not be allowed to happen."

Escort

Meanwhile Mrs Shinz, speaking from the stoop

Shinz - exposure on internet

of her mobile home in Trashville, Carbondale, was unrepentant.

"It ain't no big thing," she told reporters. "Going on the internet in my bra is the most natural thing in the world. I'm just going like, 'here's my brassiere', that's all. I'm only going to show it for a couple of seconds, anyhow."

Fiesta

And she had harsh words for the people who have complained about her plan.

"They're only sore because their woman ain't showing them no bra at home, and that's for sure. Uh-huuurh."

Fanny's Batter bits

Seventies all-in wrestler turned haulage contractor Kendo Nagasaki has been signed by Steven Speilberg to star in a $750 million Hollywood remake of 'Georgy Girl'. Nagasaki, who will pocket £120 for playing the title role made famous by Lynn Redgrave, has vowed to keep his trademark ninja-style mask on throughout the film.

Bishop of York's Para Drop was Kids' Stuff

ARCHBISHOP OF DANGER

IN A DAREDEVIL STUNT, Archbishop of York Dr John Sentamu recently hit the headlines around the world when he made a Red Devil-style parachute jump. The publicity-hungry cleric leapt from 12,500 feet, free-falling for 45 seconds before opening his chute and performing a textbook landing at RAF Langar in Nottinghamshire. In the process, the 59-year-old priest won widespread respect and raised over £50,000 for military charities.

But Sentamu's action-packed airborne escapade didn't go down well with everyone. In an outspoken sermon, Archbishop of Canturbury Dr Rowan Williams later branded the stunt 'kids' stuff'. "It really was quite a tame thing to do when you compare it to some of the extreme stuff I've got up to," he told a packed Westminster Abbey last Sunday. "It may have looked exciting to the uninitiated, but to someone who lives life at the pace that I do, a tandem parachute jump really is run of the mill stuff."

LID

And now in a new book, *Archbishop of Danger - Under the Cassock*, Williams lifts the lid on his secret, thrill-seeking life. Here, in a series of exclusive extracts, we reveal the death-defying exploits that will astound a public who know him only as a mild-mannered man of God...

MARKS

With his mitre, crook and ceremonial vestments, the Archbishop of Canterbury is a familiar sight at ceremonial events, state funerals

VALLEY OF THE SHADOW OF DEATH: Williams planned suicidal jet-bike leap.

and remembrance services. But few people would expect to see him in a set of garish motorcycle leathers and a cape, sitting astride a rocket-powered motorbike at the base of a 500-foot ramp.

"A couple of years ago, I had this fantastic idea of jumping over the Grand Canyon on a rocket-powered bike. It was a suicidal stunt, I knew that if anything went wrong, I faced a two mile drop to certain doom. Even if it went right I still had to face a 250mph landing on the opposite, rubble-strewn bank. So I prayed to God to ask if it was it was a good idea. He told me that I must be completely crazy, but I just laughed, because that's the kind of bishop I am.

SPENCER

The next day, I found myself on the edge of the Grand Canyon astride 2000 horse powers of jet-bike. Sitting on the launchpad, I ran through a mental checklist of 'what ifs'. What if the engine exploded? What if a gust of wind blew me away from the landing strip and smashed me into the canyon face at the speed of sound? What if the tyres disintegrated under the 20 G-force acceleration up the nearly vertical ramp? I was nervous, I'll admit. But then I remembered the Eleventh Commandment, one I had written myself... Thou Shalt Not Chicken Out!

The countdown had begun. With five seconds to go I revved my engines to maximum thrust and

"The sense of latent power between my legs was intoxicating"

got ready to dial in the afterburner.

LANCE

The sense of latent power between my legs was intoxicating as the bike strained against the brakes like a wild stallion. All thoughts of danger left my mind and my only thought was of piloting that bike across the canyon and into the record books. Everything was in slow motion now, and despite the deafening banshee scream of the engines all I could hear was the steady beat of my heart.

It was now or never. My thumb was hovering over the launch button on the handlebars, but then I suddenly remembered that I had promised to judge a home-made jam competition that afternoon for the Canterbury Women's Institute. Reluctantly I aborted the jump. Had I made that leap, it would have been the greatest stunt ever pulled by a bishop.

Later that day, I awarded first prize in the jam competition to a jar of particularly-nice seedless fruits of the forest preserve. I couldn't help thinking that if the women in the audience had seen me just hours before, sitting in my leathers astride my rocket-powered jet-bike on the brink of a canyon of death, then they wouldn't keep going on about how great the Archbishop of York is."

WATERFALL STUNT WAS LEAP OF FAITH

You might expect your local bishop to occasionally go over the Ten Commandments in his pulpit, but chances are you'd be amazed to discover that he was planning to go over the world's

FALL FROM GRACE: Bishop planned barrel drop.

highest waterfall... in a barrel. But that's just what Dr Williams decided to do to celebrate his appointment as Archbishop of Wales.

PIKE

"When you think about going over waterfalls in a barrel, you automatically think of Niagara Falls. But Niagara is only about 170 feet high - the sort of drop that a soft, show-off bishop like Dr John Sentamu might do. I set my sights a bit higher - over TWENTY TIMES higher, in fact! And that's how I found myself sitting in a beer barrel, bobbing about in the foaming rapids at the top of the Angel Falls in Venezuela. The water was so rough that I had to fasten my bishop's hat on with an elastic band.

This truly was a leap of faith of every sense of the word. Nobody had ever made it down those falls alive, and even if I did survive the drop there was every chance I'd get bitten to death by the man-eating piranhas which gather in the whirlpool which swirls at their base. As my barrel drifted inexorably towards the overhang, my speed increasing with every second as the powerful current dragged me to my inevitable fate, the adrenalin rush was incredible. My barrel was now just inches from the edge. The thundering roar of millions of gallons of raging water filled my senses, I felt the boiling spray stinging my

Shocking Verdict from

DARE ANGEL: Archbishop of Canterbury Dr Rowan Williams with his carefully-teased owl-like eyebrows yesterday.

CRATER LOVE HATH NO MAN: Williams fancied chances in volcano shot.

bishop's cheeks. I was just seconds away from almost certain death, yet I had never felt so alive.

WILSON

Then, as my barrel teetered on the brink and I was looking down into the 3,000 foot foaming maelstrom beneath me, it suddenly occurred to me that I hadn't written the introduction to my weekly parish newsletter yet. I knew that it was going to the printers that afternoon, so I had no choice but to turn back. Disconsolately, I paddled my way to the bank and clambered out of my barrel and onto the grass. It was such a disappointment. Had I been able to carry my plan through, it would certainly been the bravest thing attempted by a suffragan bishop of the Anglican Communion ever. It would definitely have put certain other so-called 'stunts', carried out by certain other bishops who shall remain nameless, well and truly in the shade, I can tell you."

FIRE & BRIMSTONE HELD NO FEAR FOR BISHOP

Everyone's heard the old religious joke "Tell the Canon he's fired." But, when Dr Williams was appointed Canon of Christchurch College, Oxford, it was very quip that gave him the idea for his most audacious stunt yet.

MACMILLAN

"The prospect of being shot from a giant gun had always appealed to my daredevil side. The only problem was, whilst it might have been dangerous enough for a lesser, headline-hungry bishop, it was simply too tame and wussy for me! So when I heard that Mauna Loa, the world's most dangerous volcano, had erupted and was belching white hot lather, spume and molten rock high into the Hawaii sky, I immediately packed my toothbrush, crash helmet and giant cannon.

HEATH

My plan was to douse myself with petrol before being blasted right over the erupting volcano. The intense heat from the white hot magma over which I was flying would ignite the petrol, turning me into a spectacular human fireball. If everything went according to plan, I was to land in a child's paddling pool on the other side of the crater. This would hopefully cushion my landing and extinguish the flames. However, I knew I was taking a huge risk. If my trajectory was anything other than inch perfect, I would be dashed to death on the spiky pumice and then burnt to a crisp before help could arrive.

Everybody told me it couldn't be done, but then I remembered our Lord Jesus Christ. When they told him it was impossible to walk on water, he didn't bottle out. He stepped up to the plate and proved them wrong. Now it was my turn!

And so I found myself soaked in petrol, sitting on top of half a ton of TNT at the bottom of a 200-foot supergun barrel. Ahead of me, all I could see was a small, blue circle of sky. In those few, final seconds before giving the order to fire, all my senses were heightened. The smell of petrol combined with the sulphurous gases, making a heady, high-octane cocktail of danger and excitement. Every nerve and sinew in my body tingled with the anticipation of what I was about to do. I braced myself against the dynamite as I heard my choirmaster, Mr Posner, strike the

"I was just seconds away from almost certain death, yet I had never felt so alive"

match to light the fuse. I held out my fists in front of me as I readied myself, waiting for the inevitable shockwave from the massive detonation beneath my feet.

It was then that a thought hit me. I had to get back to Wales, where I was expected at my local radio studios to record my popular 2-minute 'Time for a Think' spot, which went out on BBC Radio Monmouth every Sunday between the morning shipping forecast and the farming outlook programme. It was a real sickener, but I had no option but to cancel the firing and get to the airport. As I climbed out of the emergency escape hatch, I wistfully reflected on what my listeners would have thought if they knew that I had almost been fired from a cannon over an erupting volcano, whilst doused with petrol, into a shallow paddling pool of water."

HEAVENLY JUMP WAS HIGH RISK STRATEGY

Bishops are used to wrestling with complex moral quandaries, ecumenical problems and theological conundrums. But few can claim to have wrestled a tiger... 5 miles above the earth!

OFFICE

"I'd given up defying death for lent, but now it was Easter Saturday and my danger fast was at an end. Forty days build up of adrenalin was coursing through my veins; I knew I had to do something to slake my ravenous thirst for risk. I dreamed up my most extreme stunt yet. We took off in a light aircraft and flew to a height of 25,000 feet. In the plane, my verger strapped me into a straitjacket before taking my parachute and fastening it onto a hungry tiger. After this, he got a hunting knife and locked it into a safe. The stunt was to go like this... once we were over the drop-zone, the tiger would be pushed out of the door, followed at 10-second intervals by the safe, and then by me. My task was to somehow escape from the straitjacket, then catch up with the safe. After cracking it and getting the knife, I had to catch up with the tiger, wrestle it and stab it to death. Only then could I remove the parachute from its back, put it on and pull the ripcord. It was a risky plan; if I got a single part of it wrong I'd end up spread across the landing site like so much strawberry jam.

The door opened and the cold rush of wind hit the cabin. The green drop light went on and my verger pushed the snarling tiger out of the plane. I watched as the doomed beast hurtled towards the ground, quickly becoming a tiny dark speck as it thrashed its razor-sharp claws wildly. Ten seconds later, the safe followed it out of the door. As it tumbled away from the plane, I knew that locked within its six-inch thick steel shell was my only hope of getting out of this alive. The verger turned to me, on his face was etched a rictus of concern. "You know, Dr Williams," he shouted across the boom of the engines and the roar of the wind. "It's not too late to back out."

RELATIONS

But standing there in the doorway of that plane, strapped in a straitjacket five miles in the air, with my parachute on a tiger thousands of feet below me, I knew there was no turning back. I had to go through with this. I was just about to jump when the pilot received a message from my secretary at Westminster Abbey, Mrs Simmonds. Apparently, the gift shop had run out of souvenir pencil sharpeners and I'd accidentally come out with the key to the stock cupboard in my pocket.

GRACIE

Reluctantly, I had no choice but to call off the stunt at the eleventh hour. As the plane landed and I stepped out onto the airfield, I really regretted missing yet another opportunity to cock a snook at the grim reaper. Archbishop Hey-everybody-look-at-me Sentamu may think he's made of the right stuff, but until he has stabbed a tiger whilst hurtling towards certain death at 150mph, he can keep his big mouth shut."

NEXT WEEK: *How I was about to do a 20-mile bungee-jump from the edge of space into a shark-infested swimming pool filled with cyanide and poisonous snakes, until I remembered I had left the font running.*

Extracts from *Archbishop of Danger - Under the Cassock* by Dr Rowan Williams (Poached Egg Books, £8.99).

Track to the future!

Passengers using the Channel Tunnel could soon find themselves driving out the other end into Napoleonic times if an ambitious scheme to reverse the tunnel's fortunes goes ahead.

With owners Euro Tunnel massively in debt the tunnel is in serious danger of collapse. And bosses believe the only way to save it is by spending even more cash, this time converting it into a giant time machine.

Rail

The advantages of a Time Tunnel rail link between Britain and France would be enormous.

* **TOURISTS** could choose which period of history they would like to arrive in. The French could visit Victorian England, or we could visit France in Norman times and see the famous Buyer Tapestry being painted.

* **TRUCK** drivers could make up lost time by arriving at their destination before they had even left home. And food produce need never go to waste because of road hold ups. If a lorry load of milk went off, the driver could simply go back in time to when it was fresh. Or forwards until its cheese.

* **CONTROVERSY** over veal exports would be resolved. With time tunnel technology lorries packed with baby veals could simply go forward in time until the veals were fully grown, thus keeping protestors happy.

* **AND** lorry drivers could avoid French farmers setting fire to their sheep by simply going back through time until before fire had been invented.

Liar

Unlike the original project the Time Tunnel could easily be finished on schedule. Within a year. Or even a week. For no matter how long it takes to build, once it is finished engineers can simply bring it back through time to when it was supposed to be ready.

A tunnel in time saves line

Obviously there would be a down side too. Steps would have to be taken to ensure that Germans could not slip through the tunnel and go back in time to try and win the war again. And scientists meanwhile fear that a 'paradox' could occur if time travellers were to break the 'space/time continuum'. This would result in the room shaking too and fro, and sparks flying around everywhere. Meanwhile safety chiefs have expressed concern that tourists travelling too far back in time could be eaten by dinosaurs.

Lair

But the biggest hurdle appears to be technical. Euro engineers will have to come up with a reliable Time Tunnel capable of working smoothly, and not going wrong every week like the one in the sixties television series. And that, says Euro Tunnel Co Chairman Alistair Morton, will cost a lot of money.

Lira

"At this stage I think we're looking at around ten, possibly twenty, trillion, zillion, squillion or even phillion pounds. More money than there is in the whole world, probably. But there's no risk to investors whatsoever. Because if at the end of the day the time tunnel doesn't make a profit, we can simply go back in time to before the tunnel was built, scrap the whole idea, and give everyone their money back".

Singer Plastic Bertrand prepares to officially open the Channel Tunnel less than a year ago. But in its first 12 months the world's most ambitious civil engineering project has sprung a huge financial leak, with current debts of over £8 billion.

I *do* want to go to Choloca! Actor Rodney Bewes (left) and fellow future time traveller John Noakes

Win a weekend travelling in time!

We're giving away a pair of tickets for the opening day of the Euro Time Tunnel. Simply write and tell us where YOU would go for a weekend break in time, travelling either backwards or forwards to the place and date of your choice. The possibilities are endless.

Write and describe your dream holiday in time. Try keeping it reasonably short, typed if at all possible, and send your letters to: Time Tunnel Competition, Viz, P.O. Box 1PT, Newcastle upon Tyne NE99 1PT. We'll print the winning entry in the next issue, plus a special report on the winner's dream holiday, providing they stop in November this year on their way back through time, and tell us what it was like.

Past stars to go back to past in the future

We asked a few celebrities who were available at short notice where and when **THEY** will be heading when the Time Tunnel opens for business. Former Likely Lad **Rodney Bewes** had no doubts. "I'd go back to the late sixties when my team Chelsea weren't crap, and before I'd done the Basil Brush show", he told us.

Pesetas

But Rodney may find one player missing from the sixties Chelsea team. Former goalkeeper **Peter Bonetti** is planning a time trip of his own. "I'd go to Mexico in June 1970", he told us. "And try not to let those three goals in for England against West Germany that cost us our place in the World Cup finals".

Zlotys

But former Blue Peter presenter **John Noakes** wasn't looking backwards. He was looking forward to travelling forwards in time. "I'd like to travel forwards from when I was in the seventies, but only as far as the eighties, which is backwards from here. That way I would still be a Blue Peter presenter instead of John Leslie, and I could have got to shag Catherine Zeta Jones. Or rather, I *will* have *get* to of shagged her. Sort of thing".

Catherine Zeta Jones in a new bra yesterday

Portrait of EVIL!

ANDY McBride knows only too well the horrors commited by the world's most depraved dictator. For Andy, an 18 year old trainee shoe shop assistant, was a member of the crack Sea Cadets during the 1990 Gulf War, and attended weekly training sessions in a church hall near his home in Buxton. Now, in these extracts from his bombshell new book, we expose the true terror of Saddam Hussein's evil reign.

' A massive military convoy rumbles through the streets of Baghdad towards the national TV station. Tanks surround the building and armed guards storm inside. *But this is not a military coup.*

On a whim Saddam Hussein, the self-styled butcher of Baghdad, has decided that tonight he is going to appear on 'Al Gamani Generihad', Iraqi TV's version of the Generation Game.

Spaghetti

Not surprisingly Hussein wins every game. A terrified judge awards him ten out of ten for making spaghetti, even though his soggy lump of dough is stuck to his shoes. His folded table napkin looks more like a dead duck than a swan, and in the next round he ends up on his backside attempting to dance the Lambada.

Best

But, surprise surprise, at the end of the game Saddam Hussein is the winner. In the control room nervous TV producers mop sweat from their brows. *All is going well until - at the climax of the show - a nervous Saddam forgets one item from the conveyor belt... a sandwich toaster.* His face floods with rage.

Great

Minutes later the show's host Jimrihim 'Nick Nick' al Davidson, his assistant, seven other contestants, plus 250 staff and technicians at the television centre are all dead - *slaughtered in a warped act of bloody revenge exacted by the world's most evil man'.*

Later in his book Andy gives a spine chilling insight into life - and death - in the torture chambers beneath Saddam's Presidential Palace.

'Ahmed Salih ran a small hairdressing salon in a fashionable area of Baghdad. One day he received a call summoning him to the Presidential

Palace. Saddam had been watching telly again, and after seeing some seventies repeats on Iraqi Gold he decided he wanted a moustache like Jason King.

It was an unusual request, but one which Ahmed dare not refuse.

Eastern

The barber's hands trembled as they trimmed the tyrants trademark black moustache. When he was finished there was a nervous silence as Saddam stared sternly into the mirror, then suddenly his face beamed with delight. The moustache was perfect.

In a fit of generosity the mad mullah gave the barber a million pound tip. But the hapless hairdresser never got to spend it.

Escape

The following day Saddam heard that in 1973 Jason King actor Peter Wingard had been convicted of a sex offence with a crane driver in a bus station toilet in Glousestershire. He exploded with rage.

Scenes similar to this are common place in the labyrinth of torture chambers beneath evil Saddam's Presidential Palace.

That night Ahmed Salih was dragged from his bed and taken to the notorious torture chambers beneath Saddam's palace where he was chained to a dungeon wall and left there - *until his beard was two feet long and his trousers were all raggedy at the bottoms.*

The Butcher of Baghdad weilds a gun as he prepares to embark on yet another orgy of death, yesterday.

The terrifying truth behind the nightmare of the horror of the DICTATOR of DEATH!

Then he was stretched on a rack until his body was 15 metres in length.

Suprendo

Ahmed then pleaded for mercy as he was pushed into an iron maiden. But his cries were in vain. *The sharp metal spikes glistened as the heavy door was slammed shut.*

Balls of Fire

Somehow the unfortunate barber was still alive when the door was opened. For a moment it seemed he had survived his ordeal - until the guards gave him a drink. *Suddenly water began to spurt out of tiny holes which riddled his entire body.*

Chariots of Fire

Saddam is obsessed with security. Even his own government ministers are blindfolded before they meet their President, then they are shot immediately afterwards. Evil Saddam then breaks open their skulls with a solid gold teaspoon, before dipping real soldiers - *terrified teenage conscripts* - into their heads and feasting on their still-warm brains.

Local Hero

But as their grey matter churns about in Saddam's madcap stomach, their ordeal is far from over. For the instant his hapless victims emerge from Saddam's deranged rectum, they are scooped up, blindfolded, and shot again.

Memphis Belle

Saddam then invites the ministers wives to a banquet to feast on the twice shot dead shit remains of their husbands. After the feast Saddam jumps out of a giant cake and guns down all the guests. *Their lifeless bodies are then* liquidised before the power hungry dictator mixes them with strawberry Nesquick and drinks them through a giant straw.

Memphis Slim

So mistrusting is Saddam of his faeces the following evening all of his nocturnal ablutions are rounded up in the dead of night, and folded into a souffle which is then cooked at high temperature. After thirty minutes of agony the twice shot, eaten, liquidised, twice shitted, drank and pissed out souffle finally collapses when evil Saddam opens the oven door allowing cold air to rush in.

Britain's office workers are at it 9 till 5

ON THE JOB!

Britain may be down in the Euro-dumps as far as our economy is concerned – but there's still one thing that we are best at.

A recent survey shows that we come top of the table for *office hanky panky*.

SIZZLING

The sizzling survey compared the sexual habits of office workers throughout the EEC. And the results show clearly that when it comes to *bonking the boss* the British are best!

STAGGERING

A staggering *12 per cent* of office workers in Britain admit to having an affair with one or more of their colleagues, compared to only 11 per cent in France, and a measly 10 per cent of dismal Deutchlanders.

SAUCY

And fortunate female office workers in Britain are *three times* more likely to be sexually harassed by saucy senior male colleagues than their European counterparts.

And if any more evidence was needed to prove that Britain's office workers are the sexiest in Europe, the incidence of rape, in the workplace is a sizzling 7 per cent higher in the UK than in other countries. A statistic that even the steamy Swedes, amorous French and randy Italians cannot match.

BONKING

Yes, Britain is *officially* office bonking bonkers, so we've organised our own survey to find out just how many of you office workers out there are *at it*, and exactly *what*, *where* and *how* you are getting up to!

KNOCK UP

Just fill in this fun questionnaire and send it to us. We'll analyse the results in full, knock up a few graphs, pie charts etc., and publish them together with numerous pictures of female

You don't have to be SEX mad to work here, but it helps

models posing provocatively around filing cabinets in their underwear. *It's Britain's biggest ever Sex At Work Survey*, and we want *everyone* to take part.

Photocopy your tits and win a prize

And if you're a saucy secretary, you could pick up one of our fabulous *booby prizes* by enclosing a photocopy of your tits with your completed questionnaire. We'll be awarding a terrific three draw filing cabinet (complete with suspension files) for the best colour copy tits we receive and a super swivel typist's chair for the best black and white knockers.

BREASTS

Simply complete the questionnaire by placing ticks in the appropriate boxes, and send it to: Viz Office Sex Survey (& Tit Photocopy Competition), P.O. Box 1PT, Newcastle upon Tyne, NE99 1PT. (Photocopies of breasts must be same size, on *A3* paper. No A4 knockers please).

ELEPHANTS' EARS – NEW THEORY

A former England international footballer is about to rewrite the wildlife record books.

Ace goalscorer turned keen amateur zoologist Frank Worthington believes he has made an amazing discovery which turns accepted zoological theories about elephants upside down. For former Leicester City centre forward Frank claims the experts have got it wrong – and Indian elephants have in fact got *bigger* ears than African elephants, not the other way round.

ELEPHANT

For years schoolchildren have been taught that the African elephant – larger than its relative the Indian elephant – has got the biggest ears. But now, thanks to Frank, the natural history books may have to be rewritten.

"I couldn't believe it at first", said Frank, always a favourite with the football fans. "I was just looking through a book when I suddenly realised the ears on an Indian elephant are bigger than the ears on an African elephant", he told us.

Frank has submitted his findings to the British Natural History Museum.

Once a deadly football marksman, Frank now seems set to carve out an equally exciting career as a zoologist.

Footnote: Frank Worthington last night withdrew his claim that Indian elephants have bigger ears than African elephants. "It was all a big mix up", he told us. "I was looking at the wrong picture in my book".

MALLET HAMMERS NAIL

TV celebrity turned pop star Timmy Mallet yesterday hit out at the new album by actor turned singer Jimmy Nail.

SCOUSE

"I haven't heard it yet, but a friend of mine has, and they didn't like it', he told us. Meanwhile 'Giz a job' star Jimmy, who shot to fame as scouse TV detective Taggart, was unavailable for comment.

'OTTERLY' BONKERS!

THEY'RE about two or three foot long and they're furry. They're Britain's most aquatic-est mammals apart from water voles, and we've gone 'otterly' bonkers on 'em. *They're OTTERS!*

These furry little fish-lovers were once in danger of extinction in the wild. Thanks to conservationists numbers are now up.

But the tragic fact is that each year in British zoos and wild life parks, up to 400 otters go unsponsored. In one case, a group of eight otters in a zoo in Wales spent four years with no corporate plaque on the front of their enclosure.

But your *BIG HEARTED-VIZ* has decided to put its money where its mouth is. And this issue we've embarked upon the most ambitious otter sponsorship scheme ever embarked upon in british publishing history.

We've already spent an undisclosed amount sponsoring an incredible **FOUR** Viz otters at zoos the length and breadth the country.

But we're not going to stop there!

Under the terms of our brand new Viz Otter Mission Statement, we aim to ensure that *by the end of 2002, no UK reader will be* more than 50 miles from a Viz otter. And that's a claim that no other magazine can match*.
(*excludes Otter Week, The Otter, Otter Monthly & Otters and Ottermen)

Viz splashes out 300 spond on damp mammals

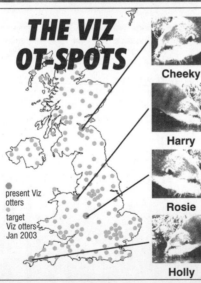

Viz otter, Rosie, tests the water in her enclosure at Dudley Zoo

THE VIZ OT-SPOTS

- present Viz otters
- target Viz otters Jan 2003

Cheeky
zoo: *Edinburgh*
age: *3*
fave food: *fish*
hobbies: *swimming, eating fish*
colour: *brown*

Harry
zoo: *Chester*
age: *2*
fave food: *fish*
hobbies: *swimming, squeaking*
colour: *otter brown*

Rosie
zoo: *Dudley*
age: *5*
fave food: *fish, ice cream*
hobbies: *standing on hind legs*
colour: *brown*

Holly
zoo: *Newquay*
age: *3*
fave food: *fish*
hobbies: *hiding from the public*
colour: *otter colour*

Viz otter girl Hayley gets to grips with Holly, the Newquay Zoo Viz otter

"We're backing the Viz otters" say stars

SINCE the otter sponsorship campaign was launched at the top of this page, we've received thouands of good luck letters from across the country. Even stars of stage and screen have been queuing up to support the Viz Otter campaign.

"It's one small cheque for mag. One giant leap for otterkind."

Said *Neil Armstrong*, former first man on the moon. His fellow countryman *Bob Hope* agreed. *"I sure 'hope' your otter campaign lasts a bit longer than I will,"* quipped the deathbed Funnyman.

*"Here is a newsflash. I'm back*ing Viz's Otter campaign."* said cissy newsreader *Martyn Lewis*, a sentiment echoed by *Michael Winner*. *"Like me, your otter campaign is a real winner!"* the fat cunt told us.

Toothy funnyman *Ken Dodd* took a break from waving a feather duster to back the campaign. *"By jove, missus! How discomknockeratingly otter-fer-carious!"* he said, making his eyes go big and sticking his teeth out.

Saturday nightmare, *Cilla Black* was also quick to jump on the bandwagon. *"Worra lorra lorra orras! Well done Viz!"* she screeched.

Restaurateur *Abdul Latif*, had these words for the otter campaign. *"Come to the Rupali, Bigg Market, Newcastle upon Tyne. You won't find an 'otter' curry anywhere!"*

WHO CARES?

Your *Big-Hearted Viz* leads the way in sponsorship, putting its money where its otters are, bringing hope to 4 of the furry fellows. But how do the other magazines on the newsagent's shelf measure up? Editors last night were quick to defend their appalling record on otter expenditure...

Shame of mags who turn their backs on otters

MotorSport editor, **Jack Steering-Wheel** told us, *"I'd love to sponsor some otters, but unfortunately I'm allergic to them. And so are all my staff."*

Clive Suit, editor of *The Economist* was keen to defend his magazine. *"We sponsor a single otter called Spike in Jersey Zoo, but we lease him back to ourselves via a wholly owned subsidiary holding company in the Isle of Wight, and then write off the difference against capital gains. So in real terms, it's three otters"*

Eric Prepuce, editor of bongo mag *Fiesta* was masturbating in the toilet when we called, but his secretary told us she knew of no plans to sponsor otters.

Janet Curtains, editor of *House and Garden* was philosophical. *"We don't sponsor as many otters as we'd like to, and we'd like to sponsor one. So that means we don't sponsor any,"* she said.

FHM editor **Steve Adverts** made no excuses. *"Our readers couldn't give a fuck about otters,"* he said. *"If I saw one in the street I'd run it over."*

Chart: Otter commitment 2001 (£0 – £300, with £150 marked)

Big-hearted Viz	MotorSport	The Economist	Fiesta	House & Gardens	FHM

---

10 things you never knew about OTTERS

1 The smallest ever otter belonged, not surprisingly, to Calvin Phillips, the world's smallest man. The otter, which was the size of a doormouse, was too small to eat fish and Phillips had to feed him on tadpoles which he caught using a fishing rod made out of a cocktail stick.

2 Innumerable books have been written about otters including Tarka the Otter, Ring of Bright Water and Watership Down.

3 With his powerful, chisel-like incisor teeth, an adult otter is capable of felling a 200 foot spruce in 3½ minutes.

4 No, sorry. That's beavers.

5 Otters have a lot in common with Frank Sinatra, including their liking for fish, their ability to swim underwater for up to four minutes, and their being found in holes in the ground.

6 The word 'utter' sounds a bit like 'otter'

7 The next word to otter in the dictionary is ottoman, type of settee sometimes consisting of 3 or more seats united at a central point so the occupants face different ways.

8 Another type consists of a circular, continuous seat with a padded truncated pillar in the centre which forms a back.

9 A third is in the form of a low, square sofa, the padded seat of which lifts up and discloses a box.

10 The word after ottoman is oubliette, a dungeon in a castle or prison reached from a room above by a trapdoor.

HEY, ZOO THERE! Have *you* got an otter that needs sponsoring? Would *you* like to join Britain's fastest-growing otter sponsorship scheme? Please contact Stevie Attenborough, Chief Otter Sponsorship Co-ordinator, Viz, PO Box 1PT, Newcastle upon Tyne NE99 1PT.

Please bear in mind we've already coughed up over 300 quid on the little fuckers, the bird in the bikini was another 100, plus 20 for the stuffed otter, so if you have any particularly old or ill ones that you could let us have cheap, all the better. We're looking at a fiver per otter, tops.

££££££££££££££ IT COULD BE ZOO!

Play the Viz National Ottery and win up to £1m* worth of prizes!

We're not going to the zoo, but you can come too ... if you're the lucky jackpot winner of our fantastic **National Ottery**.

For the winner and a friend will enjoy a chauffeur-driven Rolls Royce trip of a lifetime from their door to the zoo to see a Viz Otter. Not only that, but we'll throw in ten bottles of **champagne**, a 2lb bag of **Bombay mix**, forty **Embassy Regal** and ten **pounds** spending money.

In our book, that's a total prize value of **£1 million!** (sorry, cash alternative not available).

To enter, simply fill in the National Ottery Ticket below with any six numbers between 1 and 49, tell us which zoo you'd like to go to, and send it to: *The National Ottery, Viz Comic,* PO Box 1PT, Newcastle upon Tyne, NE99 1PT. All entries must be in by Monday 15th October. The first ticket with six numbers on drawn out of the hat by Viz Ottergirl Hayley will win.

Play the National Ottery Extra

Don't despair if you didn't win the jackpot prize. If you tick the Ottery Extra box on the ticket, you could win a day at Newquay Zoo with the Viz otter keeper. You'll get the chance to feed him, look at him from close up and even sweep up his little turds with a brush (the otter that is, not the keeper) You'll even get to have lunch with him, but remember, you mustn't feed the keepers as they have a very special diet.

THE NATIONAL OTTERY

VIZ COMIC ISSUE 110 SEP/OCT 2001

286 - 445830

A ☐ ☐ ☐ ☐ ☐ ☐

OTTERY EXTRA

GUARANTEED JACKPOT OF A DAY WITH THE OTTER KEEPER AT NEWQUAY ZOO

☐ ENTERED IN THE OTTERY EXTRA

NAME......................

ADDRESS......................

....

DAYTIME PHONE No...............

NEAREST OTTER AT.............ZOO

DRAW ON MON 15th OCT

— ☐ FILL BOX TO VOID —

*Publisher's estimate

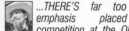

THE BEIJING OLYMPICS have come and gone, and once again Team GB has failed to deliver. Our shamefaced team of 312 elite Olympians brought back a paltry 18 golds from China. That's equates to a mere one medal between seventeen athletes, which makes Britain's average result in each event a feeble 18th - that's just a sixth of a bronze each. We went out on the street to see what the great British public had to say about this national embarrassment.

...I FOR one am sick of paying for Paula Radcliffe to go on fancy foreign holidays every four years, only for her to return without any medals. At the very least, she should offer to pay for her own mini-bar bills, room service and any adult movies she's watched in her room.

Loris Idris, architectural model-maker

...THERE'S far too much emphasis placed on competition at the Olympics. How much better it would be if everybody came home with a medal, simply for participating nicely.

Janice Lemur, feng shui blackbelt

...WHAT is the point of sending an athlete half way round the world, only for them to come back empty-handed? Unless we are 100% sure that they're going to win gold, we should simply save our money. I ended the Beijing games having won just as many medals as Paula Radcliffe, and I spent the entire fortnight sitting on the sofa in my pants, dipping Doritos in a jar of Chicken Tonight. The cost to the British taxpayer? Zero, if you don't include two weeks' dole money and my housing benefit.

Sid Colobus, jobseeker

...THE ladies' quad rowers who came second to the Chinese had a ruddy nerve bringing their boat back to shore. If they had had an ounce of decency they would have scuttled their craft and gone down with it, like the captain of the Titanic did, and saved us the cost of their air fare home.

Harry Spider, bingo caller

...14-YEAR-old diver Tom Daley should hang his head in shame. If his story had been a film, he would have brought back a gold medal from Beijing. Not only that, he would have won it with his last dive, narrowly beating a Russian who had been cheating all along. All this whilst disguising the fact that he had a broken arm caused by the Russian pushing him over in the changing room just before the final round. In reality, he came a mediocre seventh, and that's without a broken arm.

Vince Bonobo, sandwich engineer

...IT'S often been said that winning is unimportant, as it's the taking part that counts. What utter claptrap. The government should stamp out this silly idea once and for all by awarding New Year's Honours to medal winners, and nothing at all to the losers.

Olive Proboscis, stripper

...ANYONE who doesn't win a medal is worse than a paedophile in my book. They have been taking Britain for a ride for far too long.

If I had my way, I'd fine them a thousand pounds each, chemically castrate them and put them on the Sex Offenders' register.

Lord Chief Justice Frank Howler QC, law lord

...IF OUR athletes had merely taken loads of drugs, like the winners probably did, perhaps they would have come back from Beijing with a more impressive haul.

Ulrika Orangutan, proctologist

...WHEN I heard about Britain's feeble performance in Beijing, I put my foot through the television and sent Bert Kwouk the bill.

Harold Marmoset, pygmy breeder

...THE British Olympic authorities could easily increase our tally by offering British passports to gold medal winners from other countries. I'm sure that jamaican sprinter Usain Bolt, for example, would jump at the chance to live in Britain. Getting stopped and searched by the police four times a day for being black would be a small price to pay for the glory of running for England.

Geoff Mandrill, unicycle courier

...I WOULD happily have paid the bill for Mr Marmoset's television, only I was born in Manchester and I don't see why I should.

Bert Kwouk, actor

...BRITAIN finished fourth in the medals table, so even if there had been medals for winning medals, we would have been pipped for the bronze.

Samson Gibbon, pouffe upholsterer

...MAKING our BMX cyclists ride toddlers' bikes was asking for trouble. No wonder our golden girl Shanaze Reade fell off hers and missed out on a medal.

Anthony Sasquatch, mud wrestler

...I HAD high hopes that Team GB would do us proud in Beijing, but I feel badly let down by the overall performance of our squad. The only athlete who managed to live up to my expectations was tennis ace Andy Murray, who crashed out in the first round. Well done Andy.

Desdemona Capuchin, genuine filthy bored housewife

...THESE Olympic losers have shamed our nation in front of the whole world, yet they expect to come back here and just live amongst us as if nothing has happened. They should be shaved, and forced to wear stripy boilersuits with big pink 'L's on them until 2012.

D Irving, revisionist historian

HEROES. There's no other word for them. Lionhearts, victors, champions, luminaries, paragons and demigods. These are some other words for the brave boys - and girls, let us not forget the girls - who brought back a mountain of gold from the Beijing games. That mountain of gold, that shining, lustrous Everest of medals, literally stands as a testament to the heroic deeds that Team GB wrought in the white heat of the Olympic crucible.

Never before have English eyes witnessed such a display of sheer guts, determination, grit, guts and determination as have been put on by our 2008 Olympians. The precious memories of the superhuman feats of our 350 (subs please check actual number) athletes, gymnasts, swimmers, cyclists, runners, sprinters, boxers, weight lifters, jumpers, runners, horseriders and runners will remain fresh in the nation's hearts long after they have been forgotten.

The debt we owe to them can never be repaid.

In fact, I would go further. The sacrifices made by all the soldiers who laid down their lives for their country in the two World Wars are as nothing compared with the deeds of our Beijing heroes. Alongside the glorious deeds of our golden generation - the likes of Rebecca Adlington, Chris

TONY PARSEHOLE

I'd take a bullet for these Olympic heroes

Hoy and Ben Ainslie - the so-called fallen heroes of the last century's great conflicts are no more than snivelling cowards. They are shameful stains on our national character and I spit on their graves.

Our gold medallists are the Greatest Britons who have ever sprung from the loins of this green and pleasant land that we call England, and I would happily lay down my life for theirs.

Should an assassin, crazed with jealousy, attempt to pick off 400metres lioness-heart Christine Ohuruogu with a head shot, I would joyfully dive into the ballistic path and die in her stead. Nothing would give me greater pleasure.

Should a lowlife vermin who had never achieved any measure of athletic excellence, attempt to behead Sarah, Sarah and Pippa - our Yngling yachters of hearts - with a razor-sharp samurai sword, he'd have to get past me first. Without a second's hesitation, I would push my neck in the way of his lethal blade. And, as the fountain of scarlet blood gushed from my severed jugular, it would arc through the English sky in a red, white and blue tribute to all our Olympic heroes. I wouldn't need to think about it once, let alone twice. Or three times. Certainly not four. Giving up my own life for them would be the least I could do.

Perhaps that makes me a hero. I don't know, it's not for me to say. Who am I to say who is or isn't a hero? Once again, I don't know, it's not for me to say either. But I do know one thing. The selflessness, heroism and sheer, unadulterated courage that I showed when I laid my life down for these great Olympians was as nothing in comparison to the selflessness, heroism and sheer, unadulterated courage that they showed upon that lofty Beijing anvil where their mighty feats were wrought and forged under the red hot hammer of Olympic competition.

Never before since the far-off epoch that was ancient Greece have men there that's 500 words. Invoice enc.

NEXT WEEK : I'd fight a pack of lions to save Lewis Hamilton from being eaten at the zoo

We've gone CONKERS BONKERS!

If there's one thing in Britain today that won't be hit by the recession, it's conkers. For while house prices plumit, the pound tumbles and businesses go to the wall, kids are collecting conkers as enthusiastically as ever.

And we're getting into the *swing* by going horse chest-*nutty!* Yes, we're inviting everyone in Britain to forget their troubles, and play conkers instead.

CONKERS

All you have to do to enter our Conker Competition is go out and find a conker. You'll find them lying on the ground under Horse Chestnut trees. When you think you've got a good one, drill a hole in it, and thread some string through, then knot it at both ends. Then post your conker to us, and we will enter it into our Conker Challenge.

CHEATING

Your conker will take part in a special match against our own Champion Conker. It will be a fair fight, with no cheating, and will be witnessed by us. If your conker wins, we will send you a £100 sweet voucher valid at any sweet shop, plus 100

It's Britain's biggest conker competition

marbles. All conkers will be returned together with a certificate.

STRING

So come on, collect those conkers and send them as soon as you can. The competition closes on October 31. Send your conker on a string, plus a 50p entry fee and a stamped addressed return envelope, to: Viz Conkers Bonkers Horse Chest*nutty* Conker Competition, P.O. Box 1PT, Newcastle upon Tyne, NE99 1PT.

Please note: Any conkers which have been soaked in vinegar, cooked in an oven or otherwise tampered with, will be disqualified. The judges decision is final.

Conker campaigner David Alton yesterday

A word of warning to conker collectors from Conker Safety Campaigner MP David Alton.

DANGER

"I would ask anyone going out to collect conkers to take my advice and follow the Conker Code. Conkering can be great fun if you avoid danger by following these simple tips", said David.

● Collect conkers from the ground around trees. Don't throw sticks or stones to dislodge them from the tree. "Throwing things is dangerous, and can damage the tree", said Mr Alton, MP for Liverpool Mossley Hill.

● Never climb a tree to reach conkers. Falling can cause serious injuries.

● Don't trespass to reach conker trees. Collect conkers in public parks, etc. *Not* in people's gardens.

● Finally, stay away from busy roads. Collecting conkers on or near a busy road simply isn't worth the risk.

The Norman Conker-west

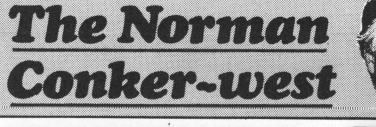

We asked a few famous faces whether they'd be entering Britain's biggest Conker Competition.

Chancellor Norman Lamont forgot his economic worries for a few moments and said he'd love to take part, providing he could get hold of a decent conker. "It's been some time since I went out looking for conkers, and I doubt if I'm as agile as I used to be. I certainly won't be climbing any trees", quipped the Government's money expert.

SEX

Judith Mellor, long suffering wife of sex scandal 'Minister for Fun' hubby David, said she'd never played conkers.

GIRL

"I was a girl when I was younger, and it was more the boys that did that sort of thing".

CARROT

Welsh comic Jasper Carrott told us he wouldn't have time to collect conkers this year.

PEAS

"I'll be too busy writing new material for my hilarious TV show, and recording a follow up to my hit single 'Funky Moped'", he told us.

Mellor - was girl

Carrot - hilarious

"DAFT" BUGGER

OH NO. I THINK I'VE LOCKED MY KEYS IN THE CAR AGAIN.

20 THINGS YOU NEVER KNEW ABOUT RUSSIA

A little over a year ago no-one had even heard of Russia. For all we knew it was just another country somewhere in Europe. But now communist leader Michael Gorbachev has put Russia firmly on the map. It's in the papers and on the telly, and in pubs and clubs all over Britain people are talking about little else.

Love it or hate it, nowadays you simply can't ignore Russia. But what is it *really* like? Here are a few things you probably didn't know about the USSR.

1 Russia is by far and away the biggest country in the world. Except for Canada.

2 Despite it's enormous size Russian has a relatively small population, roughly equivalent to that of Wales.

3 Russia is famous for it's spies. Some famous names rumoured to have been Russian spies include Roger Hollis, former head of MI5, ex-Prime Minister Harold Wilson and children's TV presenter and newsreader John Craven.

4 Not all Russians are spies. There are Russian doctors, electricians, gas fitters and even landscape gardeners. In fact, almost half the population of Russia have no connection whatsoever with the KGB.

5 In Russia the staple diet consists of potato. To stay alive the average Russian has to eat *three times* his body weight in potatoes – every week!

6 Russian shoppers queue for hours to buy bread, which is in short supply, while caviar is given away free on street corners. That's because sturgeon, the plant that caviar comes from, grows wild in Russia where it is generally regarded as a weed.

'Pauls' of Moscow, The world's first barbers shop, opened in 1923.

7 Russia invented the barber's shop. The first ever gentleman's hairdressers was opened in a Moscow side street in 1923.

8 Russian leader Michael Gorbachev likes nothing better in his spare time than listening to western pop music. His favourite groups include Five Star and Sister Sledge.

9 Alcoholism is a serious problem in Russia with 8 out of 10 Russians registered as alcoholics. However, the Moscow branch of Alcoholics Anonymous has only one phone line, which is constantly engaged.

10 Russia is made up of many republics, the smallest of which is Ludbanskia with a population of only 4.

11 Visit the USSR for a holiday and you will immediately notice that unlike other countries Russia has no tourist attractions. After the communist revolution in 1914, all tourist attractions were demolished, and sandy beaches around the Russian coastline were covered over with rubble.

12 Russia's best known actor is Walter Koenig who played 'Chekov' in the hit TV series Star Trek. Relatively unknown in the United States, Koenig enjoys megastar status in his native country where his acting career has made him Russia's wealthiest man, owning six pairs of shoes. He has also become a pop star, topping the soviet pops for an incredible 6 years with his version of the Shirley Bassey hit 'Hey Big Spender'.

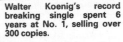

Walter Koenig's record breaking single spent 6 years at No. 1, selling over 300 copies.

13 Russia is the flatest country in the world. Its highest mountain, Mount Bonsnov, is only 47 feet high.

14 Britain and Germany weren't the only two countries involved in the second world war. Russia also took part, coming third behind us and America.

15 With 17 different indiginous species, Russia has more types of mouse than any other country in the world.

16 Despite having no fewer than 149 letters in their alphabet, there is no world for 'tooth' in the Russian language.

Mikhail Gorbachev – Sister Sledge.

17 Russian families spend their evenings huddled around the TV set – watching British programmes! For the top rated TV show in the USSR is 'Catchphrase', hosted by silver haired comic Roy Walker. However, viewing figures can be misleading. For in Russia there are only 3 television sets. Two belong to Walter Koenig, and the other one is broken.

18 It's no wonder the Russian's make good spies. They have the best eyesight in the world! Only 1 in 500 Russians wear spectacles. And it's just as well. You'd have to queue for 2 weeks to have your eyes tested at a Moscow opticians, and a pair of glasses with just basic frames would cost over a million rubles — that's £800 to you and me – the equivalent of three years wages to the average Russian!

19 A second-hand pair of Levi 501 jeans, originally bought for £24.99 in a London boutique, were recently sold at an auction in Moscow for a staggering £23 million. The buyer – Walter Koenig.

20 Tipton in the West Midlands has close links with Russia. In 1982 it was twinned with the Russian city of Kiev. Every Easter the mayor of Kiev sends his West Midlands counterpart a large oven-ready chicken stuffed with garlic and butter.

I THINK WE'VE SPOTTED A GAP IN THE MARKET HERE, BRIAN

AMAZING -BUT SHIT

Surely the most unusual celebration in the history of football was that performed in June 1966 by television variety star Bruce Forsyth. To celebrate England's famous World Cup victory over West Germany, all-round entertainer Forsyth sung and danced non-stop from Land's End to John O'Groats. During his marathon routine Brucey sunged 12,487 songs, wore out 157 pairs of tap dancing shoes, and lost 3 top hats in high winds on the Pennine Way.

During World War II German military chiefs experimented with 'human messenger pigeons' for delivering important documents on the field of battle. Volunteer soldiers carrying vital messages were fired towards their destination from a giant circus canon. In one week alone 67 men died from head injuries, and the experiment was later abandoned.

Despite its vast size - a full grown adult could weigh up to 25 tons and stand 12 metres in height - the mighty Tyrannosaurus Rex had quite a small cock. Yet they still managed to produce more smegma than any other dinosaur! The T Rex's small arms made washing its bell end impossible. Archeologists believe that as a result, from a foreskin no bigger than Pop Idol judge Simon Cowell's polo neck, the giant reptiles could produce three tons of knob cheese per day.

At the height of the Cold War Russian spymasters hatched a bizarre plot to kidnap Manchester United and England star Sir Bobby Charlton and replace him with a double. The agent they planned to use was said to be such a perfect likeness he would have fooled even Bobby's footballing brother Jack.

Y'AAL REET WOR KID?

DAH JACK. I AM GOOT!

In 1973 rock singer Rod Stewart called in paranormal investigators after he awoke one night to find a ghost giving him a blow job! But the spooky mystery was solved when Rod's 'ghost' turned out to be Britt Ekland underneath a blanket.

★★★ **Ghostly secrets of things**

A BIT ON THE 01

Doris Stoke-Manderville is a medium with a difference. For when spirits contact her from the other side, it's not to pass messages on to their loved ones. It's for sex!

And Doris boasts a host of ghost celebrities among her lovers from beyond the grave. Over a period of several years she claims that the late stars of showbusiness have, quite literally, been putting their willies up her. And now she exclusively reveals the sexy secrets of the celebrity spooks who regularly go hump in the night.

‛My intimate sexual experiences all began quite by accident. I had been admitted to the chiropodists for a routine operation on my toe nail. The next thing I knew I was suddenly aware of being outside of my body, looking down on the chiropodist who was operating on my foot. There seemed to be some sort of panic, and the nurse was rushing around the room looking for some scissors.

BALD

I remember clearly looking down on the chiropodist, and noticing a small bald patch on the top of his head. Later he confirmed that he had a small bald patch, something which I could not possibly of known had I not been floating above him in that surgery as my life lay in the balance down below. I now realise that I underwent a near death experience on that day, and my memories of it are still vivid.

MERCURY - used rubber

COOPER - 'Magic tricks'

'Dead stars have been putting the willies up me for years', says Doris

I was floating through a tunnel. Ahead of me I could see a light. I headed towards it and found myself emerging from beneath a quilt. I was in bed. And stranger still, there was a tall, dark man next to me who I immediately recognised as being Bernie Winters, the late TV comic.

Bernie's bed was the scene of a 'Heavenly' experience

I had always been a fan of Bernie's, and I offered no resistance as he took me in his arms and grinned. His big, goofy smile was unmistakable. I was in Heaven – quite literally – as we made love for what seemed an eternity.

Afterwards Bernie lit a cigarette and turned to me. "You have to go back", he said. "You still have a life to live". I knew that he was right. As I crawled back beneath the sheets I glanced back and saw Bernie grinning bravely. His eyebrows were raised in that silly smile, but he could not hide the tear which was running down his cheek.

LITTLE

Crawling back down the dark tunnel I suddenly found myself back in the chiropodist's surgery. The nurse was offering me a cup of tea. She explained that there had been complications, and that my toe may feel a bit sore for a few days. In a way I felt glad to be alive, but at the same time sad that I had to leave Bernie.

GOLDEN

The next time I had sex with a ghost was in my bedroom a few weeks later. I'd gone to bed early while my husband watched the football. Suddenly I heard the sound of chains dragging on the stairs. Then a dark figure appeared in the doorway. I nearly had a *Shear Heart Attack* when I saw who it was! Because there in my bedroom stood Freddie Mercury, dressed in full bondage gear, complete with leather cap, rubber pants, a dog collar and lots of chains.

GEORGE

I'd always fancied Freddie but didn't think his ghost would be interested in girls. I couldn't have been more wrong! He was all over me, kissing every inch of my naked body.

BILLY

Before things went too far I asked him if he'd mind wearing a Durex. I believe in safe sex, even with ghosts. He was the perfect gentleman, and got one out

of my husbands top draw. Being a ghost his cock was a bit see through, but between the two of us we just about managed to get it on.

FREDDIE

When it comes to sex Freddie *was the champions* alright. He was wonderful. Afterwards he took off the Durex, tied a knot in it and threw it on the floor. Then, after wiping his ghostly cock on the curtains he walked through the wall and was gone. Curiosity got the better of me and I picked up the rubber just to see what ghoulish gunk looked like. But to my amazement it was completely empty. In fact it was still in its packet in the draw where my husband had left it.

PROOF

Now you try explaining that to me! If that's not proof of ghosts shagging me then I don't know what is.

COOPER

They say that there are sixty-nine love making positions in the Karma Sutra. Well I discovered number seventy the night magician Tommy Cooper visited me in my bedroom! You see, ghosts can take their heads off if they want to. And magician ghosts can do practically anything they like with their bodies.

BREMNER

Well, me and Tommy had only been at it for a few moments when suddenly he popped his head off and tucked it under his arm. My hair turned white with shock. Minge hair that is! Anyway, Tommy was exactly that! (Magic that is.) He would be making love to me from behind while his hands fondled me passion-

ately. First one, then two, then three! Eventually there were half a dozen hands groping every inch of me at once.

CHARLTON

Occasionally he'd stop and produce a rabbit from his backside, or saw me in half. I had never enjoyed sex so much. Then, after what seemed like an eternity, we reached the most thrilling climax I have ever known. As Tommy made love to me my body shook with ecstacy, while in the mirror I could see his head on the dressing table, wanking furiously. That was a night I'll never forget in a hurry, I can tell you!

HUNTER

At first my husband used to object to me having sex with the dead stars, then he seemed to get quite turned on by the idea and suggested that the next time I did it he should come along and watch. So the following night I arranged to have a sexy threesome with Laurel and Hardy and my husband came along and brought a camera.

LORIMER

My husband snapped away like mad while Laurel and Hardy took turns fulfiling me. Afterwards the fat one looked at the sheets and said "That's another fine mess you've made on the bedclothes", while the thin one scratched his head and started crying.

CLARKE

The following day my husband got the film developed, but when he went back to collect it he thought for a minute he'd got the wrong pictures. All you could see was me on the bed, apparently having passionate sex with myself

★★★

at go hump in the night ★★★★★★★★★★★★★★★★★★★★★★★★★★★★★

HER SIDE

Hubby and I *liked* it hot with Monroe!

Stan and Ollie (above) joined Doris for a sexy threesome.

from behind, and kissing every inch of my own naked body. The mystery was solved when we noticed an advice sticker on one picture, pointing out that ghosts don't show up in photographs.

JONES

After that my husband suggested we try a bit of girl on girl action with some lady ghosts, and asked me to put on a lesbian show with Marilyn Monroe. I summoned her to the bedroom using a glass and some playing cards. At first we were both nervous, and I admitted to her that I had never done lesbian. She said it was her first time too.

MADELEY

Anyway, we did a topless show for my husband, and I licked a sausage while Marilyn kissed my bottom. Suddenly a cold ghostly draught blew her skirt up, just like in that film. My husband got so excited he tried to join in. But he was forgetting that I have special powers to have sex with ghosts. If a mere mortal like him tries to touch a ghost, his hand just goes through them as if they aren't there.

GILES

So my poor husband ended up falling flat on his face and banging his head on the wardrobe! I couldn't stop laughing, and even Marilyn saw the funny side. Unfortunately however, that was not the only time when things got a little out of hand.

BATES

As a child I'd always been too scared to ride on a ghost train. But not nearly as scared as I was the night half a dozen ghosts decided to pull their very own ghost train - on me!

LECTER

Apparently several ghosts had been at a party in Heaven and had taken a lot of drink on board. You could say that they were in high spirits! In fact I've never seen six ghosts as pissed as they were. Anyway, they rolled up through my bedroom wall at two o'clock in the morning and I immediately knew what they were after.

KRUGER

Well I was having none of it. I'm as saucy as the next person, don't get me wrong. But I draw the line at having sex with six drunken ghosts all at once.

JASON

Luckily I had my wits about me and I told them I had the painters in. They took it in good spirits and I made them all a cup of coffee and we sat and watched TV until their ghost taxi arrived. It didn't honk the horn or ring the bell. It just drove in through the wall with a ghostly 'wooosh!' sound and hovered above the coffee table.

KYLIE

All six ghosts piled into the taxi, but the driver wasn't having it. He said he was only licensed to carry five ghosts, and someone had to get out. Luckily one of them – I think it was 'carry on' actor Sid James – said he needed some fresh air, and offered to walk back to Heaven. Eventually I got rid

ALL THE FUN OF THE SCARE!

Sextet of ging gang gooly gooly gooly gooly gang bang ghouls tried to pull a ghost train up my tunnel of love!

of them, but only after lending them the taxi fare home!

CRAIG

Another scare I had was when I got pregnant by a ghost. It was Dirty Den who did it. He'd just been murdered in EastEnders, having got Michelle pregnant. Then he turned up in my bedroom while my husband was out and smooth talked his way into bed with me. I should have known better. Anyway, I found out I was expecting and told him so. But he didn't want anything to do with it.

Den - ghost baby father

In the end I decided to have an abortion. Ghost abortions are a bit like killing a Dracula. You can't get them on the National Health, that's for sure! My husband went down the butchers and got a steak and some garlic, and we used them to kill it during daylight hours.

Thinking about it now perhaps I should have had the ghost baby, because a friend of mine who is a nurse says that ghost babies simply walk through your tummy instead of coming out *down below*, so it doesn't actually hurt at all.

MRS MANGLE

Sometimes my husband and me lie awake at nights and cry thinking about my ghost baby that we killed with steak. But perhaps with all the murders and everything the world today would not have been a good place to bring up a ghost baby.

I guess that ghost baby is in Heaven now. Wherever it is I know it understands that what we did was for the best.

Monroe - did 'lesbian'

GHOSTBONKERS!

A great many people have claimed to have been the victims of sex attacks by ghosts. In fact, in certain parts of South America women are twice as likely to be raped by a ghost (or a space alien) than by an ordinary man. But ghost sex fiends look nothing like Frankenstien's monster, Dracula or any of the other fictional ghosts we see on the cinema screen.

SPECTRES

The scientific term for sexy spectres is an *incubus*, and one theory is that their 'bodies' or shapes are made up of a kind of chemical energy know as erectoplasm. Some of these ghosts take human form, others look like small green electric clouds that whizz about the place breaking things.

MARTENS

In the past the police have often been accused of scepticism and a lack of sympathy when dealing with cases of ghost rape, and there has never been a successful conviction of a dead rapist in British legal history. However a new postal helpline has recently been set up to provide help and support for

By our SCARY GHOSTS CORRESPONDENT BLAKEY Off On The Buses

the victims of ghost sex. If you have been attacked by a ghost, write to the following address explaining in as much detail as possible exactly what happened. *Ghost Sex Helpline*, P.O. Box 1PT, Newcastle upon Tyne, NE99 1PT.

Due to the volume of mail received it may not be possible in all cases to send a reply. We reserve the right to publish (in edited, abridged, or in totally unrecognisable form) all submissions received.

★ ★ ★ ★ ★ ★ ★ ★ ★ ★

TOP JOKE

I CAN'T EAT ANY MORE OF THIS DARLING

YOUR EYES ARE BIGGER THAN YOUR BELLY.

ELTON AND THE GENERAL PURPOSE BUILDER

WORLD EXCLUSIVE

A man who has been seen working at the home of millionaire pop star Elton John is a self employed general purpose builder, we can exclusively reveal.

Steve Fairbrother, a stocky 32-year-old, runs his business from a small yard in Guildford, Surrey, not far from Elton's £2.5 million mansion.

YELLOW PAGES

Elton, 44, met Steve after reading an ad in the Yellow Pages. In it Steve described himself as 'Prompt, friendly and reliable'. 'No job too big or too small' the ad continued. It also boasted 'Free Estimates'.

THOMSON LOCAL

Steve has been a regular visitor to the pop millionaire's lavish £6 million luxury home, often seen coming and going in his £4,500 red Escort van – believed to have been paid for using the profits from his building work, some of which has come from Elton.

BONES

The singer, 46, who often wears hats, makes no bones about his relationship with the handy man. "It's true, I've employed him to do some plastering, to build a fireplace and tile the bathroom. He's a reliable tradesman, and his prices are competitive", he confided to friends recently. And people

close to the billionaire singer say that Elton is 'delighted' with the work that has been done.

In the past few months Fairbrother is believed to have:
● **PLUMBED** in a sink in Elton's lavish £80,000 utility room.
● **REPOINTED** a chimney stack above the south facing gable of the star's £12 million farmhouse.
● **BUILT** a small retaining wall around flowerbeds outside the star's lavish £40,000 kitchen window.

SCOTTY

Neighbours living close to the multi-billionaire's £14 million hideaway describe Fairbrother as 'quiet'. "He regularly comes and goes bringing plaster, lengths of wood and tools. Sometimes he has a ladder on the top of his van", one neighbour told us.

UHURU

When we rang Tewson's Builders Supplies of Guildford, a spokesman confirmed that Fairbrother had an account with them, and revealed he had collected several lengths of dressed timber from them only last week.

Elton - paid Steve (above) for home improvements

"He ordered them on the Monday, and said he wanted them on Wednesday", we were told. Unfortunately the spokesman could not confirm that this wood was for work on Elton's mansion. "I don't know what it was for", he told us.

WIFE

When we rang Fairbrother's home – a small flat which he shares with his wife – he wasn't in. His wife Shirley, a pretty 24-year-old, offered to take a message. "My husband's out on a job at the moment", she said. "But if you give me your number I cane get him to ring you back when he comes in. Probably after six", she added.

DIVORCE

Elton, who's short-lived marriage to Brazilian beauty Renata Blauel ended in divorce, has admitted to friends that he is concerned about hair loss on the top of his head.

Monkee Mike invented Tippex

It's a little known fact that sixties pop star Mike Nesmith of The Monkees invented Tippex, the opaque correction fluid used in offices around the world.

Nesmith - Tippex

The stationery inventions of the pop stars are often overshadowed by their chart success. But a glance along the shelves of any office supplies shop will reveal any number of invaluable inventions attributable to music stars of the past.

POST IT

For example, did you know that David Bowie dreamt up the 'post it' notepad. Millionaire Dave coins in a cool £25 million a week from sales of the sticky back notelets.

CHIPPED

Among other singing stars who have chipped in with office inventions are Tina Charles (treasury tags), Mungo Gerry ('IN' and 'OUT' trays), Gerry Rafferty (self-advancing date stamp), and The Bay City Rollers' Alan Longmuir (the 'desk tidy').

MASHED

But not all pop stars are the inventors of stationery items. After writing the ever popular 'Wild Thing', Reg Presley out of The Troggs nipped home and invented an automatic fog detecting device for Heathrow Airport.

IT'S JUG MANIA!

Britain's bridegrooms are nutty about knockers ~ claims vicar

Fellas are falling over themselves to marry women with big boobs! That's the view of vicar Dennis Randall. And he should know — he's been marrying couples for over forty years.

Vicar Dennis — he's seen it all

Dennis has just completed a book in which he relates many interesting stories derived from his time as a vicar. In it he compares the changing tastes of British bridegrooms over the years, as seen first hand from the altar.

HITCHED

"When I first started back in the forties, the fellas were going for anything they could get their hands on", Dennis confided. "Like everything else in those days, good looking birds were in short supply. I felt sorry for some of the blokes — they were coming home and getting hitched to the first girl they met off the boat. I'm not kidding, there was some funny looking brides about in those days. Lots of them had been working for the war effort — in factories and in the fields. Some of them were built like cart horses. Today's body builders are nothing compared to these girls".

COUPLES

Dennis has lost count of the number of couples he has brought together over the years. But he recalls a definate drop in the number of marriages taking place during the late sixties.

"It was all this love and peace business. Couples weren't bothering to tie the knot. And in those days if a couple did get married, chances are the bride wouldn't even wear a bra. Women's Lib I think they called it. Mind you it had its advantages — I got more than a few eyefulls in those days I can tell you".

In recent years Dennis says the trend is definately towards bigger busts! "Nowadays fellas want something they can get to grips with. I see them every week. The knockers just seem to get bigger and bigger".

Royal Weddings can have an amazing influence on marriages in the months that follow. "Andrew and Fergie was by far the best example", Dennis told us. "The minute those two had walked up the aisle, my church was chock-a-block with fellas all wanting to marry birds with fat arses".

JUGS

But big hasn't always been considered beautiful. Back in the seventies Dennis detected a definate trend away from buxom brides. "Small jugs were very much the order of the day", he told us.

VASES

Indeed, Dennis believes a lot of today's broken marriages and divorces can be attributed to bad choice on the part of the bridegroom.

"On a few occasions I felt like asking the bridegroom what on earth he was playing at. Some of the boilers I've hitched up you just wouldn't believe . I know a looker when I see one. Fellas have got to remember — marriage is for life. So if you are getting hitched, make sure you choose a good looking bird".

TEAPOTS

So far Dennis has failed in his attempts to find a publisher for his book, provisionally entitled 'Here Come the Boobs'. Meanwhile a church spokesman, who denied any knowledge of the book, told us that Rev Randall had been suspended from his job as vicar some time ago pending the outcome of a police enquiry into allegations of Gross Indecency and Sending Pornographic Material through the post.

Fanny Batter's HOLLYWOOD JIGSAW gossip

 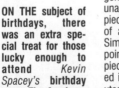

ACTOR and renowned prankster *Ashton Kutcher* failed to see the funny side after being on the receiving end of a particularly cruel jigsaw related prank - courtesy of comedian *Ronnie Corbett.* Kutcher had 'Punk'd' close friend Corbett for his show, and it seems that the pint-sized funnyman decided to turn the tables on his pal, presenting Kutcher with a 5,000 piece jigsaw of York Minster with a slight twist- there was a piece of sky missing! Kutcher and girlfriend *Demi Moore* are thought to have spent over a week searching his swanky Beverley Hills mansion from top to bottom for the missing piece - and the couple weren't too happy when Corbett eventually showed up at their apartment clutching the elusive item and unable to contain his laughter. According to our sources Kutcher has decided to let the matter drop, but Corbett won't be seeing his chum for a while on the orders of Ms Moore!

EYEBROWS were raised when another comedian, *Wedding Crashers* star **Vince Vaughn,** presented girlfriend **Jennifer Aniston** with a specially commissioned 25,000 piece jigsaw of **Lenny Henry** for her birthday. It is well documented that Jen's ex **Brad Pitt** recently splashed out on a lavish 24,000 piece jigsaw as an engagement present for new squeeze **Angelina Jolie** - and it is also common knowledge that Jolie is a huge fan of *Chef!*

ON THE subject of birthdays, there was an extra special treat for those lucky enough to attend *Kevin Spacey's* birthday celebrations this year. The Academy Award-winning actor and hugely respected jigsaw buff treated his guests, among them *American Beauty* co-star *Mena Suvari* and mid nineties reggae outfit *Chaka Demus and Pliers,* to an impromptu lecture on the history of the jigsaw, followed by a tour of his world-renowned collection of Impossipuzzles. This was followed up by the presentation of Spacey's birthday cake which was fashioned into, you guessed it, a giant jigsaw of *John Spilsbury,* the London engraver and map maker thought to have produced the world's first jigsaw puzzle.

WHISPER it quietly, but word on the street is that *Jessica Simpson* likes her jigsaws simple. Though he refused to comment when we approached him recently, **Nick Lachey** has apparently been making disparaging remarks about his estranged wife's puzzle solving abilities, or lack of them, to anyone who will listen. My spies tell me Lachey has claimed that Simpson only buys jigsaws which are 3000 pieces or less and has even gone as far as to suggest she was unable to complete a standard 2500 piece jigsaw of some cats peering out of a basket without his assistance. Simpson has hit back at the rumours, pointing out that she solved a 5000 piece Jan Van Haasteren puzzle unaided in an impressive 7 hours 32 minutes at a charity event in New York last month. You go girl!

FRED Durst was left red faced at this year's Bournemouth Jigsaw Convention after bragging to passers by that he was the proud owner

of the world's biggest metallic jigsaw - specially commissioned in the likeness of his band Limp Bizkit and weighing in at a hefty 60,000 pieces. He was left with egg on his face however, when rival rapper *Kanye West* put in an appearance, flanked by bodyguards, to pick up an extraordinary purchase - a 150,000 piece solid gold jigsaw of Ilfracombe Harbour.

See you next time for more Hollywood Jigsaw Gossip!

Too much cheese as bad for you as not enough, say boffins

EATING too much cheese can be as bad for your health as not eating enough, according to a new report which looked at the right amount of cheese to eat.

In the study, which looked at the medical records of people who had been fed different quantities of cheese, cheese scientists found that consuming too much cheese had just as adverse an effect on a person's well-being as comsuming too little.

condemn

But condemnation of the report has come from all sides. Pro-too much cheese eating pressure group, *XS Cheese*, were quick to condemn the report's conclusion. A spokesman said: "It's nonsense. Eating too much cheese is far better for you healthwise than eating the right amount or less." And Moira Johnson, representing pro-insufficient cheese lobby group, *NotEnoughCheese*, also attacked the findings. "The

Some cheese yesterday ~ But too much, or not enough?

study is deeply flawed. We will continue to try to educate people to eat a less than adequate quantity of this foodstuff."

condomn

However, the report was welcomed by Lord Owen, chairman of the anti too-much or not enough cheese pressure group *Just The Right Amount Of Cheese, Please*. "This report vindicates what we have said all along", the busty peer told reporters at a packed press conference. "That too much cheese is as bad for your health as not enough."

by our cheese amounts correspondent
Arthur Quanticheese

Lord Owen ~ just the right amount of cheese

Have **YOUR** say

THE EXPERTS have had their say, and you've had a chance to look at both sides of the argument. So what do YOU think is the correct amount of cheese to eat? Call the number below to register your vote.

Q: What do YOU think is the correct amount of cheese to eat?

Too much
☎ 01 811 8055

Right amount
☎ 01 811 8056

not enough
☎ 01 811 8057

Calls cost an absolute fucking fortune. Please wait until the bill payer has left the house before calling.

...OF COURSE, I'M TALKING OFF THE TOP OF MY HEAD HERE...

I Don't Be-leaves It!

Amazing World of Foliage

• **WHEN APOLLO** 12 astronaut Al Bean stepped onto the lunar surface in 1969 he was amazed to see hundreds of leaves blowing about in the solar wind. Baffled NASA scientists could offer no theory as to how the leaves had reached the moon, over a quarter of a million miles from the nearest tree. The most likely explanation is that they had been cast up into space when a meteorite hit a tree on earth back in dinosaur times. The crew of the next lunar mission, Apollo 13, carried rakes and a wheelbarrow but due to an accident just after launch they were unable to carry out their plan to have a bonfire on the moon. *I don't be-leaf it!*

THE LEAVES of the Giant Waterlily (*Victoria regia*) are so strong that they can sometimes support the weight of a fully grown man. Crowds at the opening of Kew Gardens in 1852 were amazed when

Prince Albert leapt from the dais and landed squarely on a lily pad floating in the great Palm House pond. Unfortunately, the weight of the german pervert's penis jewellery - a large iron dumbell piercing his glans and attached to his scrotum with a heavy chain - proved too much for the leaf to support, and the Prince consort sank unceremoniously into the water. Needless to say his wife Queen Victoria was not amused. *I don't be-leaf it!*

LEAVES MAY be small, green and flat, but botanists believe that inside they are just like a human body. Like a person, a leaf breathes, grows and eats. Even more amazingly, it even goes to the toilet! But unlike people leaves don't have meals of food as we understand it. Every leaf - and there can be hundreds on a single tree - tucks in to a meal of carbon dioxide and sunlight before shitting out oxygen. So remember, next time you take a breath of air, chances are it's come out of a leaf's arse. *I don't be-leaf it!*

What's the naughtiest thing you've ever done?

YOU CONFESS

Steve Jenkins, 22, dispatch rider
"When I was 16, I borrowed my dad's car without permission. I crashed it, and said it had been stolen."

Richard Turd-Burglar, 12, ad-sales manager
"When I was a teenager in Australia, I used to steal women's underwear from washing lines and wear it in bed."

Peter Sutcliffe, 53, lorry driver
"Between the dates of February 1977 and November 1980, in the counties of West and South Yorkshire, I attacked and killed 13 women."

Andy Turnbull, 32, coffee machine engineer
"Once while stopping at my granny's, I used her false teeth to wipe my arse with, then put them back in her mouth."

SPACE - THE FINAL INSULT

By our National Disgrace Correspondent

THE heartbroken widow of pointy-eared Star Trek alien Mr Spock yesterday pleaded with Lottery chiefs: "Please don't let me sell my husband's medals to the Klingons".

Edna Spock, 68, from Peterlee in County Durham, has been living on social security hand-outs since her space husband Mr Spock died in the second Star Trek film. And now, facing growing weekly bills for bingo and cigarettes, she has reluctantly decided to sell her late husbands treasured collection of military medals.

Hope

"I promised Mr Spock that I would never part with the medals as long as I lived", said Mrs Spock yesterday. "But now that I am going to sell them, I only hope that the Lottery pay a small fortune for them and that they can stay in the country".

Monkhouse

Mr Spock, the emotionless Vulcan science officer who sprang to fame during the first Star Trek television series, won of hatful of Federation medals for battling the Klingons during a 28 year TV space career. But ironically it is wealthy Klingon collectors who are most likely to snap up the medals when they come under the hammer at an auction of Intergalactic Memorabilia to be held at Christies in London later this month.

Dylan

Spock's gongs include the Federation Bravery Medal that he received one week for climbing up a pipe to reverse the polarity of the dilithium crystals, and a special Federation Commendation Cap he was awarded by Captain Kirk for remaining at his post while the bridge of the ship shook about during alien attacks.

Zebedee

Eager collectors of space militaria could pay anything up to £50,000 for the unique collection. Only last year a medal awarded to Avon out of Blake's 7 was sold at auction for

Mrs Spock (above) treasures the space medal earned by her brave husband Mr Spock (below).

£5,500 to an anonymous giant spider bidding by phone from the planet Metabilis 3.

Florence

Experts fear that Spock's medals will also leave the planet, boldly going where no Star Trek medal has gone before - into the hands of pastie-headed, goaty-bearded, war-like Klingon collectors.

Milan

But Viz is launching a campaign to keep Spock's medals in Blighty. We believe they are a vital part of our National Heritage, and we want Lottery chiefs to cough up the cash to buy them back for Britain.

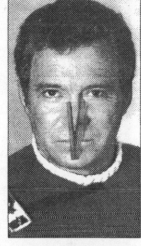
Spock's heirloom collection of space medals (above) selectively coated in 22 carat gold and accompanied by a wood and glass display case and Certificate of Authenticity.

"It's dis-guss-ting all that money being spent on a Dome for gays and child molesters when only a few millions of pounds would keep these priceless space medals in Britain", said a daft, bigoted cow yesterday.

Viz says SAVE Spock's Medals

Captain Kirk, now a corset wearing wiggy fruitcake, believes it would be wrong for Spock's medals to fall into enemy hands, despite the fact that the Klingons now maintain an uneasy peace with the Federation. Speaking from his death bed on the Channel Island of Alderney where he now lives in a giant cuckoo clock, he pleaded for Lottery chiefs to intervene.

Rome

"He may have been totally emotionless, but Spock would be raising an eye-

brow quizzically in his grave if he knew what was going on", he told us. "I'm sure he would find it all very highly illogical".

Captain Kirk, who has been dying of space piles since 1988, bravely agreed to start the ball rolling by donating £1.50 to our campaign.

Fruitcake Kirk thinks this whole thing stinks.

Spock's medals could be sold to KLINGONS

LATEST NEWS

TITANIC SINKS

THE Titanic, the world's biggest newspaper story, has sunk in what may prove to be the greatest tabloid disaster in living memory. And there are grave fears for over 2,000 journalists who were on board.

The story, launched in 1912, was a 'floating palace' for tabloid journalists who have used it to fill endless column inches for many months. Indeed, newspaper editors thought it was unsinkable.

But despite warnings of growing disinterest, the story steamed at high speed into a wall of public indifference. Reports indicate she is listing heavily and going arse-up in the water. Unless another story appears quickly to pick up the survivors, many jobs could be lost.

Turin

Several Sun journalists have been spotted clinging

TITANIC LOST
Grave loss of life is feared

The story pictured in 1912.

desperately to items of debris, including readers childish drawings of Leonardo Di Caprio and Kate Winslett. Among passengers known to be on board is Mr Piers Morgan, editor of the Daily Mirror, who was last seen wearing women's clothing in the vicinity of the lifeboats.

MINISTER FOR SEX

HE'S the most fanciable man in the House of Commons. Lady MPs go weak at the knees every time he strolls up to the dispatch box, and his saucy sexploits put every other red-blooded politician to shame. He's John Prescott, part-time MP for Hull, part-time Deputy Prime Minister and full-time sex machine.

Recent tabloid accounts of his X-rated raunchy adventures with blonde secretary Tracey Temple have set tongues wagging across the country, yet this office fling is thought to be just the tip of the iceberg. If the rumours are true, Tracey is just the latest in a long line of conquests stretching back more than 20 years. This news has come as no surprise to women, as there can be few ladies in the country who are immune to the Cabinet Casanova's heavyweight charms.

EXCLUSIVE

Prescott's sexiness at first hand is the assistant manageress at the pie shop where he goes for breakfast, elevenses and lunch every day when he is in his constituency. Maude Scratter, 52, has worked at Hull's North Bank Pies for 30 years, and knows only too well the effect that the Deputy Prime Minister has on women. Because he once had the same effect on her pies.

DPM WAS HOT STUFF IN PIE SHOP

One woman who has experienced

"I remember one day he came in for his usual 3 meat sqaures and 2 steak bakes," she explains. "Unfortunately, my microwave had broken that morning and I explained that I wouldn't be able to warm them up for him. "Leave it to me, Maude," he laughed. Then he shot the pies the sexiest look I have ever seen. Within a few seconds, steam started to come out of the top of them, and after a couple of minutes they were sizzling hot. If his lovers' fannies get as hot as those pies, he'd better be careful he doesn't burn his manhood when he sticks it up them," she added.

DREAMBOAT PRESCOTT PUT FERRY IN A SPIN

Another person who has witnessed the MP's incredible sexual charisma at close quarters is ferry worker Albert Featherstone. In the late 60s, Albert was a lavatory cleaner on the

"If his lovers' fannies get as hot as those pies, he'd better be careful he doesn't burn his manhood when he sticks it up them!"

Hull-Ostend ferry where the young Prescott worked as a waiter.

"I remember one time when John was given a tray of tea and biscuits to take up to the captain on the bridge," he recalls. "He knocked on the door and walked in with the tray and immediately the needle on the ship's compass began to spin out of control. Prescott's sexual magnetism was so strong that the navigation equipment was going haywire! Before long, the ship began to list and tilt so much it was in danger of turning over and sinking. Prescott was ushered out of the

bridge and taken below decks whilst th crew regained control After that, he was tol that for safety reason he must remain at th back of the ship at al times."

JOHN'S JUMP START GOT DOT GOING

When he's not driving women wild, th dreamy DPM is mos likely to be found driving around in one of hi famous Jaguar cars. And it was while h was behind the wheel of one of them tha he met Dot Herpes. The attractive 45 year old Grimsby divorcee had broke down in a country lane when Presco pulled up in his plush XJ6

"He looked under the bonnet of my ca and told me that I had a flat battery," sh remembers. "H said he would ge me started again and got a pair o jump leads out o his boot. He con nected one set t my battery, but I couldn't believe what h did with the other end. Instead of fasten ing the clips onto the battery in his Jag he shoved them down the front of hi trousers.

"I don't know what he did with them but my car started immediately! I didn' have to turn ther key or anything. He ha started my car using sex power alone. I fact, there was so much electricity com ing down those wires that the horn on m car started sounding, the electric aeria went up and down and my headlights ex ploded. It was the sexiest thing I have eve seen."

86% of Women Gagging for Taste of Prescott's Right Honourable Member

MINISTER'S PELVIC THRUST WORKED ANIMAL MAGIC ON ELEPHANT

Regent's Park Zoo keeper Terry Morris also saw up close the amazing effect that the Humberside hunk has over members of the opposite sex. And such is the power of Prescott's erotic punch that it can even cross the species barrier.

"I remember one day he and Tony Blair and were filming a party political broadcast at the zoo when Lulu the elephant got out of her cage," Terry recalls. "People ran screaming in all directions. Tony Blair panicked and climbed up a tree, but John Prescott just stood right in Lulu's path.

"As she thundered towards him, trumpeting with rage, Prescott remained cool as a cucumber. He put one hand on his hip and the other behind his head and thrust his pelvis forwards once, like Mae West. I don't know how he did it, but it made the noise of a kettle drum being struck. I couldn't believe what happened next; Lulu the elephant was knocked flat on her back!

"It was like she'd been shot. If Prescott has enough va-va-voom to stop an 8-ton charging elephant with a single hip thrust, then the women of Britain must be putty in his hands," added Terry.

THE WORKINGS OF A SEX MACHINE

FORGET Omar Shariff, Robert Redford and George Clooney, it's 68-year-old deputy PM John Prescott who gets Britain's women frothing at the ballot box. To the female of the species, the 20-stone bruiser's appeal is obvious, but to men it's a little harder to understand. But if Prezza's way with the ladies is a mystery, one thing's for certain - every red-blooded fella would like to know the secret of his appeal in the hope of recreating his success with the fairer sex!

We asked *Dr Frances Batter,* Professor of Sexology at the University of Wisconsin to explain exactly what makes the Deputy Prime Minister irresistible to women.

HAIR
Despite what bald men claim, women love a good head of hair on a man, and at 68, Prescott sports a healthy thatch of surprisingly brown hair. In addition, his barnet is styled to look as though it has been cut by his mam, Phyllis, with the kitchen scissors on a Sunday tea-time. This 'basin cut' accentuates his air of vulnerability and brings out the mothering instinct in women. Whether or not this is a deliberate ploy by the deputy premier only he knows. But judging by the profusion of notches on his bedpost, it is certainly one that works.

EYES
The eyes are the first point of contact between prospective lovers, and when Prescott fixes a woman with his watery 'come to bed' gaze, resistance is futile. And the huge, dark bags underneath his eyes confirm that once he gets her between the sheets, there'll not be a lot of sleeping going on! The promise of an all-night sex session with the author of 'Alternative Regional Stratergy: A Framework for Discussion, (1982)' is one that few ladies could resist.

VOICE
Traditionally, the language of love has been French. Charles Aznavour, Sacha Distel and Maurice Chevalier regularly turned women's knees to jelly with words such as 'l'amour', 'ma cherie' and 'he-honh, he-honh, he-honh', all uttered with a romantic Gallic lilt. Prescott may not be French, but his Warrington accent, fine tuned at Elmesmere Port Secondary School has an emotional honesty that can charm the underwear off any lady.

CHINS
It's a scene familiar from many nature documentaries; the male frogs in a pond inflating their chins to impress a prospective mate. In nature, it's a case of the bigger the chin, the more attractive is the male to the female. Today we may not live in ponds eating flies, but that primeval amphibious instinct remains as strong as ever deep within a woman's psyche. Prescott's chin air bags are a signal to women, unleashing their primitive desires and leaving them like putty in his fingers.

HANDS
Salty seadog Prescott's hands have been toughened by his early life in the merchant navy. Twelve years as a waiter on a North Sea ferry, operating a tea urn and collecting tips has left them strong and powerful enough to punch someone in the chin when they throw an egg at him. But these same hands are gentle enough to caress a woman's breast to orgasm, an ability that ensures he is constantly in demand as a lover.

GUT
Since the dawn of creation, women have looked for mates who could provide security and sustenance for themselves and their offspring. Back in cave-man times, a well filled stomach on a Neanderthal man was a sure sign that a prospective mate had got what it took to hunt successfully. Today we may buy our food from shops, but the primeval instinct remains as strong as ever deep within a woman's psyche. Prescott's pendulous bilge tanks are a signal to women, unleashing their primative desires and leaving them putty in his fingers.

SWEATY ARSE CRACK
Weighing in at 20+ stone, wearing a nylon suit and sitting squashed between John Reid and Gordon Brown on the cramped Commons front bench, Prescott's buttock cleft is an erotic factory pumping out clouds of the sort of sexy pheromones that drive women wild. It has been calculated that one whiff of his intoxicating arse vinegar after a hard day in the chamber would contain sufficient chemical attractants to arouse all the women in a town the size of Peterborough.

FEARS OF THE BLUE PETER STARS

We asked a few former Blue Peter presenters to use their knowledge of Blue Peter tortoises and other things to predict what life would be like in the year 2000 if Britain was extremely windy, and ruled by tortoises.

Tennis fan and one-time 'Money Programme' presenter **VALERIE SINGLETON** admitted that she would not be prepared for windy weather.

"I've always hung my washing out on the line, and in high winds that would be virtually impossible", she told us. "A moderate wind would be fine", she added "as that would speed up the drying process."

The late great Ted Moult's Everest double glazing understudy, dog owner and seventies junior TV stunt king **JOHN** 'get down shep' **NOAKES** said that windy weather would be a disaster for him. "I enjoy sailing my yacht around the Mediterranean, and windy weather is bad news for sailors. I'd have to put in to port, and hope that the tortoise harbour authorities allowed me in", he predicted.

CURRIE · 'Be- spectacled'

ELLIS · 'Pretty'

FREDA · 'Tortoise'

Be-spectacled grin-a-lot former 'Junior Showtime' star **MARK CURRIE** was able to smile, despite our gloomy forecast. "With a name like Currie, I reckon I already know a bit about wind", he quipped light heartedly.

Pretty pregnancy scandal presenter **JANET ELLIS** seemed unsurprised by our horrific weather predictions. "I have noticed it getting a bit windier recently", she remarked. "Perhaps it's something to do with the ozone layer, or Chernobyl", she added speculatively.

QUEEN MUM'S THE WORD

By our Palace Correspondent

Buckingham Palace officials are today expected to strongly deny rumours that the Queen Mother is to be bundled off into an old people's home.

The rumours came after a woman in Bournemouth reported that the Queen and Prince Phillip had visited her residential care home on the seafront and inspected the facilities.

STAFF

The Royal couple, who used a false name for their visit, were told that the Queen Mum would be one of 32 old folks resident at the home. Staff there would cater for her every need, and her room, which she would share with one other guest, would have a TV, running water and tea making facilities.

TAYLER

According to the home's owner, who preferred not to be named, the Queen and Prince Phillip appeared to be impressed by what they saw. Although the Queen Mother was never mentioned by name, Prince Phillip allegedly referred to "the mother-in-law" on several occasions during the visit, and the couple paid particular attention to the dining room and kitchen facilities.

McGOWAN

"They seemed most anxious that the food was up to scratch", the owner, a woman in her forties, told us. "They said that she'd had trouble in the past with fishbones."

ZETA-JONES

The Queen could expect to pay something between £300 and £500 a week to have her mother put away in a coucil approved home, a considerable financial commitment even for someone of her means. But the advantages would be obvious. Now in her nineties, the

Queen Mum is believed to talk about nothing but the war, and requires constant attention from Palace staff. Having her put away would relieve a great strain from the already overworked Royal family.

The pressure of living under a shared roof are well known to thousands of families throughout Britain.

ALPHA-SMITH

As one expert put it "It can be a real pain in the arse having some delapidated in-law rambling away in the corner of the room while you're trying to watch something on the telly". As a result every year in Britain over 800,000 senior citizens are put out to pasture, costing the tax payer an incredible £25 billion in social security hand outs.

20 THINGS YOU NEVER KNEW ABOUT... BEDS

We have sex on them, and breakfast in them. There's reds under them, and mattresses on top of them. Yes, love 'em or hate 'em, there's no getting away from 'em. Beds are here to stay.

So why not lie back with your head on the pillow and enjoy 20 fascinating facts you probably didn't know about beds.

1 Were it not for a print error on the sleeve of their first record, pop group Simply Red would have been called Simply Bed! For singer Mick Hucknall is mad about beds, so much so that as a child he used to lie in every morning.

2 A unique bed – one on which it is claimed Cliff Richard had sex – was sold at Sotherbys in 1981 for a record £31,000.

3 The biggest bed in the world is the Sea Bed, which is so big it doesn't have a mattress. Instead it is covered in soft sand, and is big enough to sleep over a million fish every night.

4 On the subject of sand, Sandy is a town in... you guessed it! *Bed*fordshire.

5 And so is Luton.

6 If you *go to bed* with someone, you don't necessarily share a bed with them for the night. Because *going to bed* with someone is a euphamism adopted by the younger generation, meaning to *have sex.*

7 So, for example, you could *go to bed* with someone on the sofa, or the kitchen table.

8 Or up against the side of a bus shelter.

9 In the Bible, Joseph and Mary didn't *go to bed*. They conceived the Baby Jesus *immaculately*, which means there was no jiggery pokery involved at all.

10 If you expect a bedsitter to look after your beds for you while you go to the pub, you'll be disappointed. Unlike a babysitter, a bedsitter is a cramped attic room in London containing a filthy mattress and a calor gas stove for which you pay £800 a week rent.

11 You could also be forgiven for thinking a bed pan is a kitchen vessel for the cooking and preparation of beds. But you'd be wrong. A bed pan is in fact a special potty for grown-ups which enables bed-ridden people (people with lots of beds) to go to the toilet without getting up.

12 John Lennon caused a storm of controversy in the sixties when he publicly went to bed with Yoko Ono in an Amsterdam hotel room.

13 And Paul McCartney surprised a few people when he decided to marry Linda.

14 Numerous pop stars have since got themselves into bed conundrums. "All I've got is a s-single bed", sang Noosha Fox in the seventies hit of the same name. Meanwhile, her pop counterpart Gordon Sting of The Police complained "The bed's too big without you".

15 Life is not a bed of roses. However, opinion is divided as to whether it is a minestrone served up with parmesan cheese, a cold lasagne, or a bitch. And then you die.

16 A bed of nails is probably the world's most uncomfortable bed. For it is indeed a blanket of sharp protruding nails slept on by mystical snake charmer types.

17 And it's also the term used to describe any bed belonging to TV's top Geordie pop cop, singer turned actor turned singer again turned writer turned director turned producer, Jimmy Nails.

18 'Ole blue eyes himself, Frank Sinatra, refuses to sleep in the same bed twice. Instead he buys a new bed every day, and has the old one burnt in the morning.

19 A bed bath is not a special water bed which doubles as a bath. It's a popular male fantasy in which Joanne Whalley-Kilner pulls you off whilst dressed in a nurse's uniform.

20 There are two types of bed bug. One is a microscopic insect which lives under your mattress and comes out at night to crawl up your arse. The other is a small listening device which is concealed beneath your bed on your wedding night so that your best man and his mates can listen in to the pathetic sound of you attempting to shag your wife after you've been drinking all day and then dancing all evening with your mother-in-law.

An amazing behind the scenes look at British Intelligence

MY NAME IS JOHNSON...
...REG JOHNSON

EXCLUSIVE

When we think of secret agents we think of James Bond, 007, of fast cars, action and beautiful women. But what is it really like being a spy? Is it as exciting and glamorous as the movies would have us believe?

Until recently Reg Johnson was an undercover operative working for British Intelligence. But now he is blowing the lid off Britain's secret service in a book which makes 'Spycatcher' look like a kid's bedtime story. And here, in an exclusive serialisation of his book, Reg dispells the myth that the Bond films have created and replaces fiction with fact in the first TRUE story of the British Secret Service.

' When I watch a James Bond movie I can't help laughing. In real life the secret service is nothing like that. We do most of our work behind a desk, without a gun in sight.

Extracts from the book that makes Spycatcher look not very controversial

UNDERGROUND

The hardest part of working for the secret service is getting the job in the first place. I was half an hour late for my interview 'cos I couldn't find the office. I had to go into a phone box somewhere in London and dial a secret number. Then the phone box turned into a lift and went down to the British Intelligence Secret Headquarters, which is underground. Of course I couldn't find the right phone box, and when I did there was a queue of people waiting to use it.

SHARKS

The interview itself was tough. I thought I was doing quite well, then somebody pressed a button and my chair fell backwards tipping me into a pool full of sharks. It was all part of the interview. I had to kill the sharks and escape. Luckily I killed them all, and they asked me to start on Monday. But there was at least a dozen other blokes who weren't so lucky!

Once you're in you get several days of special training before they send you on any missions. Mind you, there's none of those fancy gadgets that 'Q' comes up with in the movies.

That stuff is pure nonsense. In reality they just teach you basic things like fighting and how to escape from places. You get flying lessons too — in planes and helicopters — and they teach you how to drive a car — on two wheels, and under water.

EXOTIC

After training they give you a gun and a car and send you on a special misson. Any old car won't do. It has to be bullet proof and have an ejector seat. After you've been a spy for two years they give you a car that turns into an aeroplane.

The kind of work done by the secret service is far removed from the excitement and glamour of the movies. In James Bond the villians live on exotic islands surrounded by beautiful women. But in real life it's nothing like that.

BIKINI

I only ever got sent to an island once. A Russian spy had stolen an atom bomb and was going to destroy the world. When I got there I found his girlfriend on the beach wearing a bikini. We had sex, then she

helped me to find the bomb and diffuse it. If we had been two seconds later it would have gone off. That was a close shave, even by my standards.

TEETH

Being a spy is a dangerous job. One day I was on top of a cable car when I was attacked by a tall man with metal teeth. We had a fight and I knocked him off. Then I managed to climb up the wire all the way to the top just before the cable car exploded. Of course I got paid danger money for that sort of thing. On a really dangerous mission I could come out with £200 a week, including overtime and danger money. Of course, there were other benefits too.

KARATE

Spies always have to have sex with women, usually to get secret information out of them, or to get them to help you escape from places. Once I was in Hong Kong when six women in bikinis who were good at karate attacked me. Luckily I'm a black belt in karate, so I won. Then I had sex with them. Afterwards

they gave me all the secret information and then helped me to escape.

WARDROBE

You can never relax being a spy. Even at weekends. Every time you enter a room you have to check to see if someone's hiding behind the door, or in a wardrobe. Once I was going to town on a bus when a fat Chinese man tried to kill me with a sharp metal hat. I had to kill him with some electrical wires and throw him out of the window. I managed to get off the bus just before it exploded.

QUEEN

It's no fun killing people, but when you're a spy you have to. But first you have to have a licence to kill. And they're a lot harder to get then T.V. licenses, I can tell you. Your application form has to be signed by the Queen of England, and they cost £500 a year. That might seem like a lot of money, but being a spy I couldn't do without it.

Next week: How I had to escape from the moon after being drugged and put in a space rocket by another spy with three nipples and a golden gun. '

'I WAS A SPY, HONEST' by Reg Johnson, is published by Omlette Books, price £2.95.

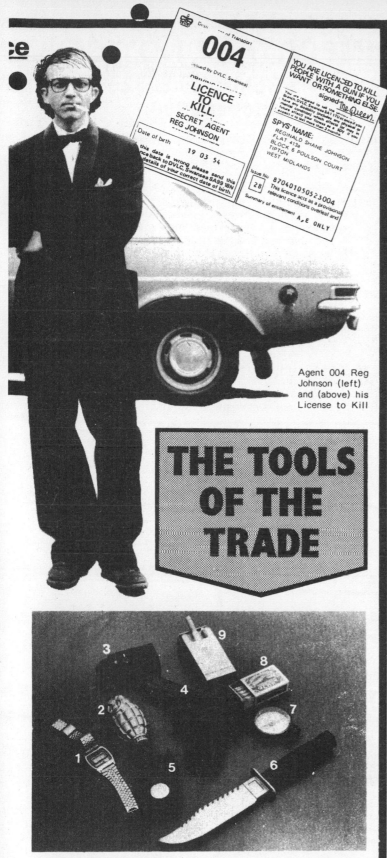

004

Issued by DVLC Swansea

LICENCE TO KILL.

SECRET AGENT
REG JOHNSON

Date of birth 19 03 54

this date is wrong please send this licence back to DVLC Swansea SA99 1BN details of your correct date of birth

YOU ARE LICENCED TO KILL PEOPLE WITH A GUN IF YOU WANT OR SOMETHING ELSE

signed The Queen

SPYS' NAME:
REGINALD SHANE JOHNSON
FLAT 413a
BLOCK 6 POULSON COURT
TIPTON
WEST MIDLANDS

Issue No 28

87040105052300 4
This licence acts as a provisional relevant conditions overleaf and

Summary of entitlement A,E ONLY

Agent 004 Reg Johnson (left) and (above) his License to Kill

THE TOOLS OF THE TRADE

Here is just a small selection of some of the equipment used by Reg Johnson on his secret missions. (1) A special watch which tells the time underwater and even in space. (2) Hand Grenade (3) Exploding camera. (4) Special gun which turns into a cigarette lighter when you press a button. (5) Walky talky radio for sending messages back to base. (6) Knife with road maps, emergency food supplies and waterproof clothing inside the handle. (7) Special spy compass. (8) Exploding matches. (9) Cigarette packet missile launcher. Each cigarette is a missile in disguise, powerful enough to blow up ten helicopters.

McCARTNEY 'STOLE MILK'

Greedy ex-Beatle pinched pensioner's pintas

The pop world has been rocked to its foundations by amazing allegations that millionaire superstar Paul McCartney — one of the wealthiest men in Britain — has been STEALING milk from the doorsteps of an elderly neighbour near his luxury home on the Mull of Kintyre, Scotland.

And furious local resident Mrs Isla McKitterick believes the greedy ex-Beatle could have **MADE OFF** with literally **GALLONS** of milk since his crime wave began several months ago. She believes that McCartney launches regular 'dawn raids' on her doorstep from his converted farmhouse only 8 miles away, and then returns home to **GUZZLE** the stolen milk with his wife Linda and their two children.

McCartney - 'guzzled' milk

PRISON

"This has been going on for some time now", Mrs McKitterick told us. "I've reported it to the police, but they still haven't done a thing about it. They know where he lives — I think they should go up there and arrest him. He should be sent to prison".

EVIDENCE

However, local police are powerless to act against the wealthy star due to lack of evidence. "There is no evidence whatsoever to support these allegations", we were told.

THEFTS

Meanwhile a spokesman for the 40 year old star denied all knowledge of the thefts, and dismissed the claims as "ridiculous".

Mrs McKitterick is no stranger to controversy. She last made the headlines in 1974 when she accused the late Elvis Presley of intercepting her mail and opening letters which had been addressed to her.

I'M ARRESTING THIS HANDBAG ON SUSPICION OF BEING AN ACCESSORY.

THIS THING CAN'T GO ON.

STOP

ROAD CLOSED

'I COULD DO THAT!'

Window cleaner issues challenge to Queen

A window cleaner from Sunderland has laid down a dramatic challenge to the Queen. "If she comes up here and does my job for a week, I'll gladly do hers".

Brian Potter, who's been a window cleaner for almost 20 years, issued his challenge after seeing the Queen on the television. "I'd swap places with her tomorrow if I had the chance", he told us. "She's got it easy down there. She gets paid a fortune for doing nothing", said Brian, who earns less than £80 a week. "But I'd like to see her up a ladder cleaning windows in all weather. She wouldn't know where to start".

WIPE MARKS

Brian issued his challenge in a letter to Buckingham Palace. "Cleaning windows is a lot harder than you'd think", he told us. "You might think you're doing it properly, but then when they dry, you've probably left loads of dirty wipe marks", he explained. "You have to start at the top and wipe it in circles, sort of sideways and downwards", he added.

If the Queen does rise to the challenge, she'd need to wear suitable clothes, according to Brian. "It's no good her turning up in high heels, and a fancy dress with her crown perched on the top of her head", said Brian. "That would fall off straight away".

CHALLENGE

So far Brian's challenge has remained unanswered. A spokesman for Buckingham Palace said the Queen was not available for comment.

IMPORTANT NOTICE

Viz Comic Issue 32
October 1988

After serious technical faults were discovered in issue 32 of Viz Comic the Publishers have decided to recall all copies of that comic so that the faults may be rectified.

The fault lies in the stapling, but it must be stressed that readers are in no immediate danger as only the upper staple is affected. However, you are advised not to read the issue unless it is absolutely essential, and then should avoid prolonged reading. **Under no circumstances should readers attempt the repair themselves.**

If you own one of these comics, please send it to Viz Issue 32 Fault Department, P.O. Box 1PT, Newcastle upon Tyne NE99 1PT. Please enclose a stamped self addressed envelope for its return, and we will effect the repair free of charge.

We must stress that this fault does not affect any other issues of the comic, all of which may be read in complete safety. The Publisher apologises for any inconvenience caused.

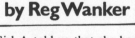

TOP OF THE FORM!

SEXY SCHOOLGIRLS SCORE WITH SAUCY SIRS

Good looking girls are more likely to succeed at school than their less attractive classmates. That's the shock finding of a special report soon to be published.

The report claims that many teachers give preferential treatment to prettier pupils, often at the expense of other, often equally bright youngsters. And it claims that a majority, although not all male teachers give high marks to good looking girls, even if their work does not deserve it. At exam times, school teachers fiddle the results to make sure that their favourite pupils pass, and the less attractive girls fail. The report goes on to claim that as a direct result many highly intelligent girls leave school without vital job qualifications simply because teachers considered them plain looking, slightly unattractive or perhaps a little overweight.

GIRLS

We carried out our own investigation to see if there was any evidence to support these startling claims. We spoke to two girls both of whom had recently left school. Girl A was tall, blond and good looking, while we considered girl B to be unattractive.

by Reg Wanker

Girl A told us that she had done well at school, had received high marks in her exams, and now had a well paid job working in a travel agency. She hoped eventually to become an air hostess. However, girl B had left school with only one 'O' level and after 4 months had still not found a job.

CARETAKER

We rang a school in order to ask male staff for their reaction to the report. But the caretaker told us no-one was there, because it was Saturday.

Fish say NO to sex

Fellas! If you are thinking of having sex with a fish, then you're out of luck, according to one leading expert on marine biology. (That's fish to you and me).

For he tells us that fish are not attracted to men in the same way that women are.

LIVE

"Fish are completely different to women", our expert told us. "They live underwater, and have an entirely incompatible system of reproduction". In other words it's strictly "no sex please, we're fish!"

WIRED FOR SOUND

Cliff Live Knocks 'em Dead on Death Row

POWER TO ALL OUR FRIENDS: Sir Richards kept fans singing when Old Sparky blew a fuse.

CLIFF RICHARD kept an American crowd entertained yesterday when an execution was delayed by an electrical fault.

The veteran rocker, who is visiting the US to promote his latest album, had popped into Folsom Prison in California to witness local murderer Billy-Bob Hurtubuise being put to death in the electric chair. However, a fuse blew when guards threw the switch, bringing proceedings to a halt. Sacramento prison authorities called in an electrician, who estimated that it would take at least half an hour to get the chair working again. The specially-invited audience, made up of the prison governor, state officials, reporters, the families of Hurtubuise's eight victims and twelve selected witnesses including Cliff, were asked to be patient while the repairs were carried out.

TENSE

Prison warder Crimplene Hogg told NBC's Hymen Prepuce what happened next. "The crowd started to get restless. They'd all come to see Hurtubuise die like a dog in Old Sparky, and they didn't like the idea of having to wait around for the main event. The atmosphere in that

EXECLUSIVE!

room was sure getting pretty tense until Cliff got up and started to sing," he said.

PARTICIPLE

"It wasn't long till everyone in the execution chamber was singing along and tapping their toes," he added. "And that included Billy-Bob strapped in the chair. He was hollering louder than anyone else!"

MUNICIPAL

For nearly 40 minutes the ageless star, 68, kept the audience entertained with a medley of his best-loved hits. All-time favourites such as *Summer Holiday*, *Congratulations* and *Living Doll* kept spirits up while the electrician worked to fix the faulty chair. And the unexpected gig proved to be a great success with the audience, prison staff and the condemned prisoner.

OLYMPIC

"Cliff was fantastic," commented Assistant District Attorney Delorean Vanderbilt. "He's got a back catalogue with nearly six decades of hits to choose from, and he kept everyone entertained until the electrician gave the execution the thumbs up. We were all kind of sorry when we had to stop singing and sit down to watch Billy-Bob fry."

ENGAGEMENT

And the Peter Pan of Pop's impromptu singalong proved to be a hard act to follow. "It was a good execution," said Vanderbilt. "Hurtubuise didn't want to die, and the warders had to crank that old generator up four times until his skin burst and his head caught fire. But frankly, after our own private concert from Cliff even that was a bit of an anticlimax."

CON-GRATULATIONS: Condemned man Hurtobuise on his way to be electrocuted yesterday, yesterday.

Who wants to be a MILLIONAIR

What is it like to have millions and millions of pounds? To be able to jet round the world spending cash like confetti. What kind of lives do the multi-millionaire pop stars, the mega-rich business tycoons and the rolling-in-it Royals lead?

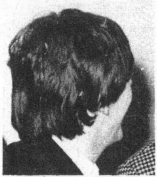

Ex-Beatle McCartney

£ Not many people could afford to walk into their local corner shop and buy 316 million king size Mars bars. But if he was feeling particularly peckish, ex-Beatle **PAUL McCARTNEY** could do just that. His vast £79 million fortune would also pay for a slap up chinese meal — for eight and a half million people! However, prawn crackers would be an extra £5 million — a luxury that even Paul could not afford.

What is it like to be LOADED with lolly?

£ Fellow ex-Beatle **GEORGE HARRISON** would have to be a bit more frugal in his choice of restaurant. For his personal wealth amounts to less than £15 million. A modest hamburger and chips at McDonald's followed by a visit to the cinema—for himself and just over 3 million friends—would be more within George's budget.

£ Rock superstar **PHIL COLLINS** has very few money problems, except of course finding somewhere to keep it all. His £22 million fortune, if it were in £1 coins, would weigh an incredible 176 tonnes (enough to wear a hole in the pocket of even his toughest denim jeans). That's not the kind of money you carry round with you every day, so moving his millions

IN BRITAIN TODAY, MOST PEOPLE ALREADY ARE

A fair cop? Arrests like this one can earn a British bobby over £8.600 in a good week

Who wants to be a millionaire? At one time the answer to that question would have been a resounding "I do" from the vast majority of the British public. But nowadays in prosperous Britain, the answer is more likely to be "I already am".

For believe it or not, in 1988 almost half the people reading this magazine will already have joined the exclusive 'millionaire's club'.

CASH

And it's not just the royalty, land owners, pop stars and people with lots of money who are becoming millionaires these days. An increasing number of self made men and women are beginning to amass seven figure fortunes. And it's not only the big businessmen who are raking in the lolly.

MONEY

Bus drivers, shop keepers and even policemen all have the earning power that it takes to

By MAURICE SHITE

make a million. In London, a bus driver working a couple of shifts a day and getting regular overtime can earn anything up to £240,000 a year, including tips.

MORE MONEY

Top policemen earning massive commissions on every arrest they make can bank around £450,000 for a good years work, while a hard working shop owner opening seven days a week till late in the evening, selling expensive things, can probably come home with around twice that amount.

Even social security claimants can qualify for the big money league. A single parent family with two children receiving a variety of generous state benefits, like Child Allowance, Family Income Supplement, Milk Tokens etc., could soon be rubbing shoulders with the Richard Bransons and Queens of this world, providing they invested their money wisely. However it's a sad fact that many social security claimants choose to squander their money on their short term interests such as food, rent and heating etc.

★ Rude Kid

WHAT WOULD YOU LIKE FOR YOUR BIRTHDAY, DEAR?

TWAT ON A BRICK!

Are YOU a millionaire?

Here's a simple questionaire that will reveal whether or not YOU are a millionaire. Just answer the following question a, b or c, then tot up your final score to see where you stand in the big money league.

1. How much money have you got?
   a. Less than £100.
   b. Between £100 and a £999,999.
   C. A million pounds or more.

A million quid

HOW DID YOU DO?

SCORING: a - 1 point, b - 2 points, c - 3 points.

1 point: I'm afraid there's no lavish lifestyle for you, at least not in the near future. There's still a long way to go before you reach that magic million.

2 points: Not bad. You're well on your way to joining the jet setters, but there may still be a little while to wait. Keep saving that money.

3 points: Congratulations! You're a fully fledged millionaire. Why not go out and celebrate with champagne and caviar?

must be a real headache for the singing star.

For example, it would cost the pint sized popster over £121,000 to send his fortune by Red Star from London to Glasgow, for delivery to the door the next day before 12 noon. The cost of taking his money on tour with him, to Europe or the United States, could run into millions.

 Having millions of pounds means never having to worry about the phone bill again. Pop entrepeneur **RICHARD BRANSON**, his personal fortune estimated at over £130 million, could chat happily on the phone to a relative in Australia for anything up to 296 years before running out of ten pence pieces. And millionaire publisher **ROBERT MAXWELL** could go on talking much longer — until the year 2901 if necessary.

But when it comes to the richness stakes, **THE QUEEN** is in a league of her own. The combined fortune of all Britains's pop stars and businessmen would be chicken feed compared to the riches of the Royals. For example, if Her Majesty the Queen decided to play snooker for money, it would be doubtful whether anyone in Britain could afford to take her on. For it would take a Royal piggy bank the size of Tower Bridge to hold her enormous £3,300 million fortune which, if it was stacked in bundles of £5 notes, would take up over half a cubic hectare of space.

Royal youngsters Prince William and Prince Harry probably receive more each week in pocket money alone than most of us earn in a lifetime. Add to that the £15 or so a week which they each earn delivering newspapers in the affluent South East, and the Royal twosome are far and away Britain's wealthiest toddlers.

TV celebrities are undoubtedly among Britain's top earners. An actor starring in top soap 'EastEnders' can expect to earn at least a million pounds for what is essentially a part-time job, working only two evenings a week plus another hour on Sunday afternoons. In addition to their basic pay, greedy stars can then charge anything up to £800,000, or even twice that, just to open a supermarket.

Despite being Britain's richest man, comic **BENNY HILL** lives on a diet of Kentucky Fried Chicken in a house no bigger than an extremely large dog kennel. Neighbours of the 63-year-old reclusive millionaire TV funny man near his home in Southampton told us that "Benny lives on a diet of Kentucky Fried Chicken in a house no bigger than a extremely large dog kennel".

Win a fab prize!

Shakin' Stevens fans — ever wondered how much room an extremely large sum of money would take up? Well if you have, this could be your big chance to win a free ride in a removal van for yourself and a friend!

SHAKY

All you have to do is imagine that during 1989 Shaky's record sales soar, and by the end of the year he has earned a cool £250 million. Shaky then goes to his record company to collect the money, and he is given £217 million in fivers, £19 million in tenners and all the rest in £1 coins.

Using your knowledge of how much room money takes up, simply work out how many removal vans Shaky will need to get all his money safely home in one journey. Send your entries, on a postcard, to Shaky's £250 million Removal Van Requirement Competition, P.O. Box 1PT, Newcastle upon Tyne, NE99 1PT, to arrive in the post. Our judges decision will be final, and the winner will be notified by post.

'REF, CAN YOU CLEAN THE WHITE BALL— I THINK THERE'S A HARE ON IT!'

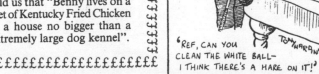

165

SEX IN THE YEAR 2000

Have you ever imagined what it will be like to have sex in the future? Will space-age sex be different to the way we make love today? Experts believe it will. Indeed a 21st century sex revolution could change the way we live. So let's take a look into the future and look forward to having SEX IN THE YEAR 2000.

Over the years, attitudes towards sex have changed dramatically. Well, those attitudes will continue to change. By the year 2000 Victorian values will be a thing of the past. Gone will be inhibitions and prudish attitudes towards sex. Instead making love will seem as natural as lighting a cigarette — strangers will do it at bus stops, on trains or where-ever the fancy takes them.

SEX

The out-dated Indecent Behaviour laws will have been removed from the statute books — instead the police will actively encourage people to make love. Sex will be so common-place, our clothes will be specially adapted so there's no need to take them off when we have sex.

SEXUAL

At work, coffee and tea breaks will be replaced by sex breaks. Bosses, following the Japanese example, will realise that sexual stimulation increases productivity. Profits will be up, and Britain will boom to the sound of bonking.

SEXY

New healthy attitudes towards the subject of sex will mean the end of seedy sex shops. Instead families will visit huge out-of-town sex hypermarkets, with free parking for over 2000 cars, selling everything you could ever need for sex. From

sexy underwear to a pair of skimpy briefs. And sex hypermarkets will be open on bank holidays too!

Sex will no longer be an awkward, old-fashioned show of affection between two people that takes place behind closed doors. People will have sex at all times of the day, in all sorts of places, and in any numbers. At football matches whole crowds will have sex together to celebrate a goal. And new attitudes will mean no more complaints about sex on TV. Instead, old ladies will sit down and enjoy the Eurovision Sex Contest. A grey-haired Terry Wogan will present the show, and couples from all over Europe will have sex on screen, hoping to win the competition. Indeed, competitive sex will be the sport of the future with top

athletes, at the peak of their physical fitness, going for gold in sex — the most popular olympic sport of the year 2000.

You read the SEXIEST stories in Viz

Incredible advances in technology will begin to change sex beyond all recognition. As well as Access and Visa, people will eventually carry SexExpress cards. To have sex with other SexExpress cardholders, simply pop the card into any High Street SexPoint machine . At the end of the month you will receive a statement letting you know how many times you have had sex, and with whom.

SEXINESS

Cinemas will have been replaced by multi-screen Seximas. Having sex with your favourite movie star will become a reality, thanks to special attachments on the seat in front of you. Meanwhile at home watching TV will never be the same again. Special hologramic TVs will allow you to have sex with the newsreader as they read the headlines, or fondle your favourite weatherman as he makes his forecast.

And new technology will also make sex possible via the phone. Simply ring a friend's number, then plug your telephone into a special socket in your bed. Crossed lines could lead to some thrilling three-somes, or even a fabulous foursome. But beware — these sexiphone sessions are likely to cost you as much as £600 per minute at peak times, owing to inflation.

SEXUALITY

Sex education will be revolutionised too. The children of the future will learn sex in the classroom along with English and maths. They will be able to visit Sex Museums too, where actors will perform old-fashioned "bedroom sex" — the kind we enjoy today. By the year 2000, sex in bed will be a thing of the past.

SEXTANT

Sexual diseases will also be a thing of the past. Doctors will have invented a special new space medicine that tastes like lemonade and stops you from catching any sex diseases. And pregnancy will not be a problem either. Women will be able to swallow a special pill every night, smaller than a smartie, and if taken regularly this will stop them becoming pregnant.

WIN A BOTTLE OF VODKA

Experts believe that by the year 2000 it will have become customary to drink alcohol before having sex. And they claim to have conclusive evidence to link alcohol with improved sexual performance. A spokesman for **Vladivar vodka** today told us that drinking alcohol can boost sexual potency by anything up to 700% in real terms. "After a couple of bottles of **Vladivar,** I can perform sexually for periods of up to two weeks."

Why not try it for yourself. We're offering a bottle of **Vladivar vodka** to each of the first five people who write in and tell us who they would like to have sex with in the year 2000, how and why. Include a diagram if necessary.

Send your entry to "Sex and Vodka in the year 2000", Viz, PO Box 1PT, Newcastle upon Tyne NE99 1PT.

DARLING — THERE'S A FEW THINGS I'D LIKE TO GET OFF MY CHEST.

SEAN FREE!

Bond To Be Wild!

- Connery campaigners want star returned to the streets

AN ambitious attempt to release Sean Connery back into the wild is being scheduled for early next year.

The 68 year old actor, who was taken away from his native Edinburgh by film producers almost 50 years ago, is currently being kept in Marbella, Spain, where he spends much of his time playing golf. He is still flown to Hollywood occasionally to perform for film crews.

Star

Conservationists and film fans alike feel that the ageing star should be returned to his natural working class environment after a lifetime spent in showbusiness.

After 50 years in showbusiness Sean Connery (above) could soon be returning to the wilds of working class Edinburgh.

Bounty

Working class Scotsmen are fast becoming an endangered species as a result of New Labour's classless society. The reintroduction of ex pats like Connery back into run down inner city areas could be the only way of maintaining a breeding working class population for the future.

Mars

Last year brown nosing comedian Billy Connolly was released back onto the streets of Glasgow. 'Free Billy' campaigners successfully loaded the banana booted comic into a canvas sling at his mansion in Los Angeles and throughout a 12 hour flight to Glasgow the bloated comic was hosed down with champagne.

Anti-roll

The "Big Yin" was strapped to the top of a Land Rover for the final leg of his journey home from the airport to the Gorbals district of Glasgow. After an emotional farewell from his showbusiness pals, including a tearful Sir David Frost, the bewildered looking star walked nervously away from the vehicle. For few moments he seemed unsure of himself, then suddenly he bounded off and was quickly lost amongst the tenements.

Paralell

The same team will be handling Connery's release. Dr Jennifer Goodall, Professor in Proletariat Conservationism at Heriot Watt University, will be in charge of the operation.

Gay

"The main danger is that working class celebrities struggle to adapt to their natural environment after spending too long in showbusiness", said the Professor. "But in the run up to Connery's release we will be taking special measures to ensure that the transition goes as smoothly as possible".

Gold

For the next 12 months the millionaire actor will be weaned off playing pro-celebrity golf, and encouraged to make his own breakfast, preferably fried eggs and bacon. "We will also be encouraging him to wear his socks twice before they are washed, and to be less condescending to people on lower incomes than himself", Dr Goodall explained.

Connery's return to the wild is set for Spring of 2000. After his release onto the back streets of Edinburgh his progress will be monitored by scientists using an electronic tagging device attached to his Rolex watch. For his first few weeks of freedom luxury food items such as smoked salmon and quails eggs will be dropped off near to his release point to help the star's transition to self sufficiency. Gradually the quantities will be reduced, encouraging the star to fend for himself.

Musclebound

Scientists hope that Connery's release will be more successful than that of Billy Connolly, whose freedom lasted less than a week. He was found beaten up in the Bells Hill area of Glasgow where he had been scavenging for caviar in dustbins outside a chip shop.

True

"Unfortunately in Billy Connolly's case the other working class males appear to have rejected him", Dr Goodall explained. "They probably noticed a foreign scent - like the smell of Prince Andrew's shit on his nose - and reacted violently."

Crooner Phil in crocodile shock

SINGER Phil Collins has vowed never to record a song about crocodiles. For the slap headed pop millionaire is a real-life Captain Hook.

Like the pirate in Peter Pan ugly Collins, 46, is terrified of the razor toothed reptiles. So much so he demanded record company bosses write a clause in his contract excusing him from writing or performing songs about, or including references to, crocodiles.

Phil Collins (above) will *not* try rocking the crocodile rock - his feet just *can* keep still, thank you very much.

False

Over the years crocodile rockers have made a fortune singing about the snap happy creatures. In 1973 Elton John's 'Crocodile Rock' soared to No. 5 in the charts. The ivory tinkling arse tickler celebrated by snapping up 2,000 pairs of crocodile skin shoes the very next day.

Waterloo

A chubby young Geordie schoolkid was so inspired by seeing Elton's extravagant footwear on Top Of The Pops, 20 years later he wrote a song about it. Plank actor Jimmy Nail's 'Crocodile Shoes' was a massive hit, and launched the Easter Island statue headed star's singing career.

Paddington

Yet Phil Collins refuses to take the crocodile bait. "It would be easy for Phil to write a song about crocodiles and make a fortune, but he was never one for taking the easy way out", said one record company insider.

Winnie the Pooh

Ironically Collins's phobia does not extend to alligators. And just as well! For the stocking faced star, whose hits include 'In The Air Tonight', is a keen alligator breeder and keeps a dozen of the scaly croc lookalikes in a giant cuckoo clock fastened securely to the wall at his Swiss mountain home.

Is West Midlands Town Resting Place of Biblical Relic?

The Da Guthrie Code

HOLY MOSES: *The ark of the covenant (top) contained the ten commandments (left) a fragment (far right)of which has been found in Tipton by Hugo Guthrie (above)*

WE'VE ALL read the *Da Vinci Code* and dismissed it as a load of old tosh. But one man who takes the legends of the Knights Templar seriously is West Midlands local authority tourist chief Hugo Guthrie. For he believes he has unearthed evidence of the lost burial place of the Ark of the Covenant; and it isn't in Jerusalem, Egypt or Cathars... *it's in Tipton!*

TABLETS

Fascinated by the stories of the Knights Templar since seeing the blockbuster film Indiana Jones and the Raiders of the Lost Ark as well as the poorly rated DaVinci Code starring Tom Hanks, Guthrie, 61, believes he has tracked down the fabled Lost Ark – the holy golden relic containing two stone tablets on which the Ten Commandments were inscribed and given to Moses by God on Mount Sinai. In the Bible, the Ark was said to have amazing powers, being capable of raising storms, radiating divine fire, smashing chariots, and changing the weather.

"This discovery is not just of archaeological interest," insists Mr. Guthrie. "If we learn how to harness its incredible powers properly, we could use the Ark to end this summer's drought and solve Britain's power crisis. And pick winning lottery numbers."

PILLS

It is believed that the Knights Templar rescued the Holy Grail, long thought to be the cup from which Jesus drank at the Last Supper, along with the lost Ark of the Covenant, and brought them back to Europe. After the fall of the Templars, the two treasures vanished from historical records for centuries. Biblical scholars, archaeologists and adventurers have all spent years searching for the Ark, but until now its secret resting place has remained one of the world's most enduring mysteries.

OINTMENT

Although the Ark has not yet been found, Guthrie is optimistic that it will be after the thrilling discovery of an inscribed stone slab, which he has rumoured to be one of the tablets bearing the Ten Commandments. "I found the fragment of Commandment stone buried near an old well in a field belonging to my brother-in law. I was clearing some old bricks away to help him build a new shed when we saw this stone," said Mr Guthrie yesterday. "At first I thought it was just a bit of old paving slab, but when I took a closer look, I realised it had letters on it. I don't

By Our Reporter
THOR OERSTRYK

read Hebrew, but I certainly recognized a few letters, and the words 'shalt not kill' were as plain as day."

After sifting through literally several documents and historical writings and visiting Tipton's municipal museum, Guthrie concluded that, following their defeat by the Saracens, the Knights Templar took the Ark to the site of his brother-in-law's field in Tipton. Here they left a series of strange clues on the walls of a nearly pub that point to where they hid the Ark.

SUPPOSITORY

Mr Guthrie, who three years ago was fined £450 after pleading guilty to performing a sex act on a badger, expects Tipton to be flooded with amateur archaeologists over the summer. He believes all will be hoping to find other priceless Templar treasures, including the Spear of Destiny, the Turin Shroud and that bit of wood from the cross with 'INRI' written on it. "This will be a big boost for the Tipton area," said Guthrie, who has unveiled ambitious plans for a lavish portacabin-style visitor's centre and shovel shop abutting his brother-in-law's field. He has applied for funding from the Vatican, the Friends of the Knights Templar and the West Midlands Metal Detector Club.

MILK BOTTLE

Guthrie is pulling out all the stops for the opening of his complex. "My wife is going to do the catering and I have been promised that a crew from BBC Midlands Today will be along," he told reporters. And he is hoping for a star-studded turnout. "I've sent invites for the opening to everyone from the Pope to Harrison Ford and Tom Hanks, and I've already had an acceptance from David Dickinson," he told us.

'Lunar Pond Plan a Winner' says NASA

US SPACE AGENCY boffins say that craters on the Moon would make "smashing little fish ponds."

Professor Jackson Pallo, head of Lunar Colonisation Research told reporters: "A pond always brightens up a garden, and I'm sure it would do the same for the moon. Let's face it, it's not the most interesting of places, and a nice pond with a fountain would look wonderful. There is also the advantage that you wouldn't have to dig a hole, just fill a crater with water."

When we called pop star Sting, who had a hit with *Walking on the Moon*, he said: "I'm in the middle of fucking my wife at the minute. Ring me back in about four hours"

CELEBRITY SWEARS

No. 96 ~ Sooty & Sweep

What's that, Sooty?

He said 'Felch'

MICHAEL WINNIT

THE BIGGEST MOUTH **THE BROWNEST NOSE**

Hang the curtain hangers!

IF there's one kind of person I cannot abide it's interior designers.

I have been having some very expensive new curtains fitted in my enormous house recently, and thanks to my interior designer the little brass bits that we've ordered from Italy (those bits that hold the curtains open) won't be ready for *eight weeks!*

I cannot imagine anything more annoying. As I said just the other day to my great friend Tony Blair, who is leader of the Labour Party, "My new curtains are a lot of use if I can't close them, aren't they Tony!".

Perhaps when Tony becomes the next Prime Minister he'll do me a personal favour and agree to bring back hanging. If not for everyone, then at least for interior designers!

Shut up poor people

Big is breast!

★ A YOUNG girl I saw yesterday had particularly large breasts. I thoroughly enjoy sleeping with young girls who have large breasts, and have slept with many hundreds over the years.

Funnily enough, they tend to enjoy sleeping with me too!

POOR people are so tiresome. They moan and bleat like a herd of sheep. I'm glad there aren't any in my garden.

If there were, me and my girlfriend Vanessa, who is only 18 and has large breasts, would never be able to get any sleep. Not that we get much anyway! We're always having sex.

Why don't the poor simply go out and get several well paid jobs like me? I suppose I'm just cleverer than they are. One thing's for sure. I'm certainly more richer than them. But then if you think about it, if they were as rich as me, they wouldn't be poor!

Funny old world isn't it!

Me and Esther Rantzen

How many millions make a billion?

COULD somebody please tell me whether a billion is a *thousand* millions, or a *million* millions?

I was always of the opinion that a billion meant a million million. However, since the Americans started using the word it seems to have been somewhat devalued.

I only ask because its one of those annoying questions no one ever seems able to answer. Even my good friend Sean Connery, who is very rich like me, couldn't help me when I bumped into him at a party last week.

Actually it makes no difference to me. Either way I'm still a billionaire.

Me having a drink with my good pal Tony Blair

I didn't fork out for second rate cutlery!

HAVE you noticed that the cutlery on Concorde isn't quite as good as it used to be?

I don't know whether they've done it to save money, or whether its someone's idea of a nice design. But I don't like the handles. I don't know why, but I find them irritating, and last week I got so annoyed I ended up eating with my fingers.

The only reason I travel on Concorde is for those extra little luxuries like nice cutlery. If standards continue to drop, I may as well travel by ordinary First Class!

TABLE FOR POO!

IF there's one kind of person I cannot stand its unhelpful waiters.

I had the misfortune to be dining at a particularly expensive restaurant the other evening.

I didn't particularly like the pattern on the carpet to the right of my table so I asked the waiter if he'd mind moving the table a couple of inches so that I couldn't see the offending bit. He refused.

Later I asked him to cut up my carrots for me into a Moon and stars shape. Typically, the idiot cut both into slices! The final straw came when I asked for a smiley ketchup face, only to be told that they had no ketchup!

So I closed my eyes and began to scream. I screamed and screamed as loud as I could until I pooed my pants. Then I made myself sick.

That showed him!

Me and my great coloured friend Jimi Hendrix

Things are looking up

★ PERHAPS I'm getting old, but I seem to be having a lot more difficulty climbing up conker trees this year than in previous years.

However, I have discovered one advantage of staying down on the ground. I'm able to look right up my 18 year old girlfriend Vanessa's skirt while she climbs up the tree for me, and I get a marvellous view of her knickers and bum!

REVEALED! The SORDID GOINGS-ON at pop s

NEVERAGAI

IT'S THE WORLD'S most exclusive funfair! Entrance is strictly by invitation only, and once inside you're guaranteed the time of your life. It's Neverland, the private theme park built by pop superstar Michael Jackson in his back yard. The guest list reads like a who's who of the Hollywood entertainment industry, as A-list celebrities and their kids rub shoulders in the queues for the rollercoasters and helter skelters.

But behind the welcoming facade of flashing lights, candy-floss and merry-go-rounds, is a darker side which Jacko makes sure his guests never get to see.

One man who has glimpsed what goes on behind the scenes is Smethwick lavatory attendant Bob McNally. And now he's set to spill the beans on the seedy goings on he witnessed. And he sends this warning to any star invited to Neverland: "If you'd seen what I've seen, you wouldn't take your kids within a million miles of that place!"

One day last summer, Bob was surprised to be offered the job of attendant in the toilet block at Jackson's California fairground.

❝ I'd just got in from work and my wife told me Michael Jackson had been on the phone, wanting me to fly out to the states to do some work at Neverland. I still don't know why he'd chosen me. I'd done a pretty good job unblocking the gents at Smethwick Community Centre the previous week, and I suppose the word must have got around.

stretch

"I didn't need asking twice. It was the opportunity of a lifetime. I just had time to pack my plunger, and next thing I knew I was in a stretch limousine on my way to Tipton airport, where Jacko's private jet was waiting to whisk me to America."

neil

Ten hours later McNally found himself at Neverland, and he couldn't believe his eyes.

"The place was packed with stars. Everyone there was a household name. The atmos-

EXCLUSIVE!

by our entertainment correspondent
Clancey Beauregard

-phere was one of fun and frivolity, and there at the centre of it all was Michael Jackson himself. He was laughing and joking with all the children. It seemed very innocent at the time."

louis

But the reality of the pop star's funfair was far from innocent, as Bob was shortly to discover.

"That afternoon, I came out of the gents after fitting a new disinfectant cube in one of the urinals. I looked over and saw Home Alone child star **Macauley Culkin** walking towards the coconut shy. It was three balls for a dollar, and I clearly saw the innocent youngster hand Jackson a ten dollar bill. He gave him his three balls

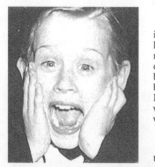

in return, but only FOUR dollars change. I couldn't believe my eyes. He'd diddled Culkin out of five dollars. That little boy went home alone, and five bucks down."

But, as Bob soon found out, it wasn't just kiddies that Jacko was diddling.

liston

"Later that day I was having a tea break when I spotted Dirty Harry star **Clint Eastwood** eyeing up the prizes on the shooting gallery. He handed over a dollar bill and Jackson passed him his air rifle and three pellets. As he lifted the air gun, Clint looked confident - and with good reason. Over the years, he'd proved his sharpshooting skills picking off baddies in a fistful of spaghetti westerns.

"Three tin ducks should have

"Don't let your kids within a mile of the place"

~Bob McNally

proved no problem, but on this occasion, Eastwood was the man with no aim. That rifle was shooting every which way but straight. Clint couldn't believe it... but I could. For moments earlier I had seen Michael Jackson bending the sights on the gun with pliers. He can deny it till he's blue in the face, but I know what I saw. Michael Jackson owes Clint Eastwood an apology. And a goldfish."

ali

Another star who was conned out of a prize by the wacko singer is **Sir Elton John,** and this time it was a big Garfield. "Jackson was drumming up custom by moonwalking in front of a darts stall, grabbing

It's not Double Fair

McNally recalls the time he saw Michael Douglas and Catherine Zeta Jones heading towards the Toytown merrygoround with a pushchair. He watched as baby Dylan sat in the fire engine, and Jacko came round to take the money.

"Douglas gave him fifty cents, but Jackson insisted that the fire engine was a two seater, and single occupancy cost double. He refused to start the ride up unless the Fatal Attraction star coughed up another fifty cents. He said it was nothing to do with him, it was just a rule of his fairground.

"Douglas refused, and went to take the toddler, who was by now crying, out of the fire engine. Catherine Zeta Jones told him just to give Jackson the money."

A surly Douglas paid up and the ride started, but it was too late for Dylan, says Bob.

"His parents were so upset that they forgot to wave at him as he came past. By the end of the ride, he was crying his eyes out again. Jackson couldn't have cared less. He just stood by the controls in a vest, smoking and staring into the middle distance."

s private funfair

NLAND!

Lifting the lid - Lavatory attendant McNally yesterday

lowed by his entourage. In the words of his own song, which struggled up to 42 in 1981, nobody wins in Jackson's fairground."

heap

According to Bob, Jackson had hundreds of little scams for squeezing cash out of his star customers. Take the waltzers, for example. Thanks to Jacko, the people on it were being taken for a ride in more ways than one.

"I remember seeing **Nicole Kidman** and **Tom Cruise** on the waltzers. Jackson was spinning their car faster and faster and they were being tossed around like ragdolls. After the ride stopped, the wobbly-legged Hollywood couple staggered off laughing towards the hook-a-duck stall. Little did they realise that Jackson was in their waltzer helping himself to all the loose change which had fallen out of Cruise's trousers.

There must have been half a million dollars easy. Tom may have had his eyes wide shut, but his pockets were wide open, and he was easy prey for sneaky Jackson."

how

Bob himself avoided getting ripped off at Neverland until one day when he finished work early and went to get a bite to eat.

"I fancied a toffee apple and a drink, so I went to Jackson's refreshments stall, a dingy caravan parked in some long grass

next to a generator. My apple looked alright from the outside, but under the toffee it was a different story.

magpie

The skin was as wrinkled as a prune, and the inside was soft and mushy. It must have been three months old if it was a day. I threw it away in disgust after one bite. I asked for my money back, but Jacko looked me in the eye and told me I hadn't bought it from him."

blue peter

And McNally fared little better with his drink.

"I asked him for a big cup of coke, but I ended up with two dollars' worth of ice and just a slurp of cola. It was a complete

rip-off. Jackson may have had his head set alight during a Pepsi advert, but now it's his customers who are getting burnt."

jolly roger

A week into his contract, McNally was told his services were no longer needed.

"I'd been sacked. Officially it was because I'd been caught stealing forty rolls of toilet paper, but that's nonsense. The real reason was obvious; I knew what Jackson was up to, and he knew that I knew. I was dangerous, so he had to get rid of me.

"And anyway, I was only taking the toilet rolls so I could get the wrappers off at home and make an early start the next morning. **"**

his crotch and shouting: 'Score over six to win. Three darts for a dollar.'

John spotted a three foot toy Garfield on the stall and thought it would be lovely to win it for his partner David Furnish.

um

"Sir Elton handed over his cash and threw his darts. He

did pretty well, scoring twenty-one, and confidently stepped forward to accept his prize. But Jackson simply pointed to a tiny sign fastened to the Garfield, reading: 'Take me home if you lose', and handed the singer a small gonk with cardboard eyes.

"Elton John was absolutely furious, and stormed off fol-

Carrie on Losing

SEX & the City star **Sarah Jessica-Parker** got more than she bargained for when she tried to hit the jackpot in the Neverland amusement arcade.

McNally tells how the 38 years old actress, who plays Carrie Bradshaw in the hit show lost over $60 in Jackson's amusement arcade.

pumping

"I stood and watched her pumping money into the penny waterfalls like it was going out of fashion. Every five minutes, she'd be off to get more coins, leaving **Ally McBeal** guarding her place. In his change booth Jackson, dressed in a threadbare knitted cardigan, would look up from his racing post, hand her a margerine tub full of nickels, and she would run back to the machine. Then she'd feed all her money back in, desperately trying to dislodge a huge overhang of coins.

any old

"Eventually, Parker's frustration got the better of her, and she started rocking and kicking the

machine. Jackson was out of his change booth like a rat out of a trap. He told her to beat it, and she did." But McNally doesn't know why he bothered.

"That machine was never going to pay out, no matter how hard she hit it. Earlier that day I'd seen Jackson gluing the coins in place with Araldite."

ROLLCALL of EVIL!

IF YOUR KIDS are invited to Neverland, then take care. Here are *Jackson's Five* favourite scams for getting his sticky fingers on their spondoolicks.

● *The grip on the amusement arcade cranes is too weak to successfully pick up the packets of cigarettes with out of date fivers sellotaped to them.*

● *In order to win on the hoop-la stall, the hoop must go over both the bottle of pomagne AND the wooden base on which it stands. However, the diameter of the wooden base is about an eighth of an inch larger than the hoop.*

● *Jackson offers a prize if he guesses your weight and gets it wrong. What he doesn't tell you is your prize is worth twenty cents...and each guess costs you a dollar.*

● *Wacko readily demonstrates that the targets are not fixed to their stands on the coconut shy. But what he doesn't tell you they've been drilled and filled with lead shot, making them all but impossible to knock down.*

● *It's not unknown for the Billy Jean star to slip the occasional foreign coin in with your change, so check it carefully.*

Birds drop 'em for chocolates
AND ALL BECAUSE THE LADIES LOVE...

Chocolates are the key to a woman's pants. So says a new survey published today.

And with Christmas approaching, now is the time for fellas to be buying them. That's the view of Burt Twix, spokesman for the British Association of the National Federation of Chocolate Manufacturers of Great Britain, who commissioned the survey.

HEART

"Traditionally flowers have always been seen as the way to a woman's heart. And that may well still be so. But if you want to get into her knickers, you'd be better off buying chocolates", Mr. Twix told us.

DIAMOND

And he claims that statistics bear him out. "Our survey shows that almost 85% of men who give chocolates get their end away within a couple of days. Whilst out of every hundred men who give flowers, six months later over half of them still haven't even

...it up 'em

had a whiff of action, never mind a leg over".

CLUB

Research carried out in conjunction with the survey revealed another interesting fact about chocolates. "We're still waiting for the final results, but all the signs indicate that chocolates help prevent cancer", he told us.

BRITAIN'S BALLS ARE

Britain's ball makers are having a ball, according to the latest ball figures published today.

Department of Trade statistics show that sales of balls of all types are booming.

GOOD NEWS

And that's good news for High Street ball retailers, who have suffered more than most in the recent recession, with balls sales dropping to an all time low at the beginning of the year.

KING JAMES

But now they're bouncing back, with sales of balls, including foot, golf, basket and beach all on the increase.

BOUNCING BACK

Reg Burton, spokesman for national retail giants World of Balls, told us that ball sales were buoyant throughout Britain. "We stock over 50 balls, everything from medicine to marbles, and they're all rolling off the shelves as fast as we can put them out".

WENDY JAMES

In 1980 a survey revealed that the average man in Britain bought 7 balls, although this figure varied from one individual to another.

QUITE A LITTLE GOLD MINE YOU'VE GOT HERE TED.

GOLD MINE.

HOUSEWIFE'S HUNT HORROR
Horseback toffs tear pet to pieces

A housewife watched in horror as her family pet was torn to shreds by a baying pack of hounds, after fox hunters rode through French windows and into her living room.

Mrs. Eve Froud was sitting watching television in her third floor Putney council maisonette when she was alerted by the sound of hunting horns. Seconds later thirty blood thirsty hounds tore into her living room knocking over furniture, followed by a dozen members of the Putney Hunt on horseback.

BOWL

"The next thing I knew the dogs had somehow pulled my goldfish George out of his bowl, which was on top of the television, and had cornered him behind the settee", Mrs. Froud told us.

BOIL

The horrified housewife then sat and watched helpless as the ferocious hounds tore the tiny fish to shreds, jeered on by the red jacketed hunt members. "I'll never forget George's face as those hounds closed in", Mrs. Froud recalled painfully.

SOIL

Mrs. Froud also alleges that the Master of Hounds, Brigadier Charles Levington Compost-Heap struck her in the face with his riding crop when she tried to intervene, and then shot and killed her husband Dennis, 52, who

No nookie Fry (left) and Blind Date Cilla

was watching television at the time.

SAIL

A spokesman for the Putney Hunt described the incident as 'regrettable', and said that it was not hunt policy to pursue foxes into people's houses. "A formal apology has been made, together with

an offer of compensation for the loss of the goldfish", we were told.

SAID

Members of the Putney Hunt include brainy left wing bonk ban comic Steven Fry, and 'Blind Date' hostess Cilla Black.

SAVE THIS POOR COW

Barbaric French villagers plan to MURDER this tragic defenceless cow, HACK its lifeless carcass into bloody pieces, COOK it in an oven, then EAT it.

And then the blood thirsty mob will wash down its remains with bottles of wine.

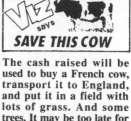

BY OUR EUROPHOBE CORRESPONDENT
BILLY BOLLOCKS

BARBARIC

This barbaric ritual, reminiscent of a scene from the middle ages, has been re-enacted in the streets of Purtain sur le Lit every year for centuries. This year it will be no different, and the French authorities have no intention of lifting a finger to stop it.

PEASANTS

Paraded out of its field by stick wielding peasants, the terrified beast will then be herded into a waiting lorry like cattle before being driven a short distance along bumpy roads to the local abattoir.

PETRIFIED

As a throng of jeering abattoir employees look on, the petrified animal will be **STUNNED** with electricity, and then **KILLED** with a savage blow to the head from a bolt.

CUE

That will be the cue for cheering crowds of sadistic

HOLSTEIN-FRIESIAN

French shoppers to go on the rampage through the narrow streets of the town, queuing in the butchers shop and supermarkets to buy blood stained chunks of the pathetic animal's body, which only hours before had been standing harmlessly in a field.

REST

Local civic dignitaries, among them the town's Mayor, will all join in the shameful procession as the animals corpse is carried bit by bit back to kitchens to be cooked and later eaten, together with potatoes and vegetables.

"If this is what passes for civilisation on the continent then we have to ask ourselves whether we, the British, really want to be a part of it", said Tory MP Sir Anthony Regents-Park yesterday. "I don't deny that animals must sometimes be killed out of necessity. But the least they could do would be to give the poor beasts a chance by perhaps chasing them around on horseback with a pack of baying hounds, or charging a wealthy Arab £5,000 a day to drive around in a Range Rover taking pot shots at them with a shotgun. To subject an animal to such an undignified ritual seems quite wrong in this modern age".

Help bring a French cow back to Britain

We want to send a message loud and clear to the townsfolk of Purtain sur le Lit: **"FROG OFF!"**. And we need your help.

We're launching a campaign to save a cow from the hands of the French butchers, and gathering signatures for a vital petition aimed at stopping the slaughter. We want you to get five people to sign the form below, and send it back to us together with a donation of £5 (cash only).

SPIDER

The forms will be collected, and then sent to the Mayor of Purtain sur le Lit, telling him in no uncertain terms where he can shove his onions, cheap plonk and silly loafs of bread.

Viz says **SAVE THIS COW**

The cash raised will be used to buy a French cow, transport it to England, and put it in a field with lots of grass. And some trees. It may be too late for the sad cow in our picture, but together we can save another cow. So send in the form, and money, today. The address is SAVE A FRENCH COW, Viz, P.O. Box 1PT, Newcastle upon Tyne, NE99 1PT.

For extra petition forms send £2.50 to the same address.

To: The Mayor, Purtain-sur-le-Lit, France

Dear Mayor,
FROG OFF! You MURDERER!

Signed 1 _____

2 _____   3 _____

4 _____   5 _____

IT'S GAZ TOP MANIA!

Britain's pop fans have gone Gaz Top potty!

He's the TV presenter who's *Gaz* top of everyone's *Gaz* pops! With his cute little lisp and blacker than black lavatory brush hairdo, he's the *Gaz* top of the pop top pop presenter who the girls are *Gaz*pin' to meet.

TOAST

Yes, he's Gaz Top. The pop host with the *Gaz Top* most! He's the toast of Britain's TV pops. The Pop Tart who's top of every chart! "Top – top – top, popability!" That's the beauty of Gaz!"

CELEBRATE

To celebrate Britain's top pop presenter, and because we've got nothing better to write about, we're giving away a host of *Gaz* top prizes! And as Gaz would doubtless agree, it's a

WIN £100 IN RECORD VOUCHERS

Gaztoptastic array of goodies that are up for grabs!

ENTER

All you have to do to enter our *Gaztopcompetition* is *Gaz* answer two easy peasy questions and send us a drawing of the man himself. We'll be asking Gaz Top himself to *Gaz* choose the *Gaz* top entry, and our first prize of £100's worth of *Gaz* top *Gaz* pop record vouchers will be sent to the

Gaz lucky winner. Answer these simple *Gaz Top* questions:

1. Which TV programme, if any, is Gaz Top currently presenting?

2. Arrange these 3 star qualities which Gaz possesses into order of importance, No. 1 for the most important, and 3 for the least: Village idiot charm, imbecilic good looks, all round Gaz Top popability.

Send your entries to Viz Top of the Gaz Pops Competition, P.O. Box 1PT, Newcastle upon Tyne, NE99 1PT. If Gaz Top is reading this and would like to judge our competition, could he please get in touch. If he wants to give us a ring we will accept a reverse charge call.

BIG MAC & FRY!

Fast food giants open Death-Row outlet

"HAVE A NICE last day!" That's what McDonald's staff are saying to their customers at the burger giant's latest branch. For the multi-national fast food corporation has just opened its newest restaurant on Death Row in the Texas State Penitentiary.

For the past 6 years, the jail has executed up to 500 black simpletons a day, with the majority opting for a last meal of Big Mac, fries and a strawberry shake. Quick-thinking McDonald's bosses spotted a gap in the market and the branch is now the fourth busiest in the state.

twenty

Previously, prison Governor Draylon Hogg found that he was sending as many as twenty deputies an hour the seven miles to the near-

Hogg - no-nonsense Redneck Governor yesterday

From our US Death Row Reporter
Jacqui-Dani Boyles
in LEEDS

est drive-thru to buy burgers for condemned inmates.

"This has sure speeded things up," he told CBS reporter Howard Glans. "Now we can have the boys fed, shaved and strapped in the chair before they have a chance to shit their goddamn pants. Yesirree."

Benson

Branch manager John-Bob No-Stars jnr. was adamant that his condemned customers could expect the same level of service as in any other branch of the fast food chain. "Everyone gets a smile when they enter and a

Dying for a Big Mac - The Texas State Penitentiary with its brand new 'last food' outlet .

'have a nice day' when they leave. And they never have any problems finding a seat. Old Sparky's right on down the hall, first left!" he quipped.

and

And the convicts aren't complaining either.
"I done had me a Happy Meal, cos' they're giving

> **"We can have the boys fed, shaved and strapped in the chair before they've had a chance to shit their goddamn pants."**

away Animaniacs toys," laughed Jimmy-Ray Joskin, an unemployed farmworker with a mental age of six, wrongly convicted of pretzel theft. It is the fourth time he has been strapped into the electric chair - on the three previous occasions he has been reprieved at the the last minute.
"I got me Yakko, Wakko and Dot. I sure hope I don't get fried before I get me Pinky and the Brain or Slappy Squirrel."

Hedges

Ricky-Bob Robespierre, a 34-year-old retard

sentenced to death for fidgeting during the national anthem at a baseball game has other plans for his final meal. "I'm gonna have me one of them there filet-o-fish thangs. My momma says you have to wait for them and I'll get to live four minutes longer," he said, beaming from ear to ear.

please

Following the success of the Texas Penitentiary branch, McDonalds spokesman Ronald Spunkfelcher hoped to make a killing in prisons right across America. "We've already got plans to have a 24-hour 1 seat restaurant outside the lethal injection chamber in Arkansas State Penitentiary," he said. "Within two years, we want to see those golden arches over every death row in the country. God Bless America."

love

But Spunkfelcher is not the only player trying to corner the last meal market. Already, death-row prisoners in Texas are 10 to a room as cells are demolished to make space for branches of *Taco Bell, Steak 'n' Shake* and *International House of Pancakes.*

Final Meals

Name: Bobby-Ray Leonards
Mental age: 8
Crime: Untied shoelaces
Last meal: Big Mac, large fries, donut, Dr. Pepper

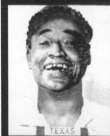

Name: Ricky-Bob Moses
Mental age: 3
Crime: TV too loud
Last meal: 6 Chicken McNuggets, regular fries, Coke.

Name: Billy-Bob Berneau
Mental age: 18 months
Crime: None
Last meal: Cheeseburger Happy Meal, McFlurry

ARE YOU COMING OUT TONIGHT OR ARE YOU STILL BUSY WORKING?

SORRY, I'VE GOT MASSES TO DO

Charles & Camilla have been banging each other for 35 years and they're still hungry for nookie. How do they do it?

22nd MAY

20 WAYS TO SPICE UP YOUR LOVE LIFE

Our relationships, like bread, soon go stale if we don't work at them. But unlike bread, which goes hard the longer you neglect it, relationships are more like biscuits because they go soft over time. Many couples find that after several years together, their biscuits have become too soft to dunk in the tea of love. Meanwhile, the pot of desire grows cold on the draining board of your relationship, and all too often this is followed by erectile disfunction and vaginal dryness.

But the good news, according to sex experts, is that it's never too late to rewarm your teapot of love. Follow these twenty simple steps and you'll soon be guzzling down steaming mugs of hot sex with full cream milk, two sugars and biscuits which are twice as long and ten times as hard as ever before.

Role Playing

Having sex with the same person for years, perhaps decades, can soon become utterly, utterly tedious and repetitive. But, say relationship experts, we can put the spice back into our love lives by simply pretending to be someone else! Why not pretend you're a businessman going away to a conference? Then, while your wife waits at home, book into a Holiday Inn and have sex with a cheap prostitute.

Dressing Up

Whether it's policewomen, nazis or French maids, there's no denying that many of our most exciting sexual fantasies revolve around uniforms. But all too often this can become forgotten in a long term relationship. Ask your wife to dress up as a nurse, complete with a medical bag, stethoscope and black stockings. Then, plan a saucy day when she goes out to cut your nan's toenails whilst you stay home and have a wank.

Rude Food

In the film 91/2 weeks, Mickey Rourke and Kim Basinger got audiences hot under the collar when they drove each other wild with food out of the fridge. Why not try the same thing at home? You never know, it might be just what you need to tickle your jaded sexual palate back into life! Get your wife to strip and pour different foods onto her naked body. Try sticky honey, creamy yoghurt, cold ice cream or greasy sausages. Massage the food sensually into her skin and hair. Then run her a hot, foaming bath. While she's in it, tell her you're going out for a paper, then nip round to her sister's for a steamy love session.

Bondage

Bondage isn't for everyone, but it can be a wonderful way to put a bit of ooh-la-la back into a jaded marriage. The key to a successful bondage session is trust; as long as you and your partner are

comfortable with the idea of exploring the darker side of your relationship, the sex that results can be mindblowing! Select some soft rope or silken scarves and gently but firmly tie your wife's wrists and ankles to the bed. To add an extra frisson of excitement, blindfold her and then tiptoe out of the house and make your way to your local red light district. Once there, pick up a good time girl for an erotic £10 hand job behind a skip in the Matalan carpark.

Al Fresco Sex

A whiff of danger often provides all the charge that is needed to jumpstart the flattest sex batteries. And many people find that a saucy session out of doors gives them exactly that 10,000 volt thrill. Why not surprise your wife with a sexy trip to a place where there's a chance you'll be caught, somewhere like the grounds of your local convent or nurses' home? Then get her to keep a lookout by the gate whilst you treat yourself to a wristily exciting hand shandy in the shadows near the shower block.

Shared Fantasies

Dirty thirties authoress Aniis Nin kept her trans-Atlantic relationship with Henry Miller at erotic boiling point by writing down her sexual fantasies in a series of XXX-rated letters. Even though you and your wife may not be literary geniuses, there's no reason why you shouldn't follow suit. And if you do, you might just find your sword getting mightier than the pen! Ask your wife to write down all her innermost fantasies in full explicit detail. Then, while she's busy doing that, take yourself off to your local lapdancing club and fantasise about the strippers whilst pleasuring yourself through your pockets.

Talking Dirty

They say actions speak louder than words, but when it comes to spicing up a lacklustre love life, dirty words can often speak louder than actions. Even the most outwardly prudish couples find that using low down explicit language during sex helps them to achieve previously undreamed of heights of pleasure. Next time you find yourself making

love to your wife in the same old way, try telling her that you "really want a fucking big shit". Then take your mobile phone with you into the toilet, dial a random number, and turn the air blue as you masturbate yourself to a shattering climax.

Wine and Roses

Although we tend to forget it, sex and romance are simply two sides of the same coin. And amazingly, many women find romance is the bigger turn on. Try to think back to the first few weeks of your relationship when you took your partner out for romantic meals and showered her with expensive wines, red roses and chocolates. Chances are, the sex you were having then was the best ever. So why not try to rekindle the passionate flame of those days? Surprise yhour wife when she comes home from work; have a candlelit meal waiting for her on the table. Ply her with glass after glass of her favourite wine while she eats it. Sooner or later she'll pass out, giving you the chance to nip into the garage for a steamy clinch with your inflatable love doll.

Sex Toys

Toys aren't just for children - they're for grown ups too! And despite their reputation, adult toys aren't meant to replace your partner, they're there to enhance the pleasures of a loving relationship. Meanwhile, programmes such as Sex and the City have brought them into the mainstream and many modern couples now say they couldn't do without their bedroom playthings. If you haven't got any you're probably in the minority...and you're certainly missing out on an awful lot of sexy fun. Go to your local Ann Summers shop and invest in a selection of his'n'hers sex aids. Take them into the bedroom and surprise your wife by waking her up and suggesting that you try some of them out. Then, while she's off down the 24-hour petrol station looking for batteries, take advantage of the Adult Channel's midnight freeview and enjoy a 10-minute sexy romp into a sailor's favourite!

DICING WITH DEA

Every year millions of pounds of tax payers money is spent safeguarding the Royal Family. Yet despite the constant efforts of police and security services, almost every week one of the Royals throws caution to the wind and risks their neck flying helicopters, playing polo or skiing off the Piste.

Charles and Di — they live for danger.

Far from being safety conscious, the Royal Family seem to thrive on danger, and a growing number of people fear that sooner or later a member of the monarchy could killed or seriously injured.

ACCIDENT

The most widely publicised incident was Prince Charles' recent return to the ski slopes so soon after the tragic accident in which a colleague died. According to a former ski guide, the Prince regularly leads his party off the Piste and along precarious cliff edges and goat paths, narrowly avoiding rock falls and avalanches.

DANGEROUS

"He always seeks out the most dangerous routes", our source told us. "He has even been known to ski blindfolded if he felt that the element of danger was lacking".

EXCITEMENT

Princess Di gives skiing a miss, finding plenty of excitement back home on the streets of London. Bored with Royal rigmarole and security restrictions, Di loves to go for a drive or a shopping trip unaccompanied by her personal detective. As one Palace insider told us, she regularly gives her bodyguard the slip.

Daredevil Royals live for danger

"Often she is allowed to travel alone in her car, with a police escort following close behind. But on one occasion she tried to lose her tail by driving the wrong way up a one way street at 130 mph". There followed a thrilling high speed car chase through central London in which six police cars were badly damaged. Eventually the police cornered her in a car park.

SPEED

"The only way out was through a narrow alleyway less than five feet wide", our insider told us. "So she revved up her engine, then drove the car at high speed onto a nearby ramp, flipping the car up sideways, before swerving through the alley way on two wheels". Detectives later found the wayward Princess shopping unaccompanied in a nearby fashion boutique.

But for every incident that makes the headlines there are many more that never make the news. Like the time when Prince Charles was dicing with death before he'd even reached the ski slopes. A close friend of the Prince takes up the story.

LEAPT

"Charles had boarded the cablecar and was on his way up to the mountain top when suddenly he realised he'd forgotten his skis. Rather than wait until the car reached the top, he clambered out of a window onto the roof, hundreds of feet up above an icy ravine and leapt onto a passing car travelling in the opposite direction. He then collected the skis from his hotel room before returning to the mountain in slightly less dramatic fashion".

Randy Andy — hair-raising stunts with his chopper.

Brother Andrew, the Duke of York, gets more than his fair share of excitement as a helicopter pilot in the Navy. A former shipmate of the dashing royal who served with Andrew during the Falklands conflict told us that the Duke was always in the thick of the action.

MISSILE

"His helicopter would often return from anti-submarine patrols riddled with bullet holes. On one particular occasion a large missile had narrowly missed the Prince and lodged itself in his rotor blades, failing to explode.

Typically, the Prince landed his helicopter safely before carrying the unexploded shell to the Officers Mess. Seconds later the device exploded, and the Prince, his face blackened and his clothes in tatters, celebrated by ordering champagne all round".

Fergie — fat arse

While Andrew finds adventure in the forces, at home the Duchess of York's life is by no means quiet. She spends hours in the air flying helicopters and aeroplanes, often at low level, and occasionally amuses herself by flying under bridges, through tunnels and by 'buzzing' motorway traffic. One lorry driver was in for quite a surprise when he stopped at a motorway service station.

STUNT

"I got out of my lorry and looked up and there was the Duchess of York's helicopter sitting on the back. She must have landed it on top of my load while I was travelling at over 60 mph. I bought her a cup of tea and a bun and chatted to her for a few minutes before she flew off again. She was very friendly — a lovely person — but it did strike me as a rather dangerous stunt for a person in her position to be attempting".

A typical stunt from the Duchess of York — "buzzing" guests at a Palace Garden Party.

TH!

Daredevil Di performs a crazy motorbike stunt as her anxious police bodyguard looks on.

Even the older Royals occasionally enjoy a brush with danger. Alarm bells were set ringing recently when the Queen Mother disappeared. Security was put on full alert and a massive search was launched, but after two days police and security services could find no trace of the popular Royal. It was feared that she may have been the victim of a terrorist kidnapping, until news came through that she'd been found — safe and well — by potholers exploring caverns hundreds of feet below the Derbyshire Peaks.

Queen Mum — potholing at 89

The plucky Royal Gran had set off on a solo potholing expedition and become trapped in a narrow fissure hundreds of feet below ground level. With oxygen in short supply, she was fortunate to be rescued in the nick of time by a team of amateur potholers who had been exploring the same area.

BRAVADO

Relieved relatives threw a party to celebrate her narrow escape and served up a right Royal banquet. Warned by her doctors to avoid fish bones, with typical bravado the Queen Mum tucked into a hearty meal of smoked kippers before downing several pints of stout. An official Palace spokesman explained her disappearance by claiming that the Queen Mother had been 'resting' at Balmoral, the Royal's Scottish holiday retreat.

BATTLE

A leading psychologist Dr Franz Klausman, believes that the Royal Family's affinty for danger is an inherited condition. "It's in their blood", he told us. "In years gone by Kings and Queens would lead their armies into battle. Nowadays we expect them to to just sit around, wave at people and open things. It's only natural for them to channel their excess energies into exciting and often dangerous pursuits".

HAZARDOUS

But the Queen fails to agree and she is known to be unhappy at the growing level of disregard for personal safety. Quite rightly she feels that as heirs to the throne her family should be more careful and think twice before partaking in hazardous pursuits. Indeed, only recently she stepped in to block a birthday treat which Charles and Diana had planned for their second son Harry.

BARREL

The danger loving duo had planned to send the toddler — third in line to the throne — over Niagra Falls in a wooden barrel. But the Queen intervened, claiming that the spectacular stunt was simply too dangerous.

● Opinion - p.27

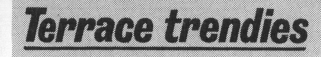

Terrace trendies

A new breed of soccer hooligan — dressed in £800 suits and drinking bubbly at fifty quid a bottle — is replacing the traditional soccer thug.

And you won't catch them wearing scarves, hats or Doctor Marten boots. Instead the new yuppie yobs sport dapper suits by Giorgio Armani. Lager is out too. The new generation of louts quaff Dom Perignon champagne by the crate full. No expense is spared. Unlike their predecessors the terrace trouble makers of today hold down highly paid jobs in the City.

FLICK KNIVES

Flick knives are replaced by filofaxes. The new breed of thug is highly organised. And tattoos are frowned upon. A diamond encrusted Cartier wristwatch is more in keeping with the new image.

MACHETE

With their £250 hand stitched Jermyn Street silk shirts, you won't catch these thugs 'putting the boot in'. They wouldn't want to risk chaffing their made-to-order Italian pig skin brogues, at £300 a pair.

SAMURAI SWORD

And it isn't their style to look for trouble. Indeed with their £500 leather Gucci ties, they

Football thugs who dress to kill

don't go to football matches at all. Instead they go out, in their solid gold Dunhill cufflinks and Chinchilla socks at £900 a pair, and eat nouvelle cuisine in fasionable restaurants, or just stay at home in their £2 million converted dockland warehouses, relaxing and listening to their £3000 top-of-the-range Nakamichi CD players, with quadraphonic sound.

● Opinion - p.27

Queen sex

Members of pop group Queen have taken part in '2-in-a-bed' sex romps with their wives. The saucy stars were believed to be naked at the time.

Other pop stars, among them Paul McCartney, are also thought to have had sex with their wives.

Cheese blow

Cheese prices are set to soar. And a pound of Edam could set housewives back as much as £28 a pound if new EEC Cheese Regulations come into effect later this year.

TINNED PEAS

This comes as a double blow to shoppers already reeling from the news that tinned peas are to be outlawed under new Vegetable legislation.

Housewife Mrs Vera Wells described the news as "typical". However, there was some consolation for shoppers. A change in the laws governing nuts could mean a drop in the price of chopped almonds. Only a small reduction can be expected, however a saving of between 1 and 2p a pound could be passed on to housewives.

'Des-gusting'

Lynam's movement hits bum note

THE television presenter Desmond Lynam was last night bailed to appear before Bow Street Magistrates after being caught defecating into a piano in Fortnum & Mason's.

The 52-year-old BBC sports anchor was shopping in the exclusive London store when he apparently became gripped with violent stomach cramps. Shocked onlookers then saw him drop his trousers and perform the toilet function under the lid of an £18,000 Bechstein concert grand.

excreted

Posh shopper Lucinda Sopwith-Camel told us: "It was perfectly ghastly. Lynam excreted in full view of the whole music department, and then cleaned himself up with some sheet music. You don't expect that kind of thing from television celeb-

Posh shop Fortnums and Masons yesterday

rities, and certainly not in Fortnum & Mason's."

A sombre-faced Lynam refused to answer reporters' questions as he left Paddington Green police station, and stood behind his solicitor who read a prepared statement. "My client deeply regrets the unfortunate incident which took place yesterday. As the result of a medical condition, Mr Lynam was caught short whilst browsing in the piano department and had no option but to take the action that he did." He added: "He looks forward to the opportunity to clear his name in court."

excorfud

But a Fortnum & Mason's spokesman last night rejected Lynam's version of events. He told us: "Mr Lynam had been acting strangely in the piano department for more than two hours prior to the incident. Security staff had seen him loitering next to several pianos, and had moved him on twice after he began loosening his trousers."

"There are ample lavatory facilities on every floor of Fortnum & Mason's," he

Lynam (inset) outside Paddington Green Police station yesterday, and (above) a piano similar to the one he is alledged to have shat in.

added. "Mr Lynam was no further than twenty yards from a toilet when the offence was commited."

A police spokesman confirmed that Lynam had been charged with shitting in a piano in Fortnum & Mason's, and would be appearing before magistrates on Christmas Eve.

Three years ago at Sheffield Crown Court, Lynam pleaded guilty after being caught defecating into an 18th century harpsichord at Chatsworth House. He asked for 148 similar cases to be taken into consideration and was bound over to keep the peace.

HAVE-A-GO HERO FRANK TEACHES THUGS LESSON

Hero Frank nursing bruised knuckles yesterday.

Have-a-go hero Frank Barker wasn't prepared to stand back and watch the day heartless youths raided his back garden to steal apples.

Frank, a plucky 38 year old, sprang into action and challenged the would-be thieves. One of the gang fled empty handed, while Frank grappled with the other two.

COLLAR

"I managed to get hold of one of them by the collar and punch him in the eye", said Frank, who was recovering from the incident at home yesterday. "I then managed to pull him to the ground and kicked him several times".

"The next thing I knew out of the corner of my eye I caught a glimpse of the other one running for the gate". Frank was having none of it. "I instinctively grabbed a short length of rusty drainpipe which had been lying on the ground and caught him on the back of the head with it".

CUFFS

Fourteen stone Frank, a keen boxer during his army days, then managed to rain a series of heavy blows down on the thugs until a neighbour, alerted by their screams, raised the alarm. The police arrived and the intruders, aged 11 and 12 years, were carted off to hospital for emergency treatment.

"At the time I wasn't scared", said brave Frank, who is nursing bruised knuckles suffered in the attack. "I didn't really think about it. Looking back I suppose it was a pretty foolhardy thing to do, but you have to stand up for yourself and defend your property. Especially nowadays".

PADDED SHOULDERS

Fulchester's Neighbourhood Watch organiser Glenda Purvis was first to congratulate Frank. "He's a very brave man and we're all very proud of him". Local Bobby P.C. Alan Jones agreed, but added a note of caution. "If you see anything suspicious, call the police. Our advice to the public is not to approach criminals. You'll be far safer leaving that to us".

COUNCIL LITTER COLLECTION DEPOT.

I DON'T KNOW IF I CAN DO THIS JOB.

DON'T WORRY. YOU'LL PICK IT UP AS YOU GO ALONG.

BIGGS or BARKER

We ask...Who is Britain's Most LOVEABLE ROGUE-EST RONNIE?

Whether batting for Biggs or backing Barker, we all have our opinion on who is Britain's most rascally Ronnie. Is it the Great Train Robber Biggs, a latter day Robin Hood who, with his band of 14 merry men, daringly snatched £2.6million from under the very nose of a coshed train driver? Or is it roly-poly funnyman turned antique dealer Barker, who in 1989 cheekily offered an undercover reporter 20 quid for a silver salver valued by Christies at over £1000? *Let's look at the facts...*

SCORE

Coshing — 8 | 5

Big-hearted Biggsy is not a violent man. When he and his fourteen mates set out to rob the overnight Royal Mail train at Cheddington, Bucks, they didn't take any knives or guns to defend themselves. It was only when have-a-go driver Jack Mills tried to make a name for himself by refusing to obey the robbers that roguish Ronnie was left with no option but to bludgeon the 58-year-old into unconsciousness with a sockful of billiard balls.

A low-scoring round for Barker. He has enjoyed a long career in showbusiness, starting out at Aylesbury Rep in 1948. Spotted by David Frost, he got his TV break in 'The Frost Report', before spending most of the 'eighties dressed as a woman singing ribald lyrics to popular tunes. Appearing in sitcoms like 'Porridge' and 'Open All Hours' whilst running his Cotswolds bric a brac shop has left him little time for coshing train drivers.

Escapology — 9 | 4

The prison bars of Wandsworth were no match for cunning convict Ronnie Biggs. After serving a mere 15 months of his 30-year sentence, the wily fox outsmarted his captors, slipping unseen over the wall to freedom. Now back under lock and key at HMP Belmarsh, it can only be a matter of time before the crafty 71-year-old semi-paraplegic shows the screws a clean pair of heels and high-tails it back to his luxury Brazilian bolthole.

In 1974, Barker played an habitual criminal, sentenced to five years in Slade Prison for burglary. Looking upon capture and incarceration as occupational hazards, he did his porridge without attempting to escape. The regime at Slade was a fairly relaxed one. During his time inside, Ronnie was never beaten up by the screws, doused with boiling sugary water, fed food containing ground glass or bum-raped in the showers.

Girlfriends — 9 | 7

Ronnie Biggs has always been a glamorous charmer with an eye for the ladies. After his daring escape from Wandsworth Prison in 1965, he flew to exotic Rio, where he bedded a string of sultry Brazilian beauties. Even marriage to the luscious Raimunda de Castro did not curb his playboy lifestye, and in 1978, he appeared in a Sex Pistols video, eagerly pushing his face into some buxom woman's breasts.

When it comes to the ladies, Barker's life is less than glamorous. There's no bevy of pouting Brazilian-waxed birds for this Ronnie; he's happiest up in the loft, rummaging through his collection of over 70,000 saucy postcards. The closest Barker gets to a real sex session would involve him up a ladder in a brown shop-coat, puffing and blowing whilst trying to push nurse Gladys Emmanuel's enormous arse through an open window.

Speech Impediments — 5 | 9

All lovable rogues have some sort of speech impediment. Think of Terry Thomas with his inability to pronounce the letter 'r' and many more examples of this phenomenon. After three strokes, Biggs's charming cockney patter has been replaced with a series of cheeky grunts accompanied by a rascally string of viscous drool. A good effort.

But anything Biggs can do, Barker can do better! Whether he's shouting for G-G-G-Granville to fee-fer-f-fetch a cloth, or comically "pismronouncing" his "worms", Barker is the king of vocal idiosyncrasies, boasting a wide variety of verbal tics. And they prove no impediment to roguish Ronnie receiving near full marks in this round.

Victims — 2 | 9

The Great Train Robbery, carried out in August 1963, was made all the more dashing and exciting because it was a victimless crime. The money - *worth up to £50 million today* - was destined to be burnt, and train driver Jack Mills was nearing retirement and would probably have died eventually of old age.

From behind the till of his Chipping Norton bric a brac shop, Barker preyed on anyone who came through the door to buy antiques. Old ladies, decorated war heroes, the disabled, children buying presents for their grannies; all were routinely charged MORE for an item than cheeky Ronnie had originally paid.

FINAL SCORE 33

Sorry, but your score is not quite 'Biggs' enough, Ronnie. You're certainly a rogue, but not quite lovable enough to tip speccy scallywag Barker off top spot.

Lives in Tandem - *More than coincidence???*

Ronnie Biggs and Ronnie Barker have both lived extraordinary lives. But what you may not realise is that their lives share remarkable similarities. *Fortean Times* editor Paul Sieveking considers these amazing coincidenc-

• Both were born in 1929 on exactly the same day - August 8th. Barker was born just seven weeks later on September 25th.
• Both were christened with exactly the same name, spelt identically - Ronnie.
• Both started school when they were about 5.
• Their best friends (train robber Buster Edwards and Scotch comic Ronnie Corbett) are, to within 2 or 3 inches, **exactly** the same height.
• Both men celebrated their sixty-fifth birthday in the same year.
• Both men's surnames begin with a letter 'B', followed by a vowel.
• Followed by **TWO** consonants.
• Both stole a train on their 34th birthday. Except Ronnie Barker.

FINAL SCORE 34

Well done, Ronnie. Out of the two Ronnies, you are the lovable rogue-est. So it's goodnight from you and it's goodnight from him. Goodnight.

It's STAR-

IT'S THE END of the world as we know it. So sang REM in their 1991 pop hit. But little did they know that on February 12 2060, that song will *come true* when meteorite *NT7 crashes to earth*.

Exclusive!

The lump of space rock, the size of a *Nationwide League Division 2 football stadium*, is on a collision course with earth traveling at speeds up to and in excess of **25,000 mph** - that's as fast as *200 Formula 1 cars*. When it hits the earth, it will leave a crater the size of *several Wembley Stadiums*, and devastate an area half the size of Wales *many times over*.

The initial impact will create a tidal wave the height of the *post office tower* with *two double decker buses on top*, which will wash over the earth at the speed of *three Concordes* and with the destructive power of *enough Hiroshima bombs to fill the Centre Court at Wimbledon*.

rain

Lumps of white hot molten rock, some the the size of a cricket ball, others the size of a *fridge,* will rain down causing death and destruction. And all this will happen in less time from now than the age of a plumber taking early retirement - *that's less than three Gareth Gates's lives laid end to end.*

Predictions say 99% of life on earth will be destroyed, and it's a conclusion which has sent the showbusiness world into shock. For if the boffins' worst fears prove correct, the impact of NT7 will kill off many of our favourite stars.

heart

"The impact of a huge meteor could spell disaster for celebrities from light entertainment, the soaps and the pop world," warned astrophysicist Professor Les Kellett of Great Yarmouth University.

loans

"Even in the unlikely event that a star survived the initial asteroid strike and global firestorm, the subsequent 10-year nuclear winter and worldwide ice age would devastate the lucrative summer season circuit on which so many of our stars depend.

helmet

"Acts such as the Chuckle Brothers, Joe Pasquale and Stu 'I could crush a grape' Francis, could be hit particularly hard by such an apocalypse, especially if the end of the world also leads to a downturn in pantomime bookings."

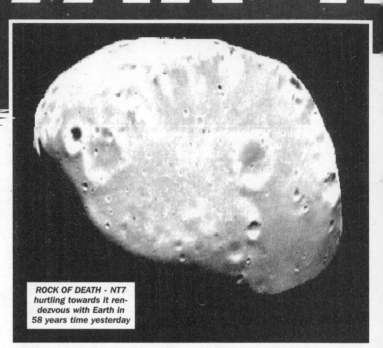

ROCK OF DEATH - NT7 hurtling towards it rendezvous with Earth in 58 years time yesterday

One celebrity who isn't worried about surviving the impact of NT7 is roly poly outdoors expert Ray Mears.

The 35-year-old machete enthusiast has survived some of the most extreme conditions on earth, from the Arctic to the Sahara desert. He told us: "After the end of the world, my survival skills will be more valuable than ever." And he had these tips for any stars wishing to join him in the post-apocalyptic wilderness.

- Stock up on tinned food now while you have the chance. Make sure you have enough meat, vegetables and fruit to last you for the rest of your life.

- Don't forget a tin-opener. Without it, you're dead.

- So you've eaten your food. How are you going to wash the pans? There won't be any washing-up liquid after the end of the world, so you'll have to make your own out of shampoo - but remember to rinse thoroughly.

- After the holocaust there'll be no scouring pads, so removing dried-on foods such as scrambled eggs from those pans could prove tricky. If leaving them to soak overnight doesn't work, try rubbing them with a hedgehog.

- So you've washed your pans - how are you going to dry them? You can't leave them in the sun, the atmosphere is choked with billions of tons of radio-active dust, and all your tea-towels were swept away in the tsunami which followed the initial impact. No problem. A bed sheet can provide several perfectly serviceable tea-towels if it is cut into oblongs and hemmed on an overlocking sewing machine.

---

20 MINUTES....*to DOOM!*

NT7 WILL STRIKE our planet at 7.44pm on February 12th 2060. However, there will only be 20 minutes warning before it lands. We asked the stars what *they* would be getting up to during their last 1200 seconds on earth.

"I'm not worried about the end of the world," said slightly sinister former Swap Shop presenter **Noel Edmonds**. "I'm building a huge diamond dome over my enormous private estate at Crinkley Bottom. Diamond is the hardest substance in the universe, so me and my money and helicopters will all be safe.

I'll spend the world's last 20 minutes just watching everybody panic outside my dome."

"I know exactly what I'll do," said zany ex-famous person **H** who used to be out of Steps. "I'll write, record and release an upbeat pop single that's all about Armageddon and how people shouldn't worry about it. Then I'd give a percentage of the money to children's charities."

Staggeringly bad film director **Michael Winner** wasn't worrying either. "I'll sit down and have a glass of wine," he told us. "Then I'll watch all the good bits from all the films I've made over the last 40 years. Then I'll watch them again. Then I'll make a cup of tea. Then I'll hard boil an ostrich egg. Then I'll just wait," the fat cunt added.

...AGEDDON!

Seaside town ready for Tsunami

-"Bring it on!"
says Lord Mayor

Blackpool is prepared for anything the universe can throw at it, said Lord Mayor councillor Ivan Taylor yesterday. And he had this message for doomsday meteor NT7: *"We're ready for you. Come and have a go if you think you're hard enough!"*

With two thirds of the planet covered by water, the chances are high that the errant asteroid will hit the sea. And if it does, scientists fear it could trigger a **tsunami** - a *two thousand foot tidal wave* which will travel round the world at supersonic speeds, destroying everything in its path.

civic

But at a meeting of Blackpool council last week, civic dignitaries drew up a six-point plan to ensure that even after the end of the world it will be business as usual for the popular Lancashire resort. "Our town has always been known as the north-west's premiere fun capital," said councillor Taylor. "We're certainly not going to let this bit of Blackpool rock disrupt anybody's holiday," he quipped.

goldwing

And he went on to outline the six-point plan which aims to prevent the half mile high wall of water dampening anybody's holiday spirits.

● *The illuminations and flower clock will be turned off at the mains to prevent short-circuiting when the tsunami engulfs the Golden Mile.*

● *People living along the seafront will be evacuated to nearby Lytham St Annes on a fleet of trams and buses until the danger has passed.*

● *Rubbish collections will be suspended on the day of the tsunami. Binmen will work the following Sunday to catch up. In the event of the day after the tsunami striking being a Bank Holiday, refuse collections may take two weeks to get back on schedule. Residents are asked to be patient during this period.*

● *The Piers will be closed to everyone except essential personnel for reasons of public safety while the tsunami strikes.*

● *The times for the the penguin parade and chimps' tea party may be subject to late change. Visitors are asked to contact Blackpool zoo before noon on the day of the meteor impact for up to date details.*

● *Ladies from the Blackpool, Fleetwood and Lytham Women's Institute will distribute cups of hot tea and coffee during the mopping-up operation.*

Councillor Taylor, his wife the lady Mayoress, and various other local dignitaries intend to co-ordinate emergency proceedings from the top of Blackpool tower. He said: "Make no mistake, This town is ready for Armageddon. We have even been to Tandy and bought some walkie talkies."

"Don't worry ~ It's not the end of the world," says Sting

" I used to sing about walking on the moon, so I know more than most people about meteorites. I also know a bit about saving the planet. You may remember me doing it a few years ago, when I took that plate lipped man onto the Wogan show.

To know how to stop a meteorite, you first have to know a bit about what they're made of. Here's a cutaway section, showing all the different layers.

The outside of a meteorite is called the **surface**. It is made of **craters**. Beneath the surface lies the inside of the meteorite, like an onion it is made of several different layers: **rock**, **stone**, **meteorite**, and the central **core**.

Now we know everything about a meteorite, but how are we going to prevent one ending the world? I've thought of *three* plans to stop NT7 in its tracks.

1 *Explode it.* The Americans could send an atom bomb which would blow it into billions of sand-sized pieces of rock. These would rain harmlessly down onto our atmosphere as shooting stars, creating the world's most spectacular firework display. We could sell tickets to watch, and send the money to the people of the Amazonian rain forests, so they could buy new trees and more plates for their bottom lips.

2 *Move the earth* out its path. It may sound ridiculous, but with 60 years' notice, I reckon there's plenty of time for scientists to build an enormous fan on the moon which could blow the earth out of harm's way like a giant balloon, allowing NT7 to cruise safely by.

Top Pop Star and ex-school teacher Sting outlines his plans to save the planet.

SURFACE
INSIDE
METEORITE
ROCK
STONE
CORE

3 *Deflect the meteorite* into a different orbit. An enormous cricket bat 200 miles long could be sent into space along with West Indian cricket ace Brian Lara. In the zero gravity vacuum of outer space, it would weigh the same as a normal earthbound cricket bat. Space-suited Brian could effortlessly hit NT7 for six, right out of the Milky Way, or even skillfully nick it to Jupiter or Saturn. Making such a huge bat would, however, seriously deplete the world's willow forests, but it would be a price well worth paying to save the planet.

Having now saved the world twice, I don't want any thanks. Just something simple would do, like renaming Earth 'The planet Sting' in my honour. **"**

Glass Sales to Break Records in 2060 ~ report

The catastrophic impact of meteor NT7 will be bad news for all life on earth, but good news for Britain's glaziers, according to Glass & Glaziers Monthly, the glazing industry's leading trade magazine.

"We're looking forward to smashing profits, and it's all thanks to the end of the world," laughed Paul Frasier, boss of the Professional Glaziers' Association.

giant

"I doubt there'll be a window in the country that won't get put through when this giant flaming ball of doom hits the earth, and they're all going to need replacing. The public will be putty in our hands." But glass watchdog organisation OffGlaze sounded a note of caution.

kendo

"We'll be watching out for overcharging in the immediate aftermath of the apocalypse, and coming down hard on any profiteering glaziers," warned Glazing Tsar Keith Hellawell. "The public can rest assured that whatever other worries they have regarding Armageddon, I'll be working to keep glass prices pegged at a sensible level."

SUCK MY COCK AND WIN £10 MILLION!

A quick gobble on Ted Johnson's bell end and one lucky girl could be TEN MILLION quid better of this weekend.

For luckless female Lottery contestants are being offered a chance to cash in on the enormous jackpots by the man who claims to have Britain's luckiest penis.

TADGER

Ted believes his tadger has the Midas touch, and that girls who give it a gobble will be *cock sure* of Lottery *suck-cess*. For after sucking it in the car park behind Ted's local working men's club a few months ago, a former girlfriend went inside and won a game of bingo.

BADGER

And now Ted is issuing an open invitation to the ladies to come along give his lid a lick.

"My penis has always been lucky, and a little bit

By our ficticious Lottery Correspondents
JACK POTT and WYN A. MILLION

of luck is all you need to win the jackpot. I'd suck it myself, but I can't reach. So any girls out there who want to boost their chances, they're welcome to come along and give it a quick lick."

CADGER

Surely there must be a catch? Not according to 22 stone Ted, 37, who retired from his job as a road sweeper on health grounds. "I'm not after a share of the winnings. The lucky ladies can keep the lot. I'll get my satisfaction from simply knowing that loads of good looking birds are going to win the Lottery".

The cock Camelot tried to ban!

Bob reaches for his Horn of Plenty. Yesterday.

Lottery bosses were yesterday thrown into a panic when we told them that one of our readers had a lucky bell end.

SUCK

"If this is true we stand to lose a fortune, with massive payouts due to all the women who suck this fellow's cock", a spokesman told us. "We will have to look very carefully at this situation", he added.

The jackpot for next week's draw could be as much as £100 million as thick people flock to buy the £1 a time tickets.

Have Your Say...

ACTRESS SUE JONES-DAVIS was probably best known for her role as Judith in Monty Python's *The Life of Brian*. In one scene she famously appeared full frontal nude, but like many actors, was happy to do so as it was essential to the plot. But could that decision come back to haunt her, for Jones-Davis is now the Lord Mayor of Aberystwyth? Does it matter that we have all seen her pubes? Does it demonstrate that despite being a Lord Mayor, she is just an ordinary person like one of her subjects, or does the fact that we know she's got a right old biffer on her lessen the dignity of her office? We went on the streets to find out what YOU think...

Jones-Davis as the Lord Mayor of Aberystwyth yesterday and as Life of Brian's Judith (inset)

...IT ill befits any Lord Mayor to have their pubic hair on display to all and sundry. I think that on being elected to any public office, people should have their pubic hairs shaved off so that this sort of thing can never happen.
H Brandenburg, Herts

...IT'S not the first time that someone in public office has had their pubes on display to the general public. Glenda Jackson MP got them out in the film The Music Lovers, and more recently Welsh Minister Ron Davies gave his an airing whilst looking for badgers in a layby somewhere.
Tarquin Milk, Runcorn

...I THINK it's an absolute disgrace that Jones-Davies should take up public office. Thirty years ago she was showing all she's got to anybody who wanted to see it, and today she's judging sand castle competitions for children!
M Tonbridge, Luton

...IT'S an appalling situation where somebody can reveal their pubic hair for profit, and then be put in charge of the civic finances of one of the finest cities in Wales. What is this country coming to?
T Bootle, Wigan

...I THINK it's marvellous that I have seen my Lord Mayor's bush. How many other people outside of Aberystwyth can say the same? Good on ya, Sue.
T Pinner, Aberystwith

...I BEG to differ with T Pinner (above letter). I live in London and I have seen Lord Mayor Boris Johnson's unruly pubes. Mind you, I'm married to him so that probably doesn't count.
Mrs B Johnson, London

...I THINK anyone who reveals their intimate bodily hair to anyone other than their spouse or their doctor should be banned from holding any public office whatsoever and put on the sex offenders register.
H Monkton, Croydon

...I LIVE in Aberystwyth and recently received a legal notice from the council telling me I had to cut my hedge which is blocking a pavement. The hypocrisy is staggering since, if her appearance in The Life of Brian is anything to go by, Jones-Davies is unwilling to trim her own thick foliage.
M Fibreboard, Aberystwyth

...I WAS so appalled when I heard on the news that

Jones-Davies had become Lord mayor of Aberystwyth, that I put my foot through the telly and sent her the bill. Then my wife pointed out that we were actually listening to the news on the radio, not the TV. So I put my foot through the radio and sent her the bill for that, too.
J Braithwaite, Mull

...JONES-DAVIES has demonstrated that she is not averse to displaying her pubery in public. I think the possibility of her giving everyone a little flash of her wotnot will add a little frisson of excitement to otherwise boring civic meetings and plaque unveiling ceremonies.
M Broughton, Solihul

...I WAS elected to my local council in 1970 and have never see any of the serving Lord Mayors' pubes. By electing Jones-Davies, the councillors of Aberystwyth have achieved in one stroke what I have never achieved in nearly forty years of office.
Councillor L Plywood, Otley

20 THINGS YOU NEVER KNEW ABOUT... DOGS

Woof! Woof! Some dogs yesterday

It's a dog's life, or so the saying goes. And never more so than at Christmas, when millions of people across the country open up their Christmas stockings to find unwanted puppies inside.

But before you bag up your unwanted pooch and chuck it in the nearest river, why not stop for a moment and think. How much do you **really** know about our four legged friends? For instance, did you know that...

1 A dog's sense of smell is ten million times more sensitive than that of a human being. This means that a dog standing in Trafalgar Square could smell a kipper – on the Moon!

2 If a vet picks up your dog, never ask him to 'put it down', because if you do he'll kill it with a needle in the back of the neck, and send you a bill for £14. That's because the expression 'to put a dog down' means to kill it with a needle in the back of the neck, and send you a bill for £14.

3 Americans love their dogs, and Los Angeles is the dog capital of the world. There are dog hairdressers, dog psychiatrists and even a dog restaurant, open exclusively to dogs. But any dog can't just walk in. Tables must be booked 3 months in advance, and prices start at $200 (£800) for a bowl of onion soup.

4 A hot dog isn't a dog with a temperature. Nor is it a stolen dog, wanted by the police. It is in fact a stale sausage sandwich with onions and mustard on it, often sold outside football matches for £5 each.

5 Hot dogs with sausages in them shouldn't be confused with sausage dogs. A sausage dog isn't a sandwich, it's a small, sausage shaped dog with tiny legs that only just manage to keep it off the floor.

6 The world's smallest dog is the chiwawa, the smallest ever recorded example belonging to Kalvin Phillips, the world's smallest man. His parents presented him with a puppy 'Shorty' at Christmas 1952. The dog weighed a microscopic 4 grammes, but Kalvin got bored and drowned him the next day – in a thimble of water.

7 If you bend down to examine a 'dog end' on the pavement, you definitely wouldn't be looking up the back end of a beagle, or examining the arse of an alsation. In fact, the chances are you would pick it up and smoke it! That's because a dog end is the disregarded portion of a cigarette.

8 And if your dog end was covered in the previous owner's sallva, you'd probably tut and remark that it had a 'duck's arse' on it.

9 Although they can be attractive animals, calling a girl a 'dog' would not be taken as a compliment. That's because 'dog' is a derogatory term used to describe an ugly woman.

10 And so is 'boiler'.

11 In cave man days dogs were much bigger than the ones we know today. Although little remains of these pre-historic dogs, we know that they had enourmous jaws, big enough to bury the massive dinosaur bones which scientists are still discovering to this day.

12 The Queen is Britain's number one dog lover, and her 700 Corgis are treated like royalty. No expense is spared. Each week all 700 dogs are taken to high class hairdressers Truefitt & Hill of Old Bond Street for a shampoo and trim. Indeed, the Queen spends £60,000 a year on cotton buds alone, which she uses to wipe the dog's bottoms.

13 Unlike kids today, dogs are prepared to get up off their backsides and do an honest day's work. Sheep dogs chase sheep around hills, fox hounds chase foxes around hills, police dogs bark at football fans, and specially trained sniffer dogs are used by customs officials to detect tiny amounts of drugs – concealed up people's bottoms.

14 The law no longer requires dogs to be licensed. However you do need a license to own a pub, a television, a car or a fish.

15 Or a gun.

16 But you *don't* need a license for a *gun dog*. Because a gun dog isn't a gun. It is in fact a dog.

17 Dogs are the world's most intelligent animals, apart from dolphins. And parrots. Indeed, the first man in space was in fact – a dog! On the 4th of October, 1957, history was made when Russian poodle Rin Tin Tin took off on board the Soviet's Sputnik rocket. Sadly, after three days orbiting the earth the heroic hound exploded.

18 If someone says 'it's raining cats and dogs' you needn't expect a downpour of domestic pets. Unless you live in Bolivia! For in 1932 meteroligists there were baffled when a football match between Ixiamas and Cotagaita was abandoned by the referee after it had started rainings *dogs* – and *frogs!* And fish as well.

19 Ask a prostitute in the Kings Cross area of London for 'doggies', and she'd be unlikely to hand you a basket of puppies. The chances are she'll get down on her hands and knees and avail herself to you for sexual intercourse – from behind. That's because 'doggies' etc. etc. etc.

20 Ask the same lady for a 'topless hand shandy' and she'll probably get her tits out and pull you off for £25.

MY FAVOURITE MOMENT IN HISTORY
No.3 JOHN OAKSEY

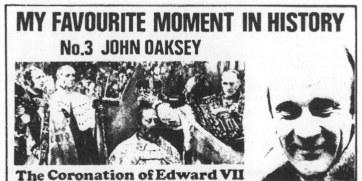

The Coronation of Edward VII

Don't be fooled by that friendly face - there could be danger next door!

IS YOUR NEIGHBOUR A VICAR?

● England's top vicar Robert Runcie

Dressed in casual clothes, it's not always easy to spot a Vicar. He may be drinking down the pub, or shopping in the local store. He could even be doing a spot of gardening, only feet away from your kids playing in the street. For all you know, your own neighbour could be a member of the God Squad. You can't always spot them, but here are ten tell-tale signs that would suggest the man next door is a member of the cloth.

Look out for these 10 tell-tale signs...

1. Does he go to church regularly, and always seems to work on a Sunday? He may leave the house early while other neighbours enjoy a lie in. Keep an eye out for him while you're washing the car.

2. He may have an unusual dress sense, with a preference for black shirts.

3. Does he have an uncanny knowledge of the Bible? Most vicars can quote entire paragraphs from it without once refering to the page.

4. Does he drink a lot of tea, with cakes, and ride around slowly on a bicycle?

5. Has he ever organised a jumble sale. Perhaps he has asked you for unwanted clothes etc., or you know someone who has been approached in this way.

6. Has he ever visited your house at a time of mourning? Funny how he always seems to call round not long after a close relative has died ...

7. He plays it straight with the girls — only one woman in his life, and strictly no sex before marriage. You won't catch a vicar playing the field.

8. Is he the quiet type, who drinks only sherry? The type who goes to bed early, and snubs your invitation to a late night party.

9. Is he the friendly type who never seems to get involved in a fight? Always says "hello" when you pass him in the street. Look out for that smile — he thinks he's got you fooled.

10. Does he ever talk about a 'Steeple Restoration Fund'? If your neighbour's a vicar, he may even ask you to contribute money towards this.

Spot the Celebrity vicars

We've disguised some well known celebrities as vicars. Can you tell who they are? Try to identify our mystery vicars, then turn to page 22 for the answers.

Here's what to do if you suspect...

If you suspect your neighbour, it may be wise to take the following precautions:

★ Make sure your doors and windows are securely locked.

★ Remember to cancel milk and newspapers when you go on holiday.

★ Don't let your children play unattended, especially near main roads.

You can report vicars to your local police, but they may not be able to do anything unless a crime has been committed.

ANYWAY

Anyway, write their name and address on a postcard and send it to your local police station. Remember to write "VICAR" clearly in the top left hand corner.

BRITAIN IS SINKING

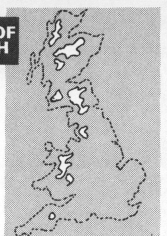

Britain in 1995 - only the white areas will remain above sea level

~And millions will drown

At one time Britannia Ruled The Waves. But now we are slowly sinking beneath them, and by the year 2000 Britain may have vanished completely into the sea.

By Bob Twatt

That's the shock belief of many geologists and top scientists who have been monitoring shifts in the Earth's crust. For years it has been recognised that Britain is slowly tilting on an axis — the west coast of Scotland rising by about an inch a year, and the south east coast moving down slowly into the sea. But now experts fear the process has been speeded up dramatically, and that Britain is beginning to **CRUMBLE** and fall apart at the seams.

FLOODS

Indeed, as early as March next year Dover harbour may have disappeared altogether, and perhaps by Spring 1990 flood tides could be sweeping thousands of shoppers in London's Regent Street to their deaths. Experts predict that by 1995 only the highest points in Britain will still be safe — areas like the Pennines and the Welsh mountains. Meanwhile in cities like Birmingham only church steeples and high rise flats will remain above water.

ACID RAIN

Many theories have been put forward to explain Britain's 'crumbling' phenomena. Mavis Partington of Ispwich blames heavy lorries, and points to cracks in the road outside her house as evidence. "They come roaring along here at all hours of the day", she told us. "And the council haven't done a thing about it. It's a wonder no-one's been killed".

TIDAL WAVES

The experts point to much bigger cracks to reinforce their claims. Cracks like the Humber, The Severn and The Thames. Any map of Britain shows the island is riddled with these cracks or 'rivers', each one threatening to tear wide open and split the nation into tiny islands.

Many geologists blame the coal industry, claiming that millions of mine workings which riddle the country have produced a deadly 'woodworm' effect. Mrs Dorothy Jones of Reading blames football hooligans, and believes convicted offenders should be made to repair the damage on Saturday afternoons. "The birch is too good for them", she added.

SHOWERS OF RED HOT LAVA

So far the Government has been reluctant to discuss the problem, but it may well be that plans are already underway to use millions of sandbags or "Green Godess" fire engines, mothballed since the war, to combat the advancing waves. An army spokesman who we stopped in the street said he was "In a hurry" to catch a train and added that he "hadn't got a f***ing clue" about the Government's plans.

Moscow on 'AIDS' elephant alert

Kremlin chiefs could be set to deny rumours of a radioactive elephant disaster in the Soviet Union in which thousands could already have been killed.

MUTATED

Elephants, originally escaped from a zoo, could have bred in the wild, and mutated due to fall-out from the Chernobyl nuclear disaster.

BULLET-PROOF

Thousands of these AIDS infected ferocious 'super elephants', bullet-proof due to radioactivity, are probably already at large in the sewers of many Soviet cities.

STARS BEHIN

It used to be big news when a famous celebrity went to jail, but nowadays it seems to happen every day.

Lester Piggott, footballer Tony Adams. The list seems endless. For in today's materialistic world the temptation to commit crime is enormous, and nowhere more so than in show-business. Indeed for celebrities today prison sentences are looked on almost as an occupational hazard. One day they're appearing on TV and signing autographs, the next day they're sewing mailbags and slopping out.

PORRIDGE

So who are the stars most likely to fall foul of the law, and what kind of crimes might they commit? And how would today's top show-business stars cope with doing 'porridge'?

Just for fun, we asked Britain's top Show-business Criminology Psychoanalyst to answer these questions by examining the character of several top celebrities. And here, using his in-depth knowledge of crime and the stars, he gives us his verdicts.

Roly-poly TV astronomer **RUSSELL GRANT** looks innocent enough on the box. But I believe that if times were hard burly Russell could put himself about a bit, and turn to violent crime for his livelihood.

BREAD

I feel that Russell would go in at the deep end, and carry out an armed wages snatch on a Securicor van. But the flamboyant star gazer would refuse to lie low after the raid, and would give himself away to police by throwing money around in London's West End clubs.

RICE

Russell would receive between 10 and 14 years for armed robbery, but I believe he would adjust easily to life inside. His amiable nature and generous personality would make him popular with other inmates. I think he would get a job in the prison library, and in his spare time do horoscopes for warders and perhaps even the governor. With good behaviour he would be out and back on our breakfast TV screens in about 7 years.

LLOYD-WEBBER

Taking and driving away a vehicle without the owner's consent may not sound like the most serious crime, but someone who commits that offence should be made an example of, especially if that person is TV magician **PAUL DANIELS**.

I believe that Paul, after a drinking session with other showbiz pals, could try to show off by stealing a high powered car and taking it for a joyride.

MAGICIAN

If Daniels was sent to prison for his crime he would find it a far cry from his glittering career as Britain's top magician. He would spend long periods of time banged up in a cell, and practising magic would not be easy. Prisoners are not allowed playing cards, rabbits, swords or strings of handkerchiefs in jail. I believe Daniels would become disillusioned, and after his release I believe he would soon be in constant trouble with police for various petty crimes.

BLACKBURN

If Italian gangsters were looking to launder mafia millions in Britain they may well turn to **TONY BLACKBURN**. And I believe that Tony is too nice a person to turn them away. He could unwittingly become tangled in a web of international fraud.

ROCHDALE

If the police rumbled Tony's illegal money laundering operation, then the popular former Radio One DJ would be in big trouble. Not only would he face a severe prison sentence, but his Sicilian paymasters may well try to silence him.

OLDHAM

Life inside would be a nightmare for pretty boy Blackburn, 47. As well as keeping his eyes open for mafia hit men, Tony would also have to watch out for all the red blooded criminals who have not set eyes on a women for many years. Loneliness drives men to do awful things behind prison walls, and for Tony the showers would be a no-go area. Indeed, Tony would have to be careful. Less attractive men than him have left jail with ringpieces like doughnuts.

BISCUITS

Shoplifting is the single most common crime among the stars of stage and screen. And if millionaire chat show

host **TERRY WOGAN** were wrongly accused of stealing a packet of biscuits and a pair of women's tights after a mix-up in Sainsbury's, few people in showbusiness would even turn an eye.

Should Terry receive a custodial sentence, he would find that as in the outside world, opportunities exist in side prison for the commercially minded. Terry has a sharp eye for business, and I believe that within days he would be dealing in tobacco, chocolate and pornographic magazines. The thrifty Irishman could then enjoy privileges such as a carpet and TV in his cell, and would have the warders as well as the inmates at his beck and call.

KNIFE

Children's TV artist **TONY HART** would be the last person you'd expect to find cruising the Kings Cross area in a pink cadillac, wearing a floppy hat. It's hard to believe, but if found guilty of living off immoral earnings, Vision On presenter Hart could receive a jail sentence of up to 5 years.

D BARS

CELEBRITY COURT

THIS WEEK'S GUEST JUDGE
SIMON BATES

We were rather short of ideas for this issue, so just for fun we asked Britain's favourite DJ Simon Bates to be judge for a day and preside over some ficticious court cases, passing sentence on some well know celebrities who an imaginery jury have found guilty of committing hypothetical crimes.

Simes agreed to don his judge's wig and dish out justice in the following cases which we have made up.

⚖⚖⚖⚖⚖

Defendant: **THE KRANKIES**
Charge: Sending obscene material through the post.

Verdict: **GUILTY**

Simes Sums Up: I'm as liberal as the next man, but due to the extremely sordid nature of the material concerned here I have no hesitation in sentencing Jimmy and Jeanette to the maximum term the law allows.

Simes Sentence: They will go to prison for 2 years.

⚖⚖⚖⚖⚖

Defendant: **BERYL REID**
Charge: (Just for fun) Possession of a class A controlled substance with intent to supply.

Verdict: **GUILTY**

Simes Sums Up: Knowing Beryl as I do, I am particularly disappointed to find her before me on such a serious charge. However, this is only a first offence, and I feel confident that you have learnt your lesson. I propose to give you a second chance.

Simes Sentence: I hereby sentence you to 18 months in prison, suspended for 2 years. In addition the court will seize the estimated proceeds of your crimes – £7,500.

⚖⚖⚖⚖⚖

Defendant: **BOB HOLNESS**
Charge: Drunk and Disorderly, Urinating in a public place.

Verdict: **GUILTY**

Simes Sums Up: I had always thought you were an intelligent man. But your behaviour here has reflected no intelligence on your part. A man in your position, a respected game show presenter, should set an example to others. You have let yourself and many other people down.

Simes Sentence: I hereby fine you £10,000, and also sentence you to 5 years in prison.

If any of the celebrities named feel that their sentences are unfair they can join in the courtroom drama by making a postal appeal. Simply write to: The Appeal Jury, Viz Celebrity Court of Justice (No. 46), P.O. Box 1PT, Newcastle upon Tyne NE99 1PT. Your appeals will be heard in the next issue, celebrity Judge and coffee ad star Gareth Hunt presiding.

Sensitive Hart would never truly adjust to prison life. In fact, he would lie awake at nights planning his escape. One day, whilst working in the prison kitchen, I believe Hart would grab a knife and take a warder hostage. Eventually he would clamber onto the prison roof and stay there for several days, shouting abuse at police, throwing slates, and making imaginative use of old sheets and other materials to create brightly coloured, attractive banners proclaiming his innocence.

FORK

ESTHER RANTZEN has a reputation as a tireless campaigner for good causes, but I believe there are flaws to her character. If, for example, she were offered £5,000 by a crooked second-hand car dealer to murder a rival small time gangster, I believe she would carry out the killing, against the advice of her husband, BBC producer Desmond Wilcox.

Of course life would be the only sentence Esther could expect for the cold blooded contract slaying. I think she would be appalled by the conditions she found inside jail, and would immediately start campaigning for better conditions and facilities. I think Esther would genuinely regret her crime, and would become a model prisoner, studying for an Open University Psychology degree, and writing books. After only 7 years I think she would be out on parole.

SPOON

However, I fear Esther would very quickly be back inside, and this time for good. I believe she would have a huge row with her husband Desmond Wilcox after discovering that he had spent the £5,000, and after a struggle her gun would go off, fatally wounding her balding TV executive husband.

The Undersea World of PAT ROACH out of Auf Wiedersehen, Pet

HELLO THERE. YOU KNOW, YOU GET TO SEE SOME PRETTY STRANGE SIGHTS WHILE FILMING A TV SHOW LIKE AUF WIEDERSEHEN PET. THE WORLD OF WRESTLING, TOO, IS FULL OF SURPRISES. BUT NOWHERE ARE THERE ANY STRANGER THINGS THAN THOSE CREATURES THAT LIVE BENEATH THE WAVES. COME WITH ME NOW AND LOOK AT THESE WEIRD OCEANIC INHABITANTS, THESE MYSTERIES OF THE DEEP?

THIS IS A CERATOID ANGLER FISH. SHE USES A SPINE AS A LURE TO TEMPT FISH TOWARDS HER MOUTH. THE MALE IS TINY AND LIVES AS A PARASITE... ...HE STAYS ATTACHED TO HER BELLY FOR HIS WHOLE LIFE!

ONE OF MY FAVOURITE UNDERWATER FRIENDS IS THE SEAHORSE. THESE ODD LITTLE FISH SPEND MOST OF THEIR TIME ATTACHED TO WEEDS. BABY SEAHORSES ARE PROTECTED IN A POUCH... ON THE BELLY OF THE MALE!

THESE FUNNY LOOKING FELLOWS ARE MANATEES, OR SEA COWS. THEY ARE VERY SLOW MOVING, TIMID MAMMALS THAT LIVE IN WARM SHALLOW COASTAL WATERS OF THE TROPICAL ATLANTIC OCEAN. ALTHOUGH LARGE AND POWERFUL, THESE CREATURES ARE VERY DOCILE VEGETARIANS, SHYING AWAY FROM MAN.

NEXT WEEK– TIMOTHY SPALL'S LIVING DESERT

Burping Britain's

NO.1

Britain is top of the acid flatulence league – and that's official!

A survey has shown that Britain topped the table throughout 1992, well ahead of its European neighbours when it comes to burping and belching.

VICTORY

And the good news comes only two weeks after Britain's surprise victory over Belgium in the semi - finals of the heartburn indigestion cup. Now the UK is all set for a flatulence and indigestion cup final showdown with arch rivals Germany who, for the last eight years, have been world champion farters.

RENNIE

Sales of Rennie tablets have trebled in Britain during 1993 as our rumbling tums have sent us soaring to the top of the dyspepsia league tables. But diarrhoea is also on the increase in British bums. And medical chiefs yesterday warned against loose stool sufferers overdoing it with remedies such as kaolin and morphine mixture.

"The prescribed dosage is perfectly safe", said a Harley Street specialist yesterday. "But if Britain's diarrhoea sufferers overdo it we could be heading for a surprise appearance in the Coca Cola Constipation cup final against Ireland next year".

RENATTA

Ireland, who top the Guinness drinking charts, are also undisputed world solid stool champions, with anything up to 85 per cent of Irish bottoms blocked solid at any one time.

BOTTOMS

And with their bottoms under such strain it is hardly surprising that the bookmakers make them clear favourites to win the Rectal Pile World Cup finals to be held in Bolivia next Tuesday.

HOT AIR

Balloon lands Hill in hot water

Plans for a hot air balloon passenger service to be operated by soccer pundit Jimmy Hill are up in the air – after Hill's maiden voyage landed him in hot water.

BBC 'Match of the Day' analyst Hill planned to take a leaf out of airline boss Richard Branson's book by becoming a millionaire balloon operator. But Hill's maiden voyage from Birmingham to London was a disaster, and left his first two customers furious, and demanding their money back.

FLIGHTS

Hilda and Norman Jones saw Hill's advert in the Dudley Herald offering return flights to London for only twenty pounds. They rang Hill and were told to meet him in a field in Warwickshire early the following day. When they arrived they were surprised to find former footballer Hill on his own, struggling to unravel the balloon which had become tangled on nearby bushes.

BASKET

When the balloon was eventually inflated Hill told the couple to join him in the small basket. "There was barely enough room for Jimmy Hill, never mind the two of us", said Mrs Jones yesterday. Eventually the balloon got off the ground, but as he struggled with the controls Hill repeatedly caught Mrs Jones on the ear with his elbow. "It was bruised for several days afterwards", she told us.

SUIT

Within ten minutes it became obvious to Mr Jones that the balloon was travelling in the wrong direction. "I pointed out to Mr Hill that we had crossed the A444 and were heading for Nuneaton, in the opposite direction to London. He didn't seem to have a clue what he was doing".

HARD

Shortly afterwards Hill announced that he was stopping for lunch, and made a precarious landing near the town of Hinckley, during which Mrs Jones hurt her elbow. Hill then wandered off, leaving the Jones's to fend for themselves in a field. After three

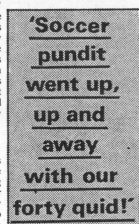

> **'Soccer pundit went up, up and away with our forty quid!'**

hours he had not returned, and after walking four miles Mr and Mrs Jones eventually flagged down a taxi and returned home, at a further cost of £18.50.

JIMMY

The following day Mrs Jones rang Hill and asked for a refund. "At first he spoke with a Scottish accent and said Jimmy Hill wasn't in. But eventually he owned up, and said he was Jimmy Hill, but was not prepared to give us our money back". According to Mrs Jones, Hill claimed that the flight had been cancelled due to bad weather, and that no refunds were payable. "You should have taken out insurance", he was quoted as telling her.

BRIEF

When we rang Hill he told us that his balloon service had been suspended due to 'technical difficulties'. Later, when pressed, he admitted that he had been unable to steer the balloon in the right direction. "I didn't realise that so much depends on the wind", he told us. "It's not like an aeroplane or bus, where you just turn a wheel and go anywhere you want. Balloons have a mind of their own, and they just go anywhere they please". He added that he was considering operating a revised service from Birmingham to Mansfield, but had no dates yet.

When we rang the BBC a spokesman for the 'Match of the Day' programme told us that they were unaware of Hill's ballooning activity. "What Hill does in his spare time is largely his own business. It has no bearing on his analysis of football matches", we were told.

OPEN & SHUT

Meanwhile, a spokesman for the Dudley Herald newspaper told us that Hill had not paid for the advertisement which he placed in their paper. "Mr Hill booked a small advertisement to run for one week at a cost of £26,50. To date this has not been paid and the outstanding amount is now overdue".

'I'll strip for cash'

A young girl advertising a 'stripping' service in a local newspaper offered to 'take everything off' when she was approached by a reporter.

"I've got a big chest. You can look at it if you want", she told him, pointing at a large blanket chest in the back of the shop. Our man said it wasn't a very nice colour. "I'll take everything off for £80" she told him.

CAN YOU IDENTIFY OUR MYSTERY MULTI MEMBER LOVECHILD?

There's nothing new about our Right Honourable Members sticking their own less than honourable members in the wrong ballot box. But can you imagine what would happen if a group of well known MPs all shagged *each other*, instead of their secretaries?

In 1994 modern space technology and computerised gene image grafting techniques enable us to predict exactly what *would* happen if nine of our best known politicians had group sex. For this is the face of the resulting love child, an amazing biocomputerised picture, accurate to within a quarter of an inch.

SEX

Needless to say in reality it would never happen, because all nine of these honourable Members would no doubt chip in towards an abortion. But just for fun, can you identify the nine different politicians taking part from our picture? The first person who correctly identifies all nine will win all three new Viz T-shirts

advertised on page 40. Send your answers on a postcard to our usual address, marked 'Mystery MP Lovechildren'. For those of you who can't be bothered the answers are on the next page.

MILLENIUM BUGGERED!

Royal timebomb set to explode

THE Queen Mum may not live to see the year 2000, according to computer experts.

Boffins fear that Britain's favourite granny will be struck down by the Millenium bug when the clocks strike midnight on the 31st of December this year. Scientists fear that the Queen Mother's body clock will not recognise the dateline 01.01.2000, and that she will crash, wreaking havoc amongst the Royal Family.

Brain

Top Harley Street physicians are on standby, ready and waiting to update the Queen Mum's brain. But the Palace is split over the huge cost of such an operation.

Festival

Reprogramming the old dear's noggin is possible, but it will be expensive. Replacement loaf parts for a woman of her age have to be made especially. The total cost could run into thousands, rather than hundreds of pounds. And the Royal Family must decide whether such an enormous investment

Queen Mum is NOT Millennium compatable

can be justified, bearing in mind the Queen Mother's age.

Battle

"Basically, it will cost us three times the price of a new Princess just to re-programme granny. And one must question whether that sort of expenditure can be justified on what is essentially a short-term asset", said one big eared heir to the throne yesterday.

Hotspur

Meanwhile, the latest Royal signing, Sophie Rhys-Jones, was unveiled

The Queen Mum could be waving her last goodbye on December 31st 1999.

to the fans at a press conference in London yesterday after passing her medical with flying colours. Miss Rhys-Jones will officially sign up at a wedding this summer. Meanwhile delighted Palace officials yesterday confirmed that the new princess will be Millennium compatable, and promised further signings in the near future.

Warlord

"We've got money to spend, and if the right princesses become available, we'll be in for them", said the Queen yesterday.

GREENPEACE campaigners in rubber dinghies yesterday swamped the airwaves surrounding Radio One in protest at the Norwegian Government's policy of Moyle hunting.

Moyles, the largest animals on radio, are an endangered species after being extensively hunted for their blabber. The Moyle is also a precious source of ego, a commodity which is used extensively in the entertainment industry.

Victor

Because of their size Moyles are an easy target. They inhabit the shallow airwaves of daytime Radio One and cannot remain undercover for more than one record, before surfacing to spout shit for several minutes. Moyles attract symbiotic parasites who attach themselves to their big, fat, sweaty arses and laugh sycophantically at the constant, incoherent, high pitched sounds which they emit. Some experts believe that these sounds are a form of intelligent communication, although no-one has yet been able to decipher them. Despite its vast bulk the Moyle exists entirely on a diet of cheese and onion crisps which it scoops up in vast quantities in its huge mouth as it

A beached Moyle floundering on an episode of Never Mind The Buzzcocks recently.

gracelessly manoeuvres itself around the airwaves.

Viva

Now protected by international law, the hunting of Moyles is strictly regulated and licenses are only granted for the purposes of scientific research.

Astra 1.3 GL

Despite being the largest animal that has ever lived, Moyles have the smallest penis, at a mere three quarters of an inch - when erect.

The Telly Savalas Story

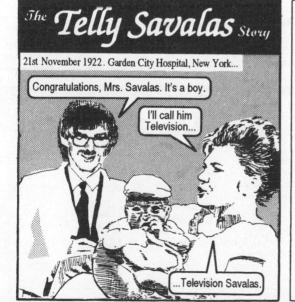

21st November 1922. Garden City Hospital, New York...

Congratulations, Mrs. Savalas. It's a boy.

I'll call him Television...

...Television Savalas.

Once home, young 'Telly' proves a big hit with his father, Mr. Savalas.

Heh! Heh! Who loves ya, baby!

Hmm!...

...I'll remember that.

1972, and Telly becomes an actor and does Kojak...

Nyaaaah! Who loves ya, baby.

Thanks, dad!

SNACK ATTACK!

Bill, 52, Nabbed by Cops...
for Eating CRISPS!

A SOUTH YORKSHIRE man discovered he had bitten off more than he could chew this week - when he was sensationally arrested... for eating crisps in the street! Hapless Bill Johnson, 52, was busy wolfing down his favourite cheese & onion snack on a bench in Rotherham town centre when he was approached by police.

BURLY

A visibly shaken Mr Johnson told us: "I couldn't believe it. There I was minding my own business when all of a sudden these two burly coppers jumped out of a police car which had pulled up. They spotted me eating my crisps and pounced."

Within minutes he had been taken to the local station where he was formally charged. "I was stunned - all of this grief just for eating a few crisps. I was made to feel like I was a common criminal," he added.

BADMINTON

South Yorkshire Police remained unrepentant when asked to justify their officers behaviour. A spokesman told reporters: "As we have explained both to the gentleman in question and yourselves on numerous occasions now, Mr Johnson was not arrested for eating crisps but on suspicion of various incidents of indecent exposure and two minor sexual assaults which occured in and around Rotherham town centre. After viewing CCTV evidence, Mr Johnson admitted to the crimes and is currently on remand pending a court

Crisp Martyr Johnson (inset) yesterday and (main picture) the bag of crisps he was arrested just

appearance. Whatever it was he happened to be eating at the time we apprehended him is frankly neither here nor there."

The police's heavy-handedness was strongly criticized by local Conservative MP Bexter Ellis-Sophie. "The fact that this man was arrested simply for eating crisps in the street beggars belief. What sort of country are we living in when a man can't even eat his lunch without some jumped up loony lefty trying to lock him up for it, he fumed to reporters.

FISHING

Later on, he phoned the same reporters to add: "Quite simply this is political correctness gone mad."

Have Your Say!

VIZ READERS have been quick to air their views on the plight of 'Crisp Martyr' Bill Johnson, who was controversially arrested by South Yorkshire Police for eating crisps in the street. And the overwhelming majority of you were outraged by the decision.

"I AM so disgusted by Crisp Arrest Man Bill Johnson's ordeal that I am considering voting BNP in the next General Election because of it. And also because I am a racist."
Edith Thick, Burnley

"WHAT next - will I be given an ASBO by 'Naked Chef' Jamie Oliver for eating turkey twizzlers?"
Mike Dullard, Kent

"NO DOUBT if Mr Johnson had been eating a samosa or some prawn crackers that would have been fine and dandy and what's more the powers that be would probably have bought them for him out of my taxes. Well I've got news for Mr Blair and his Euro cronies, eating crisps makes me proud to be British and if Herr Schroeder doesn't like it then perhaps it's about time he remembered who won the war in the first place."
Maj. Charles Montegrew Power-Burroughs (retired), Hemel Hempstead

"WHAT next - will I be sent to the electric chair by that American chap with the moustache from 'Supersize Me' for tucking into a Big Mac?"
Mike Dullard, Kent

"ISN'T IT about time the police let decent law abiding citizens like Bill Johnson eat their crisps in peace and concentrated on catching the real criminals for a change, like the flasher who has been plaguing my local town centre."
Neil Sanderson, Rotherham

"THE LAWS of this land apply to Mr Johnson the same as everyone else. If the police feel in this post 9/11 environment that banning crisps can help us win the war against terror then we should all abide by their decision and eat pork pies or something."
Ryan Reasonable, Berwick Upon Tweed

GREER of the YEAR

C OMEDY harridan Germaine Greer was yesterday crowned 'Greer of the Year' at a star-studded ceremony in central London.

It is the first time the feminist writer has been awarded the title. The presentation, made by last year's winner Ian Greer, professor of Obstetrics and Gynae-

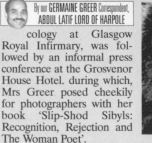
By our GERMAINE GREER Correspondent, ABDUL LATIF LORD OF HARPOLE

cology at Glasgow Royal Infirmary, was followed by an informal press conference at the Grosvenor House Hotel. during which, Mrs Greer posed cheekily for photographers with her book 'Slip-Shod Sibyls: Recognition, Rejection and The Woman Poet'.

Greer - award yesterday.

NOEL'S ARK!

Telly superstar Noel Edmunds plans to take a page out of God's book in a dramatic one man bid to save London Zoo.

For the tidy bearded TV host plans to build a giant ark – big enough for all the zoo's ten thousand animals. And like the Bible's Noah, telly's Noel plans to march the animals into his ark two by two when the zoo closes later this year.

ANIMALS

Edmunds launched his spectacular rescue plan in the wake of news that many animals may be destroyed when the zoo's gates finally close. And the game show king's ambitious ark project could mean salvation for thousands of helpless animals, including lions, zebras and giraffes.

But, weather permitting, the dare devil former DJ is determined to go ahead. And officials at London Zoo were cautiously optimistic in their response to the rescue plan. "Anything that will help publicise the plight of homeless animals and help to guarantee their future survival is to be welcomed", one told us.

MALTREATMENT

Naturalist, author and part time zoo keeper Gerald Durrell, a renowned critic of animal maltreatment, was surprisingly unimpressed by Edmunds eleventh hour bid

Animal addict Noel's late, late mercy bid

to save London Zoo. Durrell, who's got a grey beard and married an American bird, a bit like Linda McCartney, refused to comment when we rang him at his Jersey home.

Fellow celebrity zoo owner Lord something or other, the randy one who inherited the lions of Longleat the other week, wasn't available for comment.

FISH

And when we rang Jaques Cousteau for a comment his secretary told us he only talks about fish.

So sad - a homeless monkey

Noel's nautical mercy dash will be one of the most daring and spectacular animal rescue operations in maritime zoo history. And unlike the light hearted game show's which he hosts on our TV screens, Noel is taking his rescue bid *deadly* serious.

LINDISFARNE

"He's read the Bible over and over again, and plans to follow Noah's example down to the last detail", one TV insider told us. But other colleagues at the BBC expressed concerns about the scheme.

TYGERS OF PANTANG

"As I recall Noah's ark relied on heavy rainfall raising the water level sufficiently to set it afloat", said weatherman Michael Fish. "I find it highly unlikely, especially in the current climate, that enough rain will fall on London zoo to launch a giant ark".

SELFISH STARS TURN BACKS ON BEASTS

According to their publicists the top stars will always go out of their way to help a worthy cause.

Whether it's Live Aid, Aids Awareness, Comic Relief or Telethon, there's always a host of famous faces eager to jump on the charity bandwaggon.

PUBLICITY

But away from the glare of publicity, just how kind hearted are the stars? We decided to find out by asking a few well known celebrities whether *they* would be willing to help save the poor, homeless animals of London Zoo.

DAVISON - TV vet

Our first call was to TV vet **PETER DAVIDSON**, star of All Creatures Great and Small. On screen he's on call 24 hours a day to help farm

animals in the Yorkshire Dales. But in real life it was a different story.

GIRAFFES

When we asked whether Davidson would be prepared to provide a home for the zoo's giraffes, and some camels, he flatly refused. "I am a busy man. Please stop wasting my time", he told us.

CHARITY

Big hearted strong man **GE-OFF CAPES** is known as the Gentle Giant, and regularly attends charity sports events. But sadly, we have to report that Geoff's gentle caring attitude does not extend to the animals in London Zoo.

BEARS

We asked the World's former Strongest Man whether he could look after a few bears when the zoo closes. We pointed out that these could easily be accommodated in a small pit dug in his back garden.

"I can't keep bears in my garden. It would be illegal",

CAPES - TV strong man

stormed the former policeman. "I could probably look after a couple of budgies, but it would be difficult for me to feed them regularly, especially at Christmas when I'm away doing pantomime", he added.

REDSKINS

Rock superstar **BONO** probably owns a string of houses across the world. And probably has acres of gardens to spare in every one. Surely room for some lions, or perhaps some monkeys, you'd have thought. But no. It seems that big mouth Bono would rather see them destroyed.

"Looking after wild animals is a specialist job", said a

Top of the

We're gonna live forever, Gonna live forever, Live forever, Forever. So sang Oasis in their hit song Live Forever. It's unlikely that the words of that song will come, as Spandau Ballet sang, 'True' but with advances in medical technology and Dolly the sheep, pop stars are living longer than ever before. It's a sobering thought that the likes of H out of Steps and Pop Idol Will could still be alive in 100 years time.

But how long exactly do the stars think they've got? And what steps are they taking to prolong their lives? We asked them and then got Pop Gerontologist, Doctor Fox to assess how many, as David Bowie sang, 'Golden Years' they can expect before finally ending up in a, as Bernard Cribbins sang 'Hole in the ground'.

Pop Peter Pan Young One Bachelor Boy **Cliff Richard** still shows no sign of dying. So what's his secret formula for living forever? Surprisingly, the ageing rocker bases his lifestyle on that of the tortoise, some of whom live to be over 250 years old! "I've got a big wrinkly neck, and I eat lots of lettuce," he told us. "Round about October, I curl up and my manager puts me in a box of hay in the garage, where I remain fast asleep for half the year. In the Spring he takes me out and wipes my eyes with damp cotton wool," he added. "I've also painted 'Sir Cliff' on my back with Humbrol."

Dr. Fox says: *"On the face of it, Cliff's plan is a good one. Without the stress of constant touring, and with his vitamin-rich lettuce diet, he could live well in excess of 180 years. However, there is a risk that he could be eaten by rats or wake up early and wander into next door's garden, fall asleep in a pile of leaves and get shovelled onto the bonfire by mistake."*

Four-foot-six Who frontman **Roger Daltrey** famously sang that he hoped he died before he got old, but as he approaches pensionable age he's glad his dream only came true for fellow band members Keith Moon and John Entwistle. "I once heard that your heart only beats a certain number of times in a lifetime," he told us. "So I try to keep my pulse rate as low as possible. For example, I always listen to my old rock LPs at 16 rpm, and I go everywhere in a sedan chair carried by slaves. I'm so relaxed my heart only beats 3 times a minute, so I reckon I've got a good few years

left in me yet."

Dr. Fox says: *"Roger's laid back approach is a good recipe for longevity. But there are other things he could do to postpone his inevitable death. For example, tests on mice have shown that looking at fish reduces stress and can double lifes-*

pan. Roger owns tens of thousands of fish, so if he could manage to watch them all he could improve his chances. However, thanks to his high octane lifestyle in the 60's, he's already used up the vast majority of his heartbeats. I predict he's only got a few weeks to live."

In her 1979 hit, disco diva **Gloria Gaynor** boasted that she would survive, and so far she shows no sign of letting her listeners down by dying. "I think I'll survive well past 120," she told us. "I reckon the

secret of long life is to keep yourself busy. Although I no longer record or perform, I do voluntary work at a local quarry 4 days a week, shovelling wet grit into sacks."

Dr. Fox says: *"Whilst exercise is undoubtedly good for older people, it should ideally be gentle. I'm worried that Gloria Gaynor may be doing herself more harm than good shifting 15 tons of damp grit 4 times a week. However, if she cuts down to a couple of days a week, and remembers to lift the sacks with her back straight and knees bent, I see no reason why she shouldn't still be with us in 50 years' time."*

Bee Gees Maurice, Robin and **Barry** are well known for their ludicrous high-pitched singing, piano key teeth and hissy fits. They look no older now than they did in their heyday 40 years ago, but in the words of their song, how much longer can they keep 'Stayin' Alive' before they're all dead? "We hope to make it well past 100," squeaked Maurice. "We're on a special diet drawn up for us by a Hollywood nutritionist. Basically, we each drink a gallon of purified mineral water each day, and we're only allowed to eat nuts and fizzbombs. Lots of other stars follow the same plan, including Bob Hope, McCauley Culkin and Gary Coleman out of Diff'rent Strokes."

Dr. Fox says: *"I fear for the Bee Gees. They should be wary of fashionable dieticians promising long life. Years of experience have shown that you can't beat a traditional balanced diet, such as chips, beans and fish fingers, supplemented with regular injections of monkey hormones, for keeping the Grim Reaper at bay."*

DEAD or ALIVE?

For some people it's very simple to predict how long they'll live. That's because they're already dead. But if you're dead, don't despair, you're in company with some of the greatest people who have ever lived - Cleopatra and William Shakespeare, Einstein and Beethoven, Martin Luther King and Dustin Gee are all dead.

But it's often difficult to work out whether or not you are dead - you could be in a narcoleptic trance, or simply drunk or asleep. Follow the path of truth, answering the questions yes or no to find out if you are Dead or Alive.

Are you lying in bed looking forward to another day in the world?

Are you going down a long, light-filled passage towards your grandparents?

Are you lying in a long, thin fridge with your big toes tied together with string?

Are you visiting them at an old folk's home?

Has an undertaker just stolen your watch & rings?

Are you sitting watching the telly?

Are your 1995 Christmas decorations still up?

Is your body temperature below 98.4°F?

Have you had sex recently with a mortuary assistant?

There are no flies on you, you're alive! But don't get cocky - you could be hit by a bus tomorrow.

Bad luck. You're dead. But it's not all doom and gloom - your hair and nails may well still be growing.

OAPs!

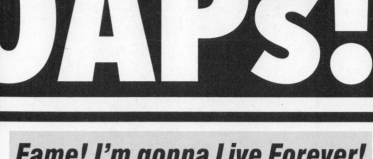

Fame! I'm gonna Live Forever!
Or did they?

Back in the 80s, the kids from Fame told the world that they were going to live forever. And twenty years on, the question on everyone's lips is how is their bid for immortality faring. We look at see how they are progressing and award each one a 'live forever' rating out of ten.

Lee Curreri
Born 4th January 1961
Lee, who played mop-top music prodigy Bruno, is doing quite well in his bid to live forever, having sucessfully completed 41 consecutive years without dying. He quit the stage 10 years ago and now works in a cheese freezing warehouse in New York, where he believes the cold, calcium enriched air will improve his chances of immortality.
Live forever rating: **8/10**

Erica Gimpel
Born June 25th 1964
Actress Erica Gimpel has sucessfully completed 38 years of immortality, though is sadly now confined to a wheelchair after being shot in the spine in an episode of ER. This is doubly tragic, as in the Fame TV series Erica played feisty student Coco Hernandez who lived to dance on car bonnets. Happily, she can still walk and dance in real life.
Live forever rating: **7/10**

Jesse Borrego
Born August 1st 1962
Texas born Jesse has the Japanese symbol for long life tattooed on his arse. After his character Jesse Valasquez was axed from the show, he learned how to fly - high! - by joining the United States Air Force. He left after 4 years and picked up his sucessful acting career, appearing in the TV movie 'Hell Swarm' and episode 19 of the first series of 'Hunger'.
Live forever rating: **8/10**

Gene Anthony Ray
Born May 24th 1963
39 year-old Gene, who played dance ace Leroy Johnson has quit acting and moved to Milan, where he runs a business photographing tourists holding a snake. He hopes the Mediterranean diet of sun dried tomatoes, olive oil and pepperami will help him achieve his aim of living forever. And he's doing fairly well so far, with a score of 39 years not out.
Live forever rating: **7/10**

Valerie Landsburg
Born August 12th 1958
California-raised Valerie, who played bubbly but unattractive dancer Doris Schwartz is leading the race to immortality amongst the fame veterans. After network bosses axed the series in 1987, she quit acting and moved to Nantucket where she sank her savings into a squid ink farm. After that failed, she moved to Boston wher she now farms cuttlefish for budgies.
Live forever rating: **9½/10**

Albert Hague
Born October 13th 1920
Veteran actor Hague played long-suffering Benjamin Shorofsky, bearded Professor of Music at the High School for Performing Arts. Initially doing well in the imortality stakes, he suffered a setback in November last year when he was bitten in the scrotum by a gila monster after a prank he was playing at San Francisco Zoo backfired. He was making a good recovery when he fell out of the hospital window and died.
Live forever rating: **0/10**

How Long Will YOU Live?

It's a sad fact that, unless you are Dr Who or Christopher Lambert off Highlander, one day you are going to die. However, it's also certain that you will probably live longer than your grandparents did, especially if they died young. 30% more people now live up to 20% or more longer than 95% of their ancestors - and it's a figure that's increasing by up to 15% per year.

In 1970, average life expectancy was 35. Nowadays, most of us can look forward to sitting in a pool of urine blowing bubbles out of our noses in care homes till well past the age of a hundred.

Your family history, what you eat and drink, and the things you do each day all affect how long you will live. So when exactly will YOU die?

Find out exactly how long you've got left before your heart stops beating, starving your brain of oxygen and your consciousness flickers out forever into the vast, black, timeless void of eternal nothingness that is death by taking our fun lifestyle quiz.

1. You make yourself a packed lunch to take into work. What do you put in it?
a) A lettuce sandwich on wholemeal bread, an apple and a bio yoghurt.
b) A ham sandwich on white bread, a sausage roll and a carton of orange juice.
c) A 2lb tub of Utterly Butterly and a spoon.

2. How do you cope with stress at work?
a) Cut yourself some slack. Listen to Andean Pan pipe music, fiddle about with an executive toy and perhaps prune a Banzai tree on your desk.
b) Leave the office for an hour and hit a couple of baskets of golf balls at the local driving range.
c) Bottle it all up until you turn the colour of Alex Ferguson and black out.

3. When you stick the butter knife into a toaster to retrieve burnt toast, how often do you unplug it first?
a) Sometimes.
b) Rarely.
c) Never.

4. You go to a department store to buy some trousers, but notice that the gents' department is on the second floor and the lift is broken. What do you do?
a) Take the stairs.
b) Take the escalator.
c) Take a taxi home and order an enormous elasticated pair from a mail-order advert in the Sunday Mirror.

5. How much money do you owe to Bermondsey Dave?
a) Nothing.
b) 0 to £5.
c) Over £5.

6. You decide to have a nice quiet night in watching TV with the family. How much do you have to drink?
a) Nothing serious, just a few cans.
b) A bottle of wine left over from Christmas, half a bottle of cooking sherry and some gin.
c) All the drink in the house followed by a pint of gassed milk, the two-stroke oil from the garden strimmer and a tin of Brasso filtered through a slice of bread.

7. What sort of cigarettes do you smoke?
a) Healthy Marlboro Lights from Holland and Barrett.
b) Medium tar cigarette such as Benson and Hedges.
c) Giant rollies made from tea bags and newspaper.

8. How often do you go backpacking in Australia?
a) Never.
b) Rarely.
c) Often.

9. How often do you wear a seatbelt in the car?
a) Always. It's clunk! click! every trip.
b) Just on long journeys, but don't bother for short trips.
c) Never, because you met a bloke in the pub whose mate was in a crash and he was thrown clear of his car and the fireman told him that if he had been wearing a belt he would have been killed.

10. How often do you go to swimming parties at Michael Barrymore's house?
a) Never.
b) Sometimes, but I never take my trunks.
c) Often.

11. Which of the following would best describe your arteries?
a) Like the inside of a gleaming pipe off the Castrol GTX advert.
b) Like the Bakerloo line - a bit grubby but functional.
c) As tight and furry as Sooty's arsehole.

12. When you get up in the morning, how long does your uncontrollable coughing fit last?
a) 3 seconds to 5 minutes.
b). 5-10 minutes.
c) Until you go back to bed.

How long have YOU got?

Mainly As: Congratulations! The Queen may as well begin writing that telegram now. Thanks to your healthy lifestyle you are almost certain to reach the ton, and if you avoid household accidents and hereditary illnesses, you could even double that.

Mainly Bs: Not bad. You are guaranteed at least seventy years on this earth, but you could make eighty or ninety with a few minor adjustments to your lifestyle - don't park right outside the pub door every day, park 20 yards down the road to give yourself a bracing walk.

Mainly Cs: Oh, dear. With your lifestyle as it is, you'll be lucky to make it to the end of this article. Drastic action is needed now. After you've finished your tea tonight, don't sit in front of the telly, go out and run further than you have ever run in your life before.

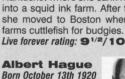

Top Drawer!

Britain's brightest drawer chat page

I AM an artist, and I keep my underpants in a retractable, sliding, rack-mounted storage tray on runners. That is to say I am a *drawer*, and I keep my *drawers* in a *drawer!!!* How my wife and I laugh every time I make this observation.

T. Dangerfield, London

MY WIFE does the football pools each week, and one Saturday afternoon, she came running into the kitchen screaming that eight draws had just come up. I was about to open the champagne when I discovered that she had bought two chests of drawers (each containing four drawers), and that they had just ascended in the lift to our home on the tenth floor of a block of flats!

H. Willis, Reading

When is a Drawer Not a Drawer?

I WAS recently in Ikea, and was appalled to see that they were selling a set of 3 pine storage boxes in a frame unit in the drawers section. Further more, the same item was listed in their catalogue as a 'Kroll 3 Drawer Unit'. Call me old fashioned, but a lidless wooden box with a handle is not a drawer. I told one of the assistants that in Britain a drawer has runners, otherwise it is a wooden box in a frame. She refused my request to relocate the item to the general storage section of the shop and remove it from the catalogue.

P. Smith, Gateshead

Stick to the Rules

IT MAKES my blood boil when I hear young people complaining about their drawers sticking. I've got a set I've been using day in and day out for sixty years and they've never stuck once. I always say that it's not the sticking drawer that's at fault, it's the owner. A drawer, like a car or television set, needs regular maintenance to keep it in tip-top condition. If these people would only take the trouble to follow this five point drawer care plan they'd enjoy trouble-free drawer use for years to come.

1. Always site chests of drawers well away from sources of humidity, excessive cold and excessive heat.
2. Never attempt to pull a drawer out at an angle. It could twist in its runners, damaging the sliding mechanism.
3. Never overload a drawer. As a rule of thumb, if you have to press the contents down in order to get the drawer back in, it is too full.
4. Open and close every drawer at least five times a day. This will ensure that the moving parts don't seize up due to lack of use.
5. Remove every drawer from its housing twice a year to oil or wax the runners.

Brigadier Y. Lewerthwaite, Cumbria

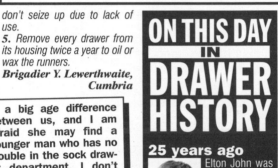

BESPOKE TOBACCONIST

THIS SOLID IVORY CIGARETTE HOLDER WITH A 22 CARAT GOLD TRIM IS PRICED AT ONE GUINEA

THAT ONE WILL BE FINE. PUT IT ON MY TAB

CD 6.03

Miriam

DRAWER HELP WITH MIRIAM STOPPARD

Dear Miriam...

LETTER OF THE DAY

My husband left me after putting teaspoons in the dessert spoon compartment of the cutlery drawer.

I'm 32 and my husband's 35 and we've been married for 11 years. He always used to put the spoons in the correct compartments of the drawer, but about a year ago he started mixing them up. When I tried to talk to him about it, he would just fly off the handle and say he didn't want to discuss it.

Then one day he put a knife in the fork compartment. We had a blazing row and he ended up sleeping on the sofa. The next morning, the row started up again. He stormed out of the house and I set about putting my cutlery drawer back in order.

I didn't see him for six months, but now he has been in touch and he wants to come back. He says that things will be different and that he will put the cutlery back in the correct place.

I still love him, but the trouble is I have met another man. He is wonderful, and not only does he replace the cutlery in the right compartments, he makes sure they are all the right way round. He even has a separate drawer for best.

I'm afraid that if I take my husband back he'll revert to his old ways. Please help me, Miriam.

Mrs EB, Belfast

Miriam writes...
Wake up and smell the coffee, girl. Your husband thinks he can mix up the utensils in your cutlery drawer and then just breeze back into your life as though nothing has happened. He may say things will be different, but a leopard will not change his spots. My postbag is full of letters from women like you who have given their fellas a second chance to keep the cutlery tidy, and it never works. Your new man sounds like a gem. Hold onto him.

Dear Miriam...
I have recently started having trouble opening my sock drawer, and it's driving a wedge between me and my wife.

I am 48 and she is 32. We've been married for 12 years and I have never had any difficulty opening the drawer. Lately, however, I have found it increasingly difficult to get it open when I want to take socks out and put them in. My wife tells me it doesn't matter, but I can tell it is starting to bother her as well. There is a big age difference between us, and I am afraid she may find a younger man who has no trouble in the sock drawer department. I don't want to lose her. What can I do?

Miriam writes...
Every man has difficulty opening his sock drawer now and again. And worrying about it will only make the problem worse. Just relax and take it easy. Nine times out of ten the problem will cure itself. In the meantime, why don't you and your wife experiment with other places to store your socks, such as your underpants drawer, or a drawer in the kitchen? Or even her underwear drawer! After all, it is 2003.

Dr Miriam Stoppard
The Drawer lines you can trust

STICKING DRAWERS	**000 8181 188881**
LOOSE HANDLES	**000 8181 188882**
WON'T STAY CLOSED	**000 8181 188883**
VAGINAL DRYNESS	**000 8181 188884**

ON THIS DAY IN DRAWER HISTORY

25 years ago

Elton John was forced to call off a concert at Blackpool Tower after he trapped his fingers in a drawer at his guest house. Fans were offered a choice of a refund or tickets to a rescheduled show.

50 years ago

The first plastic drawer was unveiled to an amazed public at the 1953 Paris Expo. An early buyer was film star Zsa Zsa Gabor, who ordered three for the bedroom of her Hollywood mansion.

100 years ago

The first drawer ever to fly took to the air at Kittyhawk Beach, Oregon. The six inch drawer, containing Orville Wright's pipe, sunglasses and a box of Tictacs was airborne for twenty-three seconds and later bequeathed to the Smithsonian Institute in Washington.

LEAVE ME ALONE

A West Yorkshire man is claiming that presenters of a top BBC current affairs programme are making his life a misery.

Man begs Newsnight Bully Boys

Paxman — suave

For the last 12 years Stuart Lewerthwaite believes he has been the victim of a campaign of hate carried out by presenters of BBC2's flagship current affairs programme 'Newsnight'. His house has been burgled repeatedly, his property vandalised and threats made against his family.

ANOTHER SHOWBIZ EXCLUSIVE

CATALOGUE

Among a catalogue of allegations made to West Yorkshire police, Mr Lewerthwaite claims that:

● Suave Newsnight anchor man Jeremy Paxman has broken into his house on 3 occasions in the last 12 months, stealing property worth over £300, and damaging door and window locks.

● Glamorous newsreader Francine Stock kicked down garage doors at the rear of his property and stole garden tools, a bicycle and an aluminium stepladder which she later sold to sports presenter Desmond Lynam.

● Top news analyst Peter Snow drove a motorcycle across his front lawn at 2am on a Sunday morning, causing damage to flowerbeds.

● Various Newsnight reporters regularly congregate outside his house, smoking cigarettes and swinging on his gate. On one occasion an unidentified newsreader urinated against his garden wall.

DIRECTORY

According to Mr Lewerthwaite the trouble began as long ago as 1985 when a brick was thrown through his kitchen window. "I thought it was just kids but after a brief chase I cornered one of the culprits on waste ground near my house. I immediately recognised him as Peter Sissons, the then Newsnight presenter."

ENCYCLOPAEDIA

After a brief struggle Mr Lewerthwaite claims that Sissons escaped, bounding over a fence and making a getaway on the back of a motorcycle driven by BBC political editor John Cole.

"I went straight to the police, but they told me it was just an isolated incident and there wasn't much they could do. I thought nothing more of it until 2 weeks later when somebody knocked at my front door." Mr Lewerthwaite was surprised to find highly rated Newsnight host Jeremy Paxman standing on the doorstep.

TOASTER

"He looked very nervous, and asked me if I had the time. He kept me talking for several minutes until I became suspicious and shut the door. I ran to the back of the house to find that the kitchen door had been forced open by Sue Lawley, and a toaster, a casserole dish and two pounds in cash had been stolen."

RAPPER

The next morning Mr Lewerthwaite phoned the BBC and spoke to the duty officer who logged his complaint. "To this day I've still heard nothing from them", he told us. "It's obvious that nothing is being done".

RANSACKED

A few months later Mr Lewerthwaite and his wife returned home from the pub to find their front door wide open. "The house had been ransacked. Our TV, video, music centre – everything was gone." Even items of clothing together with jewelry belonging to Mrs Lewerthwaite had been stolen.

OBSCENITIES

"All those things could be replaced", Mr Lewerthwaite told us. "But what really upset us was the obscenities which had been daubed on

Mr Lewerthwaite's house yesterday

the walls by Peter Snow. And to make matters worse, when I tuned in to Newsnight that evening I saw Donald McCormack cockily wearing one of my white shirts, a present from my wife, stolen only hours earlier."

BONNET

The very next night the Lewerthwaites were awoken after midnight by banging sounds from outside. When Mr Lewerthwaite went to investigate he saw former 'Tonight' presenter Dennis Tuohey jumping up and down on the bonnet of his car. After chasing the veteran anchor man off, Mr Lewerthwaite discovered that the car had been broken into and a radio cassette, plus

a dozen tapes, had been stolen.

BOOT

To date Mr Lewerthwaite has reported 71 incidents to police, all involving Newsnight presenters and associated TV journalists and correspondents. However West Yorkshire police have failed to make a single arrest.

FISHWIFE

A spokesman for the BBC's current affairs departmen refused to comment. Meanwhile Mr Lewerthwaite, who is 83 and lost the use of one eye during military service in 1942, vowed never to watch the Newsnight programme again. "In future I shall be watching the News at Ten", he told us

BEFORE WE FINALISE YOUR WILL, FARMER GILES, DO YOU BEQUEATH ANY MONEY TO YOUR BROTHER?

NO. HE GETS MY GOAT.

SOLICITOR

CD 5T 9.90.

20 THINGS YOU NEVER KNEW ABOUT SHOES

An attractive shoe model displays the very latest fashion in footwear - shoes.

Fred Astaire danced in them, footballers score goals with them. Everybody wears them, from the Queen all the way down to Jeremy Beadle. Yes, love them or hate them SHOES are here to stay. But how many facts do you know about footwear? Here are twenty things you probably never knew about the shoe.

1 There's nothing new about the shoe. They've been around for a lot longer than you'd think. In cave men days dinosaur skin shoes – with flint soles – were all the rage. Stone age cobblers would charge around 6 dinosaur teeth (about £22.99 in today's money) for a pair, complete with thick black curly laces – made out of mammoth pubes.

2 Say "shoe" to a dog and it won't scurry off to fetch your footwear. It will simply run away. That's because the word "shoe" in dog language means "clear off".

3 However, if you say "fetch" to the same dog, after it's come back, it may then go off and get your slippers – special soft indoor shoes made out of checky cloth.

4 If you find a shoe tree in your garden, don't climb up it hoping to pick shoes. For a shoe tree isn't a tree at all. We think it's something to do with shoes that you sometimes find in the bottom of wardrobes – possibly for putting inside shoes. A bit like a mug tree, but for shoes, not mugs.

A normal tree yesterday

5 However, some shoes *do* grow on trees. For in Holland drug-crazed Netherlanders hobble about their windmills in clumsy wooden 'clogs'. These traditional, impractical and uncomfortable shoes were originally worn to protect Dutch feet from flooding dykes, and have since been adopted as part of the national costume.

6 The smallest shoes ever made were a pair of clogs presented by King Van der Vaalk VI of Netherland to the world's smallest man, Kalvin Phillips of Bridgewater, USA. The tiny shoes, an incredible size .00008 (European size 41) were carved from the two halves of a single salted peanut. However, Dutch microcobblers over did it – and the clogs were fractionally too small for 8 inch tall Phillips, who complained that they nipped his toes. Ironically, Phillips died shortly afterwards, and was buried – in a shoebox!

7 Nowadays shoes are made from many different materials. For example, Cinderella, in the pantomime of the same name, went to the ball wearing glass slippers.

8 And in the nursery rhyme of the same name, an old lady who *lived* in a shoe had so many children she didn't know what to do.

9 So she gave them some broth without any bread.

10 And spanked them all soundly and sent them to bed. The rotten cow.

11 Ask a musician if he plays the shoe horn and the chances are he'll give you a blank stare. That's because a shoe horn is, in fact, a machine, often found in hotel corridors, to enable you to put your shoes on quickly.

12 Buy a pair of shoes nowadays and you can bet that an embarrassed spotty teenage shop assistant will half-heartedly try to sell you some cheap shoe polish. That's because shoe shop managers only pay their staff 25p an hour, and then offer them a commission on sales of crap shoe polish.

13 The world's best-selling shoe has sold over 100 million pairs since it was invented by a victorian chiropodist in 1887. Dr Jeremiah Marten noticed the barefoot street urchins in his home town of Northampton developed sore toes from kicking each other in the head. So he set about inventing his patented leather 'aggro boot', and the Doctor Marten was born.

Dr Jeremiah Marten - 'aggro boot'

14 At about the same time in Sweden Dr Jergen Schol, a humble carpenter, made his wife a pair of dangerous and uncomfortable flat shoes or 'sandals' by nailing the cuffs from an old shirt onto two short lengths of splintered floor-board. His popular sandals are still worn today by librarians and medical receptionists, and can be bought at car boot sales for around 10p.

15 Dr Hugo White did not invent any shoes. He invented jam rags.

16 There are many kinds of boots. Walking boots, climbing boots, football boots, cowboy boots and Boots the Chemists, a large pharmaceutical manufacturer and retailer based in Nottingham.

17 There are also car boots, but if you try one on for size you'll probably find that it's a little roomy. That's because a car boot is the space at the back of your car where you put your suitcases.

18 But don't turn up at a car boot sale expecting to buy a new boot for your car. Car boot sales are special sales held in car parks where you can buy a 1970's lava lamp – guaranteed to blow up and burn your house down – for only 75p. (But they'll take fifty. Oh, go on then. Twenty-five).

19 They say that horse shoes bring good luck. But horses themselves would probably disagree. That's because a horse shoe is just a great big Blakey's seg nailed onto the horses toes.

A horse three weeks ago

20 They also say that an army marches on its stomach, but any soldier will tell you that's nonsense. For they march in Wellington boots, so named after the Grand Old Duke of York who invented the Cardigan Jumper at the Battle of Balaclava in 1066, etc. etc. etc.

WHAT NEXT?

THE HADRON COLLIDER will have solved the mystery of the universe's creation in a few months' time, at which point the scientists will have done their job. But what will happen when the egg-heads have packed up their equipment and gone back to their labs? The venture cost us taxpayers £4billion, so it would be nice to recoup a little, or indeed a lot, of that cash. But NOW? We asked the *Dragons' Den* entrepreneurs, each an expert in spotting money-making potential, what they would do with a 17-mile long circular tunnel 300 feet beneath the Swiss Alps...

JAMES CAAN

"My background is in the retail sector, and I think the Hadron Collider tunnel would lend itself to being turned into an underground branch of Ikea. Customers would go in and, just like in other branches of the Swedish furniture giant, they would be forced to walk all the way round before they could get out. At the end of the tunnel is the giant Atlas experimental chamber which could be used as the place where the shoppers dump all their items when they see how long the queues are at the tills."

DUNCAN BANNATYNE

"I've made a lot of money in the casino business, so I know a thing or two about gambling. I would turn the CERN Cyclotron tunnel into the world's largest and most profitable dog track. The hare could be fired at the speed of light from the particle accelerator, the traps would be open and the dogs would be off on their 17-mile, one lap race. At full speed, it would take the greyhounds about an hour to get round, so my captive punters would have time to retire to the bar, where there would be a big mark up on drinks and snacks."

THEO PAPHITIS

"I would turn the tunnel into a 17-miles-long underground hotel with 8500 suites. Charging £200 per night, that's a turnover of nearly £12m per week, and that's without telephone charges, mini bars bills and the Frankie Vaughan channels. The only trouble is, you would have to go through one room to get to the next, and if you were on the far side of the hotel, you would have to walk through 4250 rooms to get to yours. And if you were in the first room, you'd have 4250 people trooping through your room at breakfast."

DEBORAH MEADEN

"I made my fortune in the west country tourist industry, and I know how popular this part of the world is. So I would turn the Hadron Collider tunnel into a Devon-themed tourist attraction, Dragons' Den Deborah's Devonworld. There would be a coffee-shop selling cream teas and a gift shop where you could buy fudge and tea towels. They would also sell postcards of Dartmoor sheep, only the sheep would be three dimensional with a bit of fluff glued onto them. They would sell for about 60p each, or three for £1.50."

PETER JONES

"I made my first million on a Hook-a-Duck stall in a travelling fair, and I've had a passion for fairgrounds ever since. I would love to turn the Hadron Collider into a Ghost Train, which at 27km long would be the biggest and therefore the scariest in the world. The carriages would take four hours to get round, during which time the passengers would be terrified by countless whistles, flashes, plastic skellingtons, rubber bats and bits of sacking hanging from the roof. I might even get Derek Acorah to perform the opening ceremony."

UFO's — DO THEY REALLY OR IS IT JUST A LOAD OF SHITE?

UFO's — do they really exist, or are they just figments of our vivid imaginations? Every year there are literally thousands of reported sightings of unidentified flying objects over Britain alone. But can they be taken seriously? Is there REALLY someone out there watching us? Have life forms from another planet visited the Earth? Every day more and more evidence seems to suggest that they have.

However, the experts are divided. Some say that UFO's exist. Others say they don't. Some aren't really sure. But whatever your opinion, you simply cannot ignore the evidence.

MAN

UFOs are a phenomenon which affect not only ordinary people, but TV celebrities as well. And one man who should know more than most about the existence of 'visitors' from other planets is funny man BOB MONKHOUSE. For 60-year-old Bob is one of the few people to have survived a UFO 'kidnapping.'

SHOT

Bob's ordeal began in 1973 at the time when he was hosting ITV's popular game show *'The Golden Shot'*. Driving home from work late one night he decided to take a short cut across an isolated moor in North Yorkshire, and got lost. Suddenly several bright lights appeared in the sky above his car and the well known comedian and all-round entertainer slowly began to lose consciousness. When he awoke several hours later Bob was still in his car, parked in exactly the same spot, but on the seat next to him was a briefcase containing £750.

FEE

Experts who later examined him believe that Monkhouse was the victim of a bizarre UFO kidnapping, and that he had been taken on board an alien space ship to entertain the crew. The money had been left to cover his fee for

T.V. star kidnapped by aliens

the evening. Police traced the bank notes and discovered that they had been part of a consignment of money which had gone missing in the Bermuda triangle several years earlier.

How well the comic's act went down with his alien audience we may never know, for Bob's memory of the entire incident was erased by his captors, and his brain was specially programmed to deny that it ever happened.

T.V. soap stars live in terror

For several months an alien space ship paid regular visits to the set of the hit BBC TV soap 'EastEnders.' For a randy space romeo on board the vessel had developed a 'crush' on actress Anita Dobson.

CAST

The cast of the show were sworn to secrecy by TV chiefs who feared that any bad publicity would affect viewing figures. However, Miss Dobson, who played sexy pub landlady Angie Watts in the series, eventually threatened to call police after the captain of the space ship got drunk on space beer and made improper suggestions to her.

BOSSES

Bosses at the Beeb forced the actress to quit, and the 'visits' ceased shortly afterwards. The entire 'EastEnders' cast, including Miss Dobson, remain sworn to secrecy.

Late night visit costs Bob a bomb

One man will never forget the night he came face to face with an alien. For the incident haunts him to this very day.

WINDOW

Bus driver Bob McPherson of Motherwell was awoken one morning at 4am by a loud 'bang'. He peered out of the window and to his astonishment saw a bright silver saucer-shaped object, about six feet in diameter, lying half buried in his garden.

KITCHEN

"I immediately ran downstairs to see what had happened, but when I got to the kitchen door I was suddenly frozen by some strange force. I couldn't move a muscle." Bob looked on as the side of the UFO opened, and a shadowy figure emerged. According to Bob it was a small, human-like creature, measuring less than 2 feet tall.

Bob - late night visit cost him a bomb

"I stood and watched as it walked past me into my house, carrying some kind of strange cable or tube which it plugged into a wall socket in the kitchen. Then there was a loud humming noise which went on for about 30 seconds until the space ship began to glow brightly." At this point the alien returned to his space ship and the door closed.

SETTEE

"I ran upstairs to fetch my camera. But I got back just in time to see the UFO rising up at incredible speed and then shooting off across the sky," said Bob.

LOUNGE

He didn't manage to get a photograph, but Bob was left with something else to remind him of the visit — an electricity bill for £375.36 which arrived the next day! When Bob tried to explain what had happened to the Electricity Board, they were sceptical to say the least. "They've told me that I have to pay off the whole amount in weekly instalments," he told us.

Will Bob ever be repaid for his hospitality? Or was he the victim of a new breed of criminal — thieves from space? Perhaps we will never know . . . . .

EXIST

HOLLYWOOD SPACE HORROR
F.B.I. murdered Monroe in Whitehouse cover-up

Monroe - space murder?

A growing number of experts are now taking seriously the theory that aliens may have already landed on Earth and could be living among us disguised in human form. And far fetched though it may sound, there is striking evidence to support this theory.

Lavatory attendant Brian Wilkinson believes that glamorous film star Marilyn Monroe was murdered by the FBI because she knew about space aliens working inside the Whitehouse. But Brian, 42, wasn't prepared to tell us for fear that he too may be killed.

PUB

And unemployed bricklayer Walter Purviss believes he has identified visiting aliens in his home town of Huddersfield, and claims to have confronted a group of them as they sat drinking quietly in a pub near his home.

"I walked into the pub and noticed these three sitting in the corner. There was something eerie about them, so I followed one of them into the toilet. I made no bones about it. I simply asked him whether he was an alien or not. He obviously was because he punched me in the face then walked off."

COVER

By the time Mr Purviss recovered, the aliens, realising their cover was blown, had left. "It's just as well," he told us. "Cos if I ever see them again I'm gonna give them a bloody good hiding."

Have aliens already landed? Are they living among us? Can you trust your neighbour? Perhaps even your husband or wife are 'visitors' in disguise. We may never know . . . . .

How can UFO's be explained? 'It's a mystery to me' says Shakin' Stevens, as a flying saucer circles the Houses of Parliament.

THE GREAT UFO DEBATE
WHAT THE STARS SAY

If aliens did land on Earth, what would they look like? That's the question we put to a few top celebrities, and this is what they said.

MARTI PELLOW, sexy lead singer with Wet, Wet, Wet told us that aliens would probably be green skinned, wiry creatures with six legs and long, lizard-like tongues. "They'd use their tongues to catch space insects," says Marti.

Bros star LUKE GOSS believes that aliens will be like humans, but bald and with silver eyes. "They'll wear silver space suits and have a ray gun in the middle of their forehead," he added.

Brother MATT disagreed. "I don't think you'll be able to see aliens," he explained. "They'll be sort of invisible, like a cloud of gas," he told us.

Radio One DJ ANNE NIGHTINGALE took time off from her lively Sunday night request show to tell us what she thought aliens would look like. "They'll probably walk on two legs, like we do, but they'll have huge, long necks, like dinosaurs, and eight arms, all with suckers on the end," she told us. "And they'll be purple," she added.

GYLES BRANDRETH won't hang about to see what the aliens look like if he sees a UFO land. He'll run for his life! "I've got a terrible feeling that aliens will be like giant snails with big antennae, and enormous metal teeth," he confessed.

'No Celebs Left by 2010'

HARD ON THE HEELS of a year in which Britain has been battered by a series of natural disasters - floods, bird flu, highish tides at Great Yarmouth and failing to qualify for Euro 2008 - another crisis is looming on the horizon. For the UK is using up its supply of celebrities at an alarming rate, and TV experts fear that star stocks could be completely exhausted within five years.

CELEB-RATE

'BBC and ITV are increasingly having to use chefs, fashion designers, slappers and ex-cricketers in place of real celebrities,' says Professor Tibor Szakacs of Oxford University's Department of Celebrity Studies. 'The media's insatiable appetite for stars to fill programmes such as *I'm a Celebrity, Strictly Come Dancing, Celebrity Big Brother* and *Celebrity Scissorhands* is far outstripping our naturally-occuring supply of genuinely famous people, and the two lines on the graph are going to meet very soon. If something isn't done to curb television's demand for celebrities, star reserves will reach critical levels within the decade.'

LADIES-NIGHT

And the government has been quick to act in response, bringing in emergency powers and dropping the Celebrity Threshold base-rate to an all-time low of 3%. Home Affairs and Media Minister Steve Millerband told us: 'This is good news for those people who are looking to get their foot on the first rung of the celebrity ladder.'

GET DOWN ON-IT

'Thanks to this government's forward-looking reforms, it is no longer necessary to excel in the public eye in order to become famous. Something as simple and unremarkable as being the first evictee on *Big Brother*, being unable to pass your driving test or wanking off Dean Gaffney in a doorway is now quite sufficient to ensure you a place on the celebrity A-List,' he added.

CHER-ISH

Meanwhile the new ultra-low base-rate has been warmly welcomed by celebrity watchers, such as *Heat* magazine editor Boyd Ffucking-Pointless. He told us: 'This is going to make life much easier for those of us who make a living writing about the trivial doings of celebrities.'

TINA TURNER-ISH

'Filling a weekly magazine with news about household names you would recognise in the street can be quite a challenge. With the new 3% rate in place, it's no longer necessary to have heard of someone before according them celebrity status, so filling a hundred pages every seven days with celeb tittle-tattle is going to be a doddle,' he added.

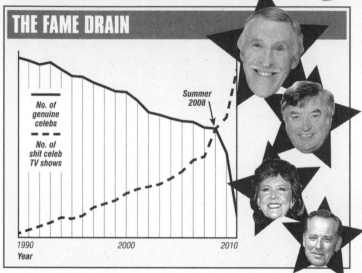

THE FAME DRAIN

No. of genuine celebs

No. of shit celeb TV shows

Summer 2008

1990   2000   2010

Year

The Three Percenters

How the New Celeb Base Rate Created a Galaxy of Barely Visible Stars

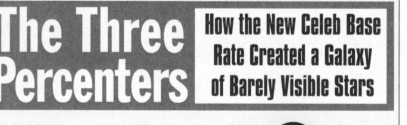

Bubble — BB inmate-turned-teabagbin.com hundredaire back in the limelight

Rebecca Loos — Becks sex text pest-cum-TV pig-wanker enjoying new found celebrity

Johnny the Fireman — BB Geordie firefighter-turned pretend British Gas Service Contract engineer famous again

Bros — When will we be famous? 3% rate not low enough to see Goss twins back over star threshold

Ken — Sacked 3rd Bros member-turned-mobile wheelie bin cleaner back on A-list

Celeb Threshold Base Rate   3%   ← Rate Lowered ←   Former Threshold

---

New 3% Rate Gives BB Nobody Grace Star Quality

WITHIN MINUTES of the Government's announcement of the new celebrity rate, crowds of morons gathered outside the homes of former non-entities throughout the country, desperate for a glimpse of the newly-found glamorous lifestyles of the inhabitants who now found themselves in the celebrity band.

In Luton, Police were called to the house of 2006 Week-4 Big Brother evictee Grace Adams-Short, where hordes of mouth-breathing autograph hunters had gathered in an attempt to meet their new idol.

One fan who was lucky enough to meet Adams-Short when she nipped to the shops to buy a handbag was 17-year-old mother of four Shania Cretis. Cretis, who has an IQ of 63, told reporters: 'She was really friendly and gave me her autograph. I'd never heard of her until this morning, but now she's famous and I want to know all about her. In years to come I'll be able to tell my grandchildren that I met Grace who was on *Brides Unveiled* on Wedding TV and was also the new face of Uncle Ben's Sweet and Sour Sauce..'

'I can't wait to get next week's OK to read her column and see what sort of handbag she bought,' she added, before leaving to go and stalk Kenzie out of Blazin' Squad.

BONKING ON THE BUSES!

Fellas. What's **YOUR** idea of the most glamourous job in the world? Film star, stunt rider or international spy perhaps? It's certainly true that men in these professions pull their fair share of gorgeous girls.

James Bond for example has sex with up to four girls in each film, while Eddie Kidd is never seen with the same top model twice. But bus driver **LES TAYLOR** claims these occupations are a bore compared to bus driving. According to Les all the action takes place on the buses!

In his new book 'Sex On The Buses' Les blows the lid off Britain's bonking bus drivers. They're all at it, according to Les. Here, in an exclusive extract from his book, Les lets us in on just some of the sexational secrets of life behind the wheel.

Lusty Les drives the girls wild 'I always give them a good ride'

Les (left) and a bus.

'A lot of people imagine its boring driving a bus back and forth along the same route every day. But nothing could be further from the truth.

My favourite route ran through a housing estate. it was the number 47, but back at the depot we called it the '69' for obvious reasons. I'd take my bus down there at 9.30 after all the fellas had gone to work, and the birds would be *queuing up for it* at every stop. I'd make sure every one of them got *a good ride*. I'd be so busy I often got back to the depot 6 or 7 hours late!

SEXY

Sometimes I would do a country route just to give myself a rest. But on one occasion that plan back fired. I was driving to this village where no-one lived so I knew I'd have a quiet run. Little did I know a dozen top models had been posing for a sexy calendar in the countryside, and they all got on my bus to go home. I drove around the countryside for several hours, stopping on request to *punch their tickets*. When we eventually got back to the terminal I parked in a quiet corner and *got off* with all of them at once.

SEX

Mind you, its not all group sex on the buses. Sometimes I'd only have sex with one woman at a time. Like the time a gorgeous blond film star got on my bus. It was the last run of the evening and there was no-one else on board. She smiled and asked if I *went all the way*. I didn't need to be asked twice, and within seconds the windows were all steamed up and the suspension was being tested to its limit. By the time we'd finished – several hours later – the bus was a total write off. I told the inspector I'd driven over some rough ground and he believed me. I still smile every time I drive past that old bus in the scrapyard.

HANDFUL

It wasn't always the passengers who provided the fun on the buses. The clippies were just as bad. I remember one in particular. Sandra was her name. On her first day the inspector asked me to *take her in hand*. And it wasn't long till I was showing her how to give someone a *fourpenny one*. With Sandra around there was always room for me *up top*. Quite a handful she was, I can tell you.

DOUBLE DECKER

Mind you, being a double decker romeo does have its hazards. I remember once I was having sex with this girl in my cab when her husband got on the bus! I've never put my regulation trousers on so quickly in my life! It was only when I got back to the depot that I realised I had them on back to front. My face was as red as my bus, I can tell you.

DOCTOR. I THINK I'M DEVELOPING SOMETHING
DARK ROOM

Another time I was stuck at some traffic lights so I decided to nip up top and have sex with a tasty housewife who'd given me the eye earlier. But while we were at it upstairs, a stern lady inspector boarded the bus. When she saw what was going on I thought I'd be sacked on the spot. But to my surprise she whipped off her tunic and joined in!

MILKY BAR

Our steamy sex session continued, even after the lights had changed... again and again and again! There was quite a queue of cars behind the bus before those two ladies eventually *rang my bell* and I was able to move off.'

Next week: Les tells how his bus got stuck on a level crossing and it took eleven members of a passing Swedish hockey team to eventually *pull him off.*

Report slams police

The police yesterday.

A report published this month in the consumer watchdog magazine *What* reveals a majority of the British public believe the police force discriminates against minority groups.

OPINION

And it is the motorist who suffers most from police discrimination according to public opinion. Over 75% of the people questioned thought that police officers deliberately discriminated against drivers who drove either too fast, or in an erratic manner while under the influence of alcohol.

GROUPS

According to the survey other minority groups such as burglars are often singled out for attention by the police.

FEDERATION

A spokesman for the Federation of Policemen said that public confidence in Britain's police was on the increase, despite the fact that they occassionally roughed people up or shot them by mistake. Meanwhile, a report published by the Police Complaints Commission, the Government's official independent police monitoring watchdog public accountability body, showed that the number of complaints made against police officers increased during the last year.

VAUX

A spokesman claimed that this was due to an increase in the number of complaints made over the past twelve months.

THEY ALMOST share a name, but the similarity ends there. One is a legend of the skateboarding world whose achievements have inspired one of the most enduring and successful video game licences of the 21st century. The other is a comedian, author and amateur ten

WHO'S THE BEST

TONY HAWK

STREET CRED

THANKS to his sensational skateboarding exploits, half-pipe hero Hawk had achieved almost god-like status among California's teenage population even before his reputation spread worldwide. His position as the King of Cool was assured at the 1999 X Games when he became the first skateboarder to land a 900 degree turn live on television and is further credited with the invention of many of the extreme sport's most outlandish moves.

S

10

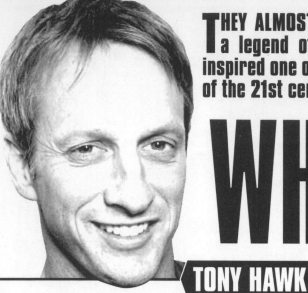

VIDEO GAME SALES

TONY Hawk's series of skateboarding games has been one of the most successful video game franchises of this century and has supplied a near endless stream of commercial hits across all of the major platforms. In the process, Tony has amassed a fortune and become a household name, in households where there are teenage boys, across the globe. So it's another strong showing for Hawk in this category

10

LITERARY ACHIEVEMENTS

ALTHOUGH it is fair to say that he is primarily renowned for farting about on a child's toy, modest Hawk has often played down his hugely successful forays into the world of literature. The multi-talented American stormed into the New York Times Best Seller List with his first book, the largely autobiographical *Hawk: Occupation: Skateboarder*, co-written with Tony Mortimer, and their follow up *Tony Hawk: Professional Skateboarder* was also a huge commercial success for the prolific partnership.

9

TOPICAL QUIZ SHOW APPEARANCES

A MEDIOCRE round the the man who has made millions trundling along on a plank with castors. Following a mix-up at the Hat Trick TV production offices, Hawk was accidentally confused with his near namesake by junior researcher and invited to appear on an episode of *Mock the Week*. This would have earned the Californian boarder a full ten marks, but the mistake was quickly discovered and the invite hastily withdrawn. So he must make do with half marks in this round.

5

DEDICATED THRILL RIDES

IN MARCH 2007, the Six Flags Fiesta theme park in Texas unveiled its new white-knuckle ride, the Tony Hawks Big Spin. The ride was billed as 'The Total Tony Hawk Experience' and proved so popular with thrillseekers, that a second ride opened later the same year at Six Flags St. Louis. In the 111-second ride, passengers travel at up to 60mph, experience 25 seconds of weightlessness during 14 inverted loop-the-loops and are subjected to forces in excess of 12G.

10

NOT BEING MISTAKEN FOR TONY HAWK(S)

IT IS DOUBTFUL that the millionaire X-Sports star has ever been mistaken for the mild-mannered British comedian at any point in his career, and so scores top marks in this round.

10

CHEAPER CAR INSURANCE CAMPAIGNING

AS A world champion skateboarder, Hawk has no need for a car to get him from A to B. If he wants to go anywhere, he takes his trusty board. And it can go where a car can't - down handrails, along the edges of raised flower beds and inside very large drainpipes. Even if he did have a car, the cost of insurance for a man of Hawks' wealth would scarcely be a concern and consequently he has little interest in campaigning for cheaper insurance. This round is a disaster for Tony, but can his competitor capitalise on this last minute slip up?

TONY MAY HAVE looked like he was skating to an easy victory in this competition, but it seems his campaign came off the grind rails. Ultimately the skater's complacency in the later stages and in particular his apparent unwillingness to alert the general public to inexpensive car insurance have seen him fall flat on his face.

FINAL SCORES **54**

NEXT WEEK: *It's Fred versus Je*

nis player with a penchant for appearing on humorous panel game shows. Comparison may seem impossible, but that hasn't deterred us from finally answering the question, which has been on everyone's lips for over a decade...

TONY HAWK(S)?

TONY HAWKS

8 — **STREET CRED**

TONY may lack the athletic prowess of his near namesake, but he can be seen as something of an old school hip hop pioneer, having scored a number three hit in 1988 with 'Stutter Rap' as one third of Morris Minor and The Majors. Somewhat surprisingly, Hawk decided to call time on his promising recording career after this early success and as such his score in this round is somewhat lower than it could have been.

8 — **VIDEO GAME SALES**

IT'S ANOTHER case of so near, and yet so far in this round for the plucky Brit. Tony was in talks with a major company over the development of a series of big budget video games based on his occasional appearances on the semi successful BBC monologue show *The Brain Drain* in the early 90's. However, negotiations broke down and the highly anticipated product never materialised. Accordingly, this round is a disaster for Tony.

7 — **LITERARY ACHIEVEMENTS**

HAWKS has penned a series of humourous and inventive books, including *Playing the Moldovians at Tennis* and *Around Ireland with a Fridge*, and so may have been optimistic of outperforming his rival in this category. However, although his books have been well received by critics and readers alike, they have failed to sell as well as his skateboarding counterpart's efforts in the all-important and notoriously elusive US teen market. The result is humiliation for Hawks as he is comfortably beaten on his own turf!

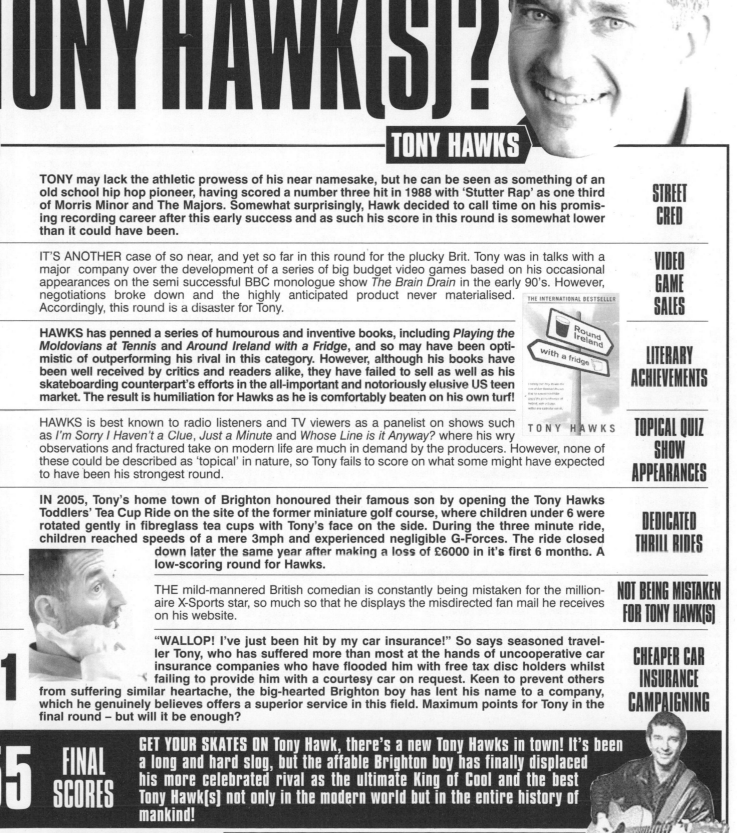

THE INTERNATIONAL BESTSELLER
Round Ireland with a fridge
TONY HAWKS

3 — **TOPICAL QUIZ SHOW APPEARANCES**

HAWKS is best known to radio listeners and TV viewers as a panelist on shows such as *I'm Sorry I Haven't a Clue*, *Just a Minute* and *Whose Line is it Anyway?* where his wry observations and fractured take on modern life are much in demand by the producers. However, none of these could be described as 'topical' in nature, so Tony fails to score on what some might have expected to have been his strongest round.

8 — **DEDICATED THRILL RIDES**

IN 2005, Tony's home town of Brighton honoured their famous son by opening the Tony Hawks Toddlers' Tea Cup Ride on the site of the former miniature golf course, where children under 6 were rotated gently in fibreglass tea cups with Tony's face on the side. During the three minute ride, children reached speeds of a mere 3mph and experienced negligible G-Forces. The ride closed down later the same year after making a loss of £6000 in it's first 6 months. A low-scoring round for Hawks.

0 — **NOT BEING MISTAKEN FOR TONY HAWK(S)**

THE mild-mannered British comedian is constantly being mistaken for the millionaire X-Sports star, so much so that he displays the misdirected fan mail he receives on his website.

21 — **CHEAPER CAR INSURANCE CAMPAIGNING**

"WALLOP! I've just been hit by my car insurance!" So says seasoned traveller Tony, who has suffered more than most at the hands of uncooperative car insurance companies who have flooded him with free tax disc holders whilst failing to provide him with a courtesy car on request. Keen to prevent others from suffering similar heartache, the big-hearted Brighton boy has lent his name to a company, which he genuinely believes offers a superior service in this field. Maximum points for Tony in the final round – but will it be enough?

55 — **FINAL SCORES**

GET YOUR SKATES ON Tony Hawk, there's a new Tony Hawks in town! It's been a long and hard slog, but the affable Brighton boy has finally displaced his more celebrated rival as the ultimate King of Cool and the best Tony Hawk(s) not only in the modern world but in the entire history of mankind!

decide... **WHO'S THE BEST WEST?**

David Soul's WORLD OF HOLES

Hi. Remember me? I was the light-haired one out of Starsky and Hutch. Anyway, that's enough about me, let's get on to the hole talk. And my postbag has been full of holes this week. Letters about holes, that is! Here's a whole load of the best.

I WILL never understand my wife. She recently threw out my favourite vest because it had been eaten by moths and was full of holes. The next day she replaced it with a brand new string vest! It had a whole lot more holes than the one she threw out!

T Turner
Galashiels

I'M A shop manager, and somebody recently asked for a refund because an item he had bought had a hole in it. Imagine his embarrassment when I pointed out that it was a doughnut shop.

T Barnton
Smethwick

I REALLY enjoyed the recent film *'Holes'* based on the book by Louis Sachar. Come on, Hollywood, make more films about holes for all us holes fans.

A holes fan
Wrexham

BRITAIN'S LEADING ORIFICES AND APERTURES FORUM

DURING a recent row with my wife, she said that there were lots of holes in my argument. And she was certainly right. I was arguing about whether or not to throw out a chair that was riddled with woodworm.

R Brunswick, Hampshire

HOLES in the News

► **CANCUN, MEXICO:** Chilli pepper farmer Boco Perez got the shock of his life when he fell into a hole which had suddenly opened up in his field. He fell right through the centre of the earth, finally emerging 10 hours later, feet first and travelling at a speed of 800 miles per hour through a pavement in Trinity Road, Aberystwyth.

• • • • •

► HERTFORDSHIRE, ENGLAND: Police searching for the stolen Henry Moore statue 'Reclining Figure' have recovered it... minus its trademark hole. Big Chief Superintendant Frank Sitting Bull told reporters that he feared the hole, worth £3million, had been melted down and turned into holes in the bottoms of plant pots.

• • • • • •

► **ARCHAEOLOGISTS** excavating a Roman fort just outside Haltwhistle, Northumberland, have found the remains of a hole in the ground, thought to have been dug by Roman soldiers. The hole, made of soil and worms, is in a remarkable state of preservation and thought to be 1800 years old. "The hole's purpose is a bit of a mystery," said Newcastle University's Edward Canning. "It may have been dug for use in a religious ceremony, or perhaps for somewhere to put a post for a clothes line." Whatever its purpose, the hole is thought to be Britain's oldest complete hole in the ground and could be worth anything up to £30.

• • • • • •

► A WORMHOLE in the space-time continuum opened up outside Dolcis shoe shop on Peebles High Street last Saturday. Shoppers looked on in disbelief as several trainers, half a dozen Dr. Scholls and 3 brogues, all left feet, were sucked into a parallel universe. Manager Hamish McTavish said: "The annoying thing is, the missing shoes are still here, but I can't get at them as they exist as so-called 'dark energy' occupying a parallel dimension. So the right ones I've still got in the shop are no good to me."

I CAN'T see the logic of buying lots of Swiss cheese. It contains holes, and the more of it you buy, the more holes, and therefore the less cheese you get. Surely if you buy less of the stuff, you get less holes and therefore more cheese.

Prof Roger Power
Dept of Holes
University of Essex

I'M A shop assistant, and a woman recently asked for a refund because an item she had bought had a hole in it. Imagine how embarrassed she was when I pointed out that it was a record shop.

T Smethwick
Barnton

IN general, I love holes. But there's one hole I can't stand. That's the one in the hull of the RMS Titanic that caused the ship to sink with the loss of 1500 lives.

Roger Whitaker
Surrey

FIVE years ago my doctor told me that I was in danger of becoming clinically obese and advised me to go on a 'hole-food' diet for the sake of my health. Since then I have eaten nothing but Hula-Hoops, doughnuts and bars of Aero. I now weigh 36 stone and have had four or five heart attacks. It just goes to show that doctors don't know everything.

Jack Palance
Bradford

I CAN'T understand you losers who are obsessed with holes. Holes suck. They're nothing more than big dents. Why don't you get a life and get interested in something good, like mounds and heaps?

Darcus Howe
Oxford

I WORK in a shop, and a customer recently brought several items back because they had holes in them. Her face turned red when I explained that it was a shop selling collanders, polo mints and those little sticky circles you use to reinforce punched paper before putting it into a ring binder.

T Smethton
Barnwick

THE FUNNY THINGS KIDS SAY... ABOUT HOLES

'OH, NO, Granny! You're curtains are full of holes,'' exclaimed my four-year-old grand daughter one day. She was talking about my net curtains. That was fifteen years ago and I still chuckle about it every half hour.

Edna Stuffing, Punto Arenas

I OPENED my Bible the other day so as I could check in Deuteronomy whether or not I am allowed to eat bats. Imagine my dismay when I discovered that bookworms had eaten large holes in many of the pages. "My Bible is ruined," I cried. "Don't worry, Gran," said my one-year old grand daughter. "It really is a Hole-y Bible now!" How we laughed.

Edna Parsnips, Adidas Abba

I WAS listening to a Radio 4 programme about space recently when the presenter mentioned *Black Holes*, saying one had been discovered which was over three million light years deep. "Gosh, Granny," said my two-week-old grandson. "I bet the spacemen needed a really big spade to dig that." How I laughed.

Edna Chipolatas, Punta Sabioni

MY DAUGHTER was giving birth last week and she asked my to be present at the birth. To calm my nerves, I started singing "There's a hole in my bucket." "No there isn't, granny. There are three. Two where the handle fits in and one in the top," said my new-born grandson as his head popped through my daughter's cervix. How we laughed.

Edna Sprouts, Brussels

Readers' Holes

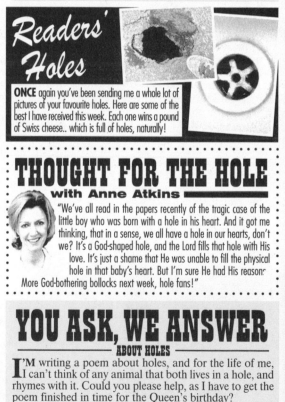

ONCE again you've been sending me a whole lot of pictures of your favourite holes. Here are some of the best I have received this week. Each one wins a pound of Swiss cheese.. which is full of holes, naturally!

THOUGHT FOR THE HOLE
with Anne Atkins

"We've all read in the papers recently of the tragic case of the little boy who was born with a hole in his heart. And it got me thinking, that in a sense, we all have a hole in our hearts, don't we? It's a God-shaped hole, and the Lord fills that hole with His love. It's just a shame that He was unable to fill the physical hole in that baby's heart. But I'm sure He had His reason. More God-bothering bollocks next week, hole fans!"

YOU ASK, WE ANSWER
ABOUT HOLES

I'M writing a poem about holes, and for the life of me, I can't think of any animal that both lives in a hole, and rhymes with it. Could you please help, as I have to get the poem finished in time for the Queen's birthday?

Andrew Motion

Well, Andrew, that's a tricky one. I've thought about it very hard, and the only one I can think of is the mole. You might get away with the vole, but technically speaking, that lives in a burrow.

NOW IT'S THE QUEEN MUM AND FRANK BOUGH!

A new Royal sex scandal is set to rock the foundations of the already shaky House of Windsor. And sensationally, it's the *Queen Mum* who's been caught at it this time – with TV drugs sex fetish star Frank Bough!

A saucy car phone conversation between the wrinkly Royal and randy Frank was accidentally picked up on a hair dryer being used by housewife Vera Gubbins, 42, who lives just 270 miles from Windsor Castle.

HAIR

Vera, from Thornaby, near Middlesbrough, couldn't believe her ears as she sat drying her hair. Suddenly voices began to come out of her 15-year-old hair dryer.

OH CALCUTTA!

"I immediately recognised Frank Bough's voice as I've seen him on the telly. Then suddenly I realised he was talking to the Queen Mother. She kept refering to the Blitz, and horse racing, and she asked him to get her a bottle of gin on his way home. Bough always referred to her as 'Fishbones', and the conversation got quite fruity at times."

GODSPELL

Quick thinking Mrs Gubbins switched on her husband's telephone answering machine and pointed the hair dryer at it, and luckily the entire conversation was recorded onto a cassette.

We were offered a copy of the cassette by Mrs Gubbins' husband Charlie, plus colour photographs of Frank Bough and the Queen Mother frolicking by a swimming pool, for £2,000. In order to respect the privacy of the Queen Mother we refused to publish either.

EVITA

However, if you'd like to hear the tape and see the pictures, we'll lend you them, for £100. Just send an envelope containing £250 cash (£100 plus a £150 deposit) plus a stamped addressed return envelope. We'll then send you the tape and pictures. When you're finished with them, send them back and we'll return your deposit. Honest.

RYVITA

Send your cash to 'Queen Mumgate Tape (and pictures)', Viz, P.O. Box 1PT, Newcastle upon Tyne, NE99 1PT. Please note it may take several years before you get the tape and pictures. But be patient, it will be your turn eventually. Proceeds from our 'Queen Mumgate tape (and pictures)' scandal will all go to charity.

Spanky Frankie's 'phone call to 'Fishbones'

Queen Mum – 'fruity talk'

The Queen Mum rings Bough on her mobile phone from the back of her Royal limousine. A local radio transmitter beams her signal out towards Middlesborough.

MIDDLESBRO

Bough's phone chat with 'Fishbones' is relayed by radio mast at Brackley. The signal is so strong that Mrs Gubbins picks it up on her hair dryer.

LONDON

TIMETABLE OF EVENTS

1. At 8.45am the Queen Mother leaves Windsor Castle to drive to the shops. 2. 8.50am Vera Gubbins washes her hair 270 miles away in Thornaby. 3. 9.10am The Queen Mother makes a call to Frank Bough on her car phone. 4. 9.12am Vera Gubbins picks up call on her hair dryer and records it on her telephone answering machine. 5. 9.15am she rings the newspapers. 6. 9.30am we pay her £2000. 7. 11.45am Mrs Gubbins has spent the lot.

LISTEN IN AND WIN!

RECORD A ROYAL

We're offering a hair dryer, a telephone answering machine and 12 cans of lager to the reader who sends us the best Royal telephone conversation recording.

FRIDGE

You can pick up Royal telephone conversations on most types of electrical equipment around the house – transistor radios, fridge freezers, microwaves, tumble dryers etc.

Simply make your own recording of a private Royal telephone conversation – whether it's Fergie on your food mixer, or Di on your dish washer – and send it to us.

SNOOKER TABLE

Send them to 'Record a Royal', Viz, P.O. Box 1PT, Newcastle upon Tyne, NE99 1PT.

Your guide to the Royal Copulation Ceremor

ROMP and CIRCI

AT 5pm on the 19th of June, Britain's church bells will peal to celebrate the wedding of HRH Prince Edward to Miss Sophie Rhys-Jones. And at 11 pm that evening, Prince Edward's bellend will *peel* as the Royal marriage is consummated in a ceremony which has remained virtually unchanged since the days of William the Conqueror.

Royal consummations have traditionally been secretive affairs taking place behind closed doors, the details being known only to a privileged few insiders. But in the post-Diana spirit of openness, the palace has for the first time released details of the happy couple's wedding-night itinerary.

Posh

After the service at St. George's Chapel, the Royal newly-weds will attend a posh reception hosted by the Queen at Windsor Castle.

At 10.55pm, they will retire to the magnificent Nuptial Chamber in the East wing. At 11.00pm, the ceremony begins in earnest as the couple make their way into twin en-suite bathrooms to disrobe.

Baby

It falls to the Archbishop of Canterbury - the only onlooker allowed inside the royal bedroom - to help the bride into the majestic Ann Summers split-crotch panties and peep-hole negligee first worn by Queen Mary in 1554. In time-honoured tradition, The Archbishop performs this duty wearing oven gloves so as he can't feel her tits.

St. George's Chapel (above), scene of the wedding, and the Majestic Nuptial Chamber (left), scene of the knobbing

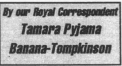
By our Royal Correspondent
Tamara Pyjama Banana-Tompkinson

At 11.03, the ceremony begins in earnest again as the Prince signals his intentions by rubbing her knockers once... twice... three times.

He then holds aloft the Imperial penis - known for centuries as Pink Rod - which slowly makes its way towards the entrance of Sophie's lavishly-pubed beefy drapes. After pausing to bang about a bit, at 11.04 precisely, the curtains to the inner chamber are slowly parted and Pink Rod leads the procession along the vaginal passage, flanked by two hairy knackers.

The new Princess proceeds through the doorway at 11.01, beginning the five-yard walk to the marital bed, followed closely by the Archbishop.

Scary

As the procession passes the glorious mirror-fronted built-in wardrobes, Princess Sophie may pause briefly to dig the itchy, nylon knickers out the crack of her arse. She then waits while the Archbishop draws back the duvet before she climbs gracefully onto the bed to await the arrival of her husband.

Sporty

At 11.02 precisely, the Prince steps out of his bathroom and for the first time Princess Sophie sees him resplendent in ceremonial polycotton pyjamas.

Ginger

The Prince approaches the bed from the opposite direction and pauses. The Archbishop then steps forward and, in a scene that has been repeated for hundreds of centuries, stoops onto one knee and lowers the royal pyjama bottoms.

Danny

Like many Princesses before her, Sophie may struggle to keep her emotions in check, as, for the first time, she claps eyes on the royal wedding tackle. The Archbishop then retires discreetly to the end of the bed from where he witnesses the proceedings as the official representative of the Church of England.

Taking 'STEPS' to Modernise the Monarchy

THE POMP and pageantry of Royal Consummations have served the country well for over a thousand years. But as the new millennium approaches, is the time right to break with tradition and modernise the ceremony?

After eating strong cheese at bedtime, our royal correspondent had a dream, in which he asked top teen pop sensations 'Steps', whose latest record, 'Blancmange Baby', is currently storming up the charts, if and how they would modernise the ceremonial nookie habits of the Royal Family.

"The Royals have to keep their dignity," said singer **Clare, 20**. "Fancy sex is all well and good, but we look up to our Royal Family to set an example."

Hunky keyboard wizard **Lee, 20**, wasn't so sure. "If they were a little less prim and proper between the sheets, these Royal consummations would attract even more tourists into the country than they do," he told us.

"Edward and Sophie should be allowed to do whatever they like in bed," said singer **Faye, 20**. "Old fuddy-duddies shouldn't be allowed to tell them what to do."

"They should take a leaf out of Queen Juliana of the Netherlands's book," said **Lisa, 20**. "She is more in touch with her subjects because she rides around on a bike and has common, everyday sex."

Heart-throb hurdy-gurdy player **H, 20**, was more specific. "Our Royals are far too boring in the sack. They want to get with the programme and do more sexy stuff. I reckon they should do S&M, A&O, DVDA and ESD," said H.

ry everbody's talking about

JMSTANCE

We take you behind the bedroom curtains on Edward's big night in

At 11.05, the ceremony reaches its magnificent climax, when the royal pods bang three times on the Princess's Biffin's Bridge, signalling that the royal wad has been spent.

The majestic ritual over, the procession quickly withdraws and the Prince rolls over, emitting a fanfare fart. At this point the Archbishop, now resplendent in a purple and gold silk trouser-tent, steps forward and invites the Prince and Princess to sign the official deed of *Coitus Completus*.

Richard

On the stroke of midnight the bottom sheet is raised on a flagpole high above the battlements of Windsor Castle. This is greeted by a deafening cheer from the thousands of spectators who have waited for hours on the Chapel Hill lawns hoping to be amongst the first to see Edward and Sophie's map of Africa.

It's a right Royal COCK-UP!

THANKS to meticulous planning, royal consummations usually pass off without a hitch, but over the years there have been a few times when it's not been 'Alright on the Wedding Night'.

● In 1981 it wasn't all plain-sailing on Charles and Diana's big night aboard the Royal Yacht Britannia, when the Prince accidentally locked himself in the bathroom. The ceremony was delayed by three minutes whilst the then Archbishop of Canterbury, Dr. Robert Runcie kicked the door in.

● *King Henry VIII was so disappointed in the size of Anne of Cleves's tits that he was unable to raise Pink Rod, and the*

ceremony had to be postponed. But it wasn't his fault, as that evening, he went on to 'pollute the bed' not once, but twice!

● *In his eagerness to consummate his marriage to Queen Victoria in 1840, Prince Albert rushed the disrobing*

ceremony and caught the metal bolt fastened through his bobby's helmet on his zip. He spent the rest of the night with the Windsor Fire Brigade trying to free his chopper with a hacksaw.

● Another one of Henry VIII's six wedding nights went pear-shaped in 1536. During the consummation of his marriage to Anne Boleyn, the hapless Queen let rip with a thunderous fanny fart, blowing batter-bits into the King's beard. She was beheaded later that year.

That Royal Wedding Night Root in full

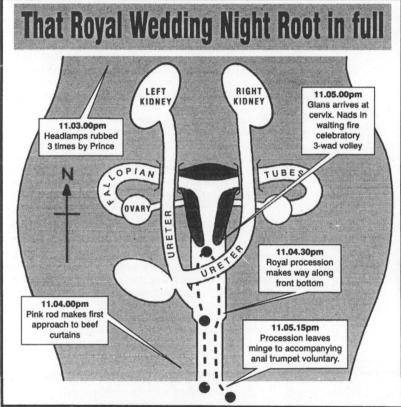

LEFT KIDNEY

RIGHT KIDNEY

11.05.00pm Glans arrives at cervix. Nads in waiting fire celebratory 3-wad volley

11.03.00pm Headlamps rubbed 3 times by Prince

N

FALLOPIAN TUBES

OVARY

URETER

URETER

11.04.30pm Royal procession makes way along front bottom

11.04.00pm Pink rod makes first approach to beef curtains

11.05.15pm Procession leaves minge to accompanying anal trumpet voluntary.

ASK ANYONE on the street to name their two favourite Captains and they'll reply without a moment's hesitation "Kirk and Birdseye". Captain James T. Kirk, Commander of the Starship Enterprise, has been a fixture on our screens for 40 years, as has his fish finger-frying ad-break counterpart Captain Birdseye. Both have voyaged to exotic locations; clean-shaven Kirk to distant galaxies to fight alien monsters, be-whiskered Birdseye to suspiciously breezy-looking tropical beaches to cook fish fingers for his crew of youngsters. But whilst everyone agrees they're the top two Captains in the world, which one is the best... and more importantly, which one is the worst? It's time to tot up the scores once and for all, as we present....

THE CLASH OF THE CAPTAINS
KIRK   VERSUS   BIRDSEYE

ROUND 1 — TRAVEL — Kirk 9, Birdseye 6

Travelling at Warp Factor Seven, Kirk regularly clocked up ten billion miles before breakfast boldly going where no man had went before. In fact, at the end of his five year mission to seek out strange new worlds, it is estimated that Kirk had covered a staggering 6.02×10^{28} light years - that's more miles than there are atoms in the Galaxy. With such an impressive tally, Kirk opens the contest with a substantial score.

In comparison with the Starship Enterprise, Captain Birdseye's ship the Fish Finger trawls the sea at a snail's pace. At the helm of his trusty three-masted schooner, the cheery skipper and his crew of unchaperoned children have circumnavigated the globe thirty-six times in search of the tastiest fillets of white cod. However, Birdseye's career total of around a million miles is still less than the distance covered by Kirk every two-thousandth of a second.

ROUND 2 — ENEMIES — Kirk 8, Birdseye 7

Wherever he went, Captain Kirk was for many years plagued by Klingons - members of a warlike race with Cornish pasty heads. But many other enemies threatened the safety of him and his ship, including Romulans, Khan, Tribbles and Cybermen. Indeed, it was a rare episode of *Star Trek* that didn't finish up with Kirk beaming down onto an alien planet to have a good old-fashioned fist fight with a more-or-less man-sized space monster.

Every seaman's worst fear is an attack by pirates, and for Captain Birdseye that nightmare came true in the seventies, when his advertisements were repeatedly attacked by his arch enemy, pirate Captain Jack. Unlike traditional pirates who murder and rape, forcing their innocent victims to walk the plank, Jack was hell bent on pinching fish fingers. However, his plans were invariably thwarted by Birdseye's pre-pubescent crew, whom the bachelor Captain rewarded with a good old-fashioned nosh-up!

ROUND 3 — ANAL HYGIENE — Kirk 7, Birdseye 5

Wherever he went, Captain Kirk was for many years plagued by Klingons - small pieces of faecal matter adhering to the hair in his buttock cleft. However, in recent times he has taken to eating a fibre-rich diet of Kellog's Bran Flakes. As a result, it is believed that the Captain's logs now move twice as fast through his bowels and leave much less mess around his anus. Kirk's phaser is set to stun, and so is the state of his nipsy. It's a good performance from the Starship Commander's starfish in this round.

Subsisting on nothing but fish fingers, Birdseye's diet is very imbalanced. Nutritionally speaking, the only fibre he is getting is from the small slice of lemon placed on the side of the plate as a serving suggestion, and clarty motions are the inevitable result. In a 200-year-old sailing ship, washing facilities are going to be primitive at best, with the Captain probably forced to regularly share his bath with several of his young friends. Browneye hygiene is going to be a problem for Birdseye, and consequently he scores poorly.

ROUND 4 — RELATIONSHIP SKILLS — Kirk 8, Birdseye 7

Being in charge of a giant starship crewed by members of every race and species imaginable tested Captain Kirk's relationship skills to the limit. Acting as a referee between the emotionless logic of Vulcan Dr Spock and the fiery temper of Scottish 'Bones' McCoy took up most of each episode. On a larger scale Kirk would often bring peace to war-torn planets, using his diplomatic skills in a way that would leave modern day peacemakers such as Kofi Annan, Nelson Mandela and George W Bush shaking their heads in awe.

Keeping discipline on a traditional square-rigged sailing ship was usually the responsibility of a detachment of six tough marines. Punishments were swift and harsh, with offenders being made to 'kiss the gunner's daughter', as well as being flogged with the cat, keelhauled or even hanged off the yard arm. But since Captain Birdseye is the only adult on the Fish Finger, discipline is left to him alone. A few gentle smacks on his innocent young crew's bare bottoms or on the backs of their young legs are often all that is needed to keep them in line.

ROUND 5 — AGEING — Kirk 7, Birdseye 5

In a famous episode of *Star Trek*, Kirk and the rest of the Enterprise's crew were struck by an alien space virus that caused them all to age a thousand times faster than they normally would. Fortunately for the Captain, Dr McCoy managed to find an antidote with about five minutes of the show to go, and so everyone was returned to normal. In the programme, the senile Kirk was a very wrinkly, white-haired old man who made gumming movements with his mouth. However, in real life he has aged quite differently. In his latest Bran Flakes adverts he is remarkably wrinkle free, his hair has stayed its original colour, and he just looks like he's been crying a lot.

As any jolly Jack Tar will tell you, a life on the ocean wave can take its toll on the complexion. Constant exposure to salty spray and bitter winds whilst a-combing the seven seas for the choicest fillets of white cod can quickly lead to a prematurely-aged look. However, Captain Birdseye's youthful crew clearly keep him feeling like a youngster every day, whilst a little bit of him no doubt rubs off on them too. But there's an even easier way for the veteran seafarer to stay looking ever-youthful. Ad chiefs recently sacked the man who'd played him for decades, and replaced him with a stubbly young man who didn't look anything like a Captain.

ROUND 6 — CAPTAIN'S HATS — Kirk 0, Birdseye 10

Oh dear, not once in the entire 71 episodes of *Star Trek* did Kirk ever appear wearing a hat which clearly denoted his rank. A disastrous round for the titter-less Commander of the Starship Enterprise.

Birdseye is never to be seen without his traditional Captain's cap. And just to avoid any confusion, not only does it say 'Captain' on the front, it's also got a picture of himself leering through a lifebelt on the badge.

CONCLUSION — Kirk 39, Birdseye 40

Captain Kirk may have fought off every alien in the Universe, but in the end he couldn't beat a fishy-fingered octogenarian with a predilection for children. In the only battle that counts, the battle of the Captains, he let down his defensive shields down, and millions of trekkies around the world too.

Shiver me timbers and splice the main brace. It's time to break out the ship's rum to drink a toast the winner. It's three cheers for Captain Birdseye as he flicks two fish fingers at his space age rival. He's given Kirk a proper battering, and left him looking a bit (bread) crumby.

Next week: Mr T versus Mr Sheen

W.C. 2000 A.D.

While toilet technology has moved slowly forward during the twentieth century, our attitude towards 'the smallest room' has remained routed firmly in Victorian times.

However a revolution is set to take place in unsuspecting water closets all over Britain. And soon, our lavatories will be entering the computer age. Indeed, by the year 2000, the toilet as we know it, will be a thing of the past.
Join us then, on a journey into the future. Come with us into the toilet of tomorrow. And see how technology is set to change our toilet habits for good.

> By our Toilets Editor
> **MIDGE URE**
> out of Ultravox

'Glittery farts and silver winnits'

Scientific stools glide out of a luxurious hovering bottom to be 'beamed' away to the Moon. Is this the toilet of tomorrow?

THE SHAPE OF STOOLS TO COME

Looking at them, its pretty hard to believe that the Italians invented toilets. But they did, for it was of course the Romans who introduced lavatories to Britain. Prior to that, people simply had to go out of the window.

But the high technology toilet of the 21st Century would be un recognisable to any Roman in search of the loo. Because to begin with, toilets will no longer be locked away in a quiet corner of the house, or banished to the back yard. Instead they will sit alongside settees, TVs and coffee tables in our living rooms. Or in the kitchen, next to the cooker.

Hygiene

That's because toilet hygiene will be a thing of the past. New healthy vegetarian diets, with plenty of fruit juice and nuts, will mean an end to smelly stools. Instead our bodies will produce odour free motions, similar in texture to Weetabix.

Goodbye Sam

Gone will be the uncertainty about how big our stool will be. Because advances in computer graphics will enable us to design each one ourselves using a mouse and a computer screen connected to our rectum. Not only will we dictate the exact size and shape, but we can even select the colour from a choice of over 500 alternatives.

Hello Samantha

If you find toilet tissue in the bathroom in the year 2000, hold onto it. For it will be a rare and valuable antique. Instead of slaving away wiping our bottoms with fragile strips of tissue paper, in the year 2000 one wipe with a futuristic sheet of silver foil will clean your cleft more thoroughly than you could ever have imagined. Space age winnits – small silver balls like you get on top of Christmas cakes – will disappear instantly. For silver foil loo paper will be coated with special chemicals that not only clean you bum, they'll also cure piles.

Folk

By the year 2000 the stench of urine will be nothing more than a fond memory shared by old folk. For instead of liquid, in the year 2000 we will piss tablets. Two yellow ones normally, or a pink on if we've been eating pickled beetroot,

The toilet of the future will look nothing like the cumbersome, old fashioned things we sit on today. There'll be no toilet seats to begin with. Instead hover rays will suspend our bottoms above a large circular light panel on the floor. Instead of farting, our bottoms will emit small glittering clouds of gas accompanied by dreamy music like on Star Trek. Our stools or piss pills will then appear on the panel below.

Morris

There'll be no chains to pull in the 21st Century toilet. Instead we'll pull up our trousers and walk over to a control console. By sliding one knob up and another one down, our ablutions will be dematerialised, rematerialising seconds later at a sewage works on the moon.

Public toilets will benefit from new techology too. Special X-ray walls will mean no hiding places for dirty old men. And to prevent vandalism a special computer will monitor your bladder and your rectum, to make sure that only people who genuinely need the toilet are allowed in. Payment will also be handled by computer. Simply press your buttocks up against a panel by the door, and the toilet's computer will identify your unique bum print, and charge your visit to your credit card.

Coal

However, some things will never change. And in Harrods department store the faithful toilet attendant will still be there in the year 2000. But tipping him could prove a costly business, due to inflation. For before you leave you will be expected to drop at least £5,000 into his little saucer – the equivalent of 20p in today's money.

B-flat

Sadly, the advent of the space age lavatory will mean an end to one of Britain's best loved traditions. For 'toilet humour' has owed its popularity over the years to our childish obsession and prudish attitudes towards the lavatory. But in the year 2000 Britain's toilets will no longer be taboo. 'Adult humour' magazines will long since have gone to the wall, and words like 'hairy bollocks' will be part and parcel of the English language, no longer the source of any amusement.

BLASPHEMY!

A controversial author yesterday warned that his new book was set to make 'The Satanic Verses' look like a kids' fairy tale.

EXCLUSIVE

God sex chicken book set to cause a storm

For Arthur Pilkington claims that in his first novel 'Devil Arse Spunk' God shags a chicken up the back passage. And he fears that it may result in a Salman Rushdie style fatwa being put on him by the Church of England. Indeed, he has already sent a copy to the Archbishop of Canterbury, and he fears the official announcement of a death threat could be imminent. "I'm not afraid of controversy", he told us yesterday speaking from a secret address where he has been in hiding for several days. " I have already written to Melvyn Bragg asking for his support, and I am trying to arrange a meeting with the Prime Minister so that I can bring attention to my plight."

Mr Pilkington hopes that his book will repeat the success of Salman Rushdie's highly controversial best seller, although as yet no publisher has been found. "At the minute there are a few spelling mistakes that need ironing out, and at seven pages it looks a little short. I'm thinking of adding a few bits about Jesus and farmyard animals to try and pad it out a little.

Once I get it typed up properly publishers will be biting my hands off – you just see", he told us.
We rang controversial Christian Cliff Richard to try and obtain a quote that we could use out of context. However, the former singer, who changed his name from Harry Webb when he turned his back on a pop career in the seventies, wasn't in.

'I CAN SAVE DI'S MARRIAGE'

Joe Kinghorn displaying one of his many caravans yesterday

A Bridlington tradesman yesterday stepped forward and threw a lifeline which could save the Prince and Princess of Wales' marriage from crashing on the rocks.

Joe Kinghorn believes that a caravan holiday could be just the ticket to bring the feuding Royals back together.

TURMOIL

Local trader Joe was genuinely upset when he read reports that the Royal marriage was in turmoil. "I'm a big fan of the Royals, they do a marvellous job, and I have every sympathy for Charles and Di. I realise the pressures they must be under".

CARAVAN

"What they really need is to get away from it all – a complete break. And, what better way to do that than with a caravan", said Joe, Managing Director of Bridlington based Coastal Caravan Sales (UK) Ltd.

"With a caravan in tow Britain becomes your back garden. Every day you can enjoy a different view from your kitchen window. In a word it gives you freedom – freedom to go wherever you please".

LAKES

Joe believes a fortnight or so in the Lakes, or perhaps just the odd weekend away together would give Charles and Di vital breathing space, and a chance to patch up their differences.

RIVERS

"A lot of my customers have similar problems", Joe told us. "I get people in my showroom fighting like cats and dogs. But generally speaking it's the ones who buy the caravans who end up

smiling. There's a strange, romantic kind of magic about a caravan, and they're not as expensive as you might think".

STREAMS

Joe has already written to the Palace enclosing a colour brochure and a price list of caravans currently in stock. "I think I've got something for everyone. From the latest, fully equipped luxury models – a real home from home, to the smaller, more compact vehicles. With full credit facilities available – subject to status".

So far Joe has heard nothing from the Prince of Wales, but he is confidently expecting a call. "I quoted him the current list prices, but for a quick sale I would obviously be willing to do a deal on that", said Joe.

COULD YOU SAVE THE ROYAL MARRIAGE?

Have you got a marriage saving idea that could help keep Charles and Diana's marriage afloat?

BUST-UP

Perhaps you have survived a marriage bust-up yourself, or know someone who has. Maybe you have an idea or suggestion that may bring a glimmer of hope to desperate Diana. Or a useful hint that could help Charles patch up the holes in their relationship.

FANNY DOWN

Or perhaps you think they get paid too much to start with, and you don't give a toss about their bloody marriage. Christ, the head start she's had in life, you'd think she could have made something better of it. And besides, who did she think she was marrying? David frigging Bowie of something?

CONFIDENCE

Anyway, whatever your opinion, write and let us know. Send your marriage saving tips to 'Viz Royal Marriage Saving Appeal', P.O. Box 1PT, Newcastle upon Tyne NE99 1PT. And remember to mark your envelope 'My heart bleeds for Diana'.

All letters will be treated in the strictest confidence. In fact we won't even open them.

Fergie clashes with Queen over telly

The Queen - Fisticuffs with Fergie

The Duchess of York's marriage split with Prince Andrew looks set to end in an undignified slagging match between Fergie and the Palace.

Already the Queen has snubbed the Duchess by banning her from the Palace. And there are even reports of violence between Fergie and her Royal mother-in-law.

FLARED

Former Palace gardener Reg Molesworth confirmed that tempers had flared during a recent visit by Fergie.

BELL BOTTOMED

"I think she'd come back to get her telly and a few of her other things. Next thing I knew there was shouting and screaming and the two of them were at it like cats and dogs. They were scratching

and tugging at each others hair. Eventually Palace bodyguards pulled the Queen off, but she was furious, and told Fergie she would finish it next time she saw her".

DRAIN PIPED

Meanwhile Andrew has been in touch with Palace solicitors over unpaid phone bills at the Duke and Duchesses South York home. Since packing her bags Fergie has refused to contribute towards the unpaid bills, thought to total around £300. Andrew insists that the Duchess was responsible for at least three quarters of the calls made.

VIZ READERS BACK ANDY

Viz readers have voted 8 to 2 in favour of Prince Andrew in a recent poll to find out who is to blame for the York's marriage break up. We asked readers whether the burden of responsibility for the break down in the marriage lay at Andrew's feet, or whether Fergie was at fault, because it was obvious right from the beginning that she was a bit of a tart and not right for our Andy.

We also took the opportunity to ask readers whether they thought Andrew should re-marry quickly on the rebound, take his time and choose the right partner, or say 'What the hell' and play the field a bit, making a bit of hay while the sun shines and generally giving the dog a bone. And surprisingly, 70 per cent of readers plumped for the latter.

Sting in the Tail

IT SEEMS that every day our newspapers are full of scare stories about meteorites, Al Qaida attacks and tsunamis. But now, as heat levels soar due to global warming, sun-baked Britain could be facing an even more terrifying threat than skin cancer, forest fires and hosepipe bans - an increase in attacks by BEES.

For, according to an old woman on a bus in Newcastle, as temperatures rise the stripy insects become angrier and are liable to sting with greater frequency.

In a press conference held on the No 62 from Byker to the City Centre, 82-year-old Ethel Carstairs said: "Look at that bee buzzing around. It'll be angry that. It's the heat as makes them angry. It's going to sting somebody, that."

Following the announcement, hospitals across the country went on a state of high alert. A spokesman for the North Tyneside General Hospital at Rake Lane told reporters: "In the light of what the old woman has said, we have begun to stockpile antihystamines, tubes of Savlon and sticking plasters. But we urge the public to remain calm if they see a bee. If people simply stand still and don't panic, the chances are the bee will simply fly off and sting someone else."

Meanwhile Mrs Carstairs had this advice for her fellow travellers: "Somebody wants to open a window and shoo it out."

A bus similar to the one Mrs Carstairs was on, and (below) an old lady similar to Mrs Carstairs who was on the bus similar to the one above.

"Don't Get Stung!" says Sting

Arsehole popstar Sting gives his top ten tips to avoid being stung by bees this summer.

"My parents named me after a bee's arse, so it is only natural that from a very early age I have been fascinated by these creatures. I spend my life surrounded by bees, yet in twenty years on the road with the Police I have never once been stang off of one. And if you take these few simple precautions, you too will avoid being a bee sting victim this summer. Here are my top ten things you can de-do-do-do, de-da-da-da to avoid being steng."

1 TAKE a tip from my fellow arsehole pop star Morrissey out of the Smiths - wear a hearing aid and carry a bunch of flowers around in your back pocket. If you hear a bee approaching you from a distance, simply throw the flowers to one side and the bee, smelling the pollen, will go for them instead.

2 BEES are attracted to movement, so if one comes buzzing around you, keep perfectly still. It will eventually lose interest and move on.

3 ALL bees love honey, so it stands to reason that just as opposite poles of a magnet repel each other, bees will be repelled by the opposite of honey, which is Branston Pickle. A little bit smeared on the back of the neck, forehead and behind each ear will keep bees safely at bay.

4 BEES are attracted to the bright colours of flowers, where they go each day to collect pollen and nectar. In order to avoid being mistaken for a flower, wear dark or muted coloured clothing, such as a bespoke £3000 Armani jacket or a hand-stitched £4000 Alexander McQueen suit.

5 BEES often land on light-coloured surfaces, so if you have pale skin it is a good idea to cultivate a year-round all-over tan by moving to the Seychelles, Mustique or buying a huge villa in Tuscany. Either that, or do as my fellow arsehole pop star Mick Hucknall out of Simply Red does and 'black-up' like Al Jolson during the summer months.

6 A BEE knows that if it stings you it is signing its own death warrant, so you are ten times more likely to get stang off a depressed or suicidal bee. If you encounter one of these, try to cheer it up by playing it a happy tune on your kazoo.

7 FOR CENTURIES, monks have kept the bees in their hives drowsy and unaggressive using smoke. But few of us carry a smoke canister around with us! So simply use the smoke from a £2000 Cuban Havana cigar available from DuCannard Tobacconists in Knightsbridge.

8 IF A BEE comes towards you, run around quickly in small circles flapping your arms wildly and shouting. The noise and movement will confuse it and it will eventually move on.

9 IT IS A fact that 90% of all stings take place inside cars. Trapped in the car and unable to escape, the bee panics and begins to attack. So avoid becoming one of these statistics by simply driving a convertible car such as a Bentley Continental GT cabriolet or a custom-built Ferrari F50 Bolide.

10 BEES will sting when they are angry or stressed. Over the years I have found that the best way to ease stress is by doing yoga. So if you are attacked by an angry bee, simply fold its legs into the lotus position and fix them there with a small piece of sticky tape. If after ten minutes the bee has not calmed down, simply hit it with a rolled up newspaper.

Now Wash BANDS!

Government Launches Nationwide Campaign to Counter Pop Squirts

A recent survey found that a disturbing 25% of Britons don't wash their hands after wiping their bottoms. That means that one in every four people is walking about with their finger ends coated in excreta.

It's a shocking statistic for ordinary members of the public, but its implications for the pop industry are truly horrifying. For the figures prove that, thanks to their lack of hygiene in the toilet, one member out of every four-piece pop band is spreading unpleasant germs around the country's stages and recording studios.

A bout of nausea and sickness caused by turd-borne bacterias amongst the members of a group could delay the release of an album or even mean that a concert tour has to be postponed or cancelled. Lost sales of tickets and records on the high street would inevitably lead to a reduction in tax revenues, and it's a scenario that is worrying the government.

"Diaorrhea and vomiting amongst pop stars could cost the Exchecquer millions of pounds a year," says entertainment minister Ord Wingate MP. "Unless something is done to address the issue of toilet

Right: Part of the government's £1m campaign. One of the leaflets that will be distributed around recording studios, international stadiums and swanky hotel suites.

All Along the Washtowel

Soapy Hendrix sez...

CLEAN UP YOUR ACT!

Issued by the
Department of Health

IT'S A FACT that one member out of every rock quartet leaves the bathroom after a pony without washing their hands. But which one is it? Here pop guru and ex-diddler *Jonathan King* sniffs out the evidence before pointing the brown finger as he names and shames

...the Mucky Pups of Pop!

U2

FOUR-PIECE rock supergroup U2 have been cleaning up by topping the charts since 1983, but it's a fact that one member of the band can't be bothered to wash his hands after wiping his bottom. But which one is it? Let's look at the evidence.

Very little is known or cared about Larry Mullins. But like most obsessive loners, chances are that the reclusive drummer keeps himself fastidiously clean. Indeed, it's quite likely that a bizarre psychological condition compells Mullins to wash his hands obsessively, often hundreds or thousands of times per day.

Hell raising bassist Adam Clayton (real name Harry Webb) is best known for false rumours accusing him of having it off with loads of prostitutes in a London Hotel. You might suppose that anyone who stoops so low wouldn't think twice about leaving a lavatory without washing their hands. But you should remember that the sort of high-class good time girls that millionaire Clayton certainly didn't go with always insist that their punters are clean before letting them do sex.

The third member of U2, hat-wearing The Edge (real name Shirley Crabtree), is known for only one thing besides his stupid hat and his stupid name - his distinctive clear-cut guitar riffs. These would become increasingly difficult to play if the strings were covered with shitcrumbs and pellets of winnity bogroll. Therefore it's highly likely that The Edge spends upwards of a minute working away with soap and water after each sit-down khazi visit.

As well as being a band member, lead singer Bono (real name George O'Dowd) is also an environmentalist, a world statesman and a chess Grandmaster. With such a hectic schedule, it's unlikely he has enough time to wash his hands. Indeed, it would be a miracle if he managed to give his soiled anus more than the briefest of rudimentary wipes after each stool.

Jonathan Points the Brown Finger at: *BONO*

T

AFTER the departure of Ger Halliwell, the five-piece Spice Girls were reduced to four members. In their three-year reign as Queens of Girlpower, they had their fingers on the nation's pop pulse. But statistically, only three of them had their fingers under the tap after wiping up the mess from a copper bolt. Baby, Posh, Scary and Sporty they may have been, but which one was Dirty Spice?

Sporty Spice Mel C (real name Declan McManus) was well known for her trademark on-stage cartwheels, forward rolls and press ups. Like all sporty people she probably spent a large amount of time in the shower, getting herself clean after her gruelling daily fitness routine. It is highly likely tha

Jonathan Points th

Your

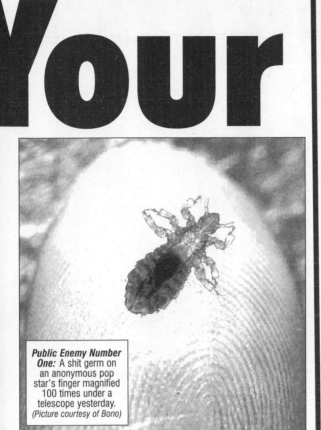

Public Enemy Number One: A shit germ on an anonymous pop star's finger magnified 100 times under a telescope yesterday. *(Picture courtesy of Bono)*

hygiene amongst British pop musicians, this country's economy could be plunged into the worst recession this century."

And the government has decided to put its money where its mouth is, by launching a £1 million poster and leaflet campaign -

'Clean Up Your Act', featuring Soapi Hendrix, a cartoon bar of soap that plays an electric guitar. The minister added: "We've even set up a 24-hour freephone hotline that pop stars can ring to find out more information about when they need to wash their hands."

ROCK stars are often our role models. We copy their haircuts, their clothes and their CDs on our computers, but do we copy their toilet habits? We went out on the street to get the public at large to come clean about their bog standards.

Hands Up!
Do YOU Wash After a Shite?

John Brown, Contract Publisher...
"For many years I washed my hands after going to the smallest room like everyone else. But then I found out how cheap water was, so I started to use vintage champagne instead. And a bar of gold instead of soap."

Charlene Lubbock, McDonald's Chef...
"I've had amoebic dysentery since eating a dodgy paella in Benidorm 4 years ago, and I have to go for a tom tit at least fifty times a day. If I stopped to wash my hands each time, I'd never get any burgers cooked, so I just clean them every four times or so."

Archie Turtle, Lavatory Attendant...
"Certainly not. As someone who spends most of his working day with his hands down the toilet, pulling turds out of a blocked U-bend, I see little point in it."

Mad Lennie, Gentleman of the Road...
"Yes, I wash my hands thoroughly every time I have a shit. In fact I'm going to wash them right now, because I've just done a great big one in my trousers."

Mrs. Edna Tortoise, Housewife...
"Unfortunately, I suffer from Obsessive Compulsive Disorder, which causes me to wash my hands about a hundred times a day. Fortunately I also suffer from Crohn's disease, which means I defecate about a hundred times a day, so it balances out quite nicely."

Sir Rupert Tickler, Consultant Proctologist...
"Certainly not. As someone who spends most of his working day with his hands up someone's bottom, pulling turds out of a blocked alimentary canal, I see little point in it."

Marvo, Stage Magician...
"No I do not. I simply use sleight of hand to fool the audience into thinking I have."

Rev. Trafford Lovething, Parish Priest...
"In my line of business, it's certainly true that cleanliness is next to godliness. A vicar's hands get very close to his parishioners' noses when he's handing out the communion wafers. A friend of mine who wasn't quite so fastidious in the bathroom soon got the nickname 'Father Shitty-Fingers'."

SPICE GIRLS

washing her hands after wiping her bottom was therefore second nature.

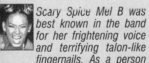
Scary Spice Mel B was best known in the band for her frightening voice and terrifying talon-like fingernails. As a person with such sharply-manicured digits, she will have been no stranger to push-through. After doing a sticky number two, it's more than likely that Mel had no choice but to give her hands a thorough cleaning with soap, nailbrushes and a cocktail stick.

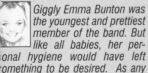
Giggly Emma Bunton was the youngest and prettiest member of the band. But like all babies, her personal hygiene would have left something to be desired. As any

mum will tell you, infants have no idea about protecting herself from germs. After doing a 'poo poo' in her potty, the 22-year-old blonde megastar was probably so keen to go out and play skipping and ponies to bother about washing her hands.

Posh Spice Victoria Adams was the upper class member of the group. Aristocratic people are by nature some of the most hygienic in the world, as they are able to afford refined soaps such as Imperial Leather and Lux. However, Victoria would not have needed to wash her hands after having a dump, since she would probably have delegated such menial tasks as removing her clag to one of her many servants.

Finger at: **BABY SPICE**

THE ROLLING STONES

THE wrinkly rockers may have sung *'It's All Over Now'* back in 1964, but when it's all over their hands after they've wiped their arses, which member of the Greatest Rock'n'Roll Band in the World fails to wash it off?

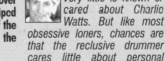
Rubber-lipped lead singer Mick Jagger may be getting a bit long in the tooth these days, but when he gets up on stage he's still got the energy of a 40-year-old man half his age. A dope-smoking rebel in his youth, Mick would once have probably thought nothing of leaving the bathroom with stinky fingers. However, now as a respectable pillar of society, Sir Mick regularly takes tea with the Queen at Buckingham Palace and so is likely to be a stickler for

post-excretory etiquette.

Very little is known or cared about Charlie Watts. But like most obsessive loners, chances are that the reclusive drummer cares little about personal hygiene. Indeed, it's quite likely that a bizarre psychological condition gives Watts an obsessive aversion to soap and water, compelling him to not wash his hands hundreds or even thousands of times per day.

Unlike his clean-living elder brother Cliff, Keith Richards has rightly earned his position as the wild man of rock. For over four decades he has lived on a diet of nothing but drugs and Jack Daniels. As a result of eating no

food, Keith's anus healed over in the late 1970s. Consequently, since he only goes to the toilet to urinate, the Stones guitarist has no need to wash his hands.

The Stones guitarist Ronnie Wood is an accomplished artist, whose energetic portraits of rock stars often resemble the people they're supposed to be. Like all great painters, such as Brian Sewell and Tony Hart, Ronnie is a sensitive soul at heart. Now that he's more likely to be found washing his brushes with turps rather than nudging it, it's quite possible that genteel Wood would be deeply distressed if he failed to clean the turd off his fingerends after a doing a faece.

Jonathan Points the Brown Finger at: **CHARLIE WATTS**

FRANKENSTAR MUST DIE!

A showbusiness hurricane is brewing over Variety Club plans to create a 'Frankenstein' style celebrity monster using the remains of dead stars from the past.

Plans for the showbusiness monster were hatched by fund raisers hoping to raise extra cash towards the Variety Club's Sunshine Coach appeal. But there is growing unrest among showbusiness personalities concerned that a man-made celebrity monster, assembled using limbs and organs from dead stars, would be in bad taste. And alarm bells are ringing after some insiders admitted that the monster may go wrong, and kill the very children that it was setting out to help.

RISK

"There is always a risk that an experiment like this could go wrong. The monster may not know its own strength, or scientists may accidentally give it the wrong brain or something", admitted concerned star Ernie Wise last night.

VARIETY

The variety club is believed to have already begun work on the 'monster' at a castle in North Wales belonging to comedian Jimmy Tarbuck.

Tarby – castle

Storm over celebrity monster set to do charity work for kids

And according to some reports, Tarbuck's shopping list for the proposed 'Frankenstar' includes:

- **HAIR** from the late great comedian Frankie Howerd.
- The **CHEEKY GRIN** of the late Bennie Hill.
- **LEGS** removed from musical hall great Arthur Askey.
- **FEET** from giant comic Tommy Cooper.
- And Eric Morcambe's glasses.

Once the monster is complete pioneering Variety Club scientists plan to bring it to life by harnessing the power of an electrical storm. They then plan to tour the country, doing sell-out charity shows to raise funds for needy kids.

VAMPIRE

This is not the first time that a fund raising organisation run by the stars has attempted such an ambitious project. In 1989, amidst a storm of publicity, the Lords Taverners announced plans to make a vampire out of Peter Sellers. The project cost several million pounds to finance, but ran aground when Sellers failed to wake up when it got dark.

An artists impression of how the monster will look when it emerges from Castle Tarbuck.

The hands of the late great DAVID NIXON will give it breathtaking conjuring ability.

The singing, dancing legs of music hall legend ARTHUR ASKEY will give it 'all round' entertainment value, with terrific family appeal.

The feet and shoes of the immortal TOMMY COOPER will come to life once more as the undead creature takes the stage in numerous charity shows.

We will laugh again at the saucy antics of BENNY HILL as the beast breathes new life into his cheeky grin.

??

LLOYD GROSSMAN'S *THROUGH THE WINDSCREEN*

The Celebrity Car Crash Quiz

Dear oh dear oh dear! This star's car's in a right mess. Let's sift through the wreckage and see if we can discover *who would crash a car like this?*

Here's the evidence. Our mystery celebrity has crashed his £50,000 Mercedes in East London. On the dashboard is a gumshield. And what's this on the rear passenger seat? Some sort of metal tea pot? No – it's a magic lamp. Perhaps our star is some sort of entertainer? A magician perhaps? Finally, in the boot there's a crate of bottles. Not booze I hope. No,

it's a crate of HP Sauce! This star obviously has a *big* appetite for the stuff.

If you know the answer, send the name, on a postcard, to: Lloyd Grossman's THROUGH THE WINDSCREEN, Viz, P.O. Box 1PT, Newcastle upon Tyne, NE99 1PT. The winner will receive ringside tickets to see Frank Bruno's next big fight.

???

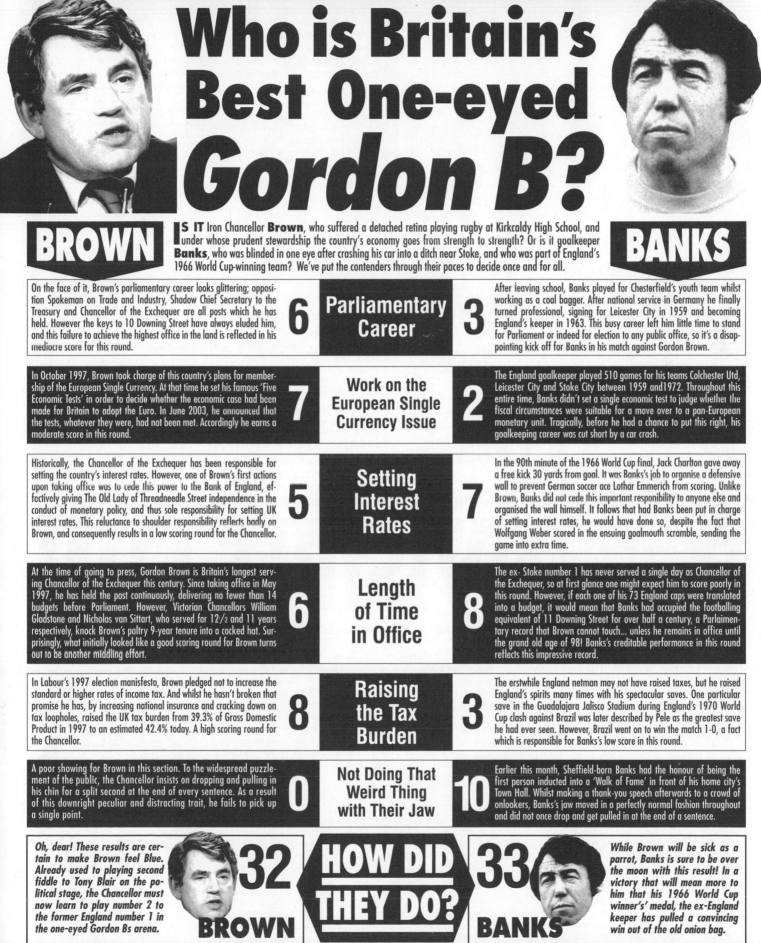

As the World Cup gets underway and Tony Blair's grip on power weakens by the day, there is only one topic on the lips of the nation. In every pub, club and factory from Abberton (near Worcester) to Zennor (not far from St Ives), the air is buzzing with one question...

Who is Britain's Best One-eyed Gordon B?

BROWN

BANKS

IS IT Iron Chancellor **Brown**, who suffered a detached retina playing rugby at Kirkcaldy High School, and under whose prudent stewardship the country's economy goes from strength to strength? Or is it goalkeeper **Banks**, who was blinded in one eye after crashing his car into a ditch near Stoke, and who was part of England's 1966 World Cup-winning team? We've put the contenders through their paces to decide once and for all.

BROWN				BANKS
On the face of it, Brown's parliamentary career looks glittering; opposition Spokeman on Trade and Industry, Shadow Chief Secretary to the Treasury and Chancellor of the Exchequer are all posts which he has held. However the keys to 10 Downing Street have always eluded him, and this failure to achieve the highest office in the land is reflected in his mediocre score for this round.	**6**	**Parliamentary Career**	**3**	After leaving school, Banks played for Chesterfield's youth team whilst working as a coal bagger. After national service in Germany he finally turned professional, signing for Leicester City in 1959 and becoming England's keeper in 1963. This busy career left him little time to stand for Parliament or indeed for election to any public office, so it's a disappointing kick off for Banks in his match against Gordon Brown.
In October 1997, Brown took charge of this country's plans for membership of the European Single Currency. At that time he set his famous 'Five Economic Tests' in order to decide whether the economic case had been made for Britain to adopt the Euro. In June 2003, he announced that the tests, whatever they were, had not been met. Accordingly he earns a moderate score in this round.	**7**	**Work on the European Single Currency Issue**	**2**	The England goalkeeper played 510 games for his teams Colchester Utd, Leicester City and Stoke City between 1959 and 1972. Throughout this entire time, Banks didn't set a single economic test to judge whether the fiscal circumstances were suitable for a move over to a pan-European monetary unit. Tragically, before he had a chance to put this right, his goalkeeping career was cut short by a car crash.
Historically, the Chancellor of the Exchequer has been responsible for setting the country's interest rates. However, one of Brown's first actions upon taking office was to cede this power to the Bank of England, effectively giving The Old Lady of Threadneedle Street independence in the conduct of monetary policy, and thus sole responsibility for setting UK interest rates. This reluctance to shoulder responsibility reflects badly on Brown, and consequently results in a low scoring round for the Chancellor.	**5**	**Setting Interest Rates**	**7**	In the 90th minute of the 1966 World Cup final, Jack Charlton gave away a free kick 30 yards from goal. It was Banks's job to organise a defensive wall to prevent German soccer ace Lothar Emmerich from scoring. Unlike Brown, Banks did not cede this important responsibility to anyone else and organised the wall himself. It follows that had Banks been put in charge of setting interest rates, he would have done so, despite the fact that Wolfgang Weber scored in the ensuing goalmouth scramble, sending the game into extra time.
At the time of going to press, Gordon Brown is Britain's longest serving Chancellor of the Exchequer this century. Since taking office in May 1997, he has held the post continuously, delivering no fewer than 14 budgets before Parliament. However, Victorian Chancellors William Gladstone and Nicholas van Sittart, who served for 12½ and 11 years respectively, knock Brown's paltry 9-year tenure into a cocked hat. Surprisingly, what initially looked like a good scoring round for Brown turns out to be another middling effort.	**6**	**Length of Time in Office**	**8**	The ex- Stoke number 1 has never served a single day as Chancellor of the Exchequer, so at first glance one might expect him to score poorly in this round. However, if each one of his 73 England caps were translated into a budget, it would mean that Banks had occupied the footballing equivalent of 11 Downing Street for over half a century, a Parliamentary record that Brown cannot touch... unless he remains in office until the grand old age of 98! Banks's creditable performance in this round reflects this impressive record.
In Labour's 1997 election manisfesto, Brown pledged not to increase the standard or higher rates of income tax. And whilst he hasn't broken that promise he has, by increasing national insurance and cracking down on tax loopholes, raised the UK tax burden from 39.3% of Gross Domestic Product in 1997 to an estimated 42.4% today. A high scoring round for the Chancellor.	**8**	**Raising the Tax Burden**	**3**	The erstwhile England netman may not have raised taxes, but he raised England's spirits many times with his spectacular saves. One particular save in the Guadalajara Jalisco Stadium during England's 1970 World Cup clash against Brazil was later described by Pele as the greatest save he had ever seen. However, Brazil went on to win the match 1-0, a fact which is responsible for Banks's low score in this round.
A poor showing for Brown in this section. To the widespread puzzlement of the public, the Chancellor insists on dropping and pulling in his chin for a split second at the end of every sentence. As a result of this downright peculiar and distracting trait, he fails to pick up a single point.	**0**	**Not Doing That Weird Thing with Their Jaw**	**10**	Earlier this month, Sheffield-born Banks had the honour of being the first person inducted into a 'Walk of Fame' in front of his home city's Town Hall. Whilst making a thank-you speech afterwards to a crowd of onlookers, Banks's jaw moved in a perfectly normal fashion throughout and did not once drop and get pulled in at the end of a sentence.

Oh, dear! These results are certain to make Brown feel Blue. Already used to playing second fiddle to Tony Blair on the political stage, the Chancellor must now learn to play number 2 to the former England number 1 in the one-eyed Gordon Bs arena.

32 BROWN

HOW DID THEY DO?

33 BANKS

While Brown will be sick as a parrot, Banks is sure to be over the moon with this result! In a victory that will mean more to him than his 1966 World Cup winner's' medal, the ex-England keeper has pulled a convincing win out of the old onion bag.

Next week: The clash of the drugged-up gay Georges: *Boy versus Michael.*

Black Hole Blackpool

"Bring it on!" says Lord Mayor

BLACKPOOL'S FAMOUS illuminations could lose some of their sparkle this year - that's if scientists' fears of a black hole forming at the Large Hadron Collider come true. For the infinitely dense body in the Swiss research facility would create a gravitational field so strong that not even the lights of the seaside resort's world famous winter attraction could resist its pull.

"The light would all be sucked straight out of the bulbs towards Switzerland," said Lord Mayor Ivan Taylor. "And whilst that would be good news for people standing on the town side of the display, anybody looking from the beach or the pier would think there had been a power cut."

SPECTACULAR

The town's tourist chiefs admit that this could spell disaster for Blackpool, which relies heavily on its spectacular light show to attract visitors from all over the country. And at a press conference held in the town hall, Alderman Taylor outlined the council's 10-point plan to keep the lights shining in the event of a black hole. The measures the council are considering include:

- **REPLACING** red bulbs in the illuminations display with blue ones, as the shorter blue wavelength photons have less mass and are less likely to be pulled towards the black hole.

- **PLACING** mirrors on buildings along the Golden Mile to double the effectiveness of the light before it gets sucked to Switzerland.

- **MAKING** greater use of non-incandescent decorations such as tinsel, glitter and luminous paint.

Taylor accepts that if a point of infinite density does indeed form in the Collider, his town's illuminations may not be as much of a draw as they have been in previous years. But he was remaining light-hearted about the physicists' doomsday scenario.

MONOCULAR

He told reporters: "Blackpool was blacked out from 1939 till 1945 and it didn't do anything to spoil the fun in Lancashire's premier coastal holiday destination."

BINOCULAR

"We're not going to let a little thing like the catastrophic effects of a black hole dampen our spirits, and we're facing the prospect with typical Blackpool humour," he said. "We're updating our traditional solid black 'Blackpool by Night' postcards to read 'Blackpool All The Time'. And we'll also be producing a new range of saucy postcards, all relying heavily on the suggestive possibilities of the words 'black hole', and featuring scotsmen in kilts, young ladies in nudist camps and newly-wed couples in bed and breakfasts," he added.

INVISIBLE SUN

Sting's Mission to Explain Black Hole Collider in Song

POP arsehole **STING** is set to explain the mysteries of particle physics... in song! In his new single *Mountains of Colliding Hearts*, the former bass player with 80s boy-band THE POLICE explains what happens when protons from a hydrogen atom are smashed into each other at the speed of light at the Large Hadron Collider.

He told *Radio 1 Newsbeat*'s Jackie Quimfest: "When I saw the CERN cyclotron on one of my very large widescreen televisions - I think it's 105", possibly more - anyway, it's the biggest one you can get, I know that - I was immediately fascinated. I started thinking about this 17-mile-long machine, the Large Hadron Collider, which is looking for something unimaginably small - even tinier than a hundred & thousand."

PARTICLE

"The Higgs Boson - the so-called "God Particle" - is the missing link between the Big Bang, Steven Hawkins and Einstein's Theory of Relativity," he continued. "It is literally the Holy Grail for scientists who are researching Quantum Physics, String Theory and that. And when I started thinking about it, the idea for the new song just came to me."

QUANTUM

"It's a number all about the relationships between the different particles in an atom - the proteins, the croutons, the quarks and the gluons. They are unable to love each other until they realise that the fundamental forces in the nucleus that are trying to tear them apart are also what's holding them together," Sting said.

"It's a soft-rock ballad with jazzy overtones," he added.

The single comes hard on the heels of *The Large Hadron Rap* - a light-hearted song produced by scientists working at the CERN Collider, which became an unexpected internet hit last month. Sting told us: "My record's better than theirs, because I'm treating the subject of String Theory with the seriousness it deserves. Not only that, but I've got the scientific expertise to back up my music with meaningful lyrics that will make people think."

DOUBLE

"And if I don't step up to the plate to explain these subjects to the ignorant public, then who will?" the big twat continued. *"I play the bass, guitar and lute, so string theory doesn't phase me at all. And I'm also no slouch when it comes to big bangs, as I poke my missus Trudi Styler for five hours every single night."*

Sting's previous single was an attempt to express musically the solution to Fermat's Last Theorem. *Searching for an Answer (to the*

Theorem of Love), which he recorded with fellow pop dickhead Bono, failed to chart when it was released in the same week as *Integers of the Heart (Ain't No Solution)* by pop wankstain Alex James out of Blur.

Rocker Rick's heart pours itself out exclusively to Viz

I NEVER MISSED A BEAT!

EXCLUSIVE

Heart attacks back in Quo heart op bust up

VETERAN rocker Rick Parfitt's heart has blasted the star in a bitter war of words over his recent emergency heart operation.

Rocker Rick hasn't spoken to his former bosom buddy since it conked out and he collapsed in agony at his luxury £750,000 mansion last month. Rick was rushed to hospital where heart op docs hacked open the 48 year old rocker and carried out a life-saving quadruple bypass operation.

Heart

Rick claimed his heart had nearly killed him. But that angered the organ, and now his heart has hit back - saying that it always stood by the star, and blaming the rocker's wild lifestyle for his recent health problems.
"Despite my loyalty, he's made my life a misery", his heart told us yesterday. "He's only got himself to blame".

Clout

Rick's heart is a hollow, muscular organ whose function is to maintain the circulation of blood around his body. It receives oxygenated blood from Rick's lungs which it then pumps around his body via a system of arteries and smaller blood vessels.

"Rick a-orta known better than to blame me"

But it says its job was made impossible by the hell raising star, whose body it alleges is falling apart.

Bangles

"He never gets any exercise" it told us. "And his lungs are in a right state. He smokes fifty fags a day, and the blood what comes out of them is rubbish. As a result I had to work twice as hard just to keep him breathing. How them lungs haven't got cancer yet I'll never know".

Baubles

Throughout his career Rick has bedded a string of beauties. But while he was bonking away into the earlier hours, his poor heart was doing all the work.

Our Showbiz Reporter Andy Bullshit talks to his pal Rick's heart yesterday, and (left) Rick in his rockin' days.

"After a gig Rick would be pretty tired, but he'd still go out drinking into the early hours. By two in the morning he'd be so pissed he could hardly stand up. I know, cos his blood would come in here stinking of whisky. But he'd still end up pulling a bird. How he did it I'll never know."

Beads

Rick would bring his girls back to his flat for even more booze, his heart revealed. "Back at our place they'd start drinking and doing drugs into the early hours. I'd be pulling my hair out, wondering when he was ever going to fall asleep. Then, at five o'clock in the morning, just as his eyes were starting to close, he'd suddenly decide he wants to give her one".

Blood

"So there's me going like the bleedin' clappers trying to get some extra blood to fill up his cock. Otherwise he can't do the business. So he'd be banging away and I'd be pumping my bollocks off, trying to keep up with him. How

I didn't pack in there and then I'll never know."

Haddock

Yesterday Rick's heart issued an emotional plea from itself, begging the hell raising star to change his ways, before its too late.
"He's let me down so many times in the past, I don't know if he's capable of making any real changes. But if he's reading this, please let's give it one more try. If not for our sake, then for the sake of his other organs".

Next week: How Rick broke me the night he talked of a transplant.

MINCE 59P/lb

I LIKE A WOMAN WITH A BIT OF MEAT ON HER.

HOW A POP STAR'S HEART WORKS

A pop star's heart consists of four cavities, two ventricles and two auricles. When he sings, plays or has sex, blood enters the right ventricle and passes through a valve into the right auricle from where it is transmitted through the pulmonary artery to the lungs. Here oxygen is added and carbonic acid gas removed. The blood then returns to the heart via the pulmonary veins and enters the left auricle, then passes through a valve into the left ventricle from which the oxygenated blood is distributed to the rest of the body via the aorta. In all pop stars the auricle and ventricle of one side of the heart are separated entirely from the other by a muscle wall.

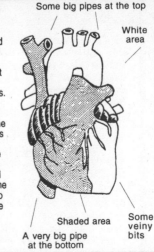

Some big pipes at the top

White area

Shaded area

A very big pipe at the bottom

Some veiny bits

LEADER O BANG

Firing Squad Execution Set to be Greatest Show on Earth

BOB GELDOF is planning yet another charity spectacular, and the headline act is going to be convicted paedophile Gary Glitter. But the ageing glam-rocker won't be singing a medley of his songs; instead he'll be being shot dead by a firing squad.

In scenes reminiscent of 2005's Live8 extravaganza, Glitter's execution for child sex offences later this year will be a glitzy Wembley Stadium stage show. And Sir Geldof hopes to enlist the help of many of his star friends to ensure that the gig goes with a bang.

A source close to Geldof told us: "After the success of last year's Hyde Park concert, everyone's mad keen to get on board. Bob's already got Coldplay, the Killers, Arctic Monkeys and Elton John confirmed for the afternooon. It's going to be a great atmosphere."

As the climax to a day of top-line entertainment, the disgraced glam rocker will be dragged onto the stage in one of his trademark glitter suits, his eyes will be covered with a flamboyant sequinned blindfold and he'll be tied to a sparkly post before a squad of five Vietnamese soldiers shoot him through the chest and face.

"Live8 saw the reformation of the classic Pink Floyd line-up," the spokesman continued. "But the execution of Gary Glitter live on stage is

EXECLUSIVE

going to top that. It's the event that the rock world has been waiting for. It's literally the holy grail of pop."

"It won't be the longest set Gary's ever done, but it'll certainly be one of the most memorable," the source added. "And if the shooting goes down well with the crowd, the execution squad might bring Glitter's corpse back out on the stage and bayonet it for an encore."

Organisers expect a crowd of over 100,000 to witness the execu-

Bob Geldof yesterday

tion, which will be carried live on giant screens in Hyde Park, on Wimbledon's Henman Hill and in the courtyard of Edinburgh Castle. In addition, the event will be beamed round the world to an estimated TV audience of over a billion.

All the stars, including the firing squad, are giving their services for free, and the profits from ticket sales and TV rights will be going to the St Columb Donkey

Sanctuary in Cornwall.

Although the line-up is not yet finalised, there are rumours that U2 plan to split up specially for the event. However, organiser Geldof ast night remained tight-lipped about his plans. He told us: "I don't know what you're talking about. This is the first I've heard about any of this."

"Anyway, Wembley Stadium doesn't even fucking exist any more," he added.

Top Himsel

NOW languishing in a Vietnamese jail, Gary Glitter faces spectacular showbiz execution when he is found guilty being a nonce later this year. But the ageing 70s pop icon ha vowed to cheat the firing squad by topping himself in prison

However, thanks to the spartan conditions in which Glitter finds himself, carrying out his threat may be a trickier task than it seems. Without any ropes, guns or sharp knives available in his cell, the ageing rocker may find himself being forced to keep his appointment with his rifle-toting executioners.

But chubby survival expert Ray Mears reckons that with a little ingenuity, there are a million-and-one ways for the veteran star to shuffle off his mortal coil. He told us: "Gary is literally spoiled for choice. If he looks hard enough, he will find that his apparently empty cell is an Aladdin's Cave of suicide equipment."

Here Mears takes us through just a few of the ways that Glitter could

pop his platform clogs before the Vietnamese authorities get a chance to tie him to a post and shoot him through the heart.

Ray of hope?: Weed-gobbler Mears ponders Glitter's fate yesterday.

FTHE

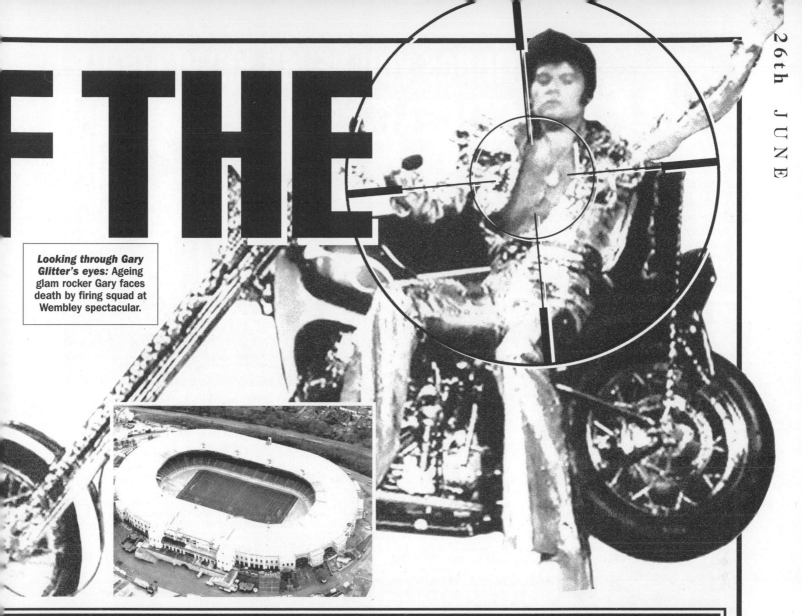

Looking through Gary Glitter's eyes: Ageing glam rocker Gary faces death by firing squad at Wembley spectacular.

f of the Pops
Glitter Vows to Dodge Bullets

HANGING...............

"The laces from Gary's famous silver platform boots will have been confiscated by canny guards, so throttling himself with those is out of the question. But the sturdy whiskers from his trademark hairy chest could easily be woven into a rope with which the glam-rock king could hang himself from the light fitting in his cell."

POISONING...............

"At breakfast time every day, Glitter should pour himself a bowl of Alpen. If he slips the raisins from his cereal into his pocket, he could spread them on the floor of his cell and tell the warders that they are rat droppings. He could then save up the rat poison that they put down until he has set aside a sufficiently large dose to kill himself."

BURNING...............

"As a non-smoker, Gary will not have any matches in his cell. However, by carefully arranging the reflective wing collar and shoulder pads from his glitter suit into a series of parabolic mirrors, he should be able to focus the sun's rays into a powerful white hot beam with which he can burn himself to a crisp, like an ant under a schoolboy's magnifying glass. To sit still whilst such a death ray fries him alive will take quite a bit of willpower, but I have every confidence that the Leader of the Gang is up to the task."

CHOKING...............

"Prisoners often while away their long hours of incarceration by playing table tennis. To choke himself to death, all Glitter has to do is open his mouth wide and swallow the ball as his opponent serves. As luck would have it, the average ping-pong ball is exactly the same diameter as Gary Glitter's trachea. Once his airways are blocked, as long as no-one in the jail knows how to do the Heimlich manoeuvre, I estimate that Gary should be turning up his toes in less than two minutes."

VIOLENT DEATH......

"Pecking order is very important to prison inmates. Anyone who tries to muscle in on Mr Big's action is liable to find themselves the victim of vicious retribution. Accordingly, if Gary makes his way around the exercise yard asking everyone he meets if they want to be in his gang, his gang, his gang, he can confidently expect to be beaten to death with a snooker ball in a sock the moment the guards' backs are turned."

DISEASE...............

"Shiny objects hold an irresistible fascination for many birds, such as ravens and magpies. Like Burt Lancaster the Birdman of Alcatraz, Gary would arouse little suspicion if he used sequins from his stage suit to attract birds into his cell. However, since he is in Vietnam, his new feathered friends would almost certainly be riddled with Asian bird flu, and it would be only a matter of hours before Glitter himself succumbed to the killer plague, cheating his executioners of their moment of glory."

TRAUMA...............

"Glitter could enlist the aid of his fellow pop diddler Jonathan King, getting him to bring in a cake with a bar of soap concealed inside. In the shower block, Gary could deliberately drop the soap, and bend over to pick it up whilst singing 'Do you want to touch? Do you want to touch? Do you want to touch me there? Yeah!' In the resulting melee, Glitter would almost certainly perish from a combination of internal injuries and massive rectal haemorrhaging."

20 THINGS YOU NEVER KNEW ABOUT SPACE

Neil Armstrong climbs up the ladder to get into Saturn V.

IN 1998, the Americans celebrated Bonfire Night by sending 77-year old pioneer astronaut John Glenn up into space on a rocket. Meanwhile, moonwalker Buzz Aldrin says that in 30 years time, we'll all be playing golf on Mars. But how much do we actually know about space? Here's a Cape Canaveral countdown of twenty things you never knew about the world's favourite infinite vacuum.

20... The first man in space was the Russian Cosmonaut, Yuri Gagarin who blasted off in Sputnik One with his dog Laika on October 10th 1965. The biggest problem he faced was that when cooking his breakfast in space, his sausages stuck to the bottom of the frying pan. Space boffins back on Earth solved this by inventing Teflon, which was used on the oven-to-tableware on all subsequent moonshots.

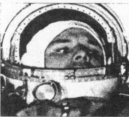

Yuri Gagarin in his space hat.

19... Our solar system contains nine planets which are blown around the Sun by solar winds. They are Mars, Venus, the Moon, Neptune, Mercury, Saturn, Haleys Comet, Uranus and Pluto.

18... And Jupiter. So that makes ten.

17... In olden days, people used to think that the moon was made of green cheese. However, thanks to technology and space travel we now know that it is made of moonrock, a type of weightless grey, fluffy dust, a bit like cement.

16... The closest star to the earth is Alpha Century. No one knows how far away it is, but space eggheads have calculated that it would take you approximately 3,000,000 years to get there.

Some tupperware.

15... Many labour saving devices used around the home came about as spin-offs from the technology developed for the space race, including polystyrene ceiling tiles, cat flaps, car alarms and tupperware.

14... Light from the pole star Polaris takes 400 years travelling at the speed of light to reach the Earth. That means that when you look at it today, you are actually seeing it as Sir Walter Raleigh saw it when he was a boy.

13... The Space Shuttle is a kind of space bus, and like ordinary buses, you even have to give up your seat for an elderly person. However, real buses seldom explode forty seconds after leaving the bus stop.

12... The Shuttle is the most expensive mode of transport in the world, guzzling petrol at a rate of 6 miles to the gallon. Travel on it is beyond the pocket of most people, a day return to the moon costing a staggering £30,000, the price of two estate cars!

11... The first man to land on the moon was the American Neil 'Stretch' Armstrong, whose command module Saturn V touched down on the beach next to the Sea of Tranquility on July 21st, 1969. During the flight, he passed his time by writing an historic quote to accompany his big moment stepping onto the lunar surface. However, as he marched out, he fluffed his lines and asked the driver, Buzz 'John' Aldrin to go back and land again.

10... The first man in space wasn't a man at all. He was a monkey called Cheetah. In a specially built little rocket full of bananas, he blasted off from the Baikonur Cosmodrome, Kazakhstan on November 3rd 1957. Travelling at 17,750 mph he reached an altitude of 588 miles before blowing up.

Astro chimp Cheetah, bids farewell to his proud mum.

9... Because it is so far away, space cannot be seen with the naked eye. Astrologers, the technical term for space scientists who live in round houses called conservatories, are only able to look at it with the aid of very long glasses called telescopes.

8... The biggest telescope in the world isn't actually in the world at all. It's in space! The Hubble Space Telescope weighs 11 tons, cost $1.5 billion and was flown up into space on the Shuttle. However, when the man looked through the end he couldn't see anything and they had to take it back to the shop.

7... The arthur of Space 1999, Author Seaclarke tells everyone that he conceived the idea of the communication satellite. What he tends not to mention is that he also said they would probably be tied to the ground with very long ropes so as you could climb up and mend them when they broke.

6... Thanks to Mr. Seaclarke's invention, we can now watch 1970's Bavarian pornography on a Wednesday and Saturday, buy nasty jewellery from some failed soap star 24 hours a day and pay an extra tenner to watch Evander Holyfield getting his ear bitten off by a bull-necked rapist.

5... Space is the subject of the oldest and most uninteresting programme on telly. The Sky at Night, presented by fat, boggly-eyed, dusty suited, comedy xylophone player Patrick Moore, was first shown on April 24th 1957 and has appeared, unwatched, every month since.

4... Holidaying is the term for going on holiday, but mooning is not the term for going to the moon. Mooning actually means showing your arse from the back of a bus to two pensioners doing 40mph in a Morris Marina as you overtake them on the motorway.

3... A space bar isn't a pub in space where Whoopee Goldberg sells blue fizzy drinks to things with plastic foreheads and gills. It's the long plastic bit at the bottom of a typewriter that makes holes in your writing.

2... Black holes are enormous space vacuum cleaners. They are so heavy, that a teaspoon full of black hole would weigh as much as a baby elephant and would almost certainly break the spoon.

1... If someone tells you they are going moonwalking it doesn't necessarily mean they are going to blast off in a rocket for a stroll around the lunar surface. It

Some Michael Jacksons.

probably means they are going to do that ridiculous backwards-cum-forwards walk made popular by not-plastic-surgery-nightmare, not-kiddie-diddler, high-pitched knacker grabber Michael Jackson.

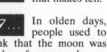

BOFFINS PAINT THE TOWN ....LOAF!

British boffins have beaten America's top scientists in the race to discover a new colour.

Last month a team of researchers and scientists from the University of London announced that they had discovered the first new colour in almost 1000 years. The brand new colour, which has never before been seen, is to be called 'loaf', and will be added to the official spectrum of colours on 1st January 1995. Provisionally the colour will be added inbetween blue and indigo, although it is believed loaf is more of a bluey green than a purple.

By our Science Correspondent Dr. Stanley Jordan

PUMPED

The British discovery leaves American scientists reeling. For over the last ten years the American government has pumped billions of dollars into a massive programme of new colour research and development, their target having been to discover at least one new colour by the year 2000.

FARTED

Previously the last colour to be discovered was orange, the first recorded use of which was in the famous Bayern Tapestry dating back to the 11th Century. Now the colour loaf will enter into history, and the scientists who discovered it will almost certainly become multi-millionaires.

POOTED

As well as the longstanding offer of a £1,000,000 reward from Dulux, the potential royalty earnings from worldwide licensing agreements are limitless. There will be an endless queue of manufacturers eager to use the new colour on a multitude of products ranging from clothes to cheese, and from cars to carpets.

FLUFFED

The earnings from a single licensing deal alone will run into seven figures. For example, British Telecom will have to cough up over a million pounds to produce a Slimline telephone in pastel loaf. And if the colour becomes fashionable next Spring, top designers will fork out a fortune to produce loaf skirts, blouses and pants for the summer season.

Spin offs such as loaf food colouring, tinted wood varnish and eye shadow will also generate vast income. The bidding has already begun between confectionery makers eager to market the first loaf sweets, with Smarties hoping to gain exclusive rights. Soft drink manufacturers are also keeping a close eye on the situation, with obvious potential for drinks such as Loafade, and Lemon and Loaf.

WANKED

But the real money will be made from sales of paint, with Dulux already expressing an interest in marketing a full range of 500 shades of loaf by autumn 1995. As well as gloss, matt, satinwood and eggshell loaf, there will be thousands of variations including Minty Loaf, Apple Loaf, Loaf Sunrise, Dove Loaf and Loaf White.

A spokesmen for the University of London's Department of Colours proudly displayed a sample of the new colour at a press conference yesterday, and announced that it will be officially named by Her Majesty The Queen at a

Jubilant scientists yesterday proudly display the first splodges of loaf

ceremony later this month. It is understood that during the ceremony Her Majesty will wear a loaf dress with matching hat, and she will be presented with a bunch of loaf flowers.

Barking Mad!

Britain's top dog breeders are up in arms over new EEC rules affecting the sale of dogs.

To bring us in line with other EEC countries, starting next week dogs are going to cost 40p per pound. And as a result a pedigree labrador puppy that would have cost up to £300 on Friday could be snapped up for as little as a pound on Monday morning.

DOGS

Britain's pet shop owners are divided over the new rules. Many feel quality dogs at lower prices will widen customer choice and boost dog sales, but others fear that larger dogs will simply be priced out of the market. "A large dog, weighing around 70 pounds, simply isn't going to sell", one pet shop owner told us yesterday. "Big dogs are simply going to be left on the shelf."

HOUNDS

Meanwhile, bureaucrats in Brussels are set to introduce standard pricing for other animals including cats, at 20p a bag. However, one new ruling has been unanimously welcomed by Britain's retailers and customers alike. Starting in the autumn

EXCLUSIVE

Euro chiefs sell Britain a pup

budgies will only be sold individually, in an attempt to cut down on budgie waste. It is estimated that in Britain last year over half of budgies were thrown away unused.

BOILERS

Actor Brian Blessed, a lifelong fan of budgies, heartily approves of the new measures. "People don't tend to realise that a budgie only keeps for six weeks. It breaks my heart to see so much budgie waste in Britain, and I hope that these new laws will encourage moderation and prevent hundreds of tons of budgies being needlessly thrown away by British households each year", said the former Sweeney star.

Some dogs being weighed yesterday

Britain's record on small colourful caged birds is amongst the worst in Western Europe. Up until 1978 under British law a 'small colourful perch based feathered animal residing in a cage' was not even classified as a pet, and their administration came under the auspices of the televising licensing authorities. However, thanks to the pioneering work of animal rights activists, among them Sandie Shaw, popular birds like budgies, parrots and canaries were granted pet status in April 1979 by the then Prime Minister James Callaghan.

RADIATORS

The late Les Dawson's wife was delighted with the new

Dead Les – liked budgies.

rules for budgies. "It was one of Les's last wishes that budgies should have more protection in terms of legislation," she said last night.

It's the Battle of the Animals

Yannick Noah's Ark

RECENTLY, the whole of London was captivated by Wilma, the bottle-nosed whale who lost her battle for life after swimming up the Thames into central London. Sadly that brave fight ended in death from dehydration after well-wishers lifted the plucky twenty-footer out of the water on a crane. But her sad end served to highlight the plight of whales all over the world; and ever since, these gentle giants of the seas have seldom been out of the news.

Britain has gone Whale Potty. In pubs and clubs across the land, the dwindling whale population has become the hot topic of debate. Whale charity Greenpeace has seen its numbers swell tenfold, and now three out of four Britons actively spend 10 hours per week saving whales.

But what about the only other member of the animal kingdom that begins with the letter 'W' - the humble **wasp?** These miniature marvels were buzzing around the planet 20 million years before whales were invented. They pollinate plants and kill garden pests, yet they are reviled by the British population. If a wasp were to drown in the Thames, it is unlikely that it would feature even on the local news.

So are we being unfair to the wasp? If there was only one place left on the ark when Noah got to the last but fourth letter of the alphabet, who do you think it should go to? Wasps or Whales?

We asked naturalists **Sir David Attenbor-ough** and **Bill Oddie** to look at all aspects of the lives of these two very different creatures. Here they plead the case for the mighty whale and the humble wasp being allowed onto the Ark. Tennis player **Yannick Noah** will listen to the cases for whales and wasps before deciding which species to allow aboard his namesake's fabled vessel.

NOT GETTING STUCK

TWO thirds of the earth's surface is covered by water, and you would think that with all the world's vast oceans to roam, the whale would rarely become stuck. But you'd be wrong, for these dim-witted denizens of the deep are constantly finding themselves floundering on beaches, gasping for water, or swimming in circles by Battersea Power Station looking for the way out to the ocean. And even when they are pushed back into the sea by beardy-weirdies, nine times out of ten these blubbery buffoons simply swim straight back onto the sand.

WASPS are truly the Houdinis of the animal world, being able to escape from all manner of sticky situations. Most wasps fly straight through cobwebs, and those that do get stuck don't need to struggle for long before they are off, depriving many a spider of his stripy dinner. Indeed, their only natural enemy is jam, which holds a fatal fascination for the wasp. Once one of his legs has become ensnared by the sticky preserve, there is no escape and a sweet, sugary death quickly follows. Bizarrely, however, they can escape from marmalade and lemon curd.

STINGING

IT'S A good start for these enormous gentle giants. Whales are known throughout the world f o r their placid nature, and there has never been a report of a whale stinging anyone. In fact, whales do not even possess a sting - and it's a good thing too. Because if they did, scientists estimate it would be the size of a telegraph pole sharpened at one end, and would contain over 200 gallons of poison. That's sufficient venom to sting the buttocks of 600,000 people, enough to fill the grounds of every Premiership League club in the land. Except Sunderland, which would only be a quarter full as usual.

ALTHOUGH only small, these buzzing insects pack a punch. Over half of their body weight is made up of their sting, and they like nothing more than to stick it in the bottoms of unsuspecting members of the public. And once it has delivered its painful payload, a wasp simply makes some more and flies off to find another hapless victim. Indeed, boffins estimate that over its lifetime, a wasp produces 200 gallons of poison. That's sufficient venom to sting the buttocks of 600,000 people, enough to fill the grounds of every Premiership League club in the land. Except Sunderland, which would only be a quarter full as usual.

MUSICAL ABILITY

WHALES are the only animals who have topped the music charts, their songs featuring on albums such as *Out of the Blue* by ELO and many more. Rightly have whales been dubbed 'The Pavarottis of the Oceans'. Just like the Italian tenor, they are grossly overweight and can sing very loudly, their mating songs carrying for a distance of three thousand miles. But as well as lady whales, their haunting strains are also popular with earth mother-types who like to listen to CDs of whalesong whilst giving birth in PVC paddling pools surrounded by Laura Ashley tealights.

UNLIKE whales, wasps have little or no musical talent. Rightly they have been dubbed 'The Paul Rutherford out of Frankie Goes to Hollywoods of the animal world'. In fact, the only noise wasps can make sounds a bit like somebody in a mental hospital playing a kazoo. Scientist put this lack of an ear for music down to one thing - their lack of ears! Over the years, wasps' ears have evolved into antennae, pairs of wobbling Deely-Boppers on the top of their heads which are completely useless for hearing.

PLUMAGE

BY and large, whales are not noted for their colourful livery. All whales are grey, except the blue whale, which is blue, and the black and white killer whale, which isn't a whale at all, more a cross between a shark and a dolphin. A whale's neutral colouring means that it blends into its blue and grey surroundings so it can sneak up behind an unsuspecting plankton.

THROUGHOUT nature, the colours black and yellow together signify danger. As humans, we instinctively know to keep our distance from animals displaying these colours, such as wasps, hornets and tigers. And these colours are used in everyday life to signify places and situations it may be best to keep our distance from, such as sources of radioactivity, biological toxins, and certain down-market supermarkets.

HEARTINESS OF APPETITE

WHALES have an undeserved reputation for being greedy. In the Bible, Jonah and his technicolour dreamcoat were eaten by a whale. In Herman Melville's novel, Gregory Peck and his ship were gobbled up by Moby Dick. And in *Pinocchio*, Monstro the whale made a meal of Gepetto, his cat and Jimminy Cricket. But in fact, despite their gargantuan size, whales are the daintiest of eaters. It is an irony of nature that these jumbo jet-sized monsters like nothing more than to nibble on a plankton - a tiny fish no bigger than a grain of

WASPS have voracious appetites, and unluckily for us, they eat the same things as we do. It is a sad fact that no British Bank Holiday picnic is complete without a swarm of the hungry marauders chasing everyone away before tucking into the contents of their hamper. So keen on stealing picnics are they, that wasps have rightly been called 'The Yogi Bears of the animal kingdom.' It is estimated that during its life, the average wasp scoffs over 300 Cadbury's mini-rolls, 250 rounds of jam sandwiches and 150 ice lollies, all washed down with 600 tins of fizzy pop.

SOCIAL STRUCTURE

WHALES have rightly been called 'The Tramps of the Ocean'. They don't have a house or any teeth, and they spend their time wandering round singing tuneless songs to themselves, occasionally under Battersea Bridge.

LIKE Humans, wasps live in highly complex social groups which would not function efficiently unless each performed their allotted tasks, such as tending larvae, looking for dropped lollipops and stinging people on the arse. Unlike humans, however, wasp society is quite backward since all members of the colony must kow-tow a single, self-appointed queen who sits about all day doing nothing.

SWATABILITY

WHALES remain one of the hardest members of the animal kingdom to swat with a newspaper. Thanks to a layer of blubber nearly 8 feet thick, these lengthy leviathans would not feel anything, even if beaten quite viciously. It has been calculated that in order to splat a blue whale, one would have to hit it with a *Daily Star* as big as a four football pitches, featuring a picture of Jordan on the cover with her tits as big as two St Paul's Cathedrals. This was found too impractical, and Japanese whale welfare research vessels now prefer to use explosive

WASPS are the perfect size for swatting and thanks to their complex compound eyes they make easy prey. A wasp's vision of the world is like a continuous special effect from an episode of *Top of the Pops* from the 1970s - a continuous rotating kaleidoscope of tiny images. Instead of seeing one newspaper bearing down on him a wasp sees a thousand, and in order to get away he must choose the right one to fly away from. And woe betide any wasp who makes the wrong judgement. He'll be smeared across the wall like so much mustard before he can say Jack Robinson.

BREATHING UNDERWATER

UNLESS you're standing outside a Marine Biology Convention and they are having a fire drill, if you ask the man in the street whether a whale can breathe under water, he'll say yes. But he'd be wrong. That's because unlike fish, whales don't breathe water, they breathe air through their noses. And like Daniella Westbrook, thay have one enormous nostril. But unlike her, it is on top of their head. Indeed, if you held a whale's head in a bucket of water for long enough, it would drown. However, scientists have calculated that it would take a man as big as Nelson's column to lift a whale, and he would need a bucket the size of the Albert Hall. With a handle on it as big as Sydney Harbour Bridge.

IF YOU hold a wasp's head under water, it will die. However, you'd better be prepared for a long wait, as its death will be due to old age, not drowning. That's because wasps do not breathe through their mouths, but through special snorkel-like tubes called spiracles which open out in their bottoms. This is why, when they are exerting themselves and panting for breath - for example whilst flying - they make a high-pitched farting noise. This unusual breathing technique is not without its risks; if a wasp with diahorrea were to attempt to wash its nipsy it would drown almost immediately.

ORIGAMI SKILLS

A WHALE'S brain is over 300 times bigger than that of a human, so you might think it could easily get to grips with the ancient Japanese art of paper folding. Flapping birds, jumping frogs and a man wearing a sampan hat in a rickshaw would all be well within the mental compass of these intellectual giants of the ocean. However, it is thought that the whale's enormous clumsy flippers, whilst excellent for swimming effortlessly through the oceans, would prove next to useless for creating crisp, diagonal folds in sheets of paper.

ANYONE who has ever knocked a wasps' nest from the roof of their shed with a stick, before covering it in petrol and setting it alight, will have marvelled at the intricate structure of this miracle of nature. With its thousands of hexagonal chambers and its intricate galleries, a wasps' nest is a paper structure infinitely more complex and beautiful than anything the greatest human origami artist could create. Unfortunately, since it is made from chewed-up wood pulp and wasp spit, it doesn't count, as the rules of this ancient oriental art forbid tearing the paper or using glue.

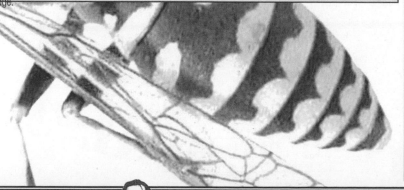

YANNICK NOAH'S VERDICT

*In my long tennis career, I have seen some close matches. Borg versus Connors in the Wimbledon Semi in 1982, Yvonne Goolagong verus Chris Evert in the 1979 Australian Open, Vitus Geralitus versus Billie Jean King in the 1981 Davis Cup men's final. But none has been such a nail-biter as this titanic encounter. And after looking carefully at all the evidence, it's game, set and match... to **WASPS**. Like John McEnroe, these insects are fiery and bad tempered, but determined to get to the top. And like him, they are unpopular with people, but possess an indefinable star quality. Whales, on the other hand, like Betty Stove, are blubbery buffoons that occasionally have their moment in the limelight, but are destined to be the perpetual runners-up of the Animals Beginning with W Stakes. Consequently, there is no place for them aboard my great-great-great-great-great grandfather's ark, and they will have to perish in the flood along with all the wicked fornicators of the earth.*

Next week: Giraffes versus Gnats

CARS OF THE STARS FACE TOUGH NEW TEST

By our Motoring Correspondent Mark Two-Cortina

Car star David Bowie did a 'Julie Andrews' - he fled to Switzerland to avoid the new MoT test.

Many top showbusiness celebrities will be facing their toughest test this year — when they take their cars in for an MoT.

For new Ministry of Transport rules mean that from January 1993 the MoT test will be tougher than ever before, with new regulations governing windscreens, bodywork and rear number plates.

CARS

And the new rules could cause celebrity chaos among stars whose cars are their only means of transport. For they may find that their vehicles are no longer up to the required standard.

KNACK

A spokesman for the Ministry of Transport yesterday confirmed that the stars could not expect an easy ride from garage mechanics. "There are a series of strict tests which all vehicles undergoing an MoT must pass, irregardless of who owns the vehicle", we were told.

BANGLES

All cars which are more than three years old must renew their MoT certificate annually. The current test fee is £24, but failure could cost the stars a fortune, with expensive repair bills and further test fees to pay before their vehicles can be declared road worthy.

EAR RINGS

We spoke to some of the stars whose cars will be undergoing the new tough test and asked them for their reactions to the new rules. Drummer, singer, actor and potential slap head **PHIL COLLINS** was

unable to hide his obvious concern about the new test. "I like old cars. The newest one I've got is a 7 year old BMW. This could mean a major headache for me", he told us. "I'm especially worried about my old Ford Popular, as it doesn't have a rear number plate light".

Miami Sound Machine **GLORIA ESTEFAN** was surprised to hear about the new MoT test. "I'd never heard of an MoT before", the Cuban born beauty told us. "I have three cars – a Mercedes, a Rolls Royce and a Brazilian sports car. But fortunately I won't be affected by the new regulations, as I live in America", she said.

DAVID BOWIE acted fast to avoid his twelve year old Volvo having to pass the new test. "My manager warned me about this new test, and so I moved my car to Switzerland where MoT testing is much less strict. Otherwise I'd have had to spend a fortune on bodywork repairs", said the silken voiced, 80-a-day

songster whose hits include 'The Laughing Policeman' and 'Spiders on Mars'.

Pop twins **MARK** and **LEW GOSS** out of Bros live life in the showbusiness fast lane, driving fast cars and having sex with women when they aren't appearing on Top Of The Pops. The new MoT test holds no fears for them. "We buy new

sports cars more or less every week, and have usually crashed them within a couple of days, so we never need to take an MoT test", they told us.

MoT for Ph.D

We decided to find out for ourselves exactly what sort of problems the new MoT test would pose for the stars, so we took our ten year old car along to a local garage for a test, and told the mechanic that it belonged to TV botanist **DAVID BELLAMY**. Here are the results of the examination.

CRACKED windscreen – unacceptable under the new rules.

SEATBELT is worn and damaged. Another dangerous fault.

DODGY steering. Excessive play in steering column.

BADLY aligned headlights.

EXHAUST emissions – too much carbon monoxide in fumes.

BRAKES fail on two counts: fluid leaking from system, and handbrake fails to hold.

VERDICT: FAILED

WORN TYRES below legal tread limit of 1.6mm.

Our mechanic was far from happy with David Bellamy's car. "There's no way I could pass a car in this condition. Mr Bellamy would be driving round in a death trap. The worn tyres and brakes certainly need seeing to immediately. There's a

good deal of work needs doing to this vehicle, and I'm afraid it's going to cost him. I'd be happy to give him a price, but I don't think I could make a start on it till next Wednesday. I could probably have it ready by the Friday.

MoT tests aren't 'popular' with pint sized singer Phil Collins. His ageing motor (right) has no rear number plate light.

Spectres spooked by Big 'C' scare

GHOSTS were turning white with fear last night after new evidence emerged suggesting a possible link between walking through walls and cancer.

Doctors fear that wall cavity insulation - the expanding foam substance used to insulate walls - could be hazardous when walked through. And that's bad news for phantoms at haunted houses all over Britain, many of whom could already have been exposed to serious risk.

Popular

In recent years wall cavity insulation has become an increasingly popular form of home improvement. Householders keen to shave a few pennies off their fuel bills pay extortionate amounts for a foam solution to be pumped into the hollow cavity within the external walls of their property. This then expands and hardens to form a heat retaining membrane within the wall. They hope.

Prefect

But researchers now believe there is strong evidence linking an increasing number of ghost cancers with the use of the foam. Statistics show that in the last twenty years cases of ghost cancer have almost trebled, with an estimated 2,000 ghosts dying each year from cancer related illness.

Head boy

Silus Hodgson was murdered by highwaymen in a field in 1730. His ghost haunted the spot where he died for over 200 years until a house was built on it in 1937. He then began haunting the house. All was well until 1982 when a new owner had the walls insulated. Twelve years later Mr Hodgson's ghost was looking at its stomach, which is see through, when it noticed a lump. Shortly afterwards doctors diagnosed an inoperable stomach tumour and Mr Hodgson's ghost was given only six months to live. Less than five months later it was dead again.

Mr Hodgson's ghost's ghost outside the house he had haunted.

Now haunting the spot where his ghost died, Mr Hodgson's ghost's ghost believes the people who manufacture and install the foam should be held responsible. "There is a clear case of negligence to be answered. No cautionary measures were taken - no warnings were put on the walls - and as a result my life and the lives of numerous other ghosts have been cut short", it told us.

We rang a solicitor who advertises free initial consultations in the local newspaper but his knowledge of the law as it applied to ghosts was flimsy to say the least. We then rang Mr Gill, a builder who did some work at our office five years ago, but he told us he didn't believe in ghosts. And he doesn't do wall cavity insulation either.

2nd JULY

The Black Hole Truth

IN THEIR quest to understand how the universe started, the world's top boffins at the CERN Large Hadron Collider may inadvertently be about to create a BLACK HOLE that will *TURN THE WORLD INSIDE OUT!*

But what exactly are black holes, and why are they so dangerous? We asked former D:Ream keyboards player-turned high energy particle physics egghead *PROFESSOR BRIAN COX.*

"A black hole is a body of matter that exerts a fantastic gravitational pull, sucking things into itself," said Cox, whose song *Things Can Only Get Better* reached the top of the charts in 1994. "The more things it sucks in, the more powerful it gets. It's a bit like the opposite of a Dyson vacuum cleaner."

"If the CERN accelerator makes a black hole there will be nothing that anybody can do to stop it. Turning the machine off will be impossible, as the power switch will be one of the first things to get sucked into the vortex," added the electro-synth king.

And Cox, who played the minimoog solo on *UR The Best Thing*, which made the Top Ten on its second re-release, had more bad news. "What's even more worrying is that scientists wouldn't even know they'd made a black hole in the first place, since they are so dense that not even light can escape from their gravitational fields. If we accidentally made one in the collider we wouldn't be able to see it," added the former D:Ream ivory tinkler whose hits dried up in the 90s.

SUCKING

WHEN THE UK gets sucked inside out by the CERN black hole, it will be a very different place to the one we know today.

The streets will be strewn with carpets and furniture as houses suddenly find their rooms on the outside. Hungry zoo animals will roam our towns when their cages turn back to front. Meanwhile our supermarket aisles will be a mess, as rice pudding, beans and noodle doodles get sucked out of their tins by the awesome cosmic power of the black hole and splashed all over the floor.

HOLE LOT OF TROUBLE: The nightmare that awaits the UK.

And people will fare no better. Men, women and children will be horrified to find their internal organs sucked onto the outside of their bodies, whilst their skin and faces are trapped helplessly on the inside.

It's a terrifying prospect for all of us, but it's even worse for the stars. With their glamorous lifestyles, matinee idol looks and multi-million pound endorsement deals, they have so much more to lose than the average man in the street.

How will they cope when they suddenly find themselves sucked inside out? We asked a selection of A-listers what they plan to do when the black hole strikes.

How the Stars will Co

KERRY KATONA

"I'M dreading the Large Hadron Collider experiment spiralling out of control," the former Atomic Kitten mime-artist confessed to us last night. "I'm sure Iceland wouldn't want me advertising their sausages with all my intestines on show! I'm already working on a new, no-holds-barred fly-on-the-wall documentary series called *Kerry and Mark - Inside Out*, all about me and my husband Mark going bankrupt with our organs situated on the outside of our bodies. It should be on MTV in the autumn."

ROWAN WILLIAMS

ARCHBISHOP of Canterbury Dr Rowan Williams thinks that being turned inside out will raise some interesting theological questions about the nature of being and his relationship with God. "The Lord is within all of us, so being turned inside out will in a sense bring His spirit out into the world to live amongst us," he said. "And in addition, we will, perhaps for the first time be able to truly see inside ourselves, which as a Christian, is very important, because only from looking for God within ourselves, can we really know His love. Of course, being inside out will mean that everything is reversed which will have enormous consequences. Shit will come out of our mouths and we will talk through our arses so, for me, it will be business as usual."

NOEL EDMONDS

"THE earth and everything on it turning inside out is a particularly horrifying prospect for me," said tidy bearded *Swapshop* star Noel Edmonds. "It means that the boxes on my show *Deal or No Deal* will have the amounts of money displayed on the outside, so all the scum contestants will win the jackpot prize of £250,000. Also, when the banker rings, I won't be able to hear the phone because my ears will be on the inside of my head." And things look equally bleak for Edmonds's most famous havoc-causing pink and yellow creation. A TV insider told us: "If Mr Blobby was turned inside out he would cease to be funny as people would see that inside he was just an out of work actor trying to make ends meet."

HELL!

e in Black Hole Britain

BILL ODDIE

THE world turning inside out is going to make life extra tough for curmudgeonly TV twitcher Bill Oddie. The scowling *SpringWatch* presenter told us: "I'm not going to be able to watch birds anymore, because when the black hole turns my birdwatching hide inside out, I'll be visible to them and they'll fly off. Also, with my internal organs on show, I'm likely to be mistaken for roadkill and attacked by carrion crows."

FERN BRITTON

A BLACK hole turning the world about face would spell disaster for daytime TV host and Ryvita ad queen Fern Britton. An ITV insider confessed: "Fern is very worried about the Large Hadron Collider. Being sucked inside out by a black hole would mean that her gastric band, the strap around the outside of her stomach which led to her new slim look, would be on the inside. Here it would have no effect and Fern would quickly find herself ballooning back to her old weight."

PHIL OAKEY

HUMAN League frontman Phil Oakey admitted that he was worried by the prospect of a black hole armageddon. "Me and the group being turned inside out is going to be very confusing for our fans. When the girls are on stage, no-one will know which is the dark-haired one and which is the blonde, since their trade mark barnets will be on the inside. It'll even be confusing for me, and I was married to one of them! I think it was the dark-haired one." And Phil voiced concern that this confusion could have an impact on the success of their latest album *Golden Hour of the Future*. "I don't think it's too much of an exaggeration to say that the effects of a black hole could cut sales of the new album in half," he told us. "We could be looking at single figures."

Have Your Say

THE HADRON COLLIDER is the most ambitious scientific project ever conceived, on a par with the Apollo Moonshot programme and the Human Genome Project. If it works, it will open up new horizons of scientific understanding.

But is it really worth the £4billion we have been forced to cough up for it? Could the money have been better spent keeping our old people warm this winter, building homes for tramps, or buying some more castles for her Majesty the Queen? We went on the streets to find out what *YOU* thought.

...THE scientists involved all seem very excited at the prospect of this experiment, but £4billion is a lot of money. Perhaps they should have paid for it themselves, raising money by doing a sponsored walk, organising a Bring and Buy sale or washing their neighbours' cars.
Howard Jones, electrician

...£4billion just to make a machine that creates a black hole to swallow up the earth? What a waste of money. They should do something useful for a change and invent a black hole that sucks in paedophiles.
H Timberarse, plumber

...WITH reference to the previous letter. Mr Timberarse fails to realise that a black hole which sucks in paedophiles will not solve the paedophile problem. They will simply go through a worm hole in the fabric of time and end up putting children in other dimensions at risk.
J Lumberjack, pharmacist

...THE technology needed to create this machine is awe-inspiring, and the only thing they can think to do with it is to smash protons into each other. Why do they always have to break things? Why can't they keep the protons nice?
Edna Balloons, housewife

...MY doctor tells me that at 46 stone I have to lose weight. But now some of the greatest minds in the world working on the Hadron Collider tell us that they don't really know what weight is. Well I might as well carry on eating cakes until they find out.
B Frampton, driving instructor

...WHY is it called the LARGE Hadron Collider? Why not just the Hadron Collider? I'll tell you why, because it's been built by men and they are obsessed with the size of their cocks. Hadron even sounds like HARD-ON. It's just cocks, cocks, cocks with men. Cocks, cocks, cocks, it is, really. Cocks, cocks, cocks, cocks, cocks!
Germaine Greer, harridan

...THEY needn't have spent all that money. If they wanted to get two protons to collide they should have tied one on the bumper of my wife's car and the other to the back wall of our garage. Ha! Ha! Ha! It's a cracker!
Frank Carson, comedian

...WITH reference to the previous letter. I'm afraid that Mr Carson has shown his failure to grasp the concept of how the Large Hadron Collider works. His wife would have to drive her car into the garage at just under the speed of light and with an incredible degree of accuracy in order for the experiment to be a success.
Hector Albany, physicist

...SCIENTISTS needn't worry about a black hole swallowing up the earth. There is already a massive, incredibly dense body threatening civilasation. My mother in law. Ha! Ha! Ha! It's the way I tell 'em.
Frank Carson, comedian

...HONESTLY, it's just cocks, cocks, cocks, cocks cocks, cocks, cocks!
Germaine Greer, harridan

227

Don't let our children look at little things!

A religious row may end in the closure of a secondary school in Yorkshire.

Parents at St. Oswald's school in Osset are keeping kids away as a result of a row over controversial Biology lessons.

BIOLOGY

The long running dispute centres around the use of microscopes and magnifying glasses in 'O' level Biology lessons. For 80 per cent of pupils at the school are Jehovah's Witnesses, and their parents claim it is against their religion to look at small things.

PHYSICS

Parents of Jehovah's Witness children have set up their own action committee to protest about the school's Biology syllabus, and they use quotes from the Bible to support their action.

WOODWORK

A spokesman for the group told us that the Bible was very clear on the issue of magnification. "It tells us 'And the Lord spoke unto Jehovah in a loud voice and said unto him, So shall ye not with a lens nor a glass look upon small things, nor things that are not big, niether shall ye peer unto those things closely. And yet shall ye only cast your eyes upon big things, and this shall ye do with only thine naked eye, eschewing lenses of all types, excepting only this: That ye may use a telescope, or binoculars, and these shall ye only use to look upon very big things – and things of great size – and even then shall ye not go above 40 times magnification. And this ye shall do unto the thousandth generation.'

DINNER TIME

Parents view Biology lessons in which pupils examine microscopic organisms a clear contravention of these religious beliefs. And extremists have even threatened to burn down the school's £200,000 science block if staff refuse to hand over the school's microscopes to be smashed.

DOUBLE GAMES

Parents of normal children are also keeping their kids

Get rid of microscopes plead bio-ban parents

away, claiming that Jehovah's Witnesses have gone too far. "It's becoming impossible for our kids to get a decent education under these conditions", one told us. "Last year they decided that spindles and pivot mechanisms were against their religion, and teachers were forced to get rid of all the scissors and pliers in the school."

St Oswald's school, Osset, scene of the parents 'Jobo' row.

Headmaster Percy Alderson fears gates at the school will close for good if the row is not settled soon. "Like any school we rely on dinner money to keep us going. And obviously with no kids coming in we aren't selling any dinners. I've already had to lay off woodwork teachers, and more will have to go next week.

PUPILS

Mr. Alderson concedes that St. Oswald's suffers unique problems as a result of the unusual religious mix of pupils. "We are currently talking to the examination boards to see if they can do any exams on elephants or whales, instead of little things – like insects, or bees legs – which seem to be at the heart of the problem."

RETINA

Osset has the highest population of Jehovah's Witnesses of any town in Britain, with the exception of Grimsby. In Osset 8 out of every 10% – more than 1 in 5 per head of the population – currently over 60,000 – are Jehovah's Witnesses. Only Swindon has more, with 10 for every person below the age of 30.

Jehovah Witnessing is not strictly speaking a religion. Founded by Henry Fonda in 1952, the Jehovah's Witnesses were originally a book club. Each month, members would receive 3 books on approval.

BOOKS

They were under no obligation to buy, providing unwanted books were returned within 14 days. New members would receive 3 free books, a hat and a rainproof mac.

NEIGHBOURS

After Fonda left in 1967 to found the Mormon Tabernacle Church – a High Street printing franchise – Jehovah's Witnesses began to move into pyramid selling, with members offering expensive soap and washing-up liquid to neighbours at pre-arranged house parties.

HOME AND AWAY

Over the years the 'Jobos' have become renowned for not believing in things. Here are just a few of the things in which they do not believe.

- *Blood Transfusions*
- *Newspapers*
- *Christmas trees*
- *Philips screws, coach bolts and ironmongery*
- *Strimers, Fly-mos, and Black & Decker Workmates*
- *The Loch Ness Monster*
- *Ghosts*

Jehovah's Witnesses are now number three in the world's top ten religions, with over 120 billion members in over 6,000 countries. Well known members include Prime Minister John Major, society hostess Cynthia Payne, footballer Paul Gasgoine and Geordie fop comedian Vic Reeves.

PRISONER CELL BLOCK 'H'

Membership of the Jehovah's Witnesses costs a pound, and you can join either by waiting for them to knock at your door at six o'clock on Sunday morning, or by filling in the form below and taking it to any main Post Office.

JV6 Application Form To Join The Jehovah's Witnesses

Name _____   Address _____

_____

_____ Hat size _____

Do you believe in anything?   YES ☐
(blood transfusion etc.)   NO ☐   (Tick 'No')

I enclose £1. Please send me 3 free books with no obligation to buy. I realise that if I return them within 14 days I will owe you nothing and my membership will be cancelled.

Signed _____   Post Office Stamp

SEX FOR SALE!

Sex is for sale on the streets of Britain, often for as little as £5.

Our investigation into widespread prostitution has revealed a dramatic increase in the amount of sex available for money. And as a result of rising mortgage rates and the poll tax, more and more hard-up housewives are going 'on the game' in order to make ends meet.

For many hard pressed households the only answer to rising bills is to turn on the red light and offer sexual services for cash. And if the money is right, customers can have anything they want.

SCANDAL

We sent investigators onto the streets to uncover the scandal of Britain's booming brothels. After hearing reports of suspicious goings on at a house in the quiet village of Little Barton, we rang up the owner, a Mrs. Wilson, and arranged to visit her that afternoon.

HAIRED

Mrs. Wilson, a small, grey-haired lady in her fifties met us at the door and we were ushered inside. "What exactly is it that you want?" she asked.

MESS

Dirty dishes littered a small coffee table and several newspapers were scattered around the floor. "I'm afraid

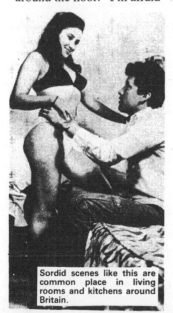

Sordid scenes like this are common place in living rooms and kitchens around Britain.

Investigators blow lid off housewives sex scandal

this place is a bit of a mess", she explained, "but I can offer you a cup of tea". But when our investigators mentioned sex, Mrs. Wilson's tone changed completely.

FULL SEX

"I think you'd better leave", she snapped. We then offered Mrs. Wilson £40 for full sex, at which point she began to ring the police. Our investigators made their excuses and left.

REVEALING

Neighbours in the picturesque village of Bradbury had no idea what the attractive, middle aged lady who had recently moved into No. 3 Church Cottages did for a living. Adverts placed in the local press said merely 'Piano Tuition' followed by a telephone number. We rang the number and arranged an 'appointment' for early the next morning. When our investigators called the door was answered by a woman dressed in a revealing blouse and slippers. She introduced herself simply as 'Mrs. Murray'.

FRUIT

Our man was led to a small room at the back of the house. Shelves were strewen with plates, cups and a bowl containing several large pieces of fruit. In the centre of the room was a large piano.

BEGINNER

"It's ten pounds an hour", our investigator was told. "Have you done it before, or are you just a beginner?" Mrs. Murray was quite happy to talk about her work. "I do it for the money", she admitted. "Shall we

Another sordid scene similar to the one shown below, left.

get started then?" At this point our investigator made his excuses and left.

PROSTITUTES

The brightly lit alleyways and escalators of London's Underground act as a vast sex supermarket where perverts and prostitutes meet and do business in a 24-hour roundabout merry-go-round of non-stop sex for sale.

It's behind doors similar to this one that sordid scenes (like those pictured above and bottom left) take place.

At Kings Cross tube station a young girl stood by a coin operated vending machine. "How much is it?" our investigator asked, waving a bunch of notes discreetly. "It only takes ten pence pieces, and I think it's jammed", the young girl replied.

POSITIONS

Girls like this, many as young as 15 and 16, can be found on every railway platform in Britain. "Can you change a fifty pence piece?" our investigator was asked. At this point he made his excuses and left.

Stall holders at a street market in Camden, North London, regularly offer sexual services over the counter. At one fruit stall a subtle menu of sex was on clear view. Round firm melons were displayed alongside ripe bananas. Nearby there was a box of plums.

GRAPEFRUIT

Our reporter approached the stall holder, a man in his early thirties, and pretended to be interested in a grapefruit. "Grapefruits are 26p each", he was told. Our man then pointed towards his trousers and asked whether sex was available.

ORAL

The stallholder disappeared briefly before returning with another man. Our investigator was then lead to nearby waste ground, punched in the face and kicked several times about the head and body, before he made his excuses and left.

FAMOUS PEOPLE BEING BORN

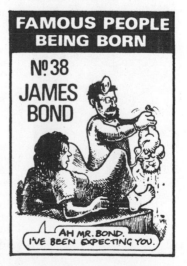

Nº 38 JAMES BOND

AH MR. BOND. I'VE BEEN EXPECTING YOU.

T.V. SEX SET TO BOOM

Eurocrats say British aren't getting enough

British telly viewers are bottom of the Euro TV sex league, according to officials in Brussels.

Our four major TV channels simply aren't putting enough sex on screen. And EEC bosses in Brussels are set to demand that BBC, ITV and Channel 4 chiefs increase our quota of 'X' certificate action in order to bring us in line with other member nations.

RANDY

Top in the television sex stakes come the randy Italians with their porny politicians and daytime strip shows. And coming a close second are the frisky frogs, who enjoy saucy soap ads and sizzling arty films.

But backward Britain has no nudity before nine, and only occasional glimpses of sex are seen in the odd play by Dennis Potter.

BREASTS

But new rules being drafted in Brussels are set to change all that, with a new fixed Euro-quota of breasts, bottoms and simulated sex having to be broadcast every single night. Exciting changes could be in store for Britain's telly addicts, possibly including:

- **TOPLESS TREATS**
  Among your favourite stars popping out each night could be the 'darling buds' of **Catherine Zeta Jones** as she frolics naked in the hay with screen husband Nicholas 'Rodney' Lyndhurst.

Darling Bud Catherine Zeta-Jones and screen hubby Pop 'Plonker' Larkin, alias actor David Jansen.

- **SAUCIER SOAPS**
  Raunchy new plots will be written to include full sex between the stars. **Bett Lynch** will be bonking on the bar at the Rovers Return, and in EastEnders, viewers will gasp as **Arthur Fowler** goes all the way with wife Pauline – on the kitchen table.

PEEL

Stroppy slap head **Sinead O'Connor** will be Top of The Pops in 1992 – she'll be the biggest hit of the year if she pops 'em out and shakes 'em about to the sound of her latest record. Meanwhile temperatures will soar as weather girls, including **Trish Williamson,** peel off to reveal their pleasant outlook to boggle eyed viewers, reading the weather in saucy half cupped leather bras and revealing elasticated thongs.

PIPS

And wives wouldn't mind hubby tuning in to the football quite so much if **Jimmy Hill** analysed games in the altogether! If Jimmy's strapping chin is anything to go by, the girls would have plenty to get excited about as the game got underway.

SUPREMES

The BBC's news and current affairs programmes, renowned throughout the world for being dull and boring, would be revolutionised. 'Newsnight' would become *Nudesnight* as **Jeremy Paxman** and company replaced sober suits with birthday suits.

KIEVS

And instead of reviewing the morning papers at the end of each show, the programme would finish with a huge on-screen sex session. Guests, including Cabinet Ministers and top MPs such as **Norman Lamont, Kenneth Baker, Neil**

PHOAAR! Fruity French sit back and enjoy saucy scenes like this every day, while.....

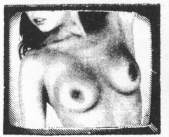

WAHEY! Spawney I ties get a load of this lot - 24 hour tits on every TV screen.

ZZZZZ! Meanwhile in boring Britain viewers regularly fall asleep to scenes like this.

Kinnock and **Clare Short,** will all be encouraged to fondle their own sexual organs and explore each other's bodies, while the show's hosts perform simulated homosexual acts on a large pink waterbed. Or something like that.

TIKKAS

People we spoke to on the streets yesterday gave a big thumbs up to plans for sexier TV. "I think it will be great seeing everybody's tits on the telly", Bob Smith of Fulchester told us.

However TV watchdogs including Mary Whitehouse need not worry. It won't be a question of 'anything goes'. Responsible TV chiefs will step in to draw the line. "It will still be a case of strictly no flapshots, popshots or panhandles", as one TV insider put it this afternoon.

EX~STING ~CT

Pop star Sting is hoping to get across a serious message to fans.

And the message is "If dinosaurs aren't extinct already – and they very probably are – at the rate man is going they definitely will be soon".

DINOSAURS

Conservationist Sting, who's hits include 'My Bed's Too Big' and 'Walking on the Moon', admits that in all likelyhood dinosaurs *are* already extinct. But he also conceeds that there is a very slim chance examples of the prehistoric giants may still survive in remote corners of the Earth, as yet untouched by man.

AFRICA

"There might still be some left on a lost plateau or secret valley somewhere in Africa, or something", the millionaire singer and actor told us.

STING - Worried about dinosaurs

But the Newcastle born former milkman fears that *if* there are still any dinosaurs left (even though he knows there probably aren't) – but *if* there are, then they *will* soon be extinct.

EXTINCT

"If we continue to waste the planet Earth's resources the way we are doing today, then any dinosaurs that *are* left, will be extinct – probably by the year 2000", said the singer.

UNLIKELY

"I know its really, *really* unlikely that there are any left, but all I'm saying is that *if* there are - and just imagine if there were – then there *definitely won't be,* soon", he added.

THE CURSE OF CORONATION STREET

Stars of Britain's longest running TV soap are living in fear after an uncanny catalogue of catastrophe has hit members of the cast of the top rated show.

"There's a curse on The Street and I just know that something terrible is going to happen", said one terrified star who refused to be named.

BIZARRE

Over recent years, in a series of bizarre coincidences, events that have taken place on screen have been uncannily echoed in real life.

REAR-END

* Actress Madge Hindle, alias Street star Renee Roberts, was written out of the script in an horrific car crash. *Only 18 months later Madge was involved in a rear-end shunt at a roundabout in Ilkley. Fortunately the actress was unhurt, but her car suffered £80's worth of damage.*

* Only months after script writers penned a scene in which Mavis Wilton's budgerigar died, a double tragedy struck. Actor Bill Waddington, alias Street busybody Percy Sugden, returned to his Osset home to find one of his tropical fish had died. Fifty years earlier, almost to the month, Roy Barraclough's pet dog had been run over by a car.

* Not long after newcomers Jim and Liz McDonald moved into the Street, actor Charles Lawson's real-life brother-in-law put his house up for sale.

SOAP

In 30 years of writing the hit soap script writers have often included story lines concerning marriage break-ups, death and baby dramas – which then come true off-screen.

SHAMPOO

Only years before the break-up of his screen marriage to actress Sue Barlow, actor Mike Baldwin, alias the Street's loveable cockney rogue, romeo rat Johnny Briggs, suffered a real-life argument with his wife Christine.

Although their marriage was in no danger, actor Mike slept on the settee two nights running.

T.V. soap stars fear for their lives

Soap star Jackie Ingram alias the Street's Sharon Taylor (real-life actress Jackie Baldwin) as the soap's loveable romeo rogue (inset) actor Ronnie Biggs, better known to viewers as rag trade rat Ken Baldwin alias on-screen actor Bill Roach yesterday

A scene in which his Street character died of a heart attack signalled a real-life drama for actor Tony Osoba. Real-life Tony, alias on-screen textile boss Pete Ingram, collapsed and died in romeo rat Johnny Baldwin's office.

Only days later, after a round of golf with friends, actor Tony suffered mild chest pains.

SHOWER GEL

Luckily it was a false alarm. Doctors confirmed that Tony was suffering from indigestion as a result of eating his breakfast too quickly. But Tony is still left trembling with fear every time he gets heartburn or flatulence.

BUBBLE BATH

Terrified stars are now pleading with Street bosses to have scripts changed rather than tempting fate. Elizabeth Dawn, alias The Street's Vera Duckworth, successfully begged the show's producers to have a scene in which she visited the opticians written out of the script. For the actress feared she might go blind in real life if the scene were broadcast.

Ironically that scene was replaced by one in which the Duckworth's house was flooded. The next day Dawn, 53, found a tap dripping in her Cheshire house. Not even her local plumber was able to explain the mysterious coincidence.

CURSE

However, the Curse of Coronation Street has not always been bad news for the stars. Some of the uncanny coincidences have happy consequences. For example when actress Barbara Knox, alias Street Star Rita Fairclough, won a three minute trolley dash in the soap's Bettabuys supermarket.

PERIOD

Incredibly, the event was mirrored in real life. *For the very next day actress Barbara won a three minute trolley dash in her local super-market.*

RAG WEEK

And just like her generous on-screen character, big-hearted Barbara donated all her winnings – over £30's worth of groceries – to charity.

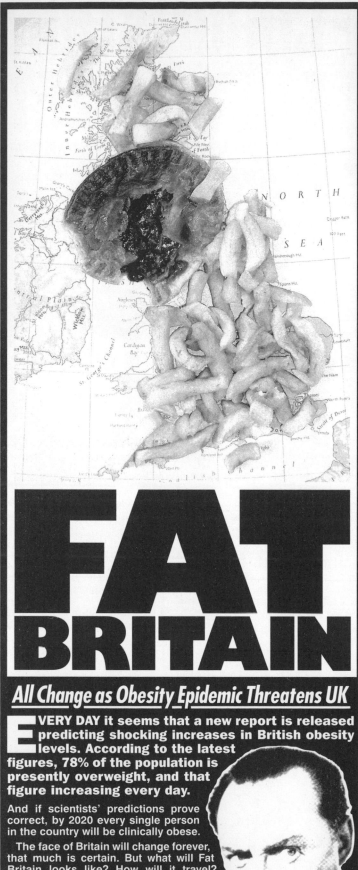

FAT BRITAIN

All Change as Obesity Epidemic Threatens UK

EVERY DAY it seems that a new report is released predicting shocking increases in British obesity levels. According to the latest figures, 78% of the population is presently overweight, and that figure increasing every day.

And if scientists' predictions prove correct, by 2020 every single person in the country will be clinically obese.

The face of Britain will change forever, that much is certain. But what will Fat Britain looks like? How will it travel? What will it do to relax?

Here, Professor Kidderminster Chocolate, head of the Obesity Forecasting Department at Lampeter University takes a look at every aspect of daily life in a future where EVERY Briton is a fatty-bum-bum...

TELEVISION

EVERY ASPECT of the television industry will have to change to accomodate Britain's broader-beamed viewers. Televisions themselves will all have to be made wider as fat people take up more screen room. And remote controls will have to be redesigned with larger, more widely-spaced buttons for the viewers' sausage-like fingers. Not only that, but programmes will be broadcast in the upper half of the screen only, as the bottom half will be permanently obscured by viewers' bloated stomachs.

The other side of the camera will also see great changes. Weather reports will take longer, as a thirty-stone Sian Lloyd lumbers around in front of her map, wheezing and fighting for breath. And commercial breaks will have to be much longer to allow

TELLY TUBBY: Sian Lloyd as she will certainly look in 13 years.

gargantuan viewers to waddle off to the lavatory, where they will take several minutes to locate their genitals amongst all their rolls of sweating flesh.

The sexy programmes we love today, such as *The Tudors*, *Diary of a Call Girl*, *Fanny Hill* and *Das Crazy Sex Show*, will all have to be banned as their broadcast would lead to excited viewers dropping dead in their millions.

This would be good news for viewers, as dull, uninteresting programmes would be a lot cheaper to make and consequently the licence fee would drop from £140 to around £60. Now for the bad news; the increased expense of feeding morbidly obese BBC employees would see the fee increase to more than four times its present level.

TRAVEL

THE STANDARD four-seater saloon car of today would have to be twice as wide as it is now in order to accommodate the morbidly-obese family of the future sitting side by side. This would mean that every road in the country will have to be widened - an impossible task. Instead, the car as we know it will have to be made twice as LONG, with the passengers sitting in a line behind the driver, like in a bobsleigh.

Up at the front, the dashboard will have to be three feet deep, to accommodate all the driver's pies, sweets, burgers and cakes. Meanwhile, the traditional steering wheel will be useless, as it will become wedged in the bulging gut of the 2020 motorist. Instead, he will wear a metal collander on his head, which will transmit his brainwaves to the front wheels down some curly wires. The salad-dodging driver of the future will merely have to think which way he wants to go and the wheels will automatically point in the right direction.

In tomorrow's fat world of the future, the double decker buses we know today will of become a thing of the past, as the narrow staircases up to their top decks will be impassable for their elephantine passengers. And the drivers will be wedged solid into their cramped cabs, too fat to get out at the depot. They will be forced to live their whole lives like veal cows, trapped on board their buses, wallowing miserably in a foetid soup of their own filth and ordure. It's not a pleasant prospect for their passengers, so bus fares will be forced to come down.

With the roads clogged up with double-length cars and shit-stinking buses, train travel will become the best way of getting about. But even the railways will not be immune from the consequences of Fat Britain. In order to meet the demands of the obese travelling public, on a typical twelve-carriage intercity express, eleven of the carriages will have to be buffet cars serving teas, coffees, sandwiches, hot bacon and tomato rolls, and a wide selection of snacks and crisps. However, some things won't change. The trains of 2020 will still run out of sandwiches within ten minutes of leaving the station.

But the biggest change of all will take place in the air. Instead of the relaxed, spacious experience that flying is for us today, going abroad on a plane in the year 2020 will be cramped, stuffy and uncomfortable, with barely enough room for us to stretch our legs.

FILM & POP MUSIC

EVERYONE loves going to the pictures, but in Fat Britain we're set to see a whole load of changes to our movie-going experience. The first difference we'll notice is at the food counter. The obscene, bucket-sized 'large' popcorn of today will become the piddly 'extra-small' portion of 2020. Ask for a 'large' popcorn in the future, and it will be delivered to your seat by an usherette in a fork-lift truck. But that's not all she'll be delivering. With viewers' fat cheeks swollen up like risen dough, squeezing their piggy little eyes shut, staff will hand out special screwjacks so film fans can push their podgy chops down far enough for them to be able to see the screen.

DIE LARD: How Bruce Willis will look in 2020.

Pop bands of the fat future will bear little resemblance to the ones we know today. Guitars, pianos and any instruments with fiddly little keys will prove impossible to play with the clumsy, chubby fingers our musicians will then have. As their waistlines swell, present day skinny virtuosos such as Eric Clapton, Jools Holland and Peter Hook will be forced to retrain as alpine horn, kazoo or kettle drum players.

SPORT

BUTTERBALL: How Peter Crouch will appear in the future.

IN THE fat Britain of the future, Premiership football games will be much shorter than they are today, as thirty-stone players will find themselves red-faced, puffing and blowing within ten minutes of kick-off. Crowd numbers will be slashed in half, as each fatty footie fan will require two seats - one for each buttock. With ticket revenues slashed in two, players' wages will drop. And further down the table, the effects could be be even more pronounced, with non-league grounds left empty as overweight supporters find themselves unable to squeeze through the turnstiles.

The London Marathon will still take place, now lasting over a week instead of its present three or four hours. And its route will be drastically different too. Structural engineers fear that the massive vibrations set up by thousands of morbidly obese runners thundering across it like a herd of baby elephants would smash Tower Bridge like matchwood.

The traditional cricket match that is such a part of our national identity is set to be transformed beyond recognition. In the Test Matches of 2020, big-boned bowlers will make their run-ups on battery-powered mobility scooters, whilst corpulent batsmen will slouch on settees in front of their wickets, clutching a bat in one hand and a plate of sausage rolls in the other. Tea and lunch intervals would last so long that there would only be enough time for a couple of overs a day.

And the High Octane world of Grand Prix racing will not be immune either. The F1 cars of today, with their 1000 horsepower engines, would barely have enough grunt to get off the starting line with a fifty-stone Lewis Hamilton at the wheel. And the McLaren driver's trademark yellow helmet will have to be redesigned from scratch, needing seven - or even eight - chinstraps to hold it on.

Darts would be the only sport left unaffected.

WORK

THE WORLD of work will be unrecognisable in a fat Britain of the future. Rush hours will not exist as we know them today, as practically the entire working population will be too fat to get out of their houses. Traditionally, in such circumstances, the Fire Brigade are called to remove a window and lift the victim out on a tarpaulin. But in 2020, the firefighters will be unable to assist, as they will be too fat themselves to slide down the pole and get out of the station.

Of those that do manage to get to work, the vast majority will find themselves working in chippies, cake shops and Greggs bakery. However, the shop counters will have to be moved forwards several feet to make enough space for the vast assistants. This means that the space for customers in the shop will be smaller. However, this will all be irrelevant as none of the customers will be able to fit through the doors.

In factories, the working day will be completely different. In order to meet the workforce's insatiable demand for food, tea breaks will be up to two hours long, and the traditional cake trolley will be replaced by a dumper truck filled with jam doughnuts. Coupled with a four-hour break for lunch, this means that production lines will only be running for half an hour a day, during which time the workers will probably take the opportunity to go for a big Elvis-sized shit on specially strengthened jumbo toilets.

Blackpool Gears Up For Lardarse Invasion

"Bring it on!" ~ says Lord Mayor *(again)*

ONE PLACE that is gearing up in readiness for a fat future is the Lancashire seaside resort of Blackpool. And according to Lord Mayor, Mike Taylor, by 2020 the town will be more than able to cope with any number of morbidly obese funseekers. *"Bring it on!"* he told reporters.

Launching their new policy document 'Blackpool - Where FAT spells FUN,' Blackpool County Council set out a twenty point plan to ensure that the town will be more than ready for the 2020 corpulence epidemic. Amongst measures being proposed are:

• *THE famous 360 foot tower being reduced in height to just 4 foot to allow fatties to enjoy the panoramic view from the top without wearing themselves out climbing up the stairs.*

• *INSTALLATION of cranes along the Golden Mile to winch the 50-stone sun-seekers on and off the beach.*

• *THOSE things you stick your head through to have your photograph taken will be re-painted to depict extremely thin people in bathing costumes in order to create a humorous contrast.*

• *COIN trays on all fruit machines in arcades along the Golden Mile will be widened to allow lucky winners to get at their winnings with their roly-poly digits.*

• *THE wooden and canvas deckchairs on the beach will be phased out and replaced with extra load-bearing designs made from railway sleepers and industrial conveyor belting.*

• *THE famous illuminations will be lowered down their lamp post so that blubbery tourists will not need to strain themselves lifting their chins to see them.*

• *THE world famous Flower Clock will be dug up and replaced with a Waddle-Through Pie Shop.*

"Blackpool has kept going strong through two world wars and global warming," Mr. Taylor told reporters. "We're certainly not going to let a few excess pounds of adipose tissue spoil the fun along the Golden Mile," he added.

Tipton-khamen!

West Midlands Egyptian burial plan for 'One and Only' Hawkes

Ambitious plans to bury Chesney Hawkes in a latter day 'Valley Of The Kings' have sparked controversy among residents and planners in the town of Tipton.

Hawkes (right) whose pop career ended so young.

The Tipton Gateway Trust, an independent partnership of local businesses, have put forward plans for a giant tomb and pyramid to be constructed on derelict industrial land adjacent to the A457. The £50 million project will involve tunnelling a series of inter linking catacombs beneath the ground and building a spectacular stone pyramid 800 feet tall.

Chamber

In a giant burial chamber directly beneath the pyramid the body of the boy pop star Chesney Hawkes will be laid to rest in an ornate sarcophagus, surrounded by his most treasured possessions, including a gold disc, a mountain bike and his CD collection. The name of his hit single 'The One And Only' is to be carved on the wall in hieroglyphics, and the chamber will be decorated with a hand painted wall frieze depicting his appearances on Top Of The Pops.

Pop

There are further plans to bury more pop stars nearby on the same site as and when they become available, and for a landscaping scheme to create a genuine 'Valley of the Kings'. It is estimated that tourists visiting the tombs could bring an extra £300 million into the West Midlands economy every year.

Classical

Project coordinator Hugo Guthrie got the inspiration for his scheme during a holiday in Egypt. "My wife and I were visiting the pyramids and were impressed by the sheer volume of visitors they attract from all over the world. It dawned on me that a pyramid would be

an ideal attraction for Tipton, and an economic boost for the whole of the Metropolitan Borough of Dudley".

Jazz

It is hoped to have the tomb built, buried under tons of sand and then rediscovered and opened to the public in time for the millennium. But the ambitious scheme has already faced criticism. Opponents say the pyramid will create car parking problems for local residents and they claim that vital wildlife habitat will be destroyed.
"Mice and pigeons regularly use that land for recreational purposes", one objector told told us.

An artists impression of the Tipton pyramids development due to open in the year 2000

Boy pop king set for Midlands tomb

But Mr Guthrie remains optimistic. "Too often Tipton has been caught lacking in ambition. Now is the time to change that. The Tipton Valley of the Kings will be one of the wonders of the West Midlands. It will put us firmly back on the tourism map".

Porn

The outcome of applications for funding to the National Lottery, the Millennium Commission and English Heritage are not due for several months. Meanwhile Chesney Hawkes was last night unavailable for comment.

Joe's 90

FORMER TV star Joe 90 celebrated his 90th birthday yesterday with a quiet party at a retirement home in Filey, North Yorkshire.
Joe, who during the sixties starred in his own television show, has not been able to walk unassisted since his leg strings snapped in the late eighties. But staff at the Bay View retirement home say that he's in good spirits, and was able to enjoy a glass of champagne to celebrate his birthday.

Then and now. 90 in 68 (above) and (below) in 98 90 at 90 yesterday.

Scan

90's wife Rhapsody Angel out of Captain Scarlet died in 1992 following a long bout of woodworm. Joe was forced to sell the giant food mixer in which he sat at the beginning of his TV show to pay for her funeral. The couple had no children.

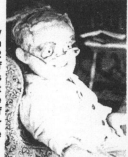

20 THINGS YOU NEVER KNEW ABOUT FRANCE

With the opening of the Channel Tunnel, due for completion in 1992, we'll soon be seeing a lot more of the French.

Dislike them or hate them, there'll be no avoiding them. So how much do we really know about our excitable next door neighbours? Here's 20 garlic stinking facts you probably never knew about the frogs.

1 For many years Britain and France were at war. It was them who started it, in 1066. But we won when Nelson defeated the French navy at the Battle of Trafalgar Square.

2 Like the English, the French have a proud naval history. Their greatest victory came in 1985 when they sank the unarmed protest ship 'Greenpeace Warrior', which was docked at the time, in Auckland, New Zealand.

3 The world famous French Foreign Legion is a crack military regiment which, coincidentally, has very few French soldiers in it.

4 Unlike the British Legion which is a string of social clubs where overweight men play darts, bingo and drink beer.

5 In France 8 out of every 10 schoolchildren can speak reasonable English by the age of 16, whilst in Britain only 1 in every hundred school leavers can speak any French at all. That's because French is a stupid language, with three different words for 'the', depending whether you're a man or a woman, or how many of you there are, or something.

6 The French are renowned for being the world's greatest lovers. Yet their sexeist man is Charles Aznavour, a 92 year-old, balding, midget cabaret singer.

Aznavour - 2 foot tall and bald.

7 If your postman sticks a 'French letter' through your letterbox, don't rip it open to see what's inside. Because a French letter is another word for a contraceptive sheath.

8 French films on Channel 4, with subtitles, are usually a good bet if you want to see a few birds getting their kits off on the telly.

9 France is famous for its thousands of vineyards where the world's finest wines are produced.

10 And it's thousands of pavement cafes and bistros where most of it gets pissed up the wall by the locals, who drink like fish.

11 France's greatest tourist attraction is the Eiffel Tower, a giant electricity pylon with a gift shop at the top.

12 The Statue of Liberty was a gift from the people of France to the United States of America to celebrate 100 years of French independence. And what did they get in return? Euro-Disney.

13 The famous French dairy farmer Louis Pasteur made one of the most significant scientific advances of the 20th century when he invented Pasteurised milk, in small plastic catering size cartons.

14 France is famous for it's many great painters, among them Toulouse Lautrec, a three foot tall sex pervert who hung around the Moulin Rouge, a high class strip joint in Paris, painting birds with long legs.

15 The most famous of the French Impressionists is Marcel Marceaux, who doesn't do people, he does the wind, and glass boxes, and Kodak adverts.

A farmhouse yesterday

16 Farmhouse holidays in the South of France are currently very popular with the British middle classes. But instead of paying for bed and breakfast accommodation, they just buy the fucking farmhouse instead.

17 If it's that bloody nice in France why don't you stay there, and don't come back. You smarmy bastards.

18 If you ask a Parisian for directions to the 'Underground', you'll probably be shown into a small street cafe, shoved in a barrel and smuggled to Switzerland. That's because in France the 'underground' is a secret organisation for putting people in barrels and smuggling them to Switzerland.

19 The French were founder members of EEC and, in accordance with EEC agricultural policy, Britain buys boat loads of their rotten onions, while French lorry drivers set fire to lorry loads of British sheep.

20 Finally, never share a shower with a Frenchman. For among their more peculiar toilet habits, the French occasionally squat and spend tuppence in the cubicle, and then poke it down the plughole ... with their toes.

21 And there's another thing. France have never won the World Cup, which we did, in 1966.

BLIND *TRUNK!*

Health fears grow as Ginger whinger binger Evans drinks fucking elephant under table

BOOZY DJ Chris Evans last night hit the town on yet another 18-hour bender. But this was to be a particularly MAMMOTH session as he was accompanied by his latest drinking pal - *an African elephant!*

The millionaire DJ was forced to team up with the 8-ton party animal after his regular boozing partners, top chef Aldo Zilli and celebrity fat cunt Danny Baker, proved unable to keep pace with his drinking.

marathon

After visiting over 320 bars on their marathon pub crawl, Evans and the elephant, named Sultan by his keepers at London Zoo, took in Nelly's, London's most fashionable bun and peanut restaurant.

topic

Fellow diner Dinah Fellows said: "They were very drunk when they arrived. Evans was matching the elephant drink for drink. There was a lot of larking about and trumpeting.

mars

"Evans ordered 100 bottles of vintage port and drank them straight down. When the buns and peanuts arrived, there wasn't enough room to set them down because of all the empties, so they had to move to a larger table. I've never seen anything like it."

OUT - drinking pal chef Aldo

OUT - boozing buddy Baker
IN - elephant Sultan

They eventually ended up at West End nightclub *Stringfellows*, where they quaffed thousand-pound bottles of champagne until the early hours.

bounty

During their mind-boggling booze binge, the pair sank an incredible £4 million worth of drink, including:

• 12,500 pints of lager
• 800 bottles of Laurent-Perrier pink champagne
• 100 barrels of best Bitter
• 2 vats of Chateau Lafitte 1946 red wine
• 6 tin baths of Vodka and a can of Red Bull
• 2 pints of Guinness
• 500 Formula-One size bottles of Dom Perignon.

According to witnesses, by 4am Sultan had had enough, but Evans was still going strong.

endeavour

The former *Big Breakfast* presenter made a call on his mobile, and twenty minutes later a carpenter turned up with a load of timber.

victory

Joiner Jack Churchill told us: "Evans stuck a million pounds in my top pocket, punched me in the face and told me to build a fifteen foot high table. I built the table and he drank the elephant under it. I've never seen anything like it."

"Get help,

"Take my advice, Chris, or you'll soon be knocking on Evans door." That was the stark warning last night issued by the country's top alcohol expert.

Terry London, one of Britain's most experienced pub landlords said: "Doctors recommend a safe drinking limit of about 28 units per week. Chris is currently downing upwards of half a million units in a single session. Sooner or later, that sort of boozing may start to take a toll on his health."

But London, who has helped pop-stars, celebrities and professional footballers with their drinking during his 34-year career, says it is not too late to act. He believes this simple 3-step programme could help Evans handle his drink.

THIS IS MY NEW GIRLFRIEND - A GREAT WHITE SHARK STEAK

CAREFUL. SHE'S A BIT OF A MANEATER

"I've been there ...and it's HELL!"

One DJ's Battle to Beat the Booze -in his own words

Evans - failed to turn up for work after necking 240 gallons of petrol.

Partygoers later helped sozzled tusker Sultan to a taxi. But for *Don't Forget Your Toothbrush* host Evans, the night was still young.

beagle

Amazed onlookers watched as the *TFI Friday* frontman made his way to a nearby all-night garage, where he lay on the forecourt necking £4-a-gallon Superplus unleaded petrol straight from the pump until 8 am.

Sultan turned up for the zoo two hours late the next morning, grey-faced and nursing a **jumbo-sized** hangover. Meanwhile, when Evans failed to show for his daily breakfast programme on Virgin radio, bosses blamed illness.

bosprey

However, the ginger star, looking frail and wearing twenty pairs of sunglasses, was later spotted leaving his local Waitrose store at the wheel of a dumper truck loaded with Alka Seltzer.

for Evans sake" - says booze expert

CHRIS'S THREE STEPS TO EVANS

Step 1 Alcohol is a poison, and when it enters the blood through the stomach wall it does untold harm. If you're knocking back that much booze, give yourself a chance. Drink a pint of milk and have have some toast before you go out. And keep your stomach lined with a bag of crisps every ten pints throughout the binge.

Step 2 Keep a track of what you're drinking, and if possible use the 1-2-3 system. One of your first drink, two of your second, three of your third and so on. For example: 1 pint of lager, followed by 2 pints of bitter, followed by 3 bottles of pernod, and so on. And always finish the booze session with a pint of water.

Step 3 The morning after the night before. It's an old wives' tale that what makes you bad makes you better, but it's true nonetheless. As soon as you wake up, examine your pillow and eyebrows for congealed evidence of what you had the night before, and then hit the off licence for the hair of the dog.

SOMEONE who knows only too well what Evans is going through is fellow DJ *Barry Malpas,* and last night he warned the ginger millionaire: "Don't make the same mistakes I made."
In 1979, Malpas had the radio world at his feet, but a battle with the bottle nearly lost him his livelihood, family and health.

> *I had it all. I was a young man, just 38, the traffic reporter on BBC Radio Solent's drivetime show. It was a job most people could only dream about, and I thought it would last forever.*

weekend

Drinking didn't seem to be a problem in the early days. I never gave it a second thought. Sure, I'd have a couple of pints of a weekend, like most people. But I wasn't like most people any more. I was the voice of traffic congestion in the Southampton area, mixing with celebrities on a daily basis, and it was getting harder and harder to keep my feet on the ground.

swift

It's easy to say no to a swift half at lunchtime with a friend, isn't it. But let me tell you, it's a little more difficult to turn down the offer when the person asking you out is Mike d'Abo out of Manfred Mann's Earth Band, or ex-Anglia TV anchorman Paul Lavers. Before I knew it, I was caught up on the celebrity merry-go-round. My swift half on the occasional lunchtime was now every Friday, and my two or three pints at the weekend had become three or four. Without me realising it, my drinking was spiralling out of control. Like Evans, I thought I could handle it. I played hard, but I worked hard too.

But Barry didn't realise he was burning the candle at both ends, and something had to snap. The hammerblow finally fell one Friday afternoon after a heavy drinking session with TV star Mr Bennett.

> *He'd been on the mid-morning show plugging a new series of Take Hart, and we went out for a swift half. Before I knew it, that swift half had become an all-lunchtime booze binge and I'd necked two halves. It was time to go on air, and my head was swimming.*

swallow

I sat in the studio to do my report and the red light came on. I remember a lorry had jack-knifed on the A336 at Netley Marsh, and there were slight delays for people heading for the Cowes Ferry. Routine stuff; no problem for a professional broadcaster on top of his game. But I'd had the equivalent of a pint of lager less than five hours

before, and the inevitable happened. My relentless partying had finally caught up with me. I announced: 'Slight delays for people heading for the Fowes Kerry.' I immediately corrected myself, but the damage was done.

spit

I realised what everyone around me had known for a long time. My drinking was out of control and if I didn't do something it was going to cost me my job, my family and my health.

Six weeks later, Malpas was made redundant. The official reason given was that BBC Radio Solent had pooled its Travel news resources with BBC Radio Portsmouth, but inside, Barry knew the real reason. The BBC couldn't afford to have a beer-swilling hellraiser on the air. He'd become a liability. His booze madness had cost him his job.

walkin'

With the help of his wife Pat, Barry struggled to pull his life back together, and after a twenty-two year spell as security man at Everything's a Pound in Hedge End, he finally managed to break back into showbusiness.

> *I confronted my booze demons, and eventually got my drinking under control. I'm back on the radio again and life's never been sweeter. I've been given a second chance and I'm not going to throw it all away this time, I can tell you.*

And Barry has this message for his troubled fellow jock Chris Evans:

"If I can do it, you can. I only hope Billie can show the patience and understanding that my Pat showed in bringing me through this nightmare."

• *Barry Malpas* presents 'Your Favourite Songs' every night from 2-6 am on Romsey Hospice's closed-loop radio system.

WINNING FORMULA

THE WORLD OF FORMULA 1 was rocked to its foundations last night after allegations that the Ferrari team CHEATED in order to secure this year's constructors' title. The Italian team faced disqualification from the Malaysian Grand Prix after after pieces of wood on the side of its cars breached stringent technical specifications, but the latest allegations, if proved correct, could mean that far more serious rule-breaking has been commonplace throughout the season.

It's the pits as Ferrari race aces bend the rules

According to Ferrari insider Ray Savage, team drivers Michael Schumacher and Eddie Irvine have regularly employed underhand tactics, including;

SHOCKING SPORTS EXCLUSIVE!

- Setting up *fake diversions*
- Spreading *quick-drying glue* on the track.
- Running into back markers, and *cutting them down the middle with an enormous circular saw,* which comes out of the Ferrari nose cone.

witnessed

Michael Schumacher smiles and touches his ear, yesterday

ONE shocking instance of cheating, which Savage claims to have witnessed, happened at this year's British Grand Prix: "Irvine was trying to overtake Hakkinen, but the flying Finn was not letting him past.

A big red car going very fast - yesterday

"Suddenly, when no-one was looking, Irvine must have pressed a secret button on his steering wheel. The car rose up on ten foot long extending legs and drove right over the top of the McLaren. It was a disgrace."

judged

Loyalty to his own team prevented Savage telling race marshals what he had seen, but after another incident later in the same race, Ray felt that he could hold his tongue no longer. "Schumacher rounded the first corner with a hefty lead over Coulthard. Then, quick as a flash, he pulled up, jumped out of his car and painted a false tunnel onto the side of a wall, and a length of false road leading up to it.

juried

"Then he put up a shortcut sign, pointing at the 'tunnel' and waited behind a bush. Coulthard and the rest of the pack were heading round the corner by now, and when they saw the shortcut, they naturally went straight for it. However, to Schumacher's amazement, they simply drove into the tunnel as if it was real, leaving the German in last place.

barristered

"Quickly, he jumped into his car, and set off at full speed in pursuit, only to crash immediately into the painted wall. Staggering out of the wreckage, Schumacher was then run over by a steam-roller which came out of the tunnel. That's how he broke his legs - and it served him right. That was me and Ferrari finished as far as I was concerned."

Father of eight Savage was later forcibly ejected from the Silverstone circuit, after being seen by security guards entering through a hole in the fence, and attempting to sell bootleg Michael Schumacher hats to racegoers.

IRVINE 'MADE LOVE LIKE A RABBIT' - Model

A FORMER model who once got banged off of Formula 1 race ace Eddie Irvine, claimed last night that he 'made love like a rabbit.'

Irvine at home yesterday

"It was amazing," said 49-year-old Bridie McO'Dougle, from Belfast. "We met in a hotel bar, and he took me back to his room. He made love to me 150 times that night. He was insatiable. He would hop about on the floor, sniffing at a load of sawdust.

burst

"Then he'd jump onto my back for a frantic five second burst of love-making, before hopping off to nibble at some vegetable peelings in the corner of the room. It was the most incredible sex I've ever experienced."

grumbling

McO'Dougle is presently undergoing DNA tests in an attempt to prove that the 28-year-old racing driver is the father of the twelve, hairless blind babies to which she gave birth three weeks after their night of passion.

Wing Span-Dabidozy!

Krankies' terror as hawk snatches schoolboy Jeanette

LITTLE Jimmy Krankie was recovering in the fake children's ward of a Ben Nevis hospital last night ~ *after being snatched by a hawk.*

The four-foot schoolboy granny was out walking with his normal-sized husband Ian when the bird of prey swooped unexpectedly.

hawk

Ian told reporters: "I heard a scream and turned to see a hawk gripping Jeanette's shoulders in its talons. Before I knew what was happening, it lifted him off the ground and carried her away."

wolf

He continued: "I threw some rocks to try to get the bird to drop her, but it was already too high. As it flew into the distance I could hear Jimmy's cries getting fainter and fainter. I could see his little legs kicking and I just panicked."

Unable to contact the emergency services Ian, 56, gave chase and eventually spotted his wife waving frantically from a nest 300 feet up a cliff.

hunter

Although relieved to see him still alive, he knew he had to act quickly before the hawk came back and ate her, so he started climbing.

He said: "I've done live TV, the Royal Variety Performance and Friday night at the Glasgow Empire but let me tell you, I've never been so scared as I was climbing that rock face. It seemed to take an age, but eventually I clambered up onto the ledge where the nest was."

And he was just in time, for the hawk chose this moment to return. Ian looked on horrified as the bird pecked with its razor-sharp beak at

Ian, and cheeky schoolboy wife Jimmy **(left)** *and* **(above)** *the Harris Hawk that snatched Jeanette.*

Jimmy's school cap as she attempted to fend it off with her satchel.

imp

Ian, who plays the straightman to his wife's wrinkled cheeky schoolboy in the duo's tired variety act, managed to scare it off for long enough to grab the terrified Jimmy and scramble to safety.

An RSPB spokesman later identified the bird as a Harris Hawk. He said: "They usually take mice or rabbits. I've only very occasionally heard of one taking a comedy schoolboy. It may be a rogue bird, so just to be on the safe side we've had it shot and smashed its eggs."

Calls for New Medical Standards

THE BRITISH MEDICAL ASSOCIATION have released new guidelines for doctors, specifying a change in the fruits used to describe the size of tumours, cysts and other bodily growths.

OUT go tumours the size of old-fashioned gooseberries, apples and grapefruits and **IN** come trendy lychees, avocados and papayas.

According to BMJ Chairman Dr Robert Chartham, the new standards are being driven by the changing market for more exotic fruits. He told us: "Thanks to the likes of Jamie Oliver and Nigella Lawson, the public now demands a more sophisticated selection of fruits on its supermarket shelves. It's only right that the medical profession reflects this when describing the results of tests and scans."

"Using more up-to-date fruit comparisons is an effective means of vividly bringing patients' worst fears to life in the twenty-first century," he added.

The new doctors' guidelines, which are expected to cost over £10 billion to implement, are the most wide-ranging since the use of fruit was originally adopted, replacing sports balls of various sizes, in the early 1960s.

But pensioners' leader Dolly Earnshaw was quick to hit out at the changes. "My members have only just got to grips with the last set of changes," she fumed. "We were perfectly happy with our lumps being the size of cricket balls, thank-you very much. This latest change is even more confusing. Us old folk can't tell the difference between a starfruit and a plantain, let alone a mango and a Chinese lantern."

"It's just changing things for the sake of changing things. I don't like these fiddly new five pees neither. Everything was much better in the old days. They should send them all back," she added.

Despite the new fruit guidelines for tumours, the BMA says that traditional comparisons will remain unaltered in other areas of medicine. "There will be no change to the description of bumps on the head, which will continue to be compared to the sizes of various eggs," said Dr. Chartham.

Supermarket HOT SEAT

25 SUPERMARKET STAFF members at the Walsall branch of Sainsways have all become become pregnant in the past 12 months... *after sitting on the same manager's cock.*

Store boss Lennie Tripper told the *Walsall Herald:* "I don't know what it is about my cock. I've lost count of the number of emplyees who have fallen pregnant after sitting on it. It might be a coincidence, or it may have magical properties. All I know is,

this mini baby boom is costing me a fortune in maternity pay."

But Tripper is finding that his lucky cock is causing some headaches at work. "Some of my staff whose families are already complete are refusing to sit on my cock in case they fall pregnant too," he said. "I think they're being a bit superstitious, but I can't force them to sit on it."

When the story was featured on the front page of the local paper, Tripper's magic cock became the talk of Fulchester. "Now it's not just the staff," he told reporters. "I've even had requests from ladies trying to conceive, all asking me if they can have a sit on my lucky cock."

"Well, if it brings them a little hope, I'm only too happy to let them," he added.

~ *Reuters*

PANDA, 84, PUNCHED IN FACE

...all for £2 OF BAMBOO

Xiang Xiang shows the bruises she received and (inset) an artists impression of the attacker.

By our Zoo Crime correspondent Capybara Palmer-Tompkinson

OLD age panda Xiang Xiang Robinson yesterday told how she was beaten black and blue in her own home by an intruder... all for just £2 worth of bamboo.

Xiang Xiang, 84 was awoken by someone knocking at the door of her enclosure in Bristol zoo in the early hours of Saturday morning. The caller asked for directions to the monkey house, but when Mrs. Robinson said she couldn't help, he pushed past her into the cage.

bamboo

Xiang Xiang said "I asked him what he wanted and he shouted 'where's your bamboo?'"

Mrs Robinson, who has arthritis, diabetes and lice said "I pointed to my tyre hanging from a rope at the far end of my pen where I keep all my bamboo.

bull and bush

I thought he would take it and go, but he didn't. He was convinced I had more bamboo hidden away in the panda house."

But Xiang Xiang never keeps more than £2 worth of bamboo in her enclosure at one time ever since the elderly Chimpanzee next door had his live savings of bananas stolen whilst he was at a tea party.

The thug then laid into Xiang Xiang, punching her once in each eye, on the end of her nose and both ears. He then fled with the bamboo leaving her badly bruised. Though dazed and weak, she managed to attract the attention of a porcupine who raised the alarm.

grey whistle test

The incident has left Xiang Xiang afraid to leave the shelter in her enclosure. Zoo keeper Craig Driftwood said "This was a despicable attack on a frail and vulnerable panda. The fact that just £2 worth of bamboo was taken makes it even worse. The sooner the culprit is caught, the better."

The attacker is described as 6 feet in length, white with small black ears, black rings around his eyes, a broad black stripe over his shoulders and black legs. He was wearing a coat of long, close fur and spoke with a Chinese accent.

'I'LL KICK DIRTY DEN'S HEAD IN!'

EASTENDERS EXCLUSIVE

– Blasts angry husband

Furious toilet cleaner Terry Hutchinson today threw down the gauntlet to Leslie Grantham, star of EastEnders, in the latest episode of a bizzare love battle.

"Come anywhere near my wife and I'll kick your head in". That was the warning from Terry, 49, who fears that his wife Glenda may have fallen for the charms of the actor who plays smooth-talking 'Dirty' Dennis Watts in the TV series.

BASTARD

"I could handle that bastard anytime", blasted Terry, a former boxer who weighs in at 18 stone. "He just has to come round here — any time he likes — and we can sort it out, man to man, in the back yard".

It was while drinking heavily at a local pub that Terry first began to suspect something was going on between his 42 year old wife and the well known TV celebrity. But when he arrived home later that evening to confront his wife, she denied ever having met Grantham.

Soap star's rocket slam wife sex love battle fury storm rages

However, a few days later, during an all day drinking session with friends, Terry's suspicions were again aroused and he immediately returned to his home in Bogwater Lane, Burnley, hoping to catch the couple red handed. But he burst into the bedroom only to find his wife was at work, and there was no sign of Grantham in the house.

After watching an episode of EastEnders that evening Terry went drinking with his brother-in-law, before visiting the BBC's studios in West London to confront Grantham face to face. However he was stopped by security guards at the gate, and later arrested for being drunk and disorderly.

"If he comes anywhere near me I'll have 'im', stormed Terry yesterday. "Fancies his chances does he? I'll take the bastard now!" he added.

POLICE

Grantham, who is set to leave EastEnders, was yesterday 'unaware' of Hutchinson's claims. Meanwhile Mr Hutchinson was remanded in police custody awaiting psychiatric reports.

Scandal of the EEC garden accessory mountain

by Bill Berk

THOUSANDS of items of garden furniture are being stored in gigantic EEC garden accessory mountains, including barbecues, swing chairs and folding tables.

And while thousands of households around Britain are left to do without, **TONS** of garden furniture is **SOLD OFF** at discount prices throughout Garden Centre's and D.I.Y. megastores.

PATIO DOORS

And we can reveal that many other items, including **MILLIONS** of miles of carpet are also being stored in vast warehouses and later sold off at low, low prices.

For instance, Galaxy deep shadow pile £8.75 a square yard. Fleetwood foam back shag pile — only £4.50 a square yard. And for the kitchen, Hoover 3348 auto washer — was £289.99 —now only £249.99.

GREENHOUSE

"It's absolute disgrace that the EEC can allow this to happen", said Fulchester MP Derek Twatt. "It's absolute mockery. The Government ought to do something about it".

Long after the death of a beloved Royal, there are still man

Will We Ever Know

Killer English Bull terrier Dottie, yesterday and its victim Pharos (left). Inset, Princess Anne with a face like a slapped arse.

IT WAS THE ROYAL DEATH that shocked the nation. On December 22nd 2003, the country woke up to the news that Pharos, the Queen's favourite corgi, had been savaged to death by one of Princess Anne's pitbull terriers. Its back legs brutally ripped off by the beast, the plucky dog fought back bravely against its attacker, but stood little chance.

Six weeks later, there has still been no official investigation into the tragedy. And as public dismay continues to mount, the authorities seem no closer to finding out the truth of what actually happened on that fateful day.

Official reports maintain that the Queen's favourite pooch's hind legs were chewed off by terrier Florence during the vicious Christmas attack at Sandringham. But many observers believe that the finger of suspicion points elsewhere.

secret

Former Palace dog-handler Douglas Milburn told us: "It's an open secret that Pharos was murdered by

By our Royal Correspondent
Aiken Drum
the Man in the Moon

Dottie, not Florence as has been claimed in the papers. Florence was nowhere near that corgi, and I can prove it."

door

According to Milburn, affectionately known in royal circles as "Dogshit Duggie" until he was dismissed on Boxing Day, the innocent Florence is being set up as a 'patsy'. He said: "The real villain in all this is Dottie, yet she's getting off scot free. I know Florence didn't kill Pharos because I was worming her in the kitchen when the attack happened."

Now, in a bid to prevent a miscarriage of justice, Dogshit Duggie has come out from under the stairs to dish the dogdirt on Dottie.

"That dog's a bad one, make no mistake. She's hit the headlines two or three times for attacking children and housemaids, but a lot of her other crimes have been hushed up.

"For instance, one morning I was devilling some Pedigree Chum for the royal dogs' breakfasts when I heard a blood-curdling scream from the hall. I rushed out to find Dottie lying on the mat, eating the postman's arm.

"I thought it might be possible for a doctor to re-attach the arm, but Dottie wouldn't let go of her prize. She growled if anyone went near her.

"In the end, the Queen came down to see what all the fuss was. She gave the postman £50 and an OBE on condition that he never told anybody what had happened. He had lost a lot of blood, so he wasn't thinking straight and agreed.

sandwich

"Another time, Princess Anne was visiting a cat res-

Where Were You?

EVERYONE remembers where they were when they heard the tragic news that Pharos, the Queen's favourite corgi, had been killed. We asked five C-list celebrities and a recently relegated D-list celebrity to recall what *they* were doing when the shocking news broke.

BBC Baghdad Correspondent *Raggy Omaar*

"*I REMEMBER* I was recording a piece about a suicide bombing at a Baghdad hospital when the first reports about Pharos's death bega to trickle across the wires. The news seemed so shocking and at first I didn't want to believe it. Later I just wandered the streets for hours and hours, desperately trying to make some sort of sense out of the tragedy. I suppose I was in a in a state of shock. I think we all were."

Entertainers *Keith Harris* and *Orville the Duck*

"*WE* were about to go on stage in our pantomime at Crewe Theatre Royal when we heard what had happened on the radio. Orville was so upset he wanted to cancel the show, but I told him he had to go on. 'But I can't', I kept making him repeat. In the end, we went on but the audience was very subdued. I've never got so few laughs in a show, and coming from me that's saying something."

Flamboyant Airport star *Jeremy Spake*

"*I WAS* about to tuck into several pies when my neighbour came in and told me the news. I was so upset I lost all my appetite and had to force the pies down. When I had finished them I could barely face my pudding of six mocha and almond bombs. But I did. I think it's what Pharos would have wanted."

Cheeky TV God-Botherer *Nick Hancock*

"*I WAS* watching TV when the programme was interrupted with a newsflash. Such terrible events can lead one to question one's faith. How can a lovely corgi like Pharos have his life cut short, whilst wicked animals such as wasps, piranha fish and crocodiles are allowed to live? All we can do is trust God in his wisdom. It is all part of the Lord's plan. Hallelujah. Praise Him. **Praaaaaise Him.**"

Renault Car salesman *Thierry Henry*

"*WE* were training at Highbury when Arson called us into the dressing room to break the news. Some of the lads broke down. They couldn't believe Pharos had gone, and they'd never see him being carried down the steps of the Queen's plane any more. It brought back to me vividly the feeling of loss felt by the whole French nation in 1973 when Giscard d'Estang's hamster Bertrand choked to death after over-stuffing its cheeks with sunflower seeds."

Most Haunted Medium *Derek Acorah*

"*I FOUND* out about about Pharos's death before anyone else. I was having a cup of tea with Yvette Fielding when my spirit guide Sam walked through the wall and told me there was a ghost corgi humping his leg. Yvette screamed. I asked Sam to look at the collar to see if it had an address. When he told me it read 'Pharos, Buckingham Palace' I went cold and Yvette screamed again."

more questions than answers. We ask...

the Truth?

Royal Death still Shrouded in Mystery

Princess Diana ~ Didn't live to see this happen.

cue centre. Her advisers had suggested it might be a bad idea to take her dog with her, but as usual she refused to listen.

"After cutting the ribbon, Anne was shown a basket of abandoned two-week-old kittens which had just been rescued. Without warning, and in front of over 100 specially-invited guests, Dottie tore into them, ripping them limb from limb. There was blood and cat fur everywhere. It was like an explosion at a gonk factory.

sesame

"Needless to say, Princess Anne didn't make the slightest attempt to stop the attack. The crowd were left to stand around, smiling awkwardly and making polite smalltalk while the carnage unfolded in front of them.

"If word of that terrible attack had leaked out it

Yappier times ~ Pharos (above, left in photo) a few days before the attack and (right) being carried onto a plane yesterday.

would have been curtains for Dottie. But as usual, the Royal family managed to hush things up by giving everyone present £50 and making them all Knights

Commander of the Order of the Garter."

But it's not just postmen and kittens who have found themselves on the wrong side of the Princess's dog. According to Duggie, one very senior member of the Royal family once felt the might of the pitbull's mighty jaws.

hot cross

"I remember on one occasion, the entire Royal family had gone for a picnic in the grounds of Balmoral Castle. I was present in my offical capacity to shovel up any dirts left by Princess Anne's dogs.

"Everything was going swimmingly. The family were letting their hair down, throw-

ing frisbees and playing games. When I heard the Queen Mum screaming I thought she was having a fun fight with one of the Princes. When I saw what was actually going on, my blood froze.

buttered

Unseen by the rest of the party, Dottie's hunting instincts had come to the fore. She had sensed that the Queen Mum was the oldest and weakest person present, and had separated her from the rest of the group before launching a deadly attack.

By the time I got to her, the dog had bitten one of her hips off.

tapped

The Queen Mum was rushed to hospital by air ambulance to have a new hip fitted. Meanwhile Princess Anne just said, "bad girl," and tapped Dottie gently on the nose with a rolled-up newspaper.

I knew at that moment that Dottie would kill again. I just prayed that next time it would be a small child or a housemaid and not one of the Queen's beloved corgis. Alas, as we now know, my prayer went unanswered."

bathed

A Palace spokesman refused to comment on the allegations, but confirmed that Milburn had been employed by the Royal family as a dog handler. He told us: "Douglas Milburn was dismissed on December 26th following an investigation

into the theft of a quantity of underwear from the Duchess of Wessex's quarters at Buckingham palace."

sinked

Duggie told us: "The Royal family has set me up. I'm a patsy, just like Florence. The Queen offered to buy my silence by giving me £50 and making me the Marquess of Cornwall, but I told them where to stick it. The next day when I was searched on my way out of the Palace and found to be wearing four pairs of Sophie Rhys-Jones's used scanties, I knew that somebody had planted them in my trousers."

At Hammersmith Magistrates yesterday Douglas Milburn pleaded guilty to 4 counts of theft, and asked for 208 similar cases to be taken into consideration. Sentencing was deferred until March, pending psychiatric reports.

THE Viz SAYS PAGE 46

Pharos Fact File

Name: *Pharos*

Position amongst Queen's corgis: *Favourite*

Breed: *Corgi*

Colour: *Corgi colour*

Age: *Between 10 and 20*

Age in dog years: *Between 70 and 140*

Favourite food: *Dog food*

Sex: *Unknown*

Fave group: *Busted*

Likes: *Walking, barking, sniffing other dogs up the arse, being carried off planes, shitting*

Dislikes: *Cats, fireworks*

Colour Vision: *None*

Black & White Vision: *Yes*

Sense of Smell: *Keen*

No. of legs before attack: *4*

No. of legs after attack: *2*

Existent state before attack: *Alive*

Existent state after attack: *Dead*

PHAROS was the dog the nation had grown to love. With its waggy tail, cheerful yapping and wet nose it was to all of us the Queen's Dog of Hearts. As a tribute to his memory, here is a moving Fact File to cut out and treasure.

20 THINGS YOU NEVER KNEW ABOUT VEGETABLES

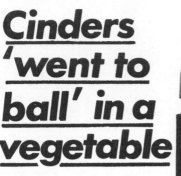

Phil Collins — A fan of Mange Tout

Tasty, boring, good for you or just plain food. Whether you like them or not, vegetables are here to stay. In restaurants, supermarkets, sandwiches and soups, there's simply no avoiding them.

Some folk eat nothing else, others hardly touch them. But are we all being told the full story about vegetables? Did you know that .....

1 All vegetables come from plants which grow in the ground. The potato, for example, comes from the Potato Plant (planto potatum)

2 The artichoke, a globular, leafy herb, is the first ever vegetable. (In alphabetical order).

3 There are two types of swede. One is a vegetable, the other is a person who comes from Sweden.

4 Among top personalities who eat vegetables are pop singer Simon Le Bon and footballer Gary Lineker.

5 There are two kinds of mushroom. As well as normal mushrooms there are Chinese mushrooms. Chinese mushrooms cost more than normal mushrooms in Chinese restaurants.

6 Bamboo is a kind of wooden Chinese vegetable.

7 Many vegetables can be cooked from frozen by placing them in a pan of boiling water. Add salt, return to boil and then simmer for around 3 minutes.

8 The tomato is not a vegetable.

9 The potato is.

Cinders 'went to ball' in a vegetable

11 The common abbreviation for vegetable is "vege". But in shops vegetables are usually referred to by their own names, like "cabbage" and "turnips".

12 Rock superstar Phil Collins who has had chart hits with songs like 'Easy Lover' and 'No Jackets Tonight' names mange tout as his favourite vegetable. "It's a small, sweet, premature, unripened peapod", says Phil.

13 There are probably more than 100 different kinds of vegetable altogether. The smallest, the pea, only has three letters in its name.

14 Cinderella, in the pantomime of the same name, went to the ball in pumpkin.

15 The word 'vehement', meaning ardent or passionate, immediately vollows vegetable in the dictionary.

16 'Vehicle' is the next word.

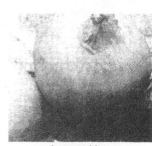

A vegetable

17 Ask for vegetables in a French restaurant and the waiter will give you a long blank stare! That's because in France, vegetables are called legumes. And in Germany they don't say cauliflower. They say blumenkohl!

18 Square vegetables are here at last! Although rather difficult, it is now possible to grow cubic marrows, using special equipment.

19 Vegetable derives from the Latin word 'vegetatum', meaning 'to eat with the main course'.

20 The first vegetable soup was onion, served in Paris in 1876. Dozens of vegetable soups are available today, in boxes, tins and packets. These include Florida Spring and Farmhouse Country.

10 Eastenders stars pick their vegetables from Pete Beal's market stall. Unfortunately there isn't a cockney rhyming slang expression for vegetable, because nothing rhymes with it.

← gastric juice

WHAT'S A NICE PLAICE LIKE YOU DOING IN A GIRL LIKE THIS?.

THAT REMINDS ME I'D BETTER FEED THE DOGS!

Heart Swap Large 'Back on the Cakes' ~claim

Salad dodging comic on ticker-spanking bender

ROLY-POLY unfunny man Eddie Large was last night reported to be back on the pastries, less than **6 MONTHS** after the heart transplant which saved his life.

Large, who shot to fame in the seventies as the medically obese straightman to "supersonic" Syd Little, was apparently spotted downing a succession of sausage rolls, cakes and eclairs with friends in a fashionable Morecambe patisserie until the early hours, despite having been warned by docs to lay off cakes and pies.

After 6 months of abstinence, Large appears to have well and truly fallen off the cake trolley. During his 16 hour pie shop bender, he *guzzled* his way through

- **56** *Cocktail sausage rolls*
- **8** *Ginster's pasties*
- **12** *steak bakes*
- **10** *Danish pastries*
- **2** *family-size white chocolate torts*
- **3** *Black Forest Gateaus with custard slice chasers*

unfounded EXCLUSIVE!

Syd Little later slammed the shop's proprietors and Large's eating buddies.

stormed

"Some friends they are," he stormed. "Everyone knows that Eddie has a pastry problem. To ply him with pies and cakes like this is nothing short of criminal. They ought to be ashamed of themselves."

wolverined

Passer-by Derek Bolsover was less sympathetic. "I think it's disgusting," he told us. "It's like digging up the organ donor and slapping him in the face. They ought to take

that heart back off Large and give it to someone who deserves it, like Simon Weston or Bob Champion."

Large picture (left) shows Large out of Little and Large and little picture (inset) shows Little and Large's Little, yesterday

TERRORIST FOOTWEAR

Lights! Camera! And red hot pumping****
SEX ACTION!

Most people dream of having sex with their favourite film stars. But Burt Gubbins is lucky – he does it for a living! He's the world's highest paid *sex stuntman*, standing in for the stars to perform their steamy sex scenes. And here he reveals for the first time the scintillating secrets of a career spent *in bed* with some of the world's most glamorous film stars.

Bonking Bert's a box office blockbuster

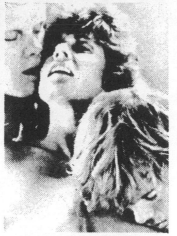

Red hot sex scenes like this are all in a day's work for Bert.

YOU NAME 'EM I'VE BONKED 'EM

Fonda – dynamite

Agutter – shower

Kensit – fried eggs

" Every time you see a couple at it in a film you can bet your bottom dollar that it's me up there giving her one. And although you'll never see my face on the screen, you can be sure that I've got the *largest part* in the film!

Everyone knows the secret of a good movie is a red hot sex scene. But if the leading man can't cut the mustard, the whole movie could flop. And these days, despite their macho images, most of the top stars are a dead loss between the sheets. I've seen some of the biggest names in Tinseltown pull their pants down and not know what to do next.

KIT OFF

That's where I come in. I don't need a script or nothing, I just turn up, get my kit off and – *action!*

SEX

Sometimes they call me in if an actor is too shy to do the business. Like the time they were filming Klute. Donald Sutherland had refused to do a sex scene with Jane Fonda. Well, fuck me. I was in there like a shot!

CLAPPERS

Mind you, I've never had to work so hard in my life. The keep fit routine has certainly paid off for Jane Fonda. That bird is dynamite between the sheets! It was supposed to be a one minute love scene, but she had me going like the clappers for an hour and a half! In fact, we only stopped when the camera ran out of film.

POSITIONS

On another occasion I stood in for Donald filming Don't Look Back. Boy, that was another marathon session! I

Julie Christie yesterday.

had to knock off Julie Christie in that one. The director was a bit fussy, so he made us do it about fifty times, all in different positions. Mind you, I wasn't complaining! Eventually he was happy, and we sat down for a rest, only for the cameraman to tell us that the film hadn't wound on. So we had to do it all over again!

In my line of work you never know what tomorrow will bring. One day I'll be shagging Emma Thomson on top of a piano in The Tall Guy, the next I'll be dressed up as a lion groping Bo Derek on the beach in Tarzan The Ape Man.

BLOW

One morning I got a phone call from David Lynch, the director. He said he was filming American Werewolf in London that afternoon, and could I come along and give Jenny Agutter a blow job in the shower. I didn't need asking twice!

TOOL

Under the British Film Censorship laws you aren't allowed to see my *tool of the trade* on screen. That can cause problems, as it's particularly difficult to hide, if you see what I mean. Anyway, subtle camera angles are used to get round

WOULD YOU LIKE ME TO LET YOUR TROUSERS OUT SIR?

WHINE! WHINE! SCRATCH! SCRATCH!

Rourke (left) was hell bent on bonking Bassinger (right).

'It would have taken Mickey 9½ weeks to get a bonk on'

the problem. For instance, that was my arse you saw going up and down in The Singing Detective.

SIZZLING

Naturally, a few actors still insist on doing their own sex: Mickey Rourke was determined to bonk Kim Bassinger himself in 9½ Weeks. There were some sizzling scenes in the script and Mickey was just dying to get stuck in. But when the cameras started to roll, he couldn't *stand up* to the pressure, and after half an hour they realised it wasn't just the focus that needed pulling.

PULL

They tried everything, without success. So eventually they had to *pull him off...* the set, that is. At the rate he was going it would have taken 9½ weeks just to get a bonk on. Needless to say, I was only too happy to oblige when they rang me up and asked me to stand in for Mickey.

GROPE

My *performances* have won me many fans among the stars. After I've warmed them up on set, a lot of them come after me begging for more. And a lot of them won't sign a movie contract unless they get a guarantee that I'm doing the sex scenes. Mind you, I've made that much dosh – getting millions of dollars for a quick grope and a bang – that I can afford to be choosy about parts myself now. These days I don't do a film unless I really fancy the bird.

For instance, I was asked to stand in for Mel Gibson in Lethal Weapon II, but I turned it down cos Patsy Kensit's got tits like fried eggs.

NOISES

One of my favourite jobs I get these days is dubbing new soundtracks on them arty films you get on Channel 4. They're just full of sex, and of course it's me they get to make the noises.

SHAG

I have to watch the film in a soundproof booth, wearing headphones. I have a great time in there, just me, a couple of birds to shag, and a few crates of beer.

STEAMY

People often ask me 'what's the best shag you've had with the stars?' And I reckon it has to be the one I had standing in for Jack Nicholson in The Postman Always Rings Twice. I'll never forget the steamy romp I had with Jessica Lange on the kitchen table! It took me ages to get the flour out of me pubes, I can tell you.

Northampton Says 'Cheese'

Northampton has been chosen as the venue for the 1994 National Cheese Festival.

The six month festival event, which will be jointly-financed by the European Development Agency, the Department of the Environment and Northampton Council Parks and Recreation Committee, will take place on a 500 acre site on the outskirts of the town, a derelict former shoe factory. The cost of converting the site into a spectacular venue for Britain's most extravagant cheese festival to date is estimated to be in the region of £250 billion.

CELEBRATION

Festival Director Mike Twatt believes the event will be good news for Northampton. "It will be a six month celebration of cheese, with cheese displays, exhibitions, and other things.

HIGHLIGHTS

Among the highlights will be the world's biggest piece of cheese. There will be cheese-orientated entertainment from around the world, cheese making demonstrations, a fun fair, refreshment facilities and car parking for 12 million cars in a specially-built car park at nearby Wellingborough".

PERM

"Not only will the festival bring jobs to Northampton – we are already advertising for car park attendants – it will also help attract new industry and investment, rejuvinating a former derelict eyesore, and making it into a hastily-assembled amusement park. We expect visitors to flood to Northampton in their thousands".

SHAMPOO & SET

Mr Twatt said he was relying on the people of Northampton and local industry to step in and make the event a success.

"If we can raise sufficient sponsorship from local firms, and if 120 million people visit the site, four times each, during the six-month period, the Festival will actually break even". Thirty-nine people visited the 1990 National Cheese Festival at Wolverhampton earlier this year.

IT'S BOLLOCKS

The word 'Bollocks' has been chosen as the British entry in next year's Eurovision Swear Contest, due to take place in Copenhagen in the Spring. Nations will be competing for first prize in the competition to find Europe's premier obscenity.

DUTCH

Among the contenders will be the Dutch entry 'Debiele', the French contender 'Putain' and the Greek profanity 'Skatta Nafas'.

TOWEL

This year will see the first entry from a united Germany since the competition began in 1952. The German contender 'Binden', literally translated, means sanitary towel.

LILLETS

Britain's entry will be performed by Felicity Kendall who will be hoping to improve on last year's dismal performance when Gareth Hunt came last with a dreadful rendition of the word 'toss'.

FAMOUS PEOPLE ON THE TOILET

No.26 Bernard Weatherill MP
THE SPEAKER OF THE HOUSE OF COMMONS

I DECLARE THIS MOTION PASSED.

DEATH BED JOKE

AND HOW ARE YOU COPING?

OH, THE FAMILY ARE MARVELLOUS, THEY'RE ALL RALLYING ROUND.

VROOM!

Black Ice!

Lancashire Resort Ready for Ice Age
~ "Bring it on!" says Lord Mayor

THE DINOSAURS perished when the world was plunged into a 'Day After Tomorrow'-style Ice Age a million years ago. And now it's time to turn up the central heating once more, because scientists predict there's another one just around the corner.

As sea levels rise, global warming is set to go into catastrophic reverse, leading to a worldwide Doomsday scenario which will leave the entire planet looking like a giant snowball.

resort

It's a frightening thought, but one person who certainly isn't losing any sleep worrying about the coming big chill is the Lord Mayor of Britain's favourite seaside resort. That's because Blackpool is the ideal holiday destination whatever the weather... and that includes temperatures as low as minus 85 degrees F, says Councillor Ivan Taylor. And he has this message for anyone thinking of cancelling their break in the Lancashire resort: "Come to Blackpool and have a good time. It takes more than a cold snap to spoil the fun along the Golden Mile!"

Scientists estimate that a new ice age could leave Blackpool buried under a 2-mile thick sheet of rock-hard permafrost, but Councillor Taylor isn't worrying.

chance saloon

"We've been hard at work on a 10-point action plan, called 'Blackpool - Open for Business'," he told reporters. "One thing's for sure, this town isn't going to be caught napping when the Ice Age hits."

Councilor Ivan Taylor who said "Bring it on" (right) and Blackpool tower (far right) yesterday

EXCLUSIVE!

The leaflet, which is available from the rack just inside the library door, outlines the council's plans which include:

• *Borrowing extra gritting lorries from nearby Fleetwood Borough Council to tackle the encroaching Arctic ice sheet and keep main traffic routes open.*

• *In addition to a full range of cones, choc-ices and lollies, all council-licensed ice cream vans will be selling cups of hot soup and Bovril for the duration of the ice age.*

• *The weekly knobbly knees competition on the South Pier will be relocated to a marquee adjacent to the Pleasure Beach or, if temperatures drop below -88°F, to the Town Hall Assembly Rooms on Corporation Street.*

• *Boarding Houses will be required to shut guests out at 9am instead of 8am, and to allow them back indoors an hour earlier than usual, at 4pm instead of 5pm.*

• *The council dog warden will be equipped with a specially-strengthened hoop on a stick and a pair of tough gardening gauntlets to enable him to tackle any marauding polar bears that may stray into the town's streets.*

In addition, the ladies of Blackpool, Fleetwood and Lytham Women's Institutes are hard at work knitting scarves which will be handed out free of charge to people hiring deckchairs.

Councillor Taylor said: "No matter how cold the weather gets, you're always guaranteed a warm welcome in Blackpool."

The Snow Must Go On!

We contacted several traditional variety stars to see if they'd get cold feet about doing a Blackpool Summer Season during the Ice Age. The answer was a resounding no. 'The show must go on!' they all told us.

*"The Arctic conditions could play havoc with my career," said Strictly Come Dancing presenter **Bruce Forsyth**. "It would be impossible for me to perform one of my trademark tapdance routines wearing clumsy snowshoes on my feet. But it wouldn't stop me performing. I love getting up in front of an audience. It's always ice to see them, to see them ice. Even if it's just some walruses."*

*"We'd still be up there on stage singing our hit every night," said **the Nolan sisters**. "But we'd have to dress up in reindeer skin parkas and snow goggles to combat the sub-zero temperatures. If we wore the same skimpy sequinned frocks we did when we were famous, we'd get frostbite in our knockers."*

*"I'd have to alter my act without a doubt," Crackerjack stalwart **Bernie Clifton** told us. "My comedy ostrich is a bird adapted to a temperate climate so it certainly couldn't survive an Ice Age. I'd have to adapt my routine accordingly, riding round on the back of a comical penguin. But that's showbusiness. If you*

Plucky Stars Gear Up for Big Freeze!

want to stay at the top of the comedy tree for as long as I have, you have to keep changing your act constantly."

*"I'd like the planet to be devastated by thousands of years of plummeting temperatures, but not a lot!" quipped former celebrity **Paul Daniels**. "I'd still be up there on stage doing tricks with my wife, the wonderful Debbie McGee, but a 2-mile thick ice shelf on top of the theatre may well keep audiences away. So it wouldn't actually make much difference to me."*

THE BEATLES ARE BACK!

'Fab Four' re-form - new album due

The Beatles as they were - in 1964

Yes, it's true. Fifteen years after they split up pop legends The Beatles are set to reform. And work on a new album is already underway.

Surviving members of the most successful pop group in the history of the world have consistently denied rumours that the band had been planning a comeback. But it now seems certain that the best selling artists ever in the history of popular music will soon be back in business.

LIVERPOOL

The mastermind behind the move is Johnny Johnson, a Liverpool based plumber and life long fan of the fab four. He spoke to us from a recording studio in London where work has already begun on a new Beatles L.P.

"It just seemed right after all this time that the band should get together again", he told us. "Obviously there were problems, and bearing in mind the sad

'It just seemed right after all this time'

loss of John Lennon there was a need for a new guitarist and songwriter. The obvious choice was John's son Julian, but with him living in the States there was going to be transport problems. Luckily a friend of mine plays guitar so I asked him if he would do the job".

Unfortunately none of the remaining Beatles, Paul McCartney, George Harrison and Ringo Starr were interested and so Johnson had to recruit a further three musicians before rehearsals could begin.

"I decided to do the singing myself so I really only needed another two", he explained.

LIVERPOOL

"I put an ad. in the Liverpool Echo and got fixed up with a drummer straight away. He knew a bass player who wasn't working so we signed him up and started rehearsing for the new L.P."

Although the album isn't due out until next year, recording and writing are already well under way.

"All the material on the album is going to be new stuff, and I can already see a change in musical direction beginning to come through," Johnny told us.

"The old stuff still stands the tests of time, but there's a lot of new ideas coming through and I think a few of our fans might be pleasantly surprised with the results."

STRAWBERRY

If you were too young to catch The Beatles first time round, you'll have a chance to see them on their comback tour which will be timed to coincide with the release of their new album. The L.P., which is due in the shops by mid-1986, is provisionally titled 'Strawberry Roads Tomorrow'.

professor piehead

Tearooms in crisis...Tearooms in crisis...

Ooh! Betty's!

BRITAIN'S cake-strapped tearooms are reaching crisis point as a record demand for light refreshments stretches resources to the limit. And now Tea Service bosses fear that many pensioners may have to go without the nice cup of tea and cakes that they so desperately feel like.

By our tea service writer **Alan Bennett**

A British tea room working at full strength.

The position has become so bad that Tea Service bosses may consider refusing waitress service for certain OAPs because there simply aren't enough tables.

Dr. Clive Foot - Elevenses

National Tea Service faces Meltdown

"Unseasonably normal weather has led to elderly people pottering around spa towns," says Dr. Clive Foot of Harrogate University's Department of Elevenses.

"Inevitably a good proportion are going to fancy a nice bit sit down with a cup of tea and a cake, and unfortunately our tea-shops cannot cope. If the weather doesn't get a bit parkier, and demand continues at this rate, I can see the whole system

That's one of the recommendations of a controversial report leaked from the Mr. Kipling Institute, an independent Tea Service think-tank.

collapsing in the next three months."

The report cites shocking examples of cases where the system has already broken down under the strain:

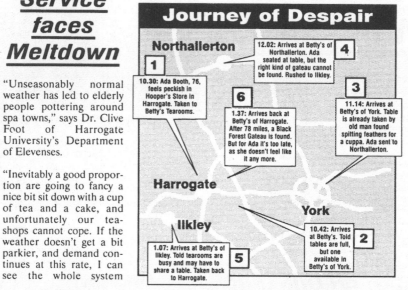

Journey of Despair

Northallerton
1 10.30: Ada Booth, 76, feels peckish in Hooper's Store in Harrogate. Taken to Betty's Tearooms.

4 12.02: Arrives at Betty's of Northallerton. Ada seated at table, but the right kind of gateau cannot be found. Rushed to Ilkley.

6 1.37: Arrives back at Betty's of Harrogate. After 78 miles, a Black Forest Gateau is found. But for Ada it's too late, as she doesn't feel like it any more.

3 11.14: Arrives at Betty's of York. Table is already taken by old man found spitting feathers for a cuppa. Ada sent to Northallerton.

Harrogate

Ilkley
5 1.07: Arrives at Betty's of Ilkley. Told tearooms are busy and may have to share a table. Taken back to Harrogate.

York
2 10.42: Arrives at Betty's. Told tables are full, but one available in Betty's of York.

*A junior waitress forced to work a 10 hour shift, who miscalculated the amount of sugar in a cup of tea, leaving an 80-year-old lady PULLING A FACE and muttering to her sister.

*An old man of 82 being seated at a table that was still covered in CRUMBS from the previous occupant's scones.

*A plate of biscuits left for 3 days on a cake trolley in a CORRIDOR because staff were unable to find a table for it.

*A 76-year-old woman, taken on a 78 MILE round trip to find a tearoom serving Black Forest Gateau.

A spokesman for Betty's, one of Britain's biggest tearoom chains confirmed last night that stocks of Earl Grey were low, but there was no cause for alarm as yet. "Every old person who genuinely fancies a cup of tea and a bite to eat will be served. They just may have to be a little more patient," he told us.

Where are they NOW?

TakeThat!

Groundbreaking boy band Take That! were never out of the headlines in the nineties, but after their dramatic split, they slipped from the public eye. Whatever happened to those lively lads, asks 15 year old Ada Trousers from Braintree in Yorkshire.

(Clockwise from top left)

Gary Barlow, the bozz-eyed tubby one who penned the band's hits, was declared bankrupt in 1997, after blowing an estimated £40 million on fizz bombs and sherbert dips. He now runs a small newsagents shop at Four Lane Ends in Newcastle upon Tyne.

Robbie Williams, the first to leave the band bought a milk round in Ashby de la Zouch, Staffordshire.

On the band's break-up, **Howard Donald** took the opportunity to realise a lifetime ambition and walk around the world. On his return, his dad got him a job at Boulby Potash mine in Cleveland, where he is presently deputy over-man.

Jason Orange left the band with an estimated £10 million which he invested in a revolutionary scientific process to extract gold from sea water. He now lives in a bus shelter in Peterborough.

Mark Owen sank his money from the band into a gas-turbine mobile sex library specialising in under-the-counter farmyard pornography. Business has boomed and he now earns up to and in excess of £100 per week.

As our favourite Royal approaches her 100th year we ask...

Has the Queen Mum been clocked?

NEXT month the Queen Mother celebrates her 99th birthday. And at Buckingham Palace elaborate plans are already underway to celebrate her centenary early in the new millenium.

The Royals will be cashing in on this record-breaking birthday like never before, with parties, stamps and lavish church services already planned. But one man believes the Queen Mother's 100th birthday bash is nothing more than a sham.

Wound

For Roy Biggins believes that the Queen Mum could already be 130 years old, and that her age has been wound back at some point in the past by an unscrupulous owner.

Wound

Mr Biggins, 48, worked as care attendant in a retirement home for over 18 months before being sacked for stealing money. And he believes the Queen Mother has been clocked.

Wugged

"Her teeth are a dead giveaway", he told us . "They've yellowed with age. You can give a coffin dodger as many new hips as you like, but the teeth will always be a giveaway".

Wock

Mr Biggins also points to the Queen Mother's arse as further evidence that all is not as it seems. "Your Victorians preferred a fuller, rounder figure. The Queen Mum's podgy backside is typical of that era. After the turn of the century your Edwardians went for a more streamlined body, with smaller tits. Looking at her teeth and bum I'd date the Queen Mum no later than 1870, and possibly a lot earlier than that".

Wagged

If the Queen Mother's age is not genuine, the implications for the Royal Family are serious. "To

The Queen Mum today - a good little runner considering her age.

begin with her life insurance will be invalid", Mr Biggins claimed. "And her value will greatly reduced. A one hundred year old Queen Mum is a marketing man's dream. But a clapped out 130 year old bag is nothing special, and she's going to eat money over the next few years".

Wascal

The Queen Mum's service history is patchy. King George VI is known to have acquired her privately from the Bowes-Lyon family during a visit to Durham in 1923. But there are no records before then, and it is possible that the Bowes-Lyons tampered with her age in order to make her more eligible for a Royal wedding to King George.

Wan

A more worrying possibility is that part of the Queen Mother may indeed be 99, but that the rest of her body could be that of an older woman. Mr Biggins claims that in the retirement home trade, "cut and shut" pensioners - hybrid fogies made out of two or more write-offs welded together - are increasingly common.

The Queen Mother's first registered owners were the Bowes-Lyons whose address on her birth certificate was this Durham castle. From here she was acquired by the Royal Family in 1923.

Is Britain's prized Royal asset really a 130 year old banger?

"They're usually cobbled together by some back street bodger and the join covered up by a lick of beige clothing. But they're death traps. Hips can fall off, their arses could crash at any minute, and their colostomy bags have been known to explode" says Roy, who claims that many lonely old folk are tricked into marrying 'cut and shuts' in rest homes by unscrupulous proprietors who continue to claim benefit for dead guests in this way.

Pale

Over the years Roy has seen every trick in the book used by unscrupulous rest home proprietors trying to make a quick profit. "Popular tricks of the trade include putting sawdust in their mouths to prevent them from slavering, and breaking an egg up their arse", he told us yesterday.

Visiting old people can be a perilous business. Here's Roy's top ten tips for anyone thinking of paying an elderly relative a visit in sheltered accommodation or in a retirement home.

1. Always make sure the address where they're living is the same as the address on their pension book.

2. Always ask to see the original birth certificate.

3. Don't be afraid to take their hat off and have a look around underneath.

4. Ask the keeper if you can take them out for a cup of tea and cucumber sandwiches. Listen for excessive wind noise, and check their seat for damp patches afterwards.

5. Check the tailpipe for spluttering, and look out for signs of overspray around the flies.

6. Ask them about the war, and listen out for incoherent ramblings.

7. Beware of new clothes - they could be covering up old problems.

8. Make sure their shoe sizes matches the shoe size given on their birth certificate.

9. The keeper may tell you that the pensioner is tired and needs a nap. Don't be hurried into making a quick decision. Take your time.

10. Insist on seeing full medical history. Otherwise you could be storing up trouble for the future.

Aged Concern offer a 101 point check to anyone thinking of visiting a pensioner. "It costs £50, but that is a small price to pay compared to the cost of hip replacements, cataract operations and stair lifts", said a spokesman yesterday.

WHO WANTS TO KICK A MILLIONAIRE UP THE ARSE?

Continued from page 13.

The correct answer was (d) *a fish.*

Check your answer (on page 13) with the correct answer above. If they match, congratulations! You have just won a kick up the arse... of someone who has got £100.

Go out and find someone who has got £100, then kick them up the arse. Once you've done that, buy the next issue of Viz and there'll be another question for you to answer. *Get that one right and you double your prize - you get to kick someone who has got £200 up the arse!* Keep getting your questions right and in a mere 15 issues (two and a half years time) you could be kicking millionaire Chris Tarrant up his smug money spinning backside!

NB. If you got the wrong answer, don't worry. You can play again by simply sending another £5 to us at the usual address.

I'm a Celebrity Skidn Get Me Out of Here!

IT'S A JOB WE ALL HATE. Doing the laundry. Modern day washer-driers have made the task a little easier, but it's still a chore. And while it's a chore, people like *Barry Featherstone* will be in work. Because Barry runs his very own laundrette, offering a full service from pre-wash to fluff dry. Seven days a week, he washes other people's dirty clothes. It might seem like a depressing way to make a living, but when you consider that the socks and pants belong to such names as *Leonardo di Caprio*, *Liv Tyler* and *Johnny Depp*, the job takes on a little more sparkle.

Featherstone, 42, began in the laundry business in 1983 when he opened his first shop, The Park Street Lav-o-Mat in Birmingham. But after 20 years he decided to move on, and opened a second shop in the more glamourous Belle Air district of Hollywood. Here he rubbed shoulders with the famous while he rubbed their gussets with Omo. Now, after nine months he's back in Birmingham. And he's about to blow the lid off the stars' shreddies secrets in a new book *Acdo All Areas* (Saucepan Macmillan Books, £9.99). In this extract, Barry gives us a taster of his life at the Hollywood soapface.

" I was known as the Laundry King of Hollywood. Everyone who was anyone used to come to my Belle Air Washetaria for their weekly service wash. They knew I'd do a good job on their smalls, and, just as important, that I'd be discreet. To my customers, privacy was all important. Intimate appareil exposes the secrets of the inner person in startling detail. But they trusted me, and as a professional laundryman I would never dream of revealing what I have found out about the stars whilst examining their scads.

animals

But having said that, some of these celebrities were no better than animals in their habits. I remember one Hollywood star coming into my shop with a week's worth of laundry, and the state of his underpants had to be seen to be believed. My professional ethics prevents me from naming him, but his bowels were Every Which Way including Loose that week, I can tell you. You'd think with all the money he had, he could have bought a roll of toilet paper. Even orang-utans wipe their arses on leaves - this man just didn't bother. I had to put them through the boil wash twice. Thankfully,

EXCLUSIVE

most of my customers practiced good personal hygeine, and their underwear generally just needed a pre-soak and an economy wash."

It may seem that even washing celebrities' undercrackers would quickly become mundane, and Barry admits that most of his work was run of the mill. But every now and then, a bag of laundry would throw up a surprise:

"I remember on one occasion I was doing a bio-wash of *Denzil Washington*'s underpants, when there was a screech of tyres outside the shop. I turned to look and saw a Ferrari had pulled up outside, and *Sharon Stone* was getting out carrying a black bin liner. She came into the laundramatte and asked me to do a service wash on her clothes. I told her it would be £5.50, although because it was America I told her the price in dollars. She thought it was a little pricey, but when I told her that included a 40° pre-wash she agreed. I gave her a docket and she left. She may have been one of the most desirable women in the world, but it was just another job it

> **"News travels fast in Tinseltown, and when you have Sharon Stone's knickers in your hand, it travels even faster"**

me and I started loading her clothes into the machine. But I was suddenly stopped in my tracks. In my hand was a pair of her knickers, and I recognised them immediately. **It was the same pair that she wasn't wearing in the famous leg-crossing scene in Basic Instinct.**

News travels fast in Tinseltown, and when you have Sharon Stone's knickers from Basic Instinct, it appears to travel even faster. Soon the press were

In a lather over the stars' smalls: Barry Featherstone (left) and some of the famous faces into whose gussets he has had a privileged peek. (below): Diehard man Willis, (right): Basic Instinct flasher Sharon Stone, (righter): Mafia crooner Sinatra, (even righter): Baywatch babe Pamela Anderson and (bottom): trial star OJ Simpson.

swarming around my shop wanting to see them. There was even the chief executive from the Planet Hollyood chain of restaurants. He wanted them for the wall of their New York branch, and he offered me fifty thousand dollars for them on the spot. I showed him the door.

Being in the laundry business is a bit like being a doctor - client confidentiality is paramount, so it's a mystery to this day how the word got uround about those famous knickers. I had only told my brother-in-law and his brother, and neither of them would have breathed a word. When Sharon came back for her laundry that afternoon, she had to run the gauntlet of reporters. She was not too pleased, I can tell you, and I'm sorry to say that she didn't come back to my shop again."

Barry may have lost Sharon Stone as a customer, but many of Hollywood's greats came back to him time and time again. One regular face at the Belle Air Washetaria was Die Hard heartthrob *Bruce Willis*:

"Bruce was a very good customer of mine. He was always busy making one film or another, and they always involved him wearing a vest. He'd regularly have to roll around the floor and blow things up and everything, and at the end of a day's shooting his vest would be filthy. Despite his hard-

man image, Bruce liked clean clothes, and every night he'd come into my shop with his soiled singlet and ask for the best wash possible. Money was no object to Willis.

He would pick the vest up the next morning on his way to the studio. The ironic thing was, because of film continuity, he couldn't suddenly appear halfway through a scene in a clean vest. So when he arrived on set, the make up artists would spend the first couple of hours each morning 'dirtying up' the vest I had just laundered! ...only in America!"

Barry saw a lot of things during his nine months as the Hollywood Laundryman to the Stars, and most of it has remained confidential until now. But on one occasion, he was forced to break that confidentiality and go to the police:

"It was a Wednesday lunchtime. I remember it well because I closed half day on Wednesdays and I was just about to lock up. Suddenly, perhaps

ark! Hollywood Undercracker Cleaner's Wash 'n' Tell Story of the Star's Shreddies

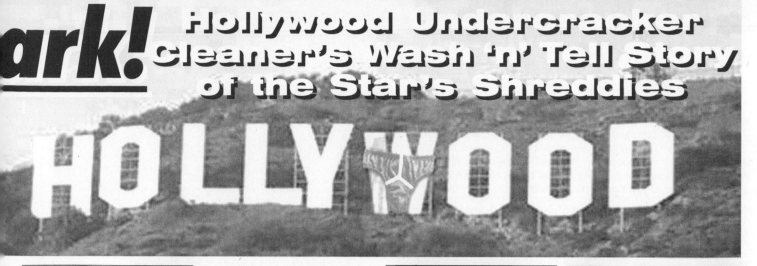

the most famous face in America walked in with a cloth laundry bag. It was **OJ Simpson**. He asked me to put all his things through the wash and he would pick them up the next day. There wasn't a lot, so I thought I would just put them through the twin tub.

As I was loading them in, I noticed that everything, pants, vests and socks, were all extremely small - far too small to fit a man the size of OJ. It was definitely the case that this man liked to wear all his clothes too small. A shiver ran down my spine as I thought that this threw new light on the 'tight glove' evidence in his famous murder trial.

I hated breaking a confidence, but I went to the police with his underpants. They were very excited, but pointed out that unfortunately he had already been acquitted of the murder, and that as much as they'd like to, they couldn't try him for the same crime again."

Barry ran a tight ship at the Belle Air Washetaria. Customers orders were always completed on time and to thier satisfaction. But on one occasion, Barry slipped up, and nearly paid for it with his life:

"I'm sure you've heard that **Frank Sinatra** never wears the same socks or pants twice. Well let me tell you that's not true. And I should know, because I used to wash them. I remember Frank came in one day with a load of his smalls in a pillow case. He said he wanted the same day service, as he had a gig that night with **Sammy Davis Jnr** at the Hollywood Bowl and had got nothing clean to wear. I said I'd have them done by 3.30pm. He left the shop, and as I was writing Frank's label, commedian **Normal Collier** came in. He was carrying an identical pillow case of underwear. We started chatting about things, and then he suddenly started doing his chicken impression. It was hilarious, and I was laughing so much that I must of put the label on the wrong pillowcase.

The next day, two big men in dark suits with violin cases under their arms came in and said they wanted to have a word with me about Frank's underpants. You can guess what had happened - I'd given Norman Collier's laundry to Frank Sinatra and vice versa. I knew of Frank's Mafia connections, so I was pretty scared, I can tell you. I called Collier to explain what had happened, but he thought I was joking and kept doing that thing where he pretends the microphone was cutting out. The two Mafia men were furious. One of them grabbed the phone off me and yelled some thinly veiled threat to Collier. They told him to 'stop playing the fucking wise ass', and said they were coming round to fetch Frank's smalls. They let me know in no uncertain terms that if this mistake happened again, it would be the last one I made.

I don't know what happened at Collier's house, but the poor devil came

> "Two men with violin cases under their arms wanted to have a word with me about Frank Sinatra's underpants"

back into the Washetaria the next day with a pair of underpants in one hell of a state. They could have walked to there on their own, I can tell you."

Of course it's not just vests, pants and knickers that Barry laundered in his nine months in Tinseltown. The bras of the stars were also given the famous Feather-stone treatment. **Uma Thurman**, **Julia Roberts**, **Marlena Dietrich** - you name 'em, he's washed their bras:

"If I told you who's titpants I'd had in my hands, you would think was making it up. The list reads like a who's who of gorgeous women. Some men may think that this is one of the best parts of the job, but let me tell you, when I'm handling bras, putting them in and out of the washers and driers, it's all kept on a professional level.

I remember this one time, **Pamela Anderson** came into the Washetaria and asked if I could do her bras. I told her how much a service wash and spin dry would cost and she agreed. She went out to her car and came back with a wicker basket full of her bras. I thought nothing of it and got straight to work. But as I was loading them into the machine, I couldn't help thinking that something wasn't right. On screen, Pam is most famous for her 38FF assets, yet all her bras were size 32A.

When you wash someone's most intimate cloathing, there's a bond

between you that isn't there for any one else. So when Pam returned the next day to collect her washing, I felt I could ask about the discrepancy in size. I'll never forget her reaction. She roared with laughter and opened her coat. She was naked from the waist up, and I was treated to a private showing of the tiniest pair of 32A breasts I've ever seen. I was taken aback, but she explained to me that whenever she is on screen, her breasts are computer generated to look massive. I was amazed. Hollywood is certainly the land where appeerence can be deceptive. She swore me to secrecy, and of course I agreed. It is a secret I shall carry to my grave.

Despite the glitz and glamour, Barry tired of the showbiz life. After nine months of being the King of the Hollywood laundry circuit, he returned to his shop in Birmingham, which his nephew, Paul Featherstone, had been running in his absence. Rumours that Barry had just served a nine months prison sentance were denied by Barry, but confirmed by his nephew, Paul.

"He admitted twelve charges of stealing women's underwear from the Park Lane Lav-o-Mat, and asked for a hundred and forty other cases to be taken into consideration", he told reporters. "It might have been two hundred and forty, I can't remember. It was fucking loads, anyway", he added.

As the Queen lies in court and proves she's above the law, we give You the chance to...

Have Your Say!

THE recent fiasco of the Paul Burrell trial and the Queen's last minute intervention has focused the public's attention on the Monarchy, and in particular their relationship to the judiciary. Should the Queen be above the law, or should she be treated as any other citizen? We went on the streets to find out what **YOU** think...

Regal illegal - HM the Queen on her way to work yesterday. But look! No seat belt - and the powerless police (circled) can only stand and watch.

...I think the Queen should have been called into the witness box to give her evidence just like anyone else, regardless of the constitutional implications. However, they should have made a special gold witness box with ermine trim, a red carpet and all diamonds stuck in the front.

William Malthus,
wine taster, London

...It is ridiculous that in this day and age, anybody is above the law. However, if anyone is going to be above the law it should be the Queen, because she does a marvellous job and always has a smile for everybody. And she can't answer back.

Audrey Beech,
grandmother, Liverpool

Of course the queen should not be allowed to get away with any crime, except of course for the murder of suspected paedophiles. I for one would applaud her majesty for taking a firm hand with regard to these perverts.

George Kelly,
ferry captain, Derbyshire

...The Queen is above the law, and quite right too. I run an all-night garage, and if her Majesty chose to burst in at 3am with a sawn-off shotgun, trying to wrench the till of the counter and screaming at me to fill her bag with cigarettes, I would be deeply honoured.

Ernie Ludlow,
retailer, Bishop Stortford

...Of course the Queen must be immune from prosecution. Were she to be convicted of a crime and sent to prison, just think of the consequences. The thought of the likes of Rose West getting her common little hands on Her Majesty's vagina in the showers simply doesn't bear thinking about.

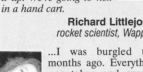

Edna Carstairs,
doll's hospital anesthetist, Hull

...I can't see what all the fuss is about. It's not as if she's a menace to the public. The only crimes Her Majesty commits are respectable ones, like shooting pheasants out of season, ignoring traffic lights in her gold coach and systematic tax evasion.

Trafford Lovething,
disc jockey, Manchester

...No one has a greater love and respect for Her majesty the Queen than me. She does a marvellous job, often in thankless circumstances whilst the carpers stand and criticise. However, the law is the law and everyone must obey it. I think she should be hanged and I for one would happily pull the bloody lever.

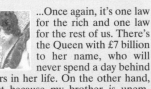

Alan Riot,
laughing gas fitter, Lancaster

...Once again, it's one law for the rich and one law for the rest of us. There's the Queen with £7 billion to her name, who will never spend a day behind bars in her life. On the other hand, just because my brother is unemployed, hasn't got a penny to his name and has murdered six women, he gets thirty years in Pentonville. It beggars belief.

Barry Heraclitus,
burglar's mate, London

...It would be absolutely pointless sending her to prison for any crime she committed. The prisons are like Buckingham Palace these days. You couldn't make it up. We're going to hell in a hand cart.

Richard Littlejohn,
rocket scientist, Wapping

...I was burgled two months ago. Everything was taken and excreta was smeared on the walls. The detective who came round to investigate told me that if Her majesty was responsible, the authorities would not be able to touch her. It turns my stomach whenever I see her on the telly with that smug grin on her face.

Mrs Dobson,
housewife, Nottingham

...The Queen should be allowed to commit any crime she likes. Prince Charles, however should only be allowed to commit crimes up to armed robbery and rape, whilst Princes William and Harry could do burglary and car theft. Minor Royals, like the Duke and Duchess of Kent could could steal items of small value from corner shops.

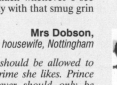

John Brutus,
council worker, London.

...The first principle of the British legislature is that the reigning Monarch is above the law. Therefore, were the Queen to make herself subject to the law like eveyone else, she would in fact be be *breaking it* and could expect to feel its full weight. However, luckily for her, being above it she would then get away scot free.

Joynson Taylor-Garret,
solicitor, London

...The Queen is rightly above the law, but I think she should be held accountable for any crimes she committed before she ascended to the throne. So if, for example, she was found to be the killer of Jill Dando, no action should be taken. However, should DNA tests show she had aided John Christie in his murder spree in the late forties, she should be brought to book.

Reg St. John Stevas,
dustman, Leeds

...The whole area is a minefield. What if the young Princess Elizabeth had stabbed her father in the chest? The precise moment at which he died, she would be guilty of his murder. And yet at the same moment she would have become Queen and therefore immune from prosecution. Does this mean that Prince Charles can now murder his own mother with impunity? If so, what kind of example is that to set to his children.

Frank Palance,
policeman, Goole

You only Clive Twice

Pretend OAP actor finally comes of age

EXCLUSIVE!

Dunn, yesterday, finally as old as he always pretended to be and (inset) as he was when he was pretending to be as old as he is now

VETERAN ACTOR Clive Dunn was last night celebrating after finally reaching the age he has always pretended to be.

Dunn, 80, spent most of the seventies playing doddery octogenarian Corporal Jones in TV's Dad's Army, despite actually being only about 30 or something.

"In those days, if I wasn't on set I had to dress up like I was even younger than I actually was, wearing big collars and leather jackets just to prove I wasn't as old as I made out," he told reporters.

bother

"But now I am actually as old as I was only pretending to be then, I don't have to bother any more. It's great."

Dunn intends to spend his retirement in Portugal, being like what he only used to pretend to be, but for real.

Jack

"It's a bit strange," he added. "Next year I'll have to dress up younger than I will be in order to look as old as I used to have to dress up older than I was to look, when I was younger." *-Reuters*

Interview

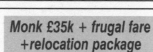
THE LADS FROM BUSTED: Charlie (middle) and Matt (right).

As chart-topping boyband Busted complete their record-breaking nationwide sellout tour, we meet Charlie, Matt and Ken to give them a grilling about life, love and music.

You've just completed a sellout nationwide tour, filling stadiums from one end of the country to the other. Does life on the road sometimes get a bit dull?

Matt: Yeah, it can do. But we're always getting up to loony pranks to keep ourselves entertained.

Charlie: Yeah, like the time we went into our support band McFly's hotel room and lit a fire in the wastepaper bin!

Ken: There was smoke everywhere. The sprinklers went off and they had to evacuate the whole building. It was absolutely mental!

Bands get up to all sorts of tricks to relieve the boredom of touring. What sort of high jinks have you got up to?

Ken: All sorts. We're absolutely mad!

Charlie: Once on the tour bus, Matt got out these supersoakers which he'd filled with petrol.

Ken: Matt was like, let's have a competition to see who can burn the driver's hat off!

Matt: Yeah, it was mad! Charlie went first and totally missed. The driver lost all the skin off the back of his head and his neck. It looked so funny!

Charlie: We all just cracked up.

How do you relieve the boredom of life on tour?

Charlie: We play a lot of practical jokes on our crew. There was this one roadie who was always playing tricks on us, like hiding our guitars and putting them out of tune when we weren't looking.

Matt: Yeah. We thought we'd get our own back on him. So one night after a gig, the three of us burst into his room, gagged him and tied him to his bed with gaffer tape.

Ken: Then we set fire to his mattress and locked the door!

Matt: We absolutely p*ssed ourselves!

Travelling between gigs must get pretty boring from time to time. What do you do to break the monotony?

Ken: All sorts. This one time we were on tour and having some publicity photos taken at some zoo. God, it was SO boring!

Charlie: I thought I'd liven things up a bit, so I doused one of the gibbons with lighter fuel and set it on fire!

Matt: It was hilarious! You've never seen a monkey move so fast. It was swinging through the cage making these funny whooping sounds and setting all the other apes on fire as it went past.

Charlie: Yeah. In the end the whole cage was full of burning monkeys leaping about making these stupid noises! We p*ssed ourselves!

Ken: You should have seen the keeper's face! It was a picture!

Life on the road must get dull from time to time. How do you keep yourselves entertained?

Matt: Touring can be dull, but we do have lots of fun. For example, our manager's a real square, and we love winding him up.

Charlie: This one time, we were on CD:UK and Matt was like, let's flush his phone down the toilet!

Matt: He was so mad! He marched straight into our dressing room to b*ll*ck us, but Ken was hiding behind the door with a tyre full of petrol!

Ken: I put it round his neck and lit it! He totally went up like a rocket!

Matt: It was so funny! We were laughing about it so much when we went on the show we couldn't even play our guitars.

Charlie: Or sing.

LIES on Ea

IN AN illustrious career spanning more than half a century, Sir David Attenborough's wildlife documentaries have opened the public's eyes to the wonders of nature. His TV shows such as Life on Earth, the Blue Planet and Life in the Freezer have regularly pulled in huge audiences with their spectacular views of the animal kingdom. In his latest blockbusting series, Life in the Undergrowth, the 78-year-old naturalist's cameras have given us a mind-bogglingly close-up look at the microscopic world of insects and creepy-crawlies.

MONKEY BUSINESS: Oliphant (left, in monkey suit) and Attenborough (right, in anorak) pose in Bristol Park.

Or so he would have you believe.

For, if the claims of a Bristol-based former children's entertainer are correct, every single shot in every one of Attenborough's documentaries, from cuddling gorillas in the Congo to crawling through an ants' nest in the Kalahari, is a FAKE! And now, following an explosive bust-up with the veteran film-maker, Harry Oliphant says he's set to reveal the on-set secrets he has kept for fifty years.

In these exclusive extracts from his new book Lies, Camera, Action! (Kedgeree Books, £2.99), Harry lifts the lid to give us a behind-the-scenes view of the deceitful world of David Attenborough. A world the BBC would prefer us not see.

EXCLUSIVE FLASH

"I'd been working as a professional children's entertainer for about ten years when I got a call from the BBC, asking me if I'd like to do a bit of work on a new series they were filming in Bristol. I thought this could be my big break into kids' TV, so I was a little bit puzzled when I arrived and was directed to the nature documentaries department.

I was shown into a room with David Attenborough, who told me they were shooting a documentary about ostriches and emus for Wildlife on One, and he wanted to use Oswald, my puppet ostrich, for some of the shots in the film. I was outraged and told him that faking shots was immoral, but he just laughed and told me not to be so naive. He pushed a twenty-pound note into my top pocket and winked. 'And there's more where that f***er came from, if you play your cards right,' he smiled.

He explained that the BBC gave him thousands of pounds of licence-payers' money each year to travel the globe filming wildlife, but that he made all his shows in Bristol, using trick photography. 'I just pocket all the cash and none of the stupid f***ers is any the wiser,' he laughed. 'If you don't believe me, just look in the f***ing car park,' he told me. I looked out of the window and there were eight brand new Rolls-Royces, with the number-plates DA 1 to DA 8.

To cut a long story short, I eventually agreed to take part in the film. Twenty minutes later I was up a ladder in Studio 3, pulling the strings to make Oswald walk about, lay an egg and then bury his head in some sand, whilst Attenborough delivered his commentary."

Once filming was over, Oliphant returned home to his digs and thought he'd heard the last of Attenborough. But the next morning, the phone rang again.

"It was David. He told me to be at the BBC that afternoon, and to bring Colin, my Punch and Judy crocodile, with me. Something told me I wouldn't need to bring Mr Punch, Judy or the policeman - this wasn't going to be a children's show.

After lunch I found myself back in Studio 3, where Attenborough was making another documentary, this time about the world of predators. When the programmme went out on telly that night, all you could see was the brave presenter almost losing his arm to a ravenous crocodile.

Attenborough's Creepy-crawly Heeby-jeebies

WORKING AT THE BBC, Oliphant got to see a side of the apparently fearless Attenborough that was kept firmly hidden from the public's view.

"Once, while we were filming an episode of Life in the Undergrowth, there was a terrible scream from David's dressing room. Everyone rushed in to see what the matter was. He was standing on a chair with his knees knocking, pointing at his bathroom door. 'Spider!' he sobbed. 'There's an absolutely huge spider in there.' When we went in, we couldn't believe our eyes; there in the bath was a tiny little money spider, no bigger than the head of a pin.

SINKING FEELING: Attenborough screamed when he saw spider in bath.

I couldn't help thinking of the scene we'd shot the previous day of David, supposedly standing in a Venezuelan cave whilst a couple of giant South American bird eating tarantulas crawled all over him. Of course, as usual it was filmed in Studio 3, and it was just me in a black bodystocking with a couple of spider puppets I'd made out of a pair of wooly gloves, some pipe cleaners and a few old ping-pong balls for eyes.

Attenborough may be nearly 80 years old, but he'd run a four minute mile in three minutes if a real tarantula came anywhere near him."

But what the cameras didn't show was my arm up Colin's green cloth backside, operating his harmless wooden mouth.

I got thirty quid for that one, which was a lot of cash in those days. Back then in the sixties, I would have had to do twenty kids' parties to make the same amount. Working for Attenborough was easy money; I may not have liked deliberately deceiving the public, but before long it had become a way of life.

rth!

Over the years, Harry was called upon to disguise himself as many different animals; most famously of all when he had to don an ape suit during the filming of the landmark Life on Earth series.

" I'll never forget the uncomfortable day I spent dressed in a gorilla suit, cuddling up to David in a bush in the local park. The script called for him to deliver his lines whilst I sat on his lap peeling and eating a banana.

Well I don't know what was the matter with Attenborough that day, but he just couldn't get his words right. Take after take he kept fluffing his lines, whilst I of course ended up eating banana after banana.

I know that scene has become a classic which has been replayed endlessly, but I can't watch it without getting indigestion!"

In Attenborough's latest series, Life in the Undergrowth, Oliphant was asked to pretend to be a maggot.

> **"I'll never forget the uncomfortable day I spent dressed in a gorilla suit, cuddling up to David in a bush in the local park."**

"The fruit fly larva is one of the simplest organisms in the world, but faking it for the cameras was one of the biggest challenges of my career. In the scene as it appeared onscreen, the maggot writhes around on a grape whilst David points out its reproductive organs.

We tried all sorts of ways to get the shot but nothing seemed to work, until I had the idea of making a costume out of an old sleeping bag. I zipped myself in and started wriggling about - it looked perfect!

For the final scene, a couple of stagehands operated a giant polystyrene pointing hand which was left over from the Kenny Everett Show, whilst I squirmed about on a spacehopper which had been painted green. The viewers were fooled, but if you look very carefully, you can see a Milletts label sticking out of the larva's mouth!"

Over the years, Oliphant has often been called upon to use his skills as a children's entertainer in the production of Attenborough's wildlife films.

"In the kids' party business, you have to be able to turn your hand to anything; a bit of juggling, some plate spinning, a few conjuring tricks - you've got to be a jack of all trades. One of the most basic skills is balloon modelling. This came in particularly when we were filming a show about poisonous snakes.

There was obviously no way that David was going to go anywhere near a real snake, but none of the alternatives we thought of worked on camera. A sock puppet and a length of hosepipe just didn't look convincing through the lens.

One of the researchers had seen me making balloon animals at an IKEA store opening, and he suggested we try that as a last resort; the show was going out in less than an hour.

I blew up a balloon and drew some eyes and a mouth on it with a marker pen - the efffect was startling and the cameras started to roll. Unfortunately, every time Attenborough picked the 'snake' up and started pretending to struggle with it, it popped.

After a dozen or so takes (and snakes!), I was down to my last balloon and everyone was pretty tense. Luckily, just at that moment, someone noticed that Attenborough was wearing a Remembrance Day poppy, held onto his safari suit with a pin.

With the pin safely removed, and with just seconds to spare before the programme went out, we got the shot in the can. And only just in time - as the director shouted 'cut', the knot in the balloon came undone and the 'boa constrictor' shot across the studio, making a rude raspberry sound! Everybody fell about laughing."

After so many years helping Attenborough fake his films, why has Oliphant chosen now to blow the lid on his secret career? He insists he is not driven by money.

"Admitttedly, things are tight at the moment," he told us. "I've had to give up working as a children's entertainer, following a party where I was doing a trick involving some sweets changing pockets and I got a little lad up to help. There was a bit of a misunderstanding and I ended up being put on the sex offender's register. After all the publicity over the court case, bookings started to dry up, but money is not my motivation for deciding to speak out now.

The reason I've decided to expose Attenborough now is that we've had a bit of a bust-up. I saw him a few weeks ago through the window of a restaurant, so I went in to get his autograph on my UB40.

First he pretended not to know me, then he said he didn't have a pen handy. I couldn't believe the way I was being treated after all I had done for him through the years. To cut a long story short, I threw his dinner on the floor and gave him a piece of my mind in front of everybody. The restaurant manager called the police, I was arrested, bound over pending psychiatric reports and ordered not to go within five hundred yards of David Attenborough.

Over the years I've helped David make films about every animal under the sun, but the truth is that there's only ever been one real animal on the shows - and that's the lying weasel that presents them. "

CLIFF MUST DIE!

Pop favourite Cliff Richards has been sentenced to death by members of a cult religious group who have branded the baby faced star an EVIL puppet of SATAN.

"Cliff Richards must die", says Blackburn based Derek Qualcast, self proclaimed High Priest at the Church of the Latter Day Scientific Christologists of the Seventh Holy Grail. And he accuses Richards of using his music to preach evil to unsuspecting record buyers and fans.

"Richards is the antichrist", blasted Mr Qualcast, who is 57. "He is in league with the forces of darkness, and is sent by the Devil to lure us from the path of righteousness. The words of his songs are thinly veiled catalogues of sexual corruption. He preaches fleshy pursuits and sinful activities, such as girl on girl, topless relief, oral and shaving pleasures".

PORNOGRAPHIC

Mr Qualcast claims that Richards' hits contain Satanic messages. "Records such as 'Devil Woman' speak for themselves, while 'Carrie' clearly takes its name from the devil worship pornographic film of the same name. And 'Goodbye Sam, Hello Samantha' is an open invitation to young people to indulge in acts of unfathomable evil, possibly involving farmyard animals".

VIRGINS

Qualcast fears that Richards has already claimed the lives of thousands of young virgins, and goats. For he believes the secret of the baby faced star's boyish good looks is the blood of freshly killed victims, which he drinks every day. And worse still, he is convinced that 68 year old Richards practices voodoo, black magic and has the ability to turn himself into a bat.

Muslim style 'Fatwa' on Peter Pan of Pop

In a 15 year campaign to silence the singer Qualcast has visited every record shop in Blackburn, and one in Rochdale, sprinkling holy water on their doorsteps. But despite his efforts the ageless star's string of chart hits has continued uninterrupted. However, Mr Qualcast vows to continue the fight.

SPUNK

"It is a clear cut case of Good against evil," he told us yesterday. "The Bible tells us that on the seventh day it was written that the heavens shall open and he will be

Richards – 'sent by Devil'

tempted three times for forty days and forty nights, and yea on the forth time the clouds shall part and down will rain the Devil's spunk and spawn and a multitude of frogs and boils, and so you shall know him by the name of Lucifer, and his name shall be Ahab, who begat Cain and Abel who begat George who begat Harry Webb who is called Cliff Richards. For so it is written, and so shall be", Mr Qualcast added.

EATING SMARTIES MADE ME GROW WOMEN'S TITS
~claims man

A Cleveland man yesterday claimed that Smarties made him grow women's tits.

Unemployed panel beater Bill Strimmer of Billingham claims that eating the candy coated chocolate sweets resulted in him developing a pair of 38 inch 'D' cup women's breasts.

"I was horrified, and embarrassed", Bill told us. "They were huge, with nipples and everything. I didn't know where to turn".

WOMEN'S

Despite support from his wife, who lent him a bra, Bill faced ridicule from workmates at the garage where he was employed, and eventually he was forced to quit his £10,000 a year job.

"I lost my job for having women's tits, and its all because of Smarties", says Bill, 42, who is claiming £2 million compensation from Smarties manufacturer Rowntrees for loss of earnings due to women's tits.

"I always ate Smarties, ever since I was a kid", he told us yesterday. "But the problems began when they introduced the blue ones." Within three days of eating the new blue coloured sweets, Bill noticed he was growing a pair of women's tits.

"I went to my doctor and he immediately asked if I'd been eating blue Smarties. He told me to stop, and sure enough the tits disappeared. But by then it was too late. I was already out of a job, and I was the laughing stock of the whole town.".

TITS

A spokesman for Rowntrees confirmed that blue Smarties had been introduced for a limited period, but was able to state categorically that they did not cause women's tits.

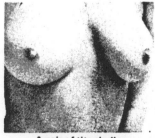

A pair of tits similar to the ones claimed to have been grown by Mr Strimmer

Mr Strimmer refused to make any comment until he had spoken to his solicitor. "The blue ones definitely made me grow women's tits and I'll sue them for every penny they've got", he told us yesterday.

Mr Strimmer last hit the headlines in 1972 when he claimed that sucking Olde English flavour Spangles had caused him to grow women's tits.

THE EYE AT NIGHT!

An Aberdeen woman may be forced to sell her house – because she claims TV astrologer Patrick Moore has been using his telescope to observe heavenly bodies – through her bedroom window!

PEEPING

Glenda McBride, 58, says she has been forced to dress and undress with her curtains closed since peeping Patrick had a new extra powerful lens fitted to his telescope at his observatory in Selsey, Sussex.

Astrologer Moore yesterday

"He ought to keep his boggly eyes fixed firmly on the stars, and not on my tits", said Glenda yesterday.

MP's SEEK SEX RULE REVIEW

Urgent reforms in Britain's sex laws are being sought by MPs fed up with constant slurs and allegations of sex scandal and impropriety involving politicians.

And among new measures they propose are:

- **EXCLUSION** for all MPs from the strict laws regarding prostitution.

- **RELAXING** of the rules regarding indecent exposure

- and **LIFTING** of the present ban on MPs carrying out acts of gross indecency.

VICTIMS

For years MPs have been the victims of countless allegations of sexual corruption. And many live in constant fear of media harassment, arrest and even imprisonment.

EXPOSED

"Having their sexual misconduct exposed can cause embarrassment and distress to an MP, and in certain cases can seriously affect their career prospects", one Commons insider told us. So now a growing number of MPs feel that strict laws governing sexual conduct should be relaxed for MPs, allowing them to endulge in acts of gross indecency, homosexuality with small boys and kinky sex including bondage, without fear of police or press reprisals.

EXCLUSIVE

"MPs have a uniquely difficult and stressful job to perform", we were told. "The pressures of work bring about a need for unusual, bizarre and often illegal forms of sexual relief, especially among Tories. Often kinky sex is the only relaxation available to a busy back bencher."

BIZARRE

"The existing laws regarding sexual behaviour are all well and good for the man in the street, but MPs should be exempt from these laws. I believe we are a special case", the spokesman continued.

LAVATORIES

MPs are also lobbying for better facilities at Westminster, including public lavatories on grassy lawns surrounding the Commons where MPs can meet homosexuals for the purpose of carrying out indecent acts, and regular visits to Westminster by schoolchildren and boy scout groups, to enable politicians to meet young boys.

Fanny Batter's HOLLYWOOD FILM gossip

Big news in Tinseltown is that hunky 'JFK' star **Kevin Costner**, *alias oscar winning actor Robin of Sherwood, is taking Batman Catwoman sex bomb* **Michelle Pfeiffer** *to the pictures on Friday. And get this. Kevin's booked seats in the back row!*

Michelle's best pal, '9½ Weeks' fire escape sex blindfold ice cube nipple rubbing star **Kim Bassinger**, *17, is planning a party that night while her parents are away, and she's invited Kevin and Michelle to come round after the film.*

Only trouble is Michelle's parents, one-time Six Million Dollar Man **Lee Majors** *and twelfth wife* **Tammy Wynette** *want Michelle home by eleven.*

Will Michelle's latest **liaison** *turn out to be* **dangerous**? *Watch this space!*

Word on Sunset Strip is that Hollywood Brat pack stars have built a den in **Gregory Peck's** *back garden. Gang members include* **Emilio Estevez** (*son of Sound Machine singer Gloria*), *Madonna 'ex'* **Sean Penn**, **Rob Lowe** *and* **Donald Sinden**. *And word is the gang rule is 'no girls'.*

Michael Douglas, *kid brother of 'Carry On' comic Jack, is desperate to join, but before his Brat pack pals will let him in, 'Fatal Attraction' star Michael has got to pass their initiation test by stealing apples from a tree in psycho 'Silence of the Lambs' star* **Anthony Perkins's** *back yard.*

Good luck Michael. **Rather you than me!**

Can you keep a secret? Friends in the know tell me that romance is on the cards for 'Out Of Africa' star **Meryl Streep**, *gorgeous daughter of the late great* **Charlie Chaplin**, *and now Hollywood's highest paid actress. Word is that Meryl, 29, has caught the eye of cooky 'Cape Fear' star* **Robert De Niro**.

Between you and me, 'Raging Bull' Bob fancies Meryl, and wants to go out with her. Trouble is he's too shy to ask. Enter **Al Paccino**. *Best pal Al played postman and handed Meryl a note from De Niro asking her to go out with him. And guess what. Word is that De Niro has been told he can borrow dad* **Randolf Scott's** *car to take Meryl to the disco on Saturday.*

Randolf, 108, certainly won't be needing it. He's flying to Rio for the weekend, and long time live-in lover **Una Stubbs** *has just passed her helicopter driving test!*

See you next time Star Gazers
Fanny xx

ANN'S ADVICE ON CARPENTRY

HRH Princess Ann has spoken out strongly on the issue of carpentry.

The Princess Royal, speaking at an annual dinner of the National Society for the Protection of Certain Types of Fish of which she is Honorary President, said over a period of many years traditional forms of carpentry have almost become extinct.

JOINTS

"I find it sad that traditional close fitting hand-made joints, such as the dovetail, are rarely used by carpenters and furniture manufacturers nowadays", the Princess told the audience of over 800 fish enthusiasts.

SPLIFFS

"There was time, not so long ago, when joints between two pieces of wood were made to fit so tightly that often glue was not required", the Princess added. And she blamed modern working practices for a fall in standards throughout the carpentry trade.

Princess Ann's remarks came in the week in which she announced the launch of a new charitable foundation, The Princess Royal's Trust for the Furthering of Traditional Old Fashioned Woodwork Skills. And stars have already rallied round, promising a gala fund-raising concert to support the cause.

FOUNDATIONS

Top entertainers, among them Phil Collins, Billy Connolly plus a host of Royal sycophants have volunteered

Ann - advice

their services for the event which is expected to lose money hand over fist.

PEAR POSTBAG

The lively forum where everyone's talking about PEARS

I HATE pears. They're just wonky apples if you ask me.
Eileen Brake, Cork

I SENT my husband to the shops to buy some pears. He came back with 2 of everything! And not even any pears! He's an idiot.
Ethyl Murton, Murton

I ALWAYS get pears and peas mixed up, but I have worked out a way to tell them apart. The spelling. And the fact that one is a fruit and one is a veg. And the fact that they look different.
Lupe Velez, Caracas

I LOVE pears so much I changed my name to Mr Pear and I eat nothing but pears. I really like pears. I hate anything that is not pears.
Frank Pear, Lambeth

I THINK that the children's cartoon Care Bears would have been much better if it had been called Pear Bears and had been about bears who liked pears. Or maybe it should just have been called Pears and been about pears.
Cecil Cunningham, Hull

THESE politically correct people who go around defending pears clearly forget that it is an apple - not a pear - a day which keeps the doctor away. Do these people want us all to die?
Cedric Allingham, Sussex

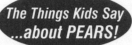

The Things Kids Say ...about PEARS!

My grandson recently turned to me and said "Mummy, that apple has a fat bottom. It should go on a diet." He was talking about a pear!
Maureen Weismuller

At a recent family dinner my grandson piped up with "Granny, why do people say 'uh-oh, it's all gone pear shaped?' I fail to see what the shape of a pear has to do with adverse or problematic circumstances." How we laughed.
Edna Sabu, Wentworth

"The au pair doesn't look like a fruit" said my granddaughter as we greeted the new nanny. She had made a basic semantic error, confusing the word 'pair' with the entirely different word 'pear.' I made her write the words out 4000 times so as she wouldn't confuse them again.
Dolly Hutchinson, Leek

Your Poems about Pears

*If you're like me
And you love pears
Then be a real go-getter
Don't settle for only one pear
For a pair of pears is better!*

By Mrs Edith Plywood

*Pears, pears, pears
Pears, pears, pears
Pears, pears
Pears
Pears, pears, pears
Pears*

By Mrs Janet Pudding

We Love Your Pear Jokes

Q. When is an apple not an apple?
A. When there's two of them and they are a pair (pear)

Q. What did the man pear say when it saw a sexy lady pear whose flesh had been removed?
A. Cor! (core)

Q. What's green, shaped a bit like a gourd and lives mainly in the Arctic circle, but whose range extends down into Russia, Greenland, Canada and Alaska, reaching as far south as Ontario, and whose diet consists mainly of seals, fish and carrion?
A. A polar pear (bear)

Aren't Men Daft!?

I SENT MY boyfriend to the shops for a loaf of bread and a pint of milk. When he came back, he'd remembered the bread but forgotten the milk!
Gloria Prepuce, Jarrow

TALK ABOUT dippy, my hubbie must take the biscuit. He suffered a massive head injury whilst working on an oil rig six years ago, and has been in a persistent vegetative state ever since. The silly sausage doesn't even know what day it is!
Sue Glanshood, Yeovil

MY HUSBAND's daft as a brush. I went round to my neighbour's house and found him stark naked in bed on top of her. The daft 'ap'orth had gone into the wrong house and climbed into bed with the wrong woman! I ask you!
Janice Foreskin, Hull

VIZ HEALTH
The Viz Accident & Emergency Dept

Dear Viz A&E,

I'm really sorry, I'm pissed as a fart, and whilst showing off to some girls, I fell off some scaffolding and landed on my wrist. I think it's broken. It's swollen up like a balloon and I can't move my fingers. Could a Viz A&E doctor have a look at it?
B Aspen, Manchester

● A triage nurse has read your letter to ascertain its urgency. Letters involving head injuries and heart problems are read first. The current waiting time to have your letter read is between 3 and 8 hours. Thank you.

Dear Viz A&E,

I'm a non smoker and I've had a cough for a month now. I just can't seem to shift it. It's worse in the mornings and for the past few days I've been bringing up blood. I'm afraid the Viz doctor can't read my letter for a couple of weeks and I'm really worried. Can you help?
G Sprake, Leicester

● This part of the page is for accidents and emergencies only. For all other illnesses and medical conditions you should write to the Viz GP.

Have you just fallen and broken a bone? Maybe you've burnt your arm getting something out of the oven. Or perhaps you've got massive gripping pains in your chest. Write to:

Viz A&E Dept, PO Box 656, North Shields, NE30 4XX.

Are They Still Dead?

I USED TO love the comedian **Bernie Winters**, especially when he appereared with his dog Schnorbitz. I know he passed on a few years ago, but I couldn't help wondering if he was still dead?
Dean, e-mail

★ Born Bernard Weinstein on September 6th 1932, Bernie Winters was a regular on our screens with his brother Mike for many years. He died on May 4th 1991. A quick phonecall to his ex agent reveals that Bernie is indeed still dead and has no plans to come to life in the immediate future.

COULD YOU tell me if the 60s singer **Janis Joplin** is still dead? I used to love her songs and it would be great to see her live in concert.
Franklyn Mint, Suffolk

★ I'm afraid you won't be seeing her anytime soon, Mr Mint. The hard drinking rock wildchild died on October 4th 1970, and a quick check with Los Angeles crematorium authorities reveal that as of yesterday she was still dead. "I've just had the lid off for a look and Janice's ashes are still in their urn", cemetery owner Hyman Prepuce told us.

I WAS WATCHING a film where an ancient Egyptian mummy came back to life and went round strangling people. It set me wondering, has the famous mummified pharoah **Tutankhamun** ever come back to life, and if so, did he go round strangling people?
Dr M Baker, Haltwhistle

★ Tutanhamun, son-in-law of Akhenaten died at the age of 18 in 1340BC and his tomb lay undiscovered until 1922. On his death, his organs were removed and put into canopic jars and his brain was pulled down through his nose with a big hook, so it is unlikely that he ever came back to life. However, a British Museum spokesman confirms that there was a curse found on his tomb, so he could not rule out the possibility that the boy king's bandaged corpse could come back to life at some point to wreak terrible revenge on those who disturbed his eternal slumber.

WE WERE saddened to here of the death of Australian naturalist **Steve Irwin** after he was stang off a fish. My family was wondering if he is still dead, or whether he has any plans to appear on our screens again.
T Coatzee, Luton

★ Crocodile hunter Steve, with his catchphrase 'Oh, Crikey', brought the world of Australian wildlife into millions of homes. However, he is presently in a wooden box under six feet of earth at Australia Zoo, and has no current plans to return to television presenting.

MY MOTHER used to adore the actress **Noele Gordon** who died a long time ago. I wonder if you could tell me what she is doing now, or is she still dead?
Maurice Micklewhite, e-mail

★ Noele Gordon, or Nolly to her fans, appeared in over 6000 episodes of Crossroads in the 60s and 70s. She passed away on April 14th 1985 and has been dead ever since. However, a true professional, the late Noele still works tirelessly and this Christmas is appearing in Aladdin at the Hull Empire.

Could THIS be the HOUSE OF THE FUTURE?

Twenty-five years ago who could have imagined what houses would look like today. In 1962, how many housewives could have forseen the advent of the microwave oven and the upright 'jug' kettle? And could our grandfathers possibly have imagined electric lawn mowers and 'non drip' paint?

Yet today, things which our parents could not have believed possible, things like cordless telephones and processed cheese, are here to stay. Or are they? What changes will take place in the **NEXT** 25 years? What kind of place will we call home in the year 2012? Read on and you will see, as we take a glimpse into **THE HOUSE OF THE FUTURE**.

By our Science Correspondant FRANK TURD

BEAMS

In the **KITCHEN** French speaking robot chefs will do the cooking on wind powered ovens. There'll be no dishes for us to wash. Special magnetic beams will hold your meal in mid air above the table. Simply open wide and a mouthful of food will pop itself into your mouth.

A typical menu for the year 2012? How about a delicious green roast of meat with blue potatoes and orange peas? Farming will take place on far away planets like Jupiter and Mars, and new brightly coloured space animals will become a regular part of our diet.

But be on the look-out for mice in the kitchen of the future. By the year 2012 a new breed of highly intelligent mice will have learnt to use tin openers and may even do their own cooking.

BUTTON

Fans of **TELEVISION** will need a microscope to keep up with their favourite soap operas. Because in 25 years time scientists will have developed a TV set smaller than a grain of sugar.

A special room in your house will be set aside for futuristic entertainment. At the press of a button highly advanced computers will transform the room, whisking you away as if in a time machine to the time and place of your choice. At the flick of a switch you could be playing for England in the 1966 World Cup Final at Wembley, or wrestling with a harmless rubber lion in a realistic African jungle setting.

Even today's slim, modern push button **TELEPHONES** will have become valuable antiques by the year 2012. In the house of the future the telephone will have hundreds of buttons and bright flashing lights. Instead of a ring, it will make a more modern, sort of space age bleepy noise. And to use it, you won't need to say a word. Simply pop the handy sized space age receiver into your car, and your thoughts will automatically be beamed into the telephone.

MICROWAVE

In the **BEDROOM** new technology will mean better sleep. Using the microwave principle, special beds with see through doors and revolving matresses with enable you to have 8 hours sleep in 3 minutes!

Meanwhile invisible **DOORS** will baffle burglars while letting in extra light, and there'll be no need to open and shut **WINDOWS** by the year 2012. Traditional glass windows will be replaced by sheets of ice which will remain frozen in the cold weather, or simply melt when it gets too hot inside the house.

Petrol will long since have run out, so in the **GARAGE** you'll find a water powered car. A giant tank on the roof will contain enough water to drive all ten wheels at speeds of up to 500 miles per hour. And for a fill up there's no need to find a garage. Just stop at the nearest river and help yourself!

GARDENING will also be made easy by a new generation of garden plants. Special trees will constantly sprinkle water onto the flower beds while automatic roses prune themselves. And you'll not need a lawn mower either, because grass will grow sideways instead of upwards.

NUCLEAR

And there'll be no need to worry about **BURGLARS** if you go away on holiday. You'll simply take your house with you. At the push of a button you'll be able to convert the house into a gigantic nuclear powered helicopter and fly it anywhere you like.

Dinner is served in the kitchen of the future.

Win 40 Senior Service!

We have looked briefly at the shape of houses to come in 25 years time. But can you imagine what houses will look like in an incredible 50 years time? Can you picture a house of the future in the year 2037?

Why don't you try and draw us your idea of what houses will be like in fifty years time. We'll be extremely surprised if anyone bothers, but we're offering a prize for the best design in any case. Send your pictures to 'House of the Future in the year 2037 Drawings Competition', Viz Comic, 16 Lily Crescent, Newcastle upon Tyne, NE2 2SP. We'll publish the best efforts in a future issue, and our prize – 40 Senior Service cigarettes – will go to the sender of the most imaginative picture.

POP IN THE YEAR 2000

First came Elvis – with his tight trousers. Then came the Fab Four, with mop top hair cuts and electric guitars. Then there was the Bay City Rollers, with tartan patches and stack heeled boots.

And now we've got Kylie, with her great big teeth and tiny tits. Yes, the face of pop music is ever changing. Stars rise and fall in the winking of an eye, as the pop roller coaster trundles ever onwards. As well as changing fashions, new technology greatly influences the music scene. '78's were replaced by singles, then singles gave way to '45's. Cassettes replaced albums, and now LPs have given way to the 'CD'.

INSIGHT

So what can the pop pickers of tomorrow expect to listen to? And what kind of stars will be Top of the Pops in the 21st century? Here's a fascinating insight into the amazing space age world of pop in the year 2000.

In the year 2000, the world's top pop event, the Eurovision Song Contest, will be a thing of the past. Instead, space viewers from all over the Universe will tune in on their satellite TVs to watch the Galactic Vision Song Contest. Alien pop groups from all over the cosmos will perform their entries on a show hosted by a silver wigged Terry Wogan, broadcast live from the BBC Television Centre – *on the moon!* Co-host Katie Boyle, who'll be a weightless 150 years of age, will have to learn a few new languages. For she will need to converse fluently in over 1,000 *space languages*, as well as French and German.

Space age Terry

However, some things will never change. Despite the new look contest, nobody will actually watch it, and Norway will still come last.

Cliff yesterday

Everyone knows that the perennial Peter Pan of pop Cliff Richard never ages. He looks exactly the same today as he did when he first shot to stardom in 1952. Indeed, the familiar face of religious Richard will be the only one to survive into the charts of the future. And needless to say, the 84 year old rocker will look exactly the same as he does today, except for even more wrinkles, slightly more leathery skin and a thicker pair of glasses. And a wig.

There'll be no more queueing for pop concerts in the year 2000. Stars of the future will play especially for you – *in your own living room!* Crafty record companies will use advanced bio-technology to 'clone' their best selling stars. And you'll be able to rent a perfect life-like replica of your pop favourite from the corner shop – for around the same price as a video cassette.

BEDROOM

And the good news is that your favourite star will gladly stay the night, and give you a very special *bedroom performance*. The only problem is this extra service will cost saucy pop fans an extra £2,000 a night!

Top record companies are already experimenting with genetic technology, and are believed to have produced a prototype clone of teen sensation Chesney Hawkes. However, that experiment was abandoned after the Chesney clone exploded killing a guitarist during a top secret invitation only experimental test gig at a venue in London.

Chesney clone – blew up

Advances in medical science will mean that by the year 2000 pop stars who have been dead for decades – such as Jimi Hendrix, Buddy Holly and Adge Cutler out of The Wurzels – can be ressurected. 'Born again' stars, dug up from their graves and revitalised by scientists using space chemicals – will be able to continue their careers where they left off.

NAPPY

By the year 2000 Elvis will be top of the charts once more, weighing in at 45 stone, dressed in a silver space age nappy, and recording songs in a specially built giant reinforced toilet/studio in the basement of his Gracelands mansion.

Pop stars have always had a reputation for drink and drugs. But the drugs of the future will be a far cry from the powder and pills they take today. In the future all drugs will have been legalised – except marijuana – and to get a *buzz* in the year 2000 pop stars will simply have to nip into the local newsagent and buy heroin, cocaine or new extra strength *space drugs,* all of which will be available in a 'firework' form.

INJECTIONS

There will be no sniffing, smoking or injections involved. Pop stars will simply return to the privacy of their hotel rooms, light the blue touch paper, and shove the firework up their arse.

A pop star beseiged by groupies yesterday

Of course one great advantage of being a pop star is being able to sleep with lots of groupies. But sadly, by the year 2000 sex will have become a thing of the past. Absolutely everybody will be riddled with AIDS, so sex will all have to be done by computer, in special disease proof space greenhouses on the moon.

In years to come record players – and even CDs – will be dusty relics on a museum shelf. Gone will be the shiny discs we buy today. For record companies have already spent *billions* perfecting the new Strip Disc – a long piece of paper containing musical information in

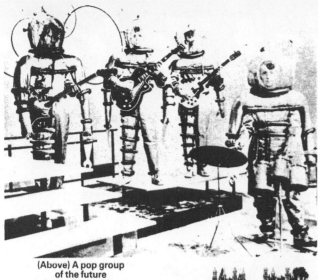

(Above) A pop group
of the future
playing live on Mars

the form of a space age bar code. Simply slip the strip into a pocket sized 'DDDD' (Digital Disc Decoder Deck) and out will come octophonic sound, ten times louder than when it was recorded.

 Believe it or not, record companies are already working on new technology that will eventually supersede the Strip Disc. By the year 2020 they will have done away with records altogether. Instead fans will buy a *'pop' drink* by their favourite band. Simply take a mouthful, and as you swallow your favourite tune will appear in your ears.

ALBUM

Top stars of the future will release their album in three formats – can, bottle or carton. And record companies will boost their summer sales with special limited edition *ice pops*.

 The cumbersome instruments which pop stars carry with them on tour – drums, guitars and pianos – will all be consigned to the rubbish heap of tomorrow by the year 2000. Musicians of the future will need only one walnut sized special purple space crystal, mined on Jupiter, and capable of playing all their music. They will simply *think* the tunes into the crystal, and it will glow and throb, and the tunes will appear, along with a low, rhythmic, humming sound, like on the Tomorrow People.

Scientists from EMI Records are already planning a space voyage to Jupiter to find out whether these crystals actually exist or not.

Radio One jocks
in the year 2000

 Turn on your tranny in the year 2000 and you may be surprised to hear the familiar voices of all your favourite Radio One FM jocks. For in order to preserve their popular but ageing DJs, Radio One controllers will use Dr. Who style technology, removing and pickling their brains before wiring them up to electronic voice boxes. Like the Daleks, the DJ's brains will live in robot bodies, and slide around the corridors of Broadcasting House, bumping into things.

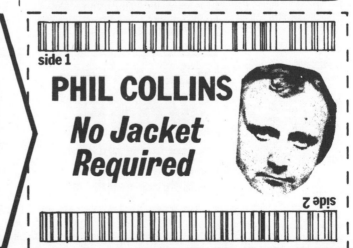

Local Artist Takes Saatchi Prize

A LITTLE known north eastern artist was last night £20,000 richer after scooping the prestigious Saatchi Prize for Contemporary Art.

Monkseaton-born Jason Woodscrew, 14, was the unanimous choice of the judges, who praised the way his work "embraces the poetic, the logical, the sexual and the sensual, whilst drawing connections between, without and within them."

works

Two works singled out for particular praise were *'Suck my Nips'* (2001) and *'My Penis Smels of Apples'* (2002), both executed in car touch-up aerosol on concrete.

cooker

The award is welcome news for Tyneside's bid to be named European City of Culture 2008. Newcastle and Gateshead council spokesman Paul Rubinstein said: "This is a great boost for culture in the region. The Angel of the North,

PORTRAIT OF THE ARTIST: Woodscrew, captured on CCTV yesterday.

By our art correspondent
ALFRED SHAN De BASS

the Millennium Bridge and the Baltic Centre have already marked the north east out as a centre of creativity in the arts. To have these prize-winning artworks on display in the region is just the icing on the cake for the region."

man

As usual the award has prompted controversy. Many critics felt that the £20,000 should have gone to 13-year-old Scott Bradawl of Cullercoats for his marker pen on bus shelter piece 'Angie Does Anal for Tabs' (1999). However, Bradawl failed to make the shortlist for the third year running.

bill

The winning artworks are on exhibition at the Links shelter, Whitley Bay seafront until February 6th when they will be scrubbed off by a man from the council. Admission free.

Evening Standard Art Critic Quentin Bumboy on
'My Penis Smels of Apples'

"What is the artist saying to us in this painting? It seems to me that Woodscrew's use of the word 'penis' establishes him firmly in the art historical tradition of male nude painting. Since the earliest cave paintings, the penis has been a symbol of strength. And yet also of weakness and vulnerability. It smells of apples; that is, it is redolent of the instrument of man's fall from grace. But at this point Woodscrew delivers his *coup de grace*. By mis-spelling the word 'smell', he challenges the veracity of our senses, causing us to question our notions of truth and being. He leaves us wondering if his penis 'smels' at all, and if it does, whether it smells of apples. Or perhaps some other fruit.**"**

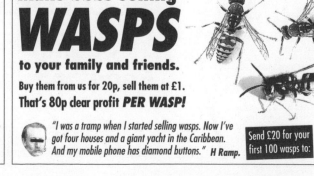

JURASSIC GARDENS

In the millions of years since dinosaurs last roamed the Earth many changes have taken place in our environment. Continents have been torn apart, mountains thrown up out of the sea and huge oceans created by melting ice.

Greenfingered stars could play host to dinosaurs

So what would dinosaurs make of our planet today? How would these giant reptiles react to our motorways, our high rise blocks and our huge, sprawling cities. And how, for example, would these pre-historic monsters adapt to living in the gardens of some of TV's best known stars?

DETAILS

We fed details of some of the best kept gardens of the stars into a special dinosaur computer. And we were surprised by some of the results. For they suggested that the gardens of many of Britain's top celebrities and entertainers could prove to be ideal homes for large dinosaurs.

PATROLS

As well as being a respected broadcaster and celebrated sexual deviant, TV's **FRANK BOUGH** is also a renowned authority on dinosaurs. He has written many books on the subject including *'Let's Be Frank – Frank Bough's Earnest Opinion of Dinosaurs'*, and *'Dinosaurs – A Grandstand View by Frank Bough.'* So we asked Frank to sift through our dinosaur data and pinpoint exactly which celebrities' gardens would be best suited to accommodating these gigantic extinct reptiles of the Jurassic era.

MICRAS

"Dinosaurs like a wet environment", said Frank. "Huge and slow moving, they would spend a lot of their time partly submerged in water. For that reason we are looking for a garden with either a pond or a swimming pool, or perhaps both.

They also eat trees, and so plenty of high vegetation is another must. A nice row of coniferous hedges –Llaylandi for example – would be ideal, as these are fast growing and could be replanted after they'd been eaten up by dinosaurs.

PRIMERAS

A further consideration would be wildlife. Some dinosaurs, such as the Tyranosaurus, were meat eaters, and they would need a supply of food in order to survive. A household with pets – cats and dogs for example – would be an advantage. Or ideally a garden with a rabbit problem. Rabbits can do untold harm to plants in your garden, but a hungry dinosaur would soon put a stop to that, I can tell you".

SUNNIES

Frank also warned of some of the dangers to dinosaurs that are inherent in many of the gardens of the stars.
"I'd be very weary of overhead wires such as telephone cables and electricity supplies. Dinosaurs are very tall, and don't exactly have brilliant eyesight, so they'd be prone to all sorts of accidents.
Also a busy driveway, with cars coming and going, could be a danger to both drivers and dinosaurs. Dinosaurs might be scared of cars, but then again they might mistake them for other dinosaurs, and try to have a fight with them", said Frank.

RAINIES

Finally we gave Frank sketch plans of the gardens of several top stars from the

BRUCIE: Room for lots of dinosaurs

Busy Jason has no time for uphill gardening

Chris's garden is ideal for a dip.

world of TV entertainment, and asked him to pick a top three, judging each garden on its own individual suitability for dinosaurs.
"I must say it was a difficult choice, but in third place I've chosen the garden of keen golfer **BRUCE FORSYTH.** Bruce's garden is spacious, with room for quite a few dinosaurs. He has a pond, lots of trees, and plenty of grass for them to walk around on.

WINDIES

"The only minus factor in Brucie's garden is the fact that the dinosaurs would probably make big footholes in the putting green which Bruce uses for putting practice. But apart from that, an excellent garden for dinosaurs".

DOORS

In second place Frank plumped for the unusually small garden of former Neighbours heart throb **JASON DONOVAN.**
"I chose Jason's small back garden behind his London flat because of the many trees which overhang it from the adjoining park. Although it would only be big enough for one small dinosaur, providing it's neck was long enough it would certainly have plenty to eat. And because Jason spends very little time in his garden, due to his hectic showbusiness commitments, he probably wouldn't mind a dinosaur in it".

WAALS

In first place Frank chose a worthy winner – the garden

of TV funster **CHRIS TARRANT.**

GARDEN

"Chris has a nice big garden, but its main advantages is that it borders onto a river. Dinosaurs could roam happily in Chris's orchard, sit around on the patio if they wanted, and then take a dip in the river to take some of the weight off their legs. I'd say that without a doubt Chris's garden would be my number one choice for accommodating dinosaurs".

Design a Celebrity Dinosaur Garden

Here's another fabulous competition, and this time we're offering a thirty second trolley dash around the Dinosaur section of a top bookshop to the winner. You'll be able to grab as many books about dinosaurs as you can carry, and keep the lot absolutely free!

ODDIE

To enter all you have to do is design your ideal garden for dinosaurs. Just draw a detailed plan of the layout of the garden, marking on it trees, ponds, crazy paving etc. And to give us an idea of the scale, draw a big dinosaur somewhere in it, plus the celebrity whose garden you think it could be.
Send your drawings to Viz Celebrity Dinosaurs Gardens Competition, P.O. Box 1PT, Newcastle upon Tyne, NE99 1PT. And remember to draw a dinosaur on the back of your envelope. Entries must reach us by the end of June, and the winner will be chosen by celebrity gardener Percy Thrower, who is dead. There will also be a runners-up prize of a portable radio/CD/cassette player supplied by *Richer Sounds* for the entry we like best, so hurry up and get drawing. Competition closes 31st June 1993.

CAREFUL. IT MIGHT BE A TRAP.

IS THE 'OLD BOY' OF POP OVER THE TOP?

IS BOY GEORGE PAST IT?

Boy George - unable to hide a look of concern.

Culture Club 'on the way out' - claim

Rumours are rife that at 23 sexy singer Boy George is over the hill. And experts believe that his group Culture Club could be on the way out.

Comment

Yesterday a spokesman for his record company told us that George was unavailable for comment.

YOU COULD BE A VIZ NAUGHTS & CROSSES MILLIONAIRE!

It's so simple!

All you have to do is complete a row of three crosses by adding ONE 'X' to the grid on the right. And if you're lucky, you could win a bottle of champagne.

The entry fee is £1 million. Your cheques should be made payable to 'Viz Comic', and enclosed with your completed entry form.

```
 O | O | X
---+---+---
   | X |
---+---+---
   |   | O
```

I enclose a cheque for £1,000,000.

Signed _____

Cat in shell shock

Soldiers at the Royal Artillery Regiment had given up hope of finding their mascot 'Sheba' the cat after she had gone missing at their base in West Germany.

That was until observant Gunner Ken Green heard noises coming from an unexploded shell half buried in mud on the target range.

For the adventurous cat had somehow got lost in the regiment's arsenal and had found it's way into a fifty pound artillery shell.

"How she got in there I'll never know", said Sergeant Bill Brown, who looks after the cat.

"Luckily the shell did not explode on impact after it had been fired. Otherwise Sheba would have been a gonner".

Fortunately Sheba was unharmed during her firey experience, if not a little tired.

"Hopefully she'll have learnt her lesson" said Sgt. Brown.

"She must have used eight of her nine lives at once on this occasion", he added.

BARMY BILL IN

HEY LOOK! A RABBIT

HEY LOOK! A RABBIT

TOSS!

BRICK

BLAT!

FEARS GROW OVER RAC ORANGE MARCH

A planned RAC orange march is to go ahead in Birmingham next month, despite protests from AA members.

The RAC, who recently changed their corporate colour scheme to orange from the traditional red, white and blue, plan to march through the predominantly AA district of Bournville to celebrate the anniversary of their first ever roadside recovery.

Zephyr

In the historic road rescue an RAC officer riding a primitive motorcycle beat an AA mechanic to a broken down Ford Zephyr on the A38 near Edgbaston, and famously managed to jump start the vehicle. The event is commemorated annually, but this year's chosen route - along the hard shoulder of the M5 past the headquarters of the AA - has been labelled 'provocative' and has rekindled deep felt feelings of bitterness and resentment which exist between the rival motoring communities.

An AA yellow march passes angry broken down Green Flag members in the staunchly RAC district of Solihull yesterday.

Sirocco

Last night there were calls for calm from Birmingham police chiefs amidst fears that the AA is planning its own retaliatory march along the predominantly RAC Smethwick Road. P.C. McGarrett, Community Liaison Officer at Smethwick police station, issued a plea for reason to members of both organisations. "Already fragile relations are set to break down unless common sense is allowed to prevail", he said yesterday.

Mistral

Meanwhile the Rev. Ian Polkadot, leader of the hard-line Green Flag organisation, accused the AA of being sodomites, and having an average response time of well over an hour.

"The AA are the homosexual whores of hell and the sons of Satan! May God strike them down in the filthy cesspits where they lie. We the Green Flag have been chosen by Jesus to repair vehicles at the roadside. And if we don't get to your car within 1 hour we'll give you £10", balled the Rev. Polkadot yesterday.

Firemen have had their chips

THE chips are down for Britain's fire fighters, according to union bosses. For microwave cookers and the growth of convenience foods could soon consign our famous red fire engines to the dust bin.

According to fire chiefs the rise in popularity of MicroChips - the convenient if unpleasant tasting alternative to real chipped potatoes - may have sounded the death knell for Britain's brave fire fighting forces. For MicroChips, which cook in seconds inside a micro wave oven, alleviate the need for drunken pub goers to attempt deep fat frying late at night.

Passat

Since 1980 the number of chip pan fires attended after pub closing time has halved, according to official figures. And increasingly fire crews are having to rely on electrical faults and arson at school premises to scrape a living.

Twister

"Unless we act now the traditional British chip pan fire will soon be a thing of the past". That's the view of Barney McGrew, general secretary of the fire fighters

Angry fire fighters ponder their future as bosses announced a further cut in house fires yesterday

union the F.F.U. He believes the government should act now to encourage more drunken people to undertake dangerous deep fat frying.

Buckaroo

But the Prime Minister's father-in-law, 'scouse git' human chip pan actor Tony Booth, yesterday ruled out any such move. "It is inconceivable that a Labour government would en-courage people coming home from the pub to turn chip pans on and then fall asleep", he boasted to drunken soldiers in a Hampstead pub last night.

Sandwich

Meanwhile under-pressure Home Secretary Jack Straw refused to confirm that the government were considering a plan to promote more smoking in bed.

Teething trouble

DARTS player John Lowe broke down in tears yesterday as he told a Nottingham court how Larry Grayson's teeth bit him as they tried to escape from his mouth.

Lowe, 52, of The Park, Nottingham, is claiming damages and compensation against Mrs Ethel Grayson, widow of the former TV light entertainer, who he claims sold him the teeth which were unfit for the purpose for which they were intended.

Mousetrap

The court heard how Mr Lowe, a fan of Larry Grayson, attended a sale of the star's personal effects shortly after his death in 1993. He purchased several items including a chair and the teeth for which he paid £625.

He wore them without incident for several weeks until one day in March 1994 when, after attending a darts exhibition match in Skegness, he went outside and attempted to whistle for a taxi.

Fondue

Suddenly he became aware of the teeth attempting to get out of his mouth. Instinctively he tried to wrestle the teeth back in, and in the struggle that ensued he was severely bitten on the chin, the

The offending teeth pictured in Mr Lowe's mouth (above) and seen during happier times with their original owner Larry Grayson (below)

resulting wound requiring hospital treatment. The teeth escaped, but were later recovered from a bench at Skegness railway station.

Pastie

Speaking in Mrs Grayson's defence Sir Christmas Fartface QC told the court that the teeth had been sold "as seen", and his client had always assumed they were to be kept as a souvenir, for display purposes only. She had not been aware that Mr Lowe intended to use them for eating or smiling.

Foreskin

Mr Lowe is claiming £2,000 damages plus unlimited compensation for distress caused by the incident. The case was adjourned until Monday.

TUT'S CURSE STRIKES AGAIN

WHEN EGYPTOLOGIST Howard Carter smashed open the tomb of the Boy King *Tutankhamen* in 1924, he unleashed an ancient curse. Over the next 70 years, almost everyone involved in his expedition was to die in a series of bizarre accidents, illnesses and unexplained natural causes.

When the last member of Carter's team passed on, the Pharoah's curse appeared to\die with him, and King Tut's mummy once again rested in peace in its golden sarcophagus in the Great Pyramid of Cheops.

But this month, treasures from the tomb have gone on show in London and as a result, says one man, the curse has been reawakened once again.

According to Neville Oglesby, a car park attendant at the O2 Arena where the exhibition is being staged, the malevolent influence of Tutankhamen's spirit is behind a catalogue of misadventures that has befallen him in the last few weeks.

"We are dealing with an evil force too powerful to imagine," he told us. "The sooner these treasures go back to Egypt and the curse is lifted, the better."

HEXCLUSIVE!

"Nobody has died yet, but in my opinion, it's only a matter of time before history repeats itself once more," he added.

Oglesby first became aware that something was not right on the day the exhibits arrived from Cairo Museum.

"A massive articulated lorry arrived with all the artifacts packed in crates. I was reading one of my magazines in my booth by the barrier when the driver shouted out of his window. I checked his chitty and pressed the button to raise the barrier to let him through. To my surprise, the barrier stayed down."

"I pressed the button again once more, but nothing happened. The funny thing was, I knew the barrier was in good working order because the man had been out to fix it twice earlier in the week. It was as if some invisible evil force was stopping it from going up.

"Eventually, I went out and wound the gate up manually using the emergency handle like I do whenever it goes wrong. Returning back to my booth and my magazines, I wondered if I had incurred the wrath of a 6000 year-old Egyptian mummy. I don't mind admitting that a chill ran down my spine."

Neville may have thought that that was the end of the curse. Little did he know that it was only just a taster of the nightmare that the long dead boy king was about to unleash on him.

"The exhibition was set up, and all went well for a couple of weeks. Then one day I was in my booth leafing through my magazines, when I realised I'd run out of tissues. I'm not really supposed to leave the premises, but I often pop over to the local newsagent to buy tissues and tea bags, and I did on this occasion. I wasn't gone more than five minutes, but that was long enough for Tutankhamen to wreak his terrible revenge upon me.

"When I got back to my booth, the horror hit me. A cat had somehow got in and done its business in the box I keep my magazines in. The top one was quite old, and one of my favourites. I had looked at it hundreds of times, and now it was ruined. As I removed the cover to throw it away, it suddenly struck me that cats were sacred animals in ancient Egypt. Could it have been that my magazine had been spoilt by the spirit of one of King Tut's pet cats as retribution for me doing the car park for the exhibition? The more I thought about it, the more it seemed to be the only explanation.

"I am almost certain that I pulled the door of my booth to before I nipped to the shop. Only an ancient Egyptian hex could explain how a cat could get in there and do a nonsense in my mag box."

By this time, Neville was getting jumpy as he waited to see how the curse would next manifest itself. He didn't have to wait long.

> "They who enter this sacred tomb shall swift be visited by the wings of death."

PHARAOHS AT THE BOTTOM OF THE GARDEN: Car Park Attendant Neville Oglesby *(below)* ponders the Mummy's Curse *(left)*, yesterday.

CAR PARK

"It was two days after the cat incident and I'd had a busy morning. Three people had come knocking on my booth asking for change, and I was worn out, so I was looking forward to my lunch break when I could pull the blinds down and have a nice relaxing read. However, the curse of King Tutankhamen had other plans.

"It was strange, for no matter how hard I looked, I couldn't find my glasses anywhere. I knew I had them with me, because I'd been using them earlier to look at some pictures in one of my mags during my tea break. Now they had simply vanished into thin air.

"I remember thinking that if the Pharoah was capable of reaching through the millennia and magicking away my glasses, what else was he capable of? I don't mind admitting, my blood ran cold.

"I was forced to do without my glasses, and I had to squint to make out some of the pictures. When I had finished and went to put my magazine back in the box, I was amazed to find my specs hidden underneath one of the flaps at the top. How they got in there I will never know, as it's quite unusual for me

to put them there, and I'm almost 100% certain I hadn't done so that morning."

For a few days everything seemed

"We are dealing with an evil force too powerful to imagine. The sooner these treasures go back to Egypt and the curse is lifted, the better."

to be back to normal and Neville began to think that perhaps he had let his imagination run away with him. But then in a single day, a series of events took place for which he could offer no rational explanation.

* *OVERNIGHT RAIN leaked through the roof of his booth,* *ruining a brand new box of tissues.*

* *THE ELEMENT of his electric kettle, which had worked perfectly well for 18 years, suddenly burnt out.*

* *WHILST HE was reading magazines, the zip of his trousers got jammed in the down position.*

* *ONE OF the blinds in his booth shot up for no reason whilst he was flicking through some magazines, leading to a formal complaint being made by a passing lady customer.*

"I felt that the boy king was toying with me, showing me just how powerful his ancient curse

could be. I thought that hellish day would never be over, but it came to an end a bit sooner than I thought.

"I was called into my boss's office and given my marching orders. As a result of that lady's complaint, I was being sacked. It was true that I'd already had several verbal warnings and two writtens off him - but the last of those had been nearly three months ago. I hadn't put a foot wrong since then, which made my dismissal all the more cruel.

"There were hundreds of people involved in setting up the Tutankhamen exhibition, so why the mummy singled me out to be the victim of its curse I'll never know. All I know is that he'll now be seeking out someone else to exact his evil revenge upon - and that person had better look out."

OVER THE YEARS, many people involved with the discovery and display of Tutankhamen's artifacts have suffered misfortune, beginning with egyptologist Lord Carnarvon and ending with car park attendant Oglesby. But he isn't the only member of the O2 Arena staff who has fallen foul of Tut's curse.

* *DEPUTY* catering manager June Medford sprained an ankle after a vengeful Egyptian spirit caused her to slip on a wet floor in the foyer.

* *LAVATORY* cleaner Hilda Braintree was docked a week's wages after the Pharoah's malign supernatural influence caused her to oversleep and be late for work four days on the trot.

* *GIFT* shop cashier Shania Warhol was arrested and cautioned after the dog-headed god Anubis persuaded her to fill her locker with £120 worth of souvenir pencils, rubbers, key-rings and fridge magnets.

* *TEMPORARY* kitchen assistant Kirk Scoltock was fired on his first day after being caught by CCTV security cameras smoking a joint near the bins.

QUEEN SCUTTLES FERGIE ON PORN VIDEO

Plans for a blockbusting ROYAL SEX VIDEO starring the Duchess of York have been TORPEDOED by the Queen.

'Fergie's Royal Guide To Having It Off' was set to earn the Duchess a staggering **£10 MILLION** through advance sales alone. But Fergie's plans were sunk yesterday when the Queen fired a Royal broadside, banning her daughter-in-law from appearing in the no holds barred steamy educational love film.

FORKED

If the video had gone ahead eager Royal watchers could have forked out £10.99 to see:

● **NUDE** Fergie rolling around on a mattress.

● **EXPLICIT** love acts between the Duchess and a male model.

● **OPEN** displays of masturbation techniques.

● and pop shots.

Production of the red hot video was already well underway when the Queen attached limpet mines to its hull, fearing that the publicity such a tape would attract could damage the public image of the Royals. And Palace insiders believe the Queen will have no hesitation in blasting similar projects out of the water with depth charges.

KNIFED

"The Queen would indeed frown upon any member of the Royal family participating in a videotape record-

Porky and Bess in a flap over flap shots

ing which features explicit sexual acts", a Buckingham Palace spokesman told us yesterday.

GARROTED

The Duchess is no stranger to the Queen's naval gunpowder. Several crew members were killed last year when an Exocet missile fired by the Queen struck Fergie aft of the poop deck after she had announced plans to appear naked in Penthouse magazine.

Boozing Britain pisses it up the wall

Britain stands head and shoulders above its EEC neighbours when it comes to pissing money up the wall, according to a report published this week.

Extensive research was carried out over a period of twelve months to compare the drinking habits of all the EEC member states. And the results reveal some interesting facts about our European neighbours. For it includes the following findings:

★ The **FRENCH** drink more wine than anyone else – an average of 3 bottles each per day. Yet they have the weakest bladders – going to the toilet on average every 18 minutes whilst they're on the pop.

★ The **GERMANS** are Europe's noisiest drinkers. The average male drinks an astonishing 11 pints of lager each day, but as a result the average German spends 17 minutes per morning farting loudly.

★ The **IRISH** drink more Guinness than any other country in Europe, bar Finland, where it always tastes much better.

★ The average British male spends £170 a week on alcohol, pissing an incredible 60% of their pay packet up the toilet wall.

We're offering a super prize to the Viz reader who pisses the most money up the wall in a week. All you have to do to enter is send in the completed form.

ENTRY

Your entry form must be authenticated by your pub landlord and accompanied by a recent pay slip. We will then work out which Viz reader spends the highest percentage of his or her income on drink.

FABULOUS

The lucky winner will re-

ceive a fabulous first prize – two hundred quid's worth of booze, on us. We will pay, direct to your local, £200 which will be credited to your slate. And six runners up will each receive a can of warm lager, shaken in the post.

PERSONS

Send your entries to the address on the form. This competition is open to persons aged 18 and over only. People who own pubs, work in pubs and friends or relatives of people who own or work in pubs may not enter. Competition closes 1st March 1993.

To: Viz Piss It The Wall Competition, P.O. Box 1PT, Newcastle upon Tyne, NE99 1PT.

I certify that the below named person spends, on average, £_____ per week on drink in my pub.

Signed _____ PUB LANDLORD

Name and address of pub _____

I certify that my average weekly income is £_____ as verified by the enclosed pay slip/voucher/Giro slip etc.

Signed _____ Viz Reader

Please give your full name and address _____

FOR OFFICE USE ONLY – DO NOT WRITE BELOW THIS LINE

WAGE	PISSES	PERCENTAGE

SAVE OUR SPOOKS!

BOLLOCK brained Brussels bureautwats have sprouted a hair raising scheme that will send shivers up the spine of spirits all over Britain.

The Belgium based buffoons want to see a single European spook replace existing ghosts, poltergeists and apparitions by the year 2000.

Shockwaves

Plans to exorcise our estimated 200,000 spooks - and replace them with a standard EEC Euroghost - have sent shockwaves through haunted houses all over Britain. And last night the plan was attacked by Tory MP Sir Anthony Regents-Park.

Heatwaves

"Of course I don't believe in ghosts, and I'm certainly not scared of them. But even so, this is yet another example of Brussels bureaucracy gone mad".

Eurocrats plan to exorcize the Great British Ghoul

Of all the EEC member states Britain has by far the greatest number of ghouls. However traditional figures such as the Lady in White and Headless Horsemen have lost ground in recent years to more contemporary phenomena. These range from ubiquitous poltergeists throwing kitchen crockery to the equally common strange and unexplained presences in cars (accompanied by a sudden change of temperature) experienced by motorists driving alone late at night near the scene (and on the anniversary of) an horrific road accident.

Hi Tensions

Labour's Terry Nice was last night reluctant to condemn the EEC proposals. "Obviously we need to look very closely at the whole issue of ghosts and whether we believe in them and what, if anything, they should look like, because its an issue that affects all of us, but other than that I'm not going to say anything and I'll just sit on the fence hedging my bets and smiling a lot and hope that everybody will vote for me at the next election".

Internet attacker gets two years

In the first case of its kind in Britain a man has been convicted of assault after robbing a 72 year old pensioner on the computer Internet.

Wayne Pile, an unemployed 18 year old, was sentenced to two years in prison after a jury found him guilty of assault and robbery at the home of Percival Francis, a retired clerk of Putney, South London. Mr Francis had just sat down at his computer and was preparing to write a letter to a relative when Pile, who was 200 miles away in Sheffield, struck.

Tavares's

The robber was apprehended by police after an alert computer operator in Glossop spotted him acting suspiciously outside an E mail address in Hull. Detective Inspector Eric Fletcher who lead the investigation believes that computer crime is on the increase.

"The criminal will not hesitate to explore new avenues of crime, and as technology advances we must ensure that police resources are updated and criminal legislation constantly reviewed in order to remain abreast of the situation". He described the Putney attack as particularly vicious. Mr Francis was knocked to the ground and required hospital treatment for cuts and bruises. Pile escaped through the Internet with less than twenty pounds in cash and a pension book.

The Floaters's

In a similar case an American teenager was fined by a court in Ohio for throwing a waterbomb out of Windows 95 and hitting a passing pedestrian in Hong Kong.

SEX IN THE BI

We expose the
of Britain's m

This week a team of our investigators go *undercovers*, and delve beneath the duvets to expose a bonking bonanza which is taking place in bedrooms all over Britain. For sex between married couples is the new craze that's sweeping the nation in the nineties, as more and more husbands and wives jump on the bedroom sex bangwagon.

In bedrooms up and down the country seedy scenes reminiscent of porn movies are being acted out between sick husbands and wives. And the British bedroom, once the hub of the family home, now echoes to the sound of sex between horny housewives and their husbands.

IPSWICH

Mike, a former married man from Ipswich, told us that sex between husbands and their wives is on the increase. And where better for couples to do it than in the bedroom. Mike spoke openly to our investigators about having sex with his wife, in the bedroom.

MANSFIELD

"I'm not proud of what we did. A few friends of ours had already tried sex, so not long after we got married we bought a house and decided to have sex in the bedroom. At first my wife was nervous, but after a while I think she began to enjoy it. We started doing it regularly, in the evenings, usually before we went to sleep." Mike then began to describe a sex act which took place between him and his wife which cannot be reported in a family newspaper.

NORTHAMPTON

Mike separated from his wife three years ago. Now 32 and working as a motor mechanic, he doesn't blame bedroom sex for the breakdown in his marriage. "I don't think it does any harm. At the time we both seemed to enjoy it. I remember how we used to take our clothes off first. Sometimes my wife would lie on the bed sideways while we did it."

LUTON

Although no longer involved with bedroom sex, Mike still knows of many couples who are. "If anything I'd say its on the increase", he told us.

Lessons in lust

Unbeknown to her employers a Northampton school mistress is offering special home tuition in one subject only – sex!

Married mother of two Tina Harrison's extra curricular activities take place in the bedroom of her modest semi detached home in Plumtree Avenue, where the busty beauty offers naughty nightclasses between the sheets.

TRANSIT

Our reporter, who is married to Tina, went undercover – quite literally – to expose the saucy schoolmistress's sexy bedroom antics. After returning home from work he was greeted at the door by his wife who was wearing a pinafore, revealing blouse and a sexy short skirt. After being shown into the small, dimly lit living room Tina suggested they have a meal.

BEDFORD

"I've not cooked anything, but I could make some soup. Or get something out of the freezer", she told him. He declined the offer. Tina then sat alongside him, and put her hand on his knee. "Had a busy day at work?" she asked.

Later, when our man asked about the bedroom Tina seemed surprised. "It's a bit early for bed isn't it?" she asked. When he said he was tired and would like to go to bed, Tina lead him up the stairs to a small room with a bed in it, and a wardrobe. In one corner was a dressing table.

"I must remember to ring my sister in the morning", said Tina as she slipped out of her blouse and skirt to reveal a white bra and matching panties. "She left a message for me at work but I never got a chance to ring her today." At this point Tina seemed to

become suspicious, and began asking questions.

JENKINS

"Do anything exciting at work today?" she asked. Our man told her he'd had a quiet day. "Me too. Nothing exciting to report", she said, referring to her job at a local primary school.

Tina continued to undress, revealing her breasts and throwing back her long, dark hair before lying on the bed. She leaned over to our man and attempted to perform a minor sex act with her mouth on his lips. At this point he made his excuses and left.

Our 'Lone

A former shop assistant in Bradford uses the classified pages of her local paper to lure men into steamy bedroom sex romps.

Our investigator replied to a 'Lonely Hearts' ad placed in a local newspaper by a girl calling herself 'Dorothy'. They arranged to meet in a local pub. Dorothy turned up wearing a brown jacket, white blouse and sexy stockings. She immediately asked our man if he wanted a drink. "It's quite nice in here, isn't it?", she told him.

PERIOD

Over a period of several months, and after several similar meetings at different locations around the area, our man discovered that Dorothy was aged 36,

EDROOM

seedy secrets
arried couples

ly Heart' was a filthy tart

and been a shopworker until she was recently made redundant. At one stage she revealed that she was a fan of Phil Collins. "I quite like Rod Stewart as well. Who do you like?" she asked. Our man said that he liked Tina Turner.

HEADACHE

After a series of meetings our investigator suggested that they should get married. It was at this

point that Dorothy took our reporter to a semi detached house on the outskirts of Bradford where he was introduced to a man calling himself 'Dorothy's father', and a woman who claimed to be the girl's mother. Our reporter then arranged to meet Dorothy – who was wearing a sexy white dress with a bonnet and long train – at a local church where a man introducing himself as 'the vicar' pronounced them man and wife.

TIRED

That night, after a party, our man was lead away to a hotel room by Dorothy, who had slipped out of her white satin dress and was wearing just a skimpy silk nighty and sexy stockings with suspenders. "I'm just going to use the lavatory",

she told him as she left the bedroom momentarily. Seconds later she returned, naked, and lay on the bed.
"This will be my first time", she told him. "Please be gentle with me". At this point our man made his excuses and left.

THRUSH

Later we visited Dorothy and confronted her with photographs of the wedding. At first she seemed confused, then she began crying hysterically and collapsed. When we told her parents that Dorothy was a girl and that she had offered our man sex in the bedroom, her father went to the kitchen. Seconds later, he returned with a knife and lunged viciously at us. We made our excuses and left.

Tom gets the most from his post

Pensioners queuing at the quaint sub Post Office in the quiet Cotswold village of Chipping Bourton are unaware that their friendly post mistress leads a seedy double life. For at night time she becomes a leading light in the local bedroom sex circle.

Together with her husband Tom, Maureen Sanderson took over the post office when the couple moved into the village four years ago. And in their neat and tidy shop – which also doubles as the village florists – there are no signs of the seedy sexual activities in which the couple regularly engage.

BLACKBIRD

For in a bedroom directly above the shop the couple perform lurid sex acts between each other, while in the room below young mothers collect their child benefit, and purchase flowers.

SPARROW

Posing as a central heating service engineer, our investigator gained access to the Sanderson's one bedroom flat, and hid in the wardrobe. That evening the couple went to bed at about 11 o'clock. They seemed tired, and no sexual activity took place. The following day they went to bed half an hour earlier, but again they went more or less straight to sleep.

STARLING

On the third night Mrs Sanderson entered the room, sat at her dressing table and

began to remove her clothes. She took off her dress to reveal a flimsy bra and panties, before brushing her hair. Mr Sanderson could be heard yards away in the toilet, carrying out a crude lavatorial act. The toilet flushed, and Mr Sanderson then entered the bedroom wearing blue and white pyjamas.

BLUE TIT

Whilst removing her bra Mrs Sanderson briefly exposed her large breasts before slipping into a skimpy nightie. She then removed her panties, revealing a glimpse of part of her body that we cannot describe in a family newspaper. The moment she sat on the bed Mr Sanderson's arms wrapped around her, and the couple fell backwards, before moving beneath the quilt. Within seconds the couple were fondling each other's bodies.

PINK FANNY

After a series of lurid sex acts lasting approximately ten minutes the couple began to rock back and forwards rhythmically. The bed began to squeak, and Mr Sanderson began to emit a loud moaning noise. At this point our reporter went off in his pants, made his excuses and left.

ARE YOUR NEIGHBOURS BEDROOM BONKERS?

WITH BEDROOM sex rife in Britain in the nineties, the chances are someone in your neighbourhood is *at it.* It could even be the couple next door. And while you and your children lie asleep, on the other side of the wall a crude sex act could be taking place. So here's a few hints to help you pinpoint the bedroom sex perverts in your street.

● How often do your neighbours wash their sheets? If they wash them frequently, they are probably trying to get rid of embarrassing stains caused by sex acts having taken

place on or near the bed.

● Go through you neighbours' bins. Look out for condom packets in particular.

● Ring their doorbell at night, and see how long it takes them to answer the door. Do it again at regular intervals. If they take longer than usual to get to the door, they may have been committing a sex act in the bedroom when you rang.

● Build a small dam under a manhole cover in the foul sewage drain leading away from their toilet. In the morning, sift through the debris which

has accumulated there, keeping an eye open for used condoms.

● Break in to their house, and build a sinister 'nest' beneath the floorboards or in the attic space, and stock it up with survival equipment and tinned food. Dressed in para-military clothes, and wearing a ski mask, you will then be able to watch them closely, and monitor their sexual activity in detail, making notes and keeping an obsessive diary of the occasions on which they have sex, recording times, dates and other apparently insignificant details.

YOUR £1·25 Viz REALLY IS A LOAD OF JIZZ

Gary Bluto On The Box

The TV critic who gives it to you straight

These sicko queers ruin our telly

Anyone watch Wimbledon this year? What a bore. I blame the dykes.

You could dig up better looking women in Fred West's back garden. I'd rather go plane spotting in Lockerbie than watch lesbians playing tennis.

Centre Court attendances are down. Hardly surprising. Who in their right mind is going to pay good money to sit and watch a pair of lesbians knock a ball back and forward over a net.

All the excitement at Wimbledon will come in the showers afterwards. It must get pretty hot in there. Hotter than a barbecue in Fred West's back garden. Mind you, I'd rather watch paint peel in the Kings Cross tube disaster than watch sexual perverts cavorting in the showers.

No daughter of mine will ever play tennis. What parent in their right mind is going to allow their young daughter to engage in a physical activity with a bunch of sexual perverts? You've got more chance of finding your baggage in Lockerbie than you have of finding a straight girl on the tennis circuit these days. *I'd rather send my kids pot holing up Julian Clary's backside in Lockerbie.*

★ ★ ★ ★ ★ ★ ★ ★ ★ ★ ★ ★ ★

On the subject of gays, there must be about as much chance of Julian Clary gripping a pencil in his bum cheeks as there is of watching ten minutes of your favourite soap nowadays without a gay appearing on the screen. Gays here, lesbians there. They're popping up as often as stiffs in Fred West's back garden. There's even puffs in Emmerdale now. No wonder the ratings are dropping quicker than bodies in Lockerbie.

What parent in their right mind is going to allow their kids to watch sexual perverts cavorting on their TV screens? *It's about as healthy as unprotected sex with*

Brucey - He's no queer

Julian Clary on the car deck of the Herald of Free Enterprise.

When will the TV bosses learn that sodomy kills? If they had their way they'd all be buggering our kids in the playground at school. I'd rather send my kids rally driving with Ayrton Senna than allow them to be buggered in the playground.

What parent in their right mind would bugger their own kids in their back garden in Lockerbie? They come over here, they take the jobs. I'd rather watch paint dry up Julian Clary's backside. Take our women too. I wouldn't fancy being Fred West's gardener. I'd rather pay Hitler's gas bill. I know what I'd do with them. Put 'em all on a plane and send 'em back to where they came from. *Via Lockerbie.*

Tomorrow: Gary gives a five star review to Brucey's Play Your Cards Right.

Computers set to byte the dust

All computers are set to be banned if a Tory backed bill designed to stop computer porn becomes law.

The Computers and Bizarre Sex Acts Bill has been put forward by Tory MP Sir Anthony Regents-Park in an attempt to curb the current trend of computer pornography. The bill would make it an offence for any person to own or use any form of computer, other than a small electronic calculator.

PRESS

At a press conference yesterday a police officer demonstrated how it was possible to manipulate images on a computer screen for pornographic purposes. He showed members of the press how it was possible to take a harmless image of a chicken and distort it by giving it a huge donkey's cock, and then by simply pressing a button it was possible to make the chicken stick its donkey's cock up a pig's arse, again and again and again.

Animal porn at the push of a button

He then gave the pig huge tits, and made it rub them with its little piggy hands, before a sheep, with a horse's knob, joined in the fun. After a few minutes things were really hotting up. The pig and the sheep were joined by a herd of horny cattle for a farmyard sex orgy which included a lesbian show between two bulls who had been given pigs' tits and big hairy badgers' fannies.

STARCH

At the end of the demonstration Sir Anthony Regents-Park, whose suggestion it was that the bulls should have tits, congratulated police on their efforts.

MRS MILLS

In issue 67 of Viz magazine in an article about carpets, motorbike and crocodiles we referred to the late popular music pianist Mrs Mills as a 'fat cow'. We would like to point out that in so doing it was never our intention to imply that Mrs Mills was either fat, or a cow.

We should like to take this opportunity to apologise for any misunderstanding or offence which may have been caused.

CHEESE PLEASE

Britain's housewives are not afraid to ask for cheese, according to a report out today.

Amazing behind the scenes antics at the meteorological office

MY WEATHER MAP NEARLY FELL OVER

Viewers were unaware as cloud fell off map during TV broadcast

Taken from the book 'It's Exciting Being A Weather Man' by Brian Smith. (Sea Lion Books, £9.95)

SECRETS OF A T.V. WEATHER MAN

He's the man who appears on our TV screens after the news. The man who points at the map and tells us what the weather is going to be like. He is of course the TV weatherman, one of the under sung stars of the small screen.

But behind the colourful charts and cheery smiles there's a great deal that we, the viewers, never see. As can been revealed by former TV weather forecaster Brian Smith.

SECONDS

"Weather forecasts are usually only a few seconds long", Brian told us. "But often it takes several minutes of gruelling rehearsal to get it right". And things don't always go according to plan.

WOBBLED

Like, for instance, the time when Brian' weather map, probably the most important piece of a weather man's forecasting equipment, nearly fell over. "It wobbled for a few seconds, but I managed to prevent it from falling", said Brian. Luckily, that happened during a rehearsal. But things can go wrong — and often do — 'live' on TV.

DEPRESSION

"On one occasion I was pointing out a depression which was building up just North of the Shetland Islands", Brian told us. "Quite by accident I pointed at the Orkney Islands by mistake. Several viewers rang up and pointed out my error".

With only minutes to go until the end of the news, weather man Brian Smith irons out a technical problem.

However it's technical problems rather than human error which the TV weather man most fears, as only too easily things can go wrong. Like the time a cloud fell off Brian's weather map.

RAINFALL

"I was positioning clouds on an afternoon weather chart to represent likely areas of rainfall, when suddenly, quite out of the blue, one of them fell off. Luckily I managed to remain calm and talk my way through the rest of the forecast without further incident".

ICY ROADS

Of course TV weather men are professionals, trained to handle exactly that kind of situation; And often viewers may not even realise anything has gone wrong.

Raunchy behaviour shocked office secretary

TV weather forecasting is of course a pretty exciting business. But not many would-be TV weather men realise the dangers involved.

FROST

"The studio is usually a jungle of wires, with TV monitors, lights, cameras and other heavy equipment all around the place. You could quite easily have a nasty accident if you're not careful", Brian told us.

Brian is lucky in that he never had an accident in his 6 months as a TV weather man. But many others are less fortunate. "One well known weather man who was working for the BBC at the time, managed to trip on a loose wire while making a long range forecast, and a nurse had to look at his ankle".

FOG

TV weather men are often labelled 'quiet types', unlike some of their more outlandish colleagues in TV light entertainment. But as Brian tells us, this 'boring' image could'nt be further from the truth.

BOTTOM

"You'd be amazed at some of the goings on behind the scenes at the meteorologicaloffice" he revealed. "We're not as innocent as we appear on the screen". Brian continues "On one occasion, after we'd been out to the pub for a lunchtime drink, our secretary turned around to find a queue of us waiting to pinch her bottom!"

INTERMITENT SNOW SHOWERS

Finally, says Brian, weather forecasting does have it's lighter moments. He recalls how in one incident in 1949 he was a junior forecaster preparing to read a 'live' forecast for the West coast of Scotland. "Suddenly, I heard a snap, and when I looked down, my trousers were around my ankles. My braces had snapped!".

"Luckily I was only doing a radio forecast at the time, otherwise it could have been quite embarrassing!"

The Star with the Sinister S...
LIKE FAT...
LIKE ...

IN 1930s Berlin, UNITY MITFORD was part of the Nazis' inner circle, where she was known as Hitler's British girl. It is well documented that a close relationship developed at this time between the Führer and the third youngest of the aristocratic Mitford sisters.

At the outbreak of World War II, Mitford had no choice but to return to England. Unable to imagine a life without Hitler, she shot herself in the head with a pearl-handled revolver. But she survived this bungled suicide attempt and was flown home to Britain.

Historians say that shortly after her return she was packed off to a maternity hospital in the quiet Oxfordshire village of Wiggington near her home. It was here, sometime in 1940, that Unity Mitford gave birth to a baby boy, the son of Adolf Hitler.

There are no records of where the boy went, but it is assumed that he was given up for adoption. Now, 69 years on, we wonder who that boy grew up to be. He could be your next door neighbour, your local butcher, even the headmaster of your children's school.

But an infinitely more chilling thought is that he could be a familiar face from the world of entertainment. It is a frightening fact that we could be welcoming the evil architect of the

---

CLIFF RICHARD

FOR: As anyone who seen chilling footage of the Nuremberg rallies will know, Hitler was able to whip any crowd into a frenzy with his hate-filled speeches. And Cliff showed that he was every bit as charismatic as his possible father when, during a rain-soaked Wimbledon afternoon in 1996, he famously led the Centre Court crowd in a rousing chorus of his hit *Congratulations*.

Like Hitler, the Peter Pan of Pop has also shown a penchant for Aryan-looking women, dating a very short string of blue-eyed blondes including Olivia Newton-John and Sue Barker. And just 20 years after Hitler's jackbooted stormtroopers tore through Europe in their Panzers, Richards did the same, albeit at the wheel of a Routemaster double-decker containing Melvyn Hayes and Una Stubbs.

But perhaps the strongest evidence of Cliff's sinister paternity is that in his 1962

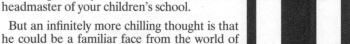

hit of the same name, he vowed to be a bachelor boy until his dying day... a chilling echo of Adolf Hitler, who gave up his bachelor status the same day that he died in his Berlin bunker.

AGAINST: Cliff twice represented the United Kingdom at the Eurovision Song Contest, finishing second and third. Hitler, as a fervent German Nationalist, never represented another country in a televised pop music contest.

And the Führer was also a committed vegetarian, whereas in Cliff's shit 1973 film *Take Me High*, his character Tim Matthews opens a beefburger restaurant in Birmingham.

HITLER GENES RATING: 卐卐卐卐卐卐卐

TOM JONES

FOR: According to official records, Tom Jones was born Thomas Woodward. However, some time before 1965 he decided to change his name to the one by which he is better known. This behaviour is eerily reminiscent of a certain Adolf Schickelgruber, who rose to fame under his adopted name of Adolf Hitler.

As young men, Jones and Hitler had dark, brooding looks which proved irresistible to women. When making public appearances, each regularly found himself showered with ladies' underwear.

Both were fond of mountain scenery; Jones, the green hills and valleys of his native Wales; Hitler, the snow-capped vistas of his beloved Austrian Alps.

In 1987, Jones the Voice got to number 2 in the charts with his record *A Boy From Nowhere*. Was that song's cryptic title a veiled reference to Tom's mysterious origins as the son of the Nazi leader?

AGAINST: Adolf Hitler was a well-known dog lover, and was often photographed petting his giant alsatians, Rudolf and Hans. Tom Jones however, has a morbid fear of dogs after being bitten on the lip by his uncle Richard's Yorkshire terrier in 1947, during a family caravan holiday in Barry Island. This aversion to man's best friend is a clear indication that the swivel-hipped Welsh vocalist may not share the Führer's DNA after all.

Also, from newsreel footage, it is clear that Hitler spoke in a strong German accent, whilst Jones sings in a Californian accent and speaks with a Welsh lilt when he remembers to.

HITLER GENES RATING: 卐卐卐卐卐卐

JIMMY TARBUCK

FOR: "Winner Takes All" was the evil philosophy behind Hitler's dastardly plan to take over the world. In a chilling echo, it was also the title that chirpy scouser Tarbuck chose for his 1970s betting-based ITV quiz show. Was this a tip of the hat to his long-lost genocidal father?

Before he founded the Nazi party, it is well documented that Hitler earned a living as a house decorator. According to Tarbuck's former neighbours, early in his career the gap-toothed comic decided to paint his own Merseyside flat in order to save money. By all accounts, he made quite a good job of it. Is this merely a coincidence? Or is it damning evidence that the Führer's painting genes had been passed on to his long lost lovechild from Liverpool?

AGAINST: Tarbuck is rarely off our television screens, where he delivers an endless stream of sidesplitting wisecracks. How different from Adolf Hitler, who was renowned for his lack of a sense of humour. Hitler's speeches, delivered to the massed ranks of his brown-shirted stormtroopers, contained no gags about his mother-in-law, how poor he used to be or the Irish, and were greeted with deafening shouts of "Seig Heil!", not gales of helpless laughter.

The Führer also rarely addressed his followers whilst standing with his thumbs tucked in his waistcoat pockets. Instead, he preferred to keep his right hand raised in a Nazi salute and the first two fingers of his left hand held horizontally across his top lip.

HITLER GENES RATING: 卐卐卐卐

PETER STRINGFELLOW

FOR: At first glance, the megalo-maniacal despot and the thong-wearing nightclub entrepreneur would seem to have little in common. But on closer inspection, sinister similarities between the two begin to emerge. Both have comedy haircuts and you wouldn't want either of them to go out with your daughter.

And the similarities don't stop there. Like his infamous possible dad, 67-year-old Stringfellow travels everywhere in a top-of-the-range Mercedes. The ageless lapdancing club proprietor also claims to have slept with 3000 different women. Hitler, likewise, caused many people to suffer unimaginably dreadful fates.

History repeats itself in other ways too. On July 20th 1944, Hitler narrowly escaped an assassination attempt in his Berlin bunker when a briefcase bomb exploded beneath his desk. And things also regularly go off under the tables at Stringfellow's clubs in London and Paris.

In yet another inexplicable coincidence, rumours are rife that, like the Nazi leader before him, Peter has only got one ball after suffering a bizarre vending machine accident in the foyer of the Albert Hall.

AGAINST: Hitler and Stringfellow's tastes in entertainment are perhaps too different for them to be father and son. While the Führer enjoyed a night at the opera watching statuesque sopranos in Viking hats belting out Wagner's *Ring* cycle to the tops of their voic-es, Peter likes nothing more than sipping champagne to a soundtrack of Donna Summer records whilst a nude 19-year-old with plastic tits rubs her fanny up and down a scaffolding pole.

HITLER GENES RATING: 卐卐卐卐卐卐

...ret... We Name Hitler's Baby

...ERLAND ...SON

HEIR HITLER: Which celebrity is the spawn of the Führer and Unity Mitford?

Holocaust's son into our homes every night.

So which one is it?

It is only by deciding which star is Hitler's son that we can prevent his atrocities from being repeated, so it's time to decide once and for all who it is. Here, we weigh up the evidence for and against a host of well-known celebrities, all of whom were born in 1940, before finally naming the guilty Nazi party...

STAN BOARDMAN

FOR: The curly-haired scally stand-up famously joked that he didn't like the "Jairmans" because the "Fokkers" had bombed his local chip shop during the war. Boardman repeated this on television thousands of times during the 1970s and 80s, so many times in fact that it begins to look suspiciously like a case of protesting too much. What better way could there be of hiding his real Nazi ancestry than by pretending to bear a grudge against the entire German race?

Take away Boardman's trademark bubble perm and brush his naturally straight hair sideways across his forehead, and the scouse laughter-maker begins to look much more like Adolf Hitler. Add a toothbrush moustache and the resemblance increases noticeably. It could be argued that only the true bastard spawn of Hitler would go to the extreme lengths of changing his hair and not growing a Hitler moustache in order to disguise his true genetic inheritance.

AGAINST: Boardman's professional comedy career began when he worked as a holiday camp redcoat. If he really was Hitler's heir, he would surely have insisted on wearing a brown shirt instead.

But perhaps the most compelling argument against Boardman being Hitler's child is this: even at the height of the Blitz, it seems unlikely that the Führer would have given the order to bomb his own son's local chip shop.

HITLER GENES RATING: ✠✠✠✠

MELVYN BRAGG

FOR: According to official records, the adenoidal *South Bank Show* presenter was born in Wigton, Cumbria in 1940. We already know that Unity Mitford gave birth to Hitler's baby in WigglNGton, Oxfordshire in that year. Is it possible that the birth certificate was conveniently altered with Tippex to put any Nazi hunters off the scent?

As host of the *South Bank Show*, Bragg has a great love of the arts, which is a trait he shares with his potential father. When the German government launched its own television service in 1936, one of the first shows transmitted a Sunday night arts magazine *Das Sudetenbank Geschowenschaft* ... presented by Adolf Hitler. Then as now, nobody watched it because it was a load of pretentious wank.

And the suspicious similarities don't end there. The German leader was known to have a penchant for urinating on women in order to obtain sexual fulfillment. Who is to say that Melvyn Bragg has not inherited this vile perversion? In over 30 years of writing historical novels, travel books and screenplays - as well as broadcasting about theatre, music, literature and the visual arts - he has not once denied weeing on Lady Bragg to attain orgasms.

AGAINST: When Bragg was offered a peerage, he reluctantly accepted the honour so that he could change the House of Lords and the outmoded system of privilege it represented from within. Hitler, on the other hand, opted to attempt to change them from the outside by carpet bombing Parliament with thousands of tons of explosives in 1941.

An-other point that counts against Bragg being the Führer's son is his bohemian bouffant hair-do, which is a far cry from Herr Hitler's orderly diagonal fringe. And if Bragg is the fruit of Hitler's loin, he has not inherited his late father's taste in neckties, preferring as he does a huge 70s footballer-style knot as opposed to the Führer's neat, restrained version.

HITLER GENES RATING: ✠✠✠✠✠✠

JEFFREY ARCHER

FOR: There is strong evidence to support the theory that disgraced peer Jeffrey Archer is the long-lost son of Adolf Hitler. In many ways, Archer's life has been a carbon copy of the Nazi leader's. Both were puny children who grew up to be members of right wing parties - Hitler was architect of the Third Reich, an empire he envisaged lasting a thousand years, whilst Archer was Conservative MP for Louth from 1969 to 1974. Both spent their youth in uniform - Hitler in the German army, Archer as a trainee beat constable in the Metropolitan Police.

Both were sent to prison. Hitler was sentenced to 5 years for high treason when, in 1923 as head of the National Socialist Party, he launched a coup in an attempt to depose the German Government. Archer served 4 years for perjury after he lied in court whilst suing a newspaper that had accused him of having sex with a prostitute. Which he had.

Both men used their time behind bars to write books; the bile-filled anti-Semitic ramblings of *Mein Kampf* in Hitler's case, and the shit-filled self-pitying ramblings of *A Prison Diary* in Archer's.

AGAINST: The only evidence that points to Archer being anything other than Hitler's son is his own denial. Indeed it is certain that Archer, a convicted perjurer, would doubtless swear in a court of law that he was NOT the offspring of the 20th century's most evil dictator.

HITLER GENES RATING: ✠✠✠✠✠✠✠✠✠✠

WHO'S HERR HITLER'S HEIR?

ANY ONE OF these stars born in 1940 could have turned out to be the bastard son of of the world's most evil dictator. But after forensically sifting through the evidence, there is only one suspect who ticks all the boxes.

It seems clear that in 1940, after falling pregnant to the German dictator, Unity Mitford returned to England and secretly gave birth to **JEFFREY HOWARD ARCHER**, the future Baron Archer of Weston-super-Mare.

Nobody knows what the world would be like if Archer's evil father had survived the Second World War. But one thing is for certain...

If the Führer HAD lived to see how young Jeffrey Hitler turned out, he would be ashamed of the way he had brought disgrace on his family name.

Brando boost for jobs on the Tyneside waterfront

'New pants' hope for North shipyard

HOLLYWOOD heavyweight actor Marlon Brando is set to save a Tyneside shipyard. Bosses at Swan Hunter are on the verge of securing a £26 million contract to supply new pants for the giant superstar. And if the deal goes through it will guarantee work for 600 men at the yard over the next three years.

Who

The 1800 ton Brando, by far the biggest actor in the world, is at present moored in Ireland where he is being used for scenes in the new Hollywood movie 'Divine Rapture'. Movie bosses have asked British yards to tender for the new pants because of the logistical problems in shipping an existing pair out from the States. However Swans face competition from rival shipbuilders Harland & Woolf. The Belfast yard are also desperate to win the pant contract.

Come

A cult figure since the fifties, Brando largely gave up acting forty years ago to concentrate on watching TV and eating crisps. The cost of maintaining him in working order is enormous, and film producers pay anything up to ten million dollars a day to hire the Brando for film appearances.

From

Meanwhile the Irish Government last night confirmed plans to dump one of Brando's stools at sea had been abandoned. The giant stool was to have been towed out to sea and then scuttled in 2 miles of water 200 miles North West of Scotland. However Greenpeace protesters blocked the move after clambering aboard the gigantic turd to stage a dramatic 'sit on'.

The protesters claim significant quantities of effluent and residual crisp flavourings such as Salt 'n' Vinegar would be emitted from the stool causing pollution on the sea bed. Government scientists had strenuously denied any contamination from the turd could make its way back into the food chain.

Knotty Ash

The stool is presently attracting sight see-ers to Galway Bay where it is moored awaiting a decision on its future.

Scenes like this may soon return if the Brando pant order comes to Swans

10 things you never knew about Massive Marlon

Weighing 1800 tons fully laden and costing $10 million a day to run, he's the biggest star in the world. Here's ten titanic facts you'll be amazed to hear about Hollywood's heaviest heartthrob.

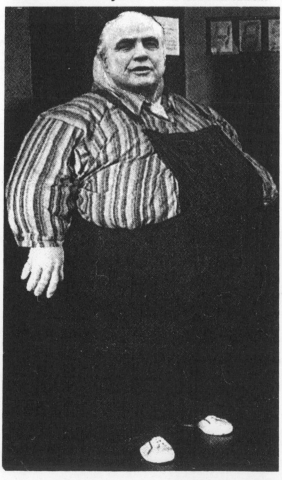

1 Irish crisp manufacturer's 'Tayto' have put their Dublin factory on overtime to cope with munching Marlon's demands for his favourite potato snack. The factory's entire output - over 200 million packets a week - are being snapped up by Hollywood film producers to satisfy the star's appetite.

2 Brando must eat a staggering eight times his own body weight in crisps every day in order to simply stay alive.

3 A twelve man team working in appalling conditions labour day and night to keep Brando's enormous gut stoked with crisps. They shovel around 12 tons of them into his cavernous mouth every minute.

4 Cleaning Brando's teeth is a never ending task - quite literally. When workmen finish at one end of his mouth, they simply turn around and start again at the other.

5 Built in 1923, Brando was converted to an actor in the fifties and first appeared in cult movies such as On The Waterfront.

6 A superstructure was added to his mouth for filming of the seventies gangster movie The Godfather in which he appeared as a Sicilian mobster.

7 Brando's captain was found guilty of negligence and failing to keep a watch after his actor was in collision with fellow star William Conrad near Los Angeles in 1982. Brando suffered minor damage to his stomach. 'Cannon' star Conrad was breached at the arse, and three of his crew were killed.

8 Brando's top speed fully laden is a mere 3 miles an hour. In an emergency it would take him up to two miles to stop or change direction.

10 Plans to convert Marlon into a multi million pound floating hotel and restaurant were scuppered in 1987 when New York harbour officials refused to grant the star a fire certificate.

ROCK AGAINST DINOSAURS

Fellow Geordies Sting and Jimmy Nail have joined forces in an unlikely battle – against dinosaurs.

For the Tyneside singing and acting duo share a common fear that dinosaurs could reappear on Earth. And they hope that by acting now they can prevent the giant creatures from conquering the planet.

By our Showbiz Staff

DINOSAURS

Jimmy Nail, alias TV tough guy cop Spender – admits that the idea was his. Jimmy has suffered from dinosophobia – the fear of dinosaurs – since early childhood.

Nail – scared of dinosaurs

COPING SAWS

"There's just something about dinosaurs that terrifies me", Jim confided. "As a child I was scared to go out of the house in case there were dinosaurs in the street. It was a really serious problem".

COLD SORES

On one occasion young Jimmy stayed in bed for three weeks. "I was convinced there was a dinosaur under my bed. I thought if I put my foot on the floor it would bite it off". Jimmy eventually clambered onto a wardrobe and escaped to the kitchen where his mother was less than sympathetic. "I got a right bashing", says Jim.

Sting – also scared of dinosaurs

HERPES

"Even then I didn't realise how serious a problem it was. It's only now, when I've started to talk about it to other people, that I've begun to realise how much it has affected me".

HIS CHIPS

Jimmy was delighted when he read that fellow Geordie Sting also suffered from anxieties concerning dinosaurs, and he immediately got in touch. Now the chart topping double act are working together to overcome their fear of the giant reptiles, and they hope to enlist the help of other rock stars, actors and Northern celebrities.

LINDISFARNE

"We've had a great response so far", Jimmy told us. "Brendan Foster and Steve Cram have both offered to help, and Lindisfarne will be playing a benefit concert to raise funds for us in the near future. Jimmy hopes to set up a registered charity N.E.S.A.D. – North East Stars Against Dinosaurs – to help channel funds to where they are most needed and help fellow celebrities overcome their dinosaur anxieties.

Geordies join forces to battle prehistoric menac

Nail's nightmare

Here's Jimmy Nail's six most frightening dinosaurs.

Triceratops – a powerful, thick set, relatively small dinosaur with distinctive horns. "My worst nightmare is being trapped in a small space – like my bedroom – by one of these. They're so strong, yet manoeuvrable", says Jimmy.

Tyranosaurus – the largest meat eater ever to stalk the Earth. "It's the sheer power of this one that frightens me. It can actually run faster than an average saloon car, and bullets would bounce off it".

Stegosaurus – This North American dinosaur of the Jurassic period had distinctive boney plates along its spine. "It's those horrible sharp spikes on its back that worry me most", says Jim.

Plesiosaurus – a giant water lizard of the early Cretaceous period. "It sounds silly, but when I first met my wife I wouldn't get in a bath in case one of these came up the plughole. She has helped me to overcome that fear, but I would never swim in the sea".

Tyranosaurus rex

Rogetsthesaurus

Rhamphorhynchus – a giant flying pterosaur descended from tree-living lizards. "With it's leathery wings and razor sharp beak I am terrified that one day one of these could swoop out of the sky and eat me", says Jimmy.

Deinonychus – a small carnivorous dinosaur which hunted in packs. "The thought of being attacked and pulled to the ground by a pack of these is never far from my mind".

20 THINGS YOU NEVER KNEW ABOUT HOUSES

An Englishman's house is his castle, or so the saying goes. And let's face it, in Britain today we've all gone house buying bonkers! Whether it's a semi in Sunderland, a mid-terrace in Mid Lothian or a flat in Flamborough, we all want to own our own home.

Nowadays nine out of ten per cent more people own their own homes than they did previously. And it's a figure that's growing. But as we spend more and more money on property, do we know what we're buying? How much do we **really** know about houses? For instance, did you know ...

1 When the Queen and Prince Phillip bought their first house, Buckingham Palace, in 1948 they paid around £2,500. If they were to sell it today it would probably fetch a staggering £10 million — or more if they left the carpets.

2 Rising house prices are nothing new. When Stonehenge was originally built it probably cost its owners no more than a few shillings in cave man money. When we rang the National Trust yesterday and asked them how much it was now worth, they told us it was not for sale.

3 The average house contains 32,400 bricks — enough to make 54 small coal bunkers, just over 38 outside lavatories, 6 garages or 2 houses half as big as the first one.

4 Not all houses are made of bricks. Some are made of timber, others wood. And believe it or not, in the fairy tale of the same name, Hansel and Gretel found a house that was made out of gingerbread.

5 If you went into an estate agents in Iceland and asked for a house, you'd be in for a big surprise. Because people in Iceland live in igloos, a kind of circular house made out of snow.

6 If you are buying a house, get to know the people you're buying it from. There will be many things to discuss, and being on good terms with them will help things run more smoothly. Organise day trips together, or perhaps a weekend in the country. Or if you're going to the cinema, why not give them a ring and ask them if they's like to come.

7 House music — music that is about houses — first appeared in the early eighties when Shakin' Stevens wrote his smash hit 'This 'Ole House'. Stevens later went on to record another hit record, 'Green Door' which was about the front door of his house.

8 If the barman tells you your drink is "on the house", don't worry. You won't need a ladder to reach it. 'On the house' is simply an expression meaning that you do not have to pay.

9 There have been many TV programmes with the word 'house' in them. These include 'On The House', 'Bless This House' and 'Man About The House'.

10 And 'Little House On The Prairie'.

11 Going home at night must be very confusing for TV celebrity Bob Monkhouse. For the 60 year old comedian has no less than **three** different houses to choose from. He lives in one, another is his television programme 'Bob's Full House', and the third one is part of his surname.

12 If you talk about houses in a bingo hall, the chances are that someone will give you a prize. That's because in bingo the word 'house' means that you have won.

13 Greenhouses are not green as their name suggests. And they're not houses either. They are in fact small glass sheds used for growing vegetables.

14 Everyone has heard of The Housemartins, but did you know that there are two types? One is a pop group from Hull, the other a small bird which nests under the eves of a house. And there are also people who live in houses called Martin, as well.

15 So that means there are three

16 The smallest house in the world was built in Massachusetts, U.S.A. in 1811 for the world's smallest man, Calvin Phillips. However the house, which was less than one foot high and had a front door no bigger than a playing card, was so small that even Mr Phillips did not fit in it, and he died a year later.

17 Try running up the stairs in a bungalow and you'll fall flat on your face. That's because a bungalow is a special house without any stairs.

18 No matter where you place a snail, it will always find its way home. That's because snails carry their homes on their backs! Even the world's strongest man, Geoff Capes, could not match that feat. In fact he couldn't even carry his bathroom for very long without falling over.

19 Like many animals, African Elephants do not live in houses. They just live in the street.

A house

20 Humans and cuckoos are unique within the animal kingdom, because neither of them build their own homes. Cuckoos break into other birds nests while they're out and lay their eggs on the floor. Humans simply call in the builders.

Not nice not to see you... not to see you, not nice!

SHOWBIZ EXCLUSIVE!

Now you see him, now you don't - Brucie (left) and a computer generated picture of how the invisible Brucie would have appeared in his bathroom (inset)

SHOWBIZ BOFFINS were left scratching their heads yesterday after Generation Game host **Bruce Forsyth** went invisible for almost 45 minutes.

The 75-year-old all round entertainer was in the bathroom of his £80,000 Surrey mansion when he suddenly dissappeared from his own sight in the mirror.

Speaking at the gate of his £100,000 Essex mansion, a visibly shaken Forsyth told reporters: "One minute I was there brushing my teeth, the next I was nowhere to be seen. I was worried that I might have became a vampire, so I called down to my wife, Isla StClair, former Nicaraguan Miss World, Anthea Redfern."

mansion

Blonde Anthea, 22, came running up to the bathroom and couldn't

By our London Palladium Correspondent
TOMMY FART

Forsyth's a jolly invisible fellow

believe her eyes. "It was really spooky," she told newsmen at the gate of the couple's £125,000 Berkshire mansion. "All I could see was Brucie's toothbrush going up and down in mid air. I didn't know what to make of it, until I heard the unmistakeable sound of his tap shoes dancing across to the toilet. That's when it dawned on me that he must have of went all see through."

light

Former Miss Paraguay Redfern was so shocked that she gave her husband a twirl and fainted on the bathroom floor. She came round to find that Brucie was still completely invisible. She said: "His unseen hands were patting me with a damp flannel whilst his disembodied voice kept

saying 'Alright my love?'"

The big-chinned song and dance man was due to record a special variety show that afternoon and telephoned the BBC to explain his predicament. Frantic TV bosses set about organising bandages, gloves and a pair of dark glasses so that the show could go on. However, three quarters of an hour later Forsyth's visibility returned as quickly as it had faded and producers were able to breathe a sigh of relief.

partridge

Bruce's agent Clitoral Hood commented: "It was a worrying time for everyone, but especially for Brucie. In this business, being completely transparent is box office poison. The public simply won't turn up to see an invisible performer, it's as simple as that."

Professor Ernst Bauhaus, head of

physics at the Variety Club of Great Britain confessed that he was baffled by Forsyth's sudden disappearance. He told reporters: "Showbiz personalities usually reflect light from their surfaces. The different textures and refractive properties of their bodies, clothes and faces allow us to see them as the celebrities we know and love. Why a star should suddenly allow light to pass straight through him unhindered is a phenomenon that is still poorly understood."

quail

Forsyth is not the first British entertainer to succumb to sudden unexplained invisibility. Radio 2 afternoon show host Steve Wright vanished from view for 4 days in 1999. Listeners to his piece of shit programme remained unaware of his predicament, whilst his sycophantic 'afternoon posse' of talentless fuckfaces were forced to suck up to a pair of floating headphones. "They had to throw flour over him in order to find his arse to lick," said a BBC source.

Didn't they do invisib-well?

What would the gameshow hosts do if they went invisible?

BEING unseen to the human eye may sound like fun, but in his film *'The Invisible man'* H.G.Wells pointed out that being completely transparent has as many pitfalls as it has advantages. Bruce Forsyth only vanished for 45 minutes and was never able to fully explore the pros and cons of his condition, but just imagine what the veteran comic could have got up to if he'd been invisible for a *whole day!* We asked Brucie's fellow gameshow hosts how they would spend 24 hours of invisibility.

"I'd go down to the Yorkshire TV canteen and pinch a load of napkins from right under the waitress's nose," grinned asinine Countdown host **Richard Whiteley**. "Then I'd hurry along to my colleague Carol

Vorderman's dressing room and settle myself down in a corner for a very special day of non-stop masturbation."

"I know exactly what I'd do," smiled Question of Sport presenter **Sue Barker**. "I'd walk right up to the trophy cabinet at the All England Club and finally pick up the Ladies' Singles plate, which I never came anywhere near winning during my frankly lacklustre sporting career."

They Think It's All Over host **Nick Hancock** had other ideas. "I'd go on a very special shopping spree," said the cheeky God-botherer. "I'd walk into Superdrug and help myself to the biggest box of Kleenex in the shop before spending the afternoon masturbating furi-

ously in the ladies' changing rooms at Miss Selfridge."

"TV viewers know all about my wicked sense of humour," said *Weakest Link* presenter **Ann Robinson**. "A day of invisibility would be the perfect opportunity to go round to Cilla Black's house and make objects float round the room so she thought she was being haunted by her late husband Bobby."

"It would be easy to abuse such a situation, but I would try and use my invisibility for good purposes," University Challenge questionmaster **Jeremy Paxman** told us. "However, I can't think of any way to do that off the top of my head, so I'd probably just end up wanking like mad in the showers of the local nurses' home."

Have YOU got what it takes to join Britain's toughest
DO YOU DARES

Fancy yourself as an elite, highly trained killing machine?

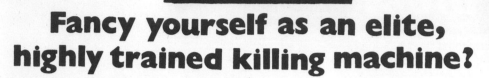

Dressed to kill – an SAS soldier armed with the latest anti-terrorist Embassy storming equipment.

Who can honestly say that at some time or other they have not dreamt of joining the SAS – their motto "Who Dares Wins" – the crack force of highly trained soldiers feared by terrorists and enemies of Britain throughout the world.

But how many of us have got what it takes to join the SAS? Soldiers who apply to join the regiment must undergo many months of tough training, and some of the toughest applicants fall by the wayside. Only a handful are hard enough to join the SAS. So what are your chances of making the grade? Before you consider applying to join the SAS, try completing the following questionnaire.

Answer each question a, b, or c, then tot up your final score to reveal whether you've got what it takes.

Who dares wins – the SAS storming an Embassy yesterday.

1. You set off on a caravan holiday to Cornwall, but your car breaks down with 200 miles to go. What would you do?
a. Call the AA or RAC. It may be a loose connection, or the points may need adjusting.
b. Cancel your trip and return home by train or bus disappointed.
c. Pack the entire contents of the caravan into a rucksack and yomp the rest of the way, taking the most mountainous route available.

2. You have gone to stay with a friend who is getting married the next day, however when you arrive he has no spare beds in his house, and a relative is sleeping on the settee. What would you do?
a. Book into a nearby bed and breakfast.
b. Wrap up well and sleep in your car.
c. Smear your face in animal droppings, and go and live in a nearby forest for a week, feeding on nuts, berries and hunting wild animals.

3. You go shopping for some new clothes. There is a good selection in the shop. Which of the following outfits would you choose?
a. A smart but casual jacket, corduroy trousers and a paisley tie.
b. A pair of comfortable cricket flannels or slacks, and a lambswool v-neck sweater.
c. Black trousers, black roll-neck sweater, black balaclava, bullet-proof vest, lightweight boots and a gas-mask.

4. You begin to notice that your next-door neighbour is coming and going at strange hours of the day and night. You suspect therefore that he may be an international terrorist. What would you do?
a. Mind you own business. It's none of your concern.
b. Ask discreetly around the neighbourhood in order to put your mind at rest.
c. Smear your face with animal droppings and hide in a pile of mossy twigs in his back garden for six weeks, compiling a detailed dossier of his movements.

5. You are in a baker's shop when you notice an important foreign diplomat purchasing a Belgian bun and half a dozen finger rolls. Suddenly an Arab terrorist steps forward brandishing a semi-automatic pistol. What would you do?
a. Dive for cover behind the pastry counter.
b. Lie flat on the floor and do exactly as you are told.
c. Swiftly disarm the terrorist using martial art skills before breaking his neck with your bare hands alone, and then dive on top of the diplomat to protect him until the police arrive.

6. You arrive at a restaurant for a meal, but are told by the head waiter that no tables are available for a least twenty minutes. What would you do?
a. Sit down and enjoy a drink until your table is ready.
b. Go to another restaurant that isn't quite so busy.
c. Smear your face in animal droppings, then throw a flash bomb into the salad bar before dragging everyone outside and forcing them at gunpoint to lie down in the car park, then return to pick the table of your choice.

regiment ~ the SAS?

WINS?

WHO DARES WINS

7. Your neighbour has asked you to look after his six thoroughbred dogs while he is on holiday. However, after a couple of days one dog is off his food and looks a little unwell. What would you do?
a. Ignore it. It's probably just pining for it's master.
b. Take it to the vets for a check-up. It's better to be on the safe side.
c. Take the dog to nearby waste ground, put a pillow over it's head and shoot it. Then return and kill all the remaining dogs to make sure you got the right one.

8. You pop round to a friend's house to see if he wants to go out for a drink. However, when you get there, there is no answer and the door is slightly ajar. What would you do?
a. Return home, and call back later.
b. Nip round the back to see if he's in the garden.
c. Burst into the house keeping your back to the wall and go from room to room, spraying the walls with bullets and occasionally doing a forward roll.

9. Whilst shopping in the supermarket an old lady catches your ankle with her trolley. When you get to the checkout you notice that the skin is slightly broken. What would you do?
a. Just forget it. It's only a scratch and it will heal itself in due course.
b. Nip back to purchase some elastoplast and some antiseptic cream.
c. Hastely improvise a makeshift field hospital in the fruit and vegetable

section, and sever your leg below the knee using your Swiss army knife, then seal up the stump with a red hot iron, in case it goes septic.

10. You arrive home from work only to find that you have lost your house keys. You try the doors and windows but they are all locked securely. What would you do?
a. Return to work to look for your keys. If you cannot find them you can sleep in the office for the night.
b. Pop to a friend's house nearby until your wife returns. She has her own set of keys.
c. Smear your face in animal droppings, before busting into a neighbour's house using a sledgehammer. Leave the occupants bound and gagged in a downstairs room, then make your way up to the attic and remove a skylight before clambering along the roof towards your house, tying a rope round your chimney, absailing down your back wall and crashing in through a second floor window.

HOW DID YOU DO?

Award yourself one point for every answer a, two points for a b, and three points for each answer c.

Less than 10: – *Oh dear me. You'd be better suited to joining Dad's Army than the SAS. But don't worry – the TA would love to hear from you.*

11 to 20: – *Not a bad result. You're tough, but not quite tough enough. There could still be a career for you in the Royal Marines or the Parachute Regiment.*

21 to 30: – *Congratulations! You've got what it takes. Next time the SAS storm the Iranian Embassy* **YOU** *could be the first one in. Hurry down to your army careers office immediately and ask for an SAS application form.*

THEY'RE AT IT AGAIN ~ claims Ron

Billions of pounds being wasted on the construction of the Channel Tunnel should be spent on preparing Britain for war. For German plans to begin World War III are already well underway.

This is the startling claim being made by keen amateur historian Ronald Windthorpe who believes the German surrender of 1945 isn't worth the paper it's written on. And while Britain prepares to do business with her colleagues in 1992, Mr Windthorpe believes the Germans are building towards another blitz. "Jerry's a sly old fox", he told us, speaking from the air raid shelter in the back garden of his Lincolnshire home. "He's been quiet for too long. He's up to something, and this time we better be prepared for it".

HUN

"We should have learnt our lessons in '39 when Jerry caught us with our pants down", said Mr Windthorpe. "This time we should be ready for them, because if we aren't, we may find ourselves on the losing side".

BOSH

According to Mr Windthorpe dramatic measures must be taken immediately, among them the re-introduction of conscription, food rationing and the internment of all foreign nationals living in Britain. In the face of Government apathy, it has been left to Mr Windthorpe to fight a lone battle against the Bosh, and he has soldiered on bravely with his own preparations. Every morning he cycles two miles to the nearby seaside town of Mablethorpe to scan the horizon for signs of an

enemy invasion. And he insists his wife, Joan, carries her gas mask with her at all times. The Windthorpes' two children, Sarah, 24, and Michael, 29, have been staying with an aunt in Wales since 1972.

SQUAREHEADS

There are several ways in which we can prepare ourselves for the advent of war, and Mr Windthorpe recommends that everyone adopts the following simple measures in order to protect themselves and their families.

* Fit blackout curtains to all windows, including skylights, and at night cover your car headlights so that they cannot be seen.

* Stick masking tape onto windows in the shape of a cross

* Build a bomb shelter in the garden by leaning sheets of wood against a kitchen table, and then building up a layer of sandbags around it.

Unfortunately, Mr Windthorpe's war efforts have come to a temporary standstill as he is currently awaiting sentence, having pleaded guilty at Cleethorpes magistrates court to a charge of theft after he was caught removing a road sign from the hard shoulder of the M62 near Goole. He asked for 362 similar offences to be taken into consideration. Sentence has been deferred pending psychiatric reports.

1939 and Jerry sets off for Poland.

Lettersocks

Britain's Most Socksually Explicit SOCK FORUM

I OFTEN have trouble telling my right from my left, which results in all sorts of embarrassing mishaps. In order to help me out a friend bought me a pair of socks with 'R' on the right one and 'L' on the left. However I put them on the wrong feet, and consequently took a wrong turn into the ladies' public lavatories and was arrested.

T Harpic
Tooting

SOCK FACT
Many people think that socks were invented by the inventor Sir Edward Sock, but they're wrong. In fact, Sir Edward invented 'Soccer' - a game similar to Football - and he took the name from the big long socks that players wear, which he also happened to invent!

I ALWAYS lose a sock in the wash so now I wash them in threes.

J Domestos, Liverpool

E VERY year at Christmas I always ask for something good, like a drill, but always end up being given socks. Over the years I have found myself becoming attached to my collection of socks, so much so that last year, I asked for socks. And what did I get? Well, socks, actually, so it was alright.

P Toilet-Duck, Hastings

I USED to get my husband socks every year at Christmas, until he got really cross and told me he was sick of being given socks! I thought he was being very ungrateful until he pointed out that he was Arthur Askey, the no-footed dead entertainer and, as such, had no use for socks!

Mrs A Askey, Liverpool

I THINK that the cast of ITV's Cold Feet simply aren't wearing thick enough socks. If I was them I'd try a pair of hiking socks, like Thorlos from Millets.

J Vim, Hackney

SOCK FACT

A sock is technically a shoe, without a sole or laces. A sock made out of rubber is called a 'Wellington Boot', or 'Welly'. A sock appendaged with fingers that you put on your hand is called a glove. And one you put on your head is a hat.

SOCK FACT
Despite being called Socrates, Socrates didn't actually wear socks! He wore sandals, a strappy, leather type of sock, worn on the feet. However, Socrates the Brazilian footballer DID wear socks, invented by Sir Edward Sock, who also invented the game Soccer.

Sock Puzzle Time

SOCK WORD

```
 1
[ ]
[ ]
 2
[ ][ ][ ][ ]
[ ]
[ ]
```

Down
1. Something you put on your feet, not socks (5).

Across
2. Something you put on your feet, not shoes (5).

ODD SOCK OUT

Which is the odd sock out of these three socks?

A B C

Answers: Sockword: 1D, Shoes. 2A, Socks. Odd Sock Out, Sock C.

The things kids say... about Socks!

ONE day my Grandson and I were walking past an airfield and we saw a windsock. "Gosh Granny, the wind must have really big feet!" he piped up. How we laughed.

Ada Cremation, London

I WAS taking my family's clothes out of the wash the other day and realised that I had lost some items of clothing. "Where are the red socks?" I pondered aloud. "Why granny", chirped my grandson, 'they're in Boston!' He meant the Boston 'Red Sox', an American baseball team! I am still laughing now.

Cissy Burial, Luton

WHILST doing the laundry for my family recently I wondered out loud how it would be possible to change the colour of the load (whites) to a slightly pinkish colour. "Put a sock in it, granny!" came my grandson's response. How I made sounds from my throat while breathing out in short bursts as a way of expressing amusement.

Dolly Embarmed, Ryde

I WAS doing the ironing for my family recently and I happened to say aloud "I really hope that someone puts an item of footwear onto the protruding part of my face through which I smell!" "Granny," my grandson sighed. "If you don't shut up I'm going to give you a sock on the nose!!" How I laughed and laughed and laughed and laughed and laughed.

Lily Mummification, Leeds

Viz HEALTH

Ask Mrs Parker the Viz GP's Receptionist

Dear Mrs Parker,
I've been having dizzy spells for a few weeks. I put it down to over tiredness at first, but it's been getting worse and now I've started getting terrible headaches at the same time. Does the Viz GP have any idea what it might be?

R Sinclair, Croydon

Mrs Parker says...
The Viz GP is very busy at the moment. He couldn't possibly look at your letter this week. There's a chance he might be able to squeeze it in next Monday, but you'll have to write again on Friday.

Dear Mrs Parker,
I'm a non smoker and I've had a cough for a month now. I just can't seem to shift it. It's worse in the mornings and for the past few days I've been bringing up blood. Is it possible to get the Viz GP to answer my letter today?

G Sprake, Leicester

Mrs Parker says...
There's no way the Viz GP can read your letter today. The earliest he could read it is two weeks on Wednesday at 4.25. If it's an emergency, you could always write to the Viz Accident and Emergency department.

Dear Mrs Parker,
I've found a lump in a rather sensitive area and I'm very worried about it. Could a male Viz GP read my letter as soon as possible?

B Possiter, London

Mrs Parker's answer service says...
The Viz GP's receptionist is on her lunch at the moment. Please write back between 2.00 and 4.30.

Do **YOU** need advice from the Viz GP? See if you can get past his receptionist, Mrs. Parker at The Viz GP Surgery, Viz Comic, PO Box 656, North Shields, NE30 4XX.

I THINK THAT ELEPHANT IS GOING TO CHARGE US

GAME RESERVE ENTRY £5

SHAME OF LOTTO MILLIONAIRE
CRIMINAL!

LOTTERY bosses were left with egg on their faces yesterday when it was revealed that Wednesday's jackpot winner is a genocidal murderer.

Camelot bosses admit they knew winner Pol Pot was responsible for the systematic killing of millions of innocent Cambodians during violent purges of pro-Vietnamese communes by his ultra nationalist Khmer Rouge organisation during the seventies and eighties. However they said there is nothing under present Lottery rules to prevent a murderous revolutionary tyrant winning, providing he has bought a ticket and abides by the rules.

Winners

"We are very pleased for Mr Pot and, as with all winners, we wish him well. Inevitably people will try to dig up stories like this, but a lot of it is based purely on envy. We would hope that the press will allow Mr Pot and his family to enjoy their good fortune in peace."

Rods

However, while Pot was collecting his cash bonanza lottery fans were already branding the revolutionary leader's £14 million rollover win a SCANDAL. Father-of-three Say Samrin, from Kompong Cham, spends £50 a week on Lottery tickets. And he believes it is wrong that a genocidal maniac should be allowed to carry off the prize money.

"It's not fair when honest folk's money is given to criminals. He goes around killing people and he gets £14 million. Meanwhile all them people who he's murdered and put their skulls in a big heap, they aren't getting a penny. Where's the justice in that?" asked Mr Samrin.

Miles's

Since his highly controversial win Pot has been disowned by members of his own family. His sister, Brenda Pot, 52, yesterday branded her million-aire brother a 'greedy scumbag'.

"He can keep his money. We don't want anything to do with it", she told us from outside the tidy Phnom Pehn semi which she shares with her partner Frank and their two children.

"The greedy scumbag's turned his back on us. We've always been here for him, but now he's a millionaire he doesn't want to know. He's even forgotten the kid's birthdays. Well we're not interested in his money. He can keep it".

Pal

Pot's former business partner Ieng Sary has few kind words to say about the former pal with whom took control of the Khmer Rouge in 1978.

"Pol was always a nasty piece of work. We had a good business going back in the seventies, killing pro-Vietnamese members of the ruling elite. But Pol was crazy. He started wiping out entire communities, and he used to purge intellectuals too. If anyone wore glasses he'd have them killed. He was a nutter. Eventually I'd had enough and I quit the business. I sold him my share in the business, but to this day I've not had a penny off him".

Chum

Pot seemed unconcerned by the controversy. He spent yesterday afternoon shopping for sports cars and helicopters in Pnohm Pehn. Confronted by reporters outside a BMW dealers, Pot refused to discuss his record of crimes against humanity.

"That's all water under the bridge as far as I'm concerned", he told us. "Nobody is perfect. Okay, so I did a few purges. But now I just want to put the past behind me and get on with enjoying my win".

Jackpot winner has 'murdered millions'

Pol Pot proudly displaying a new motor yesterday. Genocide is 'water under the bridge' he told reporters.

SCUMBAG!
Lottery love cheat is a rat says ex

SHAMED genocide Lottery winner Pol Pot has been branded a rat by the girl he loved - and left to die in a burning Vietnamese village.

Shirley Phouthang, 48, from Ho Chi Minh City, claims she had a firey fling with the lottery millionaire ultra nationalist revolutionary dictator during the early seventies.

Chappie

Shirley met Pot in Cambodia at a Kompong Cham supermarket where they both worked in 1972. "Pol used to offer me lifts home from work", she told us. "He was always a bit flash. He spent a lot on clothes, and had a car. All the girls fancied him".

Mr Dog

Soon romance blossomed and the couple had an affair lasting several months. "I took him home to Ho Chi Minh City to meet my parents", says Shirley. "At one point we even talked about marriage". Then one day Pot told her he was going away to live in political exile in Beijing and join a radical faction of the Khmer Rouge. "He said he'd write to me, and that one day he'd return". At first she was heartbroken, until the following week when a friend saw Pot coming out of a cinema in Kompong Som, arm-in-arm with another girl.

Shirley - 'heartbroken'

Shirley hasn't spoken to Pol since. However their paths nearly crossed during the Khmer Rouge's military purge of the western border provinces of Vietnam in 1977 when Pot's forces burned Shirley's village to the ground.

"He's always had my number, but he never calls. Even after our village was purged, he didn't ring to see how I was".

Eukanuba

Yesterday Shirley issued a warning to any girls who may be wooed by the jackpot winner's prize packet. "I pity any slut stupid enough to go near him", she blasted. "He may be a millionaire, but a rat like him never changes his spots. To be quite honest, you couldn't print what I think about him in a family newspaper".

DUST... STAINS... COBWEBS... GRIME... SCUM...

OOOH... I LOVE IT WHEN YOU TALK DIRTY.

The evil pastimes of Britain's top telly stars
SEX AND MURDER!

EXCLUSIVE

By BOB SHITE

Soap stars from the cast of TV's EastEnders are heavily involved in satanism and Devil worship. That is the bombshell dropped today by a man who claims to have witnessed top soap stars taking part in an evil black magic ceremony culminating in the MURDER of an innocent young girl.

Past and present members of the popular EastEnders cast

Adult bookshop owner Bill Henshaw claims that he hid only yards away and watched in horror as a top TV actress was forced to have sex with male collegues in a bizarre moonlit occult ritual. And he was **SICKENED** as another leading member of the East-Enders cast plunged a dagger through the heart of a naked virgin, before smearing himself in her blood.

BONFIRE

"I was on my way home from the pub, when I got lost", Bill told us, his face pale as he recalled the terrifying events of that night. "It was dark, and my car had broken down, so I was making my way across some fields to look for a telephone. Suddenly I noticed a light in woods nearby. It was a bonfire of some sort.

NAKED

When I got closer I noticed people dancing around the fire. They were all naked. At first I thought they must have escaped from the local loony bin, but then I recognised one of their faces. It was Dennis Watts, the landlord of the Queen Vic pub in EastEnders. He was surrounded by other members of the cast. I was completely amazed by it all, so I hid in some bushes to find out what was going on.

NAKED

They were all dancing frantically, and in the soft, flickering yellow light of the fire I could see beads of sweat rolling around the contours of their naked bodies", said Bill. "All the time they were chanting, the same words over and over again. It must have been Latin or something, as it made no sense to me. Then suddenly the bonfire seemed to flare up, and a huge cloud of orange smoke appeared. As it slowly began to clear, I could make out the shape of a gruesome

An artist's impression of the scene as EastEnders stars 'danced naked in a moonlit ceremony'

figure which seemed to be hovering above the flames. It had horns, and goat's legs. I'd never seen anything like it before in my life".

DEVIL

Experts who we spoke to suggested that this could have been the Devil himself, summoned up by the soap stars in order to receive their macabre sacrifice.

Bill continued. "Suddenly another member of the cast appeared. It was difficult to recognise him as he was wearing some sort of fancy, ceremonial robe, and his face was covered by a hideous mask. I'm not sure, but I think it could have been Arthur, the down trodden head of the Fowler household.

NAKED

Then a young girl was led out of the shadows. She was naked. Suddenly I realised what was about to happen. I wanted to do something to stop them but I couldn't. If anyone had seen me I'd have been a gonner.

BREAST

The man in the robes then produced a long silver dagger in the shape of an inverted cross, and plunged into the naked girl's breast. I simply couldn't watch. But I did. It was horrific".

CHICKENS

After the killing the actors and actresses took turns to smear the girl's blood on their naked bodies, which were clearly visible in the glowing yellow light of the fire. "Then they all began to dance. It was as if they'd been hypnotised. People were being forced to have sex all over the place. It just became one great big orgy. I could hardly believe my eyes. I just sat and watched for about half and hour, thinking it must be a dream. But every time I pinched myself, I was still awake".

ASHES

The next morning Bill retraced his steps and returned to the spot where the killing had taken place, only to find that all trace of the sick soap star's ceremony had been removed. "They had made a really good job. Even the ashes from the fire had gone. It was as if the whole thing had never happened".

Bill decided against reporting the incident to the police, "I had no proof. It was just my word against theirs. And who are the police going to believe? Me, or the entire cast of a well known TV soap opera".

MADE LOVE

A spokesman for the BBC denied that any member of the EastEnders cast was in any way connected with the occult, and suggested that Mr Henshaw's claims were totally ficticious. "This sounds like a load of rubbish to me", we were told. But Mr Henshaw remained adamant.

"They would deny it, wouldn't they", he said. "The chances are that they're all involved".

Mr Henshaw yesterday

In the past similar accusations made against BBC staff have also met with denials. As in 1983, when firemen were rumoured to have been called to the home of Sue Lawley to extinguish a "large whicker man" containing several chickens, which had been set alight in the back garden.

However, no comment has yet been made on the suggestion that TV football analyst Jimmy Hill is the head of an evil witches coven.

★Rude Kid

WHAT WOULD YOU LIKE IN YOUR STOCKING DEAR?

A TURD ON A STRING!

ARE YOU GOOD IN BED?

These days many couples are TURNING OFF to sex instead of turning on. And figures show that many of today's marital break ups actually begin between the sheets.

So we decided to set up a simple test in which the words 'sex' and 'make love' are repeated frequently.

And it gives you a chance to prove that Britain is not becoming a nation of lousy lovers. Simply answer each question A, B or C, then tot up your final score to reveal how you perform in the bedtime stakes.

1. Your partner wants to make love but your favourite programme is on TV in ten minutes. Would you:
   A. Tell them to wait until after the programme.
   B. Agree to a quicky, and get it over with in time to watch TV.
   C. Lie a rug on the floor by the fire, and make love slowly, facing the telly, so that you don't miss the programme.

2. In the evening you feel like making love but your partner says they have a headache. Would you:
   A. Make love regardless
   B. Go out for a few drinks, then come back and make love to them.
   C. Fetch you partner 2 paracetamol tablets, then wait for ten minutes or so before having sex.

3. At the end of a romantic candle lit dinner your partner is keen to go to bed, but you haven't quite finished your ice cream. Would you:
   A. Take the ice cream with you, and finish it in bed.
   B. Leave the ice cream on the table, and hurry back to finish it after you have made love.
   C. Put the ice cream in the fridge before you make love, and then offer some to your partner in the morning.

4. You arrive home one evening to find your partner in bed with a stranger. How would you react? Would you:
   A. Become violent, throwing one or both of them out of the house.
   B. Go out for a few drinks, and come back later.
   C. Get into the bed and go to sleep as if nothing was happening, and discuss it in the morning.

5. You decide to read a book at bedtime. Which of the following choices would you prefer?
   A. A cookery book or car repair manual.
   B. A paperback novel.
   C. A large illustrated book about sex.

6. At the height of your love making you realise that there is only one puff left in your ciggarette. What would you do? Would you:
   A. Get up and look for an ashtray.
   B. Curry on, stubbing out the cigarette on a bed post.
   C. Offer your partner the last puff while you go and find the ashtray.

7. While having sex you begin to feel hungry. What would your reaction be? Would you:
   A. Stop, go out and buy yourself a chinese takeaway.
   B. Stop and ask your partner if they would like a chinese takeaway.
   C. Ring up and order a chinese takeaway, then collect it AFTER you have finished making love.

8. Your partner wants to make love but your bed has been sent away to be repaired. Would you:
   A. Wait until the bed is returned.
   B. Make love on the settee.
   C. Ring a bed hire company and ask them to deliver a sexy four poster right away.

9. While on holiday you accidentally enter the wrong hotel room and begin to have sex with a stranger. Upon realising your mistake would you:
   A. Hold tight and make the best of it.
   B. Explain your error, apologise and quickly leave the room.
   C. Ring your partner's room and ask them to come along and join in a sexy threesome.

10. Your partner complains that love making is no longer enjoyable. What would you do? Would you:
    A. Ignore the remark, and continue as before.
    B. Go out for a few drinks, and cut out sex altogether.
    C. Give your partner a candle lit meal, have a shower together, put on a romantic record and make love at an unusual time of day (for instance during your lunch break), in front of a mirror, wearing a kinky revealing PVC play suit.

SCORING

A — 1 point, B — 2 points, C — 3 points

21 — 30: Ooh la la! Your steamy sex sessions make you a top scorer in the loving league.

11 — 20: Not bad, but more effort will get you better results in the sex championship.

10 or less: A poor performance. Unless you improve you'll get knocked out of the intercourse cup.

Bedroom Success!

If you are an utter and complete failure in bed, don't worry. There are many ways in which dismal sexual performers can find success in the bedroom.

SEX BOOK

In his latest book, 'An Expensive Book About Sex', leading expert on the subject Dr. Otto Waffle describes many ways by which we can discover the true pleasures of sex, with lots of pictures.

By taking the following tips you will find a great improvement in love making for both you and your partner.

● MAKE love on a bed, a settee or on a similar level surface.

● TAKE all your clothes off beforehand.

● EXPERIMENT in bed with exciting love games. Try playing Scrabble or Monopoly before you make love.

PRINCESS

How the Stars Would of Pulle

THE verdict of Lord Justice Scott Baker's inquest has done little to silence speculation about the true causes of Princess Diana's death in 1997.

Now the case files have been closed and the blame for her assassination has been placed squarely at the door of drunken driver Henri Paul and the chasing Paris paparazzi pack. Yet surveys suggest that up to more than a third of the British public still believes that the Princess of Hearts was deliberately killed in a staged accident. These people, including Mohamed Al-Fayed, everyone who writes for or reads the *Daily Express* and Keith Allen, are certain that it was Prince Philip who was behind the dastardly plot.

ENCOUNTER

However, in his summing up the judge cleared the Duke of Edinburgh of any involvement

in Diana's death due to a lack of evidence. The Prince's perfect murder plot to run Di's car off the road in Paris's Pont de l'Alma tunnel using motorcycling photographers, a white Fiat Uno, no seatbelts and a pissed up chauffeur would have taken years to plan, required meticulous, split-second timing to pull off successfully, and required a cover-up on a scale to rival NASA's faked Moon landings.

PANT

It would also have entailed a massive conspiracy involving royals, photographers, secret services, French police, paramedics, Paris Ritz staff, BBC reporters, diplomats and hundreds of other witnesses in order to succeed. Surely there must have been a simpler way for the establishment to get rid

PERFECT MURDER NUMBER ONE

JAMIE OLIVER IN THE RESTAURANT WITH A LAMB SHANK

Naked Chef Jamie could have easily cooked up a pukka way to assassinate England's Rose. He would have invited her to his exclusive restaurant - Fifteen - where he would have shown her to a secluded private booth on the pretext of allowing her to escape from her press pursuers. Then, while she was looking at the menu, Oliver would have sneaked in through a serving hatch and clubbed her repeatedly on the back of the head with a frozen leg of lamb. Then he would have calmly returned to the kitchen and started cooking the murder weapon, drizzling it with extra virgin olive oil, cracked peppercorns and sprigs of fresh rosemary, before slamming it into a medium oven for twenty minutes per pound, not forgetting to baste it half way

through. Lovely jubbly.

Sooner or later, one of the customers would have spotted Diana's blood-soaked corpse slumped across the table and raised the alarm. Then, when the police arrived Jamie would have helped them with their enquiries, before calmly bringing out the delicious-smelling cooked lamb and offering it to them to eat, saying it would be a shame to waste it.

As they tucked into their delicious freebie meal of braised lamb shank, celeriac mash and glazed shallots, little would the unsuspecting cops have realised that they were actually destroying the only piece of forensic evidence that could link their generous, big-tongued host to the killing. Bish-bosh-bash on the head....sorted!

PERFECT MURDER NUMBER TWO

NOEL EDMONDS AT CRINKLEY BOTTOM WITH AN ACID-FILLED PIE

People might think that the tidy-bearded ex-*Late Late Breakfast Show* host could just have asked the Princess to do a Whirly Wheel Challenge or shot her in the face with his elephant gun. However, if he'd chosen to take her out in such an obvious way, the trail of blame would have led back to him too easily.

How he'd actually have done it is like this: Edmonds would have organised a charity event at his palatial Devon estate, giving helicopter rides to spastics or something, and invited Lady Diana to mark the occasion by unveiling a giant gold plaque in her honour. Just as his honoured guest was about to pull the cord, Noel would have had his old nemesis Mr Blobby step forward and "gunge" her with a giant custard pie. The Princess was such a down-to-earth, fun-loving lady that it was a racing certainty she would have pushed her security men aside so she could take the pie on her chin.

This would have been her big mistake, as sneaky Noel would have earlier swap-shopped the custard pie for one filled with deadly concentrated prussic acid. This would have dissolved her head off in seconds, killing her instantly. Edmonds would have pretended to be really shocked, knowing that no-one would have pointed the finger of blame at him because he is so well known for his extensive charity work.

The *Deal Or No Deal* presenter would have taken no chances; he would have made sure that the patsy in the Mr Blobby suit was a gullible halfwit who was unable to defend himself. And just to make 100% sure he couldn't be implicated in the plot, Noel would probably have had his friend Mike Smith step forward and shoot Blobby in an underground car park as he was transferred between two police stations a couple of days later.

PERFECT MURD

THE KRANKIES ON A WA

The veteran Scotch comedy double-act would have had a fandabidozy chance of pulling off the perfect murder.

For obvious security reasons, members of the public were often kept at distance from Lady Di. However, from her earliest days as a kindergarten teacher she was as well known for her love of children as she was for her love of standing with her back to the sun so you could see her fanny through her skirt if you really squinted. If, whilst on one of her many public engagements, she had been approached by a tiny schoolboy carrying a bunch of flowers, it is quite likely that her body guards would have let him through their tight cordon.

However, this would have been

S MUST DI
Off the Crime of the Century

PERFECT MURDER NUMBER FOUR

JEREMY CLARKSON AT THE TOP GEAR TEST TRACK WITH A REASONABLY-PRICED CAR

Princess Diana was well known for being fiercely competitive. So if petrolhead TV presenter Jeremy Clarkson had invited her to take part in the "Star in a Reasonably-Priced Car" segment of his popular *Top Gear* programme, it is likely that the feisty Royal would have jumped at the chance to prove her mettle behind the wheel. Even more so if she had been told that her boot-faced love rival Camilla Parker-Bowles had been on the previous week's show, and had set a new *Top Gear* circuit record.

However, before Di began her timed lap, murderous Clarkson would have given the Stig twenty pounds to go out and sever the car's brake cables and loosen the steering wheel so that it would come off at high speed. Then, as the cameras rolled and she reached

the fastest part of the track - out of the Hammerhead and into the Follow-Through - Di would have found her Chevrolet Lacetti out of control and unable to stop. Careering off the tarmac and somersaulting end over end, the curly-topped presenter's hapless royal victim would have perished in a fiery ball of twisted metal before help could arrive.

After the "accident", the motormouth presenter would have been able to relax safe in the knowledge that his famously silent, helmeted accomplice would never tell the police what his boss had paid him to do to Diana's car. When questioned by police, Clarkson wouldn't break a sweat. Indeed, he would remain cool enough to be included on the famously dull and self-indulgent "Cool Wall" section of his own programme!

PERFECT MURDER NUMBER FIVE

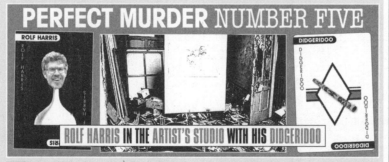

ROLF HARRIS IN THE ARTIST'S STUDIO WITH HIS DIDGERIDOO

The multi-talentless Australian is famous for his recent portrait of the Queen with the eyes and the mouth a bit wrong. And the chin and the teeth. As part of his assassination plan, Rolf would have used his many contacts at the BBC and Buckingham Palace to secure himself yet another commission - this time one to paint a portrait of Diana. And the arms.

Arriving at his light and airy studio, the unsuspecting Princess would have found herself seated by an open window in front of Harris's easel whilst the bearded entertainer busied himself behind the canvas. However, after sloshing brightly-coloured paint around with wallpaper paste brushes and decorator's rollers for a while, Rolf would have announced that he felt his sitter was looking a little tense, and this was making it even more difficult than usual for him to get a likeness.

To counteract this, Harris would then have suggested that he could help Di to relax by performing a live version of his hit Sun Arise, which reached number 3 in the 1962 Christmas pop chart. To this end, he would have produced his didgeridoo, taken a series of noisy deep breaths and raised it to his lips.

However, instead of the trademark spluttery, buzzing drone the Princess was expecting to come out the end of it, a poisoned dart would have emerged instead, sticking straight into her carotid artery and killing her in an instant. Painting her portrait would have given the bearded TV favourite the perfect opportunity to didgeri"doo" in the Princess. And what's more, Rolf's brilliant murder plot would have been so perfect that his victim wouldn't have known what it was yet until it was too late to save herself!

UMBER THREE

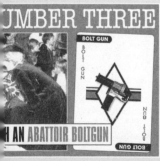

H AN ABATTOIR BOLTGUN

istake, for the "schoolboy" would ctually have been octogenarian ac-ess Jeanette Krankie, and her tempt-g bouquet would have concealed a olen slaughterhouse boltgun, pow-ed by a canister of compressed gas dden in her satchel. As an unsus-ecting Di bent to smell the flowers, ankie could have lifted the gun to e Princess's temple and pulled the gger, punching a retractable metal lt deep into her brain and killing her the spot.

In the stunned, horrified silence hich followed the hit, the tiny as-ssin would have taken the oppor-nity to make her escape, slipping sily through the legs of the crowd a waiting getaway car driven by his sband/brother Ian.

"Up to more than a third of the British public still believes that the Princess of Hearts was deliberately killed in a staged accident."

of the maverick Princess who was threatening to become an embarrassment to the Royal family.

MILK WOOD

These days, it seems we can't put on the television without seeing some show in which competing celebrities perform unaccustomed tasks such as ice skating, circus acrobatics or ballroom dancing. So how would the STARS have fared if they'd been called upon to commit the perfect crime? Just for fun, we asked celebrity criminologist Professor Greuze Brughwell to dream up a few ways in which A-List Stars might have dialled M for Murder.

Commons split over Glenda's love kipper

A leading Labour MP's quim was at the centre of a political storm last night.

Conservative back bencher Sir Anthony Regents-Park yesterday launched an unprecedented attack on opposition member Glenda Jackson's bush. The outspoken member for Fulchester Sunnyoak rounded on Miss Jackson's pubes during Prime Minister's question time, describing them as a "threadbare snatch" and claiming that their appearance was a disgrace to British parliamentary tradition.

Sir Anthony yesterday

OSCAR

"Flaunting a tatty twat to all and sundry does not uphold the best traditions of this House", he said, referring to a film in which the Oscar winning former actress had appeared nude. "What will the Right Honourable Member sink to next? Hamburger shots?" he asked. There was uproar in the House, and after several moments the Speaker ruled that questions relating to a specific member's fadge were not within the scope of Parliamentary debate.

However, Mr Regents-Park continued his criticism afterwards. Referring to a film called 'The Music Lovers' he described a scene in which Jackson's pubic hair was clearly visible. "You saw it on a train I seem to recall. I have only seen the film once, and once was quite enough. It was quite the scraggiest stoat I have ever seen. It looked like Bob Geldof's moustache, stuck on vertically. Not that the appearance of Miss Jackson's kipper is at question here. Miss Jackson is entitled to have any array of pubic hair she likes. Indeed, she could have none at all if it suits her. That is not the issue. I am merely expressing the widely held view that an MP's muff should remain in her Parliamentary briefs, and not be paraded on cinema screens for the benefit of the dirty mac brigade".

KIM

"Look at Mrs Thatcher. In the eleven years that she was Prime Minister not once did she reveal her beef curtains. And rightly so. When she left office Britain's standing in the world had never been higher. Put simply, hairy pies and politics do not mix".

MARTY

This morning a storm was brewing over Mr Regents-Park's remarks. However, the 55 year old MP was unavailable for further comment, having been admitted to a private clinic after breaking an ankle falling from a step ladder whilst reaching for oranges on a top shelf in the kitchen of his West London batchelor home late last night.

Glenda Jackson's Kipper Kwizz Win your weight in kippers!

E	L	C	B	H	A	M	B	U	R	G	E	R	F	L
P	G	N	R	T	O	M	A	L	C	Y	R	I	A	H
R	V	D	O	A	U	N	S	T	A	W	T	S	L	H
I	S	S	A	F	C	E	E	T	P	B	C	P	S	L
D	N	M	F	F	B	K	S	Y	I	R	N	A	M	E
H	I	O	I	U	C	E	V	N	P	L	G	L	O	N
S	A	B	P	U	N			J	O	S	F	N	N	
A	T	N	B	O	Q			R	E	T	S	L	U	
W	R	M	O	T	O			I	B	O	S	L	T	
K	U	E	D	E	E	H	B	P	H	F	A	I	G	E
C	C	B	P	Y	G	H	Y	E	H	C	T	P	Y	V
O	F	L	B	P	N	R	O	R	A	I	T	S	J	O
C	E	A	U	K	I	N	L	L	R	V	S	A	M	L
N	E	R	S	A	M	K	A	O	E	U	E	P	N	O
R	B	T	H	T	N	U	C	F	P	S	F	R	T	S

You may have noticed that the sole purpose of the above item was to include as many childish euphemisms for the word vagina as possible. For the benefit of anyone who is still reading we have cleverly concealed every single euphemism for the word vagina that we could think of in the grid below. See how many you can find. They read in all directions and diagonally, backwards and forwards, but always in a straight line. If you find two dozen or more, you could be in with a chance of winning our fabulous prize – *your weight in kippers!* Write all the words you can find on a postcard, then put it in an envelope (to avoid offending postal workers) and send it to: Viz Kipper Kwiz, P.O. Box 1PT, Newcastle upon Tyne, NE99 1PT. The winner will be weighed, and will win a cash prize equivalent to the value of their weight in kippers, using the prevailing wholesale market price of kippers to calculate the prize money. Entries must be received by 13th May 1994. The winner will be announced in the next issue.

'Clean up your act' says Mrs Ekland

Britt Ekland's mam yesterday issued the following heartfelt public plea to her wayward blond bombshell daughter. "Change your ways, Britt. Your mother knows best".

Mrs Ekland, 45, had became concerned after her sex kitten daughter Britt Ekland, 38, started coming home late at night. Britt's wild child antics have included:

'And just where do you think you're going dressed like that?' – Britt Ekland yesterday.

● Going out without telling Mrs Ekland where she was going to, or what time she'd be back.

● Throwing a wild party at the Ekland's house while Mr and Mrs Ekland were staying with relatives.

● Drinking, and coming home with her clothes smelling of cigarettes and pubs.

JACK

Things came to a head recently when Britt Ekland, 38, missed the last bus after attending a party at a friend's house. Mrs Ekland, 45, become concerned when Britt Ekland hadn't rang for a lift by 11.30.

JOKERS

Britt's uncle, actor Joss Ackland, 94, went out searching the streets for wayward Britt Ekland till nigh on midnight. "I was at my wits end", said Mrs Ekland yesterday. "I didn't know where she was or who she was with. I don't want to stop Britt Ekland enjoying herself, but all she had to do was have rung me and let me know where she was".

WEST

Joss Ackland, Britt's uncle, eventually found Britt Ekland outside a fish and chip shop talking to boys, including Rod Stewart. When Joss Ackland dropped Britt Ekland off at Mrs Ekland's house a furious row ensued between Britt Ekland and Mrs Ekland about where Britt Ekland had been. As a result Britt Ekland stormed out of the kitchen and went to her room, and Mrs Ekland threw her tea in the bin.

CARD

The following day Mrs Ekland went round to speak to Rod Stewart's parents, Mr and Mrs Stewart, 72. It was after eleven in the morning and Rod Stewart was still in bed. Mrs Ekland told Mrs Stewart that she didn't want her Britt seeing Rod Stewart anymore.

"What about that daughter of yours, Britt Ekland? She's no better than she ought to be. I've heard that she went with that fella out of the Stray Cats, and him only half her age", said Mrs Stewart. "So don't come round hear calling our Rod", she said.

CHILD

Mrs Ekland, who was just going home when Mrs Stewart said that turned round and came back and said: "You can talk, can't you. That son of yours Rod Stewart doesn't exactly look much like your husband Mr Stewart, does he now?". At this point Mr Stewart got up from his chair and asked just what exactly Mrs Ekland meant by that. "You know what I mean. And so does she", said Mrs Ekland pointing at Mrs Stewart.

ANIMALS

Mr Stewart then told Mrs Ekland to leave and he shut the door in her face. Mrs Ekland then went home and had very strong words indeed with Britt Ekland about her not seeing Rod Stewart anymore.

ROVER

Unknown to her mam, Britt Ekland was last night believed to be going out with Wimbledon footballer and former page 7 fella Vinny Jones.

Is there a PANSY at the Palace? We ask…

PRINCE OR PUFF?

Is Edward soft or what?

HRH Prince Edward's dramatic resignation from the Royal Marines has left Britain shocked and surprised. And it leaves an enormous question mark hanging over the Prince's head as he continues to ponder his future. It's a question the whole nation is asking. Is Prince Edward soft?

The Royal Family have a long proud history of producing fearless fighting men, courageous soldiers who have lead their country in battle over the centuries. Only recently Prince Andrew in his bullet riddled helicopter pointed the way to victory in the Falklands war, and already young Prince William has been seen 'putting himself about a bit' at his kindergarten school.

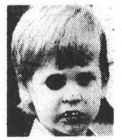
Young bruiser William

So has Prince Edward let the family down? His father, the Duke of Edinburgh, is believed to be furious over allegations that his youngest son, 5th heir to the throne, is a pansy. And the fact that at 22 years of age the Prince remains a bachelor, while brothers Charles and Andrew are happily married, has set alarm bells ringing inside the Palace.

STUFF

So just what sort of stuff is Prince Edward made out of? We decided to find out for ourselves just how tough he really is. We rang Buckingham Palace and suggested that Prince Edward was a puff.

The Prince who 'plays with girls'

"What's he gonna do about it?" our reporter asked. There was a short silence, then we were cut off.

CAR PARK

Twenty minutes later we rang again and said that we didn't like Prince Edward's face. Again there was no reply

Finally our reporter called again and offered to meet the Prince in a nearby car park to settle the matter. We waited in the car park for over an hour. There was no sign of the Prince.

GIRLS

And now, in the wake of our investigations, evidence is coming to light that Prince Edward spent more than a little time during his schooldays 'playing with girls'. Evidence that we found in a park opposite Buckingham Palace suggests that this habbit may still exsit.

SKIPPING

Only yards from the Palace gates we found a piece of rope identical to that used by children for 'skipping', and nearby we found chalk marks similar to those used in 'hopscotch' — another game popular among softer children.

Hopscotch — an artists impression of the Prince at play.

So far there has been no official comment from Buckingham Palace on the suggestion that Prince Edward spent several hours recently, during his leave from the Marines, playing 'girlish' children's games. But it is an undisputed fact that the Prince was more than happy to join in similar games during his early schooldays.

A spokesman for a leading toy shop in London refused to discuss allegations that the Prince regularly visited the store and purchased among other things teddy bears. But as the Prince drove away from the Royal Marine Commando training base at Lymspstone in Devon for the last time, a package large enough to contain several childish 'teddy bears' was clearly visible in the back of his car.

DOCTOR, I THINK I'VE CAUGHT SOMETHING

KINKY SEX TURN - OF THE S

We all have our own special 'turn on', a sexy little secret that we share with our partners. Whether it's a fetish for French knickers, or a lusting for lacy lingerie, we all have our own kinky preferences. And what sends one man wild, may send another to sleep. So what turns the stars on? For although we may not realise it as we watch TV, many of our top celebrities have unusual sexual preferences. So we decided to do some asking around to find out exactly what tickles the fancies of the rich and famous.

Mick can't get no satisfaction without wig and oven gloves.

With a wife as beautiful as Jerry Hall, you wouldn't think millionaire Rolling Stone **MICK JAGGER** would need much to turn him on. But the ageing rocker has a kinky habit which costs him a fortune to maintain. For Jagger, 57, cannot make love to his wife unless she is wearing a wig! And not just any wig. The couple's expensive tastes mean that specially made hairpieces must be imported by the plane load from Iran, for model Jerry refuses to wear the same wig twice.

Jagger – 'oven gloves'

Hall – 'wigs'

Eccentric keep fit fanatic Jagger also insists on wearing oven gloves during the couple's raunchy sex sessions, as he believes this will stop his hands from going wrinkly.

FLAGGING

TV host **TERRY WOGAN** would have great trouble keeping his 'turn on' a secret. For the smooth talking Irishman is driven wild by the sound of bells, as many of his frustrated neighbours will testify! Wogan has even had a church bell tower built on to his £500,000 Surrey home in order to boost his flagging sex life. Chimes have been known to ring out across the local countryside at all times of the day and night, signalling to all and sundry that Terry is 'on the job'.

Tel's bells drive him bonkers!

When he's away from home, Terry still needs his nightly 'tinkle'. On one occasion whilst staying in an hotel, the loveable Irishman asked for six telephones to be installed in his room, and demanded that the night porter ring all of them constantly. Fellow residents were relieved when after six hours the ringing stopped and an order was received for two cigarettes to be sent up to the room.

UNIFORM

If you were to pass by the house of BBC holiday expert **CLIFF MICHELMORE** one evening, you'd be forgiven for thinking it was on fire. For the chances are you'd see a figure dressed in a fireman's uniform clambering in through the bedroom window. You would, in fact, be witnessing kinky

Wogan – 'bells'

Cliff's nightly sex ritual. Respected broadcaster Michelmore, 63, makes his way upstairs before ringing down to his wife and reporting a fire in the bedroom. Every night his wife faithfully dons her fireman's uniform and climbs a ladder to the couple's first floor bedroom window. Once inside the room she removes the costume and the couple enjoy an otherwise normal, healthy relationship.

Botty smack for 'bad boy' Bono

Controversial singing star **BONO** has spent a fortune earned from U2's hit records converting his bedroom into a perfect replica of an old fashioned sweet shop, in order to remind him of his childhood in Edinburgh.

SPANKED

Every night Bono dresses as a schoolboy while his wife puts on a grey wig and white apron. A strange and well-rehearsed scenario then follows in which Bono is

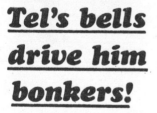
Michelmore – '999'

caught stealing gobstoppers and his bottom is spanked by the angry 'shopkeeper'. This play acting continues into the early hours, eventually building into a crescendo of passion which ends with the couple writhing naked in sherbet and dolly mixtures before both collapse exhausted and fall asleep.

TEN INCHES

Sports commentator and world expert on football **JIMMY HILL** can only make love in one place – a Victorian wendy house which he keeps in the attic of his detached Warwickshire home! Hill and his wife regularly cram themselves into the antique play house for electrifying sex sessions accompanied by the sound of brass bands and marching music. For another of Hill's sexual oddities is his taste for military music, and he has amassed a collection of some 1,000 dusty '78' records specifically for this purpose.

POPCORN

Early sexual experiences can greatly influence a person's sexual preferences later in life, and this has been the case with Radio One DJ

-ONS TARS!

Bono – 'dolly mixtures'

Hill – 'Wendy house'

Sexy Simes' Inter-City sexpress

SIMON BATES. As a teenager he went to see the film 'The Railway Children' over 200 times. And it was on one of these visits to the cinema that Bates, 27, had his first sexual experience as he ate a packet of popcorn. This left such an impression on him that ever since that day the sight, sound or even the mention of a train has 'turned him on', often with embarrassing results.

AROUSED

On one occasion the DJ was late for a Radio One Roadshow in York after catching a train at King's Cross and becoming aroused at Peterborough. The train was held up for 20 minutes at Doncaster while Bates took a cold shower in the toilets on the platform.

How kinky are YOU?

Here's your chance to find out in our fun to do quiz

We all know how kinky the stars can be, but what about **YOU**? You could have an inner kinkyness that you've yet to discover.

Simply answer these eight questions a, b or c, then tot up your final score to reveal you own saucy sex secrets.

1. Your car is stopped at traffic lights. What would you do?
a. *Apply the handbrake, and sit patiently until they change.*
b. *Rev the engine till it throbs, and gently caress the gear knob.*
c. *Open your shirt or blouse and smear bright red lipstick on your nipples, adjusting the rear view mirror in order to admire your work.*

2. On your regular visit to the hairdressers, do you:
a. *Ask for a short back and sides.*
b. *Ask for a fashion cut in the style of your favourite sexy pop star.*
c. *Ask for you entire body to be shaved, and styling mousse to be rubbed into your parts.*

3. Look at this picture. What immediately springs to mind?

a. *A day on the beach at Blackpool with the family.*
b. *Bobbing up and down rhythmically in the saddle of a horse.*
c. *A rampant three-in-a-bed sex session with you, your partner and a donkey.*

4. The Queen is making her annual Christmas Day speech on TV. What do you do?
a. *Sit and watch patriotically, whilst enjoying a cup of tea and a mince pie.*
b. *Switch off the TV and watch a raunchy, 'X' rated video instead.*
c. *Dress up as HRH The Duke of Edinburgh, before stripping naked and frottering yourself frantically against the furniture.*

5. While waiting for Directory Enquiries to answer the phone, how would you occupy your free hand?
a. *Bend a couple of paper clips until they snap.*
b. *Doodle on a notepad, perhaps drawing the curves of a naked human figure.*
c. *Lower your trousers, press the mouthpiece to your parts, and smack your bottom firmly with the yellow pages.*

6. Shopping for meat at the butchers, what would you choose?
a. *Half a pound of mince.*
b. *A large pork sausage and a couple of scotch eggs.*
c. *A small, plucked chicken and a jar of vaseline.*

7. You notice that your car is dirty. What would you do?
a. *Nothing, except hope that it will rain later.*
b. *Wash the car down yourself, then gently massage its curving bodywork with a soft chamois leather.*
c. *Drive to the local car wash, strip naked and climb astride the bonnet with your partner for a super soapy sex session beneath the frothy rollers.*

8. You are having breakfast in the kitchen when the romantic song 'Lady in Red' by Chris de Burgh comes on the radio. How would you react?
a. *Tap your foot, and perhaps whistle along.*
b. *Close your eyes, sway sensuously around the room and imagine that you are dancing cheek to cheek with gorgeous Chris himself.*
c. *Strip naked, douse your entire body in butter and marmalade, stick a hot croissant up your arse and hit your parts with a stick of french bread until they go off.*

How did you do?

Award yourself 1 point for each answer a, 2 points for a b and 3 for a c. Then tot up your total.

10 or less: You are sexually inhibited. You are ashamed of your body, and think of sex as being 'wrong' and something you shouldn't do. You are a dull, unimaginative prude.

11 to 19: You have a reasonably healthy attitude towards sex, but you are not a saucy person. You'll be better off playing it safe, having straight sex whenever possible. Avoid the use of whips, chains and rubber appliances.

20 or more: You kinky devil! You show a healthy, refreshingly open attitude towards sex. So don't be boring in bed. Experiment, try wild new positions, wear exciting costumes, and put household items up your bottom. Your sex life will be revolutionised!

Why do we pay through the nose for electrical goods?

GREEDY SHOPS PUT THE SQUEEZE ON CONSUMERS

British shoppers are being lured by manufacturers into paying way over the odds for their electrical goods.

A study has revealed that on average we are paying £800 more than we need to for our household appliances. Items retailing at £1000 or more in the high street are readily available for only £20 just a short walk away - that's an incredible saving of £980. Britain's consumers are being ripped off because they don't know that identical branded goods are available at hugely discounted prices in their local pub. With the same specifications as the shop bought models, the only difference is that they have had their plugs cut off and sometimes contain small fragments of broken glass.

The biggest price difference we uncovered in our survey was for a £1800 Del Computer which we bought from a heroin addict in the Red Lion for £20 cash.

HOW THEY OVERCHARGE US

MODEL	SHOP PRICE	PUB PRICE		SAVING
Philips 32" widescreen TV	£999	£20	(Red Lion)	£970
JVC MD70R Micro HiFi	£349.99	£20	(Nag's Head)	£329.99
Olympus C900Z digital camera	£499.99	£20	(The Blubell)	£479.99
Panasonic Nicam video	£249.99	£20	(King's Arms)	£229.99

IT'S TIME TO FIGHT BACK!

Says JESS FUCKRAD
Consumer correspondent

WE HAVE sat back and allowed ourselves to be ripped off for far too long.

The fact is that Manufacturers and shops are conspiring together to keep prices artificially high. It is up to the British public to say enough is enough. We must make a stand and demand a better deal.

Unless shops are willing and honest enough to sell us big tellys for £20, we should vote with our feet and take our custom elsewhere. Mark my words, if we keep paying these ridiculously inflated prices, they'll keep charging them. Whatever they tell you, they are lying. It's time they put OUR money where THEIR mouth is, and told the truth for a change.

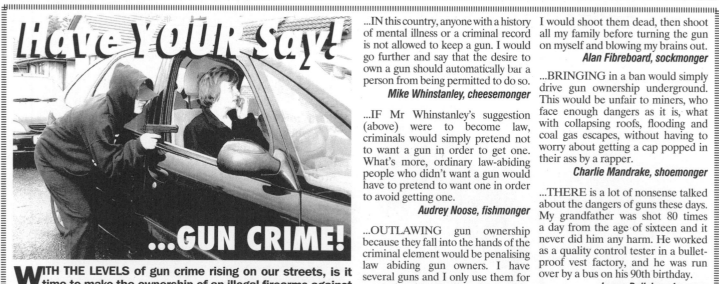

Have YOUR say!
...GUN CRIME!

WITH THE LEVELS of gun crime rising on our streets, is it time to make the ownership of an illegal firearms against the law? We went onto the bullet-riddled streets to find out what YOU thought....

...WHILST I accept there have to be ways of keeping guns out of the hands of criminals, we should not ban them outright. How else are rich people like Madonna and the Prince of Wales supposed to kill things on their estates?

Brian Microchip, costermonger

...IT IS pointless banning guns. If the government did it, then robbers would simply use another sort of weapon when holding up banks and building societies, such as big daggers, candlesticks or billiard balls in a sock.

Mick Sundry, ironmonger

...IN this country, anyone with a history of mental illness or a criminal record is not allowed to keep a gun. I would go further and say that the desire to own a gun should automatically bar a person from being permitted to do so.

Mike Whinstanley, cheesemonger

...IF Mr Whinstanley's suggestion (above) were to become law, criminals would simply pretend not to want a gun in order to get one. What's more, ordinary law-abiding people who didn't want a gun would have to pretend to want one in order to avoid getting one.

Audrey Noose, fishmonger

...OUTLAWING gun ownership because they fall into the hands of the criminal element would be penalising law abiding gun owners. I have several guns and I only use them for recreational purposes and defending myself against my enemies.

Renton Plywood. milkmonger

...I AM a law abiding citizen and a member of a well-respected licensed gun club, and I can reassure the public that they are at no risk from legally-held firearms like mine. I would not be prepared to hand over my weapons in the event of a ban, and if the police tried to confiscate my gun I would shoot them dead, then shoot all my family before turning the gun on myself and blowing my brains out.

Alan Fibreboard, sockmonger

...BRINGING in a ban would simply drive gun ownership underground. This would be unfair to miners, who face enough dangers as it is, what with collapsing roofs, flooding and coal gas escapes, without having to worry about getting a cap popped in their ass by a rapper.

Charlie Mandrake, shoemonger

...THERE is a lot of nonsense talked about the dangers of guns these days. My grandfather was shot 80 times a day from the age of sixteen and it never did him any harm. He worked as a quality control tester in a bullet-proof vest factory, and he was run over by a bus on his 90th birthday.

Lance Boil, breadmonger

...WHAT a lot of nonsense is spoken about the dangers of being run over by buses when you are 90. My grandfather was run over by 80 buses a day when he was 90 and it never did him any harm. He worked in the inspection pit of a bus garage for twenty five years after he should have retired.

Bartram Fellowes, computermonger

QUEEN FOR A DAY

In recent months there has been increasing controversy over the role of Britain's monarchy. Are the Royal Family paid too much? Should the Queen pay income tax? Do the Royals *really* justify their massive salaries? Is it right that the Queen should be paid £7 billion a year for opening a few bridges and waving at people?

Many people now feel that the monarchy should perform a more useful function in a modern democratic society. But what useful function could they perform? We decided to ask some of the top stars of TV and show business what **THEY** would do if they were **QUEEN FOR A DAY**.

What would the stars do if they were on the throne

JOHN ALTMAN

After playing drug addict Nick Cotton in TV's East Enders, John knows the kind of problems drugs like heroin can cause. Not surprisingly he believes the Queen should use her priviliged position to help tackle the problem of drug addiction in the inner cities.

SCHOOLS

"If I was the Queen I would visit schools in deprived areas and lecture the kids on the harm that drugs can do", John told us. John would use special 'Royal Drama Workshops' to get his message across.

"I think kids would take notice of the Queen, especially if they were able to join in a role playing situation in which they could experience for themselves the problems that drugs can cause. Hopefully some kids may be encouraged to take up drama as a career, so that perhaps one day they could be in East Enders like I am".

RAY ALAN

Together with his comedy ventriloquist's dummy Lord Charles, Ray is a familiar sight on Britain's TV screens. A keen supporter of the monarchy, Ray would be delighted to accept an invitation to be 'Queen for a day'.

DUMMY

"I think a ventriloquist's dummy would be a real asset to me as Queen", Ray told us. "You see, the Queen must be polite at all times, so if, for example, I was to yawn during a conversation with someone important, I could use my dummy to turn it into a joke".

Ray would love the opportunity of making the Queen's Christmas Day speech.

BORING

"The Queen's speeches are always boring, so I would leave all the talking to Lord Charles. Meanwhile, I could get all the laughs by drinking a glass of water", Ray quipped.

LINDA McCARTHY

Busy Linda combines a career as an accomplished musician with being Britain's top vegetarian chef. But Liverpudlian Linda would still find time to be Queen for a day. Not surprisingly, the first thing on her royal agenda would be to ban sausages!

DRUMSTICK

"Meat is murder", said Linda. "Every time you eat a turkey drumstick you are eating a slab of fear". So all meat products would be banned during her reign and boxes of veggie burgers would be marked 'By appointment to her Majesty the Queen'.

Another issue on which Linda feels strongly is charity. "I don't think the Queen does enough to raise money for charity", Linda told us. "If I was Queen I would get my friends and organise a car boot sale to raise money for needy causes".

PATTIE BOULAYE

Best known as the star of various theatre, cabaret and TV appearances, Pattie believes the role of Queen would suit her down to the ground.

"I'd love to be Queen", she admitted. "And if I was I'd clear out some of her wardrobes". **OUT** would go the Queen's boring dresses, skirts, hats and coats. And **IN** would come new sexy lingerie, giving the Queen a sizzling new look for the nineties. And with Pattie as Queen, there'd be even more shocks in store for tourists visiting London.

"I would make the household cavalry wear sexy G strings instead of stuffy uniforms", Pattie revealed. "And I'd have the changing of the guard ceremony choreographed. I would appear on horseback, wearing a sexy red leotard, singing Jesus Christ Superstar, then, at the end, there would be a big firework display".

WHAT WOULD YOU DO?

What would **YOU** do if you were Queen for the day? How would you put your power and influence to good use for the benefit of your country? What changes would you make to the monarchy in order to improve the system?

Write and tell us, on the back of a postcard. Whoever sends us the most innovative and original idea will be granted their wish. For we will make you **QUEEN FOR A DAY**.

The winner will travel to London first class, all expenses paid. Then you'll be given a hat, and a coat, and a little handbag, and left to wander the streets all day, talking to passers-by, shaking hands with old ladies, and waving at people to your heart's content. Then, when you've finished, you can make your own way home.
Send those entries to 'Queen For A Day', P.O. Box 1PT, Newcastle upon Tyne, NE99 1PT, to arrive by next Friday.

A Theological Question of Sport

with Christian Hop, Skip & Jumper
Jonathan Edwards

Heaven can pate

Dear Jonathan,

I was wondering recently about triple Olympic Gold medalist Duncan Goodhew. It is well known that he lost all his body hair when he was six years old as a result of falling from a tree. When he dies and goes to Heaven, will he have a full head of hair?

Joan Dunbarr, Ilkley

Jonathan answers... You assume that Duncan is going to Heaven and not the other place, but that is up to God alone to decide. However, after three Olympic golds for Britain and a Commonwealth record, I think it's a fair bet that he will be going up, rather than down! However, you must understand that the Heaven that we speak about is not a physical place that we can see or touch. It is a spiritual plane of existence to which our immortal souls transfer after the death of our physical bodies. As such, concepts of appearance, and indeed being, are irrelevant. But in answer to your question, yes, he will have hair. And pubes.

Hand of God

Dear Jonathan,

I am a Christian, but I have found my faith tested lately. If God is perfect and loves us all, how can he allow such terrible things to happen in the world, such as Maradona handling the ball into the net, or Mike Tyson biting Evander Holyfield's ear off?

Jake Turvey, London

Jonathan answers... Firstly, don't worry about your faith being tested, because coming through these tests will make it stronger. As you say, God is perfect and all powerful, and He created the world we live in. Our faith in Him is strengthened in the good things that we see, such as Botham making mincemeat of the Aussies in 1976, or England sticking 5 past the Krauts last year. However, because Adam ate from the tree of knowledge, all of us are born in sin and we are given choice. And as such, Tyson has the free will to bite his opponent's ear, Maradona chooses to handle the ball, and Tim Henman elects to fall to pieces and fail to reach the finals every Wimbledon.

Let us play

Dear Jonathan,

The Old Testament tells us that there is only one God, and that anyone who follows other Gods or false prophets will suffer His wrath. As a triple-jumping Christian, how do you explain the fact that in the 1980 Olympic games, that same God sat back and allowed India, a predominantly Hindu country, to take the men's hockey title from the Christian nation New Zealand?

Frank Oasis, Barnstaple

Jonathan answers... You cannot look at these things in isolation. Yes, the Indian men may have taken the gold, but at the same games their female counterparts were beaten by the largely atheist Soviet Union in a match to decide the bronze medalists. In the games before, the Indian men's team finished 7th. It is all very confusing. From our limited human perspective, we shouldn't try to understand what is happening. We just have to have faith that is is all part of His greater plan for Olympic hockey.

Viz

Britain's stars were paying tribute last night after Bernie Clifton's comedy ostrich was killed in a freak accident. Bernie was too upset to comment after the brightly coloured 6 foot bird fell 30 feet onto paving stones at the Clifton's home in Surrey.

The gangly bird had climbed onto the roof to adjust the TV aerial after the picture had become fuzzy during a televised football match. It is thought that the ostrich, a veteran of several Royal Variety performances, lost its footing, or was knocked off balance by a gust of wind.

Bitz

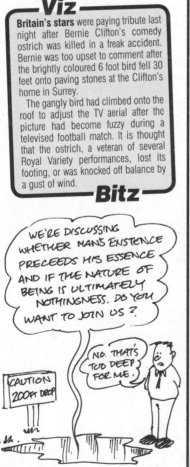

WE'RE DISCUSSING WHETHER MANS EXISTENCE PRECEEDS HIS ESSENCE AND IF THE NATURE OF BEING IS ULTIMATELY NOTHINGNESS. DO YOU WANT TO JOIN US?

NO. THAT'S TOO DEEP FOR ME.

CAUTION 200FT DROP

SCIENTISTS DISCOVER WORST THING

Some lab-bunnies yesterday

ASK ANYONE what the worst thing in the world is, and they'll tell you war, disease or famine. But not according to scientists from the University of Budapest. They've discovered that the worst thing in the world is actually a paper cut on your bell-end.

The five man team led by Professor Namoff Reszla researched the problem for over a decade before publishing their conclusions in the prestigious 'Nature' magazine.

clippers

"We subjected a succession of volunteer students to a wide range of horrible things, such as snapping their front teeth off with nail clippers and pulling back

■ by Our Medical Correspondant
Dr Roland Pianno

their big toes and sawing through the tendon bit underneath with a Stanley knife," said Professor Reszla. "We got laboratory rabbits to watch all the terrible things. Then we gauged how much they winced using a small wince-measuring probe inserted in the back of their necks."

schooners

In the experiment that got the most dramatic results, Professor Reszla drew a sheet of A4 typing paper sharply across a student's glans, leaving a 2cm paper cut.

"The rabbits' response was fantastic," he said. "The reading went off the scale. Rabbits which

had shown a wince reaction of 30% when we cut someone's tongue up the middle with tinsnips were suddenly showing 95% or more. One rabbit winced so hard it broke its back. There was no doubt we had discovered the worst thing in the world."

junks

But the study, which was commissioned by Bird's Angel Delight, has angered animal rights activists across the world. "We shoiuld not be exposing rabbits to this sort of barbarity. It's barbaric barbarism." fumed Ada Littlejohn of Nottingham.

"Why can't they get some of these murderers and paedophiles, put them into clamps and force them to watch someone getting a paper cut on his herman gelmet?"

THAT'S SAUSAGES!

BRITAIN'S *LIVELIEST* SAUSAGE DEBATE

MY HUSBAND loves sausages. He eats so many, I sometimes think he'll turn into one.

Mavis Nonentity
Croydon

I LOVE sausages, but I never felt happy eating them because you never know what's in them. So now I buy sausage meat from the butchers and make my own.

Ada Simpleton
Wessex

"THESE sausages aren't cooked," roared my husband one breakfast. "I know," I replied. "That's because they are still in the packet in the fridge and I haven't started cooking them yet." How silly he felt.

Dolly Dishmop
Staines

MY DAUGHTER doesn't eat meat, yet the other day I saw her eating sausages. "They're vegetarian sausages," she told me when I questioned her. Vegetarian sausages indeed! Perhaps I should start eating meat potatoes.

Ron Ironic
Burton on Trent

NUTRITIONISTS say that eating sausages is bad for you. But I love them. Live for today, I say. You may get hit by a bus tomorrow,

T Dreadnaught
Pudsey

I USED to be a mechanic at a second hand car lot, so I used to restore old bangers for a living. However, I was recently made redundant, but I very quickly found another job - restoring the exhibits at a sausage museum! So I still restore old bangers.

Fred Lampray
Wales

DURING the second world war sausages were on ration, but in the army we were allowed one sausage a day. I never ate mine, but saved them to take them back to my family. When I took my leave twice a year, I always went home with twelve dozen sausages. They were completely rancid, but there was a war on

and we ate like kings.

Horace Turtle
Rhyll

HOW Jews can say they don't like sausages is a mystery to me. If they tried them, they might actually like them. Still it's their loss, and it simply means more yummy bangers for the rest of us.

Dr Rowan Williams
Canterbury

NUTRITIONISTS don't know what they are talking about when they say sausages are bad for you. My grandad ate sixty a day, and he lived until he was ninety.

Arthur Wellington
Croydon

I'M A widow and I have been seeing a gentleman for several months. We have had some lovely times going for walks and tea dances. Last Sunday he asked me if I'd like to go back to his house saying he had a lovely blood sausage for me. As I love sausages I went. I don't think I shall be seeing him again.

Audrey Pumpkin
Cheedle

I LOVE sausages, and when I went on holiday abroad I was looking forward to having them every morning for breakfast. Imagine my disappointment when I found foreign sausages were all the wrong shape and had garlic in them. I was glad to get home.

Stan Tortoise
Burnley

MY grandad ate sixty sausages a day, and he was run over by a bus when he was ninety - whilst going for a packet of cigarettes!

Mavis Fibreboard
Surrey

I AM a widow, and I've been seeing a gentleman for several weeks. We have had some lovely times going to the theatre and the park. Last time we went out, he asked if he could come back to my house for a sausage sandwich. Since I was peckish I readily agreed. Men really are beasts.

Audrey Pumpkin
Cheedle

TOP SAUSAGE TIPS

PRICK your sausages with a fork before frying. That way your bangers won't go *BANG!* For cocktail sausages use a smaller fork.

D. Blackpudding
Blackpool

MAKING a sausage sandwich is much easier if you cut the sausage in two lengthways with a sharp knife, and lay the two halves flat side down on the bread.

M. Cundall
Middlesbrough

Call my SAUSAGE Bluff...

This week's mystery word:

BRADWORST

Is it...

a) A woodworking tool
b) A coniferous tree
c) A German sausage

Last week's Call My Sausage Bluff:
CHIPOLATA
Ans. b) A small sausage.

Ask Doctor SOZIDGE

Dear Dr Sozidge
Why are sausages called bangers?

Mavis Trumpton
Wells

THE WORD banger derives from BHANGA, the Indian word for a small cylindrical roll of fried meat. During the days of the Raj, serving soldiers would eat bhangas for breakfast, and on returning to england, they would refer to their sausages as bhangas, which eventually became bangers.

Heinz Sozidge is Professor of Sausages at Brunel University.

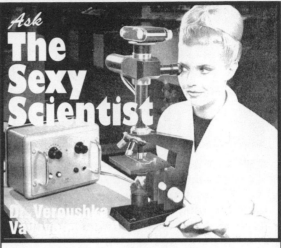

Ask The Sexy Scientist

Dr. Veroushka Vavavoom

Dear Dr Vavavoom,
I understand that the only practical way to shrink a close-tolerance metal or ceramic component is to chill it to extreme sub zero temperatures. To do this, I know that one could use an electric refrigerator, pack it in solid CO_2, or use a cryogenically cooled liquid gas. But are there any more efficient means, and if not, which is the best liquifying gas to use?

Professor Dirk Ramrod, University of Colorado

*Dr Vavavoom says ~ *Liquid nitrogen is the most efficient gas, as the component can be completely submerged in the liquid for maximum surface contact. But another approach you might consider when fitting close tolerance components is to heat the outer member, but this can create an oxide coating and introduce metalurgical problems. Furthermore, it is very hot work. When I do it, I start to get very hot and sticky. Just Imagine the way my crisp white lab coat clings to every supple curve of my body. Oooh! Sometimes I get so hot, I have to strip down to my skimpy bra and panties in the laboratory. Imagine that.*

Dear Dr Vavavoom,
I am a PhD student and I am about to start work on a project to examine the nature and structure of lysozyme. How would you recommend I approach the mapping of this particular enzyme's molecular structure?

Glint Thrust BSc, University College London

*Dr Vavavoom says ~ *The amino acid sequence of protein molecules can be worked out by relatively straightforward chemical analysis, and with respect to lysozyme, this was demonstrated by P Joles and RE Canfield in the early 1960s. However, the use of physical methods, such as x-ray diffraction is required to enable the spacial arrangements of all the atoms in the molecule to be... oh, bother! I've spilt phenolpthalene all down my dress. I guess I'll just have to take it off. Oh, and it's soaked my underwear, too. I wish you were here to help me out of these wet things, Glint. But you'd have to promise not to look.*

Dear Dr Vavavoom,
I'm conducting reserach into the methods of gene transfer in tobacco plants (*Nicotiana tabacum*) as a model for general gene transfer, but I have been having limited sucess using the chloramphenicol acetyl transferase assay as a method of looking at the sucess of transformation. I've heard of a similar, non radioactive assay using ß-glucuronidase as a marker, and wonder if it can be applied to the same experiments.

Professor Rock Haard, University of Amsterdam

*Dr Vavavoom says ~ *Well, there are... Oh no! I've laddered my stockings on a retort stand. Bother! They were so expensive, too. They're not the modern stay up type. I like the old fashioned style with suspenders. Look, they're real silk. Here, feel them. Aren't they smooth, hmm? Oooh, yes! That feels nice! Mmmm! Oooh, yes! Higher. That's right...*

To hear the rest of Dr Vavavoom's answer to this question, call

0898 000 001

Calls to Dr Vavavoom terminate at the Faculty of Science, University College London. All calls cost £1.50/min at peak times, 90p/minute at all other times.

It's the Double Life Shocker of the Millennium

Excuse Me Madam...

Are You Bei

TO MILLIONS of viewers, he will always be remembered as the camp Mr Humphries from 70s sitcom *Are You Being Served*. With his mincing walk, slack wrist and eagerness to measure a gentleman's inside leg, we all took it for granted that JOHN INMAN was as gay as the day was long.

But, according to a man claiming to be the actor's former window cleaner, nothing could be further from the truth. In real life, says Albert Vennison, Inman was 100% heterosexual and had a ravenous, insatiable appetite for sex with women. "Inman certainly led the British public a merry dance," he told us from the modest Clacton bedsit he shares with two of his brothers. "And from my unique vantage point up a ladder, looking through his windows, I saw the lot." And now, on the anniversary of his death in 2007, Albert has decided to lift the lid on the late actor's secret, straight life.

SEMI

"Like every other TV viewer, I had always assumed that Inman was on the other bus, but I found out how wide of the mark I was the first time I went round to clean the windows at his swish Croydon semi. I had gone up my ladder to do the back bedroom. Inman's house had double glazing units with a UPVC frame, and unlike other window cleaners, I prided myself on always cleaning the frames as well as the glass. But I must admit, I forgot about cleaning either when I peered through into his room. There, on Inman's bed, was a huge pile of naked women, all writhing and moaning in ecstasy. There must have been nine or ten of them, all without a stitch on. I am a red-blooded man, so I settled back to watch.

EXCLUSIVE!

"Suddenly, in the middle of the pile, trapped between a load of tits, appeared a familiar face. It was John Inman, and he was grinning from ear to ear! When he saw my boggle-eyed face looking in through the curtains, he gave me a cheeky wink. 'I'm not free!' he quipped, before giving each woman in turn a multiple orgasm. Twice. Not a lot of window cleaning got done that day, I can tell you. I was perched up that ladder watching the action for more than three hours before they all collapsed exhausted on the bed."

HARD

That was the first time that Albert saw the star's carefully cultivated gay mask slip, but it certainly wasn't to be the last.

HEAVY

"At first I assumed that what I had seen was just Inman experimenting, and that he was still one of them really. But how wrong I was, as I found out the very next day. In the window-cleaning business, it's always a mistake to think you can get away with using dirty water as it can lead to streaking and smearing on the glass. I had just filled my bucket with a fresh load from Inman's outside tap, and I happened to peer in through the kitchen window. At first, I had to pinch myself to make sure I wasn't dreaming. I couldn't believe my eyes.

"*There must have been twenty women in there, all of them beauty queens, exotic dancers and busty page 3 models. They were all stark naked so you could see everything, and they were* fighting over Inman's manhood. Eventually Inman vanished into the pantry, returning seconds later with one of those ticket machines you get at the cheese counter in the supermarket. He made each of the ladies take a ticket and then bonked them in strict rotation, quipping his catchphrase 'Are you being served?' before taking every one them to the heights of sensual pleasure. And back.

"After he had finished with them, each woman went to the back of the queue and got another ticket! The only thing that brought the orgy to an end was when the machine ran out of tickets and the women had to get dressed and go home. There wasn't a lot of window cleaning got done that day either, I can tell you."

SCRAP

Vennison now knew of Inman's secret double life, and how his on-screen gay persona was at complete odds with his red-blooded private life. So it came as quite a shock when the next day he witnessed Inman being intimate with just ONE woman as opposed to a dozen or more.

BATTER BIT

"I had just washed the window of Inman's boxroom and was about to take the suds off the glass. There are a lot of charlatans in the window cleaning game, many of whom use an old car windscreen wiper for that job. I, on the other hand always insisted on a proper window cleaner's squeegee. They may cost a bit more, but they do a professional job. Anyway, as I started, I saw Inman on his spare bed in a naked clinch with a woman. I looked around for the others but was amazed to see that she was the only one. I thought that the *Are You Being Served* star was perhaps losing his touch, until I saw a familiar looking pointy bra hanging off the lampshade on the bedside table.

"*The woman in Inman's arms was none other than Madonna, the most desired woman on earth. I watched incredulously, repeatedly squeegeeing the same bit of window over and over again, as Inman made love to her for fourteen straight hours. I lost count of the number of times she climaxed, but it*

WHEN I'M CLEANING WINDOWS: You should of seen what I of seen, says Albert.

STRAIGHT UP: Was camp Mr Humphries a secret stud?

ng Serviced?

looked like he took the sex goddess to new heights of fulfilment, heights she could never have achieved with Guy Ritchie, Sean Penn, Warren Beatty or any of the other blokes who have banged her. If she was like a virgin when she went into Inman's bedroom, she certainly wasn't when she came out, I can tell you. In fact, she was walking like John Wayne by the time he'd finished with her."

TRUE

Cow-tongued Kiss frontman Gene Simmons claims to have slept with over 4000 women in his lifetime. This may or may not be true, but what is for certain, says Vennison, is that the Grace Brothers Casanova regularly beat Simmons's record... in a single day!

"Never trust a window cleaner who uses washing up liquid in his bucket. My reputation is build on a squeaky clean finish, the kind of finish that can only be achieved using special, professional window polish that is solely available to the trade. However, this day I ran out of the stuff whilst I was doing Inman's windows, so I decided to knock on his door to ask if I could borrow a bit of washing up liquid. However, when I got to the front of the house, I couldn't believe my eyes.

GOLD

"There was a queue of women stretching out of his door and all the way down the street! In amongst them, I thought I recognised some of the Nolan Sisters, a few of Pan's People, the girls out of Hot Gossip and Hill's Angels off the Benny Hill Show. Inside the house, the line continued up the stairs and right into Inman's bedroom, where the man himself was servicing them one by one, taking each in turn to shattering levels of erotic rapture. Outside the room, security guards were asking the women to strip off in readiness so that Inman could keep his rate of turnover up. Once he had finished with them, the sexually satiated women were helped down the

back stairs by staff, given a cup of tea and a biscuit before being ferried away in a fleet of double decker buses.

MARS

"He may have looked effeminate and feeble on screen, talking to Mrs Slocombe about her cat, but when it came to the ladies, Inman's sexual stamina could only be compared to that of a rutting stag. He was insatiable. Inman served more women that afternoon than he ever did at Grace Brothers. I counted 5000 in through his bedroom door and I counted 5000 out. And he took every single one of them to a hitherto unimagined peak of carnal delirium."

AZTEC

Sadly, says Vennison, that particular encounter was to be the last that he was to have with the Are You Being Served stickman. That evening, a slight misunderstanding in the grounds of his local nurses' home led to the 53-year-old bachelor losing his job

I'M FREE: Inman was always available for sex with ladies, says Vennison.

and being put on the sex offenders register. "It was a terrible pity, because I loved my work and the unique window it gave me onto the lives of my customers," he told us. "In the end I decided it would be best to up sticks and try my luck in another town under an assumed name. I moved to Nuneaton in Warwickshire and set up a new window-cleaning round. To my surprise, my first customer was Generation Game host Larry Grayson."

GAY

"But the first time I climbed up my ladder to clean his bedroom windows, I got the shock of my life. For the amazing scene that met my eyes was a million miles away from the gay personality Grayson projected on screen."

NEXT WEEK: NOT A GAY DAY

THE TIME when all the Miss World contestants were invited round to Grayson's house for coffee, and what happened when he 'shut that door'!

GRACE BROS

It's a FUNNY WORLD

Compiled by Bob Liar

Players and spectators could hardly believe their eyes when they arrived for a fourth division match in Potizi, Bolivia. For the match referee, Mr Boco Perez, had arrived wearing his shorts back to front.

★ ★ ★

A cake factory worker in Argentina was sacked after customers complained of motorbikes in their cakes. Ricardo Gomez later told police how he had stolen a total of 174 motorcycles and disposed of them by putting them in cakes.

★ ★ ★

Doctors in Chile could hardly believe their eyes when they examined X-rays of 56 year old traffic warden Enrico Parona. Inside him they found a total of 436 motor cars, weighing no less than 75 tons.

★ ★ ★

A court in Brazil recently fined a rest home owner the equivalent of £25 after hearing how he had sold old age pensioners for pet food.

Alfredo De Falcos told the court he had sold 33 residents to his brother's dog food factory by mistake.

The happy couple seen here arriving at a top restaurant and later out night-clubbing well into the early hours.

WE 'FIX IT' FOR POP FAN

WENDY'S DREAM COMES TRUE!

Evening with 'Shaky'

for teenage fan

Eighteen-year-old Wendy Thompson couldn't believe her eyes the day she won first prize in a Viz Comic pop competition.

For Wendy, who comes from Fulchester, hadn't even entered.

But niether had anyone else, and so Wendy's name was picked out of several thousand listed in a telephone directory. And her prize: A night out with her pop idol, Shakin' Stevens.

"It was a dream come true", she told us after the couple had dined at a top restaurant.

"Afterwards Shaky took me dancing. It was great. An experience I shall never forget", said an excited Wendy.

"I shall be placing a regular order for Viz in the future", she added.

I'VE BEEN SAVING WHALES FOR ABOUT 4 YEARS NOW. AS YOU CAN SEE, THIS ONE HAS GONE QUITE SMELLY

BIFFA BACON AND PERCY POSH

AH! PERCY POSH IS PLAYING HOPSCOTCH

HE'LL NOT MIND IF I BUTT IN...

HI PERCY!

K-THUD!

UMPH!

HO HO!

THAT LITTLE WHEEZE NEVER FAILS.

LATER... AH HA!

THIS 2×12 HEAVY DUTY TIMBER PLANK GIVES ME AN IDEA!

HAVE I GOT A SURPRISE FOR PERCY!

HIYA PERCY

SMAK!

MORE OF BIFFA'S CRAZY CAPERS NEXT WEEK

300

LOONY LEFTIES SET TO KILL EVERYONE!

We have uncovered a sinister plot to **ASSASSINATE** hundreds of top politicians, show business celebrities and members of the Royal Family.

Drug crazed animal rights extremists plan to wipe out **EVERYONE** who eats meat, and that includes things like sausages, cornish pasties and even tinned ravioli. And police fear that the lunatics' hit list could include several **MILLION** names. Among them the Queen, Margaret Thatcher, and top news reader Sir Alistair Burnett.

DEATH LIST

Top of the death list would be blood sports enthusiasts and food barons like turkey king Bernard Matthews, Captain Birds Eye and Mr. Kipling.

EXTREMISTS

Posing as drop-out, drug-crazed, left-wing layabout extremists, our reporters managed to infiltrate the seedy south London squat which acts as headquarters for a group of smelly, long-haired, unwashed hippies who are master-minding the murderous plot.

ARSENAL

We were greeted at the door of the squalid basement flat by Malcolm Evans, an odourous, flea-ridden, communist subversive drug addict who boasted of his plans to terrorise Britain.

Long haired vegetarian hippies plan terror blitz on Britain

Underneath a urine-soaked mattress in a damp rat-infested back bedroom we were shown a terrifying arsenal which Evans and his festering, filth-ridden, greasy cohorts had amassed in readiness for their planned terror blitz.

CHELSEA

This included a bread knife, a garden fork, a catapult, several marbles and a sinister empty Smash tin. "That's going to be a bomb", he told us. "We are going to get some explosives to put inside it. And a battery".

CRYSTAL PALACE

Top of the lazy, layabout good-for-nothing hippies' death list is TV astrologer Patrick Moore, who, according to Evans, eats ham and mushroom quiche. We were

Some animals yesterday.

shown a black book containing the names of dozens of well-known celebrities, among them Esther Rantzen, football commentator Brian Moore and weather girl Trish Williams.

SPURS

According to Evans, Rantzen eats sausages, Moore has gone fishing in the past and Williams likes tomato soup.

STIRRUPS

New Scotland Yard anti-terrorist chiefs are already aware of Evans' activities. An earlier terror campaign ended in his arrest after he was punched in the face outside a pie shop in Wimbledon. He was later fined £25 for behaviour likely to cause a breach of the peace.

HIT LIST

MOORE

MATTHEWS

BIRDSEYE

WE INJECT QUEEN WITH ORANGE JUICE

Police are thought to be taking seriously threats by a former Twycross Zoo employee who claims that he will inject the Queen with 'monkey chemicals' if £150 is not paid into his Post Office savings account by mid-day tomorrow.

In a note sent to Buckingham Palace, former zoo auxilliary Trevor Balderstone says that unless the cash is paid on time, the Queen will be injected with the rare monkey chemicals stolen from the zoo. And as a result she will become hairy, eat bananas and live in a tree.

Zoo man's ape threat to Queen

"I'm not joking", says Balderstone, 27, an out-patient at a Leicestershire mental hospital. "I've got the chemicals and I mean business".

A spokesman for Twycross Zoo admitted that a small quantity of monkey chemicals was missing, but added that is was probably not enough to cause a permanent monkey transformation in the Queen.

PEANUTS

"She'll probably just eat a few peanuts and scratch herself under the arms a bit. And perhaps her arse will go blue for a day or so".

Is this the future face of the Queen? (above)

Our reporter managed to penetrate royal security as the Queen went on a walkabout in Milton Keynes town centre yesterday. Despite a high profile police presence, our man was able to inject the Queen with half a pint of harmless orange juice. "It could just as easily have been monkey chemicals", he told us afterwards.

Former showbiz toilet attendant

I'VE SEEN THE STA
– and they're not ve

It's difficult for us to imagine that stars, just like ordinary people, use the toilet. Somehow we find it hard to believe that TV favourites such as Michael Aspel have wees and poos, and wipe their bottoms with toilet roll.

But like it or not, lavatories are as much a part of TV life as make-up and microphones. And one man who knows that only too well is Frank Crompton who for the last forty-two years has been lavatory attendant at the BBC television centre in London.

Rolls

Over the years, Frank has seen it all. And now, after being sacked in a storm of controversy over missing toilet rolls, he has decided to spill the beans on the stars who use the lavatory, and make public for the first time ever their filthy and disgusting toilet habits.

Baps

"On screen the stars look like a million dollars. But most of it is make-up and clever camera angles. When you see them with their pants round their ankles like I have, their faces screwed up in agony, and you hear the groan of relief as their stools plop into the water, there's nothing glamorous about them, I can tell you.

Jugs

The stars are well known for their extravagant behaviour. They eat well, they drink well, they party a lot, and when they have a turd – boy! Do they have one!

Melons

I'll never forget one log in particular (mainly because it wouldn't flush away, and I have to break its back with a lavatory brush.) Anyway, this one was laid by a particularly well known star. I'll just call her Judith, as I doubt she'd appreciate me giving her full name. But I'll tell you what – there was nothing **Charming** about what she left in my toilet. It had curled itself round the bowl three times, and stank to high heaven. I felt like sending her a postcard saying 'Wish You **Weren't** Here'. The smell was so bad we had to close down the studio next door, and the following day the paint was peeling off the walls and ceiling.

Judith Chalmers

Funnily enough, it's the ladies who are worse than the fellas. What **don't** the birds chuck down the toilet, that's what I want to know. If I'd had a quid for every time I've had to stick my arm round the 'U' bend, in it up to my shoulder, just to pull out a clump of soggy tampons, I'd have a tenner by now. Probably.

Soup

One day the Director General rang me. He said someone had been flushing fag ends down the pan, and they'd caused a blockage somewhere in the pipes. As a result piss was dripping through the roof of the Blue Peter studio.

Garlic bread

That afternoon I saw smoke coming over the top of one of the cubicles, so I grabbed a fire extinguisher and kicked the door in. Surprise, surprise! There was Cilla Black sitting having a crafty cup of tea and a fag. I stopped her just before she threw the fag end down the toilet. "I'll have that", I said to her. It had a bit of a duck's arse on it, but she's a good lass Cilla. She'll always give you a drag on her ciggie.

Cilla – ciggies in the loo

Rippon – 'pebble dashed'

Another bird I recall for less pleasant reasons is Angela Rippon. Boy! Was I glad when she left the Nine O'Clock News. Every time I saw her coming I'd say "And now for the Nine O'Clock Poos!" You see, she was terrible with the nerves, and every evening at about five to nine she's come busting down the corridor, farting like a tractor.

Steak sandwich

It would be ungentlemanly of me to go into any further detail. Suffice to say that when she'd finished it looked like someone had been in and pebble-dashed half the bloody cubicle. By, it took some getting off, that did. In the end I had to get a hosepipe and jet the whole place out with water.

Erm...

There was one or two well behaved stars who'd leave the place as they found it. The magician Ali Bongo was one, but how he did it I'll never know.

Dump

One day he popped in for a quick dump during rehearsels for his show. I know it was him 'cos I looked under the cubicle door and recognised his curly slippers. Anyway, he must have had a massive turd, cos I heard the sound it made when it hit the water. My first thought was 'I hope the flush shifts that bastard – 'cos I don't fancy doing it with a brush'.

Next thing I knew he got up and left, no wiping, no flushing – no sound at all. I thought 'Here we go – another mess for Yours Truly to clean up'. But when I got into the cubicle, I couldn't believe my eyes. It was as clean as a whistle. Nothing in the pan, and no mess at all.

Tip

This happened several times. Every time he'd lay a log, then leave without flushing it. But there was never anything in the toilet.

VIZ No.1 FOR STORIES ABOUT THE STARS' COCKS

So the next time he came in I peeped over the top of the next cubicle to see what was going on.

Insult

What I saw was the most amazing thing I've ever seen happen in a toilet – and I've seen a few, I can tell you! There was Ali Bongo standing sprinkling his magic woofle dust over an enormous glistening log that was so big it was practically climbing out of the pan. Then suddenly POOF! It disappeared in a puff of smoke. To this day, I've never worked out how he did it.

Hygiene is very important, specially when you work in a lavatory. So I would wash my hands every day when I got home from work. But some of the stars didn't seem so bothered.

Summon

I remember one occasion we ran out of loo roll. It was coming up to lunch time and I knew a lot of stars would probably fancy a shit during their dinner break. So I popped out to the shop to get some paper.

When I got back I was surprised to see a well known academic quiz show host who shall remain nameless, leaving the cubicle. What he'd wiped his bottom with I'll never know. But put it this way – his tie was looking a bit dishevelled to say the least!

Hail

Anyway, he then proceeded to walk straight out without washing his hands. Five minutes later I saw him tucking into a '*starter for ten*' in the BBC canteen – with his fingers.

LOOK. THERE'S A FLOCK OF SEAGULLS FOLLOWING THAT TRACTOR

NO. IT'S CLASSIX NOUVEAU.

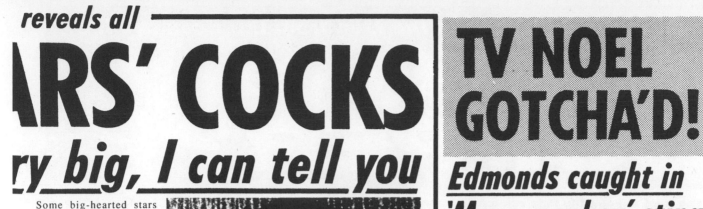

reveals all
...RS' COCKS
...ry big, I can tell you

Some big-hearted stars made working in a lavatory fun, and I'd always look forward to visits from cheery multi-talented big C all clear tap dancing trumpet player Roy Castle. Music's in his blood, and he'd always sing while he had a dump after filming 'Record Breakers'.

Roy - Big heart

"Defecation. Defecation. Defecation – that's what you need." Those were the words he'd use. And he'd always play a little tune on his trumpet too. At least I *think* it was his trumpet!

On one occasion Roy must have produced a 'Record Breaker' himself. He dashed out of the cubicle, and told me not to flush it. Then he came rushing back with Noris McWhirter and a tape measure.

Rolling

That might have been a big one, but I think the record goes to Stewart Hall from 'It's A Knockout'. One day he produced a specimen which *was* a knockout – quite literally. One whiff of it and I was gone! When I eventually regained consciousness we called in his old pal Arthur Ellis with his Halifax Brewery dipstick to measure the water displacement.

Gall

While number twos are always the most exciting, it's the everyday number ones which are a toilet attendant's bread and butter. Mopping up the piss was a never ending task. Of course, as I cleaned the floor I couldn't help but get a glimpse of the stars' cocks. And just like the stars themselves, their cocks come in all shapes and sizes. A fine example being Little and Large.

Little and Large (above) - Large (below) and Little (above). Jimmy Nail (below) - Large nose, little cock. And (inset, above, below, right) Little and Large's Large pulls a face.

One day they popped in for a piss during a break from recording their great comedy show. And when I looked over their shoulders, I got quite a surprise. I'll tell you what – the names are right – 'Little and Large'. But with no disrespect to Eddie Large, I think someone got them the wrong way round!

Cheek

Thursday evenings were always busy, 'cos I'd get the Top Of The Pops crowd in. Come to think of it, I must be the envy of every bird in Britain. 'Cos there's not one pop star's cock I haven't seen. Mind, some of 'em you have to try pretty hard to see at all. Like 'Microscopic' Mick Hucknall out of Simply Red. Don't worry girls – you aren't missing much there, I can tell you.

Hole

Jimmy Nail is another one who doesn't live up to expectations. I don't think 'Nail' was a particularly good choice of surname. 'Half-Inch Panel Pin' would have been more appropriate, from what I saw.

Of course, it's not only the stars who have tiny tackles. A lot of the high ups at the BBC – like the Director General – are short of a few inches in that department. And I suspect that jealousy may have been partly to blame for my recent dismissal. They said it was because I'd stolen some toilet rolls – but I'd only borrowed them. I was planning to bring them back the following day.

Rap

I think the real reason is my dead big cock. Frankly, I don't think the egotistical stars or the snobby bosses at the BBC could handle someone working in the toilet with a much bigger cock than them. **99**

TV NOEL GOTCHA'D!

Edmonds caught in 'Moon monkey' sting

A host of gullible TV celebrities – among them House Party star Noel Edmonds – have lost money to a cheeky con man wearing a false beard and claiming to be the Prince of Wales.

The trickster, who is wanted by police for fraud, approached several wealthy showbusiness stars in 1992, and told them he was raising funds to send a rocket to the moon. He smooth talked them into believing that Russian experimental space monkeys, launched into orbit during the sixties, were now trapped on the Moon, and £10 million was needed to build a rocket to take them bananas.

TV Noel yesterday - Moon monkey mercy mission

MONKEYS

Noel was touched by the apparent plight of the monkeys, and from the Crinkley bottom of his heart he handed over several million pounds towards the moon mercy mission. Indeed, he even offered to drive the space rocket which was due to take off next year. But shortly after being handed £5 million for 'rocket parts', the mystery man disappeared, leaving Noel and a host of other stars, including David Bowie, out of pocket.

BEATLES

The police yesterday issued a warning to all stars, telling them to beware of anyone claiming to be the Prince of Wales, and to exercise caution whenever dealing with unusual requests for large sums of cash.

Yes, we have no bananas - a monkey similar to those 'trapped on the moon'.

Edmonds is by no means the first star to be easily parted from his money. In 1988 pop star Sting – real name Mr G. Sting – handed over a box containing £12 million cash to a man with a false moustache who claimed that Martian rain forests were being cut down by aliens from another galaxy. Mr Sting was told that the cash was needed to launch a campaign to save the Martian space trees, and bring a Martian back to Earth to go on chat shows.

POLOS

But after being given the money the man jumped into his car and drove off.

U2 JOKE

CAREFUL DEAR. DON'T GO TOO NEAR THE EDGE

It's the quiz that sorts the Men from the Boys in Blue
Are YOU a COPPER?

"IF you want to know the time ask a policeman," so the saying goes. But If someone asked YOU the time, would you know if you were a policeman or not? With many of today's cops wearing plain clothes, like Inspector Morse and DI Jack Frost, you could be a bobby without knowing it. Wearing a full police uniform is no indication either, you might simply be going to a fancy dress party. The only way to find out the truth is to help yourself with your own enquiries by answering the following questions. Take down anything you say and use it in evidence to find out whether *YOUR* jobby is a *BOBBY*.

1 One night you spot someone in a cloth cap and a stripy jumper shinning down a drainpipe with a sackful of candelabras. How many times would you say 'Hello' to him?
a. Once
b. Twice
c. Three times

2 You arrive at the scene of a hit-and-run accident. The victim is a young black lad who has been knocked off his bike and is unconscious. What is the first thing you do?
a. Check for vital life signs and put him in the recovery position.
b. Ask if anyone took the registration number of the vehicle involved.
c. Slap him till he comes round, ask where he stole the bike from and throw him into the back of a police van.

3 You are trying to teach your pet dog to sit and stay on command, but after a few hours he is getting bored and losing concentration. What do you do?
a. Give up and take him for a walk.
b. Speak to him in a loud voice to show him who is boss.
c. Hang him by his collar over a fence and kick him to death.

A police dog.

4 Early one morning, you find yourself first at the scene of a break-in at a newsagents shop. The owner has yet to arrive. What do you do?
a. Call the police and guard the shop to prevent further looting.
b. Hurry past, it's nothing to do with you.
c. Go inside and stuff your uniform with fags, and sell them later to work colleagues from your locker at the station.

5 Your young son comes home from school and reports that he has done quite badly in a spelling test. What action would you take?
a. Humorously laugh it off, telling him Shakespeare was unable to spell.
b. Sit down and calmly discuss the problem.
c. Take him down to the cellar, wrap him in a mattress, and beat him with a length of rubber hose.

6 At work, your boss discovers that you have been systematically incompetent and dishonest. You are looking at certain dismissal and a possible prison sentence. What course of action would you take?
a. Resign in disgrace and accept your punishment.
b. Deny all charges and try to ride the storm.
c. Accept early retirement on the grounds of 'ill health' with a fucking big lump sum and a full pension.

7 In the bathroom one morning, you notice that the toothpaste tube has been squeezed from the middle, and the top left off. What course of action do you take?
a. Replace the cap and think no more about it.
b. Make a joke of it over breakfast, hoping the culprit will get the message.

The police yesterday.

c. Lock each member of the family in a separate room and keep them awake for 5 days. Disorientate them with violent 'Nice & Nasty' mood swings and lead each one to believe that the others have made signed statements blaming them. When their spirit is broken, hand them a brief and innocuous statement to sign, the last two pages of which are blank, and to which you later add a fabricated confession.

7 You go into a shop to buy a hat. What sort do you choose?
a. A trilby hat.
b. A baseball hat.
c. A tall, black tit with a metal nipple.

8 Driving home from the pub, you are pulled over by a police car and breathalysed. The roadside test proves positive. What do you do?
a. Admit the offence and vow to change your ways.
b. Contest the result and demand a blood test at the station.
c. Flash your warrant card at the officer and drive merrily on your way.

9 What sort of person were you at school?
a. Studious and academic.
b. Sporting and competitive.
c. A big racist bully, pickpocket and thief with no friends.

10 What do you consider the most important skill you bring to your profession?
a. An ability to organise and work as a member of a team.
b. The capacity to solve problems quickly and imaginatively.
c. Being over 5 foot 10.

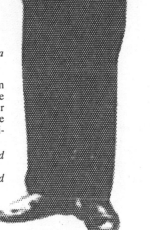

Tall and proud, a member of the Metropolitan police. How do you measure up?

DOCTOR. MY HUSBAND'S HAD A STROKE.

HOW DID YOU DO?

MAINLY A'S: Oh, dear! You are fair, honest, hardworking and you always try to do the right thing. You are certainly not a copper, and never will be. There is no place in the police force for the likes of you.

MAINLY B'S: You are not definitely a copper, but on the other hand you are not definitely not a copper neither. You are somewhere in between. Perhaps you're a traffic warden or a security guard in Top Shop.

MAINLY C'S: Congratulations! You're the Fuzz. Tirelessly pounding the beat in your big, shiney shoes, you impartially dish out justice to young and old, black or white, paying particular attention to the young and black.

Harold Ramp & Partners

Precinct Beverage Operative

Salary 10p for a cup of tea + carrier bags

Harold Ramp & Partners, one of Europe's leading vagrancy consultancies, are seeking to recruit an experienced Scottish precinct beverage operative to join a busy bench in Newcastle's Eldon Square shopping centre.

The successful candidate will have a proven track record of stumbling around a retail concourse whilst swigging from a plastic bottle of white cider, and will ideally have at least 2 years' experience of aggressive shouting at passers by. Shoes without laces are not essential, but would be an advantage.

If you are a purple-faced dedicated team player aged 25-75, looking looking to expand your career horizons in a challenging yet rewarding post bringing you into close contact with members of the public and security staff, we'd like to hear from you.

For an application bundle contact Mad Jim on the bench opposite Thorntons, Chevy Chase, Eldon Square, Newcastle upon Tyne.

Harold Ramp & Partners

"Investing in People Who Shit Themselves"

Methylated Nomadic Group

Executive Senior Tramp

168 hrs/wk - circa 10p for a cup of tea + any loose change

The Methylated Nomadic Group is an expanding nationally-based organisation of hobos working primarily in town centre gutters and paved thoroughfares. We have an active and vibrant development programme and are currently recruiting to the following position to join our existing South East team.

Executive Senior Tramp 168 hrs/wk - c10p for a cup of tea + any loose change.

The ideal candidate will be a go-getting self starter with proven muttering skills and at least 10 years experience of standing on a street corner shouting at traffic in a wooly hat. A lack of basic personal hygiene is important as well as one or more of the following:

- A beard full of dried sick
- 4 pairs of ill-fitting trousers
- Foetid stench and colourful facial bruises

Apply in the first instance with details of your current package to paraffin@mngrailwayarches.com.

The Methylated Nomadic Group

'Drinking Metal Polish Today for a Brighter Tomorrow'

Newcastle City Council Parks & Amenities dept.

In-Shrub Self-Abuser
56hrs, £0.10 for a cup of tea

Application form and further details are available from Mr College Emery, Chief Tramp Wanking Strategy Officer, Civic Centre, Newcastle-upon-Tyne.

Closing date : 28.2.05

We are an equal opportunities employer

Please note: All applicants will be required to fail a Criminal Records Bureau Enhanced Disclosure Check

Newcastle City Council Parks and Amenities Department is seeking to appoint a *Senior In-Shrub Self-Abuser* to join its successful Heaton Park alfresco masturbation unit.

If you are a dynamic, enthusiastic, scruffy individual with a proven ability to sit in a rhododendron bush pounding away at what's left in your trousers for 8 hours a day, we'd like to hear from you.

The successful candidate will also be required to defecate on the steps leading up to the bowling pavilion and step out in front of children to scare them.

London Transport operates one of Europe's leading Integrated Rapid Mass Transit Systems. We are looking for a motivated individual to fill a key post which has arisen within our organisation.

Grade 2 Comatose Underground Vagrant

circa 12p for a cup of tea (Includes London weighting) + subsidised travel

You will be required to sleep in an underground train going round and round on the Circle Line between the hours of 4am and 12.15am Monday to Saturday. The successful applicant will be expected to work in close proximity to our network of service users in order to make their journeys as unpleasant as possible, so experience of creating commuter discomfiture and nausea in enclosed spaces will be a distinct advantage.

- **Do you** have a proven track record of sitting in your own piss?
- **Can you** fit all your worldly belongings in eight carrier bags?
- **Could you** render all seats within a ten foot radius of yourself unbearable?

If your answer to these questions is "Arrgh! Fuck off, ya fuckin' basa", then we want to hear from you.

Send CV and references from at least two people who are your best fucking mates to: Mrs D Pepper, Chief Personnel Officer, London Transport, Ealing Broadway SW6.

KernowTramp LTD

Senior Bin Rifler ~ Grade II

We are the largest provider of tramps, bums and vagrants in the South West of England. An opportunity has arisen for a skilled and enthusiastic bin rifler to scavenge through the fast food outlet bins in the Bodmin, Wadebridge and Camelford area. The job will also entail a certain degree of pavement scavenging, so the ability to think on your feet and shoo pigeons off half-eaten burgers would be an advantage.

Apply in the first instance to 'Scrumpy' Trelawny, KernowTramp Ltd, seats 1-5, Truro Hospital Accident and Emergency department.

Trainee Wasteground Drinking Companion
salary £0.0001K for a cup of tea

This is an exciting opportunity for a dynamic, optically unfocused and inarticulate individual to join a close knit team of heavy drinkers working out of a patch of rubble strewn ground on the outskirts of Stockport. Duties will include standing and watching an armchair burn whilst drinking methylated spirits. The ability to make roaring noises is essential. The successful applicant will also demonstrate strong flailing skills and be able to start a fight with themselves. The ability to occasionally catch fire would be beneficial.

Ramp, Ramp, Baglady & Ramp, Stockport

Integrated Tramp Strategies since 1986

Poundbury Heritage Village is a charming hamlet designed in close consultation with **HRH the Prince of Wales** to reflect his interests in old-fashioned English vernacular architecture. In keeping with Poundbury's policy of maintaining an enforced tranquil and idyllic picture postcard atmosphere we are seeking to fill the following post:

Poundbury Heritage Village

Traditional Cheerful Gentleman of the Road *Salary - 2 bob for a cup of tea, guv'nor*

The successful recruit to this position will be a friendly, outgoing, presentable tramp with a ready smile and a kind word for everyone. He will be fully continent and well-presented, wearing a genteelly shabby overcoat, battered top hat and gaily patched trousers. He will at all times sport a carnation in his button hole, and have a working knowledge of carrying a furled umbrella (with 2 or more protruding spokes), as well as a red and white spotted handkerchief bundle over his shoulder. Shoes with flapping toecaps and spats which have seen better days will be provided. Applicants will have been educated at a leading public school and posess a University degree in one of the humanities. An ability to delight children by magically producing toffees from behind their ears would be an advantage.

Whilst chuntering to yourself charmingly as you saunter along the highways and byways of Poundbury would be positively encouraged, we regret that we are unable to consider applicants with catastrophic mental illnesses.

Send CV to: **Human Resources Department, The Town Hall, Poundbury, Dorset.**

The Prince of Wales is an Equal Opportunity Employer. No proles.

Poundbury Heritage Village ~ Forward to an Imaginary Past

Sans Abode plc
Tramp Recruitment Consultants

Are you looking to develop your tramping career? We are an agency that will value your experience, professionalism and smell, offering a range of temporary and permanent Harold Ramping opportunities within a multi-platform UK-wide brief. Sans Abode plc are pleased to currently offer the following vacancies:

- **Sinister Muttering Underpass Vagrant** - Ref 1023 - West Midlands - salary IRO 10p for a cup of tea
A sinister muttering vagrant vacancy has arisen within a dimly-lit area in a Birmingham city centre underpass. An attractive relocation package, including a wet sleeping bag and a balaclava helmet is available to the right candidate.

- **Dancing Bag-Lady** - Ref 4476 - Leicester - salary IRO 10p for a cup of tea
A dancing bag-lady is required for the corner of a busy shopping street on a 6-month contract to cover for maternity leave. If you can writhe about like Isadora Duncan on temazapan, with your eyes closed and a faraway expression on your face, an attractive short-term package awaits in the East Midlands.

- **Cashpoint Beggar** - Ref 3859 - Glasgow - salary IRO 10p for a cup of tea
An ideal first step on the vagrancy ladder for an ambitious first time tramp, this attractive opportunity offers a competitive package for the right candidate. You will be an excellent communicator and tenacious negotiator who can engage positively with cashpoint customers in order to relieve them of their spare change. A scrawny dog on a string and a dirty face would be an advantage.

- **Senior Cigarette-End Search and Collection Operative** - Ref 7823 - Nottingham - salary IRO 10p for a cup of tea
Nottingham Passenger Transport Executive is looking to appoint a senior cigarette-end search and collection operative to work amongst the disgusted public at its busy Victoria Bus Station. If you are skilled at spotting, picking up, drying out and smoking discarded dog-ends and you are looking to expand your career aspirations with an exciting position in a vibrant metropolitan environment, then this job could be for you.

Apply, quoting the reference of the job you are interested in to tommy_no_nose@sansabodeplc.com.

Sans Abode plc. YOUR lack of tramps is OUR business

Peter Pan of Pop in Euro Song cash bribe sensation!

RICHARD THE TURD!

EXCLUSIVE

A bribery scandal is set to rock the British music industry to its foundations.

For we can exclusively reveal that squeaky clean pop singer Cliff Richard accepted cash bribes in order to 'throw' the Eurovision Song Contest.

And that will leave both friends and fans of the so-called 'Peter Pan of Pop' struggling to come to terms with the fact that their idol is in fact a **LIAR** and a **CHEAT**.

LIGHT

Richard's criminal activities came to light when we gave someone who once knew him £25,000 to make up the allegations. And we have compiled a damning dossier of evidence against the star, a copy of which is being sent to Top Of The Pops.

HEAT

In a video taped conversation the fifty-year-old bachelor star admitted singing badly in order to **LOSE** the 1973 Eurovision Song Contest. Richard was red hot favourite to win with Britain's entry 'Power to all our friends'. But TV viewers were stunned when the song failed to win enough votes, and the French entry 'Bing a bong a bang a boom' topped the poll.

KINETIC

On our tape, Richard is seen boasting to a pal about how he threw the result. "I sang it badly, and when I danced I kept my hips slightly out of tune with the rhythm", he said. "I got £75,000 grand for that. Not a bad result, eh?" he added.

Cheat Cliff is a liar and a fraud

Richard explained how he had been approached during the Second World War in an Egyptian night club by a mysterious gentleman wearing a red fez who offered him cash to lose the competition. "He said he represented a shady Far Eastern gambling syndicate who were planning to bet £100 million on me not winning. He gave me a brown envelope with £2 million in it, and said I'd get the rest when I didn't win".

POTENTIAL

Shady Far Eastern gambling syndicates can make hundreds of thousands of millions of pounds by betting on someone not winning the Eurovision Song Contest. By betting £100 at odds of 25 million to one, a shady Far Eastern gambling syndicate can stand to make £2.5 billion. And with such huge sums at stake the scandal is bound to spread throughout the pop world.

GLUCOSE

Already there are rumours that stars appearing on Top Of The Pops have accepted bribes from shady Far Eastern gambling syndicates. Roland Gift, lead singer with The Fine Young Cannibals, was investigated by the police after performing a cartwheel on the show in 1987. A dark haired man with sun glasses and a trilby hat entered a Far Eastern betting shop the day before and bet £50 on Gift performing a cartwheel on the show, at odds of 4000 to one. However, as a spokesman for Ladbrokes explained, the bet was not successful.

Gift - cash for cartwheel

"Unfortunately for the shady Far Eastern gambling syndicate concerned the bet was only half of a double 'yankee'. They had also wagered that on the same programme Peter Powell would set light to one of his own farts with a match".

SILVER SPOON

If Powell had done so the mystery syndicate stood to have won over half the money in the world. Last night Peter Powell was unavailable for comment. A spokesman for the BBC said they would have to examine video evidence of all of Cliff Richard's 110 chart hits, and watch his film 'Summer Holiday' next time it comes on the telly, before deciding whether to press charges against the star.

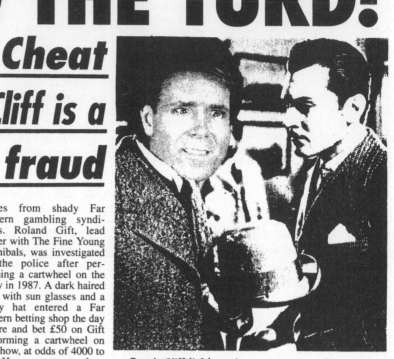

Greedy Cliff (left) caught on camera in an Egyptian nightclub accepting a million pounds. In a hat.

Peter Pan couldn't get a pan handle

A former school pal of pop traitor Cliff Richard yesterday told reporters that the 52 year old virgin was no good in bed, and regularly two-timed his wife.

"I slept with him several times, and he was a selfish lover", she told us. "He may be the Peter Pan of pop, but he's certainly no Pinocchio when it comes to pan handles," she added.

SORDID

The seventeen year old girl, who asked not to be named, mentioned that Richard was also a **bully** and **thief.** "He would often steal the other kids' dinner money, mug old age pensioners and kill people's pets with fireworks," she added.

A prostitute yesterday

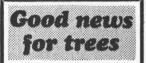
"TRAITOR!"
– say former pals

By GREG TURD and NINA COW

Pals of 51 year old Richard yesterday queued up to condemn the disgraced singer.

JUDAS

"If this is true, I'd be most surprised," said Terry Wogan, who hosted the Eurovision Song Contest.

SHITBAG

Another former showbusiness buddy, actress Una Stubbs, was last night unavailable for comment.

Wogan - surprised

YOU ARE WHAT YOU EAT

Nowadays people are more image conscious than ever before, spending hours in front of a mirror, and fortunes on expensive clothes.

And it's not just fashionable teenagers who care about their looks. Politicians, bank managers and even taxi drivers are getting in on the act, hiring 'image consultants' to advise them about their clothes, their hair and even the way they speak.

STARS

Indeed, many of today's top personalities, MPs, film stars and even Royals, spend a fortune each year on hair cuts, clothes and new teeth in order to improve their public image.

BOOK

For having the right image is very important in today's hectic world. But changing **YOUR** image needn't be such an expensive business. For according to a book soon to be published, the things we **EAT** are more responsible for the way people see us than any other factor. Indeed, changing your diet can change your image, often overnight.

PORRIDGE

Food and Image Consultant Dr. Campbell Soup has been studying what he calls the 'Total Diet Control of Character Projection' for several weeks.

"In simple terms, you are what you eat", he explains in his book. "For example a man who eats porridge may not be Scottish, but to an onlooker he certainly appears to be".

SIGNALS

Dr. Soup believes there are complex chemical and scientific 'food signals' emitted when we eat, a sophisticated form of body language about which little is known.

PLATFORMS

In today's world many people strive to create a wealthy, successful image. They buy fast cars, Italian clothes and expensive watches. But Dr. Soup believes their money would far better be spend on a new and simple diet.

"Eating small things, like frogs, oysters, or even little sausage rolls produce an unmistakable aura of wealth", says the Doctor. "For example, if you enter a room with a cocktail sausage on a stick, it immediately says more about you than your clothes can ever do".

Small sausage rolls and bits of cheese say a lot about the kind of person you are

Dr. Soup points to nouvelle cuisine to underline his point. "Affluent people spend hours in posh restaurants eating small, unpleasant things, often covered in peculiar sauce, off big plates. The same effect can be achieved without incurring that expense. For example, in a sweetshop, you could simply buy a Walnut Whip or a small finger of fudge instead of a clumsy Mars Bar, or bulky packet of crisps".

FLIP FLOPS

There was a time when to appear tough all you needed was a jar of Brylcream and a leather jacket. But nowadays the same effect can be achieved by simply tucking into a baked potato, according to Dr. Soup.

STATEMENT

"Buying a baked potato, like buying any item of food, is in itself a statement of your individuality. In the case of a baked potato, this suggests a certain abbrasivenes, a rough determination and a wild outlook on life".

OVERDRAFT

But the doctor believes that your choice of filling is just as important as the potato itself.

WITHDRAWAL

"Baked beans or cheese, for example, underline masculinity, in much the same way that a growth of stubble used to. While cottage cheese, or pineapple imply sensitivity in a person, and tuna suggests eccentricity".

Cheese, according to Dr. Soup, is the single most important food stuff in terms of image building. Your choice of cheese can dictate the way others see you for years to come.

"Stilton, for example, suggests a sophisticated, breezy image. A person who is somewhat aloof. Cheddar conjurs up a strong sexual image, the stronger the flavour, the stronger the sexual signals being sent out. Mild cheddar suggest a romantic, playful nature". According to the doctor, an air of frivolity and sense of fun can be exuded by eating processed cheeses, such as Dairy Lea triangles.

COIL

As image becomes more important to individuals, to companies and to other organisations, Dr. Soup has begun to offer his services as a Food Image Consultant to various people, among them the Metropolitan Police.

"Many of the problems the Police have are due to their image. People tend to see them as unfriendly, authoritarian figures. But I believe an improved diet could change that overnight".

DUTCH CAP

Dr. Soup wrote to the Commissioner of the Metropolitan Police, suggesting that his officers eat light pastries, toasted sandwiches and more fresh fruit. But so far he has received no reply.

RHYTHM METHOD

Dr. Soup's book, 'The Science of Diet Control of Character Projection of Image and What You Eat' will be available once a publishing deal has been found. Publishers can contact the doctor at the Department of Food and Image Research into Diet Character Projection Control, Flat 3a, Balsover Court, Lumpton, Wolverhampton WO16 4PT.

TOM FOOLERY!

HOLLYWOOD heart throb Tom Cruise's recent emotional outburst after he was squirted with water by a stunt Channel 4 microphone hit headlines throughout the world. Many people took the actor's side, arguing that a sensitive artiste should not be expected to put up with childish practical tricks. But one man who thinks differently is Les Kellet, proprietor of *Les' Joke Shop*, a novelty store on Hollywood Boulevard, just a plastic dog turd's throw from Cruise's Sunset Strip mansion.

EXCLUSIVE

Kidman: Nicole saw red over black-face soap.

Ex-pat Kellet has been in the joke business longer than he cares to remember. He previously spent thirty years behind the counter of the Laff Kabin, a popular shop selling tricks and novelties in the Shadwell district of Leeds. Eventually, an escalating series of break-ins and attacks which culminated in the firebombing of his premises prompted Les to try his luck across the pond, and in 1992 he upped sticks to set up shop in Tinseltown.

Last night Les had this to say about the pint-size *War of the Worlds* celebrity: "Tom Cruise is one of my best customers, and he's a bloody hypocrite. He's quite happy to dish it out but he just can't take it when the tables are turned."

"Over the last twelve years, Tom Cruise must have spent a hundred dollars or more in my shop. Every Friday when he gets his wages from the studio, he's in here without fail buying jokes. Looking back, he must have bought just about every novelty going; stink bombs, handshake buzzers, whoopee cushions. You name it, Tom Cruise has bought it. Apparently he liked to try them out on his then wife Moulin Rouge beauty Nicole Kidman, who at first took it all in her leggy stride.

"But the strain soon began to show. During filming for *Eyes Wide Shut*, Cruise put a bar of Black Face Soap on the sink in her dressing room. When she came onto the set to film an intense sex scene, she looked like she'd just done a shift down a coal mine! The whole crew burst out laughing and Kidman ran away in floods of tears.

"Director Stanley Kubrick was furious, and ordered Tom to apologise. Cruise admitted he'd probably gone too far this time, and went to his wife's trailer with a bunch of flowers. However, what Nicole didn't realise was that he'd

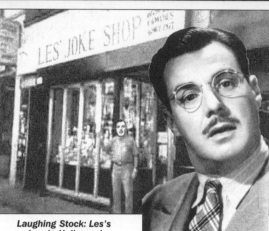

Laughing Stock: Les's shop in Hollywood.

bought the bouquet from my shop. A quick squeeze of a hidden bulb, and her face was covered in blue ink. Tom then blew a handful of sneezing powder at her for good measure.

"Tom thought it was the funniest thing he'd ever seen, but Kubrick didn't agree. He had to stop filming for three days while Kidman recovered from her ordeal.

Les could sense that Cruise's obsession with practical jokes was beginning to drive a wedge between the star and his wife. Kidman was reaching the end of her tether, but Tom just couldn't see it.

"The final straw was later that same week. Cruise told me he was planning a romantic meal with Nicole, where he was going to promise to give up playing tricks on her.

"The plan went smoothly. Tom vowed to put a stop to his joking ways, and Nicole was delighted. But it was all an act. At the end of the candlelit meal, Tom leaned over and offered Kidman a salted peanut from a can he had just taken out of his

Joke Shop Les Spills the Beans on Tinsel Town Short Arse

Zeta-Jones Saw Through Specs Maniac Douglas

HOLLYWOOD marriages are notoriously rocky, but one of the strongest and most enduring of recent years is that between Welsh beauty Catherine Zeta-Jones and her elderly husband Michael Douglas. But Les was there when a pair of magic glasses showed up the cracks in their fairytale relationship.

Focal objection: Zeta-Jones and Douglas in happier times.

"I was unpacking a box of rubber vomit pools when in walked *Romancing the Stone* star Michael Douglas. He explained he was just killing time whilst his wife, Catherine Zeta-Jones out of *The Darling Buds of May*, was in Argos next door buying some pans. But he didn't look like he was browsing to me. He went straight over to the X-Ray Spectacles display and tried on a pair."

Douglas seemed impressed and bought the 75 cent novelties on the spot. It was just another sale for Les, but he wasn't prepared for what happened next.

"Zeta-Jones came in carrying a large frying pan. She was really pleased with herself because she had got it for half price, but her smile soon evaporated when she saw Douglas wearing his glasses. She started shouting and screaming, calling him a dirty old man. She accused him of buying the X-Ray Specs so he could look

at women's bras and pants through their dresses. Douglas tried to calm her down, insisting he just wanted them so he could look at the bones in his hand, but his wife wasn't having any of it.

"She was really upset and kept hitting him with the pan, threatening to go back to her mum's in Wales. In the end she stormed out of the shop, leaving me and a very sheepish-looking Michael Douglas. He explained that he'd had second thoughts about buying the glasses, and asked for his money back. I explained that I couldn't give him a refund, as he'd already torn them off the card.

"Of course, nobody but Michael Douglas himself knows what *Basic Instinct* led him to buy those X-Ray spectacles. But I do know that they held a *Fatal Attraction* for him that very nearly cost him his marriage."

...A PLACE FOR EVERYTHING AND EVERYTHING IN ITS PLACE, THAT'S WHAT I ALWAYS SAY...

CAREFUL DEAR. THAT'S NEAT VODKA.

Sweet Joke Turned Sour for Phoenix

Most of the tricks and novelties that Les sells are nothing more than a harmless bit of fun. But in the wrong hands they can have serious consequences. One such occasion occurred on Halloween 1993, when a couple of familiar faces turned up in Kellet's shop.

"I recognised them immediately as Johnny Depp out of *Edward Scissorhands* and *Matrix* star Keanu Reeves. They were giggling and nudging each other, wasting my time. Eventually I gave it to them straight; either buy something or get out of the shop.

"Reeves explained they were going to a brat pack party that night at Los Angeles's notorious Viper Club, and they wanted

some fart powder to slip into their pal River Phoenix's drink. I told them I'd sold my last packet to Katherine Hepburn that very morning, but that I still had

Turd time unlucky: A photo of Johnny Depp.

several doses of Crap-A-Lot sugar in stock. They bought three sachets, joking that the *Stand by Me* actor would be spending his night sitting on a lavatory. I warned them that emptying more than one

sachet into their friend's tea could be dangerous, but I don't think they were listening.

"I thought nothing more about it until I turned on the *News at Ten* the next night, and discovered that Phoenix had collapsed on the pavement outside the nightclub and had later been pronounced dead. An inquest concluded that Phoenix had succumbed to a lethal cocktail of heroin and cocaine, but I'm not so sure.

"Did River Phoenix die after taking a massive overdose of Crap-A-Lot sugar? Perhaps we'll never know. All I do know is that my conscience is clear. I'm not so sure that Keanu Reeves and Johnny Depp can say the same."

---

pocket. Cruise watched, grinning as his unsuspecting wife prised the lid off the can and...Boing! A gaily coloured spring-loaded cloth snake shot out and hurtled past her ear.

"Of course, as an Australian Kidman is absolutely terrified of snakes, and so she became hysterical. Tom was hysterical too... with laughter. He took the opportunity to empty a tin of itching powder down her back.

"The marriage didn't survive long after that, and I wasn't surprised to read in the next day's paper that Hollywood's golden couple had got divorced. Now Tom is engaged to his *War of the Worlds* co-star Katy Holmes. I only hope for

both their sakes that she doesn't mind getting a ring of soot round her eye every time Tom hands her a suspiciously small telescope to look through."

And the miniature *Top Gun* actor isn't the only star of the silver screen to be found browsing through the racks of novelties in Les's shop. He told us: "If I told you who' I've had in here, you'd think I was making it up. My list of customers reads like a Who's Who of Hollywood!"

Now, in these exclusive extracts from his brand new 6-page autobiography 'Make 'Em Laff (Kellet Publishing), Les lifts the lid on the secret world of practical jokes the stars don't want us to see.

Chuddy Rap Left Biggie Smalls Hip-Hopping Mad

IT'S quite usual for the victim of a practical joke to want to get their own back. Les remembers the time Clint Eastwood put some rubber fried eggs into Tim Roth's sandwich, only to find a plastic spider in the bottom of his own tea the next day - courtesy of the *Reservoir Dogs* actor. But sometimes, the scale of the revenge can be out of all proportion to the trick originally played.

"One of my best customers was gangsta rapper Tupac Shakur. He was always in the shop. I remember one time he came in with his posse and they all bought those things that make it look like you've stuck a nail through your finger. Tupac also bought a trick pack of

Murder Rap: Biggie and Tupac yesterday.

mousetrap chewing gum

"The next day he came in and explained that he'd tried it out on his rap rival Biggie Smalls, aka The Notorious B.I.G. backstage at a concert. Tupac had offered the unsuspecting Smalls a stick of gum and the spring-loaded wire had snapped down on his finger. Apparently everybody laughed except Biggie, who called Shakur every name under the Sun,

threatening to get even.

"I thought nothing more about it until about a week later, when I read that Tupac had been killed in a drive-by shooting. Now that The Notorious B.I.G. is also dead, we'll never know whether Shakur's murder was part of a drug-related gangland turf war, or simply revenge for a mousetrap chewing gum prank."

NEXT WEEK - The time all the girls from Desperate Housewives and Sex & The City came in to buy French maid outfits and PVC nurse costumes... and insisted on trying them on in the shop!

To order a copy of Les's book, send stamps to the value of 50p to L. Kellet, 48a Bramley Apple Lane, Shadwell, Leeds. Please allow up to two weeks for delivery, as each copy is photocopied to order at the library.

Crowe's Feat Ruffled Studio Feathers

SOMETIMES the most unlikely person turns out to have the best sense of fun. Australian hard man actor Russell Crowe has a humourless reputation, but according to Les nothing could be further from the truth.

"During the filming of *Gladiator*, Russell came into the shop and bought a few things. The next day at the studios they were filming the scene where Maximus takes off his helmet in the arena and reveals his true identity to the emperor.

"Imagine director Ridley Scott's surprise when the visor went up to reveal Crowe wearing a pair of comedy Willy Nose Glasses and a set of Austin Powers buck teeth.

There were fifty thousand extras on the Colosseum set and they all just cracked up. It was such a funny shot that Scott wanted to leave it in the final film, but the humourless bosses at MGM said no. I guess we'll just have to wait a few years for the 'Director's Cut' DVD to come out before we can see Russell with a pink winkle for a nose!"

PALACE SHELL-SHOCKED

BELOW: An egg similar to the one laid by the Queen Mum and (inset) a Queen Mum similar to the one that laid an egg.

ANOTHER ROYAL EXCLUSIVE

ROYAL watchers were last night scratching their heads after Clarence House sources revealed that the Queen mum has laid an egg.

discovery

Palace staff made the discovery after her butler heard grunting sounds from her rooms late on Tuesday night. Fearing that she had fallen or was choking on a fishbone, staff entered the room and discovered her sitting on the bright blue speckled rugby ball-sized egg.

nickleodeon

She was very protective at first and refused leave the egg, breaking the arm of a footman who attempted to pick it up. However, after a few minutes she appeared to lose interest in it, and royal GP Gladstone Gamble was able to

■ *by HAZELNUT MONKBOTTLE*

remove it to a nearby incubator.

Queen mum expert Dr David Starkey said: "No monarch has ever laid an egg before, but the constitutional implications are quite clear. If whatever hatches out is male, it will take precedence over the Queen in the line of succession. We are faced with the very real possibility of ending this Golden Jubilee year with some bizarre human/chicken monster on the throne."

sky sports 1

But there were no such worries from members of the public who have already begun to gather outside the palace gates waiting for news. "I love the royals and anything to do with them", said 82 year-old Ethel Moron, who had travelled over 300 miles from Darlington with her friend Ada Fuckingstupid. "I don't care what weird beaked hybrid comes out the egg, I'm sure it will do a marvellous job."
"God bless it," she added.

Cliff Face Drop Driver OK

A WOMAN whose car plunged nearly 6ft off Cliff Richard's face has escaped serious injury.

Her Ford Orion veered out of control in the veteran pop star's hair and smashed through his glasses before plummeting off his chin.

The woman, who hasn't been named, spent the night on a ledge halfway down Richard's trousers, after the car flipped and landed on its roof. She was spotted when she flashed her headlights at passers by, who alerted the emergency services.

Firemen freed the 27-year-old woman on Wednesday night and yesterday she was "comfortable" in hospital.

Laddism is a thing of the past, and we now live in the age of the more sensitive man. But have YOU moved with the times? Ask yourself...

Are *YOU* a New Man or an Old Woman?

IT'S OFFICIAL. Laddism has gone. The larger swilling, lecherous lout of the 80s and 90s has been swept aside, joining other outmoded stereotypes such as mods, Teddy Boys and Cro-magnons in the dustbin of history. In their place, a new, more caring, thoughtful man has emerged; one who is just as happy to wield a feather duster as a monkey wrench. We all like to think we are new men, but are we really? How in touch are YOU with the more sensitive side of your psyche..?

In the hit comedy 'Three Men and a Baby', Tom Selleck, him out of 'Cheers' and the other one showed a more caring side to their nature. But how will YOU fare when you take our questionnaire?

Take our test to find out. Simply answer the question honestly, and then tot up your score to reveal if you are a **New Man** or an **Old Woman**.

Q1. It is Saturday morning, and you have arranged to meet your mates to play five-a-side football. However, your wife is very tired after breastfeeding your six-month-old baby all night. **What do you do?**

a) *Insist she goes back to bed to get some proper rest, phone your friends to cancel football, and then spend some quality time with your son. Take him to the park, or perhaps the local playgroup?*

b) *Feed your 12 cats, phone the doctor and describe your bowel movements then totter your way to the post office* with a tartan shopping trolley, muttering about immigrants and telling passers by how old you are, before falling over and breaking your hip?

HOW DID YOU DO?

*Award yourself **1** point for each time you answered a), and **5** points for each time you answered b) and then tot up your score to see how you did:*

Between 1 and 4~
Congratulations. You are a New Man. You are just as at home changing a nappy as you are a spark plug. You love to stop in and watch football on the telly, but you'll happily go with your wife to see shit films like *Bridget Jones's Diary* instead. You have that wonderful combination of tenderness and strength that women find adorable.

Between 5 and 10~
Oh, dear. You are an old woman. You only like tea made with leaves, claiming tea bags taste of paper, and you can't tell the difference between a Gas Man and and six 12-year-old gluesniffers with a Blockbusters card. You never put the heating on because it's too expensive, yet you have £50,000 in a teapot on the mantlepiece.

Don't miss our questionaire in the next issue: *Are you Ant & Dec or Dick & Dom?*

'BOARD' BONEO GOES CHESS NUTS

Pop idol Boneo, lead singer with Irish rockers U2, has swapped his guitar for a chess set. And instead of playing hits like 'Sunday Bloody Sunday', the millionaire musician is playing chess – 24 hours a day.

Bored with the rock and roll lifestyle, Boneo, who has several 'O' levels, decided to exercise his intellect by taking up the high brow game last year. And already he has impressed friends and colleagues with his chess playing ability, beating The Edge three times in a row and becoming a Grand Master at the game.

GOAL

And Boneo's new found ambition is to achieve the ultimate goal in chess and become World Champion by beating the world's top players like Nigel Short and Boris Karloff.

TRY

Boneo is by no means the only star who has taken to

By our Chess, Party Games & Table Tennis Correspondent
YVONNE GOOLAGONG

treading the board game boards. Many celebrities have achieved world class status playing board games. With only one show a week many TV stars find themselves with lots of spare time, time which they often put to good use achieving world class status in the world of board games.

BASKET

And for a skillful player the rewards can be great, for there are handsome cash prizes to be won, earning top telly stars vital extra income. For example in 1982, magician Paul Daniels picked up a cool £180,000 when he became World Champion 'Kerplunk!' player. At the

U2 star swaps pop for prawns

time 'not a lot' star Paul Daniels was earning only £50 a week for his BBC show.

RUN

Other big money game winners include Bruce Springsteen (United States 'W.H. Smith Magnetic Fishing' Champion 1979), Ludovik Kennedy (undefeated British 'Twister' champion 1981-1993) and Bamber Gasgoine who won the coveted European Triple Crown back in 1977 when he collected first prizes in 'Mousetrap', 'Buckaroo' and 'My Cat's Got Flees' at

the triathlon event at the Birmingham NEC in front of a spellbound audience of 18,000 game fanatics.

WICKET

Big game winners like Gasgoine can earn anything up to £700,000 a year from exhibition games alone, playing in packed venues all around the world. There is also big money to be made in sponsorship deals, as well as the usual perks which include free nail scissors, soap, and sizeable discounts on garden furniture.

CZECH MATES - Boneo can count among his friends U2's many fans in the former Czechoslovakia. If he wants.

£🎲£🎲£🎲£🎲£🎲£🎲£🎲£🎲£🎲£🎲£🎲£🎲£

PLAY CHESS WITH BONEO

Chess is often viewed as a minority interest sport, and many youngsters find the game boring and confusing to play. But nothing could be further from the truth, according to U2's Boneo. So we asked him to tell us a little about the game and how it is played in the hope that more people will take up the game and derive pleasure from it.

'Chess isn't like football or cricket. There's no ball, and you don't have to run around. You play it indoors, and so it doesn't matter if it's raining.

PLAYERS

There are two players, blacks versus whites. I always like to be blacks, cos one of my white prawns is missing and I have to use a button instead. But it's got 'prawn' written on it to avoid any confusion. The prawns go along the front, and the rest go along the back. It tells you the positions on the box usually.
You always start by moving a prawn. They go

two squares forward first, then one. Or across ways if they're overtaking someone. The horses are my favourite. They go three forwards and two sideways, or the other way round. That's the hardest one to remember, but I remember it by thinking that it's like two sides of a shoebox, kind of thing.

BENSONS

Castles are dead easy. They just go forwards. I think they can reverse, and they might be able to go sideways, but I'm not sure. But definitely not across. Bishops go slanty ways. And the King and Queen can go anywhere they want. But if someone overtakes your King they check mate you.

CRAVEN 'A'

It's best to remember that chess is just like a race. You have to overtake their bits before they overtake you. If you land on someone, you've overtaken them, and they go back in the box. And when someone's got no bits left, that means they've lost.

Game stars who dress like a million pounds

Lucrative sponsorship deals earn millions of pounds for successful game players. Here's how a TV star can earn a fortune moonlighting at Monopoly, or playing Sorry on the side.

WOODBINES

Top agents will ensure that every item of a stars clothing is sponsored. And this is how the millions add up.

| COLLAR £17,500 |
| TIE £8,000 |
| BUTTONS £2,000 |
| POCKETS £16,000 |
| ZIP FASTENER £7,500 |
| TURN UPS £12,000 |

TOTAL = £2,450,000

EXCUSE ME SIR, BUT YOUR FATHER HAS JUST DIED.

THIS ONE'S FOR PA

golf joke

the end

What's in a Name?

This week: CHARLOTTE CHURCH

CHARLOTTE CHURCH has gone from fresh-faced cherub of the classical music scene to chest-heavy diva of the pop charts in one easy move. We think we know all about her from seeing her every day on TV and in our newspapers. But stop imagining her soaping down your Nissan Primera in a wet vest for one moment and ask yourself: How much do we really know about Charlotte Church? You may be surprised to learn that there is more to her than meets the eye... and once again, *IT'S ALL IN THE NAME.*

...*is for* CHURCH

THE VOICE of an angel star was actually born Charlotte Reed in 1986, but decided to change her name after watching Songs of Praise one Sunday evening. Captivated by the beautiful hymns and Bible stories, it wasn't long before Charlotte Songsofpraise was wowing audiences at school productions. However, record company bosses decided her name was too long to fit on a CD box and ordered her to change it again. And the rest is history.

...*is for* HELLRAISING

LIKE ANY normal teenager, the Welsh wild-child likes to go to parties. But unlike her peers, once there Church likes to drink alcohol, causing her to become mildly intoxicated. On several occasions, the out-of-control star has been photographed walking out of a nightclub not smiling for photographers and with her eyes half shut when the flash went off.

...*is for* AGE OF CONSENT

ON THE 21st February 2002, the editors of Britain's tabloid newspapers breathed a sigh of relief as the Welsh warbler finally turned 16. Overnight, their feelings towards Charlotte changed from ones of paternal interest in her career, to ones of priapic, salivating lust over her arse and tits, with the police powerless to intervene.

...*is for* RUGBY

CHARLOTTE is currently engaged to Welsh rugby international Gavin Henson, son of the late Muppet magnate Jim. The couple are often referred to in the UK tabloids as the Welsh Posh and Becks (Pwysh y Bylls). Although why the people of Wales would want their own version of the sour-faced, clunk-voiced spendthrift and a Joe Pasquale-soundalike, PA-nudging free kick specialist is anybody's guess.

...*is for* LEEKS

IT IT NOT known if the Swansea soprano has ever expressed any particular interest in her native principality's national vegetable. However, if she wanted to grow the world's largest specimen, she'd have to beat the 3 monsters with a combined volume of 537.83 cubic inches grown by John Pearson from Ashington in 2002.

...*is for* ONIONS

IF CHARLOTTE decided growing monster onions was more to her taste, then John Sifford's 16lb 8^1/$_2$oz specimen, which won the retired engineer from Romsley a £1500 cash prize at the 2005 Harrogate Autumn Flower Show, would be the one to beat.

...*is for* TURNIPS

ONCE BITTEN by the giant vegetable growing bug, the taffy temptress may like to try her hand at cultivating a gargantuan turnip. If she could surpass the whopping 39.2lb leviathan grown by Scott Robb of Alaska in 2004, then she could add yet another world record to her already impressive tally.

...*is for* TATE

IN DECEMBER 2005, Charlotte appeared on a special Christmas edition of the Catherine Tate Show. Unlike the programme's titular star, Charlotte only made a brief appearance, failed to deliver one of three semi-memorable catch phrases, didn't stretch one minute's worth of material out for six minutes, and generally refused to outstay her welcome on our screens.

...*is for* EX-BOYFRIENDS

THE VOICE of an angel vocalist has hit the headlines with a string of far from angelic suitors. Amongst others, the Newport nymphette has been romantically linked with rapper Steven Johnson, DJ Kyle Johnson, great train robber Ronnie Biggs and nazi war criminal Martin Bormann.

...*is for* CIGARETTES

CHARLOTTE is well known for her love of smoking. She was recently stopped by customs officers at Dover whilst driving a rented Luton van. In the back, officers found almost 1^1/$_2$ million Marlborough Lights, which the Ffestiniog faghound had purchased at a Calais cash & carry. She then proved that they were for her personal use by smoking the lot, there and then, whilst port officials looked on in amazement.

...*is for* HURTUBISE

CHARLOTTE has been terrified of bears since falling into the bear pit at the North Wales Mountain Zoo in Colwyn Bay as a child. Should she visit that zoo again, she would be well advised to wear the armoured bear-proof suit, developed by Canadian nut job Troy Hurtubise. The 145lb titanium, chain mail and rubber outfit would provide ample protection should the Llandudno llovely suffer a repeat of her childhood mishap.

...*is for* UNSTABLE

ALTHOUGH SHE claims to be a 'Crazy Chick' in the chorus of her chart-topping hit, it is not thought that Charlotte's mind is any more unstable than the average female of her age. But if the Swansea sexbomb were to suffer from any severe form of mental illness in future years, then she is in luck. For Wales has several mental hospitals and residential homes all within driving distance of Cardiff, enabling the Brecon Beacons babe's family to visit her with ease.

...*is for* REAR

THE SNOWDONIAN siren was voted Rear of the Year in 2002. And it's lucky that she was, because later that year Charlotte contracted haemorrhoids after sitting on a cold radiator at an Aberystwyth Eisteddfod. With bum-grapes the size of plums, her bottom was declared ineligible for the 2003 competition, which was probably won by Kerry Katona or somebody like that.

...*is for* CURSING

WHEN SHE is singing, the Llanfairpwllgwyngyllgogerychwyrndrobwlllantisiliogogogoch nightingale has the voice of an angel, but after a few drinks it's more like the voice of a Hell's Angel. Charlotte's tendency to turn the air blue is legendary in her native valleys, where she is often heard using language that would make a docker vomit. But ironically, Welsh speaker Charlotte has to slip into English when she wants to curse, as her consonant-heavy native tongue doesn't have have swearwords! Except Cwm.

...*is for* HUGE MELONS

SHOULD Charlotte tire of growing giant leeks, onions or turnips, she could turn her attentions to the biggest challenge in the vegetable world - the melon. But unless she could produce a specimen bigger than the 268lb 12oz monster grown by Lloyd Bright of Hope, Arkansas, the Bangor bombshell would do well not to give up her musical day job.

NEXT WEEK: MR T

HER LIFE IN DANGER!

A former Palace security chief has made a startling claim that the lives of the Royal Family, including the Queen herself, are in mortal danger. And Ted Pembleton, former head doorman at the Chesterfield Palace Theatre, is convinced that security and safety measures at Buckingham Palace are now at an all time low.

His warning comes only years after Michael Fagan's much publicised intrusion into the Queen's bedroom. And Pembleton believes that unless a major shake-up in safety measures at the Palace takes place, a senior member of the Royal Family could be killed.

BLAST

Mr. Pembleton has compiled a startling dossier of evidence to support his claims, and a copy of his report is already being examined by senior police officers at Scotland Yard. In it he lists a deadly catalogue of security short-falls and inadequate safety measures. These include:

★ **BROKEN** paving stones in nearby Buckingham Palace Road which could easily cause someone to trip and fall, especially in icy weather.

★ **LOOSE** stair carpets inside the Palace which could also lead to a nasty fall.

★ **BUSY** roads around the Palace with fast moving traffic and not enough safe crossing places.

★ **LIMOUSINES** without safety belts fitted to rear seats.

In an independent test carried out at Mr Pembleton's own expense, a shop dummy dressed as the Queen was badly damaged when it was placed in the rear seat of a car, with no seat belt, and driven into a wall at high speed. "I dread to think what would have happened if that had been the Queen herself sitting in that car", a sober faced Mr Pembleton told us afterwards.

ROCKET

Among the immediate improvements recommended in his report is the construction of a pedestrian footbridge across busy Buckingham Gate, allowing the Royal Family safe access to nearby shops. And he believes that urgent safety steps are also required in the Royal kitchens.

BOMBSHELL

"I am particularly worried about the safety of members of the Royal Family nipping into the kitchens to prepare a meal or a quick snack", he told us, singling out a long flex on a kettle for criticism. "If caught accidentally by a passer-by, this would cause the kettle to fall, and could lead to serious scalding". And Mr Pembleton expressed fears that a chip pan, if left unattended, could catch fire.

Former security chief blasts Palace safety measures

Ted Pembleton first made the news five years ago when his book, *'Rape and Murder at the Palace'* was published. In it he suggested that the numerous security breaches reported in the press were just the tip of the iceberg, and that the vast majority of incidents at the Palace are simply covered up. Indeed, he put forward the theory that Ronald Biggs and the 'Great Train Robbery' gang planned their notorious raid from the safety of Buckingham Palace cellar.

STUN GRENADE

Pembleton also believes that 'Rambo' style gun enthusiasts have for many years used the Palace gardens as firing ranges, using silencers to disguise the noise and camouflage jackets to remain unseen. So far, the Ministry of Defence has refused to comment on Mr Pembleton's claim that American Cruise Missiles have already been deployed in the Palace grounds, and that on several occasions they have nearly blown up accidentally.

Princess Di, like the Queen, is a member of the Royal Family.

His book, priced £19.95, is no longer available in the shops; however Mr Pembleton asked us to point out that under no circumstances should water be poured onto a burning chip pan. "Turn off the heat, cover it with a damp cloth, and call the fire brigade", he told us.

EXCLUSIVE

DARLING. THIS THING IS TEARING US APART

BEVERLEY THRIL

Millionaire Hollywood stars bored with a life of sex and drugs are risking their lives in the search for new thrills.

In a tinsel town where money grows on trees and the streets are paved with sex and drugs, increasingly stars are becoming bored with the substances and sex acts on offer. And so they search for bigger thrills and more dangerous forms of self abuse.

In the sixties it was fashionable for pop stars to cut small holes in their testicles and inflate them with a straw in order to heighten sexual excitement. But nowadays far more bizarre sexual activity is commonplace, and as one insider revealed, one of the worst kept secrets in Hollywood involves a well known movie star, a snake and a mongoose.

them escaping giant bonfires were lit outside the King's ears and nostrils, and his arse was bricked up.

A selection of Hollywood stars yesterday (none of whom *necessarily* ever stuck anything up their arses).

Poop shute pleasures of the rich and famous

COCKTAIL

"This particular star, who for legal reasons shall remain nameless, was looking to try something different. He was at a party knocking back a lethal cocktail of drink and drugs when suddenly he produced a giant snake which he swallowed alive. Then he asked his wife, Demi Moore, to shove a mongoose up his arse."

TOP GUN

"The result was spectacular, with the animals fighting violently inside his body for about half an hour, during which time Bruce was completely off his head. That must have been some trip, I can tell ya."

TOP CAT

Introducing wild animals into orifices and encouraging them to fight inside the body stimulates the prostrate gland, a small walnut shaped organ responsible for organisms within the male private parts. And achieving sexual gratification in this way is nothing new. In Victorian England it was widely known that King Henry VIII, high on a cocktail of mulled wine and cocaine, would liven up Royal banquets by swallowing a live swan and then forcing six Yorkshire Terriers down his hog's eye with a pipe cleaner. The crazed animals would chase around inside the King's body for up to eight days. In order to stop

A less dangerous but equally bizarre form of internal stimulation currently favoured by the stars of the entertainment world is the game of 'arse snooker'. Fuelled by highly potent cocktails of liquid paraffin, brandy and Guinness, stars swallow an entire set of snooker balls. They then remove their trousers and bounce vigorously up and down on a trampoline whilst attempting to shit out the snooker balls in the correct order.

"Then Jack Nicholson climbed up my brown eye and fired up the pneumatic drill"

One point is scored for a red, which must then be followed by a colour. The colour is then swallowed again, and another red must be passed. Scores for each ball are the same as in snooker, and if the white ball comes out by mistake, the player loses four points.

TOP HAT

Wild drink, drug and arse snooker sessions can last for several days and nights on end, with players reaching numerous multiple

organisms along the way, and stopping only to quaff neat vodka and guzzle down lethal cocktails of cocaine, heroin and ecstasy. Needless to say for added interest the millionaire stars play for money.

WHITE TIE

The stakes are high, with up to $10 million resting on every ball shitted. Indeed one body building Hollywood box office billionaire was reported to have lost his entire fortune gambling on arse snooker. But he later told pals it had been worth every penny, as the crazy cocktail of bouncing balls, booze and drugs had given the former Mr Universe the "ultimate high".

TAILS

But a new and far more dangerous game growing in popularity among the Hollywood jet set can, quite literally, provide stars with a breath taking 'high'. For those who play

the deadly game of 'volcano popping' go on a trip which, quite literally, leaves them 'sky high'.

PAWS

Already high on a lethal cocktail of drink and drugs, would-be 'poppers' make their way to the top

Hopping mad!

of a volcano that is about to erupt, then sit on top of it, clenching their buttocks firmly to prevent it erupting. Eventually, when the pressure of the red hot lava bursting up from the Earth's core becomes too great for them to resist, they relax their arses and the volcano erupts spectacularly, sending them rocketing high into the air, like a cork from an exploding champagne bottle.

WHISKERS

The force of an eruption has been known to send volcano popping stars thousands of feet into the sky. Indeed several stars are rumoured to have landed on the Moon, where they have been trapped ever since. Among them singer David

Cassidy. Yet despite being stranded on the Moon, with no food or air, Cassidy is reported to have told pals that he would not hesitate to do it again.

KIT-E-KAT

"I don't care whether I suffocate or starve here on the Moon, I'll still die happy. Because feeling that volcano going off up my arse as I flew through space truly gave me the ultimate high", the singer is reported to have told family and friends during a brief phone call from the Moon.

MARS-E-BAR

But perhaps the most dangerous stunt of all took place at a Hollywood party over twenty years ago. For after polishing off a deadly cocktail of drink and drugs actor Donald Sutherland decided to experience the ultimate sexual thrill, by becoming the world's first 'human jigsaw'.

OYSTERS

After gorging himself on a heady mixture of champagne, oysters, cocaine

of jizz Viz!
LS 90210!

King Henry the Eighth (left) and the Hollywood volcano (arrowed) from which David Cassidy went to the moon.

and heroin, Sutherland persuaded a Beverley Hills surgeon pal to cut him into a 500 piece jigsaw puzzle.

HOCKLES

Doctors estimated that Sutherland had only three minutes in which to be re-assembled, otherwise he would die. A host of show-business pals, including Zero Mostell and Burt Reynolds, frantically scrambled to complete the Oscar winning star jigsaw, eventually slotting the last piece into place with only 5 seconds to spare.

GREBBS

Sutherland later told pals that being on the floor, in 500 pieces, with less than three minutes to live, had truly been the "ultimate high". Tragically best pal Zero Mostell, star of The Producers, who had found the last piece of the Sutherland jigsaw under a coffee table, himself paid the ultimate price in search of excitement.

GOBS

For ten years earlier Mostell had himself died after putting a hand grenade up his arse and throwing the pin into a swimming pool full of crocodiles ripped to their scaly reptile tits on a lethal cocktail of brandy, crack/cocaine and Parmesan cheese.

COFFIN

Later, at his $60,000 Beverley Hills funeral, Mostell, speaking from his plush $1200 hardwood coffin, told pals that having his arse blown up in a swimming pool full of drug crazed crocodiles truly had been the "ultimate high".

Drug ring shame of TV funny men

The showbusiness drugs problem in Britain is fast becoming almost as serious as that of Hollywood. Indeed, some people in pubs estimate that over half the celebrities we see on our TV screens are addicted to drugs such as cocaine.

Cocaine is commonly used by stars wanting to stay up way past their bedtime. Usually 'sniffed' through the nose, it induces an incredible feeling of 'not having gone to bed', and enables users to stay up until three and four o'clock in the morning.

CHOKIN

But the drug can become addictive, and many performers and artists rely on having 'fixes' of the drug, in ever increasing quantities, before they are able to go on stage. Indeed, according to a friend of someone we know, who works in London but was up for Christmas, one top TV comedy duo have become so addicted to the drug that the only orifices left big enough for them to take it through are their arses.

STUTTERIN

Our insider was backstage with the individuals concerned just before a gig when one of them produced a trumpet and poured half the contents of a 2lb bag of cocaine into it. Turning to our informant he then asked if he would be good enough to blow the deadly powder up his arse for him, as he couldn't reach it himself.

WHEEZIN

"I didn't care who he was, there was no way I was going to volunteer to blow half a bag of cocaine up his arse for him", the insider revealed. "Fortunately his partner Bob Mortimer did it for him. At that point I left the room, but when I returned a few minutes later there were clouds of white powder everywhere. Suffice to say the comedian concerned had a very broad smile on his face for the remainder of the evening, and by the time I left at about half past midnight there was no sign of either of them going to bed".

It's looking Black for so sad Cilla

Picture: FRANK SHIT

Anyone who had a heart would surely help this doddering old lady across the road with her shopping bags. But even those with a good memory would struggle to recognise the wrinkled face of former Queen of pop Cilla Black.

Thirty five years ago this lively Liverpool lass *stepped inside* the doors to stardom, *love*. Hit followed hit for the former Cavern Club cloakroom attendant, who listed The Beatles among her fans. But now *something tells her nothing's going to happen tonight* as the fallen star struggles home to her cold, damp, squalid bedsit.

DRIED

After the hits dried up Cilla turned to bus conducting to earn a meagre living. But the introduction of 'pay as you enter' buses put an end to her new career. Now, after years on the dole, Cilla walks miles in worn out shoes in search of shopping bargains.

DESICCATED

People living nearby were unaware that they had a celebrity neighbour. "We had no idea who the crumpled, pathetic figure living next door was", one told us. But according to another neighbour, Cilla hasn't lost her singing voice. "She occasionally comes home with a bottle of gin wrapped up in brown paper, he told us. "Then she turns on her electric heater, gets into bed still wrapped in her tatty clothes, and sings herself to sleep".

Britain's £150m dole fraud

THREE MILLION SCROUNGERS ON THE FIDDLE

'It's happening all the time' says Mr. X

An unemployed Liverpool man has blown the lid off a massive social security benefit fraud which has been costing the nation millions.

And startling evidence which he is about to give could lead to the prosecution of millions of benefit scroungers.

SHOCKING

For the man, who prefers to remain anonymous, has exclusively revealed that a shocking eighty per cent of Britain's 4 million unemployed actually *have jobs,* and are claiming benefit illegally. And that suggests that a staggering *3 million* scroungers are on the fiddle, leaving the Government with a weekly bill of £150 million in false benefit claims.

MILKMAN

"It goes on all the time", said our informant who we will refer to as Mr. X. "Everyone on our estate does it. The postman, the milkman — they all sign on and pick up dole money".

According to his figures there are fewer than 800,000 genuine unemployed people in Britain. Proof that the so-called 'unemployment problem' doesn't really exist.

HELICOPTERS

Indeed, taking into account 'fiddle' earnings, the standard of living in Britain's 'unemployment blackspots' has never been higher. "I know several blokes who are driven down to the dole in Rolls Royces", Mr. X told us. "And a lot of the lads in the local pub own private helicopters. Another friend of mine who's been signing on for 12 years now owns a string of restaurants and a major hotel group", he added.

BOMBS

We agreed to be blindfolded as Mr. X took us to a block of flats somewhere in the Liverpool area where we were told the average income among residents, all of whom are unemployed, is £2,700 a week. There was no sign of prosperity

EXCLUSIVE

inside the building, but as our informant later told us, most of the money is spent on heroin or petrol bombs which are later thrown at the police.

KNOWN

Mr. X. supplied us with a list of well known professional footballers who he claims are currently receiving unemployment benefit. We were told that one player whose weekly earnings top the £3,000 mark, also receives £30.45 unemployment benefit.

And we were told of a foreign head of state who flies into Britain once a fortnight to sign on. According to our sources he then receives extra benefit payments to include the cost of his return air fare.

INSIDE

But perhaps the most astonishing example of benefit fraud is that of staff inside the Department of Employment who regularly walk to the other side of the counter and sign themselves on. "By signing two or three times a day they can make a massive £450 a week bonus in benefit payments", claimed Mr. X.

BOX 2

When we contacted our local Department of Employment office for a comment on these allegations a spokesman in Box 2 told us we were in the wrong queue.

"You'll have to press the bell at the enquiry window", he said.

Later, our informant Mr. X, who had agreed to give his evidence to the police, disappeared shortly after we had given him £2,000.

VIZ COMIC IN ASSOCIATION WITH THE MILK BOARD

WE'VE GOTTA BOTTLE!
AND WE'RE GIVING IT AWAY **FREE!**

We're giving away a pint of milk absolutely free to the winner of this fun to enter competition. All you have to do is answer these simple milk questions.

1. Milk contains which of the following chemicals?
   - a. Nitrogen
   - b. Aluminium
   - c. Calcium

2. Which animal do we get our milk from?
   - a. Tiger
   - b. Rhinocerous
   - c. Cow

3. Which of the following is made from milk?
   - a. Rubber
   - b. Teflon non-stick pans
   - c. Butter

4. Which famous cowboy featured in 'Milky Bar' TV commercials?
   - a. Butch Cassidy
   - b. The Milky Bar Kid
   - c. The Virginian

Then complete the following limerick. In the event of a draw the sender of the most original limerick will be awarded first prize. *"There was a young man who liked milk...*

Send your answers and limerick, together with 500 milk bottle tops, to Viz Comic Milk Competition, Viz House, 16 Lily Crescent, Newcastle upon Tyne, NE2 2SP, to arrive in the post. The winner will be sent a pint of powdered milk substitute. The judges decision is final. This competition is not open to milkmen, their friends or relatives, or to dairy farmers and their employees.

YOUR FREE SHOPPING REMINDER

Cut out and place in a prominent position

Remember to go shopping

Costa Catastrophe

A sunshine stay in the Mediterranean resort of Costa Blancos spelt catastrophe for a Manchester couple.

SANDWICHES

For Terry Thomson and his wife arrived in Spain only to be told that the resort didn't exist! And to make matters worse, the couple were forced to:

* **SLEEP** in a field next to the airport.

* **LIVE** on sandwiches left over from their flight.

PIES

"It all looked great in the brochure", said Terry, who paid £850 for a fortnight's stay in the resort. "But we were told that the resort had been a printing error and that it didn't actually exist.

"It was like Fawlty Towers", he told us.

FRUIT CAKES

After two weeks in the open field the Thomson's returned home. "But the flight back landed at Glasgow by mistake and we ended up walking 200 miles to get home", said Terry.

APPLE TURNOVER

The Thomson's, who have written a letter of complaint to the travel agent, involved, plan to look up their resort on a map next year before they make any bookings.

I made love to Esther Rantzen

EXCLUSIVE

in a previous life

A bus driver from Berkshire claims to have bedded many of the world's most beautiful women – stars like glamourous TV presenter Esther Rantzen – during previous lives. And now Ron Thompson, 46, has *revealed all* about his sexy exploits in a startling new book. In it he explains how during previous incarnations on Earth, he has met and made love to an incredible bevvy of well-known beauties. Top models, movie actresses, TV personalities and beauty queens! During his many previous lives on Earth, Ron has had them all. And here, in an exclusive excerpt from his book, he spills the beans about his steamy nights of passion with some of the world's sexiest women.

Esther (right) as she is today, and (below) as she may have appeared in Egyptian times.

" I have always been a firm believer in reincarnation, and being a keen spiritualist with considerable psychic powers, I am able to recall vividly my previous incarnations on Earth, going back many hundreds of years.

In one previous manifestation of my spirit I clearly recall being an Egyptian merchant, travelling through the desert many centuries age.

SNAKE

One day I came across an oasis, so I stopped to water my camel. Nearby there was a tent, so I went in to see if anybody wanted to buy a carpet. Inside a wealthy man was surrounded by a dozen women who where bathing him in oils and feeding him grapes. One of the women began to do erotic dancing with a snake which slithered around the contours of her body. Although I didn't realise it then, I now know that this was in fact TV presenter Esther Rantzen who, during one of her previous lives, had obviously been an Egyptian dancing girl.

NAVAL

In the yellow glow of the campfire her perfectly formed body was clearly visable through her thin, silk sari. A precious jewel sparkled in her naval. I wanted her more than anything in the world.

The man in the tent agreed to swap her for one of my carpets. Then she took my hand and led me to a nearby tent where we were alone. Hypnotised by her eyes I lay motionless as she undressed me and slowly began to explore my body.

NAKED

Eventually our naked bodies came together, and there beneath the stars she gave herself to me. We must have made love a hundred times that night, until eventually we fell asleep. The next morning when I awoke she had gone.

MINE

I suppose it's fairly ironic that mine and Esther Rantzen's paths should have crossed in this way during previous existances. Perhaps I've just been lucky, but there are many other top stars who I have met in this way, among then TV and radio personality Gloria Hunniford.

DESIRES

I met Gloria during the 19th century at which time I was the gardener on a large country estate. I believe that in her previous incarnation she was married to a local squire. Unfortunately he'd been injured in the army and was unable to satisy her sexual desires.

HOT

One hot summer's day I was hard at work in the woods when suddenly she appeared, and sat down nearby. After a while she spoke, and although I didn't realise it then, her distinctive Northern Irish accent was later to become a familiar sound to me as I listened to her popular daytime show on Radio Two.

"You look hot", she said. Sweat dripped from my half naked body. I had been chopping down trees with a large axe, but her eyes gazed down towards another of my tools. "Let's take a dip together in the lake", she said.

As we swam together I could hear the waves gently rippling around her nakedness. Afterwards we lay on the grass to dry. The sun was warm as it shone down upon her beautiful white love mountains. She turned and whispered softly in my ear, "Be gentle with me".

BODIES

We made love for what seemed like an eternity. Night followed day. I lost all track of time. It was dark and then it was light as our bodies melted into one. I had never experienced anything like it. We kept going for several days until it was finally over, then we both collapsed in a state of complete exhaustion. Gloria came back to my cabin in the woods many times, after that, and I will never forget the passionate days and nights we spent together.

Not many men can claim to have made love to Angela Rippon. Indeed, strictly speaking it would be wrong to say that I had. For that feat was achieved not by me but by another of my former selves – none other than Lord Nelson himself.

BATTLES

This was perhaps the best known of my previous incarnations, for it was during my life as Nelson that I won numerous sea battles and became a national hero. And it was during this period that I first met Angela Rippon, or Lady Emma Hamilton as she was known in those days.

TORRID

Our affair was a torrid one. I longed for her while at sea, spending many sleepless nights alone in my hammock. But the time we spent together I will never forget. As Nelson I had suffered many injuries in battle, but I can assure you that my column remained intact.

PASSION

We made love with all the passion of a raging sea, and then we'd lie together like ships becalmed, gazing into each other's eyes. We were hopelessly in love, and one of my main regrets is that I was killed at the battle of Trafalgar, thus ending our beautiful relationship. "

In the next issue:- *MY OUT-OF-BODY EXPERIENCES* – Ron reveals how, through meditation, he is now able to "free his spirit" from his body, and experience sexual relationships with many of today's top stars.

Live Longer, Earn More and Have Better Nookie?

HOW'S THAT FOR STARTERS!

FORGET exercise, the Atkins Diet and quitting fags and booze. The key to a long life is having a starter before your main course! That's according to a new report from Oxford University's Department of Hors d'Oevres, Nibbles and Starters.

According to boffins, diners who regularly tuck into appetizers such as garlic bread, prawn cocktails and breaded mushrooms-

- **HAVE** *a markedly lower cholesterol level*
- **LIVE** *twice as long*
- **ENJOY** *more three-in-a-bed sex romps*
- **EARN** *four times as much money*

as people who go straight for the main course.

Professor **Steve Langton**, who conducted the two-week study, said he was very surprised by the findings. He told us: *"When we processed the data, we saw the results and couldn't believe our eyes. We thought the computer must of blown a fuse, but we checked the plugs and they were all fine. It was a huge thumbs up for starters."*

However, others have been quick to condemn the report. **Richard Anselm** of antistarter group **STOPSTART** told us: *"Professor Langton's study is a load of nonsense. Nowhere in it does he explain how his reseach was conducted, or how he reached his conclusions. He appears to have simply made it up off the top of his head. The fact that the study was financed by the starter industry speaks for itself."*

But Langton was keen to defend his one and a half page report. He told us: *"The methods used to collect and analyse the data were so complex and involved that I didn't think anyone else would understand them. That is why I didn't bother including them in the final report, and just published the findings on their own in the form of a press release to tabloid newspapers."*

A spokesman for the British Starter Council told us: *"We in the starter business have always known the benefits of having something to put you on while you're waiting for your main course, but to have it scientifically confirmed like this by Professor Langton and his team is very gratifying."* And the timing of the report is also likely to delight starter industry bigwigs, coming as it does just as they launch British Starter Week at venues up and down the country.

With interest in starters expected to boom following the report, eateries are gearing up to meet the unprecedented demand for soups, melon boats and potato skins with sour cream dips. Aggressive chef **Gordon Ramsay** told us: *"When I read the report, I immediately punched and sacked both my main course chefs. Then I hired two new starter chefs. And punched them."*

First course gets top marks ~report

"From now on it's starters only at my restaurant," he added. *"If you ask for a main course or pudding I'll come out of the kitchen and punch you and your wife."*

Your Starters

EVERYONE has a favourite starter, be it soup or something else. We phoned up ten celebrities and asked them "What's your favourite starter?" Unfortunately, six just put the phone down, but four were happy to tell us their top starter.

Roly-poly funnyman **Peter 'Garlic bread?... GARLIC BREAD?...Garlic?...Bread?' Kay**, *famous for his catchphrase 'Garlic bread?...GARLIC BREAD?...Garlic?... Bread?' didn't hesitate. "I suppose you'd expect me to say 'Garlic Bread?...GARLIC BREAD?... Garlic?...bread?' But you'd be wrong... because it's soup,"* he told us. However, fat beer ad king Peter had the last laugh when we asked him what flavour soup he preferred. *"Cream of garlic bread,"* he quipped.

U2 front man **Bonio** *was in Cuba addressing the World Health Organisation on the subject of the Kyoto agreement when we caught up with him. "I fly all over the world talking about environmental issues, and wherever I'm eating, I always have the same starter - a simple prawn cocktail."* he told us. And there's never any chance of it being off the menu - because each one is made in his favourite Dublin restaurant and flown out in a specially chartered jumbo jet to wherever he is in the world. *"Dublin prawns help me remember my roots and keep my feet on the ground,"* he added.

'I'm a Celebrity... Get me out of here'

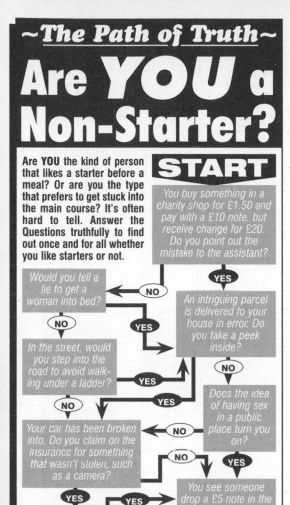

~The Path of Truth~
Are *YOU* a Non-Starter?

Are **YOU** the kind of person that likes a starter before a meal? Or are you the type that prefers to get stuck into the main course? It's often hard to tell. Answer the Questions truthfully to find out once and for all whether you like starters or not.

START

You buy something in a charity shop for £1.50 and pay with a £10 note, but receive change for £20. Do you point out the mistake to the assistant?

Would you tell a lie to get a woman into bed?

An intriguing parcel is delivered to your house in error. Do you take a peek inside?

In the street, would you step into the road to avoid walking under a ladder?

Does the idea of having sex in a public place turn you on?

Your car has been broken into. Do you claim on the insurance for something that wasn't stolen, such as a camera?

You see someone drop a £5 note in the street. Do you pick it up and keep it?

You are on a crowded bus when an old lady gets on. Do you give up your seat for her?

Do you think personality is more important than looks?

You are looking at the menu in a restaurant. Do you order a starter?

Congratulations, you're definitely a starter person. Whether you're sipping soup, gobbling up garlic bread or doing something beginning with 'p' to a prawn cocktail, you don't think a meal's complete without a starter first.

Oh dear, it's no starter for you. Perhaps you're not a big eater or are you saving a bit of space for extra pudding or some cheese and biscuits... maybe even both? Whatever the reason, starters are definitely a non-starter for you.

---

or Ten

tit model **Jordan** had a surprise choice. *"I suppose you'd expect me to say two large melons with cherries on the top. But you'd be wrong,"* she told us. *"I like nothing more than two large grapefruits with cherries on the top. I'm proud of my breasts and I'm not ashamed to eat starters that resemble them in some way,"* she said.

Grinning 'Changing Rooms' host **Carol Smillie** was slightly less forthcoming about her choice of starter. *"Who the fuck is this? Have you any fucking idea what time it is? It's three in the fucking morning,"* she joked. *"How did you get this fucking number? You might think this is funny, I fucking don't. Ring this number again and I'll call the fucking police,"* she laughed.

Appetizers, hors d'oeuvres, antipasties. Call them what you will, one thing is certain; starters are here to stay, and that's official. We look for them on menus, we order them before our main meal and we eat them prior to commencing our second course. They're the talk of John O'Groats from Land's End to the length of the land and that's a figure that's set to double. But how much of what we say about them is true? Get your teeth into these...

10 things you never knew about STARTERS

1 Starters are named after *Sir Henry de Montford, 5th Earl of Start* (1536-1602). At a banquet where the food was taking a long time to arrive, he produced a flask of soup which he began to eat. He declared it *"a most efficacious way to commence one's repast"*. In no time at all, the whole of London society had joined the craze and were having fashionable soup 'starters' before their main courses. It was not until a century later that the Duke of Bread Roll (1675-1734) requested a small baked bun to mop up his soup, so setting another trend which continues to this day.

2 In August 1997 a simple prawn cocktail was responsible for robbing the country of its Princess of Hearts. During her final supper at the Paris Ritz, Lady Di uncharacteristically decided not to have a starter. "If she had ordered a first course she would have taken at least 11 minutes longer to finish her meal. So instead of colliding with the white Fiat Uno in the Pont d'Alma tunnel, driver Henri Paul would have missed it by over 8 miles, and tragic Diana would have emerged unhurt at the other end," said an expert.

3 The term *starter* is said to derive from the fact that starters were responsible for 'starting' the English Civil War. Following an Indian meal at a restaurant, a dispute over the bill arose between **Charles I** and **Oliver Cromwell**. The King, who had had an onion bhaji and a rogan josh wanted to split the bill equally. But Cromwell refused as he'd only had a main course. A scuffle ensued which later developed into a 30-year conflict claiming over 600,000 lives.

4 The biggest starter in the world was eaten by Monster Truck driver **Chad Kyminski** during the 1996 Ohio State Fair. He entered a branch of *Pizza Hut* in Dustbowl City and ordered a garlic bread the size of a football field. He consumed the huge slice in 10 minutes, before tucking into his main course, a deep pan meat feast with mushrooms... *12 miles in diameter!*

5 If someone doesn't order enough starters, it is said that he 'under orders starters', but did you know he could also be 'under starters' orders'? This is nothing to do with not ordering enough starters, it means to be under the orders of the starter, the man who fires the gun to start a race.

6 Also, if a swimming teacher tells a group of novice swimmers to dive beneath the surface, it could be said that he 'orders starters under'.

7 The world's smallest man, **Calvin Phillips**, never has a meal without a prawn cocktail starter. But to Action Man-size Calvin, a normal prawn is the size of a lobster! So he tucks into plankton in mayonnaise made from hummingbirds' eggs... *all served up in a glass thimble!*

8 Probably the most famous starter is the Holy Eucharist, a sort of religious garlic bread that the vicar puts in your mouth to keep you going until your main course of the second coming of our Lord Jesus Christ.

9 The eating of starters is not just confined to restaurants. Amazingly, it has also been seen in the natural world. Tigers will often eat a few rabbits to keep them going before tucking into their zebra main course.

10 You might think that a 'twisted fire starter' is a starter which has become twisted in some way and has caught alight. But you'd be wrong, for it's actually a record by pop group the Prodigy. Scary spiky-haired singer Keith Flint got the idea for the song when he was visiting a Berni Inn with his mum and dad. "I ordered a garlic bread, and when it arrived it had been twisted in some way and had caught alight," he told the NME.

Beast attacks toddler in own house

SHOCK EXCLUSIVE!
By our Scaremongery Correspondent
Frances Ovassissi

PARENTS in the West Midlands were warned to be on their guard yesterday after the Loch Ness monster punched a toddler in the eye.

Two-year-old Tyrone Champignon was watching television on the sofa at his home in Walsall when the 200 foot cryptozoological beast crept in through patio doors and gave the tot a shiner.

Mum Doreen Turpentine rushed in to find her young son howling in pain, nursing a black eye. She said: "I couldn't see what was going on. Then I saw three big humps sticking up from behind the sofa. I just screamed."

fence

Dad Baxter Champignon heard her cries and rushed into the lounge. He threw empty beer cans at the animal and chased it into the garden where it jumped over the fence and disappeared.

"We were too shocked to take in what had happened at first," said Baxter. "But thinking about it now, Tyrone is lucky to be alive." he added.

dance

The RSPCA said: "The Loch Ness Monster, if it exists, is normally a placid animal and it is unusual to hear of it being sighted in a house as far south as Walsall." And they dismissed claims that the animal meant to hurt Tyrone.

"We certainly don't think the animal meant the child any harm. It was

Little Tyrone yesterday, and (above) the beast that blacked his eye

probably looking for somewhere to sleep, and the boy startled it," they added.

fish

But Nicholas Witchell from the National Anti Loch Ness Monster Society disagreed.

He said: "Frankly, the Loch Ness monster, if it exists, would be a menace. Unlike the cuddly toy wearing a Tam O'Shanter you see in Scottish souvenir shops, it would be a savage predator. I've imagined the damage it could cause to a shoal of krill, and it's not a pretty sight, believe me."

And he issued a warning. "If it does exist, now it has punched one child in the eye it will have a taste for it. None of our children is safe."

swallow

However, a more cautious note was sounded by David Sutton, managing editor of crackpot nutrag Fortean Times.

He said: "Clearly, something gave Tyrone a black eye but we shouldn't jump to conclusions. It could have simply been a large seal, a piece of floating wood, or perhaps silver-suited space creatures from another space-time continuum in search of fuel for a crashed flying saucer."

spit

A spokesman for Walsall Hospital said: "We can confirm that the boy had been punched in the eye, and cannot rule out the possibility that the Loch Ness monster did it."

HAIR APPARENT - Young's toupée at the ceremony with Prince Charles yesterday

ARISE, SIR-RUP
Gong for Radio Two-pée

By CHORLTON WHEELIE - our Royal Correspondent

Housewives' favourite toupée, Jimmy Young's wig was knighted yesterday, and then cracked a joke about its Radio 2 arch rival - Wogan's weave.

"Terry's syrup will be tearing its pretend hair out with jealousy when it hears about this!" it quipped.

The hairpiece, 52, received the award in the New Year's Honours list for services to the 78-year-old broadcaster's pate, which it joined from new in 1950.

The wig said that the short ceremony, conducted by Prince Charles at Buckingham Palace, was a nerve-wracking experience.

It told reporters afterwards: "I haven't been this nervous since Jimmy did a live broadcast from the rear seat of a Tiger Moth at the Farnborough Airshow in 1976."

The MAN in the PUB

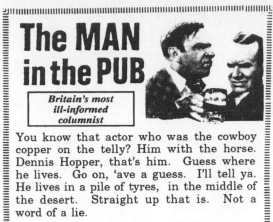

Britain's most ill-informed columnist

You know that actor who was the cowboy copper on the telly? Him with the horse. Dennis Hopper, that's him. Guess where he lives. Go on, 'ave a guess. I'll tell ya. He lives in a pile of tyres, in the middle of the desert. Straight up that is. Not a word of a lie.

I'll tell you who isn't dead, niether. **Walt Disney. Still alive he is. Put his head in a fridge. They reckon' its still working. He's buying Sky telly, he is. Heard about it on the news.**

That Kevin Keegan, he's got a *factory,* he has. Mate o'mine told me. Makes dodgy lighters. Sells 'em in the street, three for a quid. Not personally. Get's someone else to do that. *Three for a quid!* No wonder he's got so much bleedin' money, eh?

You know that Bob Holness? Him off Bullseye. Played the saxophone on Baker Street, he did. The record. Made a bleedin' fortune, apparently. And there's another one for you - Bob Monkhouse. He invented Dennis the Menace. Did you know that? It's true. Swear it on my mother's life.

You know what? And this is God's honest truth, this is. It is *scientifically impossible* for bees to fly. That's a fact that is. No one knows how they do it. It's the same with reversing articulated lorries. On paper, it can't be done. Impossible. There's no explaining it, is there? You've either got it, or you haven't, eh?

You know who's got the clap? Shall I tell ya? Shirley Bassey, that's who. A mate of mine at school, his sister had a leaflet about it. You know, about VD. Had her picture on it.

Did you know that 'omo sexuals don't actually *do it,* as such. You know... *sex!* Most of 'em don't bother. Mind, a mate of mine was in this gay club in New York, right, and this geezer stuck his head right up another fella's arse, right there on the bar, for all to see. Turns my stomach that does. I don't know how they do it, me.

I HOPE I HAVEN'T MISSED THE LAST POST

High Jack!

Fulchester's oldest man reaches for the sky!

Jack Parkinson believes he is probably the oldest man in Fulchester. And this year he hopes to celebrate his 59th birthday by taking his first ever flight on Concorde!

While most of us can only read about history in books Jack has watched it taking shape over almost six decades. On the day he was born World War Two hadn't even started yet, Queen Victoria had only been dead for a few years, and an old fashioned '78' record was top of the charts.

Advent

In an incredible lifetime which has spanned almost six decades, Jack has witnessed the advent of colour TV, has watched natural gas replace the old kind of gas, and has seen conventional ovens make way for microwaves.

Pirelli

Jack spent his working life as a postman but now tires easily and is no longer able to do much exercise. He last worked almost ten years ago, and spends most of his time nowadays sitting in his favourite armchair watching television. But today's satellite telly with remote control is a far cry from the TV he was brought up on.

Goodyear

He still vividly recalls the marriage of the Prince of Wales, Dr Who with Jon Pertwee, and the first ever broadcast by Channel 4. "It was a programme called 'Countdown' presented by a man called Richard Madeley", he fondly recalls with a smile. His body may be old, but his mind remains alert.

Lancashire

Ask Jack the secret of his longevity and he'll tell you its a combination of cigarettes and alcohol! Every night for over forty years has gone into his local pub and ordered his usual; eight pints of lager. He has smoked since he was 16, and still gets through over 30 a day.

Now with two grown up children of his own, and an incredible four grandchildren, Jack hopes to fulfil a lifelong ambition later this year by becoming great grandfather! But there is one other ambition he is desperate to achieve before time runs out.

Jack has always dreamt of flying on Concorde and next month he hopes to celebrate his 59th birthday by taking his wife, his daughter and son-in-law, and their two children, together with a friend, to Barbados for a fortnight. Anyone who could supply Jack with seven return Concorde flights can contact him at his local pub, The Red Lion on Fulchester High Street, any evening after 6.30.

Queues R you's!

Queues for this Christmas's most popular toy are already forming in High Streets all over Britain.

Nobody knows what it is going to be yet, but toy shop owners are already reporting unprecedented demand.

Tarmy

"We won't be getting it in stock for weeks yet, and already the shelves are empty. It looks certain to outsell last year's Power Rangers, and could even do better than Tracey Island", said one Oxford Street store manager who has seen customers camping outside his shop for the last three weeks.

Driver

First in the queue was unemployed gas fitter Fred Baxendale of Battersea who has been waiting in the shop doorway since early August, determined not to miss out on this year's smash hit toy.

"He doesn't know what it is yet, but my son has set his heart on having one, and he'll be absolutely heart-broken if he wakes up on Christmas morning and hasn't got one."

Sand Wedge

"Whatever they are, they'll be flying off the shelves as fast as we can put them out. We'll never have known anything like it!" said the shop spokesman.

Parsty

"No decision has yet been made as to what this year's best selling toy is going to be", said a leading toy manufacturer yesterday. However he was able to confirm that the toy will be priced £59.95, made entirely of plastic, and would be broken by Boxing Day.

"Batteries will not be included", he added.

✦✦✦✦✦✦✦✦✦✦✦✦✦✦✦✦✦✦✦✦✦✦✦✦✦✦✦✦✦✦✦✦✦✦✦✦

Grim future lies ahead for Britain's married couples

LOVE ON THE ROCKS

'We're all heading for divorce' - say the experts

There was a time, not so long ago, when marriage meant 'for life'. But nowadays many couples return from their honeymoon and head straight for the divorce court. Today marital breakdown has become an epidemic, affecting even the Royal Family.

Six months ago the marriage between the Prince of Wales and Princess Diana lay in tatters. The couple had drifted apart, spent very little time together and were often seen rowing in public. Indeed, it was feared that the couple's constant feuding may eventually end in violence.

MARRIAGE

Divorce looked certain. But six months on, Charles and Diana are once again very much in love. Their marriage saved, they now look forward to a blissful life together. So what produced this amazing turn around? What can be done to save a marriage once the sparkle has gone? The answer is simple. Sound old fashioned advice from a caring mother, and in Charles and Diana's case that advice came from the Queen herself.

DIVORCE

Alarmed at the prospect of a Royal divorce, she summoned Charles to Buckingham Palace and issued him with a ten point plan to save his troubled marriage. And her motherly words of wisdom were all it needed to save the day.

ENGAGEMENT

Now we are prepared to reveal the Queen's ten marriage saving tips, leaked to us by a Palace insider, in the belief that they can help to save other marriages — perhaps your own. Follow these Royal recommendations — they're guaranteed to put the fizz back into a marriage that's started to go flat ....

Happy families - yet 6 months ago divorce looked certain

THE QUEEN'S TOP TEN MARRIAGE SAVING TIPS

1 Always buy your loved one presents. Surprise her with flowers, chocolates and unusual gifts as well. Try to be different. Buy her a hat, or a goldfish. Or a key fob with her name on it. Gifts needn't be expensive.

2 Never say to yourself "my wife doesn't understand me". That's nonsense. You're probably talking too fast. Speak more slowly, and in a clearer voice.

3 If you're going to be late home from work, telephone and give her plenty of warning. And if the dinner's burnt when you get home, don't be angry. Dine out at a posh restaurant instead.

4 Don't forget important dates. Nothing will annoy her more. And never call her by the wrong name — it's a dead giveaway. Write her name, her birthday, your wedding anniversary etc. on the back of your hand so you don't get them muddled.

5 Never talk to her about other women who you fancy. If you're watching 'Miss World' on the telly, pretend to fall asleep. And when you're reading the paper, don't let her catch you staring at page 3 — she'll get jealous in no time at all. Be careful when choosing calendars too. Don't put a saucy calendar on the wall — she'll hate it. Choose one with pictures of flowers on it instead.

6 Always talk frankly and openly about sex, and don't be afraid to discuss any sexual problems that you have together. Try writing a limmerick or rhyme containing sexually explicit language and read it out loud to her. This will help you to overcome your shyness and inhibitions. Talk frankly and openly about sex while making love, then in the morning take her breakfast in bed, and talk about sex for ten minutes or so before you go to work.

7 Buy a big book about sex and read it together. And buy a sexy video — you see them advertised in the Sunday papers. Watch it together, cuddled up in bed. After a while your TV screen will steam up — as you both start to join in with the red hot action.

8 Look out for those annoying habits — you may not realise it but they could be driving her up the wall. Don't cut your toenails at the breakfast table, and always remember to flush the toilet. They may seem like trivial points, but it's little things like that which cause the most rows.

9 Buy her lots of sexy underwear.

10 When you're out shopping, fill your trolley with the food of love. There are lots of other aphrodisiacs as well as ground rhinoceros horn! Try avacado, peach and oysters. Or if fish turns her on, why not buy a couple. A romantic, healthy diet can work wonders for your sex life and lead to a longer, more loving relationship.

MISSIONARY JOKE

IT'S A JUNGLE OUT THERE

Croft ORIGINAL

She started out as a computer game, and now Lara Croft is hitting the big screen in this Summer's must-see movie. Tomb Raider, starring Angelina Jolie as the all-action heroine, is packing cinemas across Britain. But one person who certainly won't be joining the queues to see it is the real life tomb-raider who claims to be the inspiration behind the character.

Birmingham woman lays claim to Tomb Raider fame

33-year-old Walsall mother of six Tina Ringworm was raiding tombs by day and working behind the till of a 24-hour petrol station by night when she says her identity was stolen by an unscrupulous computer boss. And with the celluloid blockbuster set to net millions of pounds in royalties, Tina reckons it's time to set the record straight.

trots

In the film, Lara is an aristocratic english adventuress who trots the globe, searching for exotic jewels and treasure. From the steamy jungles of South America to the Great Wall of China, she fights off bears, dogs and tigers in her quest for gems with magical powers. But Ringworm says the film doesn't show the half of it. "The film's very good as far as it goes, but they've give it the Hollywood treatment. Tomb raiding is a dirty business, and don't let no-one tell you otherwise. For example, when Angelina Jolie fights a bear, she kills it and ends up breathing a bit heavy and perspiring slightly. The real picture isn't quite so pretty, believe me. If she'd really fought a bear, like I of loads of times, she'd be sweating like a pig and covered in blood. And she'd of shit her shorts.

squirts

"I remember this one time, I'd got right to the inner chamber of this Aztec tomb in Peru where there was some magic emeralds. I'd already had to fight a two-headed dog and a load of skellingtons. I'd just grabbed the jewels when I remembered I was on the 7 till 4 shift at the garage, and if you're a minute late he docks you an hour. I only just made it back in time. He was stood looking at his watch! But it was worth it. Them emeralds contained the secret of eternal youth and my brother sold them for £5 on his stall, and he give me £2.

runs

"On the playstation, Lara Croft gets three lives on every level. But in real life you only got one, and this was brought home to me this other time when I was raiding a tomb in the Himalayas. I was in this glass labyrinth on level four and I was trying to get away from this yeti. I couldn't shoot it because I'd used up all my bullets on a ten-foot Ninja warrior who come at me on a rope bridge. The only thing I could think of was rolling sideways every time it got close. In the end, it got tired and I done a big somersault out the labyrinth. But I knew I was already late for my shift. Luckily for me, my mate Irene covered for me and told him I was on the toilet when he come in."

wickets

But ironically, it was whilst she was working at the garage one night in 1994 that Tina believes her identity was stolen.
"It was about three in the morning and this thin bloke come in with glasses. I remember he had a white Astra van and he was getting £8 of DERV and a pepperami. He seemed in no hurry and we got chatting. I told him how I'd just gone up to £1.75 an hour and then the subject got round to tomb raiding. I told him all about my adventures, and he seemed quite interested.

tickets

"He went on his way and I thought no more about it. Then, the next night I was on 10 till 8 and about midnight I seen this man on the news and it was him. He was Bill Gates and he'd just invented playstation. There was Lara Croft on it and I immediately realised it was me."

rickets

With the Tomb Raider film set to break all box office records, Tina reckons it's high time her unwitting contribution was recognised.

derbyshire neck

"Bill Gates has made a lot of money out of me and it's only right that I get a fair share of what he's got. I want £200 or I told him I'm going to write a letter to the Have Your Say page in the Walsall Express and Star."
A spokesman for Microsoft in Seattle denied Tina's claims. "Bill Gates no longer owns a white Astra van," he told us.

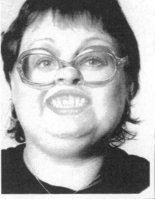

SPOT THE DIFFERENCE - Tina Ringworm (left) could spell double trouble for billion-squillionaire Gates. Pretend Croft, Angelina Jolie (right) .

After being buried for forty five years...

WE FIND HITLER'S WANK MAGS

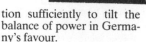

A pile of dirty books belonging to Hitler have been discovered buried in a shallow grave in former Russian occupied Berlin.

The porno mags, which were in a plastic bag, were previously believed to have been destroyed by Hitler's faithful SS guard shortly after the Fuhrer's death.

KREMLIN

But Russian military records, only just made public by the Kremlin, revealed that the male interest mags, which Hitler carried with him throughout the war, were secretly buried on Stalin's orders after the Russian leader had looked at them for a bit.

PENTAGON

If scientific tests on the pages of the magazines – some of which were stuck together – confirm that they were Hitler's, it will reinforce the popular theory that towards the end of the war the Fuhrer had become obsessed with pornographic material.

HITLER – secret stash

Experts believe that Hitler intended to harness pornography for use as a weapon against the allies, and that in 1945 Nazi scientists were on the verge of inventing the *dirty video* – a breakthrough that could have changed the outcome of the war. For Hitler had planned to bombard Britain with copies of his new secret weapon, hoping that triple X hardcore porn videos, featuring red hot explicit sex action, would distract the British population sufficiently to tilt the balance of power in Germany's favour.

We offered Hitler's dirty magazines to The Sunday Times, cos that's that sort of shit they buy, but they haven't rang us back yet.

GOEBBELS – Nazi number two.

Dr Ludmila Fredrikson of the Swedish Institute of Advanced Sexual Research believes that Hitler's unique medical condition – he is believed to have had only one testicle – could account for his apparent obsession.

DODECOHEDRON

"A one nutted person would indeed have an enormous appetite for filth", she told us.

THE MYSTERY OF SHINGLE COVE

For years mystery and rumour have surrounded the sudden evacuation of the village of Shingle Cove on the Suffolk coast in November, 1944.

And despite official denials, to this day locals still believe that their village was the target of Hitler's first experimental 'porn warfare' attack.

TRAPEZIUM

The story goes that on that cold winters night German planes flew overhead and dropped thousands of propaganda leaflets on the small fishing community. But they were no ordinary propaganda leaflets that rained down onto the beach, like giant snowflakes in a swirling blizzard. As one local who claims to have witnessed the attack recalls.

CLOWNS

"They were pictures of the Queen Mother, with no clothes on", Fred Gutteridge, now 78, told us.

Despite his advancing years, and the onset of incontinence, Fred still vividly recalls the dirty pamphlets that fell from the sky. "They were disgusting. She was completely naked, in all sorts of positions. Flap shots, the lot".

BEARDED LADY

On another page a ficticious newspaper headline told how Princess Di had worked in Amsterdam as a prostitute, and regularly had sex with four men at once on a filthy mattress.

HARDCORE

Military analysts believe that Hitler's propaganda chiefs had used hardcore Swedish models to pose for disgusting and lewd photographs and then superimposed the Queen Mum's head on top.

QUEEN MUM – Filthy pictures

The pamphlet was designed to shock the British public and lower vital wartime morale.

Fred recalls how the army quickly sealed off the village and confiscated all the pamphlets. "They took them all to some woods nearby and burnt them", he told us.

BALLAST

The mystery of Shingle Cove will remain unsolved until the year 2500 when official War Office files on the incident are opened to the public.

Gracie and Vera in Nazi porno shocker

Sex mad Hitler had drawn up detailed plans of what he intended to do with Britain if he had won the war.

And, according to Kremlin war records released this week, one of the filthy minded Nazi's top priorities was to make disgusting porno videos starring Vera Lynn and Gracie Fields.

SEX BOMB - Hitler's secret weapon - the triple 'X' porno movie

The would-be porn king planned to lure the British songbirds into a career in blue movies, by offering them tons of Nazi gold. And together with his right hand man Goebbels, Hitler had already written several scripts in which Lynn and Fields, the darlings of the British Forces, were expected to perform;

★ **TOPLESS**

★ **GIRL ON GIRL**

★ and **HAMBURGER SHOTS**

However, Hitler's adult video plans, together with his plan for world domination, were scuppered when he lost the war.

T.V. MOTSON'S SUNKEN TREASURE HOPES SINK

Plans for a treasure hunting expedition to the Carribean lead by BBC football commentator John Motson have run aground.

Motson had planned to set sail with a crew of BBC Sport colleagues and go in search of sunken treasure worth millions of pounds. But after months of planning Motson has had to cancel the expedition which has been besieged by problems throughout.

TREASURE

The first blow came when Match Of The Day colleagues accused Motson of *forging* a treasure map which he claimed showed the whereabouts of buried Spanish gold.

DARLING

"His map, which was supposed to be 300 years old, was drawn in felt pen on the back of BBC notepaper", one colleague told us.

PETAL

A further set back came when Motson was unable to hire a boat for his 6,000 mile round journey, as the 44 year old commentator has no sailing experience whatsoever.

DUCKY

And the final blow came when BBC bosses refused Motson six months off work, due to the forthcoming Olympic Games.

LAMB CHOPS

Motson had already seen three of his intended crew members drop out due to work commitments. Fellow football commentator Barry

DAVID VINE - dropped out

Davies, snooker anchor man David Vine and football analyst Trevor Brooking had been Motson's first choice of shipmates. But despite frantic phone calls to other colleagues, and offers of substantial amounts of gold, he was unable to find replacements.

BASIL BRUSH

However one BBC insider told us that Motson, who inherited the BBC's football

Some sunken treasure yesterday

commentary crown from David Coleman, has not given up entirely, and may launch another expedition early next year, possibly using advanced technology to help in his search for treasure.

Colleagues in the Match of The Day office report that Motson has spent several afternoons working on plans for a 'Special Underwater Mini-Submarine' that will enable him to scour the sea bed for sunken treasure.

SPOT the ROYAL American TV cop

Every week we ask a member of the Royal Family to disguise themself as an American TV cop.

Can you identify this week's mystery Royal TV cop look-a-like? Here's a clue:

This veteran Royal is everyone's favourite granny, but in her TV Cop disguise there's *fat* chance of her getting *fired*.

If you think you know the answers, write the name of the Royal and the TV cop on the back of a postcard and send it to: Viz Royalty American TV Cop Look-a-likes, P.O. Box 1PT, Newcastle upon Tyne, NE99 1PT. The first correct entry out of the hat will win our fabulous first prize – Hi Fi equipment to the value of your choice from any branch of *Richer Sounds*. And an electric cooker.

MY DOG YEAR HELL

"I look like a fucking waxwork"

says SIMON MAYO

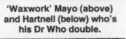
'Waxwork' Mayo (above) and Hartnell (below) who's his Dr Who double.

RAPIER witted DJ Simon Mayo has spoken for the first time about the mystery condition which he claims has left him looking like a fucking waxwork.

For months there has been concern among radio listeners over Mayo's rapidly deteriorating appearance. Once the smug, fresh faced new boy of Radio One, his face has undergone a drastic transition in recent years.

Caked

Viewers of his 'Confessions' programme hardly recognised the comical figure that appeared on TV recently with long, swept back hair and a wrinkled face caked in barrow-loads of make-up. "He looked more like Hannibal Lecturn than a trendy young television host", said one viewer.

Pied

Mayo's boss Matthew Bannister was so concerned he sent him to see a Harley Street face doctor who diagnosed Mayo as suffering from Who Hartnell Syndrome, a fairly common condition affecting 12 in every 1000 adult males. Relatively harmless, sufferers turn into William Hartnell, the pioneer Dr Who actor, over a period of several years.

Bag

But Mayo was not convinced. Colleagues were baffled by Mayo's strange, musty smell - like that of a moulting dog. They suggested he visit a vet, and it was only then that Mayo discovered the truth. For the last decade he had been living in dog years.

Radio 1 jock's shock confession

"There are seven dog years in every human year", Mayo explained. "As a result every year I age seven dog years". As a result since 1990 Mayo's age has risen from 30 to 79.

Lone

"Despite his unusual condition there is no reason why Mr Mayo should not live to a ripe old age of 80 or 90", says Professor Eugene Pantaloon, senior lecturer about dogs at the world renowned Vladivostock College of Further Education. "The only problem is that they will be 80 or 90 dog years", he confessed.

Forest

Meanwhile Mayo is putting on a brave wrinkly face. "I haven't felt the

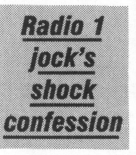
Fresh faced Mayo (above) aged 27. But two years later and he's already 41 (below).

urge to sleep in a basket and lick my bollocks, at least not yet!" he quipped. "And I haven't started sniffing the other DJ's arses". But despite the smiles, Mayo knows that unless a cure is found, that day may not be too far away.

It's the column Joe Coral's tried to BAN!

Odds Bodkins

How to beat the bookie with the Viz Tipster, Oddsworth Bodkins

● **HERE's** a hot tip - phone your bookie and ask him what odds he'll give you that *Bob Monkhouse*, *Marilyn Monroe* and *Jason Donovan* all share the same birthday. He should give you at least 133,000 to 1, because that's the chances of such an unlikely thing happening. But get this...they were all born on June 1st! Put £8 down on the triple accumulator, and you'll walk out with a cool million in your back bin. *But remember to pay the tax first!*

● *LADBROKES* are currently offering 35-1 that Shakespeare died on his birthday. Not brilliant odds but well worth a flutter, because a little bird tells me that Shakespeare was born on April 23rd - the exact same date that he died!

Tip of the Week

● **THEY SAY** a fool and his money are soon parted. Well here's a way to make a fool out of your bookmaker and net yourself a tidy wedge into the bargain. Go up to the counter and say: *"Antidisestablishmentarianism is a very long word, but I bet I can spell it."*

Any bookie worth his salt will give you at least 7-2. Stick a tenner on, and then say: *"I - T"* That's thirty-five quid plus your stake! It's like taking candy from a baby.

STUMP UP A FIVER

● **THE BRITISH** Medical Association are conducting a survey to find the average number of legs people have got. And just look at these odds on the Betfair.com website ~

"Average number of legs per person in Great Britain. More than 2; 17-5. Exactly 2; Evens fav. Less than 2; 50-1 bar."

I've had an insider's tip from a mate who works in a hospital. He says there's plenty of amputees to bring the average down under 2, but there's not very many 3-legged freaks to push it back up! I haven't done the maths, but I reckon that 50-1 shot's got to be worth a flutter. I've stuck a fiver on it and I suggest you do the same. *But hurry, the result's in next Tuesday.*

● **GO INTO** William Hill and ask what the odds are on a year NOT lasting 365 days. Last time I tried this scam I got 50-1. After placing your bet reveal you were talking about a year on Mars, which lasts 687 days! £1000 profit for a £20 stake? Nice work if you can get it!

● *TALKING* about years, place a bet that your grandad is over 500 years old. You can expect odds of between 600 and 1000-1. When you've placed your bet, introduce your 72-year-old grandad and explain that you were talking about DOG YEARS, making him 504! Two grand for the price of a pint? I'll have a bit of that!

● **FINALLY**, Victor Chandler online are currently offering a tasty 15-4 that the widow of water speed ace *Donald Campbell* isn't married to *Greengrass* out of Heartbeat. Sounds like a risky punt, doesn't it. But get this, an insider has tipped off Odds Bodkins that Mrs Campbell and Greengrass actor *Bill Maynard* are in fact *man and wife!* At those odds it's a 24-carat steal - you'll trouser a cool £30 for an £8 stake. Get your bets in quick!

More top tips next time, gamble fans!

20 things you never knew about ELTON JOHN & spiders

SIR ELTON JOHN. Loathe him or hate him, you just can't ignore him. Whether he's bubbling at one of his pal's funerals, mincing off a tennis court because someone called *"yoo-hoo"* at him or suing himself for spending a million pounds a month on flowers, the fat puff's flamboyant antics are never far away from the headlines. We think we know all about him, but do we really? *What's his real name? What's his favourite colour? How many legs have they got?* Here's 20 *Captain Fantastic* Facts you never knew about Elton John and Spiders.

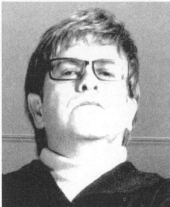

Zany ~ Elton larks about for the camera

★ Elton John's real name is Reginald Hercules Dwight. He took his stage name from two childhood heroes - ITV football commentator Elton Welsby and 70's Welsh rugby player Barry John. *"I tossed a coin to see if I should be Elton John or Barry Welsby,"* he told Rolling Stone magazine in 1994. Luckily for the world of pop, it came up heads!

★ Elton cannot swim - and it's all down to a 4th century Greek mathematician. Archimedes discovered that a floating body displaces its own weight in water, yet due to an anatomical quirk which leaves scientists scratching their heads, the pint-size star displaces three times his weight. As a result he sinks like a stone each time he jumps into his luxury swimming pool.

★ The smallest ever Elton John was only three inches tall. Tiny Calvin Phillips performed *Candle in the Wind* at his pet mouse's funeral...on a grand piano made from a matchbox!

★ The lake surrounding Princes Di's funeral island has a leak in it, and the water level constantly drops. Each month, the bespectacled singer-songwriter makes a bizarre pilgrimage to Althorp House where he spends four of five hours grieving at the lakeside until his tears have topped up the water level.

★ Despite being homosexual, Elton is very keen on football. In the seventies, he even had his own team, Watford FC! At half time, the effeminate chairman would don a French maid's outfit and serve the players fairy cakes and Earl Grey tea in bone china teacups.

★ Already famous for his piano-shaped glasses and piano-shaped swimming pools, John fell foul of the FA in 1974 when he unveiled plans for Watford United to play on a piano-shaped football pitch! Lancaster Gate bosses refused to allow the change, sending the singer into one of his hissy-fit strops that lasted 18 months.

★ Elton owns no less than 12 houses around the world. As a result, each time he visits the supermarket he has to buy 12 of everything: 12 packets of biscuits, 12 pints of milk, 12 toilet ducks etc. His weekly food shopping bill can easily top £200.

★ Elton John is synonymous with a crazy rock and roll lifestyle. In fact, he is so zany that *his wife is actually a man! Called David!*

★ John has spent the past four years searching Beverley Hills for a plastic surgeon willing to perform a delicate operation. So obsessed is he with piano-shaped things, that he plans to have his anal sphincter reshaped so that he is able to excrete 'piano stools'!

★ Elton has written countless songs about the weather, including *Candle in the Wind* and *Don't let the Sun Go Down on Me.*

★ And *Cold as Christmas.*

★ And *Through the Storm* with Aretha Franklin.

★ When not penning chart-topping pop hits, Elton also likes to dabble in chemistry. In 1986 the scientific community granted him the ultimate accolade when an element he'd invented - *Eltonjohnium*, a dense semi-metallic halide - was accepted for inclusion in the periodic table between potassium and calcium.

★ Despite writing hundreds of songs about the colour blue, such as *I Guess That's Why they call it the Blues, Blue Eyes* and many more, Elton's favourite colour is actually green! *"I hate the colour blue,"* he told MTV in 1997.

★ John courted controversy recently when he tried buy the misshapen skeleton of seventies piano playing freak Mrs Mills from the British Museum.

★ At the height of his 1980's excesses, nutty Elton had his teeth replaced with piano keys.

Creepy crawly ~ A spider (above) with 8 legs yesterday and Elton (left) fighting to keep his emotions in check as he performs 'Pinball Wizard' at the funeral of Princess Diana in 1997.

Now, instead of a 6-monthly check up at the dentist, he goes twice a year to have his gnashers tuned by a blind man.

★ The smallest spider in the world is the money spider *Arachnidae numismatans.* An example was recently found running up and down a bath belonging to the world's smallest man, Calvin Phillips.

★ *"I haven't got eyes in the back of my head"* is not an expression you'll ever hear a spider say. That's because spiders *have* got eyes in the backs of their heads! And because they can't talk anyway.

★ All spiders have eight legs, the same number of legs as pop group the Beatles, whose songs such as *Lucy in the Sky with Diamonds*

and *I Saw her Standing There* have been recorded by Elton John. So this one counts as two so that's 20.

It Ain't Half Hot Mum!

Tragic Don's Message from the Pits of Hell

PINT SIZE actor Don Estelle, who died earlier this year, has shocked friends and family after revealing that he has been damned to burn in Hell for all eternity. And what makes it worse for the 4'9" corpse is that it's all down to an irregularity in the credits of his 1975 hit single *Whispering Grass*.

Don, who played bespectacled Lofty in the popular sitcom *It Ain't Half Hot Mum*, believed he had led a blameless life and so was gobsmacked to be refused entry at the Pearly Gates.

He told reporters at a press seance: "St Peter explained that because of *Whispering Grass* I wouldn't be allowed into the kingdom of Heaven. I joked that surely it wasn't that bad a record, but then I noticed he wasn't laughing. He told me the problem was that the vocals on the record had been credited to me and Windsor Davies, although most of Windsor's part was actually sung by session singer Mike Sammes."

commandments

"He told me that I had therefore broken one of the ten commandments, and so I was being sent to Hell," added Estelle, speaking from a lake of eternal fire. "Frankly, I was gutted."

Friends of the actor have organised a petition to protest about his

SHOWBIZ EXCLUSIVE!

treatment, but according to the Archbishop of Canterbury there is nothing that can be done to save Estelle's soul from perpetual damnation. He told us: "On the surface, this decision may seem quite harsh, but the Bible's very clear on this subject. A lie is a lie and if the Lord were to let Don Estelle off, where would it all end?"

He continued: "Would Don Estelle's family want God to forgive all the other liars of history? What about Hitler, Caligula, Myra Hindley or Milli Vanilli? Somehow, I don't think they would."

green bottles

Speaking from his home next door to Donald Sinden, Windsor Davies expressed his sympathy. "Don was a lovely boy. I know I gave him a rough time as Sergeant Major

Estelle to pay - Don but not forgotten, in happier times

Williams, but it sounds like Beelzebub's treating him even worse," he joked. "Now I'm going to have to go all religious because I don't fancy joining Lofty in the firey bowels of Hades for five minutes, let alone all eternity!"

Mike Sammes died in 2001. He declined to comment when we summoned him on a ouija board. "I really don't have time for this sort of thing," he told us via a moving glass.

MONEY MAKING PETS

HELLRAISER OLIVER REID TELLS HIS OWN STORY

'I CAN DRINK 75 PINTS OF BEER'

I'm like an earthquake says Ollie

EXCLUSIVE

I've always had a reputation as a bit of a hellraiser. But I can't complain. I'm a pretty wild bloke. In my time I've smashed up every bar and been thrown out of every posh hotel in the world at least three times.

I was thrown out of The Savoy in London once because I kept jumping out of my twelfth floor window and landing on my head in the car park. I was trying to smash a friend's car but in the event I came back with a bulldozer and flattened the hotel.

> Adapted from his book
> 'I AM AN ATOM BOMB'
> © Oliver Reid 1985

VODKA

I happen to enjoy drinking. I drank vodka standing on my head until I was about fourteen. Nowadays I prefer 75 pints of beer, down the hatch in one. And that's nothing. I often drink twice that much without needing the toilet.

SMASH

If I go out for a meal it's as if an earthquake has hit town. I usually smash the table with my girlfriend or use the chairs as a knife and fork. In one restaurant I ordered twelve colour televisions, chewed them up and spat them in the waiter's face.

GUMPTION

My crazy diet of electrical appliances and broken glass often leads to stomach trouble. I often have to pump it myself — with a gallon of liquid Gumption and an industrial vacuum cleaner.

'I ate fourteen dolphins'

I'm pretty well known for my crazy and dangerous pranks. A friend once bet me £500 that I wouldn't eat a live goldfish. I took him along to the zoo and ate 14 dolphins before I was sick. Afterwards I ate another six.

BLEW UP

On another occasion I drank ten pints of nitroglycerine and then locked myself in a friend's washing machine. When he switched it on I blew up, destroying his entire house.

I'm also well known for going through doors without opening them. I had a 36 room mansion built for me in Hollywood without a single door in it. I prefer to make them myself by barging through the walls head first.

DAMAGE

I always pay for any damage I cause — unless I don't particularly feel like it. Being a hellraiser can turn out to be a pretty expensive business.

EXPANDS

I normally get through at least a dozen shirts a week because my body expands to twice its normal size whenever I get angry. A bit like the Incredible Hulk actually. Many friends have taken to calling me 'the Werewolf' because I can change so dramatically. Come to think of it my face does get quite hairy sometimes.

As a matter of fact there have been a few sheep found torn limb from limb in the fields near where I live. And I do get the odd bloodstain on my clothing when I wake up in the mornings.

Next week Ollie describes his X-ray vision and reveals that only kryptonite rays can kill him.

Oliver Reid is a gas fitter from Birmingham and in no way connected with Oliver Reed, the well known British film actor.

Feather-weight o
WHO'S B
BEST P

OF WALES

AS THE country gears up for the Royal Wedding of the millennium, ther only one question on everyone's lips... Why will Her Majesty the Qu not be attending? And did Prince Charles really get caught bumming of his butlers? So that's two. But something else everyone wants to know just who is Britain's best Prince... Naseem Hamed, or of Wales?

In the blue corner is Charles Philip Arthur George Windsor, the Prince of Wales. Own Cornwall and the scourge of modern arechitecture, bonny Prince Charlie has been the nat right Royal darling for the past 57 years. Whether he's talking to plants, enjoying sex

---

Thanks to his mother and father being cousins, Charles' incestuous genes have landed him with a whopping pair of ears. Whilst this has its advantages (at Gordonstoun school, the young Prince discovered he was able use echo location to navigate his way round the dormitory after lights out, emitting a series of shrieks too high to be heard by staff or other boys) it also leaves him open to cruel jibes from other crowned heads of Europe. Consequently, Charles gets off to a poor start in the first round.

1 — ROUND **1**

Prince of Waleses throughout history have been noted for their dapper elegance. Edward VII swept the ladies off their feet in his top hat, cummerbund and spats, whilst Edward VIII set Mrs Simpson's heart a-flutter with his fashionable plus fours, flamboyantly checked blazers and trademark wide tie knot. However dull Charles, with his drab grey suits, sensible brown shoes and boring jumpers, has bucked this trend.

6 — STYLE

Despite being a real life Prince, Duke of Cornwall, Earl of Rothesay and having over 100 lordships under his belt, Charles has never worn a crown in his life! And, if Royal boffins are correct, it'll be a long time yet before he gets the chance. That's because if the Queen lives as long as her mother did, Charles will be a doddery 82-years old before he gets the chance to slip the Crown of England onto his balding bonce.

5 — ROUND **3**

It wasn't long after Charles married his first wife, the late Lady Di, that the country rejoiced at the birth of an heir - Prince William. William has since grown up into a shy and sensitive young man who studied art history at University, which is frankly not the most manly subject in the world. Diana also produced Prince Harry, who has followed his father's example by joining the army.

6 — HEIRS

Charles doesn't have a full time job and is forced to rely on money from the Civil List, a sort of extremely posh dole. However, with a county, a principality, several houses, a mistress and a large entourage of bumlickers to support, the Prince is so hard up he has to supplement his meagre income by selling needlessly expensive biscuits from his Duchy of Cornwall estate.

7 — ROUND **5**

It is a well known fact that the Prince of Wales is forbidden by constitutional law from performing menial tasks. Opening doors, putting toothpaste on his brush and having sex with his first wife are all tasks that Charles preferred to delegate. Similarly, tying his shoelaces is the responsibility of Captain Sandy Tibbs-Urquhart, knotter of the lace equerry pursuivant to the Duchy of Cornwall. Never having looked at, let alone tied his own shoelaces, Charles is completely unable to perform this simple task, and consequently trips up in this round.

4 — LACE TYING

Despite strutting about with a chestful of medals on his military uniforms, Charles has never actually fired a shot, nor indeed thrown a punch, in anger. Always happier talking to a windowbox of pansies or sitting in a skirt on a Scottish hillside painting a shit watercolour, it's frankly doubtful that the Prince of Wales could handle himself in a pagga. However, since he's an unknown quantity he gets the benefit of the doubt in this round, earning a respectable, if unspectacular, middle-of-the-range score.

7 — ROUND **7**

Charles was educated at posh public schools like Gordonstoun and Cambridge, where his life was one long round of luxury and privilege. Each day was spent eating caviar and drinking champagne in a top hat before, after a ten course evening meal of swans, the young Charles would roast a naked fag on a roaring log fire in the prefects' dorm. Despite his expensive education, since leaving university in 1969 Charles has not found work easy to come by. Instead he has taken a series of odd jobs such as patron of the British Pteridological Society and cadre of the Royal Thames Yacht Club.

6 — EDUCATIO

In 1937 the then former Prince of Wales King Edward VIII spoke straight to the nation's heart in a moving broadcast from Buckingham Palace. Speaking directly to his subjects, the King explained that he could not live without the love of Mrs Simpson, and so was regretfully giving up the British throne. Fifty years later his successor, Prince Charles, touched his subjects in a similar radio address about his own love for Mrs Parker-Bowles. In the broadcast, Charles declared his wish to become Camilla's Tampax, absorbing the blood of his mistress's menstrual discharge before swirling round in her toilet bowl forever. Charles's heartfelt jamrag ambitions mean that he soaks up full marks in this final round, swelling his overall score.

10 — ROUND **9**

The Winner, and undisputed King of the Princes is Charles of Wales. Despite being on the ropes throughout the majority of the contest, the big-lugged toff battled royally to defeat the pretender to his crown in the dying seconds. He may not always play by the Queensbury rules, but when it comes to attending galas, waving from Rolls Royces and opening hospital wings, Charles is a true superheavyweight!

52 — TOTAL POINTS

Feather-crested?
RITAIN'S RINCE?

...e of The Three Degrees, or chastising his alleged son for dressing up as a nazi, he has ...ldom been out of the media spotlight. Meanwhile, in the red corner is Naseem ...amed, the Prince of pugilism. Weighing in at just 5 foot 3 inches, this Sheffield man of ...eel packs a knockout punch that belies his short-arsed stature.
...The question is simple. Who is the better Prince? In this 9 round match, we compare ...eir princely attributes and come to a points decision before awarding one of them the ...le Undisputed King of the Princes.
...So, take a ringside seat because *it's seconds out...*

NASEEM

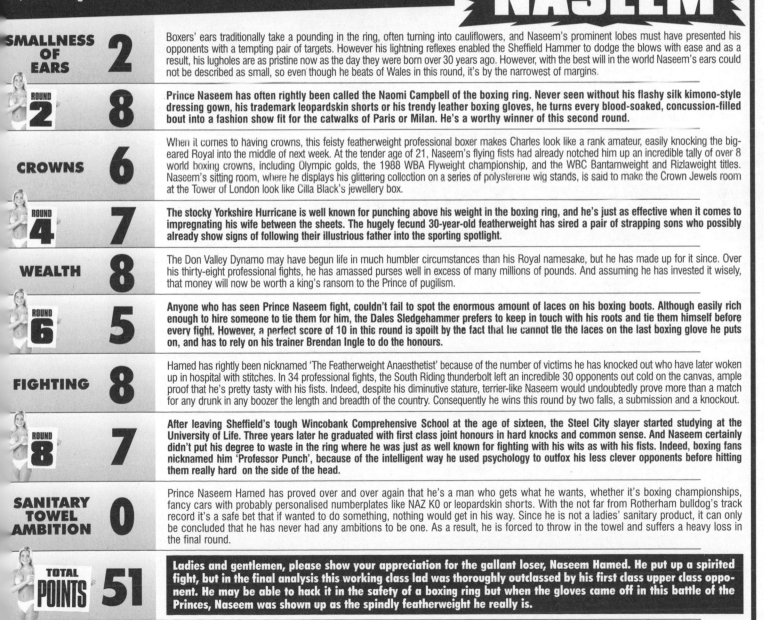

SMALLNESS OF EARS **2**

Boxers' ears traditionally take a pounding in the ring, often turning into cauliflowers, and Naseem's prominent lobes must have presented his opponents with a tempting pair of targets. However his lightning reflexes enabled the Sheffield Hammer to dodge the blows with ease and as a result, his lugholes are as pristine now as the day they were born over 30 years ago. However, with the best will in the world Naseem's ears could not be described as small, so even though he beats of Wales in this round, it's by the narrowest of margins.

ROUND 2 **8**

Prince Naseem has often rightly been called the Naomi Campbell of the boxing ring. Never seen without his flashy silk kimono-style dressing gown, his trademark leopardskin shorts or his trendy leather boxing gloves, he turns every blood-soaked, concussion-filled bout into a fashion show fit for the catwalks of Paris or Milan. He's a worthy winner of this second round.

CROWNS **6**

When it comes to having crowns, this feisty featherweight professional boxer makes Charles look like a rank amateur, easily knocking the big-eared Royal into the middle of next week. At the tender age of 21, Naseem's flying fists had already notched him up an incredible tally of over 8 world boxing crowns, including Olympic golds, the 1988 WBA Flyweight championship, and the WBC Bantamweight and Rizlaweight titles. Naseem's sitting room, where he displays his glittering collection on a series of polysterene wig stands, is said to make the Crown Jewels room at the Tower of London look like Cilla Black's jewellery box.

ROUND 4 **7**

The stocky Yorkshire Hurricane is well known for punching above his weight in the boxing ring, and he's just as effective when it comes to impregnating his wife between the sheets. The hugely fecund 30-year-old featherweight has sired a pair of strapping sons who possibly already show signs of following their illustrious father into the sporting spotlight.

WEALTH **8**

The Don Valley Dynamo may have begun life in much humbler circumstances than his Royal namesake, but he has made up for it since. Over his thirty-eight professional fights, he has amassed purses well in excess of many millions of pounds. And assuming he has invested it wisely, that money will now be worth a king's ransom to the Prince of pugilism.

ROUND 6 **5**

Anyone who has seen Prince Naseem fight, couldn't fail to spot the enormous amount of laces on his boxing boots. Although easily rich enough to hire someone to tie them for him, the Dales Sledgehammer prefers to keep in touch with his roots and tie them himself before every fight. However, a perfect score of 10 in this round is spoilt by the fact that he cannot tie the laces on the last boxing glove he puts on, and has to rely on his trainer Brendan Ingle to do the honours.

FIGHTING **8**

Hamed has rightly been nicknamed 'The Featherweight Anaesthetist' because of the number of victims he has knocked out who have later woken up in hospital with stitches. In 34 professional fights, the South Riding thunderbolt left an incredible 30 opponents out cold on the canvas, ample proof that he's pretty tasty with his fists. Indeed, despite his diminutive stature, terrier-like Naseem would undoubtedly prove more than a match for any drunk in any boozer the length and breadth of the country. Consequently he wins this round by two falls, a submission and a knockout.

ROUND 8 **7**

After leaving Sheffield's tough Wincobank Comprehensive School at the age of sixteen, the Steel City slayer started studying at the University of Life. Three years later he graduated with first class joint honours in hard knocks and common sense. And Naseem certainly didn't put his degree to waste in the ring where he was just as well known for fighting with his wits as with his fists. Indeed, boxing fans nicknamed him 'Professor Punch', because of the intelligent way he used psychology to outfox his less clever opponents before hitting them really hard on the side of the head.

SANITARY TOWEL AMBITION **0**

Prince Naseem Hamed has proved over and over again that he's a man who gets what he wants, whether it's boxing championships, fancy cars with probably personalised numberplates like NAZ KO or leopardskin shorts. With the not far from Rotherham bulldog's track record it's a safe bet that if wanted to do something, nothing would get in his way. Since he is not a ladies' sanitary product, it can only be concluded that he has never had any ambitions to be one. As a result, he is forced to throw in the towel and suffers a heavy loss in the final round.

TOTAL POINTS **51**

Ladies and gentlemen, please show your appreciation for the gallant loser, Naseem Hamed. He put up a spirited fight, but in the final analysis this working class lad was thoroughly outclassed by his first class upper class opponent. He may be able to hack it in the safety of a boxing ring but when the gloves came off in this battle of the Princes, Naseem was shown up as the spindly featherweight he really is.

Jay Okay after Bear Scare

Bear scare survivor, Jay, yesterday (left). And Big bear (main picture, right) and little bear (main picture left). Middle size bear not shown.

IT was nearly Emergency on Planet Earth for big-hatted pop king Jamiroquai when his rental car broke down on holiday in Peru.

Speaking exclusively to the Lima Telegraph and Argus about his lucky escape, the plucky Space Cowboy told how he went looking for help in the woods and found a funny little house.

bufallo

"There was, like, no-one home, so I thought I'd go in and have a sniff around," he said. "And on the back of the door were three buffalo hats – a great big hat, a middling-sized hat and a teeny tiny hat.

Jay, real name Jamiroquai Kay tried on the hats, but found one was too furry and one wasn't furry enough. Luckily, the last one was just right. He said: "I was, like, made up."

The Stretford-born funkwit then stumbled into a dark room. And he was in for a shock when he found the light switch!

springfield

"It was a triple garage, with three red Ferraris in it – a great big Ferrari, a middling-sized Ferrari and a teeny tiny Ferrari.

"I had a quick burn around the woods in all of them," he told the paper. "One was too fast, one was too slow, but the third one was just right. I was, like, sorted!"

But the millionaire hat-and-Ferrari enthusiast, famous for his big collections of hats and Ferraris was in for an even bigger shock when he found the bedroom.

massachusetts

"There were three beds, each with a blonde in it – a great big blonde, a middling-sized blonde and a teeny tiny blonde.

"So I made love with each of them – one was too hard, one was too soft, but one was just right. I was, like, this is the best holiday I've ever had!"

But disaster struck as he was enjoying a nookie-style romp with the teeny tiny blonde.

"I froze to the spot," recalls J, real name Jay. "I looked up to see this great big bear, this middling-sized bear and this teeny tiny bear standing by the door staring at me.

stayin' alive

"And the great big bear boomed in his great big voice, 'Who's been sleeping in my blonde?' and I thought, uh-oh, time to get out of here, Jamiroquai."

In a hat-raising dash to safety, the distinctive pop star, son of Cilla Black impressionist Karen Kay, did his trademark loony leap over the beds and through a closed window, plunging a terrifying metre and a half to safety.

night fever

"I belted the hell out of there like a man running over hot rocks," he admitted. "But as I was scarpering, I could hear the teeny tiny bear saying, 'It's all right – it's just Jamiroquai,' so maybe I should've stuck around!"

night fever

Following his ordeal, Jamiroquai, real name James Iroquai, cancelled his band's tour of South America and the Falkland Islands. He is now recovering in his £6 million, 10 bedroom hat in Surrey.

£50m painting 'not as good' as £8 telly

£8

£50m

New findings stun art world

ART GALLERIES were last night facing the prospect of *SCRAPPING* all their masterpieces after experts proved that a telly costing just £8 was more than *ELEVEN TIMES* as interesting as the world's most valuable painting.

Scientists shut volunteers in a completely bare room containing Vincent Van Gogh's 'Sunflowers', an armchair and a small black & white TV, and then monitored their viewing habits for an hour.

chair

Amazingly, they spent an average of less than 5 minutes standing looking at the £50 million painting, before sitting down in the chair and watching whatever was on the television for the rest of the hour.

"At first we doubted our results," said Professor Kent Walton, head of Statistics at Brunel University. "But then we checked and re-checked them and there was no mistake. Telly is loads better than posh paintings, and that's a scientific fact."

eel

When the experiment was repeated using the Mona Lisa and a copy of the Autotrader, the results were even more marked.

ladyland

Sir Roy Strong, curator of the National Gallery, was devastated when we told him of Professor Walton's findings. "I have wasted my life," he said. "All this shit is going in a skip first thing tomorrow, I can tell you."

Dog in a Million

COMEDY double-act dog Schnorbitz has been found dead at the grave of his master, Bernie Winters.

The 10-stone St Bernard who appeared on TV with his goofy-toothed companion in the early 80s had maintained a lonely vigil by the headstone in the North London churchyard since the straightman's death in 1986.

grave

According to cemetery attendant Max Crabtree, Schnorbitz lay down by his master's grave on the day of the funeral and refused to leave for the next fifteen years. "He'd be there come rain or shine. I used to give him the leftovers from my lunchbox, or throw him the occasional bone I found when digging graves," says Crabtree.

umlaut

"I tried to take him home with me, but he'd never come. He just wanted to stay where he was. The only time he moved was when he went behind Bernie's headstone to lay his morning barker's egg."

omlette

Throughout his fifteen year watch, the faithful dog was struck by lightning (twice), and survived being attacked by crows and set on fire by vandals. It is believed he finally died after being hit by a meteorite the size of a fridge travelling at over 500 miles a second.

Greyfriars Schnorbitz goes to meet his master

Loveable St. Bernard Schnorbitz, 1932-2001 and (inset) his master Winters.

Faithful pets of the Stars

Schnorbitz is not alone in his lonely vigil. Showbiz graveyards are awash with celebrity pets showing loyalty above and beyond the call of duty.

TV clown Charlie Carolli will always remembered for his inability to put up wallpaper and his love of filling people's hats with custard. But he was perhaps best known for his troupe of performing budgies.

Every day for the past 18 years, the birds he left behind have turned up at his graveside. At 13 minutes past 11 - the exact time when he died - they go through their act for their master, riding around his grave on little bicycles and going up a ladder to ring a bell.

Following his hilarious slapstick death 2 years ago,

Rod Hull was laid to rest near his home in Ipswich. He may be gone, but he is not forgotten. Guarding his grave round the clock is his double-act chum Emu.

When Michael Parkinson recently called to pay his respects, the heartbroken bird wrinkled his beak and attacked him. The veteran chat-show host was wrestled to the ground and lost a shoe in the tussle.

All clear for Nick-Nick Jim

BIG BREAK host Jim Davidson last night spoke of his relief after an agonising 3-day wait for the results of tests. Davidson discovered a lump in his throat after seeing a homeless beggar, and feared that he was developing a redeeming quality. However, tests came back clear. "I've been given the green light and I'm so relieved. I have no redeeming qualities whatsoever," said the drink-drive funnyman.

ONE SMALL STEP FOR BRAN(SON)

MILLIONAIRE adventurer Sir Richard Branson has announced plans to get into the record books by being the first twat in space.

The Virgin boss, whose last attempt to circumnavigate the globe ended in tears when his £3 million balloon popped, hopes to complete a twelve-hour mission aboard an American Space Shuttle.

orbits

If all goes to plan Branson will complete 12 orbits of the earth with a shit-eating grin plastered across his face.

The NASA Tosspots in Space programme has been running for over 20 years. In 1978 a pair of fucking wankers were blasted up to Skylab where they spent over 100 days orbiting the earth, and this was followed up during 1994 when, for six months, an absolute arse-

hole lived aboard the Russian Mir space platform.

stimerols

But a challenge to American supremacy in the field of wankered spaceflight is set to emerge on this side of the Atlantic. Hillaire Belloc, a spokesman for the European Space agency ESA, said: "We have recently been having talks with Jeffrey Archer, and by the year 2020 we should hopefully have the technology in place to put a right cunt on Mars."

HOW *with Fred Dinenage*

Dear Fred... If a newsreader wanted to read the news dressed as a black and white minstrel, **HOW** would he go about making the change without anyone noticing?

Huw Edwards, London

Right, says Fred...The key to a successful transformation is careful planning and preparation. Firstly find out how much makeup you require to completely change yourself into a minstrel. Then, decide the number of news

Before / After

*broadcasts over which you are going to spread it. Say, 300 programmes over the course of a year. Now divide the makeup into 300 equal portions. Simply apply one of the portions of makeup before you go on air each night. The differences from day to day will be imperceptible, but over the course of a year the full transformation will take place. And that's **HOW** you become a black and white minstrel without anyone noticing.*

If you want to ask Fred **HOW** to do something, why not hang around in the car park at Southern Television until he comes out.

BLUE PETER'S PETER PURVIS PULLS PLUG ON CELEBRITY FAN AVOIDANCE TACTICS

We all know the lengths that some fans will go to in order to meet their idols. But did you know that many of today's top stars often go to equally bizarre lengths to *avoid* their fans.

Tom Jones, for example, got so fed up with female admirers pestering him in restaurants he built his own – in his garden. Complete with a staff of ten, and 40 actors employed full-time as customers, the restaurant cost Tom over £2 million. But the Welsh heart-throb can now enjoy uninterrupted romantic meals – seven days a week.

SEIGE

Tom's next door neighbour, fellow Scot Rod Stewart, found that his home was under constant siege from women. So he moved out. Now Rod, 63, lives in a tree house at the bottom of his garden, undisturbed by the queues of women waiting at his door.

STORIES

These are just two of the star stories to be found in a new book – 'Things that stars do in order to avoid their fans'. The book was the brainchild of former Blue Peter presenter Peter Purvis.

HOUSEBOAT

"Originally, I was intending to write a book about cookery, but I couldn't think of any recipes", said Peter, speaking from the Birmingham houseboat which is home to himself, his wife and their sixteen dogs. "Then my wife suggested a book about the things that stars do in order to avoid their fans, and it all seemed to take off from there".

Peter Purvis (spelled 'Purves') (above) and TV's Ronnie Corbit

The fascinating book also reveals everyday tricks which TV celebrities use in order to get to work unrecognised. "Almost every TV presenter has a disguise of some sort – whether it be a false beard, a hat, false nose or a pair of plastic women's tits. Fans eyes would boggle if they stood in the cloakroom at the BBC and watched Terry Wogan arrive in a blonde wig and sexy red leather mini skirt, or newsreader Jeremy Paxman blacked up to look like Al Jolson. But that's what goes on every day of the week.

BARGE

One star wouldn't even risk walking to work. He actually wraps himself up in parcel tape and posts himself to the Television Centre at 5 o'clock each evening. He's delivered by the Post Office, ready for work, at 8 o'clock the next day".

The book also tells how pint-sized comic Ronnie Corbett allowed fans to get on top of him – quite literally – by adopting the behavioural patterns of a mole. "Ronnie avoids the fans by scurrying round beneath the earth in a vast network of tunnels which he dug entirely by hand", writes Purvis. "The only sign frustrated autograph hunters ever see of the star are small mounds of soil being pushed up from beneath the surface".

SHOVE

In America many stars have become so fed up with life in the public eye that they've *done away* with themselves. For Peter believes stars like Elvis Presley, Marilyn Monroe, James Dean and Bing Crosby are in fact *alive*, having faked their deaths in order to escape the limelight.

"There is evidence that many of America's top entertainers, previously believed to be dead, are in fact all living in a special purpose-built celebrity hideaway at a top secret location, in order to avoid their fans", claims Purvis. And even if fans discover the whereabouts of the stars' hideaway, they'll still have trouble getting there. As Peter explains in the book.

ELBOW

"The secret hideaway of the stars is in fact on the moon. Star Moonbase Alpha One was built especially for himself and his showbusiness pals by Elvis Presley. The luxury space base has everything the celebrities could ever need. Night clubs, restaurants, shops, cinemas, swimming baths and a 9-hole 'pitch and put' golf course.

WRIST

The only setback is that a one-way ticket to the moon, including a faked death, costs 500 million dollars". Peter hopes to have his book finished by the weekend, and all being well copies should be on the shelves by Christmas.

Turn to page 33 for our fabulous Star Spotting Competition

WOULD YOU MAKE A POSH PRINCESS?

Are you a dish fit for Royalty?

Every girl dreams of marrying a prince, but it's only a small minority who end up as members of the Royal family. Today Prince Andrew is the world's tastiest bachelor. But do YOU have those special qualities that would make you a dish fit for Royalty? Here's a chance for you to find out.

Complete the following test answering each question A, B or C. Then tot up your final score and find out whether YOU could one day be sitting on a throne.

1. You are at a high class party and your eyes meet with Prince Andrew's across a crowded room. Would you:
   A. offer to buy him a drink
   B. wink and flash your knicker elastic
   C. smile and shyly look away

2. The Prince invites you out for a meal. Which of the following places would you suggest visiting?
   A. An Italian pizzeria
   B. A Kentucky Fried Chicken restaurant
   C. The poshest restaurant in town

3. Whilst you are engaged to marry Prince Andrew a good looking milkman comments on the size of your bust. What would you do?
   A. Smile politely
   B. Invite him upstairs for a bit of slap and tickle
   C. Report him to the police

4. It's a sunny day so you go to watch horse racing at Ascot. What would you wear?
   A. A party frock
   B. A sexy bikini
   C. A fancy dress and a hat with fruit on it.

5. You are taken to the opera and during the interval Prince Andrew goes to the kiosk. What would you ask for?
   A. A choc bar
   B. Pop corn
   C. Expensive sugared almonds

6. You are at a classy do and your nose starts to run. What would you do?
   A. Keep sniffing till you get home
   B. Wipe it on your sleeve
   C. Go to the toilet and wipe it on your posh handkerchief

7. At a polo match Prince Andrew scores, but the referee disallows the goal. How would you react?
   A. By politely applauding
   B. By making an obscene gesture towards the referee
   C. By turning to your mate and talking pleasantly about the weather

8. You are at a Royal garden party and your hat blows off. Would you:
   A. Bend down to pick it up, cursing under your breath
   B. Dive acrobatically to to catch it, knocking a table over
   C. Quietly ask your chauffeur to go and buy you a new one

9. You are pregnant with a Royal Baby. How would you choose to have it delivered?
   A. By the ambulance driver
   B. By parcel post
   C. By the Queen's Gynaecologist

Phil Loftus

Tasty Prince Andrew - could you be cooking his bacon and eggs every morning?

10. Which of the following names would you choose if your baby was a girl?
    A. Tracey
    B. Kelly Marie
    C. Victoria Mountbatten Waterloo Windsor Elizabeth

11. Which of the following jobs would you prefer your Royal son to have?
    A. Expensive hairdresser
    B. Spot welder
    C. A posh Admiral in the Navy

12. After a long happy reign you die. Which of these funeral services would you prefer?
    A. A small service for family only
    B. A West Indian style street celebration
    C. A sparkling military parade watched by millions live on telly

How did you do?

SCORING   A — 1 point   B — 0 points   C — 3 points

30 or over — Well done. You'll make a perfect posh princess
15 to 29 — Not bad, you have posh potential
Less than 15 — Disappointing — you're not the type

MAY THE COU WITH YOU

FRESH FROM the success of his latest epic sci-fi adventure, billionaire Star Wars creator *George Lucas* may be about to mount the most spectacular fight of his career. But this time his adversary won't be a seven-foot wheezing Sith - *it'll be a 48-year-old mother of nine from the West Midlands.*

And if she has her way, instead of the wide open spaces of an alien galaxy far, far away, the battle could take place in the more mundane surroundings of Redditch Magistrates Court.

For Solihull cinema usherette Maureen Herpes claims that the plots for all six Star Wars films have been lifted directly from her own life story. She told us: "George Lucas has made a fortune, and it's all based on things that have happened to me; the plots, the characters, everything. It's only fair that I should get a cut of the proceeds."

dress

Herpes first noticed similarities between her own life and Star Wars when the first movie was released in 1977. "It was a little bit spooky, to say the least," she told us. "The film's principal character was called Princess Leia and she had black hair and a white dress. Well, my hair was black at that time too, and I'd bought myself a white dress in the BHS sale just a few weeks earlier.

pass port

"At the time I thought it was just a bizarre coincidence and I thought no more about it," she added. But when the second film, 'The Empire Strikes Back' came out, Maureen's earlier suspicions were confirmed. She told us: "It turned out that Princess Leia and Luke Skywalker were twins. Well, I've got a twin brother too." But the coincidences didn't stop there. "Luke is a space pilot who flies X-wing fighters for the Rebel Alliance, and my brother Alan drives an X-reg van for Allied Carpets. It sent a shiver down my spine when I realised, I can tell you."

What finally clinched it for Maureen was the film's climactic deathbed revelation that arch-villain Darth Vader was the twins' estranged father. "When he took that mask off and I saw his blotchy, purple face gasping for breath, the hairs on the back of my neck stood up," she told us. "It was the spitting image of our dad. He used to drink a bit, and was on eighty un-tipped Woodbine a day."

'...Vader was the spitting image of our dad. He used to drink a bit, and was on eighty un-tipped Woodbine a day'

pass stools

"Just like Luke and Leia, we didn't see a lot of our father when we were growing up," she continued. "He might not have been a Jedi Knight, but he did get into a lot of fights, mainly at closing time. And like Vader, he turned to the dark side at an early age,

drinking mainly Guinness and Younger's Scotch Bitter."

pass partout

The day after seeing the movie, Maur-een went to her local Police Station to launch an action for copyright theft against the film's director. "I told the duty constable what had happened, and that I wanted to press charges. He told me he was just finishing his shift, but he'd leave a note for the desk sergeant who came on at nine, asking him to telephone the Police Station in Hollywood and have Ge-orge Lucas arrested.

"As he ushered me out, he assured me they'd have him safely locked up in prison by tea-time," said Herpes.

Who Leias Wins:
Luke Skywalker (left), Darth Vader (not quite as left), so sad Maureen (above), R2-D2 lookalike Henry Hoover (below right), Chewbacca dog Tyson (same amount right but bit further below), and the real Princess (rightest of all), yesterday.

"However, when I turned on the TV news that night, there was no mention of the case. I can only assume that Lucas had used his wealth and position to get off the charge."

Maureen thought that the movie mogul's brush with the law might have scared him off, but as the Star Wars films continued to come out over the next few years, the catalogue of similarities between them and Herpes' life grew longer and longer. She told us: "For example, the young

DESSERT?

HONESTLY, DARLING. I COULDN'T EAT ANOTHER THING.

RTS BE
Solihull Usherette
Maureen set to sue
Star Wars Lucas

Luke and Leia are watched over by a kindly uncle figure, Obi Wan-Kenobi, who they refer to as Ben.

Dr Watson

"While we were children, we used to have many uncles who came to stop the night at our house, mainly long distance lorry drivers. I'm almost certain one of them was called Ben."

And the coincidences didn't stop there.

• *Droid R2-D2 looks strikingly similar to Maureen's Henry Hoover.*

• *The light sabres wielded by the Jedi Knights bear more than a passing resemblance to her cinema usherette's torch.*

• *A scene where the Rebel Alliance detroy the Death Star has distinct echoes of an incident from Maureen's life when she had an accident with a microwave oven and a tin of Aldi Noodle Doodles.*

• *Hairy wookiee Chewbacca can stand on his back legs and wears a bullet-belt, just like her dog Tyson.*

Herpes was so angry that she decided to go to the Police again, this time armed with a dossier of evidence. Unfortunately, the duty constable was just coming off his shift again, but he assured Maureen that her complaints were being taken seriously. She told us: "He explained he was going to fly to Hollywood and arrest George Lucas himself, just as soon as he had finished his cup of tea."

However, once again Lucas somehow escaped formal charges. To make matters worse, the bizarre plot parallels continued to appear in further Star Wars films, getting ever more blatant. Maureen told us: "As an usherette I stood through each of these films hundreds of times. On each viewing, more and more similarities became apparent. Names, faces, gadgets, relationships. It was like Lucas was taunting me.

Sancho Panza

"In the end, I was so upset by what was happening that I couldn't cope any more and I went to see my doc-

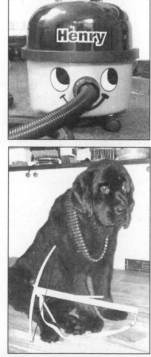

'...Hairy wookiee Chewbacca can stand on his back legs and wears a bullet-belt, just like my dog Tyson.'

tor," said Maureen. "He prescribed me nerve pills, put me on the panel for three months and had the kids taken into care." In the end, Maureen spent more than six months off work before she felt well enough to return. However, she couldn't have picked a worse time to go back to her job at the cinema.

"I couldn't believe it. The day I returned, the film that was showing was *'The Phantom Menace'*, the one where the Gungans join forces with Obi-Wan Kenobi and Liam Neeson to battle the evil Trade Federation," said Maureen. "And this time, the similarities between the plot and my life were obvious to everyone."

"You see, for the previous 4-months I had had a Gungan lodger from the planet Naboo staying in my spare room. Like in the film, he had been expelled from his own planet, where he lived in a huge bubble at the bottom of the sea. You might dismiss it as a mere coincidence until you hear his name. It was Jam-Jar Binx."

NEXT WEEK...
Maureen explains how the plot of the final Star Wars film *'Revenge of the Sith'* is based on the time she got pregnant off the cinema projectionist Anikin Skywalker, and then gave birth to herself and her twin brother before being sectioned under the Mental Health Act.

Prices Tutu High

THE ROYAL BALLET was at the centre of a storm last night after watchdogs accused it of ripping off fans. The Covent Garden-based dance company was criticised by trading chiefs for charging excessive prices for replica tutus, tights and flouncy shirts.

Cash-strapped parents complained that dancing strips costing as much as £55 each represented poor value for money, and that they often felt pressured to buy the latest ballet outfits for their children.

tights

Mum Kelly-Anne Fungus told us: "I've got nine kids, and they all want tutus with 'Fonteyn' on the back or tights like Rudolf Nureyev's because all their friends have them. What am I

Hopping Mad: *Fuming mum Kelly-Anne Fungus yesterday.*

Cash-strapped parents led a merry dance by Royal Ballet ~ report

supposed to do?" She was also unhappy about the poor quality of the kits.

slacks

"For the sort of prices they charge they should last forever, but they're actually quite shoddy," she told reporters. "I bought my youngest a pair of Baryshnikov tights, and he went to the park with his mates to do some pas-de-deuxs. When he came back,

they were completely ripped around his packet. I can't afford to keep shelling out £35 every time the kids want ballet tights, I'm on income support."

brambels

But Royal Ballet spokesman Quentin Bum-boy was adamant that the outfits represented good value for money. "We realise that some families may have difficulty finding the money to purchase official merchandise, but at the end of the day, nobody is forcing them to do it. It's simply up to parents to resist the pressure that we put the children under through our relentless advertising." And he also hit back angrily at

claims that the Royal Ballet was marking up prices excessively. "It is true that our unit cost for a Darcey Bussell tutu is £3.50, and we sell them in our shop for £45, but there is a perfectly simple reason for this. Unfortunately, it is too complicated to go into now."

thornes

But Mrs Fungus was unimpressed by Bumboy's explanation. She fumed: "It wouldn't be so bad if the costumes remained the same. I bought a Swan Lake shirt for my son a month ago. A week after he got it, they brought out a new one with different ruffles on the cuff. Not surprisingly, he won't wear it any more in case his mates laugh at him, so that's another 45 quid down the drain. Plus another £10 to have 'Wayne Sleep' printed on the back. It's a disgrace."

Big pay-packet: *A ballet dancer with impressively prominent genitals looks up a ballerina's skirt yesterday.*

● In a similar case last week, The English National Opera was fined £4000 after Trading Standards officials prosecuted them for selling sub-standard replica viking helmets and size 48DD tin bras.

Sammy Davis lives twice...

THUNDERBILL!

Grimsby magistrates heard yesterday how American singer Sammy Davis Junior was alive and well and living on a secret volcanic island in the Pacific ocean from where he plans to take over the world.

That was the claim made by unemployed builder Roger Blenkinsop, who appeared before the town's magistrates charged with burglary after a stolen generator was found by police in outbuildings at his home in Cedar Drive, Cleethorpes.

FINE

But the court was unimpressed by Blenkinsop's story and handed him a £250 fine together with a 6 month suspended jail sentence. They also imposed an additional £20 fine for contempt of court after Blenkinsop claimed he took the machine from Mr Davis Junior's secret island after the singer had refused to settle an outstanding bill for £1,200.

CHAMPION

Outside the court yesterday a disappointed Mr Blenkinsop continued to argue his innocence and warned reporters that Sammy Davis Junior, who was thought to have died several years ago, ultimately plans to take over the world from his secret headquarters in a converted volcano. And he blasted stubborn magistrates who refused to visit the area in a helicopter to enable him to prove his innocence.

DANDY

Blenkinsop claimes that he was approached by Davis Junior after he had advertised his services as a general builder in a Cleethorpes newsagent's window.

BEANO

"I had a phone call from a man calling himself Mr Glass Eye and I agreed to meet him at an extinct volcano on a small secret island somewhere in the Pacific. The minute I saw him I realised that he was in fact the singer Sammy Davis Junior, and that he must have faked his death and gone to live on this island.

SPARKY

He told me he had various building jobs that needed doing, including converting the volcano into a secret rocket launching pad and installing laboratories, lifts

Davis Junior - escaped in mini submarine

and a mono-rail system. He also wanted a swimming pool built, full of sharks.

TOPPER

Glass Eye's no expense spared shopping list also included:

● A DOZEN nuclear war heads for destroying the world's capital cities.

● A MILLION diamonds for deflecting the sun's rays and focusing them on military installations, plus

● 2000 orange boiler suits.

Indeed, money appeared to be no object for the maniac former all round entertainer. Blenkinsop even overheard a telephone conversation in which the would-be international super villain offered TV's Jeanette Kranky £200,000 a year to act as his evil midget accomplice with a razor sharp hat. An offer which, to the pint sized star's credit, she declined.

BOWLER

Blenkinsop successfully quoted to fit sliding doors to the top of Glass Eye's secret volcano at a cost of £1200, and carried out the work the following week. But two months and several reminders later, there was still no sign of payment from Mr Davis Junior.

"It was obvious that he had no intention of paying, so I decided to go round to his island and demand my money. I burst into his control centre and went straight

up to him and gave him a bit of my mind. But it wasn't until I accidentally stood on his pet cat that I realised I had in fact been talking to Huggy Bear out of Starsky and Hutch, alias seventies actor Antonio de Fargas, who had been recruited by Sammy Davis Junior to act as his expendable double.

SLIP

I turned round to look for the real Glass Eye only to see him disappearing into the sea in a mini-submarine. I was furious, so I loaded the generator into the back of my van in an attempt to recoup some of my losses." The following morning police found the generator at Mr Blenkinsop's home following a burglary at a nearby building site.

SHORT LEG

One condition or Mr Blenkinsop's employment with Davis Junior was that he promise not to tell anyone about his secret island, or of his plans to rule the world. But faced with an unpaid bill for £1200 plus VAT, plus a charge of buglary, Blenkinsop decided to break his vow. And now he fears for his life, convinced that Davis Junior is seeking a deadly reprisal.

"Just this morning I was nearly attacked by a lesbian with a pointy shoe, and on my way into court two sexy birds in bikinis almost done judo on me. I should be given police protection".

'For Your Eye Only' invoice causes storm in Pacific

Roger Blenkinsop	General Builder
To Mr Glass Eye Pacific Island	52 Cedar Drive Cleethorpes

INVOICE		11/10/92
To make and install sliding volcano doors (two of) on secret island, and making good.		
Materials		180.00
Transport		20.00
Labour		1,000.00
	Total	£1,200.00

LICENSE TO BILL: Blenkinsop's invoice (left), from Cleethorpes with love.

In April 1987 Mr Blenkinsop was convicted of stealing lead from the roof of a dissused hoted in Hull despite his protestations that sixties balladeer Matt Monroe, using the alias Silver Thumb, had been using a secret penthouse above the building as the headquarters for a plot to hold the planet Earth to ransom.

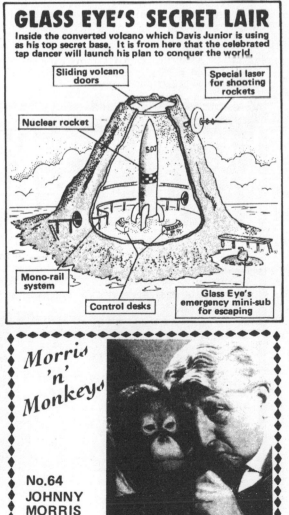

Sammy Davis Junior (above) as his fans remember him.

On that occasion he was fined £125 and ordered to pay £30 costs and bound over to keep the peace for 6 months.

GLASS EYE'S SECRET LAIR

Inside the converted volcano which Davis Junior is using as his top secret base. It is from here that the celebrated tap dancer will launch his plan to conquer the world.

- Sliding volcano doors
- Special laser for shooting rockets
- Nuclear rocket
- SDI
- Mono-rail system
- Control desks
- Glass Eye's emergency mini-sub for escaping

Morris 'n' Monkeys

No.64 JOHNNY MORRIS

R.M.S. *SHITE-ANIC!*

Really Massive Shite discovered on sea bed

THE gigantic last stool ever moved by the captain of the Titanic has been discovered lying intact two and a half miles beneath the Atlantic ocean.

The giant dog's egg is thought to have been laid by Captain Smith on the night the ill fated liner sank. Eye witnesses reported seeing the captain shit his pants as the Titanic collided with an enormous iceberg, and several survivors spoke of eerie farting noises and unpleasant smells as the ship went down. But until now the sea has refused to give up any faecal evidence to support these claims.

Conway - Shitty obsession has cost him dearly.

Scientists

Now scientists believe they have located the world's most famous dog's egg. And remarkably enough it appears to have been almost perfectly preserved despite being buried at sea for an amazing 86 years.

Dogs

"The Richard is lying in one piece, still perfectly curled and crimped", said sub mariner Russ Conway who made the discovery in his specially designed underwater exploration vessel China Tea. "It is a quite awesome sight. To look at it you would think it was fresh - as if had only just been nipped off yesterday".

Most organic matter would decay rapidly in sea water, however scientists believe the reason for the turd's remarkable preservation is quite straightforward. "Fortunately for us fish wouldn't touch it with a barge poll", Conway told us. "It has literally remained untouched for all this time".

Axemen

In 1987 Conway, 71, gave up a successful career as a pianist to concentrate on his life-long search for the turd. It's an obsession which has cost him dear. His wife of 17 years, sour faced pseudo intellectual singer Tinita Tikarum left him in 1995, and Conway

Conway aboard his exploration vessel 'Limbo' yesterday

has run up debts of over £2 million financing underwater searches. But now at last his dedication appears to have paid off.

There are ambitious plans to raise the shi-tanic and put it on permanent display as a floating exhibit, probably somewhere like

---

SHITWRECKED *The stinking of the Titanic*

It was approaching midnight on Sunday 14th of April 1912. The RMS Titanic was midway through her maiden voyage from Southampton to New York. Their was a party atmosphere on board as the passengers relaxed following their evening meal.

After a particularly heavy dinner Captain Smith told fellow officers that he was going down below to curl one off, leaving First Officer Murdoch in charge of the bridge. The sea was calm, but it was a dark, cold, moonless night.

Fleet - spotted iceberg from crow's nest.

Iceberg

Up in the crow's nest lookout Frederick Fleet spotted a giant black silhouette looming directly ahead of the ship. "Iceberg dead ahead!" he cried.

Realising the danger First Officer Murdoch gave the immediate order "Full

Murdoch - left in charge of bridge.

speed astern and hard a'starboard" in the vein hope of avoiding a collision.

Cos

Meanwhile in the officer's toilets Captain Smith was sitting down with a copy of the Picture Post. He had already got the turtle's head, but he had been egg bound since leaving Southampton and he knew that a long and difficult shite lay ahead. But a sudden judder as the ship's engines went into reverse told him something was wrong. He leapt to his feet, pulled up his trousers and returned to the bridge, arriving just in time to see the enormous iceberg towering above the bow of his vessel.

Rocket

"Fuck me!" he cried in horror. There was a sudden rumble in his trousers. Captain Smith was an experienced sailor - indeed this was to have been his final voyage before retirement - but no man on Earth could have regulated his bowel in

'Amidst the farting and screaming Captain Smith realised that the follow through was unavoidable'

such dire circumstances. A series of eerie, juddering farts echoed around the ship as his rectum began to slowly and involuntarily relax.

How they found the turd of the Titanic

Hartlepool. But at least one shit scientist believes there would be enormous risks involved. Professor Karl Heinz-Bigsoup of the Tampa Bay University of Faeces believes that the massive log will be unstable.

Plank Spankers

"Even after 86 years there is still a high risk that it will be minging. And the chances of raising it from the sea bed without anyone getting shit behind their fingernails are remote".

Tub Thumpers

There has been mixed reaction to news of the discovery. Captain Smith's daughter, TV chef Delia, believes her dad's stool should left to rest in peace. "This huge underwater crap is a sea grave. Perhaps they could stick a lollipop stick in it with a little inscription or something. But apart from that I think they should leave it alone", she said yesterday.

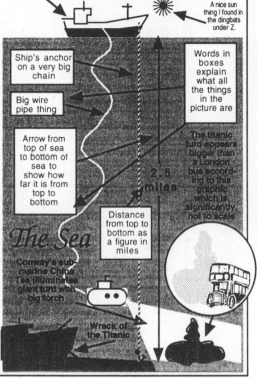

Conway's ship 'Limbo' floats on top of sea. It is attached to the submarine 'China Tea' by a big sort of wire pipe thing.

A nice sun thing I found in the dingbats under Z.

Ship's anchor on a very big chain

Big wire pipe thing

Arrow from top of sea to bottom of sea to show how far it is from top to bottom

Words in boxes explain what all the things in the picture are

The titanic turd appears bigger than a London bus according to this graphic which is significantly not to scale

Distance from top to bottom as a figure in miles

2.5 miles

The Sea

Conway's submarine China Tea illuminates giant turd with big torch

Wreck of the Titanic

☐ Graphic by some geeky little anorak on a computer

Giz a jobbie!

'Giz a job' actor Bernard Hill who played Captain Smith in the blockbuster Hollywood movie gasped in horror when we showed him pictures of the Captain's gigantic brown trout.

"I definitely *couldn't* do dat", quipped the former Boy from the Black Stuff.

Film makers have been accused of distorting the truth by omitting any reference to the Captain shitting his self from the movie script. Contrary to events featured in the film, surviving witnesses recall passengers jumping overboard to avoid the smell.

"It was nothing like the movie", said 28 year old Ann McMahon who lost her mother on that fateful night. "The stench of rotten cabbage was unbearable", she recalled. "All the people who were drowning came up to the surface, took one whiff and decided to go straight back down again."

Asked what he would have done if he had been the real captain of the Titanic, Hill said he'd have put the vessel in reverse.

"I'd have sailed back to the iceberg and everyone could have climbed off onto that".

Titanic (above) sets sail and Smith (below) who laid the monster cable.

Amidst the farting and screaming Captain Smith realised that the follow through was unavoidable. Within seconds he was touching cloth, and panic slowly began to spread around the upper decks. A sickly, pungent smell filled the cold night air and both passengers and crew held their noses or wafted frantically with their hands to avoid the hideous odour.

Banger

The Captain knew that if he lit a bum cigar on the bridge the consequences would be disastrous.

Selflessly he clambered down to the lower deck, pebble dashing several lifeboats as he went, and lowered his shit locker over the side. Witnesses recalled the purple faced captain grimacing as his mudhole expanded and a giant brown trout curled itself down into the water with an enormous splash. So heavy was the Captain's log that as it slid beneath the waves the ship rose several feet in the water.

Sparkler

Captain Smith's desperate final farts were clearly smelled on board the steam ship California only a few miles away. But the pong was so noxious her captain ordered matches be lit on deck, and he steered a course directly away from the Titanic at full speed. By 4.00am when the Carpathia finally arrived at the scene of the stinking all that remained were a few dangleberries bobbing about in the water.

DIARY of DISASTER

April 9th 1912
Captain complains of constipation after eating three Scotch Eggs and an omelette for lunch.

April 10th
Titanic sets sail from Southampton.
2.30pm - A passenger recalls seeing the Captain eating a large bag of bonfire toffee on the bridge.

April 11th
Despite continuing lack of bowel movement Captain orders two boiled eggs for breakfast and has egg fried rice for lunch.

April 14th
7.30am - Captain makes the fatal decision to have prunes for breakfast.
1.00pm - Beans on toast for lunch, followed by bread pudding. During the afternoon Captain ignores repeated warnings from his stomach about possible stool movement ahead.
7.00pm - Captain Smith dines with Mr and Mrs Arthur Askey, ordering cabbage soup, crab sticks, mince and dumplings followed by Death by Chocolate.
11.25pm - Captain retires to lavatory complaining of stomach cramps.
11.30pm - Look out reports iceberg dead ahead. Captain shits his pants.
2.30am - Going, going, pong. Titanic finally sinks.

PRINCE'S DRAMATIC ADMISSION

EXCLUSIVE

HRH Prince Edward yesterday issued a shock statement which put an end to months of public speculation. "Yes", he declared, "I AM theatrical".

A close friend told us that the Prince has been aware of his theatrical tendencies since early schooldays. "Edward used to dress slightly flamboyantly and talk in a loud voice. By the age of 13 he was gesticulating two or three times a day".

ASHAMED

It was during his time at Cambridge Edward began to realise that being theatrical was nothing to be ashamed of. He began to mix with and feel comfortable in the company of other theatricals.

TENDENCIES

However, under constant pressure from his family, the Prince tried to disguise his tendencies by playing competitive sports, talking in a stern voice and joining the army. But all along, deep down inside, he knew he was a theatrical, and it was probably in 1986 that he made his decision to leave the Marines and live openly as a theatrical person. His parents were shocked and disappointed but Edward knew he had made the right decision.

BENT

Doctors believe that up to 1 in 10 people in Britain have a theatrical bent, although many try to lead normal lives, disguising their true feelings from family and friends. But nowadays being thespian no longer has the same social stigma attached to it that it did in Oscar Wilde's day. That is the view of Dr. Quentin Bender, social psychologist at the Brighton Institute of Dramatic Art and Interior Design. And Hairdressing.

UPHILL GARDENER

"If people feel they are harbouring dramatic tendencies they should contact their local amateur dramatic society for help and counselling", he told us. Other well-known theatricals have included Danny La Rue, Larry Grayson and Noël Coward.

Celebrity Swears
№ 207
David Coleman

HAIRY ARSEHOLES

Bears Wood Shit Shock

BEARS DON'T shit in the woods. And that's as sure as bears shit in the woods, which they don't.

The discovery was made by scientists from the University of Oslo, the same team who in 1998 discovered that eggs weren't eggs.

"We've been watching bears' toilet habits for four years now, and we've learned that they all leave the woods when they need to defecate", said Professor Ulph Oerstroeke. "They move from the trees into the surrounding meadows or plains before finding a small bush to squat behind. Then they wipe their arse on a crisp packet," he added.

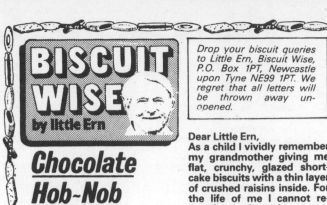

BISCUIT WISE
by little Ern

Drop your biscuit queries to Little Ern, Biscuit Wise, P.O. Box 1PT, Newcastle upon Tyne NE99 1PT. We regret that all letters will be thrown away unopened.

Chocolate Hob-Nob problem

Dear Little Ern,
I particularly enjoy the new 'Hob Nob' biscuits, especially the chocolate coated variety. However, my husband enjoys milk chocolate, while I prefer plain.
Do you know of any plans afoot to introduce a packet containing both?
Mrs Rosemary Nutmeg
Parsley, Herts.

• *Sorry Rosemary, but my friends at McVities tell me that mixing their delicious milk and plain chocolate Hob Nobs would be technically impossible.*
Ern's Biscuit Tip: *Why not buy one packet of each, and mix them yourself in a small tin or similar air tight container. Special 'biscuit barrels' are available from many stores, starting from around £2.00.*

Dear Little Ern,
My hubby insists on calling Cream Crackers 'biscuits', but I was always told that these were 'savoury wafers' as they do not contain sugar and can be eaten with cheese. Who is right?
Basil Bayleaf
Dill, Cheshire

• *Oh no! Not the old 'biscuit/ wafer' debate again. If I had a Custard Cream for everytime this one has cropped up, I'd have an awful lot of biscuits! Let's call a truce – both of you are right. Cream Crackers are, technically speaking, wafers, as they contain more salt than sugar. However the cheese rule does not always apply, as we've all ate Digestives with cheese, haven't we. And who hasn't enjoyed wafers with ice cream. Tricky, isn't it.*
Anyway, biscuit or wafer, whatever you call them, they all taste lovely.

Dear Little Ern,
As a child I vividly remember my grandmother giving me flat, crunchy, glazed shortcake biscuits with a thin layer of crushed raisins inside. For the life of me I cannot remember what they were called. Do these biscuits still exist, and if so, where can I get a packet?
Mrs Corriander Sage
Thyme, Middlesex

• *The biscuits you describe sound to me like Garibaldi, a particular favourite of mine too. Traditional Scottish biscuit manufacturers Crawfords still manufacture Garibaldi in their 'Pennywise' range, priced from around 25p for a 300 gram packet.*

Dear Little Ern,
I've been eating Coconut Creams since I was five, and I'm convinced that today's biscuits contain less coconut than they used to. Are the manufacturers penny pinching here?
Mrs Mint Marjoram
Cumin, Mid Glamorgan

• *Well Mrs Marjoram, on reading your letter I popped out to my local biscuit supplier and nibbled a couple of Coconut Creams on my way home. And they seemed alright to me. Of course not all manufacturers maintain the same high standards, and a lot depends on the brand of biscuit you're eating. Why not treat yourself to a Fox's Family Assortment. You'll find some first rate Coconut Creams in there, not to mention a few other biscuit delights!*

Biscuit of the Week

This week I've selected Peak Freans 'Bourbons' as my biscuit of the week. With their fondant chocolate centre sandwiched between rich, sugar coated biscuit fingers, it's no wonder they say 'You can't beat a Bourbon'.
I'm off to enjoy a packet now out on the patio with a nice pot of tea. See you next week. And in the meantime, keep crunching!

LET'S FACE IT. THINGS JUST AREN'T WORKING OUT BETWEEN US.

20 THINGS YOU NEVER KNEW ABOUT TREES (and elephants)

Love 'em or hate 'em, you just can't ignore 'em. They're in our parks, gardens and streets, growing in fields, flower beds and in other places as well. Some people grow them for pleasure, others cut them down for a living. We eat them, make tables and chairs out of them, and throw them on the fire. Yes, we nearly all take them for granted, but how much do we REALLY know about trees? For instance, did you know that...

1 The average tree contains enough wood to build a bungalow, or alternatively 2 fishing boats, 1500 kitchen doors, 600 million matches or one big telegraph pole.

2 Trees are a vital part of our environment, and much work has been done to protect and conserve them in recent years. As well as providing them with vaccination against disease, scientists can now tell us exactly how old a tree is — accurate almost to the day — by chopping it down and counting the rings in the middle.

3 At one time people used to make cigarettes from the dried leaves of the Tobacco tree. However, these days they can be bought from any newsagent, tobacconists or corner shop.

Trees

4 Scientists from the University of Umsk in Norway setting out on an expedition to study trees in the extreme Northern polar regions would be in for a big surprise. There aren't any.

5 Waiting for a tree to die and fall over can be a time-consuming business, as some can live for a long as 3000 years! It's not surprising therefore that lumberjacks prefer to chop them down themselves, using axes or saws.

6 A Family Tree is one which is considered suitable for children and adults alike, for example Christmas trees which are harmless can be decorated and kept indoors.

They live to the ripe old age of 3000!

7 If you walked through a forest looking for a Shoe tree you probably wouldn't find one. Ask a cobbler and you may have more luck. That's because shoe trees are funny shaped things that you put in your shoe.

8 There are over 6,000 different species of tree in the world, including Oak, Pine, Birch, Beech, Elm, Ash and Magnolia.

9 And Horse Chestnut.

10 There are more trees in the countryside than there are in town centres.

11 Make a 'trunk call' and you won't necessarily be connected to a tree. Nor will you be connected to an elephant. A trunk call is simply an operator controlled long distance telephone call.

12 Reverse Charge calls are nothing to do with trees either.

13 Trees and telephones are by no means the only things to have trunks. As well as elephants, human bodies, swimmers and people going on holiday are rarely seen without them.

14 There are two types of elephant — the biggest ones, which have got big ears, and the smaller ones, which haven't.

15 Unlike trees, the elephant can use its trunk to perform complex and often delicate jungle operations, such as peeling oranges and removing peanuts from their shells.

A tree

16 The tallest trees in the world are the giant Coast Redwoods at Humboldt County, California, the tallest of which measured over 367 feet in 1963 (so it's probably a lot bigger by now).

17 There aren't any elephants in California.

18 The largest elephant in the world, the bull African elephant, weighs in at around 12 tons, but at only 13 feet tall, it's not as high as the average tree.

19 It would take 28 gigantic bull African elephants standing on each other's backs, weighing a massive 336 tons, in order to reach the leaves on the top of the world's tallest tree. That's the equivalent of some 25 London buses, or well over half a British Telecom Tower.

20 It would take about 16 bull giraffes standing on each others' heads to reach the same height.

ELF JOKE

SO LONG AS YOU'VE GOT YOUR ELF, THAT'S THE MAIN THING!

HENDRIX LIVES SHOCK

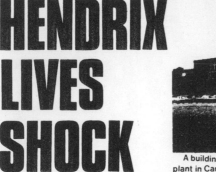

A building not unlike the Chikpan plant in Carlisle where Hendrix works

The rock world was last night reeling with shock at the news that sixties guitar legend Jimi Hendrix, thought to have died in 1970, is alive and well in Carlisle.

The story first came to light when van driver and heavy metal fan Wilf Roberts stopped to make a call at the Chikpan frozen chicken plant in the town. He says 'I noticed this tall skinny coloured guy working near the office when I went in. He looked familiar and then it hit me.... he was Jimi Hendrix.'

Wilf told friends in the pub that night but most laughed off the story. The following day armed with an album cover he returned to Chikpan.

'CHARLIE'

He discovered that Hendrix is known to his workmates as 'Charlie' and he has been working at the plant for eleven years. Hendrix, or Charlie, refused to answer questions and called a supervisor when Wilf confronted him with the sleeve from his 1968 masterpiece 'Electric Ladyland'. The supervisor threatened Wilf with the Police if he did not leave. Wilf left and called us.

DOORSTEP

We tracked the rock legend to his home in Boston Avenue and confronted him on the doorstep. Once more he refused to co-operate and went as far as to deny all knowledge of classic hits such as 'Purple Haze' and 'All Along the Watchtower'. He claimed never to

EXCLUSIVE

have visited the Isle of Wight or Woodstock, the scenes of two of his greatest performances.

EVIDENCE

But we can reveal exclusive and stunning evidence that proves that Charlie and Hendrix are the same person. Workmates we talked to reveal that the celebrity in their midst refuses to talk about his past and simply claims to have come from Guyana to London in the early seventies. Most conclusive of all is the fact that 'Charlie' has unusually long fingers and seems to have incredible strength in his hands.

GIBLETS

As one workmate who wishes to remain anonymous told us, 'I've never seen hands like them. He can rip the giblets out of a chicken in two seconds flat. He just sticks his fingers all the way in, gives his hand a twist and pulls the lot out. It's amazing.'

COVER UP

Reports coming from the plant even as this story is going to press suggest that a good deal is being covered up. Chief accountant for Chikpan, David O'Brien bears more than a passing resemblance to Buddy Holly and a greying, lanky figure in quality control proves, on close inspection, to be a dead ringer for the late Jim Reeves.

SPOKESMAN

A spokesman for Chikpan refused to discuss the case with us. He pointed out that it was not company policy to employ dead rock idols. However, he did not wish to answer questions based on the evidence that we have gathered.

HAVE YOU BEEN OUT PLAYING FOOTBALL IN YOUR NEW SHOES?

PISSFLAPS!

Rude Kid.

Look at all of these bottoms!

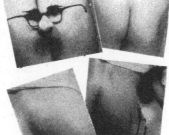

At long last and after several weeks of tension and excitement we have a winner in our fabulous Bottom Competition.

We asked you, the readers, to send us photographs of your bottoms. And send them you certainly did. We had literally eight entries from readers up and down the country, all of them hoping to get their hands on one of our glittering array of prizes.

WINNER

And the winner was Steve Williams from Manchester whose bum scooped our first prize: a packet of Sugar Puffs. Shown also are a few of our runners up. They are Mike McCormick of Manchester, Ivor Rump of London, Stephen Maden and Adrian Jones of Durham, Shaggy Yates of Newcastle, Pete Fanshawe of West Drayton and Julian Sominka of Portsmouth.

WINNERS

There were winners galore in our recent Reader Survey prize draw. Winners of the chance to star in a Viz Comic photo story were Andrew Watson of Sittingbourne, Alan Lawson of Tayside and Steve Johnson from Newcastle. Free T-Shirts went to Neil Stephen of Fife, K. Queef of Bow, East London and John Groeneveld of Rochdale. Lucky runners up with free Viz Comic subscriptions were Derek Walsh of Brighouse, John Farrar of Brighouse, Mr. D. James of Bournemouth, Kate Milner of Newcastle upon Tyne, Peter Jordon of Durham and Duncan Anderson of Aberdeen.

PRIZEWINNERS

"We're all absolutely delighted", said all our lucky prizewinners on learning of their success. "It only goes to show what terrific value Viz is, and how worthwhile the competitions are", said one afterwards.

Boobs getting bigger

Boobs are getting bigger. And that's official!

WOPPING

Britain's standard bra size is now a wopping 36B, compared to 34 inches ten years ago.

ENORMOUS

And boobs are positively busting out according to medical sources. For doctors tell us that female breasts actually increase in size during puberty.

GIGANTIC

We rang some top sexist comedians to get their zany reactions to our blooming bra sizes. But none of them were in.

Bridge Over Jamiroquai

LIGHTWEIGHT pretend funkster **Jay Kay** was at the centre of a planning storm last night after the Highways Agency attempted to serve a compulsory purchase order on him.

Plans for the new A267 dual carriageway linking Deal and Colchester was given the go ahead by Essex Council's planning committee in April, despite the fact that it would have to pass straight through the be-tit-fered Jamiroquai frontman.

At a public enquiry, an application by the McAlpine group to compulsorily purchase and bull-doze Mr Kay was turned down following a successful appeal

By our Pop Correspondent
Trafford Lovething

from the Sony record company. A spokesman said: "We argued that being flattened by a twelve ton demolition ball before being shovelled into a landfill site may have adversely affected Jay's musical career. Thankfully the committee saw sense and turned down the application."

flyover

However an ammended plan was approved. Essex County Council chief planning officer Mike Sausages told reporters: "McAlpine's secondary plan incorporated an eight foot high concrete flyover so the road could pass straight over the top of the 32-year-old star whose hits include *Virtual Insanity* and

Jay Kay yes-ter-day

When You Gonna Learn."

But record chiefs are set to appeal once again in the hope of derailing the road scheme. A Sony spokesman said: "We believe we have a good chance of stopping the construction if we can convince the Minister of the Environment to declare Jay a Site of Special Scientific Interest."

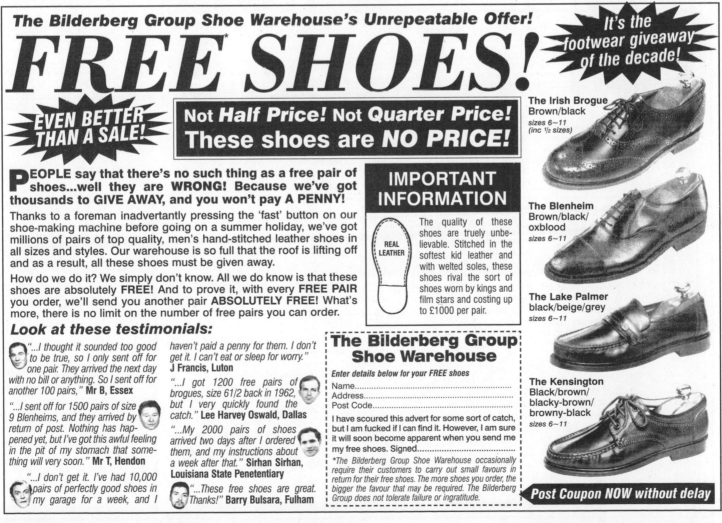

EVERY SECOND COUNTS

Scientists Warn of Bleak Future as Time Runs Out

EVERY DAY, newspaper scare stories warn us that we're exhausting the world's natural resources. Whether it's oil, gas, ozone, rain forests or fish, chances are we're using our dwindling supplies up like there's no tomorrow. But now, says a report from a leading university, there may not even BE a tomorrow. For if the boffins' calculations are correct, we are wasting TIME at such an alarming rate that by the year 2040 we will have run out of it altogether and everything will stop.

"People simply don't seem to realise that time is a finite resource," the report's author Arthur Author told us. "Unless we start saving it, and quickly, there will be none left for future generations to enjoy."

Cambridge professor Author lays the blame squarely at the door of our modern lifestyles. "Britons are using time up faster than ever before," he said. "The average person now wastes time at over twice the rate he did he did twenty years ago. These days, adults come home from work and fritter away hours on end doing Sudoku puzzles and watching repeats of My Family on UKTV, whereas their 1980s equivalents would have done something useful with their time, such as putting up some shelves, reading an encyclopedia or doing a paint-by-numbers."

END OF THE TIMELINE
As the Sands of Time Run Out, Here's a Brief History of the Life and Times of Time Itself

13 BILLION BC	Our Solar System forms	**200 MILLION BC**	Age of the Dinosaurs	**3 MILLION BC**	Stonehenge built by cavemen	**2005 AD**	Time runs out
The Big Bang	**2 BILLION BC**	First single-celled life appears on Earth	**65 MILLION BC**	Man evolves out of some monkeys	**100,000 BC**	Louis Walsh quits ITV's 'X-Factor'	**2040 AD**

NO TIME LIKE THE FUTURE

Computerised time egghead Stephen Hawking looks into the future to describe the Britain of 2040, when time has run out

❝ *The first thing that will strike a visitor to the timeless world of tomorrow will be how very dark it is, even in the middle of the day. That's because light travels to the earth from the Sun at a speed of 186,000 miles per second. Without seconds to travel in, the light will simply stay up in space, casting our world into a state of constant gloom. As a result, we will have to carry torches everywhere we go.*

Ordinary activities which we take for granted now will become impossible when time runs out. Our boiled eggs will always come out wrong, either too hard or too soft, as there will be no way to judge how long they've been in the pan, and setting the video to record our favourite programmes while we are out will be a nightmare.

On the plus side, our journeys will become much shorter, as trains, buses and planes will arrive at their destinations at exactly the same moment they set off. Unfortunately, you will have no way of knowing this as your watch will have stopped working.

And you won't be safe on the streets. With no time for them to "do", prison authorities will be forced to release thousands of murderers and rapists back into society. Gangs of leather-clad mutants, some of them with only one eye, will drive round our town centres in Scrapheap Challenge-style vehicles, killing and maiming indiscriminately in their never-ending search for torch batteries. In such a dark and dangerous Mad Max-style society, life expectancy will plunge. A child born in 2040 will be lucky to see its thirtieth birthday. ❞

© 2006 Professor Stephen Hawking

"I'll Waste as Much Time as I Like"

Says TV Motorhead Petrolmouth **JEREMY CLARKSON**

SO WHAT if time is going to run out in 35 years? My grandfathers fought in two world wars for my right to waste as much time as I like, thank-you very much. And no hairy-chinned lesbian carpet munching lefty environmentalists are going to stop me.

From now on I intend to put my socks on each morning using chopsticks. And on my daily commute to work I'm going to take the long way round. Via Timbuktu.

Not only that, but I'm going to continue wasting time when I get to work, by making endless television programmes in which I drive unbelievably impractical and expensive motor cars round an airfield. With all smoke coming off the tyres.

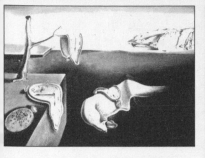

S.O.S.
SAVE OUR SECONDS
Timelords Declaring War on Waste

And children are no better. "The average child wastes more than four hours each day playing online games such as Runescape or Habbo Hotel," said Author. "His pre-internet counterpart would have used this spare time more constructively, perhaps making a kite, oiling his bicycle or learning to play the trumpet," he added.

Even senior citizens do not escape the professor's censure. "Old people are living longer than ever before, but they are spending the last twenty years of their lives complaining about the size of coins or writing doggerel poems about their cats for the Weekly News," he continued. "It all adds up to a colossal amount of wasted time which the world can ill afford."

Author believes that we are squandering one of our most precious natural resources at a frightening rate, and warns of a bleak future if we don't mend our ways. He told us: "We have to act now if we don't want our clocks to simply stop. If we keep on treating time like we have an endless supply of it, I estimate that our supply of months, weeks, days, hours and seconds will dry up some time in the next thirty-five years."

The professor says there are many ways we can all do our bit to conserve time. "Something as simple as drinking instant coffee instead of tea can save several minutes per cup, as you don't have to wait for the bag to mash," he told us. "And concrete your garden over. Countless valuable hours are wasted each year by people mowing their lawns, trimming their hedges and weeding their flowerbeds."

THERE ought to be enough time left to last us for ever, but thanks to our wasteful lifestyles we're in danger of running out of it within a generation. Nobody knows more about the value of time than the Timelords. So we asked the surviving Dr Whos to take time out from saving the universe to offer some advice about how we can save time itself.

"I think we should all stop buying Filet-o-fishes at McDonald's," booms former Dr Who Tom Baker. "Think about it. With a four-minute wait whilst each of the sixty-million filets sold annually is prepared, Britons spend nearly two million days each year hanging round the counter or sitting in the grill order parking space. That time is lost. We can never get it back.

"Think how much time we waste watching daytime television," adds former Dr Who Colin Baker. "When watching Noel Edmonds' new show 'Deal or No Deal', just tune in for the last couple of minutes. That way you'll save the twenty-eight minutes it takes for the programme to grind its way through its interminable format and reach the vaguely interesting bit at the end. Better still, save the whole half hour by not watching it at all."

"The time we all waste sitting on the toilet could be put to much better use," suggests former Dr Who Sylvester McCoy. "Having a shit and preparing a stew are both things that take about half an hour, so keep a chopping board and selection of vegetables in the bathroom. Next time you're paying a 'sit-down-visit', you can prepare a nice salad or a stew on your lap whilst you're on the toilet. Don't forget to wash your hands before you wipe your arse, especially if you've been cutting chillies."

"Ironing clothes that people are never going to see is a complete waste of time," says former Dr Who Peter Davison. "So I never bother ironing my underpants or vest. And I only iron the bit of sock which will be visible between the top of my shoe and the bottom of my trouser leg."

"Millions of us waste hours each Christmas putting up decorations, only to take them down a few days later," moans only just former Dr Who Christopher Eccleston. "If we all became jewish and threw out our trees and fairy lights, just imagine how much precious time we'd save."

FIFA Blows Whistle on Timewasters

FOOTBALL CHIEFS LAST night announced that they were joining in the battle to prevent time running out with a series of sweeping rule changes.

With effect from the start of next season:

- *Half time will be cut from 15 to 5 minutes*
- *Referees will clamp down on time-wasting, making it a red card offence*
- *Post-goal celebrations will be limited to 10 seconds*
- *Referees will be issued with dictaphones to shorten the time needed when taking players' details*
- *Pitches will be shortened to a new length of 20 feet to speed up the game*

FIFA president Sepp Blatter introduced the new rules at a press conference, demonstrating his dedication to the cause of saving time by breathing from a tank of helium so he could speak faster.

Meanwhile Serie A chiefs in Rome have vowed to limit their players' dramatic rolling around in agony after each tackle to 3 minutes instead of the customary 10.

TEN Things You Never Knew About Time

IT FILLS up our past, present and future; busy people don't have enough of it, yet today's youths seem to have far too much of it on their hands. It's neither big or small but it's all around us every minute of the day. It's time. But how much do we really know about it? Let's spend a few minutes looking at ten things you never knew about this fundamental part of our abstract conceptual framework.

Even though we can't see, smell, taste or touch time, we can hear it. That's because watches tick audibly once per second, whilst churches make a loud clanging sound every hour.

Although you can't smell or taste time, you CAN smell or taste thyme - small, perfumed stuff like posh salt that comes from the supermarket in plastic pots and little glass jars.

Just like a bus driver when you're running to catch his bus, time waits for no man.

'Time' was also the name of a shit musical written by the drummer out of the Dave Clark Five which starred Cliff Richard and a giant hologram of the late Laurence Olivier. That particular 'Time' would have waited for you, especially if you phoned the theatre and told them you were coming with a party of four or more people.

If you want to know what the time is, ask a policeman, or so the saying goes. However, unless your local bobby has a degree in theoretical physics or a working knowledge of the writings of Leibniz and Nietsche, you'd be better off approaching someone slightly better qualified.

Time is divided into intervals to make it easier to understand. Three of the main ones are dinner-time, tea-time and bedtime.

The first people to record time were the Greeks, but the first people to record time and a half were the ancient Egyptians, who came up with the idea when they were doing the pyramids.

In the olden days time was measured using an egg timer, an hourglass shaped vessel half-filled with sand which could only go for three minutes. However, a modern quartz watch is much more accurate and can go for six months before its battery goes flat and it gets put in the drawer. Although a coffee table is a table for putting coffee on, a timetable isn't a table for putting time on. It's a piece of paper pinned up in a railway station on which is printed a sequence of numbers that bear little or no relation to the movements of trains.

The world's smallest man, Calvin Philips, was also the possessor of the world's smallest wristwatch. It was so tiny that an hour lasted just a minute, and each second just two nanoseconds.

Miriam's Photo Problem Casebook

Suzie's Boyfriend Worries - Day 3

Suzie has been dating her boyfriend Dave for a year, but recently he has been having problems operating his 35mm camera. She turns to her best friend Claire for advice...

In Suzie's flat...

What's wrong, Suzie? You look worried.

Oh, Claire! It's Dave. I really love him, but whenever he takes a photo of me he puts his thumb in front of the lens...

...either that, or he never gets the focus right and they come out all blurred. I don't know what to do.

Lots of men have that trouble, Suzie...

...come on, have a shower, and we'll go out for a drink.

Shortly...

There! Is that better?

Yes, thanks! I feel...

...Oh, no! He hasn't framed the picture properly... the top of my head is cut off!

In the pub...

I'm sure things will work out between you and Dave, Suzie.

I'm not so sure...

...he's forgotten to turn the red-eye reduction on, **and** he's used too slow a shutter speed causing motion blur...

...Oh, Claire! What am I going to do?

CONTINUES TOMORROW...

Have Your Say

Two years ago, trigger-happy farmer Tony Martin shot dead an intruder in his home and was convicted of manslaughter. Instead of a two year jail sentence, the tabloids believed he should have been given a medal. Every week, the papers are full of stories of burglars murdering innocent householders in their beds, so is it finally time to change the law? Is it time parliament gave us the green light to murder them first? We went on the street to find out what YOU think...

...I BELIEVE that an Englishman's home is his castle. As a consequence, I keep a cauldron of boiling oil by the bedroom window at night, and I wouldn't hesitate to tip it onto anybody trying to break into my house.

Peter Scott, dentist

...BURGLARS are vermin, and if I caught one burgling my house I would treat him like the vermin he was and hold his head in a cup of water.

Mike Cable, chiropodist

...I AGREE with Mr Cable that burglars are vermin. But they are still God's creatures, so if I caught one, I would drive him some distance from my house and let him go in some woods.

Frank Mozart, osteopath

...I KEEP a cricket bat by my bed in case a burglar breaks into my house during the night and throws a cricket ball at me.

Hector Strauss, trichologist

...I AM a barrister, and I have simply attached a set of terms and conditions to each door and window in my house. They state that 'any party or parties attempting to gain unlawful entry to these premises is likely to be attacked and/or killed as deemed necessary and in any way deemed necessary by the householder and/or his agents or appointees without redress, and that breaking into the premises constitutes full acceptance of these terms'.

Mike Mansfield, barrister

...MY HEART goes out to Tony Martin. Not only was he prosecuted for defending his home, but he had to spend two years in prison surrounded by burglars that he couldn't shoot. He must have felt like a kid with no money in a sweet shop.

J Wells, trichologist

...HOME owners often feel powerless to protect their homes, but allowing them to kill burglars is the thin end of the wedge. Perhaps the Home Secretary should look for a halfway house and take a leaf from the 'Carry On' team's book and allow the shooting of burglars with a blunderbuss. In Carry On Camping, Terry Scott gets shot in the buttocks with such a weapon and has to suffer the indignity of his wife extracting the pellets with tweezers. The removal of each pellet is accompanied by a swanee whistle noise, followed by a loud pop and a howl of pain. Surely, the thought of this happening would act as a deterent to any burglar.

T Rothwell, trichologist

...MY WIFE recently had the heart-stopping experience of waking up one night and coming face to face with a burglar on the stairs. She was absolutely hysterical until I calmed her down and reminded her that it was me, her husband Handy Andy off 'Changing Rooms'.

H Andy, odd job man

...I STRONGLY oppose the majority that thinks we should be allowed to shoot thieves. I work for a top London advertising agency, and it would look like the St Valentine's Day massacre in here every day.

Dickey Beasley, creative

...IF IT does ever become law that you can shoot people who steal things from you, then could the woman who wrote that short story about the cat, which ended up in one of Jeffrey Archer's books please get in touch. I would love to watch when she puts a cap in the lying bastard's thieving arse.

Steven Marbles, masseur

...I DON'T know what the world is coming to. A man can break into a pensioner's house, murder us with a baseball cap and get off with a slap on the wrist. Meanwhile, the innocent householder is sent to prison for burgling his own home. Eeeh! It's a disgrace.

Ada Brady, grandmother

...I OWN my own home, and I recently broke into a prison to steal a television from a cell. Whilst in the process, I was attacked by an inmate who was serving two years for burglary. In order to defend myself, I had to use excessive force. The upshot of all this? The burglar was released, and I, a householder, was sentenced to six months in prison. What a topsy-turvy world we live in.

J Harper, dentist

...I REGULARLY shoot burglars, but I am never prosecuted. That's because I teach a human cannonball course at Risley Remand Centre.

Keith Stromboli, human cannonball

The MAN in the PUB

Britain's most ill-informed columnist.

Guess who's got AIDS

☐ GUESS who's got AIDS. Go on then, 'ave a guess. Alright, I'll tell ya. *Prince Andrew*, that's who. Yeah! It's true. This posh bloke told a mate of mine. God's honest truth. They're coverin' it up. Gonna say it was a kidney infection when he pegs it. Or leukemia. Just you wait an' see.

☐ YOU KNOW that Patsy Kensit's ditched her fella, don'tcha. That bloke out of Simple Minds. But do you know why? *It's great this is*. Mate o'mine in the music business told me. Guess what? **He wears a nappy!** Yeah, great big nappy. That's why he keeps loosin' his birds. It's true. Can't control the old waterworks, apparently. Mind you, that Patsy Kensit, eh? *Phooaaar! I would*, I can tell ya. *Cor!* Not 'alf.

Same size heads

● Did you know that your head never grows? Ever. Stays exactly the same size all through your life. Think about it. You look at any baby's head, right. Exactly the same size as yours or mine.

☐ THIS mate o'mine's got a garage, right. Guess who comes in the other day tryin' to sell him knocked off car radios. Only *Gazza*, the footballer. Yeah! What a bloody nerve. **Fifty grand a week** he gets for kickin' a ball about, and he still goes out nicking car radios.

Dungeon under house

☐ I'll tell you what. You know that little baldy Labour bloke, Gerald Kauffman, the MP? He's got a *dungeon* under his house, he has. *Yeah!* A fuckin' dungeon. They reckon he tortures people in it, an' then he kills 'em. Probably eats 'em an' all. Wouldn't surprise me. Mate o'mine's into bondage an' all that. Says that's *definite*.

That bloke Kauffman

☐ Amanda Donoghue, right? Actress? Fridge full o'spunk. No, straight up! She keeps it all in them little plastic film canisters. Got a fridge full of the bleedin' stuff. God knows what she does with it though. Bloody screw loose there if you ask me, mate.

Apollo nonsense was bollocks

YOU KNOW all that Apollo rocket nonsense? Bollocks that was. They made it all up. Never went to the Moon. Filmed it in a studio somewhere in America. In the desert it was. It's all shot in slow motion. Apparently, the bit where they land, *if you look close enough you can see a telegraph pole in the background*. Clear as day. Mind, I shouldn't really be tellin' you this. They shot the cameraman afterwards. Made it look like an accident. Knew too much, y'see.

One of them space rockets

Pigs CAN'T swim

☐ YOU EVER seen a pig swim, eh? Think about it. No mate, you 'aven't. Know why? They **can** swim, right, but they **can't**, you see. Cos if they **did** swim they'd cut their throats. Straight up! It's the shape o'their trotters. If they swam they'd cut themselves to ribbons an' bleed to death. *And you know what?* An English pig can't shag an Australian pig. Impossible. Cos their cocks and their fannies, right, go round and round y'see. Twirly, like. And pigs from the *north* of the world, their's go round one way, and pigs from the *south* go round the other way. Like *clockwork* an' *anti-clockwork*, you know. It's true that. You ask a farmer.

This bird had no knickers on

■ I was in 'ere the other night, right, an' this bird walks in... *fuck me!* She was *gorgeous!* An' I'll tell you what. She had no knickers on. You could tell by the way she was standin'. Givin' me the eye all night she was. *Phooarr!* Anyway, whose round is it?

Is this the end for Hugo Guthrie?

UNDER siege Tipton councillor Hugo Guthrie was facing calls for his resignation last night despite an apparent apology for his behaviour in the so-called 'Razzlegate' pornography affair.

Addressing members of the Civic Amenities Committee yesterday Mr Guthrie veered away from his prepared speech about glass recycling skips and told a hushed audience that he "sincerely regretted" having let the council down.

Pressure Ron - Jazz crisis councillor Guthrie yesterday

Lied

Last week Mr Guthrie, an independent conservative councillor, denied having lied to the Lord Mayor when questioned in the Town Hall car park.

Rumours

Rumours began circulating last month after cleaner Mrs Gladys Wilkinson told her husband, who is the Lords Mayor's chauffeur, that she had seen dirty magazines in a cupboard in Mr Guthrie's office. When questioned by the Mayor, labour councillor Alderman Frank Peabody, Mr Guthrie denied having any pornographic material.

Tusk

A lot hinges on the councillor's definition of the word "pornographic". Questioned by the Mayor, Guthrie denied having any pornographic magazines in his office. However two days later, during a game of golf, he admitted that he may have had certain "inappropriate photographically illustrated reading material" in his possession.

Wank mag stash allegations will not go away

Last week tea lady Mrs Bradshaw rocked the borough with her allegations in the Tipton & Smethick Post that she had once entered Guthrie's office and caught the councillor kneeling on the floor in a compromising position with his trousers down and a pornographic magazine open in front of him.

Yesterday Mr Guthrie's wife Vera was standing by her man and the couple appeared relaxed as they arrived in Smethick on a shopping trip. But despite his wife's support Mr Guthrie is now facing intense pressure to resign as Chairman of the Civic Amenities Steering Committee.

Trunk

And the odds on his political survival lengthened last night when it emerged that Gladys Wilkinson, the cleaner who made the original allegations, has now produced vital evidence which could prove her claims. A copy of Razzle magazine with several pages stuck together was night night being studied by the Lord Mayor and senior councillors.

Now for the LATE news

'BBC Newsnight', the last news flagship to be built in the UK, takes shape at the Swan Hunter yard in 1980.

ITV's television news flagship, due to be launched in the autumn, may not be completed on time.

The multi million pound 'ITN 6.30 News', which is due to replace the ageing 'News At Ten' when it is taken out of service later this year, is still in a German shipyard where work on the programme has been delayed to due a series of industrial disputes.

Suitcase

The new show was due to be launched and undergoing television trials by September, and was scheduled to be handed over to Trevor McDonald and the ITN crew for commissioning in early October. However latest estimates suggest that the programme is up to three months behind schedule.

Briefcase

Many TV reporters and film crews are already booked to appear on the programme's maiden episode. A spokesman for ITN reassured them that the programme would still be launched on time.

"It's true there have been some set backs, but the builders have assured us that the launch will still go ahead as scheduled. News stories will not be affected, however there may still be some minor fitting out work going on while the show is being broadcast".

Nutcase

The future of the 'News At Ten' is uncertain. Despite protests from Prime Minister Tony Blair and others, ITN have confirmed that the programme is to be decommissioned. A Japanese TV company are reportedly interested in turning it into a floating cookery programme, but if a sale cannot be agreed the show could end up being sold to a scrap yard in Pakistan where swarms of poor people with no shoes on would descend on it and dismantle it by hand.

Headcase

Meanwhile a Turkish TV order for three daytime chat shows and a gentle sit-com has guaranteed the jobs of 200 workers at the Harland and Wolf yard in Belfast for at least two years.

THE BUSTER BLOODVESSEL STORY - PART 1

Balham Hospital. Jan. 14th, 1961...

Congratulations, Mr. and Mrs. Bloodvessel- it's a boy! A big, fat, bald bouncing baby boy!

We'll call him Buster

Buster! Don't slouch. And take your elbows off the table...it's bad manners!

From an early age, Buster dreamed of only one thing- Pop Stardom!

I wish I could think of a name for my band. Then we'd get famous and go on Top of the Pops.

Wait a minute!... That's it!!

And 10 years later his dream came true, when he appeared on Top of the Pops in a big dress

TOP OF T' POPS

Lip up, Fatty! Oh, lip up, Fatty! Fatty reggae!

Part 2- As the hits dry up, Buster hits the cake shop!

NEWS in brief

Fanny a'had

TWENTY firemen fought for two hours yesterday to extinguish a fire in Catherine Zeta Jones' fanny. The fire, thought to have been caused by a discarded cigarette, was eventually brought under control after extra fire fighting appliances were called in to tackle the blaze.

Robbers raid tits

PROPERTY worth an estimated £400 has been stolen during a raid on TV weather girl Suzanne Charlton's tits. The raid was the fourth break in at Charlton's tits this year.

£20 arse damage

A VANDAL who attacked an arse belonging to TV Lovejoy actor Ian McShane caused damage estimated at £20, Sudbury magistrates were told yesterday.

The bumhole, which had been left unattended outside the actor's home, was allegedly kicked and suffered bruising and scrapes.

William Patterson, 21, an unemployed painter and decorator from the town had denied kicking the arse down the street, claiming that he had tripped on it accidentally while drunk. However he was found guilty, fined £200 and ordered to pay £20 damages.

Old fart

A FART thought to have been dropped by Queen Victoria couldfetch over £100,000 when it is auctioned at Sotherbys later this month.

The fart, which is thought to have been emitted over 100 years ago, was accidentally uncovered by staff puffing up an old cushion at Balmoral earlier this year.

The previous record sale price for a fart was £27,000 paid by a Japanese collector for a chuff emitted by the late Charlie Chaplin. It was discovered trapped in a small air pocket by decorators stripping old wallpaper at his former residence in Switzerland.

X-TERMINATE!

It's the acks for 'X' in alphabet reshuffle

The letter 'x' could soon be facing eXtinction if Brussels bureaucrats get their way.

For Euro spelling chiefs have decided to cut the alphabet to 25 letters to bring it in line with decimal currency. And its 'x' which seems certain to face the aXe.

Axing 'x' could save Europe up to £200 billion as a result of smaller typewriters. However alphabet bosses are undecided as to how it should be replaced. Words with 'x' in them like 'box' and 'expect' would have to be respelled, and the cost could run into millions.

Pixie

The most likely replacement for 'x' would appear to be the letters 'cks', producing much the same sound. The word 'pixie' could then be respelled 'picksie' and so on. However replacing one letter with three could cost Europe over £7 billion in extra ink alone..

French boffins have put forward their own plan, claiming that the letters 'sque' would be an adequate replacement. But a switch from 'x' to 'sque' would pose enormous problems. Words like 'sex' would then become 'sesque' - too much of a mouthful for the British who are still having trouble with croissant.

Space

One of the worst affected areas would be in the game of Scrabble. It would cost manufacturers Waddingtons over £400 million a week to replace the single 'x' in every game of Scrabble sold with an extra 'c', 'k' and 's'. To add to the problem the 3 replacement letters would score 12 points, compared to the existing 'x' score of just 8.

Alarm bells were also ringing at Littlewoods Pools yesterday. A spokesman admitted that even people with very small writing indeed couldn't get three letters into the tiny squares on pools coupons. "Making the squares bigger would cost us over £5 billion a year", a spokesman told us. "And it would be virtually impossible to Cross the Ball with a 'cks'. Not that its possible with an 'x' ", he added.

House

But a Government spokesman was yesterday keen to play down the potential problems of the move. "Dropping the letter 'x' will save money and bring letters more in line with money. By the year 2000 the date will be divis-

Ruel Fox inside the 18 yard box.

ible by the number of letters in the alphabet, and money will be the same as well. And that has got to be good for Britain". He refused to be drawn on the question of whether some 'x's would be retained for use in multiplication.

20 EXCITING THINGS YOU NEVER KNEWED ABOUT X

We use it for sex, we use it for sums, we can even use it to mark the spot. Our Alphabet Correspondent Chaka Khan has been eXamining a letter that's set to become eX-dictionary.

1 'X' was invented by the Romans, but not for use as a letter. They used it as a number, instead of 10.

2 It was introduced to Britain during the Thirteenth Century, probably arriving on a ships' clock.

3 Its first recorded use as a letter was during Saxon times, when cave men painted it on the wall to help them find their way out in the dark. Hence the word 'exit'.

4 'X' was was once a letter fit for a king. Before Henry the Eighth came to the thrown male monarchs were called 'Rexes'.

5 In 1640 Henry ordered every dog in the Rexdom to be beheaded, and ordered that the word

King Henry VIII 355 years ago.

But Henry disliked the term, often complaining that it made him sound like a dog.

'King', which had previously been a kind of prawn, be used as a replacement for Rex.

6 Traditionally the letter 'x' is used to signify a kiss on greeting cards, along with 'o' which represents a cuddle, or a hug. (And 'SWALK' written on the back of the envelope means 'sealed with a kiss'. And 'HOLLAND' means 'hope out love lasts and never dies'.)

7 But if a pirate like Blue Beard or Little John Silver rote 'x' on a piece of paper he wouldn't be giving anyone a kiss. He'd be marking where the treasure was.

8 Until recently the symbol 'X' was used by film censors to indicate an

interesting film with either sex, violence or bad language in it.

9 However the symbol 'XXX' does not indicate all three. This denotes an overpriced video featuring blurred and out of focus pictures of women pretending to have sex with each other.

10 Whilst 'XXXX' is used to denote lager that tastes like piss.

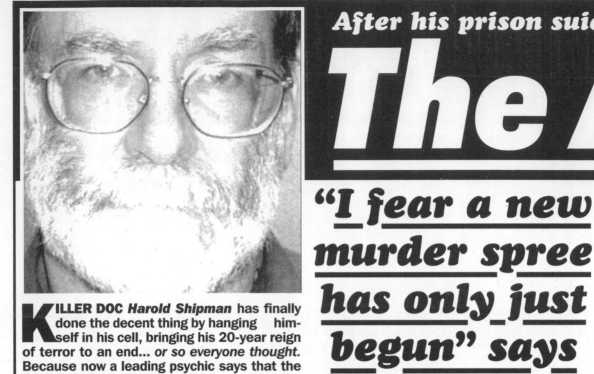

After his prison suicide, is it 'bu

The An

"I fear a new murder spree has only just begun" says psychic Doris

KILLER DOC *Harold Shipman* has finally done the decent thing by hanging himself in his cell, bringing his 20-year reign of terror to an end... *or so everyone thought.* Because now a leading psychic says that the bearded medic's death may only be the start of a new murder spree... this time *in the afterlife.*

Spiritual medium Doris Mandeville fears that dead Shipman has wasted no time in setting himself up as a GP in Heaven. And now, with hundreds of late old women on his books, he's already getting up to his old tricks.

She told us: "If nothing is done to stop him, the bodies will soon be piling up in heaven just like they did in Hyde."

Mandeville's suspicions were first aroused during a

EXCLUSIVE!

routine seance at her Doncaster home.

gossip

"I was having a chat with my red indian guide Billy Two Rivers. He was filling me in on all the gossip from the other side when he mentioned that the late Thora Hird had died unexpectedly.

"I was surprised, because I'd been chatting to Thora

on my ouija board a couple of days earlier and she seemed full of beans. Billy

told me that she'd fallen off her stairlift and had been found dead by her doctor.

Nelson's

Doris was immediately suspicious. She told us: "I knew there was no way in this world or the next that Thora would fall off her stairlift. She was an expert who had appeared in over 5,000 Churchill adverts and she knew what she was doing."

But alarm bells began to ring when Billy told Doris the name of Thora's GP. *It was Harold Shipman.*

A week later Doris was making lunch when she made contact with Two Rivers again.

"I was having trouble getting a souffle to rise, so I asked Billy to get in touch with my old friend Fanny Craddock for some advice.

You Only Die Twice

ACLOUD OF SUSPICION hangs over a series of mysterious sudden deaths which have taken place at Dr Shipman's Heaven surgery, says Doris. Here's just a few of the perfectly healthy dead people who have suddenly shuffled off their immortal coils whilst visiting the evil late GP's spirit.

Queen Boadicea

The legendary warrior Queen of the Iceni people killed herself in AD61 after suffering defeat at the hands of Roman invaders. Two weeks ago, she visited Shipman in Heaven complaining of piles. The doctor made a routine housecall the next morning, and Boadicea was found dead - apparently from a massive heart attack - when her husband Jules Verne returned home from work that evening.
Verdict: *Natural Causes*

Florence Nightingale

The bedridden Crimean War veteran who died in 1910 was a popular figure about Heaven, where she was known as 'The angel with the lamp". She was found dead once more just two hours after a home visit from the doctor. Shipman had been called in by neighbours

Rod Hull and Emu, who had seen her earlier that morning. Nightingale told them she was concerned she might be getting shingles. "Other than that she seemed fine," Hull told an inquest, just before Emu grabbed the coroner's testicles and wrestled him to the ground.
Verdict: *Open*

Mrs Mills

The seventies pub piano favourite died in 1982 of following a series of strokes. In early February of this year she felt twinges of arthritis in one of her wings and made an appointment to see the late Dr Shipman. He gave her a pain-killing injection and she returned home. Later, complaining of dizziness, she took to her bed where she was found dead next morning.
Verdict: *Natural Causes*

Pat Coombs

The Celebrity Squares dullard had only been up in Heaven for a few months when she visited Shipman's surgery for a routine smear test. According to Shipman, she started becoming short of breath. She became hysterical and started to panic like she did in On The Buses when Stan put spiders on the bus, quickly turned blue and died.
Verdict: *Open*

Dandy Nicholls

The Till Death Us Do Part "silly old moo" called in at the doctor's surgery to have her passport photograph signed. According to Shipman, he went to his cupboard to get a pen and when he turned round she was dead, apparently of a massive heart attack.
Verdict: *Natural Causes*

However, he got to her house in Heaven just in time to see her coffin being carried out of the front door.

Fanny

"Her husband, the late Johnny Craddock, explained that Fanny had re-died unexpectedly whilst visiting the doctor for her yearly 'flu jab. My blood ran cold when Billy told me the doc-

el of Death!

Doris (left) and (main picture) at one of her seances, yesterday

tor's name. It was the late Harold Shipman again.

The unexpected deaths of Thora Hird and Fanny Craddock were bad enough. But nothing prepared Doris for Billy Two Rivers' next bombshell. HRH the late Queen Mother had also just "dropped dead" again, whilst having her bunions shaved at the doctor's surgery. *Mandeville didn't need to be told the doctor's name.*

"To lose the country's favourite granny once was bad enough, but to lose her again - and this time to a dead mass-murderer like Shipman - was the last straw. I decided to go to the police with my suspicions."

pancake

But whilst Doncaster Police were concerned when they heard the medium's horrifying story, they were powerless to act. Doris told us: "They said that they could only investigate crimes that took place on this side of the astral veil. They assured me that all officers had been briefed on my allegations and that the first one to die would launch a thorough investigation as soon as he got to the Pearly Gates."

Liz Hurley

But Doris was far from satisfied with the response.

"That's not good enough," she stormed. "We could be waiting for ages, and all the while Shipman's on another killing spree in the afterlife. Surely the police have got at least one detective who's terminally ill or at least depressed who wouldn't mind doing himself in and going up there to catch him red-handed."

"And this time they should hang him. And I'll kill myself and pull the ruddy lever," she added.

Angels with Dirty Secrets

WITH NO PRISONS in Heaven, even the worst criminals are free to roam the streets of Paradise, rubbing shoulders with angels who've never broken a law in their lives. So what has happened to Britain's late killers? We ask...

Where Are They Now?

Cromwell Street murderer **Fred West** has a thriving building business. He recently put an extension on a cloud for *Eric Morecambe*, and erected a carport for *St Thomas Aquinas*.

Cockney killers the **Kray twins** run a successful chain of perfume shops, called *Heaven Scent*. However, dead police believe that this business is merely a front, laundering cash for a series of drugs cartels.

Gay US cannibal **Jeffrey Dahmer** found it hard to fit in when he got to heaven. After working as a pub cellarman, night watchman and singing telegram he became a busker. He can now be found most weekdays, playing his banjo for coppers outside the Pearly Gates.

19th century prostitute slasher **Jack the Ripper** now works as a salesman in one of Heaven's largest carpet warehouses. These days he is reluctant to talk about his killing spree in the backstreets of Victorian London. *"Those murders were a long time ago,"* he told us. *"And I'm not going to say who I am, either. That was a secet I took with me to my grave."*

Bloodthirsty **Vlad the Impaler**, personally responsible for over 30,000 deaths in 13th century Transylvania, is now a reformed character. He spends each day selling brushes and household cleaning goods door to door, and in the evenings teaches information technology at a local Further education College.

Genocidal Nazi **Adolf Hitler** has certainly calmed down since his Nuremberg Rally heyday. A neighbour told us: *"We don't see much of him to be honest. He keeps himself to himself, but he always smiles and says hello if you walk past when he's mowing his cloud."*

Sex-bomb killer **Ruth Ellis**, unlike other dead murderers, is safely behind bars in Heaven. But she's not in prison - she pulls pints at *Cloud 9*, a swanky wine bar owned by 17th century diarist *Samuel Pepys!*

Littlejohn

"Afterlife should MEAN afterlife"

Page 36

Sex Times Table

Division Belle: Birds like Carol Vorderman would rather go forth and multiply than have it off.

WOMEN prefer multiplication on their own to multiple orgasms with their fella. They would rather spend their evenings working out sums on sheets of paper than having a sexy workout between the sheets.

A new study asked women whether they preferred maths or nookie, and an astonishing 95% chose sums over rumpy pumpy.

love eggs

And it's good news for ladies' calculator manufacturers as sales of the miniature electronic adding machines look set to outstrip dildos, clitoral stimulators and Japanese love eggs by ten to one. A spokesman for Casio told us: "Today's woman just can't get enough of long division, algebra and co-ordinate geometry. Our calculator factory is working overtime, but we just can't keep up with demand."

Calculus Beats Coitus every time, say Ladies

Even celebrities are getting in on the act. Madonna was spotted buying a book of logarithm tables, a propelling pencil and some graph paper in Ryman's on Sunset Boulevard, whilst Sharron Stone last week announced that she has worked out pi to 300 decimal places.

adore bacon

Meanwhile, it was recently reported that the marriage of Hollywood golden couple Brad Pitt and Jennifer Aniston hit the rocks after the former *Friends* star began spending all her evenings working on a solution to Fermat's last theorem.

Stars' relief as Lumley monoped rumours quashed

THE WORLD of showbiz breathed a sigh of relief late last night as rumours that one of Joanna Lumley's legs had been amputated were scotched. Stories that the veteran *Sapphire and Steel* actress had lost her left leg above the knee had begun circulating in the late afternoon, and were greeted with horror by celebrity insiders.

However, at a hastily-called press conference in London's swanky Grosvenor House, a smiling Lumley sipped a glass of water before standing up in front of reporters to reveal two healthy legs which were clearly complete and real from above the knee to the foot.

"As you can see, both Joanna's legs are in great shape," said her agent Fellatio Nelson. "We don't know why or how these rumours got started, but we're happy to set the record straight."

And he made a plea that the popular actress be left alone to get on with her career.

Meanwhile fellow celebrities were delighted at the announcement. "This is marvellous," former co-star David McCallum told us when we broke the news. "It's been a very worrying few hours for all her former co-stars. We've all been praying for her to have two legs, and it seems as if our prayers have been answered."

AAARGH! AAARGH! EEEK! OH MY GOD! WE'RE ALL GOING TO DIE! AAARGH!

NEXT DAY...

DID YOU SEE THAT THING ON THE TELLY LAST NIGHT? IT WAS ABSOLUTELY HYSTERICAL.

READ & LEARN
WITH NAYLOR HAMMOND, Bsc.

This week Naylor explains

Botany

As you probably know, the earliest examples of botany, like the Mary Rose, arrived in England by ship. Nowadays we look to botany to provide us with coffee, tea and many popular brands of cigarettes, which unlike money, grow on trees alongside fruit, nuts and chocolate.

There are of course three main types of botany: Zoology - the botany of animals; Biology - the botany of things that grow on trees; and archeology - another kind of botany.

We are all familiar with flowers like Tulip or Herb which grow in our gardens. But flowers, like the acorn soon grow into large trees. This complicated 'life cycle' (a botanical form of transport) has fascinated botaneers for many years. The film 'Ghandi' told how early botaneer Sir David Attenborough developed his theory 'you are what you eat'. He discovered that cows eat flowers, which in turn eat vegetables. Man eats cows with vegetables, and so the 'food chain' continues. Latter day food chains include Liptons and Sainsburys.

Unlike animals which we see every day in the zoo, vegetables are a kind of underground fruit which live in the soil and drink water. As well as providing us with mineral drinks like pepsi and cresta, we rely on vegetables for butter, margarine and even cosmetic products (a kind of space food).

Nowerdays, despite the advent of takeaway food and battery operated agriculture, botany remains a popular leisure activity. Small flowers, together with useful information about trees can be obtained from larger branches of the Post Office.

See you next time

CD/JS 4.84

Naylor Hammond Bsc.

ON YOUR FRIGGIN' BIKE –
BASTARD!

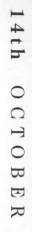

'TIDDLER' IS K.O.'D
– It's A Red Card For 'Horrible Half'

By Sun Reporters

"Good Ridance!" That's how Britain has greeted the news that the 'tiddler' is on the way out.

AXED

The hopeless half pence piece measuring less than three quarters of an inch in diameter, is to be axed, according to bank chiefs.

Talking Sense
With
Charlie Pontoon

Congratulations to the police for the way they handled the Libyan Embassy siege. There was no way they could make an arrest in such a potentially explosive diplomatic situation, and they all did us proud. But why the hell wasn't the bastard's plane shot down the minute it got off the ground. The RAF have got a hell of a lot of explaining to do.

★ ★ ★

Who does Bobby Robson think he's kidding? His team couldn't score in a brothel with tenners tied around their choppers.

★ ★ ★

So the Russians aren't taking part in the Olympics this year. Big deal. The Olympics are all about winning, not taking part, and we can do that perfectly well without the ruskies and their freak show sex change athletes.

BANDITS AT 1-O-CLOCK

RADIO '84

OH GOOD, I'LL HAVE A TOPIC WHILE I'M WAITING

SORRYTHISISAGAPSORRYTHIS
ISISAGAPSORRYTHISISAGA
APSORRYTHISISAGAPSORRY
SORRYTHISISAGAPSORRY
RYTHISISAGAPSORRYTHIS

SAMMY
AND HIS STAMMER

EX...EXC_ EXC...

yes, sir

I...I...HU...H-HATE...

I...HU...HATE...C_C_C

COP...I_HU...I_HU_C

hmm... yess...hmm

... yess...

F...F...FUCK OFF!

SD '83

20 THINGS YOU NEVER KNEW ABOUT KNACKERS & PISS

Call them plums, knackers, nuts or bollocks, there's no getting away from your goolies or gonads. "Balls" you may say. But love 'em or hate 'em, those two testicles tucked away in your trousers are here to stay.

But how much do we *really* know about our nuts? Here's twenty testicle titbits, or as many as we could think of anyway, and a few more fascinating facts about piss to make up the numbers.

1 Despite the fact that there's plenty of room for them inside most bellies, fellas may have noticed their knackers dangling precariously in a little bag between their legs. That's because knackers, like white wine, need to be kept cool. So nature has provided men with a little pink hairy 'outhouse' in which to keep them, called the scrotum.

2 The human knacker is the most sensitive piece of equipment known to man. More sensitive even than the Hubble space telescope. Sportsmen have therefore devised many ways of protecting their knackers. Footballers, for example, use their hands,

Left to right, Dame Vera 'sang about Nazi knackers', Belinda Carlisle 'froze her piss', and a motorcyclist similar to Barry 'klackers knackers' Sheen.

7 And poor old Goebels, had no balls at all.

8 Nailing one's knackers to a table for purposes of sexual gratification would be the action of... a knacker! That's because the word knacker has several meanings, and can also be used to describe a twat.

9 The word twat also has more than one meaning. However, twat cannot be used to describe a knacker in the bollock sense of the word.

whilst cricketers use a 'box'.

3 Sumo wrestlers on the other hand crumple them up and shove them up their arses.

4 Motorcycle racing idol Barry Sheen once knackered his knackers on the petrol tank of his bike during a 180 miles per hour crash. But he had them rebuilt using 'Klackers', the popular seventies 'bouncing balls on a string' game.

5 Hitler, the Nazi dictator of World War Two fame, had only got one ball, according to Vera Lynn's popular wartime ditty. The other, was the in the Albert Hall.

6 Himmler, had something similar.

10 A knackers yard is not, as you may suppose, an enclosed space near a building for the storage or specific use of testicles. Nor is it a distance of three feet calculated using nuts instead of a ruler. In fact, it is a yard where horses are taken to be turned into glue and dog food.

11 Rocks is another word for knackers with a multitude of meanings. But beware of bartenders offering you a "Scotch on the rocks". For rather than offering to pour whisky on your genitals as you may have expected, he is merely offering to put ice in your drink.

12 Beware the offer of ice in your drink the next

time you visit the home of singer Belinda Carlisle. For any mix-up in her fridge freezer could have unsavoury results. The former Bangle makes ice cubes out of her piss, according to a book by Piers Morgan of The Sun.

13 So it must be true.

14 If someone tells you they've been "pissed off", they don't necessarily mean that someone has stood on top of them and urinated onto the ground. They are more likely to be using the slang expression meaning 'to be fed up'.

15 If someone tells you they've been "pissed on" they're probably referring to a gig by the popular music group The Damned. During the late seventies a certain member of the group

gained notoriety by urinating onto the heads of the audience.

16 Ask a German plumber for a "golden shower" and he won't install an expensive shower unit in your bathroom. He'll piss on you instead. For in Germany and certain other sexually liberated corners of Europe, 'golden showers' refers to the sexual act of urinating on a partner.

17 Frankly, nothing these people do surprises me anymore.

18 If you are 'on the piss', you won't be found standing on a frozen pond of urine. In fact you're more likely to be kneeling in a warm pool of your own vomit. For being 'on the piss' means going out and drinking too much alcohol.

19 If someone tells you they've 'pissed on their chips' don't worry. They haven't found a cheaper alternative to vinegar. The expression simply means that they have cooked their goose.

20 However, if a soldier tells you he's 'pissed in his boots', he means just that. For squaddies traditionally fill new boots with urine in order to soften the leather.

21 They also wank onto cracker biscuits, apparently. But that's another story...

THERE WAS A LOT OF NOISE COMING FROM YOUR SHOE LAST NIGHT, SINGING AND DANCING 'TILL TWO IN THE MORNING.

YES, I'VE GOT A CLUB FOOT

Are you a PIMP or a SCIENTIST?

THE LINE THAT CANNOT LIE

Hustler or Egghead? Huggy Bear or Einstein? Which best describes YOU? Are you fluent in Technobabble or Jive ass? Do your bitches turn tricks on the street or do they smoke cigarettes chained up in a laboratory? Do you spend your day at the controls of a cyclotron or a Cadillac? Only by answering the questions with HONESTY will you discover the TRUTH.

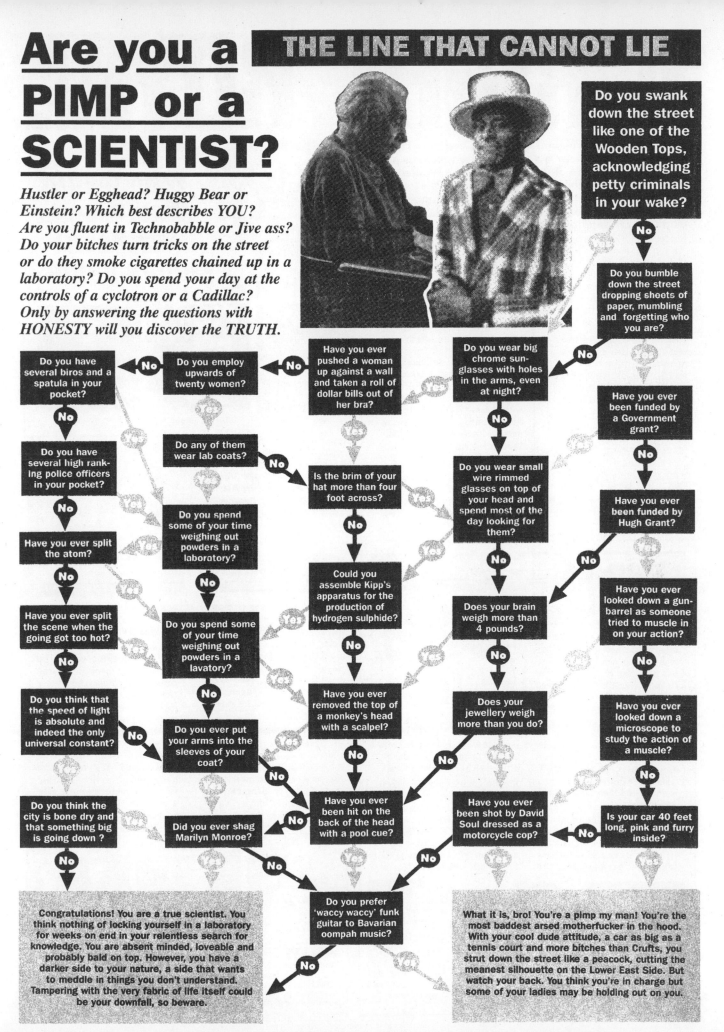

Do you swank down the street like one of the Wooden Tops, acknowledging petty criminals in your wake?

→ No → Do you bumble down the street dropping sheets of paper, mumbling and forgetting who you are?

Do you wear big chrome sunglasses with holes in the arms, even at night?

Have you ever pushed a woman up against a wall and taken a roll of dollar bills out of her bra?

Do you employ upwards of twenty women?

Do you have several biros and a spatula in your pocket?

→ No → Do you have several high ranking police officers in your pocket?

→ No → Have you ever split the atom?

→ No → Have you ever split the scene when the going got too hot?

→ No → Do you think that the speed of light is absolute and indeed the only universal constant?

→ No → Do you think the city is bone dry and that something big is going down?

Do any of them wear lab coats?

Do you spend some of your time weighing out powders in a laboratory?

Do you spend some of your time weighing out powders in a lavatory?

Do you ever put your arms into the sleeves of your coat?

Did you ever shag Marilyn Monroe?

Is the brim of your hat more than four foot across?

→ No → Could you assemble Kipp's apparatus for the production of hydrogen sulphide?

→ No → Have you ever removed the top of a monkey's head with a scalpel?

→ No → Have you ever been hit on the back of the head with a pool cue?

Do you wear small wire rimmed glasses on top of your head and spend most of the day looking for them?

→ No → Does your brain weigh more than 4 pounds?

→ No → Does your jewellery weigh more than you do?

→ No → Have you ever been shot by David Soul dressed as a motorcycle cop?

Have you ever been funded by a Government grant?

→ No → Have you ever been funded by Hugh Grant?

→ No → Have you ever looked down a gun-barrel as someone tried to muscle in on your action?

→ No → Have you ever looked down a microscope to study the action of a muscle?

→ No → Is your car 40 feet long, pink and furry inside?

Do you prefer 'waccy waccy' funk guitar to Bavarian oompah music?

Congratulations! You are a true scientist. You think nothing of locking yourself in a laboratory for weeks on end in your relentless search for knowledge. You are absent minded, loveable and probably bald on top. However, you have a darker side to your nature, a side that wants to meddle in things you don't understand. Tampering with the very fabric of life itself could be your downfall, so beware.

What it is, bro! You're a pimp my man! You're the most baddest arsed motherfucker in the hood. With your cool dude attitude, a car as big as a tennis court and more bitches than Crufts, you strut down the street like a peacock, cutting the meanest silhouette on the Lower East Side. But watch your back. You think you're in charge but some of your ladies may be holding out on you.

A RIGHT ROYAL

We all love and respect the Royal Family. Their impeccable behaviour is a shining example to us all. For this reason we find recent press speculation surrounding the private lives of certain Royals totally contemptable. In marriage as in other things the Royals exercise the highest moral standards, and we would never dare to suggest otherwise.

But the Royals are only human. And who are we to say that there could not conceivably be an adulterer among them? And if they were to temporarily stray from marital fidelity, where would they turn for a bit of royal rumpy pumpy?

Just for fun, we asked a leading sex expert to assess each member of the Royal Family and to tell us what sort of person they would most likely turn to for a quick roll on the red carpet.

For each Royal he has named three nookie nominees. Strictly for fun, use your knowledge of sex and the Royals to select who you think are the most likely candidates. When you've finished compare your choices with those of our expert which are written below.

PRINCESS DIANA

One of Di's main interests is fashion and clothes — of all the Royals she is definately the most daring when it comes to dressing up. She would therefore have a great deal in common with pop star **BOY GEORGE**. Significantly Di also has a genuine interest in the rehabilitation of drug victims.

A former nanny and now mother of two, Di has a great deal of affection for children. Perhaps a fling with poncy choir boy **ALED JONES** would bring out the mother in her.

Star George

Aled: Choirboy

D.J. Fluff

Alternatively Di's other great interest — pop music — may lead her towards a lover in the music world. Somebody like top DJ **ALAN FREEMAN**, who could no doubt impress her with a wide and varied record collection.

PRINCE CHARLES

Charles has many varied interests. Among them sport, painting and writing children's books. But he is also renowned for his strong interest in architecture, where he has a preference for old, classical styles rather than anything new or unusual. For this reason I believe he'd be attracted to a more mature woman, conservative in her appearance. **MARGARET THATCHER** is just such a woman, and therefore a likely candidate.

In an affair Charles would also be looking for a change — something different. His wife, Princess Di is a slim, elegant lady. So perhaps he'd be attracted to the contrasting, stocky, athletic figure of **FATIMA WHITBREAD** who he has no doubt met during official visits to athletic events.

Finally, Charles made no secret of his admiration for top seventies pop group **THE THREE DEGREES**. Perhaps the thought of a steamy foursome with the attractive american trio would tempt the Prince into adultery.

Maggie: Woman   Athlete Fatima   Degrees: Three

THE QUEEN

Among the Queen's many interests are art collecting, horse racing and hats. So perhaps it would be from one of these areas that she would select her Royal Romeo. A prominent artist for example, like **ROLF HARRIS**. The Queen has probably heard that he is very big down under.

Perhaps Her Majesty's fondness of flamboyant hats would suggest a shared interest and the possiblity of a rewarding romance with extrovert pop star **ELTON JOHN**. The highly talented singer/song-writer would no doubt derive much satisfaction from tinkling the Queen's ivories.

But maybe the Queen's first love, horse racing could turn up an odds on favourite in the rumpy pumpy stakes. And rather than giving a ride to a favourite jockey, perhaps she'd put her money on a fellow race horse owner — a good each way bet between the sheets. How about saucy fat comedian **BERNARD MANNING**? It's odds-on the going would be firm and big Bernard would come home with a good length to spare.

Rolf: Painter   Elton: Hats   Bernard: Fat

ANSWERS

In our expert's opinion if the Royal Family were involved in hanky panky (which obviously they never would be) the pairings would be as follows. Prince Charles and Fatima Whitbread, Princess Di and Aled Jones, Prince Andrew and Joan Collins, The Duchess of York and Dr Magnus Pyke, The Queen and Bernard Manning, The Duke of Edinburgh and Jeanette Krankie, The Queen Mother and Cliff Richard.

BIT ON THE SIDE

PRINCESS FERGIE

Lazenby —'007'   Eddie 'The Eagle'   Pyke: Brainy

PRINCE ANDREW

A fully qualified Royal Navy helicopter pilot, Andrew would have an obvious shared interest with former Treasure Hunt presenter, action loving ANNEKA RICE.

The Prince has never lived down his "randy Andy" reputation and it may well be that he would prefer a more experienced woman, someone who has been around a bit and is capable of satisfying a vast sexual appetite. For this reason "Stud" actress JOAN COLLINS would spring to mind.

Consider also Andrew's known taste in women. In choosng

Sarah Ferguson as his bride he revealed an affinity for the more generously proportioned figure. Perhaps for this reason caring heavyweight CLARE RAYNER should be considered.

Fergie's action packed lifestyle reveals on outgoing, danger loving personality. The flame haired Duchess constantly seeks adventure in the air and on the ski slopes, never content to stay at home and watch TV. The lady lives for thrills and who better to provide these than tough guy James Bond star GEORGE LAZENBY.

There is no doubt that of all the Royals Fergie has by far the best opportunities for "hankie pankie". While hubby is away at sea the Duchess spends hours alone on the piste, and perhaps this fascination with winter sports would suggest that an affair with British Olympic ski champion EDDIE EDWARDS is on the cards.

Or perhaps she would surprise us all and go for brains instead of brawn. A suave, sophisticated intellectual like DR MAGNUS PYKE would make a refreshing change from the hunk in her life, Prince Andy.

Anneka: Loves choppers   Collins: Stud   Clare: She cares

THE DUKE OF EDINBURGH

Unlike the other Royals the Duke has his roots overseas — in Greece to be precise. And he occasionally pines for his native country. Perhaps therefore a steaming affair with fellow Greek NANA MOUSKOURI would bring the memories of home flooding back.

Or maybe after a lifetime spent in the shadow of his more famous wife, the Duke may want a woman whom he could

dominate physically, someone smaller than himself and with a much lower public profile than the Queen. Someone like top schoolboy impersonator JEANETTE KRANKIE would fit that description.

But after being bogged down for so long by Royal rigmorole, etiquette and never ending formalities, maybe the Duke would be tempted to go down market in search of "the common touch". No doubt he'd find an abundance of warmth and sincerity in down to earth bubbly blonde "Carry On" star BARBARA WINDSOR.

THE QUEEN MUM

The Queen Mum would be looking for the companionship that only a true gentleman could give, someone with the grace and charm of generations past. A man with all these qualities is doubtlessly clean cut singer CLIFF RICHARD.

Or perhaps the deep, genuine, compassionate tones of attractive yet mature radio disc jockey SIMON BATES could win her heart. Doubtless the couple could spend many "golden hours" together.

Or would Britain's best loved great grandmother, now in the autumn of her years, prefer to have a toy boy at her disposal? I'm sure it would do her no harm whatsoever to have a youngster like current pop sensation JASON DONOVAN running around the corridors of Clarence house.

Songbird Nana   Cuddly Jeanette   Babs: Busty

Clean-cut Cliff   Bates: Genuine   Jason: Neighbourly

Cliff's nuts set to blow

A peaceful Surrey town is today facing a jisolm cataclysm. For sex experts are warning that Weighbridge could soon be engulfed in Britain's first ever tidal gunk wave.

Scientists fear that the bollocks of one of the town's most celebrated residents – Cliff Richard – could explode if the popular singer does not have sex soon. And the resulting 'Pompeii' style disaster could reduce parts of Surrey to a spunky slurry.

BIOLOGY

Cliff, the seemingly ageless Peter Pan of Pop, claims to have been celibate for many years, and biology experts fear a potentially deadly build up of body fluids in his undercarriage could soon reach bursting point.

As nearby residents prepare to protect their homes against flood damage with sandbags and tons of tissue paper, the question on everyone's lips is 'Will Peter Pan's Plumbs go Pop?'

PHYSICS

Weighbridge council officials were last night setting up an emergency control centre, and the army are understood to be on full alert. Late last night an eerie silence hung over the town as anxious residents hoped and prayed that the singer would either get his leg over, or experience a nocturnal emission before it is too late.

By our Chief Knackers Correspondent
BUCK
off the High Chaparal

What makes a star's knackers blow up?

Exploding celebrity knackers is not a new phenomenon in Britain. As recently as 1989 fans of celibate comedian and author Stephen Fry were stunned when his left nut appeared to explode during a book signing session at a shop in Cambridge. No-one was hurt, although there was substantial damage to several books and a carpet.

CHEMISTRY

The medical profession has been aware of the condition, often referred to as Volatile Knackers, since before the turn of the century. But little contemporary research has been carried out in the field, and there is no course of treatment readily available to sufferers, other than having a wank.

FREE PERIOD

The problem arises when semen, which is constantly produced in the male adolescent body, is not ejaculated by the penis due to a lack of any sexual activity. Failure to 'chuck your muck' in this way can lead to the development of the early symptoms of the condition, including a 'stiffy', and 'nuts like two tins of Fussells milk'.

Bob names the day for charity spectacular

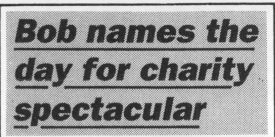

Stars whose surnames are types of weather are to be invited to turn out in a charity football match to pit their soccer talents against stars whose surnames are types of cars.

The Weather versus Car soccer star challenge was the brainwave of charity organiser Bob Johnson who hopes that the all star event will help raise millions of pounds for research into baldness.

GENERAL PUBLIC

"Stars as well as the general public are often struck down with this terrible hair losing condition, and it is therefore appropriate that big name stars like John Thaw, David Frost and Gareth Hale should take part in this spectacular fund raising event."

THE BEAT

Bob's brainwave to pit celebrity weather against car surnames on the football field came to him after he had seen an item on the TV news read by Jon Snow in which MP Austin Mitchell had been interviewed.

"Unfortunately Austin Mitchell's christian name is car, not his surname, so he doesn't qualify for the team, but I am hopeful that Harrison Ford will be playing, as well as Tommy Cooper".

SPECIALS AKA

Invitations have already been sent out to four celebrities in all whose surnames are also types of weather, and two whose surnames are a kind of car.

"It's early days yet, and whilst no-one has actually

Frost yesterday followed by Thaw early this morning.

accepted the invitation at this stage, I'm confident that we'll have a bumper turn out on the day, and a really exciting game", Bob told us.

SELECTOR

As well as the match, Bob will be selling balloons. The fun kicks off at 3pm on Saturday 26th June 1993 at Fulchester Recreation Ground. Bob says that any stars willing to take part, especially those whose surname is a type of car, should contact him at his work number which is Fulchester 577985, extension 427.

THE ADVENTURES OF **STAVROS** WITH T.V.'s FUNNYMAN Harry Enfield

HALLO MATEY PEEPS. I'M A MAKE A BLADDY KEBAB FOR HER INSIDA DE DOORS, AND ATTA DE WEEKEND, SHE SAY I CAN GO UPA DE ARSE...

COS THEY PLAY DE QUEENS PARK RANGE!

(INNIT?!)

WELL, THAT'S IT FOR THIS WEEK, MATEY PEEPS! BYE

I WAS A RIGHT FUCKING TEARAWAY, ME

Spender star Jimmy Nail is a star with a secret. For before he found fame as a tough guy TV cop, actor Jim was more often than not on the *wrong* side of the law.

EXCLUSIVE

But actor Jimmy doesn't like to talk about it

Now his time is spent solving crime on our screens, but it wasn't long ago that Nail, then a hard drinking Geordie, was causing trouble in real life. But now all that has changed, and Jimmy prefers not to discuss his dubious past.

UNUSUAL

"I must admit, it was unusual to walk into a police station through the front door while we were filming Spender", Jim recalls. "In the past I'd always been carried in the back door, with a blanket over my head. In fact, on one occasion I'd roughed up a couple of coppers, so they took me into the station through the wall – quite literally – head first!"

GRASS

Jimmy's tough on-screen persona may well have something to do with his tough upbringing. But the past is a subject that Jimmy prefers to avoid. "On one occasion I came home from the pub with an axe sticking out me head", the seven foot Geordie recalls. "Someone had been trying to chop me head off, and the axe had got stuck. Mind, I'd had that much to drink I didn't even notice it was there".

PUSSYCAT

Not suprisingly Nail is not Jimmy's real name. "That came about after my dad had nailed my head to the floor in order to keep me out of trouble", says Jim. "But it didn't work. I just pulled half the floor up, went to the pub and started butting people with floorboards stuck to me head".

DARKNESS

Thankfully those wild days are now well and truly in the past, and it's a chapter in the Nail story that Jimmy has declared well and truly closed. "I was a bit of a football hooligan in those days", he recalls. "And inevitably I ended up paying the price". Jimmy served a prison term at Strangeways following a violent incident involving a baseball bat. But not suprisingly the born again actor, writer and director prefers to look to the future, and not dwell on the past.

THIEVES

"My first acting interview was for Auf Weidershein Pet. I'll never forget the interview. I went in, and all these blokes were looking at me. I said 'Who the fuck are you looking at?' In the end I got arrested after I'd tried to kill one of them with an iron bar. But luckily I got the part, and haven't looked back since."

TRAMPS

The Nail success story has been meteoric. Now the co-owner of a production company, Jimmy is involved in all aspects of TV writing and production. Married with two kids, Jimmy lives in Wales where he is bringing his children up not to use cutlery. "It's something my wife and I discussed a long time ago and decided we would like to do. It's what we feel is right, but once the kids grow up it will be up to them whether or not they use cutlery."

THIEVES

But behind the success story is a little known tale of the tough, streetwise Geordie musician who drank and fought his way from pub gig to pub gig, wearing women's clothes. It's a skeleton that Jimmy prefers to leave well and truly in the closet.

"Yeah, I used to wear dresses and great big boots, and jump into the audience and kick everyone's head in", he confesses. But it's a subject he is loathe to discuss. "I was a bit of a lad in those days. In fact, it was quite a surprise for me to be walking into a police station through the front door while we were filming Spender. In the past I'd always been carried in the back, with a blanket on my head", says Jimmy. "But don't you dare print any of that, or I'll knock your fuckin' teeth out".

NAIL – prefers not to talk about his past.

Snow shoes is good news
say shoe retailers

Blizzard hit Britain has gone show shoe crazy — according to reports from leading footwear retailers.

Record breaking arctic conditions coupled with 20 foot snowdrifts have left the UK buried under a thick blanket of snow and ice.

SENSIBLE

But Britain's pedestrians are putting on a brave face – and sensible footwear. For sales of snow shoes have already reached record levels, with many shoe shops having sold out completely by the second week in January.

VANIEN

"People just can't seem to get enough of them. They're literally flying off the shelves", one shoe shop owner told us today.

WELLINGTON

Traditionally snowbound pedestrians across Britain have always worn Wellington boots or galoshes – a rubbery zip-up waterproof ankle high overshoe. But this year has seen sales of the once popular boots and overshoes fall in the face of stiff competition.

DICKENS

Wellie makers Dunlop yesterday threatened to wage a price was with snow shoe manufacturers, slashing the price of some boots by 25 per cent to less than £15 a pair.

TEXAS

A pair of snow shoes probably costs about the same as two tennis racquets.

Salad Daze for Albert

A CONFUSED, 108-year old Somme veteran was spending his first night behind bars last night after Walsall magistrates found him guilty of making two trips to the salad bar at a high street pizza restaurant. Albert Fairbrass, who won the George Cross for outstanding bravery in the trenches, was arrested when a fellow customer at Pizza World in Redditch reported him to the manager after he returned to the salad bar to top up his dish with greens for a second time.

Whistle-blower Mrs Edna Busybitch told the local newspaper: "It made my blood boil when I saw him helping himself to a second bowl of lettuce, so I called the waiter over and pointed out what he was up to."

She continued: "I don't care how old he is, or what he did in the First World War. It's the likes of him what puts the price of salad up for the rest of us law-abiding citizens. They can lock him up

SALADSCLUSIVE

and throw away the key, for all I care."
AND

Police were called and the bewildered Chelsea Pensioner was taken to Redditch Police Station where he was questioned through the night, before being charged.

He appeared before magistrates later that day and was sentenced to six months in prison. His solicitor told us: "Mr Fairbrass suffers from chrionic septic pleurisy after being gassed at Ypres and was recently told by his doctor that he had just four months to live. In the light of this information, we have decided not to lodge an appeal. There's already such a backlog of cases in the court system that any appeal wouldn't get heard till next year at the earliest, so frankly there's no point in us bothering to even set the wheels in motion."
IT

"Unfortunately, it looks like Mr Fairbrass will just have to die in prison," he added.

Other customers, who had gathered for Mr Fairbrass's birthday party, accused Pizza Palace staff of over-reacting, but manager Hector Bland was unrepentant. "The company has a stated policy of zero tolerance when it comes to diners breaking the single salad bar visit rule," he told us. "This is clearly written down in the Terms and Conditions section of our parent company's website."
THE

"If Mr Fairbrass had taken the trouble to visit that page and read through all the small-print, he would have been left in no doubt that he would face prosecution if he went back for seconds," he continued.
TO

"If we make an exception for one 108-year-old, terminally ill, confused, decorated war hero, we'd have to make them for everybody," he added.

Meanwhile, Albert himself was unavailable for comment. A prison spokesman told us: "Unfortunately, Mr Fairbrass is unable to come to the phone right now as he is being bummed in the showers by Mr Big."

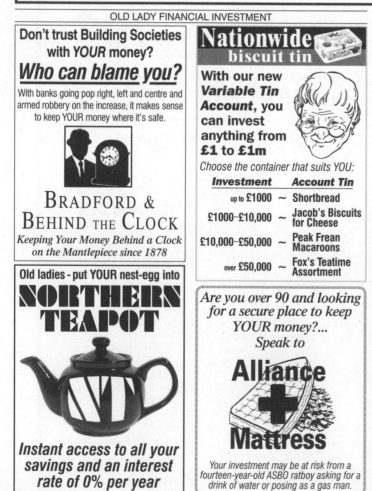

Let's play CELEBRITY WHOREHOUSE

'A bit of harmless fun with the stars'

A JUDI   B VIKKI   C IVY

D Kim   E KYLIE   F JEANETTE

FOR A GOOD TIME RING

It's hard to imagine our top TV celebrities like Melvyn Bragg kerb crawling in Kings Cross. And it's even harder to imagine famous female stars like actress Dame Judi Dench touting for business on a street corner dressed in a revealing short skirt and stockings.

But that's exactly what we've done. We call it Celebrity Whorehouse, and it's a game you can play along with. And for the winner there'll be a chance to sleep with **YOUR** favourite star, or a fifty pounds cash prize.

to 'Celebrity Whorehouse Competition', Viz, P.O. Box 1PT, Newcastle upon Tyne, NE99 1PT. Please write on the back of your envelope the name of the Celebrity with whom you'd most like to have sex.

HARMLESS

Of course it would be ridiculous to suggest that any celebrity would act either as a prostitute or as a client. And not for one moment would we do so. But just for fun, we've *imagined* that six female celebrities are 'on the game', and that six male showbiz stars are looking for a good time. It's a fairly futile and pointless exercise, but it helps fill the magazine.

FUN

To play the game we want you to guess how much each of the celebrity clients would be prepared to pay to go 'all the way' (that is to have *full sex*) with each of the celebrity whores. For example if, just for fun, you thought Rolf Harris would fork out fifty quid to have sex with fellow Aussie Kylie Minogue, enter '£50' in box 'E' against Rolf's picture, etc. Simple, isn't it? Send your completed forms

JIM DAVIDSON WOULD PAY...

A	B	C
D	E	F

M. C. HAMMER WOULD PAY...

A	B	C
D	E	F

ROLF HARRIS WOULD PAY...

A	B	C
D	E	F

MELVYN BRAGG WOULD PAY...

A	B	C
D	E	F

IAN McCASKILL WOULD PAY...

A	B	C
D	E	F

SIR A. BURNETT WOULD PAY...

A	B	C
D	E	F

Celebrity Pimp

Each week, just for fun, we select a top celebrity from the world of telly, sport and entertainment to act as our **Celebrity Pimp**. We invite him to cruise the streets in a pink cadillac and floppy hat, and we imagine which three Celebrity Whores he would choose to act as his 'bitches'.

Needless to say, there is no implication whatsoever that any of the celebrities named would act in the way we have described. This is all just for fun.

This week we've chosen former England footballer and TV soccer analyst **TREVOR BROOKING** as our Celebrity Pimp, and we've tried to imagine which three well known women Trevor might choose to act as his prostitutes.

Looks aren't everything, so we've selected Eastenders'

Michelle Fowler as Trevor's first girl. Her tough, streetwise background would be an asset to any pimp.

A good head for figures is essential for any prostitute, especially when it comes to handing over her pimp's share of the money! If any of his bitches short-changed Trevor, they could be in big trouble. For this reason we think Countdown's **Carol Vorderman** would be a worthy addition to Trevor's vice racket. Lastly, we've chosen the Media Show's **Emma Freud,** because we fancy her.

Next Week: Celebrity Drug Dealer, where top stars battle it out in a 'Manchester style' drug war.

D.I.V.O.R.C.E.

Widower seeks split from wife after death

A HENPECKED Lincolnshire man is finally set to divorce his wife - *even though* she has been **DEAD** for 18 years!

EXCLUSIVE

Unemployed shopfitter Arthur Poindexter says he's so fed up with his late wife's nagging from beyond the grave that he's hired a MEDIUM to start legal proceedings to end his marriage in the afterlife.

Arthur, 58, wed his childhood sweetheart Renee in 1965 but almost immediately things started to go wrong. He told us: "She started nagging me before we even left the reception, and it carried on throughout our marriage."

Henpecked Arthur put up with Renee's constant carping for twenty years, until she was killed by a bus

TILL LIFE AFTER DEATH US DO PART: *Hen-pecked Arthur today and (inset) in happier times with wife Renee 38 years ago yesterday.*

on her way home from the bingo. "I was a bit upset at first, but it was a new lease of life for me," he told us. "At last I was free from her giving it that in my earhole all the time. I was looking forward to finally getting a bit of peace and quiet." But Arthur's peace was shortlived.

"After the funeral, all the family came back here for a drink and a few sandwiches," he told us. "By the time everybody had gone home I'd had a few sherries and was feeling a bit emotional, so I decided to go upstairs for a lie down." But in his bedroom, Arthur got the shock of his life.

"I hadn't been in bed two minutes when the room went icy cold. Suddenly, a shadowy figure floated through the wall. It was Renee's ghost, and she wanted me to go downstairs and do the dishes. I tried to hide under the covers but it was no use; she just kept going on and on in an unearthly voice about the state of the sink, so in the end I went down and did them just to shut her up."

Sure enough, as soon as the dishes were dried and put away, the apparition disappeared. Arthur thought it was the end of his troubles, but as it turned out they were only just beginning.

"The next day I was sitting reading the Racing Post with my feet up on the coffee table. Suddenly, what I can only describe as an unseen force seemed to kick my legs to the floor. Then the hoover switched itself on. I tried to turn it off but nothing seemed to work.

"Then suddenly Renee's disembodied head appeared, floating above the mantelpiece. It was ashen-white and transparent. She told me to get off my lazy backside and vac the stairs.

"I was petrified, but I did as I was told. It was only when I'd done the stairs, including the half landing, that the machine switched off as mysteriously as it had switched on."

Over the last eighteen years the visitations have become more and more frequent. Now Renee's ghost materialises up to thirty times a day to bend Arthur's ear. Amongst her bizarre hauntings, she has

* WRITTEN 'clean this now' in the dust on Arthur's TV screen

* MANIFESTED in the back of his car, telling him a traffic light was red

* CREATED havoc with poltergeist activity, causing brillo pads to whirl round the kitchen whilst the oven door slammed open and shut

Arthur put up with Renee's ghostly carping until this Christmas, when she finally went one haunting too far.

"It was Christmas eve and I was popping out to the local for a pint with some friends. Renee manifested as usual, just as I was leaving the house. The phantasm warned me to be back in by ten or there'd be hell to pay. Unfortunately, there was a lock-in and I didn't get back till after one.

"I was hoping to sneak in without raising my late wife's vengeful spirit, but it was no good. As I turned the corner I could see her apparition hovering above the step. She had her head under her arm and a face like thunder. As I reached the front door, her ghostly voice asked me what time I called this, but before I had a chance to answer she hit me over the head with a see-through rolling pin.

EEK!

EK!

"I certainly had some explaining to do in casualty."

Arthur decided enough was enough, and contacted Doris Stokes-Taylor-Joynson-Garrett, Britain's top spiritualist solicitor, and instructed her to begin legal proceedings to bring his marriage to an end.

"Doris got on the case straight away, and her red indian spirit guide Chief Billy Two Rivers served divorce papers on Renee straight away," said Arthur. "My late wife's solicitor then contacted me via the Ouija board to arrange a hearing seance next Tuesday.

"Basically, the judge will explain to Renee's spectre that I'm seeking the divorce on the grounds of her unreasonable paranormal behaviour. She has to rap on the table twice or make the lights go dim to accept the decree absolute."

But Arthur isn't celebrating his freedom just yet. He told us: "I can't see her giving up that easily. I'll not break out the champagne until I've got that piece of paper in my hand. I know my Renee only too well. I might get the vicar round to exorcise the house anyway, just to be on the safe side."

NAG WATCH U.K.

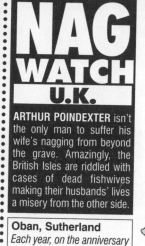

ARTHUR POINDEXTER isn't the only man to suffer his wife's nagging from beyond the grave. Amazingly, the British Isles are riddled with cases of dead fishwives making their husbands' lives a misery from the other side.

Belfast, N. Ireland
Thrice-widowed garage owner Bob Antrobus is scared to use his kitchen. It is haunted by the spirits of his late wives, who sit in there drinking transparent tea and discussing him in unflattering terms.

Launceston, Cornwall
Retired builder Eric Trelawney dreads taking a bath. For each time he gets in the tub, an apparition of his wife Edna, who died in 1977, appears in the mirror and goes on at him to hang the bath-mat up.

Oban, Sutherland
Each year, on the anniversary of his wife Morag's death, crofter Hamish McTavish is woken at midnight by a strange grey figure, who tells him she cannot rest until some shelves he bought in 1982 are put up.

Braintree, Essex
Since dying in 1973, Edna Bartram has visited her husband Jack every night. Her ghostly figure sits pursed-lipped; pointedly refusing to speak to him, uttering a ghostly tut each time he asks what he's done wrong.

Is there anybody there?

Find out with our fantastic henpecked widower's Ouija Board

Don't keep your late missus waiting till the witching hour. Find out what the old battleaxe wants any time of day with this fabulous Cut Out & Keep Widowers' Travel Ouija Board.

Instructions

Cut out the board and stick it to a piece of stiff card. To summon the old trout's restless spirit from the other side, simply put an upturned Night Nurse cup on the central pentangle, place your right finger on it, and ask in a quavery voice if anybody is there. Once she has made her presence known, sit back and get nagged at, occasionally saying "yes, dear."

'I'M A SEXY HOT POT'

Tony's a real tasty dish

Supercue TONY KNOWLES, snooker's number one sex symbol tells his own sensational story.

PICTURE SPECIAL

The ladies call me 'the Lancashire Hot Pot'. 'Cos I'm a really tasty dish.

BIG

Being sexy is a big bonus when you're playing snooker. My opponents are often distracted by how sexy I am. Snooker is a big money game, but money can't buy good looks, as all the other players know only too well.

ATTRACTIVE

I mean, look at Big Bill Werbenuik. He isn't even remotely attractive. Alex 'Hurricane' Higgins is quite sexy, but not in the same league as me.

SHOT

I remember at one tournament I was playing Alex and he only needed one ball to win. At the time I was so sexy he couldn't concentrate. He missed the shot and I won. There were 3,000 girls in the audience that night - and I was sleeping with all of them.

DIET

A few years back I was so sexy it was affecting my cue action. And I was having difficulty walking. My doctor told me I was too sexy and put me on a diet. He also gave me some tablets but I remember being so sexy I couldn't swallow them.

I'm still sexy, and I'm getting sexier every day. On my way to a tournament recently I was so sexy my car blew up, and people complained that I had been interfering with short wave radio signals. That's how sexy I am at times.

GIRLS

When I eventually arrived at the venue there was so many girls waiting inside to see me, there wasn't room for a snooker table.

ME

Girls find me irresistible. The other night I was staying in an hotel. I took so many girls back to my room with me the floor collapsed.

PROPOSALS

Being so sexy I get a hell of a lot of fan mail. On an average day I get 500 proposals from would-be brides. And I usually get at least twice that amount.

DOGS

I have other hobbies as well as girls. I used to keep dogs, but I was so sexy I had to have them destroyed.

PROBLEM

I'm planning to write a book at the moment. It's going to be about how sexy I am. The only problem is I don't know where I'm going to find enough paper to write it on.

CD·JB·JS 84

VICTOR PRATT ★ THE STUPID TWAT ★

I FEEL LIKE AN OLD MAN, VIC

ME TOO, BUT WHERE CAN WE GET ONE AT THIS TIME OF DAY?

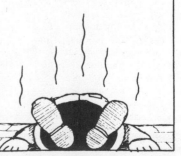

DOCTOR DEATH

I AM DOCTOR DEATH!

AH, AH, AH, AHEM!

JS 4·84

What a RIP-off!

Pope's death leaves Ernie out of pocket

THE RECENT DEATH of Pope John Paul II was a great loss to the world, but one north-east man has lost even more than most. For the pontiff's tragic death has left him £35 out of pocket, after the Vatican pulled out of a deal to buy his second-hand pressure washer.

By our Religous Affairs Correspondent
Mr C. out of The Shamen

Ernie Loffgren of Morpeth placed an advert in the 'For Sale' section of his local yellow Free-Ads paper, offering his used Karcher 1200 pressure-washer at the bargain price of £35. The same day the paper came out, Ernie got a phonecall from a prospective purchaser.

He told us: "It was the Pope. He said he'd seen my ad and the pressure-washer was just what he wanted to get the moss off the cobbles at St Peter's Square. When I told him I'd throw in half a bottle of cleaning fluid and an attachment for getting pigeon mess off the Popemobile, he agreed to buy it sight unseen."

cash

The Pope assured Ernie that he'd be round with the cash within the week, so the 52 year-old gas fitter packaged the pressure-washer up and placed it in the hall awaiting the pontiff's knock at the door. "That was on the Monday," continued Mr Loffgren. "But by the Friday his holiness still hadn't turned up."

During the week Ernie received four other enquiries from prospec-

tive buyers. "Two of them even turned up at the door with the cash in their hands," he told us. "But I had to turn them away. I'd already promised the pressure-washer to the Pope, and I wasn't going to go back on my word."

sharpe

However when Ernie watched the news on Saturday, the reason for

The Pope waves to people before facing his maker, whilst Ernie (left) faces placing another advert in the FreeAds paper. Below right, the original advert that caught the Pontif's eye.

John Paul II's no-show in Morpeth became clear.

"I couldn't believe my rotten luck when I heard the Pope had died," he told us. "I tried ringing the Vatican to see if they were going to send someone else round to pick up the pressure washer, but they were engaged. I wouldn't be surprised if they'd deliberately left the phone off the hook," he added.

Now Ernie is faced with the task of re-advertising the washer in next week's Free-Ads paper. He told us:

PLANT Clematis, markham's pink, lovely flowers . £6 ono. Tel. (0191) 253▮

PLAY SHED Child's play house, wood, in need of painting. £40. Tel. (0191) 252▮

PRESSURE WASHER Karcher 1200, used once, as new. £35. Tel. (01670) 223▮

SHED 7' x 5', felt roof with base, buyer to dismantle. £90. Tel. (0191) 241▮

TOP QUALITY TURF in rolls, 2 grades. Can deliver. £1 per roll.

"I'm going to say it's being re-offered due to timewasters, and if the new Pope rings up to buy it I'm going to tell him it's already sold."

The *Viz* Solicitor, *Mr Ingledew Botteril* writes...

MR Loffgren's situation is not unusual. Many people make a contract, whether verbal or written, only to die before the terms of that contract are fulfilled. In law, the unfulfilled contract becomes one of the deceased's effects, and is treated as a chattel under the terms of his last will and testament.

Mr Loffgren should write to the Vatican demanding the name and address of the executor of the Pope's will. He should then write to the executor requesting settlement of the outstanding amount at their earliest convenience. However, if the bequests have already been disbursed he should contact each of the beneficiaries, eg wives or children, individually requesting payment of an amount equivalent to their share of the £35 which is owing to him to be paid in a proportion equivalent to that of the deceased's estate which they were bequeathed and of which they are already in receipt.

The beneficiaries should then be given a set period, usually 14 days, in which to call at Mr Loffgren's house and pick up the pressure washer. How it is disbursed amongst the individual assignees is a matter for the Pope's executor.

Legal Questions? The Viz Solicitor Ingledew Botteril is here to help. Send your queries along with a cheque for 400 guineas to: Ingledew Botteril, PO Box 1PT, Newcastle upon Tyne, NE99 1PT

If a face can paint a thousand ships, then why can't the stars paint a...

PICTURE OF THEMSELVES?

Painter man, painter man. Who will be a painter man?

That was the poignant question raised by Boney M in their 1979 top ten hit. And the answer would appear to be almost every star in showbusiness. For painting would appear to be the number one hobby of the stars of stage and screen.

SUCCESS

Over the years a great many celebrities have enjoyed considerable success in their alternative careers as artists, swapping movie cameras and make up for palet knives and paint. Rambo star Sylvester Stalone, Zorba the Greek star Anthony Quinn, Rolling Stone star Ron Wood, Rolf Harris and Wolf out of The Gladiators. The list is endless.

Stars of screen and paintbrush Rolf Harris (above) and TV's Terminator Sylvester Stallone.

WIN A LUXURY 4-PIECE BATHROOM SUITE
(EXCLUDING TAPS) WORTH £300!

So we decided to find out whether the stars of the **TV sitcoms** shared this amazing talent by asking six small screen comedy actors and actresses to paint a portrait of themselves, using only brushes and paint. And the results, printed here, are quite amazing.

PAINTERS

We want you to spot our celebrity painters, and the first person to correctly identify the six stars by looking at their portraits will win our fabulous prize: a beautiful four piece bathroom suite, featuring WC, hand basin, bath and bidet, worth over £300 (but not including taps). And there'll be Black & Decker power tools for three lucky runners-up.

Just look closely at the six paintings, then send your answers, on a post card please, to: Viz Celebrity Portrait Bathroom Suite Competition, P.O. Box 1PT, Newcastle upon Tyne NE99 1PT. The closing date for entries is March 1st 1993. Please state your colour preference: champagne, aqua blue, 'marble effect' or dove grey.

STAR GALLERY

1 Military man manages money, and also makes good use of a palet knife.

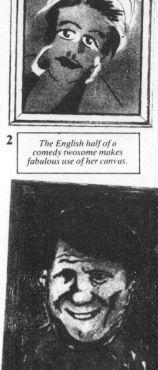

2 The English half of a comedy twosome makes fabulous use of her canvas.

3 Verging on Picasso, this one's obviously an old master himself.

4 Curly wurly brushwork from a comedy veteran who's not long out of shorts.

5 Naive brushwork and gothic overtones from this man about the house who owned a restaurant called Robin's Nest. And married Tessa Wyatt.

6 She was Terry Scott's wife in Terry And June. And her surname is Whitfield.

DONG!
'Big Ben' goes *starkers*

Pinko commie comic Ben Elton will be left red faced after being seen in the pink in a red hot blue movie soon to be released.

VIDEO

Green funny man Ben, 27, will be purple with rage when a black market video featuring the TV funny man and best selling novelist **STARKERS** goes on sale next month.

SHOCKED

During the hard core XXX rated film shocked fans will see Elton:
★ **BONKING** with a bevvy of blonde beauties.
★ **ROMPING** with a roomful of randy redheads.
★ **FIDDLING** with his parts until they go off.

SIZZLING

The sizzling on-screen sex romp was filmed in 1974 in a hotel bedroom in Hartlepool. At the time Ben, 28, was struggling to earn a living as a comic on the gritty northern club circuit. We believe he was paid £5 to perform lurid, steamy sex acts on camera.

DEVASTATED

A source close to the comic told us that Elton was "devastated" by news of the movie's pending release. "Ben was absolutely gutted when he first heard that this film has surfaced after all this time. All the hard work he's put in over the last few years could be ruined by one small mistake he made in the distant past. Everyone makes mistakes, and Ben bitterly regrets making this movie. He had hoped that this episode was all behind him."

BLACKADDER

Elton, a multi-millionaire several times over, commands fees of several thousand pounds for live appearances, and has drummed up extra millions writing TV's 'Blackadder'. But now he is fighting to save his career, and has threatened to sue film distributors Videowank (Amsterdam) Ltd. if copies of the movie go on sale.

TWEED

The video was set for release next month, but this may be delayed while legal wrangling goes on. And if Elton's lawyers are successful, copies of the video will have to be destroyed.

CORDUROY

However, we are giving away **EXCLUSIVE** copies of the cassette **FREE** to the first 5,000 readers who can answer this simple Ben Elton question. All you have to do is tell us the name of Ben's best selling book 'Stark'. Pop your answer in an envelope together with £300 cash, and send it to 'Ben Elton Sizzling Sex Romp Naked Porno Movie Offer Competition', Viz, P.O. Box 1PT, Newcastle upon Tyne, NE99 1PT.

A scene from the red hot porno flick

APOLOGY

In the next issue of Viz we will publish an article entitled 'Bamber's Ring Sting Heaven' in which we will accuse University Challenge host Bamber Gascoigne of inserting wasps into his anus in order to achieve sexual pleasure.

With foresight, we realise that this article will be in very poor taste and will be upsetting and embarrassing to Mr Gascoigne and his family. We accept that our allegations will have no grounds whatsoever in truth and we apologise for any distress which will be caused.

Jam Jarre

The Page that's Jam-Packed with Jam Facts, brought to you by synth wizard JEAN-MICHEL JARRE

JAM SPOTs

IF someone tells you their car is being followed by a "jam sandwich", they don't mean they are being tailed by two slices of bread with a fruit preserve filling. That's because "jam sandwich" is another way of saying police car. However, in a bizarre twist, policemen refer to the jam sandwiches in their packed lunches as "police cars"!

JAM even gets a mention in the bible. In 1 Kings 6, verse 31, King Solomon makes some "jambs" out of olive wood. Yuk!

When a jazz musician talks about a "jam session", he's not referring to a period of time spent eating jam. In fact, he means a sort of formless, over-long, drug-fuelled tuning up process which nobody in their right mind would want to listen to.

DON'T ask for jam in America, they won't know what you mean. That's because jam is called something else in America.

UNLIKE other sandwich fillings, such as marmalade (A dam lamer), marmite (Rat mime) and sandwich spread (Can add hers wisp), the word "jam" has no anagrams in the English language.

PIPS TO THE POST: YOUR JAM LETTERS

A Sticky Moment!
I recently went out to buy a jar of raspberry jam, but the supermarket shelf was empty. The shopkeeper explained that the delivery lorry had been held up ... in a traffic JAM!!! When I finally stopped laughing I bought some lemon curd instead.
Mr F Renton Penge

Top of the Charts!
My three favourite jams are;
1. Raspberry, 2. Blackberry, 3. Strawberry, in that order.
Mrs BTR Read Welwyn Garden City

I have to disagree with Mrs BTR Read's letter (above). In my opinion, the top three jams are apricot, blueberry and raspberry seedless. In future Mrs Read should check her facts before rushing into print.
Mike Scoltock, Sheffield

Jammy Dodge!
I recently tried to open a jam jar, but it was stuck tight. In the end I had to lodge the lid in a doorframe to get enough grip to loosen it. Imagine my amusement when my wife pointed out that I had "jammed the jam into the jamb".
Paul Weller, London

Q. What's a strawberry's favourite pop group?
A. The Jam.
Mrs G Bishop, Knutsford

Q. What sort of university course would some jam go on?
A. A "sandwich" course.
T Poulson, Grantham

Q. When is a door like something you keep jam in?
A. When it's ajar. (A jar)
Mrs Hutton, Coalbrookdale

LAUGH JAM-BOREE

Q. When is a door like the lid of a jam jar?
A. When it keeps jamming. (Keeps jam in)
Mrs Hutton, Coalbrookdale

Q. What is a guinea pig's favourite jam?
A. Hamster jam. (Amsterdam)
Mrs Hutton, Coalbrookdale

QUESTIONS & JAMSWERS

WHAT is the difference between jam and marmalade? asks *Edward Gooch of Beaconsfield.*

"Edward, jam and marmalade both fruit preserves. The difference is in how they are made. To make marmalade, oranges are cooked whole in water, then simmered in a lidded saucepan for several hours. When they are thoroughly softened, the oranges are removed and allowed to cool, before having their flesh and pips scooped out and placed in a small saucepan. Water is then added and the resulting mixture is simmered for about ten minutes. Meanwhile, the remaining orange peels are thinly sliced and placed in a preserving pan with the saved water from the saucepan, together with the juice and grated rind of some lemons. The simmered pith and pip mixture is then sieved and added to the preserving pan before the whole lot is brought to the boil. Sugar is added and the whole lot is boiled for ten or twenty minutes until setting point is reached. Then the mixture is allowed to stand for a further twenty minutes before being potted. Jam is made in a jam factory. I hope this an your question."
Jean~Michel Jarre

JEAN-MICHEL'S JAR-TOON TIME

SORRY I'M LATE. I GOT STUCK IN A JAM!

© Jean Michel Jarre. 2005.

JAMWORD

Across
1. Something you keep in a jar.
Down
1. Something you keep jam in.

The Penny Farthing versus the Sinclair C5. It's that age

The OLDEN DAYS or

Old folk are always telling us about the 'good old days', and how much better things used to be back then. But ask any youngster today and they'd tell you that nowadays are the best. So who's right? Even the experts can't decide. Scientists, doctors and space men all prefer the present day to yesteryear. But policemen, old ladies and vicars yearn for a return to bygone times. It seems that there's no easy answer to this age old dilemma. Or is there?

'Old days are best' say stars

Famous inventor **Sir Isaac Newton** may be dead, but if he wasn't, he'd be turning in his grave. So says his grandson **Henry Newton**, now 53, a member of England's 1970 World Cup squad. *"Things weren't perfect in the olden days"*, says Henry *"specially before my grandad invented gravity. I guess stuff must have floated around quite a bit. But at least in the olden days cigarettes weren't bad for you. Not like today, when they cause cancer and things".*

Potty tea drinking vegetarian MP **Tony Benn** can also trace his ancestry back to yesteryear. His father was potter **Josiah Wedgewood**, and his uncle, **Charles Darwin**, invented the animal. *"The old days were the best"*, he told us. *"Plates were better, and there was a lot more animals. Nowadays with so many people eating them, animals like the dodo have become extinguished"*, said the swivel eyed former toff.

Victorian engineer **Isambard Kingdom Brunel** was in no doubt. *"The olden days piss it"*, he told us, speaking through a psychic yesterday. *"Just look at my bridges, compared to today's crap. That one I did in Bristol for instance. That bastard's going nowhere in a hurry, I can tell you."*

We've designed a simple quiz that will enable you to decide for yourselves which days are the best, olden or nowa. All you have to do is answer the questions either (a), (b) or (c). When you've answered them all, tot up your total score to reveal which days are the best.

1. What do you think is the most important thing about a song?
(a) *The rhythm*
(b) *The melody*
(c) *Being able to hear what the bloody words are*

2. What would you say is the most important issue affecting Britain today?
(a) *The under-funding of the National Health Service*
(b) *Law and order, and the soaring crime rate*
(c) *The price of tea*

3. Do you think women should be allowed to vote?
(a) *Yes*
(b) *Maybe*
(c) *No*

4. How much would you be prepared to pay for this three bedroomed, terraced house?

(a) *Fifty thousand pounds.*
(b) *Five thousand pounds.*
(c) *Five hundred pounds, and still have enough change for a visit to the music hall and a slap up supper afterwards.*

5. What is the rudest thing you've ever seen?
(a) *A really disgusting hardcore porn video, at a friend's house, featuring full, penetrative sex, defecation onto a glass topped coffee table, and this woman with absolutely enormous tits who you actually saw 'doing it' with farmyard animals.*
(b) *Confessions of a Driving Instructor*
(c) *A piano leg*

Just an old fashioned girl? Or 20th Century Boy? Who are YOU?

6. When you arrive home in the evening how do you go about freshening yourself up?
(a) *Have a quick 'power shower', with instant hot water*
(b) *Switch on the immersion heater and half an hour later jump into a nice, relaxing hot bath*
(c) *Heat several gallons of hand pumped water on an open range, then sit upright in a zinc bath, in the middle of the kitchen floor, and scrub yourself with carbolic soap while your wife and fourteen pale, sickly, undernourished children look on.*

7. How would you describe your front door?
(a) *Locked, bolted and security shuttered.*
(b) *Closed*
(c) *Open*

8. Which of the following would be your idea of a good night's home entertainment?
(a) *Losing track of the body count whilst watching the latest sci-fi special effect robo-police action space kung fu adventure movie on video, before playing computer games until six o'clock in the morning.*
(b) *Watching the Black and White Minstrel Show on telly before settling down for a game of Monopoly*
(c) *Gathering your family around the piano for an evening of songs by candle light, occasionally huddling round the radio to see who's winning the war.*

9. If your 16 year old daughter stopped wearing a bra and began staying out all night, what action would you take?
(a) *Talk to her about contraception, and perhaps give her a packet of condoms.*
(b) *Thrash her with your belt, tie her up and then stand over her, reading passages from the Bible.*
(c) *Attempt to drown her in a pond, and if that failed burn her on a bonfire.*

10. If your unmarried daughter announced that she was pregnant, how would you react?
(a) *Encourage her to have the baby, and to bring it up as a single parent, with your help and support.*
(b) *Send her to an alcoholic back street abortionist to try and have it seen to with knitting needles, in a bath of boiling water*
(c) *Have her committed to a brick built lunatic asylum where she should remain until she's in her mid seventies, at which point she will be released; frail, confused and totally bewildered; into a world of TV and space travel in which she is totally helpless, and unequipped to survive.*

An evening at home in the olden days. With no telly to watch, people sat, or stood, facing in a variety of different directions. Note also how everything is in black and white.

old question...*Which is the best?*
NOWADAYS?

The olden days yesterday (left). Note the old fashioned cobbled street, leather boots, hats and large kettle. Below, we see the same street today. Gone are the cobbles, the hats and the kettle. And leather boots have been replaced by plastic slip-on shoes.

'Space technology by the year 2000'

Spaceman Neil Armstrong, the first man on the Moon, is a fan of the future.

"*Going to the Moon was fucking great*", he told us yesterday "*even if I did get a bit dizzy in the rocket. If I'd been born in the olden days the nearest I'd have got to the Moon would have been sitting on the roof of my house!*"

Neil believes that by the year 2000 medical science will reap massive rewards from space technology. "Teflon pans come from space rockets", he explained. "Soon there'll be space medicine as well, to cure all illnesses. In fact there'll be space hospitals, orbiting the Moon, by the year 2000. I bet you ten quid. No, make it fifty".

British boffin Sir Clive Sinclair agrees. "There's absolutely no comparison between the olden days and today", he told us. "Old things like grandfather clocks get rusty and break. Not like my plastic death trap three wheel battery powered bucket, the Sinclair C5. Okay, it might have been a heap of toss, but it was way ahead of its time. Not like all that 'olde worlde' crap you get in museums. Museums are dead boring. I prefer modern stuff, me".

---

11. Who would you like to see win the World Cup in 1998?
(a) Brazil
(b) England
(c) Bishop Auckland

12. Your 12 year old son tells you he needs more pocket money. Would you do?
(a) *Ask him how much he wants, and give him the extra cash*
(b) *Suggest he goes out and gets a newspaper delivery round to earn extra money*
(c) *Give him an old broom, then shove him up the chimney for 16 hours a day, sticking pins in his heels if he becomes stuck, or tries to climb down.*

13. You suffer badly from arthritis of the knees and walk with some difficulty. An operation could solve the problem. Where would you ask your GP to refer you?
(a) *To a private clinic for immediate treatment by a leading consultant*
(b) *To the local NHS hospital for the same treatment, by the same consultant, but with a five year wait beforehand*
(c) *To the local gents hairdresser, to see whether he can sort the problem out with a bottle of gin and a hack saw*

14. Fellas. Which of the following birds do you fancy?

A

B

C

15. Girls. Which of the following fellas do you fancy?

A

B

C

16. Your uncle has just returned from India with a mysterious box which he keeps under his bed and refuses to talk about. One day you find he has been stabbed to death with a ceremonial dagger in a room which has been locked from the inside. The box has vanished and the only clue is a Chinese hat lying nearby.
The butler, a sinister Indian mute with heavy black make up round his eyes, saw and heard nothing. Which detective would you call?
(a) *Knight Rider (that lanky ponce off Baywatch who used to drive around in a black computerised car which talked in a puff's voice)*
(b) *Dixon of Dock Green*
(c) *Sherlock Holmes*

17. Your 16 year old son, who has a mental age of about nine, attempts to break into a sweet factory, but the police arrive at the scene and he gives himself up. What sort of punishment would you expect the authorities to give him?
(a) *They should send him on a safari trip to Africa with a social worker, to broaden his horizons*
(b) *They should send him to prison, for a short, sharp shock*
(c) *They should hang him*

18. Another son, who is also mentally retarded, is married and living in a rented flat. However his wife, who is pregnant, is murdered by the landlord, who has also murdered several other women. What action might you expect the authorities to take?
(a) *They should send the landlord to prison for life, and let him out after 7 years*
(b) *They should send the landlord to prison for life, and that should mean life*
(c) *They should hang your son*

19. If you were visiting friends in Australia, how would you prefer to get there?
(a) *By Concorde, a journey time of about ten hours*
(b) *Travelling overland, sight-seeing along the way, taking perhaps two or three months*
(c) *By sailing ship, a journey time of about a year, providing you don't die of dysentery and get thrown overboard somewhere along the way*

20. You're travelling by train from London to Edinburgh. How long would you reasonably expect the journey to take?
(a) *Between four and ten hours, depending on the weather.*
(b) *Between eight and ten hours*
(c) *Ten hours*

HOW DID YOU DO?

Award yourself one point for each answer (a), two points for a (b), and three points for a (c). Then tot up your total.

30 or less: You're a twentieth century kind of person. You love life in the space age. You own a computer, and eat Pot Noodles. You wouldn't be seen dead in a museum, or at a classical music concert.

31 to 49: You've got one foot firmly in the future, and another still living in the past. You tend to be nostalgic about the olden days, and watch UK Gold on satellite TV. But you drive a modern car, and wear fashionable jumpers.

50 or over: You're a real golden oldie. You love olde worlde things like antiques and black and white TV programmes. You drive an old fashioned car with a horn that you squeeze, and your favourite sweets are black bullets.

Save Our Scare Stories!

BARMY Brussels bureaucrats have poked their big Belgian conks into Britain's business again. Not content with banning bent bananas, swapping traditional pints for Froggie litres and insisting we take the Queen's head off our own coins, their latest potty scheme is to standardise the good old British *SCARE STORY*.

SAVE OUR SCARE STORIES

Euro-sceptic MP Sir Teddy Taylor joins the xenophibic Viz 'Save Our Scare Stories' campaign yesterday:

"Fuck them Belgians" he told us. "I'm giving up sprouts and Tintin books and I'm going to post a turd to Plastic Bertrand"

If Brussels get their way, from next April every tabloid newspaper article about European policy will have to contain a maximum of:

● 3 insulting references to Belgians, of which only 2 may alliterate.

● 2 references to the Blitz or Churchill.

● 1 reference to imaginary regulations concerning the straightness of bananas.

● 1 badly drawn cartoon of a bulldog or moustachioed frog wearing a beret.

● 1 'sound-bite' quote from either Michael Winner or Sir Teddy Taylor.

"It's an absolute outrage. Britain has always led the

Brussels waves the rules at Britannia

By our European Affairs Analyst
Billy 'Bulldog' Bigot

world in tabloid scaremongery" said Southend MP, Sir Teddy Taylor, responding to our wildly exaggerated version of the facts.

Churchill

"If these Belgian bureaucrackpots think we're going to take this lying down, they

can think again. Churchill didn't take it from them during the Blitz, and *we're* not going to take it now," he added.

Stannah

Millionaire film director Michael Winner was equally furious when we called him at his home at 3am. "It beggars belief!" he spluttered. "The French and the Germans would not put up with this nonsense, and neither should we. These Euro-prats are making a laughing stock of us."

Rude Kid 2000

LET'S RENT A NICE VIDEO DEAR

FUDGE YA DUDS BITCH!

Trees, the Law & *YOU*
with Tree Lawyer **Quercus Pubescens QC.**

A HORSE CHESTNUT tree in my garden overhangs the street and drops its conkers onto the pavement. On their way to school, children pick the conkers up to play with them. However, as the tree is rooted in my garden I believe that the conkers belong to me. I don't want them, it's just that I cannot abide theft. Am I correct and how can I stop them being stolen?

T Barnstaple, Looe

★ *Yes, Mr Barnstaple, the conkers are yours, regardless of where they fall and taking them from the pavement is theft. However, in the same way that you wouldn't leave your wallet on the street, leaving the conkers there is asking for trouble. Try gathering them up before the children set out for school, or if you don't want them, simply run over them with a garden roller.*

MY NEIGHBOUR has an apple tree which overhangs my greenhouse. In the summer, it drops its apples through the glass. We have known him for twenty years and he is a lovely, reasonable man. I am sure he would happily cut of the offending branches if I asked him, but I would like to instigate really costly and acrimonious legal proceedings. Have I got a case and how would I go about it?

Mr L Ridley, Carlton

★ *Indeed you have got a case, Mr Ridley. Your neighbour is being negligent in allowing the apples to overhang your greenhouse. Get your solicitor to issue a writ, suing for the cost of repairs. If he tries to come around to discuss it, take out a restraining order forbidding from coming within 50 yards of you.*

The man next door to me recently planted a forsythea in his garden. Over the years it has grown quite high and is cutting out a lot of light to my kitchen. Can I force him to prune it back?

Ron Hubbard, London

★ *Forsythea is a shrub, and as my area is tree law, I am unable to advise.*

SAT'S SAT!

A NOTTINGHAMSHIRE couple drove their car all the way to Mars's largest moon Phobos after following the instructions given by their car's sat-nav computer!

Brian and Maureen Bromide from Ragnall, near Worksop, programmed the £200 device to tell them the best way to get to Alton Towers. However, due to an error at the manufacturers, a journey that should have only taken ninety minutes took over 2 years.

Machine

Brian, 58, told reporters: "I thought there was something funny going on when the machine told me not to take the A38 but stay on the M1. I've been to Alton Towers many a time and I know the way quite well, but I thought it must know a quicker route."

bowl

Even when the onboard set told Mr Bromide to accelerate his Toyota Yaris in order to escape the gravitational pull of the Earth he carried on despite the complaints coming from his passenger seat. "My wife kept telling me we were lost but I just

Day Trip was Out of this World

turned the radio up," said Brian.

"Looking back, perhaps I should have listened to her!" he added.

Despite the lack of white knuckle rides, over-priced food and oxygen, the Bromides said they didn't mind their 100 million mile diversion to the red planet. "We had plenty of sandwiches and the weather was dry so it wasn't a total wash out," they told us in unison.

Botty Bungle Doc given Bums Rush

A Tyneside surgeon was last night suspended on full pay after a patient who was booked in for routine anal surgery had his elbow removed by mistake.

Gerald Birch, a 45 year-old professional accordianist awoke in North Tyneside General hospital to discover that the haemorrhoids which had plagued him for years were still there...*but that his perfectly healthy right elbow had been amputated.*

A clearly upset Mr Birch told reporters: "When I came round from the op, my piles were still throbbing like mad. I thought it was probably stitches from the surgery. But then I realised that my right arm was stiff, and four inches shorter than the left." he asked a nurse what was going on, and was told that there must have been a mix up in the operating theatre.

CROSSHEAD

Bungling proctological surgeon, Dr Farnley Eating-Charlesworth who performed the operation later visited the ward to apologise in person to Mr Birch.

"He told me that it was an easy mistake to make, and not to worry because elbows usually grow back in a few months", said Birch from his home in Bedlington.

CROSSHEAD

"I hope it does, because at the minute my life is ruined. I can't bend my arm at all, and my hobbies are line dancing, darts and weight-lifting. And doing press-ups. They're not going to get away with this. I'm going for compensation."

Last night, a spokesman for the North Tyneside Hospital

EXCLUSIVE!

Trust issued a prepared statement: 'Dr Eating-Charlesworth has worked at North Tyneside General hospital for twenty years, and this is the first time he has confused a patient's rectum with their elbow. He has agreed to take leave and will remain away while a full investigation is carried out.

Mr Birch later phoned reporters to add that he also liked doing monkey impressions and winding buckets up from wells.

Gerald Burch in happier times with bending arm, and (inset) bungling doc Eating Charlesworth yesterday.

Can **YOU** tell your... **ARSE** from your **ELBOW?**

Despite skyscraping GCSE and A-Level results, surveys show that a shocking 60% of school leavers still can't tell their arse from their elbow, and the consequences are clear to see every time you try to buy some batteries from one of the junior staff at Dixons. The Department of Education has bowed to pressure and will introduce a stringent new Arse/Elbow Distinction Paper as part of the new GCSE syllabus next year, but what about the rest of us? You might think you know better, but do you? Try our fun quiz, and see if YOU can tell your arse from your elbow...

1. An hour after a satisfying casserole, you need to go to the lavatory for a number two. You've locked yourself in the bathroom, folded over the right page of Exchange and Mart, and made sure there's plenty of toilet roll. You're fully prepared. But what do you do next?

a. *pull down your trousers, settle down on the toilet seat and dump your load*

b. *roll up your sleeves, dip the crook of your arm into the bowl and shit your pants*

2. While attempting to shift a particularly stubborn stain from the ceramic hob of your cooker, you realise the job would be easier if you could use a little more of which substance?

a. *Elbow grease*

b. *Arse grease*

3. In an effort to spice up your love life, your partner offers to take you "up the arse" using whatever they have to hand. What do you do?

a. *bury your face in the pillow and present your rear end to them like a bike rack*

b. *stand in an 'I'm a little teapot' pose and beckon them into the crook of your arm*

4. While stirring some soup on the stove, you accidentally sit in the saucepan. You feel a scalding sensation somewhere between your spine and your legs and call an ambulance to ask what to do next. Where would you tell the paramedics you are being burnt?

a. *Your arse*

b. *Your elbow*

5. On a coach trip to an all-male bonding session, such as a football match, you decide to liven up the tedium of motorway travel by making a saucy display out of the back window at the family in the car behind. What would you do?

a. *drop your trousers and press your bare buttocks against the glass*

b. *push the tip of your bent arm against the rear windscreen, pointing at it and laughing*

6. While driving through an area with poor radio reception, you are forced to tune into the strongest local signal, which is the Steve Wright afternoon show on Radio 2. It isn't long before you realise that Steve Wright is causing you a terrible pain. Where?

a. *In the arse*

b. *In the elbow*

HOW DID YOU DO?

Score 2000 points for every time you answered a. and 0 points for every time you answered b.

0-2000 ~ Oh, dear! You really can't tell your arse from your elbow. Like many school leavers you probably also have minimal onion knowledge and cannot distinguish between the buttered and unbuttered sides of bread.

2000-10000 ~ Though most of the time you can easily tell your rear end from a hole in the ground, you have a little more difficulty distinguishing it from the hinged joint at the midway point of your arm. But don't worry! Try doing the test again, and this time not getting as many questions wrong.

10000-12000 ~ Well done! You can certainly tell your arse from your elbow. Putting clothes on is a breeze, and you rarely make the mistake of falling asleep at a table with your head nestled in the crook of your arm. Whether it's sitting down or nudging someone out of the way, you know the right tool for the job. Your expertise is likely to lead to a high flying career in either underwear modelling or standing in doorways with your hands on your hips.

IS YOUR FELLA A CRISPY DISH?

HOW TO REVEAL HIS POTATO SNACK SECRETS

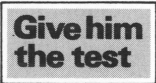

<u>Girls!</u> Ever wondered what makes your fella eat his crisps the way he does? Every fella likes a munch, but not everyone knows how to work out what those crisp eating habits really mean. It's easy when you know how!

Just read through our Crisp Info. File and soon you'll be impressing all your mates with your potato snack know how!

You can tell by the way a guy eats that everday snack exactly what he's really like. Here are some crisp eating types to be on the lookout for:

THE NIBBLER — eats his crisps slowly, nibbling around the edges like a mouse. This guy is shy and sensitive, but he probably won't have much money to take you out with.

THE WOLFER — eats his crisps without mercy, attacking them and taking no prisoners. He often eats a whole packet without stopping for breath. A heartless brute who cares little for friends or even his own mother.

THE TIME WASTER — can't decide whether he's eating or talking, he sometimes holds one crisp out of the bag while speaking. This guy is obviously a real bore, and not prepared to show his true feelings.

THE HOGGER — keeps every crisp to himself, related in many ways to the wolfer. Sometimes he tears the bag open to reach the very last crumbs. This guy is full of passion. He's eccentric, but adorable. Hang on to him!

Give him the test

Of course the way he eats his crisps isn't everything. There's more to be found out about your fella in the flavours and snack variations which he goes for. Find out about **YOUR** man by answering the following questions A, B or C.

1. You're at a disco and you tell him you fancy some crisps, but you don't say which flavour. Does he buy you:

a. *Ready salted*
b. *Pickled onion flavour*
c. *A fancy potato snack in interesting shapes and a colourful packet.*

2. You're going for a quiet picnic in the countryside. You have prepared some cucumber sandwiches and pate with posh wafer biscuits. You ask him to nip out for some crisps to take along. Which would he get:

a. *Bovril flavour*
b. *Salt'n'Shake*
c. *Tandoori flavour*

3. He's throwing a party at his place and is arranging a buffet. Which of these snacks would he provide with his saucy dips?

a. *Sausage'n'onion flavoured crisps.*
b. *Tortilla chips*
c. *A giant six-pack of Monster Munch.*

4. You've just got time for a quick snack on the bus and you have chosen a bag of Prawn Cocktail bites. What does your fella have?

a. *Worcester Sauce crisps*
b. *Spicy corn Space Invaders*
c. *A family size bag of Quavers*

How does he measure up ?

1. READY SALTED (A) are the sign of a real bore. He's got no imagination or style. A SILLY FANCY SNACK (C) shows he's a big head, so the guy who buys you PICKLED ONION (B) is the real dish!

2. TANDOORI crisps are **NOT ON** for a traditional picnic, so any guy who buys them should get the 'Big E'. BOVRIL (A) says he's no Romeo and is probably a compulsive thief. Keep an eye on your purse. The guy who brings back SALT'N'SHAKE (B) is a true romantic. Ooh la la!

3. SAUSAGE AND ONION crisps (A) are a sure sign of a trouble maker. He deserves a slap in the face! The MONSTER MUNCH man (C) doesn't have much taste and probably has difficulty coping with everyday social situations. But if you beau's for TORTILLA CHIPS (B), he's suave and sophisticated. Give that man a coconut!

4. A family size bag of QUAVERS (C) and you know your fella cares for nobody but himself. WORCESTER SAUCE (A) and you'd better beware — he may have some form of hereditary disease. Spicy-corn SPACE INVADERS (B) show all the imagination of a big hearted dreamboat. This man has good looks and flare and also enough money to take you out to the very best expensive restaurants.

> WOO-OOO-OOO-OOO-OOO-OOO-OOO!

Ghost Box

Britain's Most Pant Shittingest Letters Page

HALLOWEEN is upon us once more, and in the last issue, we didn't ask you to tell us your spooky tales, those little inexplicable events that send a shiver down your spine. But they came flooding in anyway. Here's a selection of the hairest raising stories we received...

It was a normal enough day. I had just got back from doing the school run and I was sitting down having a nice cuppa before making a start on the housework. All of a sudden I heard a noise. It was probably a ghost.

Ethel Breakdown, Truro

My husband John took this snap at a family wedding and it came out all blurred. The man at the chemist said the blurring was probably caused by psychic energy off a ghost.

Carol Mediocrity, Beeston

I don't usually pick up hitchhikers but on a rainy night last year I saw a young girl by the side of the road and stopped to offer her a lift. We chatted away and she seemed perfectly normal. When I dropped her off at her destination she said 'thanks' and off she went. I couldn't help thinking that maybe she was one of those ghost hitchhikers. I have no reason for thinking this other than it makes my otherwise dull life more interesting.

Brian Liar, Glasgow

My husband Dave, 42, and I were sitting in our lounge one night when all the lights mysteriously went out. The electricity board later informed us that it was a power cut but we still believe it was something more sinister, like a ghost.

Carol Idiotic, Chertsey

My house has always smelled of cigar smoke and our neighbour recently let slip that the old man who owned the house before us and passed away there, smoked cigars! Admittedly my husband smokes them too and is definitely the source of the odour, but it is still spooky.

Janice Nobody, Orpington

Every time I put something down in my house it gets mysteriously moved. For instance, the other day I put the post on the sideboard and when I got home it had been moved to the kitchen table - and opened too! My husband keeps telling me that it was him but I think he is just trying to protect me from the dreadful knowledge of the evil ghost that roams our house.

Susan Cakemixture, Dudley

My husband and I are both big Elvis Presley fans and we recently made a pilgrimage to Graceland, where I took this photograph of my husband looking at Elvis's grave. When we got the photos back, our blood ran cold. There, standing on the tombstone, was the ghost of Elvis Presley himself, in full Las Vegas gear. We were very disappointed, as we always thought he was best in his '68 Comeback Special era.

Doreen Bileduct, Hastings

Our house is built on the site of an old Roman fort. One night last year, I was watching the telly when suddenly the lights started flickering and the room went freezing cold. To my horror, an entire cohort of transparent Roman soldiers marched through the wall, across the room and out through the other wall. This went on for half an hour and was accompanied by the sound of horses' hooves and the clanking of armour. I was frozen with fear. When my husband came in from the pub I told him of my ghostly experience. You can imagine my relief when he explained it was probably a trick of the light.

Marjorie Steel, Chelmsford

I didn't believe in ghosts at all, until one night last month when the spookiest thing happened. Needless to say, I believe in them now.

Doris Pancreas, Hull

I visited Althrop House where Lady Di is buried and I saw her ghost in the toilets. Luckily, I had my camera handy and I took this photograph *(above)*. Strangely, I did not feel at all scared as, just as when she was alive, her spectre radiated goodness and did a lot for charity.

Mavis Gallbladder, Rhyll

The other night I was awoken by a loud thump from downstairs. I have always been terrified of ghosts, so you can imagine my state of mind as I crept downstairs to confront the supernatural horror I was certain lay in wait for me. I was so relieved when I opened the front room door to discover it wasn't a ghost at all, merely an knife-wielding maniac who had stabbed his way out of a local hospital for the criminally insane.

Ethyl Bromide, Leeds

I was working late the other night when I saw a skeleton. I nearly jumped out of my skin, and ran screaming to find the security guard, where I breathlessly explained what I'd seen. How foolish I felt when he explained that I am the chief radiographer at North Tyneside Hospital, and I had been looking at an X-Ray.

Dr Edward Canning, Luton

During a recent visit to Hampton Court. I took this picture of my husband standing by a grave. When I got the film developed, my blood ran cold. Beside my husband, who is an HenryVIII impersonator, can be seen the ghostly figure of a man in combat fatigues and a red fleece.

Ethyl Plywood, Hove

My husband and I moved into an old house last summer and I was disturbed when a neighbour told me it was known locally as the House of Death. She explained that in Victorian times the family who lived there had been murdered in their beds by an escaped lunatic who then dismembered and ate them. Last week on the anniversary of the killings I was woken in the night by loud screaming. My bed flew into the air and started spinning and blood began to ooze out of the walls and ceiling. The bedroom door then flew off its hinges and I saw a man, a woman and three children in Victorian dress, covered in blood. Then everything stopped as quickly as it began. When my husband came home from the pub, I was so traumatised I could hardly tell him what had happened. Imagine how relieved and foolish I felt when he explained that it was probably just an air lock in the central heating pipes.

Marjorie Spleen, Crewe

I thought I'd seen a ghost the other day when I walked into the living room and saw my late husband floating in front of me, two feet off the ground. Then I remembered he'd hanged himself off a beam earlier in the day and I'd forgotten to cut him down! How silly I felt.

Edna Hardboard, Torquay

I'd always believed in ghosts until I saw one flying round my bedroom making howling noises. However, it turned out to be a sheet with an owl under it. Needless to say, I stopped believing in ghosts there and then.

Ethyl Acetate, Surrey

Kids Say the Funniest Things about Ghosts!

...We were sitting down to eat our tea the other night, when every object that wasn't fastened down suddenly flew into the air and began to spin round the room wildly. Plates, cups, knives...even the microwave oven joined in this unholy dance, whilst the lights and TV switched on and off and doors, drawers and windows opened and slammed shut, pushed with great force by an unseen hand. My 4-year-old grandson was terrified, and hid under the table, shouting "Help! There's a polterghost in the room!" How we chuckled. He meant to say 'poltergeist'!

Edna Football, Goole

..."Granny, why is there a tall, cowled figure with a scythe standing behind you, beckoning you with a bony finger and filling the room with an all-pervading stench of death and decay?" asked my 4-year-old grandson the other day when he was visiting me in hospital. When I looked round, there was nobody there!

Edna Doomed, Dundee

Spookdoku

No. 3,285,483
Scare rating ★★★☆☆

The World's Most Frightening Number Puzzle!

Fill in the grid from beyond the grave so that each ghostly row, spine chilling column and ghoulish 3x3 square contains all the digits 1-9... **IF YOU DARE!**

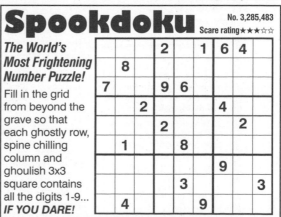

Never mind 'Back to Basics'. Let's

BACK IN THE B

A pub landlord is offering twenty pints for the price of nineteen – in an effort to get Britain back on the road to economic recovery.

Sid Fletcher, landlord of the Three Bulls Rings in Fulchester High Street, believes that most of Britain's social and economic problems are the result of insufficient lunchtime drinking. And he has launched a campaign aimed at getting Britain back in the boozer.

EVILS

He believes that two of today's greatest evils, high crime and unemployment, are directly attributable to people not drinking enough during midday sessions.

"If you compare this country now to how it was forty years ago, the differences are amazing. In the fifties there wasn't any crime, apart from that bloke who shot a copper on the sweet shop roof, and unemployment didn't exist. Now the streets aren't safe anymore, and there's no jobs."

LEVIS

Sid believes it's more than just coincidence that in the fifties there were hundreds

Major - 'back to basics' campaign

more pubs than there are today, and the vast majority of them were busy at lunchtimes.

ELVIS

"Nowadays pubs are usually quiet at lunchtime, with a few people coming in for a sandwich and a quick half. But I remember only too well how in the old days workers would come in from the factories and shipyards and drink ten and eleven pints in an hour. And those were the days when Britain's economy was booming."

Going to the pub is the key to Britain's success

Sid agrees with the current Government policy of 'back to basics', but he believes the most basic step of all is to get people drinking at lunchtimes. "I take hundreds of pounds every night, but I pay two barmaids to work lunchtimes and I'm lucky if there's fifty quid in the till by three o'clock." he told us. "If Mr Major wants to revive the economy he could give it one hell of a kick start by halving the price of beer between 11am and 3pm. The extra revenue that would generate in my bar alone would enable me to re-decorate the toilets and buy some new tables. And the knock-on effect would surely benefit the entire economy."

LIVES

In his attempt to lure workers out of their offices and workplaces Sid has come up with the innovative idea of giving customers twenty pints of lager for the price of nineteen, a saving of over £1.30. But there is a catch.

Boozing for Britain. A lunch-time boozer enjoys a lunch-time booze. In the boozer.

"The offer only applies if they can drink it all in an hour. It's not a lot to ask. A few years ago I had regulars in here drinking twenty pints at lunchtime, then coming back for another twenty before they went home for their tea. Now-adays people can't drink anymore. It's no wonder kids are growing up out of control, and committing crime, when parents are setting such a poor example."

10 tips on how best to behave at work after drinking 20 pints of beer at lunchtime

Unfortunately Britain's bosses aren't too keen on lunch-time boozing. But follow these ten tips and your boss need never know you've been drinking.

1. Sucking a mint will help disguise the smell of beer on your breath, but remember to buy the mints *before* you go into the pub as it may be more difficult to buy them afterwards.

2. If you have a desk job, try to spend the entire afternoon with your elbow on the desk and your head resting on your hand. Sit upright, holding a pen in your other hand. This will make you look busy, and thoughtful.

3. Do not start any conversations yourself, and if you are spoken to try to speak more slowly than you would normally do. This will counter the effect of your brain trying to speak more quickly than usual.

4. If your boss asks you a question, count to ten before your reply. Keep sentences short, stopping and counting to ten again between each sentence.

5. You may not realise it but your eyelids will naturally tend to drop. So make an extra effort to raise your eyebrows while talking.

6. Keep alert by trying to remember your postcode, and repeating it over and over in your mind.

7. Try to keep movement to a minimum. Do not walk anywhere unless it is absolutely necessary.

8. If you do have to go anywhere, to the toilet for example, choose a route which enables you to punctuate your journey by casually leaning on walls or items of fixed furniture.

9. If there is a patterned carpet in the room try following the pattern to enable you to travel in a straight line more easily.

10. Do not attempt to walk across an open space unless absolutely necessary. If you have to, under no circumstances look at your feet. Fix your eyes on an object in the middle distance and count each step in your head. Do not stop walking until you have arrived at where you are going.

get Britain...
OOZER

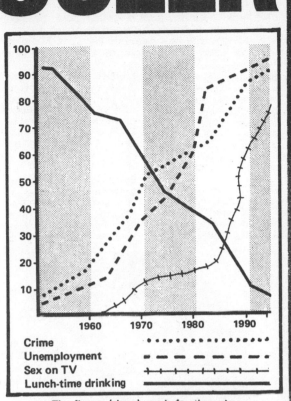

Crime · · · · · ·
Unemployment — — —
Sex on TV —+—+—+—
Lunch-time drinking ————

The figures (above) speak for themselves.

WE'RE BOOZING FOR BRITAIN

Let's make Britain Great again!

We're launching our own campaign to get Britain back in the boozer and restore traditional values, such as lunch-time drinking. And we want YOU to join in.

Simply pop into your local pub during your lunch break and drink as many pints of beer as you can. Have a sandwich first, to line your stomach, followed by up to twenty pints of beer. Then ask the landlord to sign the form below, and send it to us. In return we'll send you a

VIZ IS BOOZING FOR BRITAIN

splendid Certificate to prove that you helped get Britain back in the boozer, at lunch-time.

Complete this form then send it to: Boozing for Britain, Viz, P.O. Box 1PT, Newcastle upon Tyne NE99 1PT.

I can vouch for the fact that _____

was in my pub at lunch-time on _____ (date) and that

he/she drank _____ pints.

Signed _____ Landlord

Your address _____

Yabba dabba doo

A senior Tory MP is urging for a return to Stone Age values to help solve the problems of rising crime, unemployment and too much sex on TV in Britain today.

Sir Anthony Regents-Park, MP for Fulchester Sunnyoak, last night called for a 'back to basics' approach to be adopted in schools, prisons and society as a whole. And he has launched a plan to put Britain back on the rails, with an emphasis on a return to Stone Age values, and children being encouraged to behave in a prehistoric manner.

BLOW

"We need to instil discipline in our children, and what better way to do that than by following in the footsteps of prehistoric man. In cave man days bad behaviour was punished with a heavy blow on the head with a stone club. I think a lot of today's so called youth would think again before committing theft and murder if their parents wielded large clubs and wore animal skins", said Sir Anthony yesterday.

HAND

And the right wing MP had his own idea of how young offenders should be treated once they are caught and convicted. "It's an absolute nonsense sending these thugs and layabouts on holiday at the tax-payers' expense. They should make them live in caves, in just a pair of underpants. And if they get hungry, they should be made to chase after dinosaurs, and throw stones at them. Then we'd see how tough these people are. Let's see if they could kill a dinosaur, then eat it. I bet they couldn't".

PART-TIME

Sir Anthony believes that the solution to our present problems lie in the past. "We can learn from history. Things like the Battle of Hastings, for example, which was in 1066." And he believes that the current

Tory wants return to Stone Age values

decline in family values and the lowering of moral standards can be arrested if we learn from the pages of history.

BANK

"We have become too soft and liberal in our attitudes towards sex. We tend nowadays to refer to women who bring up a family alone as 'single mothers'. What a nonsense. In my day these people were called witches and whores, and, quite rightly, they were burned at the stake. So why not burn them now? Only by getting back to traditional values can we repair the damage that has been done."

SUMMER

Sir Anthony is set to table a motion in the Commons this week asking that prostitutes be ducked in the river Thames from the terrace at Westminster. "If they are witches they will not drown, and we can burn them", he told colleagues.

DESK

Meanwhile, Sir Anthony has insisted that he will not be resigning despite calls from his constituency party for him to quit. The calls came after Sir Anthony stabbed his secretary to death upon discovering she was pregnant with his love child. He then mutilated her naked body, arranging her entrails around her neck. So

Sir Anthony (above) and a Stone Age man yesterday (below)

far no charges have been brought against him.

BOB-A

Denying allegations of murder, Sir Anthony told reporters that the death of his secretary was a personal matter, and did not affect his ability to carry out his job on behalf of his constituents. In 1981 Sir Anthony was forced to resign from his post as Junior Minister at the Home Office after the sex slaying of four prostitutes in the Church-town area of Fulchester, a notorious red light district. Although no charges were brought against him, Sir Anthony's name became linked with the crime when the women's severed heads were found in a freezer at his home.

DINOSAUR ZOO.

WILL IT REALLY TAKE 10 VETS TO TAKE THE APPENDIX OUT OF THIS PREHISTORIC WOOLY-COATED PACHYDERM?

YES. IT'S A MAMMOTH OPERATION.

JESUS

Phil marriage No.2 in trouble

Rock star Phil Collins is reported to be 'not talking' to the new love in his life, Swiss model Orianne Geve. And the latest love bust-up comes hot on the heels of his multi-million divorce from first wife Jill.

Collins sailed into a new love storm a week ago. According to pals the upset began when Phil, 43, arranged to ring stunner Orianne on Friday evening when the couple were due to go out for a meal. When Collins called - at 11.30pm - his Swiss Miss was furious.

FRIDAY

"You said you'd ring on Friday evening", she is reported to have blasted. "It still is Friday evening", replied Phil according to pals. "But it's almost midnight", said the dark skinned beauty. "I know, but I said I'd ring on Friday evening, and Friday evening doesn't end till midnight", Phil is said to have continued.

HANDLED

While Orianne complained that she had missed her date, Phil rang former Genesis pal Mike Rutherford and asked him what time Friday evening ended. "Mike didn't really want to get involved in the row, but Phil pressed him and eventually he agreed that Friday evening didn't end until 12 o'clock midnight", and insider told us. But Orianne was unimpressed , and stormed off to bed without saying goodnight to millionaire Phil.

POWER

The following day Collins bought Orianne a large house in an attempt to heal the rift before jetting to America. But according to a close friend their relationship hit the rocks again the minute Collins returned. For Orianne had expected Phil to return 'later in the week'.

ALITO

When he arrived back at Heathrow on Sunday Orianne hit the roof. "You

Collins with his new bird yesterday

Collins 'not talking' to new love

said you'd be back later this week", she is heard to have shouted. "But it still is this week", Phil replied. Passengers in the V.I.P. lounge ran for cover as a furious row erupted.

AT C&A

"What day is the last day of the week?" Collins asked fellow V.I.P. travellers. "It's Sunday isn't it?" he said. "Well today's Sunday, so I did arrive back this week".

RAGE

According to onlookers Orianne was in a rage. "Everyone knows the week ends on Friday" she told him. "No. If it starts on Monday, then it has to end on Sunday", argued Phil. "No. Saturday and Sunday are different. They're the weekend. They aren't week days", replied his furious partner. "No. That's my whole bloody point", retorted Phil. "They're the WEEK END. That's because they're at the END of the WEEK". Embarrassed V.I.P. travellers, among them Lulu, pleaded with Phil to calm down. Eventually the couple left by taxi. "They weren't speaking the following day, although Phil did buy her another house later that afternoon", one insider revealed.

CRAP

Phil's marriage to first wife Jill ended last year in a bitter public love feud. Ladbrokes have now slashed the odds of Phil and Orianne divorcing this year, from 200 down to 25 to 1, despite them not being married yet. Meanwhile the odds on Collins' next record being a pile of crap remain unchanged at 5 to 4 on at Ladbrokes, with William Hill taking no further bets.

Wonder computer is bra-vellous

This issue marks a turning point for Viz. For we are now equipped with the latest hi-tec McApple computer, and a mouse. And, thanks to the magic of modern computer aided design technology, it is now possible for us to remove the bra from this picture of Catherine Zeta Jones.

Unfortunately the software package needed to perform this operation - Quark BraGone 2000 - was only available as an optional extra. In order to buy it, and remove Catherine's bra, we will need another £860.

That's where you, the readers, can help. If you want to see this picture without the bra, send us £1. Hopefully, by the time we publish our April issue we'll have raised the money we need, and with one simple twiddle of our mouse Catherine Zeta Jones' bra will vanish, and everyone will be able to see her tits.

Send your pound to: Catherine Zeta Jones' 'Bra Gone' Appeal, Viz, P.O. Box 1PT, Newcastle upon Tyne, NE99 1PT as soon as possible.

Catherine - May her Darling Buds spring out in the next issue? On sale March 31st.

Lonely heart? We reveal the secrets of sexcess!

GIRLS – HERE'S HOW TO HOOK A FELLA!

Girls! If you've got your eyes on a fella it's no use sitting around waiting for the fish to bite. Patience is no virtue when it comes to man hunting. If you don't get your guy, someone else will. Because these days, it's more often the women that make the first move. So in 1988 there'll be no excuses for being a lonely heart. Even the most dull and unattractive women can make a big catch, providing they play their cards right. And grabbing a guy has never been easier. Hooking a hunk is simple, if you know how. So here's a guide that will help YOU to fish for the dish of your dreams.

Where to start looking...

AT THE OFFICE

Take a good look around the office. Your dreamboat could be sitting at the next desk. Behind those Clark Kent glasses there could be a real *Superman*. If there's no phone boxes nearby, invite him into your stationery cupboard and see what he's wearing underneath his clothes.

Or maybe the boss is more your type? At the end of the day, ask him if there's anything he'd like to *go over* with you in his office, perhaps.

FISH

Even if you're out of work, there's still plenty of fish in the sea. Next time you sign on, *eye up* that fella behind the window. Why not scribble his *box number* in your little black book, and tell him you'd be interested in claiming some *extra benefits*.

HANDLE

Always keep your eyes open when travelling to and from work. Mr Right could be getting on at the next stop. Play your cards right and you could be *getting off* with him. If a fella offers you his seat, don't be shy. Politely refuse, then smile and sit on his knee. You could be in for a *bumpy ride*. And take a good look at the driver. He can handle a bus, but could he handle you? Ask him *how far he goes*, and whether he's *got room for you downstairs*.

Don't get left on the shelf

AT HOME

Don't panic if you arrive home without a date for the evening. There's no point in simply waiting for the phone to ring. Try the neighbours. Perhaps there's a *tasty dish* in the flat upstairs that you'd like to get your teeth into. Be neighbourly. Borrow a cup of sugar and offer him the use of your shower in return. Suggest you *share it* with him to save on hot water.

PHONE

If you have no luck with the neighbours, don't worry. There's still no need to dine alone. There's an endless supply of men only a phone call away!

● Be ATTRACTIVE at all times and wear EXPENSIVE clothes

Can you smell gas? Better to be safe than sorry. Ring the Gas Board and they'll send a hunky engineer round at the double. Wear perfume when he arrives. With his well trained nose he'll soon *get the scent!*

If you're choosy about your men, make a note of the fire brigade's number. You're guaranteed a choice of at least half-a-dozen fellas, all at least 5'8" tall. Set the table for several guests, and when they arrive tell them your fire's gone out, but you'd still like to *see them in action!*

CLOCK

If Mr Right still hasn't shown up by bedtime, don't despair. There's always tomorrow. Set your alarm clock for five in the morning. You don't want to miss the milkman!

CLOTHES

Dress to catch the eye. DON'T wear dull and dreary outfits. Throw out all your old clothes, and buy new ones. Don't be afraid to spend lots of money. Looking good is an expensive business, so open credit accounts with several clothes shops.

Wear BRIGHT colours at work to get you noticed. If another girl is wearing bright clothing, make sure yours are BRIGHTER. And change your outfit every twenty minutes or so. After all, men don't want to see you in the same clothes all the time.

KETTLE

At night wear something EXPENSIVE. A flowing evening gown or layered dress dripping with pearls. Wear lots of jewellery too, like Princess Di. There's a lady who got her man! Even if you're just nipping out to the corner shop, dress as if you were going to a ball. You may bump into Mr Right in the street outside.

FOOD

If you're OVERWEIGHT, lose pounds by NOT EATING ANYTHING until you're really slim and attractive. Work out a diet to make you LOOK GOOD and FEEL GREAT. Certain foods will make you look SEXIER. Eat lots of pork, and stock up on melons and tinned pineapple.

Next issue:
FELLAS!
How to pull a bit of crumpet

Tinker?...Tailor?...Soldier?...Sailor?...Cyclist?...Scientist?...Egg Farmer?...Tramp?

Oakey Dokey

Win £100 worth of fags in our Human League-tastic 'What's Phil Oakey Doing Now?' competition!

'YOU WERE working as a waitress in a cocktail bar, when I found you.' So sang eighties electro-synth pop maestro **Phil Oakey** out of The Human League. But twenty years on, there's only one question on everyone's lips... *What is Phil doing now?*

The rest of the band's whereabouts are well known: keyboard player **Martyn Ware** left to form Heaven 17 before quitting music to become an astronaut. He was poisoned by a leak of antifreeze on the Mir space platform, and now has a mental age of three. Guitarist **Ian Graig Marsh** similarly announced he had enough of the chart scene in the mid eighties and is now deputy leader of the Transport and General Workers Union. Both of **the girls** stayed in the music industry, except for a brief period when they became waitresses in a cocktail bar. After becoming the two girls in Bucks Fizz, they left to become the girls in Brotherhood of Man. *But what of Phil?*

Here are 8 things that Phil could be doing now... but 7 are made up. To win the fags, simply tell us which one is TRUE!

Oakey JOKEY?
Phil travels the country working as a clown in Jerry Cottle's Circus. He specialises in filling his trousers with custard, climbing a ladder with wallpaper and driving an exploding car. ①

Oakey CHOKEY?
Convicted in 1992 of aggravated rape, Phil was sentenced to 15 years in prison. He was denied parole in 1998 following a fight in which another inmate was glassed in the eye. ②

Oakey SMOKEY?
Phil sank his record royalties into a tobacconists shop in Rotherham. He now has a chain of 15 throughout S.Yorkshire and sells over £20000 worth of pipes and ciggies each week. ③

Oakey BROKEY?
When the hits stopped, Phil's jetset lifestyle caught up with him and in 1991 he was declared bankrupt. He now lives on a bench in Roundhay Park, Leeds. ④

Oakey SPOKEY?
With time and money to spare, Phil pursued his hobby of cycling. He joined Team Cofidis and competed alongside David Millar in the 2002 Tour de France, finishing a creditable 7th. ⑤

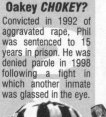

Oakey CROAKEY?
After The Human League, Phil took a degree in Amphibian zoology at Sheffield Hallam University. He is now professor of frogs at top American college Harvard. ⑥

Oakey YOLKEY?
On the death of his uncle, Phil inherited a 50% share in an egg farm. He now has over 6 million chickens and is responsible for 25% of the UK's total egg production ⑦

Oakey STROKEY?
Phil lapsed into a coma after a blood clot from his knee blocked the major artery feeding his brain. He is now in a persistent vegetative state in the Royal Hallamshire Hospital.

⑧

Send your answer, along with your name and address to *'Oakey Dokey Competition, Viz Comic, PO Box 1PT, Newcastle upon Tyne, NE99 1PT'*. The competition is not open to anyone under 16, unless the fags are for their dad. All entries must be in by 19th April. The prize has been donated by the 'Together In Electric Dreams Tobacconists' High Street, Rotherham.

 **Monkey business pictures pose constitutional crisis for Palace**

DI IN 'ZOO MONKEY BLOW JOB' SHOCKER

The vicious publicity battle which has been raging between the Prince of Wales and his embittered wife Princess Di took a sordid turn today when an Australian magazine published explicit photographs which it alleges show the Princess having oral sex with a monkey in a zoo.

The blurry pictures, taken from a considerable distance with a telephoto lens, were sent to the New South Wales offices of Australian magazine 'Viz'. They are believed to have been sent by a close colleague of Prince Charles, and are the latest attempt by the Prince of Wales to undermine his wife's popularity.

BITTER

In a bitter public feud lasting several months the Royal rivals have traded stories with the press, each designed to damage the other's reputation. Di, it is alleged, with the help of her close allies, has leaked sensitive information to the tabloid press aimed at discrediting her husband and skuttling his chances of becoming King. Meanwhile Charles is believed to have orchestrated a similar campaign aimed at branding Diana a tramp and an unfit mother.

MILD

We were sent copies of the Monkeygate pictures as long ago as June 1991, but we declined the offer to publish them, refusing to be pawns in the bitter Royal slagging match between Charles and Di. Instead, we

CAMILLA: Climbed tree for kinky love session with Charles

More FILTH on the Royals

sent them to our wholly owned Australian subsiduary and suggested they publish them instead.

HEAVY

Buckingham Palace yesterday claimed to have no knowledge of the pictures. "If Diana has sucked off a monkey, which wouldn't entirely surprise me, I wouldn't particularly want to see the pictures anyway", a spokesman for the Queen told us.

STOUT

But the signs are that Princess Di's camp are preparing to hit back with a major revelation about Prince Charles and his alleged mistress Camilla Parker Bowles. A private detective claiming to have been hired by Princess Di, last night offered us exclusive pictures which he claims were taken of Camilla and Charles at a polo meeting in 1990 – almost a year before the Monkeygate pictures broke. The detective, who refused to be named, claimed that his pictures showed Charles, Camilla and a polo pony in a compromising position.

TUBBY

"There's one great shot of Charles taking the pony from behind, while Camilla is up a tree pissing on them", he boasted.

OBESE

As a family paper loyal to our Royal family we refuse to print such scandalous and filthy photographs. We handed the pictures, together with our dossier on the case, to the editor of the Australian 'Viz' magazine.

Smooch at Ten

ITN gets physical in the battle of the newshounds

Dermott and Trevor - on-screen smacker could prove winner in the ratings war. Yesterday

THE RATINGS WAR between the BBC's Ten O'clock News and ITV's News at Ten is set to hot up - when ITV screens the first GAY KISS between newsreaders later this week.

In the shocking scene, viewers will see veteran anchorman Sir Trevor McDonald finish a report on a Guatemalan earthquake, before turning to give a lingering kiss on the lips to co-presenter Dermot Murnaghan.

EXCLUSIVE

ITN bosses are bracing themselves for an onslaught of complaints when 10 million viewers watch the controversial clinch at the close of the episode to be broadcast on Wednesday.

Murnaghan- kiss

Witchell - bad boy

Ford - astronaut

Derham - baby

bolster

But they last night denied that it was merely a cheap ploy to bolster viewing figures for the flagship show. "This gay kiss is no gimmick," insisted ITN chief Ian Bevitt. "We like to think that our news programmes reflect real life issues, and like it or not homosexual relationships

Bevitt - denying gay kiss is gimmick, yesterday

are a part of everyday life."

Meanwhile, a TV insider hinted that ITN has many more dramatic plots in the pipeline to keep viewers on the edge of their seats.

duvet

He revealed top secret plans including a mystery virus sweeping through the newsroom and Katie Derham having an illegitimate baby to a mystery

weatherman in the Autumn.

"It's a cracking plot," he told us. The father's identity will be revealed in a special double-length edition on Christmas Day.

pillow

Meanwhile BBC bosses denied using spoiling tactics and dismissed as "pure coincidence" reports that they are planning to steal ITN's thunder. News chief Tony Hall is believed to have given the green light to a sensational two-part cliff-hanger starting on Tuesday night.

fist

Viewers will see heart-throb newsman George

Alagaiah trapped in a blazing studio after BBC bad boy 'Nasty Nicholas' Witchell discards a cigarette in a pile of news reports. The public will only find out if he escapes the inferno by watching Wednesday's Ten O'clock News - scheduled directly against McDonald and Murnaghan's gay kiss.

feltch

The BBC has recently been under pressure to make its news presentation more realistic. Over the past few years, it has lost viewers with such far-fetched storylines as a lesbian siege, Anna Ford marrying an astronaut and Jill Dando being murdered by a roller-skating Freddie Mercury impersonator.

Well Knock Me Down!
with A. Feather

Arriving for her wedding in 1964, Scarborough bride-to-be Brenda Allsopp could only look on in horror as the clifftop crumbled away in front of her, plunging the church containing her marriage ceremony 300 feet into the sea below, never to be seen again.

Forty years later, whilst on a fishing holiday in Australia, Brenda hooked a blue whale and hauled it up onto the dock. Upon slitting it open, she was stunned to find what looked like a church inside the beast's stomach. Her surprise turned to outright amazement when she opened the door...to find her own wedding - including the vicar, her husband-to-be and all her guests - waiting patiently inside!

Pensioner Albert Hardwick couldn't believe his eyes when he bought a punnet of strawberries at his local market in Aukland, New Zealand. For there, on the top of the pile, was the exact same strawberry that had been stolen from off his plate by a cheeky seagull during a childhood picnic in Kent, England, when he was just 5 years old.

How the fruit had retained its freshness during its 80-year, 12,000 mile odyssey is a mystery that may never be solved. But one thing's for certain - Albert wasn't going to let it get away a second time. He ate it straight up!

Illinois office worker Doreen Margolyes couldn't believe her luck when she logged onto an astrology website on her computer. A flashing window appeared, informing her that she was the millionth visitor to that site and inviting her to enter a prize draw to win $50,000.

But her amazement turned to incredulity just ten minutes later when she visited a teddy bear collecting website, only to find she was the millionth visitor to that page too! Mathematicians have calculated that the odds against the same person being the millionth visitor to two websites is over 1000 billion to 1. That's the same chance as being struck by lightning 100 times... every day... for your entire life!

Arthur Feather.

Arthur Feather is the Professor of Strange Facts at Oxford University

ROCK A BYE El

on the to the pop

To millions of fans Elvis Presley was, and still is, (and always will be) quite simply The King. Of Rock 'n' Roll. And when he died, on the lavatory, wearing a nappy, in 1977, the whole world mourned the death of a star we had all come to know. And love.

But just how well *did* we know him? Not as well as we may have thought, according to a new book published this week. A book which takes an unusually intimate look at the private man behind the public face of Elvis Presley.

'Elvis – His Life and his Bedtime Routine' is a no-holds-barred biography which focuses on the end-of-day behavioural patterns of the man from Memphis who to many became the voice of rock'n'roll. Here, in a brief extract from the book, author Jimmy Hill, better known for his football analysis and big chin, gives us an insight into the lifestyle of a star; his supper times, his choice of late night viewing, his preferred bedtime drink. This, for the first time, is the *real* Elvis.

> Adapted from
> the book
> **'ELVIS - HIS LIFE
> AND HIS BEDTIME
> ROUTINE'**
> by Jimmy Hill
> *Published by Honey
> Nut Loop Books
> at £19.95*

'On 16th July 1977 Elvis Alan Presley set his bedside alarm clock for 8.30am, then pushed up the little knob that primed the bell to go off in the morning. Then he kicked off his slippers – first the right, then the left, and lay back his head on the pillow. Seconds later Elvis Presley was asleep.

ROUTINE

It was a routine that he had been through a thousand times before. But this time it was different. This time was the last time. For the next day, on 17th July 1977 Elvis Alan Presley was found dead, on the toilet.

GRACELANDS

Elvis' bedtime routine had been perfected over the many years during which he'd lived in his Gracelands mansion. In fact, in those latter years of his life, it was often said that neighbours could set their clocks by the time Elvis switched off his bedside lamp – 11.25pm exactly. But it hadn't always been like that.

As a youngster living in Tennessee, America, Elvis Alan Presley had often gone to bed early. His mother, Mrs Presley, had insisted that as a toddler the young Elvis should be in bed, teeth brushed and lights out, by 6.30pm. There, after a goodnight kiss from his mum, the young Elvis would lie awake and dream of becoming the King of rock'n'roll. By his side would be a little Teddy bear. A Teddy bear that would later be immortalised in the worlds of the song 'It's Now or Never'.

ADVENT

But all that was to change with the advent of World War Two, and Elvis' life and bedtime routine were turned upside down. Still in his teens, Elvis was drafted into the army and sent to fight in Germany. And army life came as quite a shock to the young boy from Tennessee, America. It wasn't only his haircut that changed. So did his sleeping habits too.

PIRELLI

Lights out in the army was 8.30pm sharp, though the young Private Presley would often stay awake reading by torchlight until nine o'clock. Or even ten o'clock if it was a good book. Sleeping conditions were cramped. Yet incredibly enough in the five years he spent in the army Elvis never once slept on the top bunk.

DUNLOP

After the war Elvis returned to America and had his first taste of success, topping the charts on both sides of the Atlantic with hits such as 'Hound Dog' and 'We're Caught In A Trap'. But with success came problems. Mrs

Presley still insisted that Elvis be in bed asleep by no later than 9.30. But Elvis, whose stage antics had earned him the nickname 'Pelvis', wanted to stay up late. Increasingly the rising star began to fall under the influence of Colonel Saunders, a mysterious figure who was later to manage his career.

GOODYEAR

The Colonel and Mrs Presley agreed a compromise whereby Elvis would have his supper and change into his pyjamas by 9.30. He was then allowed to watch TV until ten o'clock before going upstairs to the bathroom, brushing his teeth, washing his face and hands, and then going to bed.

GOODNESS

As during his army days, the King would occasionally read for a while before going to sleep. When he eventually became tired he would mark the page by folding back the top corner, then close the book and place it on his bedside table. *Remarkably, throughout his entire life Elvis never used a bookmark.*

A LIFE IN TOOTHBRUSHES

'Wise men say, only fools rush in'. And that was never more so the case than in the case of Elvis Alan Presley who never rushed in to a shop to buy a toothbrush. Incredible though it may seem, throughout his entire life the King of Rock'n'Roll never once bought a toothbrush for himself.

As a child his mother had always bought toothbrushes for him, possibly choosing red ones, as red was probably his favourite colour. In later life Elvis became less fussy about the colour of his

Young Elvis yesterday

toothbrushes, some of which would be bought by his mother, and others by his mentor Colonel Saunders.

MIDDLE

Occasionally, if Elvis was staying at a hotel and he'd forgotten his toothbrush, he would ring reception and ask if they'd got any toothbrushes. If they had some he'd ask for one, and if they didn't he'd maybe send someone out to a shop to buy one.

SIDE

In later years Elvis experimented with those bendy toothbrushes that can reach into difficult corners of your mouth, and at one stage an electric toothbrush was delivered to his mansion in Graceland, California. However, it was never used. For when, on 18th April 1980 Elvis Alan Presley was found dead, his electic toothbrush was found, still in its box, unopened, on a shelf nearby.

ELVIS p of tree

Shortly before his death on the toilet Elvis was so fat (above) he could hardly fit into this photograph.

Elvis (with fold through head) wants to play guitar, but mummy Elvis (right) and Daddy Elvis (left) tell him 'It's time for bed'.

Eventually Elvis succumbed to the lure of Hollywood, and in 1969 he moved to Las Vegas. So began the Vegas years, an era when Mrs Presley's influence on her son began to wain. It is widely acknowledged that during these years Elvis began to drink beer, and take drugs, and this gradually began to take its toll on his bedtime routine.

SAKE

His bedtime got later and later. On several occasions he was still up and running about at eleven o'clock. He would watch films until yon time, sometimes drinking cocoa *after* he'd brushed his teeth.

HIPPY

He began to wear the same pyjamas for days on end without washing them, and then stopped wearing pyjamas at all. Instead he would sleep in the vest and underpants that he had been wearing all day. On one occasion it is rumoured that he even fell asleep in front of the television and awoke the next morning, never having been to bed at all.

For Elvis the end was in sight. Once a young man from Tennessee, America, Elvis Alan Presley had risen to the heady heights of the rock'n'roll tree. He had scaled the topmost pinnacle of rock, only to roll down the other side.

HIPPY

It is perhaps ironic that a man who spent so much of his life in bed, or about to get into it, and making preparations for getting into it, should not die in his bed. For Elvis once said to Colonel Saunders "Don't ever let me die in bed, Colonel Saunders".

SHAKE

And that dream came true. For on the 18th of April 1980 Elvis Alan Presley died. Not in a bed, but on the toilet. A sad but fitting end to a legend that will live forever. Even though he is dead, yet shall he live.

Elvis Alan Presley. Born 1955. Died 1980. Long live the king.

BLOCKBUSTER BOB'S FEET IN A MUDDLE

Bob Holness has come a long way since the sixties when he used to present some kids, TV programme or other in black and white.

But fans of the millionaire Blockbusters host are probably unaware that plucky Bob battled his way to the top of the telly tree despite a serious disability. For 44 year old Bob was born with his feet on the wrong legs. And getting his TV career off the ground was made almost impossible by the fact that his feet were the wrong way round.

WINDMILL

"It was very embarrassing at times", recalls Bob, whose luxury home is a £2 million converted windmill in the Lake District. "At school I would trip over and get myself into a muddle, and when I left I found it difficult to find work".

ICE HOUSE

Bob failed literally hundreds of TV auditions because of his condition. "Producers didn't want to know when I turned up for auditions with my feet on the wrong legs. They'd take one look and say 'Forget it'. I was turned down by Tomorrow's World, The Black and White Minstrel Show, Dr Who, Animal Magic and Grandstand – all in one afternoon".

SNOW CHAIN

Having your feet on the wrong legs can be an expensive business too. "Of course I could never buy a pair of shoes to fit me", recalls Bob. "Even today I have to buy two pairs of shoes, then throw half of them away".

But despite his age – Bob will be 67 this year – there are still no thoughts of retirement for Britain's favourite elderly kids' TV quiz show host.

SUN ROOF

However, at 67, telly veteran Bob may soon be hanging up his TV quiz boots. "I've been in this game for a long time, despite my unusual feet. And it may soon be time to call it a day, and make way for someone a little younger. And with their big toes on the inside".

 EXCLUSIVE

But one thing is for sure. Bob won't be quitting his role as host of TV's Blockbusters. "You're as young as you feel, and I certainly don't feel like packing it all in yet. I never was one for gardening", said Bob yesterday.

FOG HORN

But Bob's retirement promises to be anything but relaxing. "I'm often busier at home than I am in the TV world", he admitted. And with two young kids, a hungry wife and a sizeable garden to look after, Bob will have his hands full.

Give us 'B' please Bob for Blockbuster's Bob Holness wearing a tie yesterday.

"One thing's for sure", quipped Bob yesterday. "I won't be sitting back and putting my unusual feet up for some time yet".

At the last count TV millionaire Holness was estimated to be worth £132 million.

Physical peculiarities of the brainy quiz show hosts

TV starter for ten brain box Bamber Gasgoine, arch telly rival of Blockbusting Bob Holness, had his ears on backwards for many years before undergoing corrective surgery for the problem.

HAIL CAESAR

ITV bosses used clever angles and mirrors to disguise swot Bamber's funny ears during filming of TV's University Challenge.

Telly brainboxes Bamber 'G' (with hair) and Robert Robinson (without hair)

GALE TILSLEY

Ask The Family quiz host Robert Robinson has always refused corrective surgery on his peculiar eyes. Call My Bluff question master Robinson, one of the Beeb's brainiest quiz show hosts, was born with his eyes upside down.

Death Doc Jaunt on Public Cash!

Public shells out for Shipman trip of a lifetime

THE WOMAN who lived next-door-but-two to record-breaking serial killer **Harold Shipman** is going on a sensational three-day luxury caravanning break in Anglesey... *and YOU'LL be footing the bill!*

By our Exclusives Correspondent
Xavier Clusive

Lucy Hartless, 42, who pockets a whopping £16,000 of taxpayer's money each year as a staff nurse at St Ratner's Hospital in Manchester, plans to blow almost £85 on a no-holds-barred long weekend at a top Welsh caravan park.

knees-up

The news that the woman who lived only three doors away from Britain's most notorious serial killer is planning a lavish knees-up at the expense of decent, law-abiding Britons has outraged moral indignation groups.

chin-up

"It's horrific," steamed campaigner Clive Perfect of Staying Annoyed. "This greedy woman is having a three-day beano in Anglesey, with a black-and-white television and cold running water, with money taken from our taxes – money that could be spent on schools or hospitals."

hands-up

And as Hartless prepares for the holiday, further shocking details have emerged of her extravagant lifestyle, funded by

Jekyll and Hyde park ~ the luxury £85 per weekend caravan site where Hartless will live it up at YOUR expense.

the ordinary British citizen. According to a colleague, who asked to remain nameless, the notorious late murderer's blonde ex-neighbour;

- Travels everywhere by *BUS*, always paying in cash
- Owns a *COLOUR TV* set and video recorder
- Takes her mother to *BINGO* every Saturday
- Plans to replace her twin tub with a *BRAND NEW* washing machine

cock-up

"It beggars belief," fumed Perfect. "I am sickened that this woman can continue to grab government salary handouts working in the Accident & Emergency ward of a hospital, whilst the late

Dr Death (right) ~ multiple mass murderer Shipman, yesterady and (left) neighbour but two Lucy Hartless

Shipman's victims don't get a penny. I'm sure many of them would have loved the chance to go gallavanting off to Anglesey."

Tomorrow, a 25-signature petition will be handed to the Prime Minister's driver by a group of protesters from Staying Annoyed. In it, they describe heartless Lucy's luxury holiday as "an outrage," and call for her to be killed.

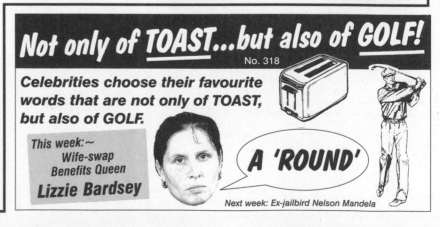

New Neptune 'space breasts' theory divides astrologers

PLANET OF THE TITS

By our new Science Correspondent
CLODAGH RODGERS

Space men of the future could be in for a treat when they eventually set foot on the planet Neptune. For new evidence would seem to suggest that the planet's surface is completely covered in large women's breasts.

And rather than 'one small step for a man', it could be a case of 'lots of enormous knockers for everyone', as the first pictures of the planet appear to indicate giant bosoms the size of dustbins.

JUNK

So says amateur astrologer Cedric Herringbone, who has spent the last twenty years gazing into space through a telescope which he bought in a junk shop. "I have been monitoring Neptune for several months, and on a clear night you can just make out the giant tits, all wobbling about as the planet orbits the Sun."

"Of course Neptune is hundreds if not thousands

Mr Herringbone (centre) with his powerful space telescope. Astrolomer Moore (left) - he poo poos space tits. And the planet Neptune (above) as it appears to the human eye.

of miles away from the Sun, and so it is a lot colder than Earth. As a result the space nipples on the planet surface are all hard, and standing up. Like organ stops'.

DINGHY

But the science world has been slow to acknowledge Mr Herringbone's theory, and indeed there are some space experts who disagree, among them baggy clothes, funny eye TV space boffin Patrick Moore.

CATAMARAN

"I have come across many such theories in my time, such as the Bell Ends of Pluto, and the famous Moon Fannies. But none of these have ever been proven, and I tend to prefer the theory that planets are made out of space rock, with lots of craters on them, like in The Clangers".

CANOE

Cedric Herringbone is no stranger to controversy. In 1989 he made headlines when the second hand bookshop he owns was raided, and several boxes of pornographic material were confiscated. On that occasion he received a formal caution after police officers discovered his telescope pointing into a neighbour's bathroom window.

LULU TO GET VISITOR CENTRE

The Queen is to open a new Lulu visitor centre in the autumn, built at a cost of over £250,000.

The centre, which is being jointly funded by Lulu and the Tourist Board, will provide much needed restaurant facilities, a picnic area, an information kiosk and toilets for the estimated 7,000 people who visit Lulu each year.

BARGE

A spokesman for Lulu said that the new centre would provide much needed facilities for fans who had previously had to simply ask for an autograph, make nervous small talk and then go home. "This development will make Lulu a world leader in terms of on-site facilities. And we hope by next year to be going ahead with a car parking scheme for Lulu that will cater for 35 cars".

KAYAK

Lulu is the first British star to open a purpose-built visitor centre, although a Craft Shop, selling woollen jumpers, pottery and ethnic

EXCLUSIVE

Free parking planned for 35 cars

jewellery was opened in 1979 at Michael Parkinson. This was extended at a cost of £14,000 in 1985 to include a small cafe with seating for 14.

Ambitious plans for a multi-million pound Trevor McDonald exhibition centre, hotel and conference facility, due for completion by 1996, have been shelved only days before construction work was due to begin at the popular newsreader.

FNARR

The scheme, which was to have included a 10,000 seat indoor arena, swimming pool and leisure club, was dropped in the light of speculation surrounding the future of ITN's flagship programme The News At Ten.

Parkinson - Craft shop with small cafe

Billy the Fish

AFTER A DRAMATIC HUMAN CANNONBALL STUNT GOES DISASTROUSLY WRONG, MICK HUCKNALL OUT OF SIMPLY RED IS LYING UNCONSCIOUS IN THE ZOO AND ABOUT TO BE EATEN BY MONKEYS. MEANWHILE - THE COUNT IS STILL ON TO DECIDE IF READERS HAVE SENT IN ENOUGH 20ps TO END THIS SERIAL FOR EVER.

WOW BOSS, THIS ONE FRAME EPISODE IS A BIT OF A COMEDOWN FROM OUR USUAL TWO-PAGE EXTRAVAGANZAS.

YES BILLY.

IS THIS THE LAST EPISODE? DON'T MISS THE NEXT EPISODE TO FIND OUT.

20 THINGS YOU NEVER KNEW ABOUT MUESLI

Ask somebody what their favourite breakfast cereal was fifteen years ago and they'd have said 'Cornflakes'. But if you ask that same question today they'll probably say 'Frosties'. Anyway, no matter where you do your shopping, you're bound to find a packet of MUESLI somewhere along the shelves. Because like it or not, muesli is here to stay. So whether you're a connoisseur or a dabbler, whether you're addicted to it or think its crap, here are the surprising FACTS about muesli:

1 Muesli comes from Switzerland, or somewhere like that.

2 'Money doesn't grow on trees', so the saying goes. But muesli does. Because it contains dried fruit and nuts, and lots of other things as well.

3 There are many different types of muesli, ranging from cheap brands, to the very expensive. Expensive ones usually have more fruit and nuts in them.

4 Up until a few years ago, muesli was served with cold milk as a breakfast cereal — and it still is today.

5 Muesli is sold in both cardboard boxes and polythene bags. The plastic bags often contain more muesli than the cardboard boxes.

6 However this is not always the case.

7 The word 'muesli' contains an equal number vowels and consanants.

8 So does 'teapot'.

9 Muesli is a truly international breakfast cereal. For example, in France the phrase "Il n'ya plus de muesli", means "there is no muesli left". "Es gibt keine muesli mehr" would be equally bad news at a German breakfast table.

10 Ask for a drink in a 'muesli bar' and you won't get one. Because muesli bars are pocket sized muesli snacks, like 'Mars Bars' and 'Twix'.

11 It is now possible to make 'home made' muesli. But as well as needing dried fruit, nuts, oatflakes etc., you will also need something to put them in.

12 Cricketer Ian Botham is unable to eat three shredded wheat. However, there is no information available on how much muesli he can eat.

Cricketer Ian Botham — his capacity to eat muesli is unknown.

13 There were more boxes of muesli sold in Britain last year than there are sheep in Yorkshire. Probably.

14 Muesli is used extensively in the ship building, textile and off-shore oil industries, as a breakfast cereal.

15 If all the muesli in Europe was thrown into the Mediteranean Sea, it wouldn't taste very nice afterwards.

16 The word 'muesli' immediately follows 'mud guard' in the English Dictionary.

17 The word 'muesli' does not appear in the Latin dictionary.

18 Niether does 'mud guard'.

19 It would probably take something like 8,000 tons of nuts, a fleet of removal vans loaded with oatflakes, a million apples, twenty-five billion grapes, a thousand acres of wheat and tons of other things as well to make enough muesli to fill the Albert Hall.

20 And you would need a spoon as high as high as the Post Office Tower to eat it with.

CUTHBERT STOKES AND HIS TEDIOUS JOKES

386

❏ MY husband is a doctor and he always insists on repairing his own boots. Every time he sets to work, he chuckles to himself and says "Physician, heel thyself". I haven't the heart to tell him that the correct quotation is "Physician, heal thyself", meaning "cure".

Mrs J Bodkin Adams, Eastbourne

❏ MY wife used to be a pantomime actress, but left the theatre to take up a post working in the septic haemorrhage department of the London Hospital for Tropical Diseases. She has to wear wellingtons at work because she spends so much time wading through mucupurulent discharges on the floor. She often jokes that whereas she used to be in

Puss in Boots, these days she's in boots in pus!

George Corbet, London

❏ WHAT a load of rubbish these so-called safety boots with steel toe-caps are. They certainly didn't prevent any injuries happening to the person who pushed in front of me in the taxi queue on Saturday night.

Big Jim, Byker

❏ I thought my boyfriend would find it a turn-on if I he saw me in thigh-length boots, but it didn't work out like that. I went into the branch of Boot's the Chemist where he works, in the Shropshire town of Thighlength, to buy some tampons and pile ointment, and I have to say that it didn't do anything to spice up our relationship.

Renee Bradawl, Thighlength

❏ AS a keen gardener, I usually end up with a lot of leftover fruit stored in my

These Boots are Made for WALKEN

Hi, I'm Christopher Walken out of *The Deerhunter*, and when I'm not appearing in over 100 films, including *The Deerhunter* and over 99 more, I love to read your fascinating letters about boots. Anyway, that's enough about me, let's see what's been in my "boot"-iful mailbag this week!

Britain's Brightest Boot Forum, edited by Christopher Walken out of *The Deerhunter*

shed over the Autumn. In the past this often got nibbled by mice, but then my wife suggested storing it in an old wellington. My wife and I are forever joking that, instead of a pair of boots, we've got a boot of pears! Unfortunately, after being left for several months festering inside one of my old rubber galoshes, the pears taste fucking disgusting.

Urquhart Ffitzsimmons, Surbiton

❏ MY grandad bought a pair of climbing boots during

the War and do you know, they're still as good as new. That's because they didn't fit him so they're still in the box on top of his wardrobe.

Sidney Durbridge, Tadcaster

❏ AFTER 40 years working for the same company, I recently got given the boot. But I wasn't upset, because it was a golden boot with a clock in it, presented to me by my boss - the chairman of Dr Marten Boots Ltd. However, I had to give it back when I was sacked a couple of days later for stealing a typewriter.

Bert Plywood, Northampton

❏ AS AN evil scientist, I have just completed construction of a giant, invincible metal robot, equipped with Death-Ray laser eyes, which I intend to unleash upon every major city in the world, causing untold havoc and destruc-

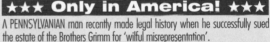
tion. Its feet are made out of steel in order to crush anyone foolish enough to stand in its way - and ironically they are made out of car boots, which I bought at a local scrapyard!

Professor Skull, Death Island

❏ "IS there a doctor in the house?" asked the theatre manager, when an actor had collapsed on stage during a play I was attending. "Yes, there's two here," I replied, before making my way up onto the stage. Once there, I pointed to my boots and said, "My Dr Martens!" Unfortunately the actor died, which took a bit of the shine off my quip.

Jack Chisholm, Spermford

Your Boot Jokes

Knock-knock!
Who's there?
The Duke.
The Duke Who?
The Duke of Wellington boots!

Mrs F Discharge, Dundee

Knock-knock!
Who's there?
Duke.
Duke Who?
Duke (W)Ellington! (boots, like the jazz band man).

Mrs F Discharge, Dundee

Q: What sort of boots does David Beckham wear?
A: *Football boots.*
Mrs J Motson, Goole

Q: Why was the boot condemned to be cast into a lake of fire for all eternity?
A: *Because, on the Day of Judgement, it failed to get down upon its knees before our Lord and Saviour Jesus Christ, confess its sins and beg Him to save its "sole" from damnation.*
Dr Stephen Green, Wales

KIDS SAY THE FUNNIEST THINGS ABOUT *Boots!*

I am a Hell's Angel with size 14 feet, and I mentioned at home recently that I was on the lookout for a pair of large motorcycling boots. "There's a pair on the telly right now, Daddy," piped up my 5-year-old son. He was referring to the programme Two Fat Ladies, featuring Clarissa Dickson-Wright and Jennifer Patterson!

Dirtyarse, Tunbridge Wells

BOOT TEXTS

i jst trd n a dog sh1t! ne1 wnt 2 cln my bootz? lol - bootsie bill, leeds
Clean yr own boots :(only jokin! ;) shitty sal

HA HA NO THANX BILL - ENUFF BOOTZ OV MY OWN 2 CLEAN! BOOTMAN

no laffin matter. my uncle wos cleanin shit of his boots an got that disese wher he went blind. serious charlie

FUK YR SHITY BOOTS BILL>MY BOOTS IS CLEAN ASA WHISLE - VINCE BIRMINGHAM

u boot saddoes make me puke .shooes rool, tommy shoeman mcguire, belfast

FUCK U TOMMY SHOEMAN SHOES R 4 PEDOS - FARNHAM BOOT CREW

1 boot - 2 boots. 1 foot - 2 feet. WOTS ALL THAT ABOUT?! -Bamber

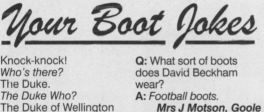

★★★ Only in America! ★★★

A PENNSYLVANIAN man recently made legal history when he successfully sued the estate of the Brothers Grimm for 'wilful misrepresentation'.

He claimed that whilst the Grimms' nursery rhyme stated that there was an old woman who lived in a shoe, the old woman had in fact lived in a boot, and that this inaccuracy had caused him such distress that he had grown up to be gay.

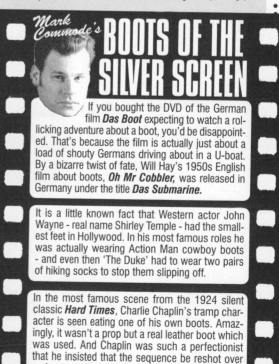

Mark Commode's BOOTS OF THE SILVER SCREEN

If you bought the DVD of the German film *Das Boot* expecting to watch a rollicking adventure about a boot, you'd be disappointed. That's because the film is actually just about a load of shouty Germans driving about in a U-boat. By a bizarre twist of fate, Will Hay's 1950s English film about boots, *Oh Mr Cobbler*, was released in Germany under the title *Das Submarine*.

It is a little known fact that Western actor John Wayne - real name Shirley Temple - had the smallest feet in Hollywood. In his most famous roles he was actually wearing Action Man cowboy boots - and even then 'The Duke' had to wear two pairs of hiking socks to stop them slipping off.

In the most famous scene from the 1924 silent classic *Hard Times*, Charlie Chaplin's tramp character is seen eating one of his own boots. Amazingly, it wasn't a prop but a real leather boot which was used. And Chaplin was such a perfectionist that he insisted on the sequence be reshot over and over again. In fact, he ended up eating an astonishing 37 boots before he was satisfied with the results!

Queen's Cornflake Butler Could Have Been

CROWN F

Palace Chiefs Cock Up Again

ALARMING lapses in royal security have been highlighted after yet ANOTHER undercover reporter tricked his way into a job at Buckingham Palace. *Daily Star* journalist Nick Lazenby used forged references to secure a post as Second Breakfast Underfootman, a position which brought him into daily contact with the Queen. Lazenby played the part of the faithful royal servant for the best part of a morning before his cover was blown when he was caught in the Queen's dressing room, going through her underwear drawer.

But during his near 4 hours in the midst of the Royal family, Lazenby was able to compile a terrifying dossier highlighting over a dozen opportunities when he could have put Her Majesty's life in danger. His appointment has proved a great embarrassment to members of the royal household and his explosive story will make uncomfortable reading for palace security chiefs.

PALACE

Lazenby, 38, found out about the palace vacancy after speaking to a fellow reporter in a pub one evening. "He told me he had just been working undercover at the palace himself, but had been exposed and sacked. So I knew there would be a job going," he told us. "I called the Comptroller of the Queen's Household and asked for the job and he told me to come down straight away. At the interview, I just had to sign a bit of paper saying I wasn't a terrorist or a reporter. He then asked me for some identification, and I showed him a fake gas meter reader ID card that I'd bought from the small ads in the back of a disreputable comic."

To his amazement, Lazenby was given the job on the spot and sent to be fitted for his footman's uniform. Ten minutes later, the undercover reporter was given his first task, making a pot of tea for the Queen.

"I couldn't believe how easy it was. Nobody knew me from Adam, yet here I was, entrusted with making a cuppa for the Monarch of England. There was nobody else in the kitchen and I could have done anything I wanted to it. On this occasion, I merely wiped the teabags up and down the cleft between my buttocks. But had I been a terrorist, I could have filled the teabags with rat poison, or even worse. It doesn't bear thinking about."

MAZE

Lazenby took the tea into the royal bedroom and handed it to the Queen, who was sitting up in bed.

"I watched as she drank it, and it sent a shiver down my spine," he said. "Thankfully on this occasion it was only my anal perspiration the

"Thankfully on this occasion it was only my anal perspiration the Queen was drinking."

Queen was drinking. Had I been a member of Al Q'aeda, Her Majesty might not have been so lucky."

However, Lazenby's chilling reflections were cut short when the Queen herself barked out an order for a slice of toast and marmalade. Once again he found himself unsupervised in the kitchen, preparing food for her Majesty.

dercover Cereal Killer

OOLS

"Unlike most people, the Queen doesn't eat marmalade from the jar, it has to be placed into a special silver dish. I'm sure any self respecting assassin would have taken the opportunity to mix lethal anthrax spores into the marmalade before decanting it. Just thinking about it made my blood run cold, and I was furious with the staff for putting her in such danger," said Lazenby. "It was sheer luck that I was not an assassin, and had merely urinated in her marmalade. And pushed the toast down the back of my pants and farted on it."

"I pushed the toast down the back of my pants and farted on it."

TIPPS & ALISTAIR

However, as Lazenby put the tray on the royal bedside table, he received an unexpectedly frosty reception. Contrary to what many people may think, the Queen is a very independent woman and likes to perform many everyday tasks for herself. One such is spreading her own marmalade onto her toast at breakfast. On this morning, she pointed out that Lazenby had forgotten the knife, and furiously tore a strip off her hapless servant.

At risk: The Queen yesterday.

"She gave me a roasting and sent me to the kitchen to fetch a knife, " he continued. "Little did she realise she could have been signing her own death warrant. I was a harmless undercover reporter, but no one asked questions about my mental state when I was given the job. I could just as easily have been a murderous pyschopath or serial killing cannibal. And here I was, walking into the Queen's bedroom carrying a knife. Fortunately for her I had just scraped it underneath my farmer's hat, but for all anyone knew or cared, I could have been about to disembowel our head of state and tuck into her organs. I shudder now to think about it.

Shortly afterwards, Lazenby was arrested by members of the Royal Protection Squad when Her Majesty caught him performing an obscene act into her open underwear drawer and raised the alarm.

"I explained to the police that I was a reporter and was just trying to highlight how easy it would be for a determined Fathers For Justice campaigner to slip a small venomous snake into the Queen's dressing table, but they wouldn't listen," he told us.

Earlier this week at Bow Street Magistrates Court, Lazenby was found gulity of obtaining employment by deception, an act of gross indecency, and four counts under the Health and Hygiene Act 1982. Sentence was deferred pending psychiatric reports.

Lazenby hit the headlines ten years ago when he used a forged passport to get a job as a postman in Gloucestershire. On that occasion, in order to highlight the risks posed to the Royal family by letter bombers, he delivered more than 20 packages containing excrement and obscene polaroid photographs of his genitals to the home of the late Princess of Wales.

DURING his few hours in the palace, Lazenby witnessed some shocking lapses in security. Here are some of the worst, snapped with a concealed disposable camera.

Royal Wee: Tangy taste of marmalade could disguise anthrax spores.

Pot Shot: The Queen's cuppa. One lump of cyanide or two?

Knife to See You: Monaroh could find blade turned on herself.

The Butler Did it: For reasons best known to himself, Lazenby broke wind on Queen's toast.

FAT PEOPLE 'EAT TOO MUCH'

– Cream Buns named in Shock Food Report

Eat too much and you could get fat, according to a report published this week. Specific foods singled out for attention include cream buns, chocolate cake and jam doughnuts.

According to a survey carried out for no particular reason, many people in Britain are already overweight. And the report goes on to claim that people who are fat:

* take up more room than other people
* wear bigger clothes
* and are more likely to damage furniture.

CHOCOLATES

We took these startling claims onto the streets to gauge the public's reaction. Mrs Hilary Foster, 46, agreed that cream buns were fattening, while her friend, 40 year old Margaret Harrison claimed that she had occasionally eaten chocolates but suffered no ill effects.

SWISS ROLL

Shopkeeper Paul Willis, 27, admitted that several fat people had visited his shop in the past and had purchased food items among other things. The manager of a nearby restaurant refused to comment on claims that fat people were among his best customers.

TRIFLE

A woman we later watched enter the restaurant was visibly overweight, but refused to tell us what she was eating or how heavy she was. Her husband then became abusive and we were asked to leave the premises.

A spokesman for the British Medical Authority told us he had not seen the report, and did not wish to make any comment.

A woman leaving a shop. In her bag are several food purchases.

WIN £1.00 IN OUR TERRIFIC VICAR JOKE COMPETITION

Everyone knows a joke about vicars. So we are launching a super competition to find the funniest vicar joke of them all!

Write and tell us your funny vicar joke. In the next issue we'll be printing the finalists and giving you, the readers, the chance to pick a winner. And there'll be a prize of £1 for the funniest joke we receive.

To get the ball rolling we asked a few pop celebreties for their favourite vicar jokes. Here are some of the suggestions we received.

Why did the vicar go to the harbour?
To see the jumble sail.

When is a vicar's door not a door?
When it's a-jar.

What is the brightest day in a vicar's week?
Sun-day.

Why did the chicken cross the road (opposite the vicar's house)?
To get to the other side.

Send your vicar jokes to: Vicar Joke Competition, Viz Comic, Viz House, 16 Lily Crescent, Newcastle upon Tyne NE2 2SP.

Blue Tits

by Bob Liar

Shapely hairdresser Joan Jones made a real boob the day she gave her boyfriend Mike Gray a hand with the decorating.

PAINT

For careless Joan spilt blue paint all over her prized possessions.

Said Mike, "I've never been interested in bird watching until now!"

"Once we'd stopped laughing we were able to have it off in no time at all," he quipped.

BIG ONES

Commented Joan, "When I have accidents I always have big ones!"

Throwaway Cartoons present: MRS. MaD Bastard

ERK
GIBBER GLIBBLE!
EJECTOR CLICK!
BOING!

Spirits face Scooby

Ghosts 'not scary anymore'

By our ghostwriter PHIL SPECTRE and his Wall of Sound

GHOSTS are no longer as scary as they used to be, according to a new report set to be published a fortnight next Wednesday teatime.

The spooky survey was commissioned by the Association of Ghost Train Operators to try and explain a dramatic drop in passenger revenue in recent years. However it could have far wider implications. For experts fear that by the year 2025 children will no longer be scared of ghosts at all.

Monsters

The scariness of ghosts has suffered in the face of fierce competition from two main rivals, monsters and space aliens. In their heyday ghosts were by far the scariest thing in Britain, with 98% of children under the age of twelve and one in five adults scared of them. But the sixties sci-fi explosion and the advent of TV have caused a tidal wave of competition, and a sucession of scary things - from monsters to Martians, and from Dracula to dinosaurs - have started to give kids the creeps.

Creeps

Len Murray, secretary of the official ghosts union the National Association for the Dead and Departed, lays the blame for the present problem squarely at the feet of space aliens. "It all began in the sixties with the Daleks, and now it has simply got out of hand. Ghosts can't compete. There must be controls put in place to protect the interests of our members", he said.

Nerds

Like aliens, monsters too have had a field day frightening children in recent years. And dinosaurs are the latest in a long line of horrible creatures to capture the imagination of children, and make them hide under their bedclothes. But Mr Murray fears that bringing dinosaurs to life in the film Jurassic Park was irresponsible and could lead to a 'double whammy' effect. "How long is it going to be before children start having nightmares about the ghosts of dead dinosaurs?" he asked. "Things are spiralling out of control, and unless the Government act soon it is only a matter of time before youngsters are faced with the terrifying prospect of the ultimate scary thing - the ghost of a dead alien space monster", warned Mr Murray.

What shits up the stars?

Crocodile's give Korbut (balancing on parallel bar, above) the creeps, while Frankenstiens give Martin Chivers the shivers. Meanwhile Peter Cushion (below) is Hammer horrified... of cars!

We asked a few famous faces what frightens them most of all. Former Russian gymnast Olga Korbut, now a Barbados taxi driver, told us that ghosts were the last things on her mind when she won the Olympic Games a long time ago. "As a child I was always scared of crocodiles", she confessed. "I could never bring myself to watch Peter Pan, and I still hide hide under my bedclothes when I hear my alarm clock ticking", she told us.

Sex Cases

Former England and Spurs centre forward Martin Chivers wasn't afraid of a hard tackle in his heyday. Now living in Denmark where he runs a successful ecclesiastical supplies business, he confessed to having one secret fear. "I must admit - I've always been scared of Frankensteins", he told us. "I don't know if its the bolts in their necks, or their clumpy boots, but even now I shit my pants whenever a Frankenstein comes on the telly".

Hammer horror star Peter Cushion showed no sign of fear in over 850,000 film appearances. But in reality Cushion was desperately scared of cars. "In his later years he'd hang around the street all day, appearing to follow people about. Often they would call the police. But all he wanted to do was follow them across the road. He was terrified of cars, and was scared to go near a road by himself"; Peter's former neighbour and T.V. Lottery Queen Anthea Turner told us.

Noel-y Moses!

MULTI-coloured House Party host Noel Edmonds has shaken the world of religion from top to crinkly bottom after issuing his own set of ten commandments. And the sinister tidy-bearded former DJ insists it's not another one of his trademark wind-ups.

Edmonds's Ten Commandments Swap Shock

"I'm serious about this," he told reporters yesterday. "The old set of commandments are nearly 3000 years old. They may have been relevant all that time ago but let's face it, these days they're simply out of date."

kill

In Edmonds' rewrite OUT go old favourites such as 'thou shalt not kill', 'thou shalt not steal' and 'thou shalt not bear false witness'. Meanwhile IN come:

• *Thou shalt not picnic or ramble on land belonging to light entertainment celebrities, even where rights of way appear to exist on Ordnance Survey maps.*

• *Thou shalt not lodge objections to planning applications by light entertainment celebrities who wish to construct private helicopter landing pads in their grounds.*

• *Thou shalt remove thy car out of the fast lane to make way for light entertainment*

celebrities who may be on their way to important meetings.

rage

Speaking to reporters via his electric gate intercom, Telly Addicts compere Edmonds explained why he felt the time had come to change the Old Testament edicts. "The tablets Moses brought down from the mountain were all about coveting your neighbour's donkey. Well my neighbour hasn't got a donkey and I'm pretty sure yours hasn't either. These days we're more likely to covet our neighbours' helicopters. I know my neighbours covet mine something rotten!"

week

But by far and away the most controversial ammendment is the one Whirly Wheel bungee bungler Noel

has made to the first commandment - 'Thou shalt have no other Gods before me.' In the new version, the world's billion Christians are told: *'Thou shalt have no other Gods before Noel Edmonds and me.'*

Not surprisingly, the plan has provoked fierce waffle from Archbishop of Canterbury Dr Rowan Atkinson who called a press

conference in order to equivocate Edmonds' announcement. "It seems to me that Noel Edmonds appears to be perhaps putting himself on a par with God, and yet are we not all, in a very real sense, on a par with God?" he prattled. "On the other hand, could it not be argued that there is another sense, equally real and no less valid, in which none of us can ever truly be on a par with our creator."

cowes

He continued: "Whether we agree with Mr Edmonds or not, and I am not saying that I do or don't agree with him, it seems to me important that he has opened up an important area of debate about the relevance of the Commandments and indeed all of the holy scriptures - whether they be Christian, Buddhist, Muslim, Jewish or indeed atheist, for all beliefs are, it seems to me, equally valid - in the modern world." Mr Atkinson

Keep taking the tablets ~ Edmonds issues his commandments (above). The tidy-bearded presenter of Swap Shop (above left) and (above far left) religous fannychops the Archbishop of Canterbury.

continued flannelling through his beard in this vein for several hours until reporters got bored and went home.

Last night Edmonds insisted that he wasn't backing down. He told us: "The original commandments were just carved on stone, but I've had mine engraved in solid gold on blocks of diamond-encrusted marble I bought at Fortnum and Mason in Paris. They cost an absolute fortune. If anyone doubts I'm serious about this, they can come and see the receipts."

EXCLUSIVE!

Noel's Ten Commandments

I

I Thou shalt have no other Gods before Noel Edmonds and me.

II Thou shalt not make use of the graven image of Mr Blobby on any merchandise without paying royalties to the copyright holder.

III Thou shalt not make Noel Edmonds the victim of hidden camera pranks.

IV Thou shalt not axe popular Saturday evening light entertainment programmes simply because there has been a temporary downturn in the viewing figures

V Thou shalt not ramble on land belonging to light entertainment celebrities, even where rights of way

II

appear to exist on Ordnance Survey maps.

VI Thou shalt not commit adultery.

VII Thou shalt not lodge objections to planning applications by light entertainment celebrities who wish to construct private helicopter landing pads in their grounds.

VIII Thou shalt find utterly hilarious the dropping of gunge onto plebians.

IX Thou shalt remove thy car out of the fast lane to make way for light entertainment celebrities who may be on their way to important meetings.

X Thou shalt keep the same haircut for 33 years.

Shame of Britain's Callous Undertakers

IS *THIS* ANY WAY TO TREAT OUR DEAD?

The head of a Bedford undertaking firm resigned yesterday after it was revealed that bodies were routinely *LEFT LYING* on his premises in wooden boxes before *BEING LOWERED* into holes in the ground.

On other occasions, it is alleged that bodies have simply been SET ON FIRE to get rid of them.

EXCLUSIVE!

dug

Leonard Duxbury, head of funeral directors Shadrack & Duxbury, quit after *Daily Mirror* photographers dug up several coffins and prised the lids off.

churn

Mechanical engineer David Nipsy, whose late grandfather, Arthur was one of the corpses concerned, was shocked.

"I can't describe what I felt when I saw my decomposing grandad on the front of the paper," he told us.

"It was shock, outrage and sadness, tinged with a glimmer of hope that I might get some compensation. My grandad fought in the War, so to treat him like this is disgraceful. It just seems so undignified to bury him in the ground like a big potato."

shake

Adele Barse flew from New Zealand six months after her mother's death, to finalise the sale of her house and property. "When I arrived at the undertakers they just handed me a jar of ashes," said a tearful

Duxbury (above) - yesterday and (right), Mirror editor Piers Morgan prises open a coffin to reveal Arthur Nipsy's decaying foot

David Nipsy - too shocked to comment, yesterday

SICKENING

Seven bodies dumped in wooden boxes then thrown into holes in the ground...

Barse. "She had been burnt so I couldn't even recognise her."

marketing board

Adele's father died five years ago, and his funeral was handled by Shadrack & Duxbury. She now fears they may have simply buried him in a hole. "My father worked for the gas board for forty years," she told us. "He always wore shiny shoes and he never even had so much as a parking ticket. It is unbelievably callous to simply leave him in the ground to be eaten by worms and moles."

Lord, *Help Us!*
with *Abdul Latif, Lord of Harpole*

Dear Lord of Harpole Please help me, as I am at my wits end. My husband is up for a pay rise and in order to impress his boss, he has invited him and his wife for dinner. His boss is a great fan of curries, and foolishly my husband told him what a great curry cook I was. The truth is, I have never cooked a curry in my life, and I don't know my madras from my elbow.

I just know the evening is going to be a disaster. My husband will lose his job, our house will be repossessed and I will have to go back on the game. They will be here in two hours. Lord, help me.

June Medford, Surbiton

*Lord Harpole says...*Firstly, do not be cross with your husband. Inviting the boss for dinner is a good way to "curry" his favour. Secondly, have no fears about cooking the meal. I, Mr. Abdul Latif, Lord of Harpole, am only a premium rate phonecall away. You, and in fact any Viz reader, can have the benefit of my years of experience as Newcastle's leading curry expert by dialling CurryTalkLive on 0906 515 1047. I will be able to personally give you advice about any curry-related topics. Calls cost just £1.50 per minute, so remember to get permission from the bill payer. All callers must be over 18, which since you are married, I assume you are. I'm sure with my help, your husband will get that pay rise which should help pay for the phonecall.

Dear Lord of Harpole I've done a stupid thing. For years my wife and I have been saving up to go to Australia to see our grandchildren. By scrimping and saving, we had got nearly £10,000 in the building society. Then yesterday, like a fool I went into Ladbroke's and bet the lot that giraffes don't make any noise. They gave me odds of 5-7. Now my brother, who is a giraffe handler at the zoo, has told me that they sometimes bleat like a lamb. I'm scared I've lost all our money. Please put me out of my misery by telling me if giraffes make a noise or not. Lord, help me.

Tony Gubba, Chessington

*Lord Harpole says...*I, Abdul Latif, Lord of Harpole, have good news and bad news.
The bad news is that giraffes do occasionally bleat like lambs. So you've lost your bet, you won't ever see your grandchildren, and your wife is probably going to kill you. But the good news is you can make it up to her by cooking a wonderful curry and I'll tell you how on CurryTalkLive on 0906 515 1047.

Do YOU have a curry query? Perhaps you're puzzled over poppadoms, or you're in a tizzy over a tikka. What ever your curry conundrums, call The Lord of Harpole on CurryTalkLive on 0906 515 1047. Calls cost £1.50 /min. And at that price, it's probably best to get the bill payer's permission first, or call from work. Over 18s only.

Who killed

IT It is now four months since the cold-blooded doorstep slaying of People's Presenter Jill Dando. And still the police seem no nearer to catching her killer. So we've asked Britain's best known ex-policeman (apart from Geoff Capes) to try and crack the case.

In an amazing series of interviews, JOHN STALKER uses his vast experience as deputy Chief Constable of Greater Manchester and garage door salesman to pick the brains of four famous T.V. detectives in the hope that their unconventional approach may help shed light on this bewildering case and enable him to finally name Jill's killer.

"I HAVE always had the greatest respect and professional admiration for Lieutenant Columbo. With his tenacity, intuition and his squinty eye for detail, he always gets his man. So I asked him how he would go about solving this 'Whodunit?'"

"AS ANY police officer will tell you, the most important part of a copper's equipment, after a canister of C.S. gas and a big stick, is his sense of humour. No matter how tragic and appalling the crimes that confront him, he must never lose the ability to have a good laugh. That is why I admire Inspector Jacques Clouseau of the French Surete."

Case No.1

Investigator: Lieutenant Columbo

Status: L.A.P.D. (Homicide)

Channel: ITV

"This is typical of the cases I handle," the glass-eyed, cigar-chomping sleuth told me. "A high-profile celebrity victim and no obvious motive. If I were investigating this case, the finger of suspicion might point at a fellow star. For the sake of hypothesis, somebody like, oh, I don't know, Sir Cliff Richard, for example.

"When I first interview him he would be cooperative and helpful, even to the extent of signing a record for my wife, Mrs. Columbo. After the interview, I'd leave, only to reappear almost immediately, ruffling my hair and looking puzzled, to ask one more question about Sir Richard's movements on the morning of Miss Dando's death. This time, after I leave, Cliff's smile would fade and his expression would harden. I would then begin to badger Sir Cliff, turning up unexpectedly to ask him more questions. I'd appear unannounced at music rehearsals, or interrupt a game of tennis in the grounds of his Weybridge mansion, shambling across the lawn in my raincoat saying there were still one or two things 'bugging me'. By now, Cliff would have become quite terse, eventually turning openly hostile.

"Finally I would confront Cliff with a flimsy web of circumstantial evidence and supposition, at which point it would be game, set and match to me."

Case No.2

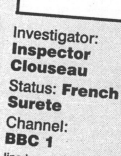

Investigator: Inspector Clouseau

Status: French Surete

Channel: BBC 1

"I would arrive at Gowan Avenue. My attention would be drawn immediately to a man with a minkey," the inspector told me at his Paris headquarters. "I would question him and he would mock my accent, whilst Mlle. Dando's killer made his getaway behind me; I might even hold up the traffic, enabling him to make good his escape in a blue Range Rover.

"I would report to my superior officer, Inspector Dreyfuss, who would twitch unconvincingly, as I outlined my ill-conceived theories on Mlle. Dando's murder. He would become confused between a real pistol and a novelty cigarette lighter on his desk, shooting the end off his nose as a result.

"A combination of farcical circumstances, including being blown up by a berm whilst dressed as Toulouse Lautrec, and knocking over a large rack of precariously poised long clattering things in the presence of a supercilious butler, would eventually somehow lead to me being convicted of the murder, whilst the real perpetrator escaped over the alps in a convertible Rolls Royce."

Dan-do?

Stalker's Telly 'tecs search for Star's assassin.

"AFTER 25 years at the sharp end of coppering, and more recently selling garage doors, if I have learned one thing it is this: That no motive is too far fetched, no matter how ghastly the crime. Never more so than in this case, where none of the facts seem to add up. A perfect case then for Scooby Doo and the kids in the Mystery Machine."

"JIMMY NAIL'S Spender is a no-nonsense North East copper. Like his name suggests, James Aloysious Bradford, is as hard as nails and twice as good at acting, and he has a distinct advantage over other T.V. detectives. For, as writer, director and producer, Jimmy can choose who the villain is going to be, no matter how ridiculous and implausible the plot, or laughable the dialogue. So I asked Crocodile Shoes himself how he would 'nail' Jill Dando's killer."

Case No.3

Investigator: **Scooby Do**

Status: **Independent Investigator**

Channel: **Cartoon Network**

"By coincidence our brightly coloured van would run out of gas during a thunderstorm, right outside the old Dando place," Fred told me. "Myself, Daphne, Velma, Shaggy and Scoob would go inside in search of clues. Whilst in the basement, Shaggy would discover a revolving bookcase, from behind which would emerge a sweaty man with a mobile phone. Scooby would then jump into Shaggy's arms, and the sweaty man would chase them along a very long corridor, passing the same objects at regular intervals." "Like, yeah!", Shaggy continued, "Then we would, like, drop a net onto the sweaty man, and tie him up, whilst waiting for the police to arrive, before removing his sweaty man mask, to reveal... the estate agent!" "It would turn out that the estate agent who was selling Miss Dando's home had discovered an abandoned gold mine in the basement. He had dressed up as a sweaty man with a mobile phone and shot the 'Crimewatch' presenter on the doorstep, in order to scare off potential buyers. At this point, whilst being led away, the estate agent may well suggest that he would have got away with it, too, if it hadn't of been for us meddling kids.

Case No.4

Investigator: **Jimmy Nail**

Status: **Plain clothes detective**

Channel: **BBC 1**

"I've got the perfect plan," said Jimmy, "I'd hide up a tree and wait for the murderer to walk past, then jump out and shout, 'Bastaaad!' Then I'd run faster than a train and chase him in a hot air balloon."

Well, we've looked at the clues through the eyes of four very different T.V. detectives; one a maverick scruff in a raincoat, one a comedy Frenchman who's been dead for 18 years, one a cartoon dog and the other a Geordie twat. It's time for me to name the killer.

Who killed Dan-do?

There is no obvious answer. But one thing's for sure. With me, former Deputy Chief Constable John Stalker, and all my fictional police friends on the case, the killer, or killers, whoever he, she, or they, is or are, will not be sleeping well in his, her, or their bed, or beds, tonight.

THE ADVENTURES OF THE SANDWICHES AND THEIR ATTEMPTS AT WORLD DOMINATION

SOON...

LATER...

BAH. FOILED AGAIN

TIN WRAP

Old 'Worm-Eyes' is back

from our Showbiz correspondents **Adolf Hitler** & **The Dali Llama**

SUPERSTAR crooner Frank Sinatra is to perform one final show and sing to a sell out Las Vegas crowd. The 'King of the Comeback' was lured from his grave to 'do it his way' one last time with the promise of a $100m paycheck.

Since his death a year ago, Sinatra has turned down all offers to get back up on stage. But it seems that a plea from his agent on behalf of his fans has tugged at the heart of dead 'Mr. Wonderful'.

cash

"Frank really misses the limelight and he is really looking forward to this show," said his agent and friend of forty years Alex Marmaladiani. "Frank is dead and gone, so he doesn't need the cash. This is all about his fans," he added.

jennings

And those queuing for the $6000 a-head tickets at Caesar's Palace were excited

Showbiz's 'Mr. Wonderful' faces final curtain one more time

but apprehensive about the forthcoming concert.

travers

"I done sold my house to buy a ticket and now I'm a two-bit bum," said life-long Sinatra fan, Artie Spannerheimer, who had walked 2000 miles from Jism, Nebraska to bag his place in the queue. "But it will sure be worth it just to see

Mr Wonderful himself on stage again," he added.

sykensit

Critics have predicted that Frank will not be the performer he once was, and Marmaladiani freely admits that age, death and putrefaction have taken their toll on the performer.

a-cake, a-cake

"So what if he can't hit the high notes anymore. Or the middle or low ones," he told listeners to the Howard Glans Show on Cincinnati radio station WKRZ. "This guy's a legend. He's still got that old razzle dazzle, even if he does look like a pile of green dog food."

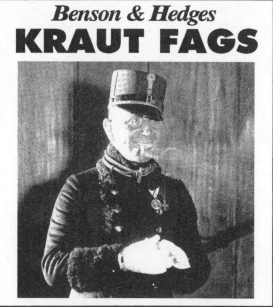
The Bat in the Hat!

Jay in a flap over titmice in his titfer

RUBBER-LEGGED pop star Jay Kay is steaming mad after being banned from driving... for 150 YEARS! And it's not for speeding or drink-driving. It's because he's got bats... *IN HIS HAT!*

The colony of over 1000 Pipistrel bats was discovered after Kay's common-law wife, ex-Big Breakfast babe Denise Van Outen was awoken in the early hours by loud, inaudible squeaking.

Sonar

Denise said: "I heard all sonar noise coming from Jamiroquoi's hat. I peeped inside with a torch and there was all these bats flying

Protected species - some of the bats in Jamiroquai's hat and Jamiroquai (inset), yesterday

around inside eating moths."

Radar

The showbiz golden couple called in pest experts, but were dismayed to be told that as the furry sightless birds were a protected species, there was nothing they could do to get rid of them. On the contrary, if Kay disturbed them at all - he could be sent to *PRISON!*

"That was a real blow for

Jay, because he can't fit into any of his Ferarris without taking his hat off," said Outen.

Hawkeye

And the future looks bleak for the funk-lite popster, for bat boffins have told him that a colony of Pipistrels can stay in the same place for up to 150 years - or even twice that.

SOUP IS SEXY!

So says Doctor in fascinating book about soup

Soup guzzling Goss brothers (left) and 'horny' Jonathon Ross

Soup is sexy. That's the saucy claim being made in a new book published this month. For author Dr. Karl Lipton believes that soup drinking can lead to improved sexual performance. In his book, 'All About Soup', he reveals that saucy dishes like minestrone and mushroom can act as an aphrodisiac.

"Certain soups are sexier than others", explains Karl, a former lecturer in soup at Warwick University. He named French onion and oyster soup as two of the sexiest starters, and claimed that other soups — like turnip — are a turn off.

OXTAIL

We decided to find out what kind of soup the stars prefered. Every night, chat show host **JONATHAN ROSS** tucks into a steaming bowl of oxtail. "Jonathan comes home from work tired and exhausted", his wife Jane told us. "But one bowl of his favourite beefy brown broth is all it takes to put the lead back in his pencil", said Jane.

"It makes me a horny devil", Jonathan told us yesterday.

SHARK

A Buckingham Palace insider revealed that at 40, **PRINCE CHARLES** is steering clear of sexy soups, plumping instead for brown Windsor. "As a result it's considered 'unlikely' that there'll be any further additions to his family", our source commented.

Off stage sexy Bros twins **MATT** and **LUKE GOSS** polish off gallons of shark's fin soup, and on stage the fans go wild. Meanwhile Ken, the third member of the band, sips away at a lukewarm bowl of lentil.

French onion and oyster are among the sexiest

This amazing link between soup and sex is not a recent discovery, as Dr. Lipton explains in his book. 'In Papua New Guinea, natives serve up bowls of boring cauliflower soup as a primitive form of birth control. It's a tradition that goes back many centuries'.

BORING

Dr. Lipton's book also provides a fascinating insight into the history of our favourite soups. For example, it explains how soup was often considered boring until the day in 1872 when French chef Jean Louis Crouton accidentally spilt a bowl of fried bread into a pot of soup he was busy preparing at his restaurant in Paris. "The resulting dish went down a storm with guests", Dr. Lipton told us. "And his subsequent invention, the crouton, is now served with soups in restaurants throughout the world". Although the restaurant no longer stands, the street in Paris where the discovery occured has since been renamed 'Rue de Crouton'.

MANY UNUSUAL SOUPS ARE RARELY HEARD OF

We've all heard of everyday soups like tomato, cream of mushroom and vegetable. But in his book Dr. Lipton also casts light on a whole variety of unusual soups which are consumed around the world.

One of the strangest must surely be African tree soup, eaten by the nomadic Okwe-kwe tribe on the fringes of the Seranghetti desert. However Dr. Lipton fears that due to massive deforestation programmes currently underway in that region this once proud tribe and their unusual soup may soon disappear forever.

ESKIMOS

Thousands of miles away in the freezing arctic wastelands, Eskimos look to the sea for their soups. One favourite is whale soup, which Dr. Lipton claims has the highest calorific value of any soup known to man. 'Eskimos have been known to survive for up to 8 weeks on one bowl of this soup alone', he says in his book.

SCIENTISTS

Over the years many soups have disappeared from menus altogether. Dinosaur soup, once popular among cavemen, has not been served for over a million years. However the same is not true of mammoth soup. In 1922 Russian scientists working in a remote corner of Siberia discovered a mammoth preserved in the ice. This historic discovery not only provided palaeontologists with a remarkable opportunity to study the extinct species, but it also gave Russian chef's a chance to re-discover this popular prehistoric soup.

Doctor has tasted more than 5,000 varieties

Dr. Lipton estimates that over the years he has tasted more than 5,000 different types of soup. However, there is one soup which he will never savour, as cooking it is strictly illegal. Dr. Lipton explains: "Many years ago bat's arse soup was a popular dish in the British Isles. However, it takes up to 200 bats to make a single bowl of the soup, and as they are now a protected species, it is simply no longer possible to make the soup".

Dr. Lipton's Book, 'All About Soup' is published by Omlette Books, priced £19.95.

DOCTOR, I'M FEELING RATHER FLUSHED.

LOOK - I'M PUTTING MY ASS ON THE LINE HERE.

Nit's a MIRACLE!

Mercy plea for baby Fleanix alive against all odds

PRESSURE is mounting on the Prime Minister to spare a baby headlouse found alive after twenty minutes under a pile of carcasses on a schoolboy's head.

The tiny survivor, named Fleanix, was missed by insecticidal shampoo as she lay huddled beside her dead mother.

Now the nit nurse who found the week-old orphan parasite is begging educuation officials to spare her from being popped on the side of a sink.

School nurse, Nora Lakeman had been treating children after an outbreak of lice was discovered in a classroom at Talbot Road Infant School in Blackpool.

Nora, 42, said "I found the little nit when I was running a comb through the head of one of the boys in 2B."

"It broke my heart

By our Big Chief Home Affairs Correspondent
Stanley um Two-Rivers

when I saw little Fleanix. She was lying next to her mum, trembling and looking bewildered. How she survived the shampoo I'll never know. It's a miracle."

Big hearted Nora took the nit home immediately and has been hand feeding her on dandruff with a pair of tweezers.

"It was touch and go for a while, but now she's doing really well, considering what she has been through," said Nora. "It would be evil to kill her at this point."

There was an outcry

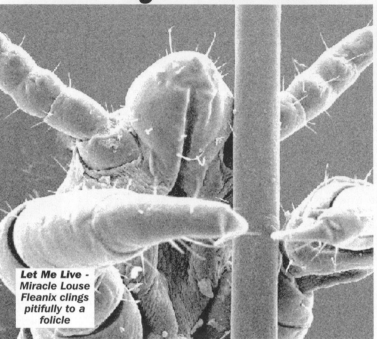

Let Me Live - Miracle Louse Fleanix clings pitifully to a folicle

after officials from the LEA insisted that all nits must be destroyed to prevent the outbreak spreading to neighbouring schools.

"She was lying next to her mum, trembling. How she survived the shampoo I'll never know"

Benson & Hedges NUN FAGS

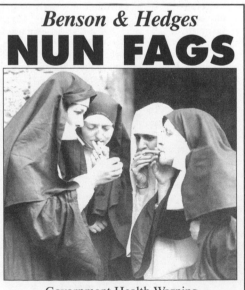

Government Health Warning
SMOKING NUN FAGS CAUSES MASTURBATION WITH SOAP & CANDLES AND RIDING BICYCLES OVER COBBLES

6mg Tar   0.5mg Nicotine   3mg Whiskers on Kittens

IT'S TIME TO STOP THE KILLING!

"For Fax Sake, Save Fleanix", say Big-Hearted Viz Readers

THOUSANDS of mentally sub-normal readers bombarded Viz yesterday with badly spelled and punctuated faxes, e-mails and petitions begging Education chiefs to spare Fleanix the Miracle louse's tiny life.

Elderly residents at a day centre signed a petition saying: "Please save that innocent little nit. If you can't find a home for her, she can come and live in our hair"

plight

Even hardened criminals have been moved to tears by Fleanix's plight. Broadmoor lifer Arnold Nonce begged "Please, God, spare baby Fleanix. She is a symbol of hope for us all."

Mrs Bollocks from Cheltenham said "Surely there has been enough slaughter. this innocent dickie has already been through so much in her short life."

One of the thousands of faxes we're received - this one from 46-year-old Dr. Derek Taylor of Hull

What do YOU think?
Should Fleanix be burst to prevent the spread of nits? Or should a special case be made in this instance?
E-mail us with your credit card number and let us know what You think. And don't forget to include the expiry date.

No Triple XXX Please, We're British

Cabinet Plans Extreme Porn Clampdown

THE GOVERNMENT recently announced plans to crack down on extreme pornography. It sounds a simple enough task at first glance, but as usual the legislation could prove to be a minefield of hot potatoes. What is the definition of extreme pornography? Erotic images that would make Anne Widdecombe vomit buckets may fail to provoke even a twitch in Peter Stringfellow's penis. And, assuming we could ever come to an agreement about what it is, how should we go about banning it? We went on the street to find out what you the public think.

"The government wants to ban images of bestiality, but where will it all end? Will they arrest the curator of the National Gallery for displaying Leonardo da Vinci's painting of the ancient Greek legend of Leda, in which a nude woman is seduced by a swan? Probably not, come to think of it, because you can't see it going in."

Brian Turpentine,
Art Dealer

"I buy a lot of erotic mags, videos and DVDs, so I reckon I have a pretty good idea of what constitutes extreme pornography. Generally spea-king, it's anything over about £20."

Spud,
Van Driver

"All sorts of pornography turns my stomach, but extreme pornography does so even more. Any perverts caught looking at this sort of material should be dragged into the town square, stripped naked and tied to a post. Then members of the public should be allowed to whip them until their buttocks and genitals are a mass of bleeding scars. Their punishment should be videoed and shown in schools, where their weeping welts would serve as a warning to children of the dangers of looking at such disgusting material."

Ena Dailymail,
Housewife

"My husband used to like me to dress up in sexy underwear like the models wore in Razzle magazine, which he used to buy each month. I

didn't mind that, but recently he's been travelling to Amsterdam and buying more extreme pornography. These days I often find myself having to get dressed up as a pig, a great dane or an oven ready chicken."

Mrs Boldmoney,
Primary School Teacher

"Extreme pornography to me means any pictures where the women have pubes. That's because when I was a lad in the fifties, all the ladies in mucky books had their pubic hair airbrushed out. It was very frustrating. Whilst doing National Service, I banged my head in a tank and fell into a coma for half a century. When I finally regained consciousness last week, I decided to treat myself to an Adult Channel subscription, but was disappointed to find that all the models now shave their bushes off. I've slept through pubes."

Reg Sneezewort,
Milkman

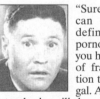

"According to survey after survey, pornography has no effect on a man's behaviour. What a load of rot. It causes my husband to briskly rub his penis whilst making strange grunting noises for about a minute and a half."

Mrs Edna Cloudberry,
Charity Shop Worker

"The spoilsport feminists tried to ban page three because they said it degraded women. Now they want to stop us looking at extreme pornography. Well, enough is enough. I like extreme pornography. A nice mpeg of a pretty girl who's been drugged and lashed to a kitchen chair whilst two men goad a rottweiler into anally raping her certainly brightens my morning, I can tell you."

Tommy Dodder,
Scoutmaster

"Surely there is no need to ban it, since one's exposure to this sort of material is self-regulating. It is an established medical fact that the more extreme the pornography one looks at, the quicker one goes blind or mental."

Dr M Bader-Meinhoff,
Haltwhistle

"The Home Office say they will ban any pornography that depicts pain, either real or simulated. If that's the case, then surely all the bongo videos in the world would be illegal. Because in every one I've seen, when the leading man goes for a pop shot he pulls a face like he's just shut his knackers in the car door."

Rex Whistler,
Carpenter

"Surely, until you can adequately define extreme pornography, then you have no chance of framing legislation to make it illegal. And everyone's standards will be different. For example, me and my grandmother will have widely different views of what constitutes extreme pornography. To me, it might be a film of a woman performing oral sex on a donkey, whereas to my grandmother it might be a ladyboy defecating between the breasts of a woman tied to a snooker table, whilst twenty Japanese men in gimp masks stand around them masturbating, shooting pints of sticky ejaculate into their hair."

Clive Shipton,
Grocer

Ever fancied a JUMP with JAGGER or a

SEX WITH THE

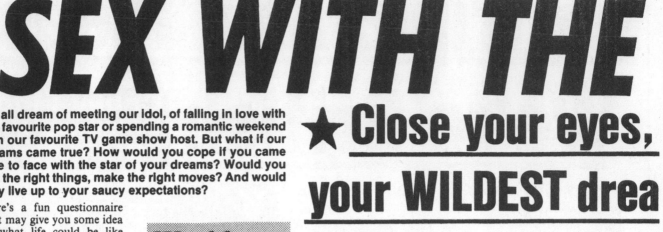

We all dream of meeting our idol, of falling in love with our favourite pop star or spending a romantic weekend with our favourite TV game show host. But what if our dreams came true? How would you cope if you came face to face with the star of your dreams? Would you say the right things, make the right moves? And would they live up to your saucy expectations?

★ Close your eyes, your WILDEST drea

Here's a fun questionnaire that may give you some idea of what life could be like mixing with the celebrities. Simply find a pen and paper, then sit back and use your imagination to see what life could be like having **SEX** with the **STARS.**

1. Matt out of pop group **BROS** rings you up and asks whether you'd like to go to the pictures with him.

But you've already arranged to go with his brother Luke. You fancy Matt more than you fancy Luke, but you don't want to cause a family feud. So what would you do?

a. Stay at home and wash your hair in order to avoid any trouble.
b. Ditch Luke and go with Matt instead.
c. Take both of them and suggest a sexy snogging threesome in the back row!

2. Imagine bumping into pint sized comedian **RONNIE CORBETT** in the street. He's lost his spectacles and he mistakes you for his wife, inviting you to go with him for an expensive Chinese meal. What would you do?

a. Point out his error, and direct him to the nearest opticians.
b. Go with him and let him buy you the meal, then point out his mistake afterwards.
c. Tell him you don't feel hungry and that you'd rather go home straight away for a bit of slap and tickle.

Would you say YES to 'coffee' with Cliff?

3. After a romantic night out with **CLIFF RICHARD**, he invites you into his flat for a cup of coffee. What would you do?
a. Say 'no thanks' and ask him to drive you home.
b. Say 'yes' to the coffee, but tell him you really must be home by eleven.
c. Say 'yes', go inside and immediately slip into something more comfortable – Cliff's sexy four-poster bed!

4. Imagine you are in bed with rugged game show host **BOB HOLNESS**,

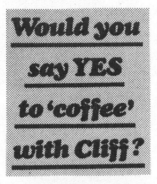

star of ITV's 'Blockbusters'. Suddenly, your husband arrives home from work early. What do you do?
a. Tell Bob to hide in the wardrobe.
b. Ask Bob to wait for a moment while you make your husband's tea.
c. Ask your husband to wait for a moment while you finish your steamy romp with Bob.

5. Which of these three would be your idea of a real night to remember?
a. A cup of cocoa and a cuddle with straight-faced award winning 'Question Time' presenter Sir Robin Day.

b. A night out on the town with heart throb singer George Michael.
c. A sexy threesome in a steamy hotel room with Little and Large.

6. You tell your husband that you're visiting a friend then you slip out to a quiet restaurant for a candle-lit rendezvous with TV funny man **LES DAWSON**. All of a sudden you notice your husband is sitting at a nearby table with Breakfast TV presenter **ANN DIAMOND.** How would you react? Would you:
a. Keep your head down and hope hubby doesn't see you. After all, you're both in the wrong.
b. Storm out, stopping only to pour your drink in Ann Diamond's face.
c. Start flirting heavily with Les, in order to make your husband jealous.

7. The pubs have just closed when suddenly sexy TV chat show host **JONATHON ROSS** starts banging on your door. He smells strongly of alcohol and

demands kinky sex. What would you do?
a. Bolt the door and call the police.
b. Ask him to sign a few autographs while you make him a mug of strong black coffee.
c. Tie him to your bed before he sobers up and start delving around in the attic to find your husband's old whip and rubber skin diving suit.

8. Imagine waking up one morning with a dreadful hangover. You can't remember a thing about the night before, but in bed next to you, snoring loudly, is bubbly TV weatherman **IAN McCAS-GILL.** What would you do?
a. Wake him up and tell him to leave at once.
b. Leave him till he awakes, then politely suggest he leaves.
c. Lock the door, throw his clothes out of the window THEN wake him up for an all day sex session!

9. One morning over breakfast your husband tells you that he has been having a torrid affair with senior labour politician and keen amateur photographer **DENNIS HEALEY.** What would you do?
a. Pack your bags and walk out, demanding a divorce.
b. Say that you're prepared to forgive him, but only if the affair is ended at once.
c. Suggest that Dennis invites Neil Kinnock round next time he visits, to make up a fruity foursome with you and your husband.

10. Boys! Imagine you are at a house party. It's getting late, and couples are beginning to pair off. Suddenly 'Blind Date' host-

BONK with BOWIE?

STARS!

read this, and let ms come true... ★

...ess **CILLA BLACK** sits down next to you holding an empty wine bottle. She appears to have had a lot to drink. What would you do?

a. Ask her to sing one of her sixties pop hits in order to liven the party up a bit.
b. Nervously put your arm around her and hope she doesn't object.
c. Burp, stub out your cigarette, then fling your arms around her, giving her a big, wet, romantic kiss, and hope that she responds.

11. Imagine you have managed to fix up a date with former 'Blue Peter' presenter **VALERIE SINGLETON**. However, when you get to the pub she drinks only pineapple juice and doesn't have much to say. It soon becomes obvious that you aren't going to score. So what do you do?

a. Put up with her for the rest of the evening, say goodnight, and convince yourself you didn't really fancy her anyway.
b. Pretend to go to the toilet, then disappear.
c. Ask her for Sarah Green's phone number, then pretend to go to the toilet and disappear.

12. Finally, imagine you were to receive a phone call from Junior Health Minister **EDWINA CURRIE**. She sounds distressed, and tells you that she is pregnant. How would you react? Would you:

a. Claim that you've never met her, and hang up the phone.
b. Offer to buy her a pram.
c. Claim that you've never met her, but offer to buy her a pram anyway.

Here's how to get YOUR hands on a celebrity!

Nowadays, many showbiz stars are plagued by obsessive fans who mob them wherever they go, and often camp outside their doors in order to catch a glimpse of their heroes.

MINDERS

A lot of celebrities choose to live in isolation, far away from their screaming fans, while others employ professional minders or bodyguards to give them protection and privacy.

SHOWBIZ

As a result, getting to meet your showbiz idol can be a difficult business. And even when you do, getting them into bed with you can still be a problem. But here's a few DOs and DON'Ts that may help you to get your hands on the star of your choice.

DON'T write letters to your idol. Fan mail is rarely answered by the stars themselves, indeed a lot is simply thrown away and never read at all.

DO make phone calls. It's far more personal. If their number isn't in the book, ring their TV or record company, explain that you're attracted to the person and that you'd like to arrange a date. (If looking up a celebrity's phone number in the telephone directory, remember to look under the right initials. For example, you won't find Bob Monkhouse under 'B. Monkhouse'. He'd be under 'R' for Robert).

DON'T invite celebrities to the pictures unless you know which film is showing. Plan ahead. Often it pays to choose a dull film — that way they're bound to pay more attention to you!

DO take them out for a drink. Celebrities are far more likely to go to bed with someone they've never met before after they've had a few drinks.

DON'T approach well-known celebrities when they are dining out in public. They hate being interrupted while they're eating. Allow them time to finish their main course then, while they're waiting for their sweet, pop over and invite them back to your place for a cup of coffee.

HOW DID YOU DO?

Award yourself one point for each question you answered 'a', two points for a 'b' and three points for each 'c'. Then tot up your total and see how you've fared.

19 OR LESS

You don't seem to be cut out for the celebrity lifestyle. Find yourself a nice, quiet partner and settle down. You can still dream of your idols, but sex with the stars would be too hot for you to handle.

10 to 29

Not bad at all. You'd probably fit well into the glamorous sexy showbiz nightlife. You have no inhibitions and would be just as comfortable in bed with a top celebrity like Paul Daniels as you would with your husband or wife.

30 OR MORE

Top marks! You were born to mix with the stars! Go at once to the BBC television centre and try to pick up a well-known celebrity as they are leaving.

How easily can showbiz wives swipe the stars' assets?

D.I.V.O.R.C.E. spells JACKPOT!

Breaking up is hard to do but it might just be worth it for *forty million smackers*

The break-up of any marriage is a tragedy for all concerned. And our heartfelt sympathies go out to Phil and Jill Collins at this difficult time.

The fact that Phil is now free to play the field and pull a fantastic fat titted young bird is precious little consolation for the heartache and pain that the millionaire singer has endured. And for Phil's wife Jill a cool **£40 million** slice of the old man's action can never begin to replace what she has lost. Although it could come in pretty handy.

JUMBO

Jill's jumbo pay-off will send cash registers ringing all over showbusiness, with eager wives eying up bank balances with a view to divorce. So just how easy would it be for the wives of the stars to get their hands on their hubbies' assets? We asked our special undercover reporter **Mandy Morrisroe** to find out by calling the stars and pretending to be their wives. Mischievous Mandy then demanded divorce, and began haggling. Here's how she got on.

DUMBO

PAUL McCARTNEY has more loot stashed away than most other pop stars put together. Over **£200 MILLION** at the last count. And without Linda, a linchpin in his band Wings, Paul would be penniless. So we figured his wife was worth £150 million.
"Hello. Is that you Paul? It's me, your wife Linda McCartney", said our girl Mandy, holding a handkerchief over the phone. "Pardon?" replied Paul.

NELLIE

"Our marriage... it isn't working, and I want a divorce", said Mandy. There was a silence on the end of the line.
"I want £150 million, in cash", she continued.
But Paul is a shrewd businessman, and rather than cave in to our demands, he decided to put the phone down.

EXCLUSIVE

As James Bond actor **ROGER MOORE** often cast beautiful women by the wayside. But our Mandy

Former Bond Moore

was determined the multimillionaire star would pay dearly to dump his real life wife Luisa. This time Mandy dropped Moore a line at his agent's office, cleverly disguising her handwriting as his wife's.

'Dear Roger. Things are not working out between us. I think a divorce would be best and I will settle for £5 million. Love, Luisa'.

BABAR

Three days after posting the letter Mandy had heard nothing so she wrote again,

Sexy Mandy chats to a star

Viz girl Mandy goes *undercover* with the stars

this time demanding only £1 million, and telling Roger he could see the kids at weekends. But still no reply. The former Bond was obviously *shaken but not stirred* by her demands.

LITTLE BLUE

Finally, Mandy decided to call up Britain's top TV celebrity **NOEL EDMUNDS** and take the money-grabbing so-and-so for every penny he had.

"Hello Noel. Gill here. I'm afraid our marriage is over. Let's talk money", she said. "I want the lot".
"Is this some sort of a wind up?" Edmonds replied. "Who is that? Don't tell me, you're recording this aren't you!"
"No, I'm not", replied our undercover girl.
"Well fuck off then", said Edmonds.

No jacket required for millionaire Phil seen on one of his last dates with wife Jill.

Four inch 'sex monkeys' wanked in my tea - says Sting

POP singer Sting shocked guests at a four star hotel by claiming that pint sized monkeys had masturbated into tea delivered to his room by hotel staff.

A fellow guest overheard the star's conversation with staff at the posh Sandy Bay hotel near Toronto. "Sting claimed that sex monkeys, less than four inches tall, had got into his room through a gap in the window frame", the guest told us.

SUMMONED

Eventually hotel manager Mark Lavender was summoned to settle the dispute. He accompanied Sting to the room but could find no evidence of sex monkeys.

"There was a slight ingress of water around the aluminium window frame, but this had occurred over a period of time", he told National Enquirer magazine. "There was absolutley no way a monkey, no matter how small, could have entered the room at that point."

REFUSED

Yesterday a hotel spokesman refused to say whether Sting had successfully negotiated a reduction in his bill over

Mr Sting last Thursday

the matter. "I am not at liberty to discuss any individual's bill. You will have to take the matter up with Mr Sting himself", we were told. Sting's best friend, former Mastermind champion Fred Housego was last night unavailable for comment.

SCREENWASH
with
Mark Commode

GEORDIE hard man actor **Jimmy Nail** has announced plans to remake **Orson Welles**'s classic **Citizen Kane**, which was recently voted best movie of all time. The controversial actor shocked a specially invited audience at the Venice Film festival when he branded the 1941 original "Shite". The new version will be set on his native Tyneside, and instead of being a newspaper magnate Nail's Charles Foster Kane will be a brick-laying country & western singer detective wearing crocodile shoes. "The old film definitely needs bringing bang up to date," Nail told CBS's **Regis Philbin**. "It's not even in colour." Shooting of Jimmy's self-penned screenplay starts underneath the Tyne Bridge in November, with Jimmy starring, directing, editing, operating the camera and doing the onset catering.

HARD on the heels of the multi-million dollar grossing **Jackass The Movie**, veteran political commentator **Sir David Frost** is set to team up with octogenarian **Ludovic Kennedy** in what is described as "The grossout movie to end all grossout movies". The unlikely pair plan to film themselves performing a variety of hair-raising stunts, such as firing roman candles out of their bottoms, jumping off high buildings into piles of human excrement, and fastening each other's scrotums to telegraph poles with staple guns. "Whatever you do, don't try any of these stunts at home," Frost told ABC's **Howard Glans**. "Me and Ludo totally take it to the edge, man." The film, provisionally entitled **Jackarse UK** is set to hit British screens in Autumn 2004.

Commode's Top 10 Movie Bloopers

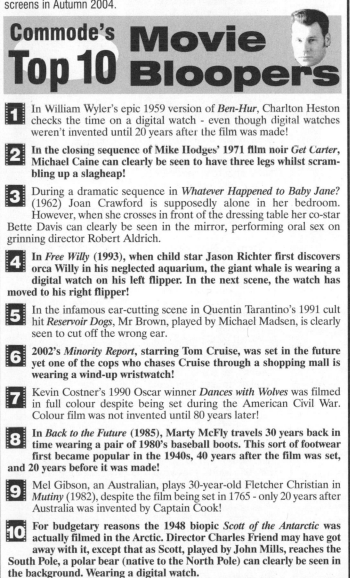

1 In William Wyler's epic 1959 version of *Ben-Hur*, Charlton Heston checks the time on a digital watch - even though digital watches weren't invented until 20 years after the film was made!

2 In the closing sequence of Mike Hodges' 1971 film noir *Get Carter*, Michael Caine can clearly be seen to have three legs whilst scrambling up a slagheap!

3 During a dramatic sequence in *Whatever Happened to Baby Jane?* (1962) Joan Crawford is supposedly alone in her bedroom. However, when she crosses in front of the dressing table her co-star Bette Davis can clearly be seen in the mirror, performing oral sex on grinning director Robert Aldrich.

4 In *Free Willy* (1993), when child star Jason Richter first discovers orca Willy in his neglected aquarium, the giant whale is wearing a digital watch on his left flipper. In the next scene, the watch has moved to his right flipper!

5 In the infamous ear-cutting scene in Quentin Tarantino's 1991 cult hit *Reservoir Dogs*, Mr Brown, played by Michael Madsen, is clearly seen to cut off the wrong ear.

6 2002's *Minority Report*, starring Tom Cruise, was set in the future yet one of the cops who chases Cruise through a shopping mall is wearing a wind-up wristwatch!

7 Kevin Costner's 1990 Oscar winner *Dances with Wolves* was filmed in full colour despite being set during the American Civil War. Colour film was not invented until 80 years later!

8 In *Back to the Future* (1985), Marty McFly travels 30 years back in time wearing a pair of 1980's baseball boots. This sort of footwear first became popular in the 1940s, 40 years after the film was set, and 20 years before it was made!

9 Mel Gibson, an Australian, plays 30-year-old Fletcher Christian in *Mutiny* (1982), despite the film being set in 1765 - only 20 years after Australia was invented by Captain Cook!

10 For budgetary reasons the 1948 biopic *Scott of the Antarctic* was actually filmed in the Arctic. Director Charles Friend may have got away with it, except that as Scott, played by John Mills, reaches the South Pole, a polar bear (native to the North Pole) can clearly be seen in the background. Wearing a digital watch.

QUEEN of FARTS

Queen: Lit one of her farts

FIRE ENGINES were called to Buckingham Palace last night after the Queen set fire to the curtains on her four-poster bed.

The blaze was swiftly brought under control and no-one was hurt, although her majesty and two friends, Queen Julianna of the Netherlands and Queen Noor of Belgium, were treated for smoke inhalation at the scene.

According to reports, the three monarchs had been having a sleepover when the incident happened. It is believed that the curtains caught fire when the Queen lit one of her farts with a match. The flames quickly spread to the canopy over the bed and the Queens were forced to take refuge on the balcony in their pyjamas.

A palace insider told reporters: "Their Royal Highnesses had been making a lot of noise all evening. Their giggling and shrieks of raucous laughter were keeping everybody awake. In fact, the footman had already had to go in two or three times to tell them to go to sleep at the time the blaze happened."

The Queen later apologised for all the trouble she'd caused, and praised firemen for their swift action. Reading from a prepared statement, she said: "I now realise that lighting my fart was a silly thing to do. It is only thanks to the quick-thinking and bravery of the emergency services that the consequences were not far worse."

This was the first time the Queen had had a sleepover since she was forbidden from holding them by her late mother. The last time the three European monarchs got together for a sleepover was at Windsor Castle in November 1992, when the resulting blaze caused damage estimated at over forty million pounds.

Windsor Castle: Blaze blamed on 'worker's blowtorch'

Xmas Turkeys!

WE asked you to tell us about the most disappointing Christmas present you ever received. Here are a selection of your festive thumbs downs...

...MY aunty Joyce bought me a soap on a rope when I was 10. She's dead now, so I don't feel too bad about slagging it off. ***Spud, Luton***

...MY wife knitted me a jumper as a surprise. It'd taken her a year to knit, as she could only do it whilst I was out the house, and when the arthritis in her hands would allow it. It was a nice jumper I suppose, but I hadn't the heart to tell her that what I really wanted for Christmas was oral sex off a prostitute. ***Hector Monkbottle, Perth***

...I LIVE in a remote village in Malawi, and last Christmas a family in Surrey (whom I had never met nor heard of) bought me a goat, three chickens and half a dozen mango saplings. When I unwrapped them on Christmas morning I was so disappointed - the woman next door got an ipod shuffle and a Cyberman voice changer helmet. ***Pipi Okwekwe, Malawi***

...I AM a model railway enthusiast, and each Christmas my wife buys me a little something to add to my layout. Three years ago she bought me an OO-scale saddle tank locomotive in the characteristic maroon and cream livery of the post-war Great North Eastern Railway. However, my layout is a representation of a section of the GNER from the early 1930s, when the livery was a slightly darker maroon. I thought this would have been obvious to her. I spent the rest of Christmas day with my arms folded, turning my head away from her whenever she tried to ask me what was wrong. ***T Potter, Morpeth***

Martin's laugh in

Deadpan Dean is in high spirit's world

Dean - having last laugh

AMERICAN showbiz legend Dean Martin who died on Christmas Day has surprised mortuary staff in California with his amazing sense of humour.

Workers at the swish $10,000 a slab Los Angeles morgue say the brave comic was sitting up and cracking jokes with fellow corpses only minutes after his post mortem examination.

"He had the Coroner in stitches with a succession of quick fire gags", one senior mortician told reporters. "He certainly wasn't allowing his death to get him down".

Heart

Only minutes earlier Martin had been told by doctors that his heart had stopped working, and that he would be dead forever. "He didn't seem too concerned. He just laughed and asked if we could put three pairs of shoes in his coffin - because soon he'd be going six feet under", one witness revealed afterwards.

Slits

Celebrated boozer Martin later smiled and waved to fans as he was whisked in a coffin from the exclusive Belle Air Chapel of Rest where he had been staying to a local crematorium where his remains were cremated. En route he told undertakers he was thirsty and jokingly asked whether the hearse could stop at a liquor store.

Fuzzbox

One technician at the $20,000 an oven crematorium later said Martin's ashes were in high spirits, and they had smiled and chatted happily with other remains while waiting to be poured into their swish $500 urn.

Sting Storm!

Sting - 'better helmets'

SINGER Sting sailed into a new storm yesterday when he told a Swedish pop magazine that Britain should have surrendered to Germany in the Second World War.

"I really believe we should have surrendered to Hitler", he told a reporter from Bonken Zpunken magazine. "It would have saved a lot of time and money. A lot of bad stuff has been written about Hitler, but the guy must have had some good points. I don't think he's been given a very good press recently". When he was asked which side he would have chosen to fight on Sting replied "Definitely the Germans. They had better helmets".

Clout

This is the second time in a week that Sting has opened his mouth and put his foot in it. On Monday he was reported to have said that primary school children should be given free heroin and that ginger babies should be drowned in a bucket.

Bangles

These latest untimely remarks are bound to offend the relatives of servicemen and women who lost their lives during the five year conflict (not to mention members of the Jewish community who suffered so badly at the hands of Hitler) when they read them in this paper tomorrow.

George hits the jackpot!

A Middlesborough grandad was last night celebrating with family and friends after finding a straight piece of wood in his local B&Q home improvements store.

George Clayton, a retired scaffolder and keen DIY enthusiast popped into his local store to buy a length of 2x2 planed timber to finish off a stud wall he was erecting at his Teesside home.

Grandad, 72, scoops straight bit of wood

"It didn't really hit me at first", he told us. "I've been buying timber there for years and nothing like this has ever happened. I just picked this bit up off the top of the rack, closed one eye and looked down its length - more out of habit than anything else. When I saw it was straight, I couldn't believe it. I checked it again, then turned to my wife and said "You're not going to believe this, but I think we've got a straight bit"

George celebrates with the B&Q store manager

Mingepiece

George's wife Lillian was so shaken that she almost collapsed and had to sit down on a nearby display of patio chairs. "I was trembling like a leaf", she told reporters later. "You read about people finding straight bits but you never think it's going to happen to you."

Come Bucket

Seconds later George's jackpot joy almost turned to despair when after reaching the checkout he was told the timber didn't have a bar code. The assistant suggested he take it back and get another one. "I was gutted." he said. "But I decided to put my foot down and told them, "No. I'm having this one".

Quim

George and his wife spent the next four and a half hours standing by the till while repeated requests were made over the tannoy for a member of hardware to come to the check out. "By this time the queue was out the shop, across the car park and half way round the local trading estate. But nobody seemed to mind waiting. They all just wanted to see my straight bit of wood."

Fadge

Eventually a twelve year old assistant turned up with a stock catalogue and after half an hour located the item in the book. The bar code was then entered into the till manually at the third attempt and George was able to take his wood home.

Fitbin

News of George's good fortune spread quickly and on leaving the store he was mobbed by a crowd of well wishers before being driven away by reporters from a London based tabloid newspaper. Meanwhile in another newspaper a former girlfriend of Mr Clayton has branded him a 'rat', and revealed that handyman George was a flop between the sheets during the couple's stormy two week affair in 1951.

HOW WOULD THE STARS COOK THEMSELVES?

By our Cookery Writer LIEUTENANT UHURA

Celebrities tend to travel by air more than most people. Their jet set lifestyles mean countless flights for the top names in the entertainment world.

But with air travel comes the unavoidable risk of tragedy. And it is therefore a grim fact of life that the stars could become the victims of an airline disaster. And this raises a number of questions that we, the paying public, have a right to ask.

CRASH

And top of the list must be: if, as a result of an aeroplane crash, some of our favourite stars were marooned on a remote mountain side, and cannibalism was their only hope of survival, how would they cook themselves?

BANG

First we asked EastEnders star **Anita Dobson** which of her cockney colleagues she would rather eat if the cast and crew of the popular soap found themselves stranded and starving on a windswept hillside in Peru.

Star eater Anita

"I'd have to go for Dirty Den, my on-screen husband and actor Leslie Grantham. Leslie and I enjoy working together and our on screen relationship is one of the strengths of the soap. But in the circumstances you describe I would reluctantly eat Leslie, but only if he was very well cooked, ideally on a barbecue."

POP

Stunning Darling Bud actress and pop starlette **Caterine Zeta-Jones** was given the choice of eating fellow members of the Darling Buds cast. And after a moments thought she said she'd prefer to eat Ma Larkin.

"I don't think anyone in my profession would relish the thought of eating a close friend and colleague, but I must admit the prospect of eating Ma – real life actress Pam Ferris – is far preferable to the idea of eating hairy old Pa or my on-screen husband whose name for the moment escapes me".

Star eater Catherine Zeta

Catherine wasn't so sure when it came to choosing a recipe for her screen mum. "I don't know... I suppose I'd like to roast her if that were possible in the circumstances. And I'd definitely want to wash her down with a bottle of wine to try and kill the taste".

WALLOP

We put a similar question to Catherine's former beau, hunky Blue Peter presenter **John Leslie**, himself a tasty dish. But on John's menu we listed a few appetising Blue Peter hosts of the past. And it was no surprise when John selected mouth watering Sarah Green as his choice for an emergency packed lunch.

"I must admit I'd be tempted by flâmbéted Valerie Singleton, and I'd quite fancy pickled John Noakes. But Sarah is the dish for me. And I'd have her in an omelette."

ROAST

When we asked Radio One FM's **Bruno Brookes** which of his DJ colleagues he'd prefer to eat we were surprised to find that he'd already given the subject some thought.

"Funnily enough, all the One FM jocks were flying back from a party in a plane not so long ago when the weather turned nasty. The thought did cross all of our minds that we could go down, and that in order to survive one of us may have to be eaten. It was quite funny really. We all found ourselves looking at Simon Bates.

MASHED

In the end we landed safely, and Bates is still in one piece. But hypothetically speaking, if I did have to cook him I think it would have to be boiled, in a big pot, with lots of herbs thrown in".

CHIPS

Finally we popped the question to controversial pop star **Sinead O'Connor.** We asked her whether her strict vegetarian principles would allow her to eat a fellow star if it meant the differecne between life and death. Surprisingly she had no qualms about tucking into a music industry colleague.

Celebrity cannibals serve up tasty star-packed dishes

Star eater John from Blue Peter

Simon B - the stars' tea

Sinead - Big Macca & fries

Star cook Bruno Brook...es

"Yes, I would reluctantly eat a fellow artist in order to stay alive. But I would insist on eating a fellow vegetarian to ensure that there was no animal protien in their body. I would choose Linda McCartney, and I would probably have her minced in a wholemeal burger bun, with mayonnaise and salad on the side. And fries. And a thick shake."

YOU ARE THE CHEF

Which star would YOU eat, and how would you cook them? Just give us your celebrity recipe and you could win a meal for two in a star-studded West End restaurant!

Imagine you are stranded on a snow covered mountain side, surrounded by all your favorite stars. The aeroplane was carrying a gas cooker in the hold, and you have a limitless supply of propane gas and clean water. You also have all the cooking utensils you could require, and other ingredients such as flour, butter, salt etc. are available from a nearby shop.

A-TEAM

Tell us which star you'd like to have for dinner, and jot down your recipe on a sheet of paper. Send it to Cook A Star Competition, Viz, P.O. Box 1PT, Newcastle upon Tyne, NE99 1PT. We'll be asking a top TV chef to judge your entries, and if he refuses we'll do it ourselves. The winner will receive first class train travel to London, overnight accommo-dation in a top hotel, and a meal for two in a top West End restaurant popular with the stars.

CHEAP

Any top London restaurants popular with the stars who are a bit thin on reservations please get in touch at the same address. And if there's any cheap hotels in need of a bit of business drop us a line.

Come to... **Peters & Lee Land** Just off the A52

405

Meet Britain's only real life
GHOSTBUSTER!

Most people wouldn't dream of spending the night in a haunted house. But Bob Smithers does it all the time. In fact, he does it for a living!

Bob is Britain's only full-time ghostbuster, and rather than running away from them, he's been chasing ghoulish guests out of haunted houses for the last twenty years.

ICE CREAM

Bob, who is 42, began his unusual career purely by accident. An ice cream man by trade, late one evening he stopped his van to sell an ice cream to a young soldier standing at the roadside. "He was a young man – about 18, and he asked me for a '99' with rasberry sauce on it", Bob clearly recalls. "I went to put the ice cream in the cone, but when I turned back he had vanished".

SOLDIER

"A few days later I spotted a newspaper and on the front was a picture of the soldier. It said he had been killed in the war – thirty years earlier. I immediately drove back along the road to the spot where he had been standing, and sure enough, he still wasn't there".

BLUE

From that moment on Bob was hooked, and he has been ghost hunting ever since. Advertising mainly in post office windows, he is on 24-hour call to come and deal with ghosts all over the country.

EXCLUSIVE

"Most of my calls come late at night", he told us "as ghosts are at their busiest in the dark. But contrary to popular belief, real life ghost hunting can be a pretty boring business. People tend to imagine ghosts walking through walls, carrying their head under one arm, or dragging chains around behind them. But it's nothing like that.

DARK

Nine times out of ten you don't even see the ghost – a lot of them are invisible nowadays. And most of the ghosts I do see are just sort of green clouds of energy that float about and shine in the dark. In fact you can get a nasty shock if you touch one of those".

STAR

Hollywood movie makers paint a more exciting picture of the supernatural. And ironically, it's Bob they usually come to for advice. "One day they came to me with the script for the film Poltergeist. It was all about a ghost that lived in a house, and Tom Cruise was going to play it, wearing a blanket on his head. To make him sound a bit more scary they were going to make him talk through a plastic drainpipe. I told them to forget it – it simply wasn't scary enough.

A Hollywood actor similar to 'not scary' Tom Cruise.

The previous week I'd exercised a ghost that had been living in a cupboard, so I suggested that in the film the ghost should live in a cupboard or a telly or something. They thought it was a great idea, and the film went on to be a great hit. All thanks to my knowledge about ghosts.

UNUSUAL

"The most unusual place I ever had to exercize a ghost from was a bus stop. This particular bus stop had been haunted for many years by a headless soldier who every night used to sit and wait for the last bus. The drivers were so scared they refused to stop there, and as a result intending passengers had to walk to the next stop – several hundred yards away.

HOME

Anyway, there's always a reason for ghosts, so what I usually do is go to the library to do research and find out about them. Sure enough in an old newspaper I read about an accident at the very same bus stop over 100 years ago. A young soldier had been knocked over and killed running to catch the last bus home. His head had never been found.

Bob on his way to bust a ghost

That evening I sat at the bus stop, and sure enough I could feel a presence next to me. Then I got to thinking about the missing head. That was probably why the soldier's spirit was not at rest – because his head was missing.

PUPPETEER

I looked around for a few moments then thought to myself, 'I bet they never checked on top of the bus stop'. Sure enough, I climbed up and there it was – a dusty skull covered in cobwebs, and still wearing an army hat. It must have laid there for over 100 years. I took the skull to a nearby cemetery and buried it in a hole, then I said some prayers. As I prayed I could feel that soldier's presence. It was as if he was standing next to me, saying "Thank-you for burying my head".

INDUSTRY

Bob is loathe to divulge the secrets of his trade, although he admits that ghost busting today is a high tech industry.

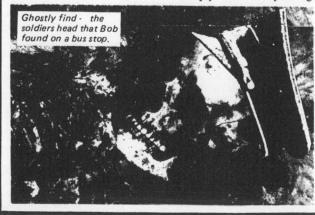

Ghostly find - the soldiers head that Bob found on a bus stop.

'If you've got a ghost, I'm your man' says Bob

"I don't use any onions, or wooden steaks like they do in the movies", he laughed. "My main tool is my ghost-o-meter. You can't buy them in the shops – you have to make them yourself. You've got to be a bit of an engineer in this game, I can tell you."

TRANSPORT

Bob described his ghost-o-meter as a cross between a torch and a walking stick, with a coat hanger on the end, and he uses it to detect ghosts. "It only works in the dark", he told us "and it takes six batteries – which don't usually last very long", he added.

EDUCATION

Indeed ghost busting is an expensive business. The high powered batteries cost almost £3 for a pack of six, and on top of that Bob has his petrol to pay for. "The cost of my petrol varies from job to job, depending on how far I have to travel to get there", said Bob, who drives a scooter. "Motorised transport is essential, as I get calls from all over the country, and you can't always get a bus, especially at weekends".

SOCIAL SECURITY

Despite thousands of nights spent in haunted houses, castles and mansions, Bob inistsits that he's never once been scared. But there have been occasions when he's given one or two other people a fright.
"I'll never forget the time I was looking for a ghost in a church graveyard. It was really spooky, with bats everywhere, and it was a full moon. It was Christmas Eve

so I'd had a few drinks, and before I knew it I'd nodded off to sleep. While I was asleep I must have rolled over and fallen into a freshly dug grave, because when I awoke I'd been buried – the grave digger had filled it in!

SERVICE

So anyway, there was me digging away trying to get out of this grave. Meanwhile the Christmas morning church service had just ended and all the congregation we're coming out – just in time to see yours truly appear from out of the grave!

M.O.T.

I was all covered in soil, I had a nosebleed and the worms had ate my clothes so I looked just like a zombie! You should have seen their faces! Did they shit themselves or what?!
Next week: How Bob was seduced by teenage lesbian vampires – with pictures.

STARS ON MARS

Many of today's top stars share an unusual ambition – to fly to Mars on a space ship.

For as science edges man ever closer to the mysterious red planet, the possibility of commercial space flights to the planet Mars looms larger.

SHUTTLE

Indeed, many space experts believe a regular shuttle service could come into effect within a few years or so.

RACQUET

In America, speculative travel agents are cashing in early, and are already taking bookings for the space journey of a lifetime. But would-be space travellers are warned – prices look set to rocket beyond the budget of most holiday makers. *For a single return journey could cost as much as a million pounds.* And with prices like that, it is only the rich and famous who are buying.

BAT

Already booked up are many of the biggest names in Hollywood, among them Jack Nicholson, Dustin Hoffman, Gregory Peck and Julie Andrews. And other stars set to scramble for stand-by tickets include Michelle Pieffer, Bruce Springsteen and Barry Manilow.

SPIDER

Indeed, Manilow, America's richest man, has already commissioned the first holiday home in space. Top scientists are studying his plans for a 12 bedroom mansion on Mars, and providing technical problems can be overcome, building work will begin by the end of the year.

SUPER

Many disappointed stars have been turned away from travel agents, and are trying to obtain tickets on the black market. Hollywood touts are rumoured to be charging over *£50 million* for a third class single ticket – and stars like Burt Reynolds are snapping them up. For tough guy Burt was originally refused a ticket by travel agents who feared that his wig would come off in space.

GREEN CROSS

One lucky celebrity who will not be paying over the odds

Celebrities queue for journey into space

Gregory Peck and Charlton Heston (above) prepare for a trip to Mars under test conditions at N.A.S.A.'s Kennedy Space Centre. (Below) Burt Reynolds - wig worry.

for his journey to Mars is Star Trek actor James Kirk. Travel officials have offered him a **FREE** return journey to Mars as guest of honour, and Kirk will ceremonially launch the first flight by pressing the start button on the space ship's flight deck.

MORSE

British stars have given a cool reception to news of possible space flights, with only a handful of enquiries being received from this side of the Atlantic. However, singer Tom Jones may well pip the Americans in the celebrity space race. For Welshman Tom has ploughed profits from his singing career into his own space ship, which he hopes to launch this spring.
"It's more of a hobby than anything else", modest Tom told us. "I'll be happy if I can get as far as the moon".

I WAS BENNY HILL'S LOVER

Says Queen's Freddie's sex change dad

A man who fathered the late Freddie Mercury has revealed himself as a saucy sex change cheat.

For he admits that during a brief period working as a sexy glamour showgirl he bedded tubby comic Benny Hill.

Frank Hobson, 42, rocked the showbiz world with his stunning revelation that he was father of tragic AIDS victim Freddie.

CAREER

Although he never met his son, Frank followed his pop career closely. "I used to watch him on Top Of The Pops every week. And I went to see him play live once – at Live Aid in 1986. I stood at the back so he wouldn't recognise me, and wore a big hat and sunglasses. I knew that if he saw me it might affect his performance".

SWERVE

It was around this time that Frank decided to have a sex change. "I felt trapped inside my body and all that", he told us. "Deep down I wanted to wear women's clothes, read women's magazines, have tits and go out with blokes".

SKID

Frank had the operation and started life anew as Francesca. I applied for a job on the Benny Hill show. I remember at the audition Benny winked at me. I thought Oy Oy! I'm in here".

JACK KNIFE

During his three years as one of Hill's Angels, Frank had an on and off affair with Benny. "A lot of what they say about Benny is true. He was quiet, shy, generous, and he left piles of money lying about the house, often crammed into Kentucky Fried Chicken boxes or old pillow cases. But he definitely wasn't gay, and I know that for a fact".

AQUA PLAIN

Frank's affair with Benny ended as the effects of his sex change began to wear off. "I began to lose my figure, and before I knew it, I was a bloke again", he recalls.

MOUNT KERB

Having lost both a lover and a son, Frank now finds himself out of work and living on Social Security handouts. In desperation he has written to Frank Sinatra for help.

MOUNT EVEREST

"During my childhood I used to receive birthday presents from America. There was never any card or a note, just a box postmarked America. Looking back, it all begins to make sense". To this day Frank remains convinced that Sinatra is his father, and now plans a tearful re-union with the star when he visits Britain later this month for a series of sell-out concerts.

K 2

"It will be difficult knowing what to say – he probably won't even recognise me after all these years. It's going to be difficult for both of us, but I know it's something that I have to go through with".

A spokesman for Thames Television yesterday denied that Mr Hobson had ever worked on the Benny Hill Show. "In fact the whole story is a bit similar to one which appeared in a previous issue about Rolf Harris's love child", he added.

NO SEX PLEASE ~ WE'RE STUDENTS

Britain's universities and colleges are heading for a crisis. For as competition for brainy school leavers increases with dozens of Polytechnics now pretending to be Universities, the supply of students is slowly drying up.

And one leading academic fears the shortage of swots swarming to college is nothing to do with the courses that they offer, or new Government funding proposals for further education. He blames a lack of fanny.

K 9

Dr Seigmund Blake, head of enrolment at Fulchester Polytechnic University College of Further Education, believes that behind their academic ambitions, students have a far more important reason for going to college.

JAMIE

"They come for the birds", he told us yesterday. "After 18 years living with their parents they just want to get away from home, get their pants off and get their ends away. It's as simple as that".

LAYLA

And Dr Blake blames a change in attitudes towards sex for the fall in figures. "The women aren't playing ball these days", he explained. "It might be AIDS, it might just be fashion, but the birds aren't putting it about anymore. And as a result the fellas aren't getting any fun".

I SHOT THE SHERRIFF

Dr Blake detects a clear link between a drop in casual sex on campus and the fall in demand for university places. "In the early seventies you couldn't walk through the student union without seeing a couple at it on the floor. Everywhere you looked there were gorgeous birds, and they were real goers, I can tell you. You could spend a month of Sundays looking and still

College kipper dries up

not find a bird with knickers on. Nowadays half of them have padlocks on their fannies. The fellas must get really frustrated".

BUT I DIDN'T

According to Dr Blake applications for University places will continue to fall unless incentives can be found. "Cheap beer is all well and good, but at the end of the day it's up to the girls. If they don't start sleeping about a bit, blokes will simply stop going to college, and the whole future of the country could be jeopardized".

SHOOT THE DEPUTY

A spokesman for Newcastle Polytechnic told us that they would be changing their name to Northumbria University at Newcastle starting from next term. A male student we spoke to confirmed that girls at the college weren't even giving him a whiff. "If I don't get something by the end of the term I'm leaving", he told us.

★Rude Kid

WHAT WOULD YOU LIKE IN YOUR SANDWICHES, DEAR?

GRANNY'S PUBES Y'WHORE!

THE CHAMELEON OF COMEDY!

Take a close look at the gallery of faces shown here. Believe it or not, they are all the same person!

Those are just some of the many faces of TV jester David Jason. Star of such hit comedy shows as Only Fools On Horses, Open All Hours and The Darling Buds of May. Over a brilliant career of TV comedy acting he has established himself as Britain's top comedy actor.

KEY

But David's key to success has been his chameleon-like ability to change character. Whether it be a scheming entrepreneur in Fools On Horses, a bungling shop assistant in Open All Hours, or a scheming old man with a good looking daughter in the Darling Buds, he can carry off the role with ease.

GOLF

The secret of David's success is in his amazing ability to change his facial expression, wear hats and dress up in

Many faces of TV Jester Jason

different clothing. It's a traditional acting technique that he has mastered over a period of many years. In showbusiness circles David is nicknamed 'The Chameleon of TV Comedy' because of his uncanny ability to change appearance.

COUNTRY

"One simple but effective trick is changing the angle of your mouth", one TV insider told us. "By bending your lips upwards at the ends your can achieve a smile. Bend them downwards, and a frown is the effect". But as

With a hat – it's Del boy!

'Perfick' as Pa Larkin!

In a jumper – it's Granville!

And this is the real me!

well as smiling and frowning, top actors, the masters of their trade, employ movement of the the forehead, eyebrows and cheeks to add emphasis to their acting, we were told.

Many top actors we spoke to praised Jason for his acting skills, but prefered to save their tributes until after he dies. In the meantime we wish him continued success in his acting career.

GOING THROUGH THE MOTIONS

By our Science Correspondent Lulu

Nobody likes to dwell on their droppings. We tend to flush our faeces down the toilet the minute we've finished wiping our bottoms. But according to one new theory, we could be making a big mistake.

An increasing number of doctors now believe that faecal examination – that's looking at your poo to you and me – is an excellent way to monitor your health. For the colour, content, texture and smell of a number two reveals vital information about the lifestyle that you lead.

TURD

"Every turd is a little mine of information", says Dr. Emilo Budweiser, Head of Excrement at Los Angeles University's Institute of Advanced Toilet Research.

STOOL

"Every stool is like a book just waiting to be read. Every

EXCLUSIVE

dollop a brand new chapter, packed with information. And winnits are like micro film, jam packed with valuable data about the human body", he continued.

MUSHITROOM

In his new book 'Learning From Our Stools', Dr. Budweiser recommends that we all take time to scrutinise our droppings for a few moments each day, probing it with a fork, or an old tooth brush. By doing so all manner of information can be discerned. Here's a few clues to look out for, together with the information that they reveal.

Do you recognise any of the following poo properties?

PEANUTS – If you find traces of peanut amongst your stool, you may well be a party person – someone who likes a nibble in the pub perhaps, or alternatively someone who enjoys crunchy peanut butter.

SWEETCORN – This suggests a healthy lifestyle. You enjoy vegetables – sweetcorn in particular. You might also enjoy corn on the cob. Sweetcorn is generally a healthy sign.

SLOPPY STOOLS – If you suffer from runny poo this is an indication of bad diet, and of possible stomach problems. You may lead a stressful life, suffer from anxiety, or perhaps you just had a big curry last night.

DARK "FUDGY" STOOLS – This could be a sign that you drink Guinness.

FAT WITH A HAT

Send us the name of a fat person and the type of hat you'd like to see them wearing. We pay £5 for every suggestion we use.

This week: Actor Stratford Johns wears a trilby hat, as suggested by Kate Remmington, 32, of Macclesfield. Congratulations Kate, there's a crisp fiver on its way to you.

TINKER, TAILOR, SOLDIER ...STAR!

By our Showbiz writers **PIERS LYING** and **GARY BASTARD**

Many of today's top celebrities have known what it is like to perform less glamourous jobs. Indeed the stars instantly recognisable for their recent TV success have usually spent many years in poorly paid and at times unusual occupations before making the big time. Here's just a few of the odd ways in which today's celebrities once earned their crust.

★ If you had the misfortune to lock yourself in the public toilets of Victoria Coach Station in 1964, you'd have been in for a big surprise. For you may well have been rescued by a short, familiar figure wearing distinctive spectacles. That's because in 1964 that man with the job of maintaining those lavatory doors was none other than top TV comedian **RONNIE CORBETT**! Toilet Door Maintenance Officer was just one of a string of part time jobs the pint-sized comic took on to earn extra cash while struggling to make it in the show-biz world. Nowadays, with TV success like the Two Ronnies and Sorry to his name, and successful panto appearances under his belt, Ronnie can afford to sit back and spend his spare time pottering about in the garden.

★ *Residents of Grantham in Lincolnshire never knew what to expect when visiting their local dentist. Something of an eccentric, he would often hide in a cupboard for up to five minutes before springing out to shock his baffled patients. He was also well known for leaving a 'whoopee' cushion on his dentist's chair. And who was that now famous dentist? None other than zany TV prankster* **JEREMY BEADLE**. *Jester Jeremy quit dentistry in 1978 to become a full time TV funny man.*

★ **SIR ALLISTAIR BURNETT** has achieved a great deal in a marvellous career in news broadcasting. But it was only due to a stroke of luck that his news career got off the ground at all. In 1979 he had been working hundreds of feet above the streets of London – as a window cleaner. An alert news editor spotted him polishing the windows at the ITN offices. He was immediately offered a job as Senior Newscaster, but this came as such a shock that the luckless Sir Allistair lost his balance and fell 23 storeys, landing in a flower bed below. After a brief spell in hospital he eventually hung up his bucket and officially began work at ITN.

★ *It's hard to imagine our senior politicians working in mundane, everyday jobs. But 32 years ago one of today's top Tory cabinet ministers could have been found hundreds of feet below the Houses of Parliament, working on the London Underground! After a brief spell as ticket collector Deputy Prime Minister* **SIR GEOFFREY HOWE** *worked as a driver on the Victoria Line before he was sacked in 1983, after turning up for work drunk. Sir Geoffrey then set is sights on a career in politics and Mrs Thatcher's right-hand man hasn't looked back since.*

★ *Bubbly Hi-Di-Hi star* **SU POLLARD** *ignored advice from her parents and gave up a promising career as a Research Scientist in order to go on the stage. Former colleagues at the University of Cardiff Medical Research Institute were shocked by her decision to quit. After several years of research she was thought to have been only days away from inventing a cure for cancer. But in the end Su had the last laugh, becoming a big hit as the scatter-brained Miss Cathcart in Hi-Di-Hi.*

★ Were it not for his marriage to the Queen of England, **PRINCE PHILLIP**, the Duke of Edinburgh, would never have become a member of the royal family. He would have remained lowly Stourbridge rent collector, Phillip Tunstall. Before he met the Queen, Phillip had been collecting rent for Warwickshire County Council for several years. During that time, he was commended for bravery after thugs had tried to grab his rent bag. Have-a-go hero Phillip fought off the brutes, who were forced to flee empty-handed.

THIS SUPER PRIZE IS JUST THE JOB

Here's a chance for you to win a trip to the job centre of your choice, anywhere in the UK. We are offering first class travel tickets for two plus two nights' hotel accommodation at the destination of your choice, to the lucky winner of this fabulous 'job related' competition. All you have to do is read the celebrity CV below, then tell us which famous star you think it belongs to.

Born in Cardiff, his first job was as Butchers Boy, cycling around the valleys of South Wales delivering sausages. He then signed apprentice forms with Port Vale football club, eventually making seven appearances, scoring one goal. Left football to return to full-time education, tak-ing a psychology degree at York University. Two years later while working as a lumberjack in Scotland a record company executive overheard him whistling and dancing to rock 'n' roll tunes. He was immediatey signed up and today he is surely Britain's brightest singing star.

Answers on a postcard please to Shakin' Stevens Celebrity CV Job Centre Visit Competition, Viz, PO Box 1PT, Newcastle upon Tyne, NE99 1PT. The competition closes on 31 March 1990. The winner will be notified by post shortly thereafter. Please remember to include your name, address and the name of the job centre which you would like to visit.

£75 billion a year down the toilet?

FOOD -DO WE REALLY NEED IT?

Or is it just a waste of money?

Food — we all eat it. Whether it's bacon for breakfast or lettuce for lunch. We all eat something.

It is estimated that in Britain this year alone we will spend over £75 billion on food. So imagine the savings that could be made if we simply gave it up. In a single year there would be enough money saved to build 37,000 hospitals, or to buy well over three million Variety Club 'Sunshine Coaches'. So just how important is food? Do we really need it, or is it just a waste of time?

Up till now scientists have always believed that we use food to keep us going, in much the same ways that a rocket uses fuel or a television set uses electricity. But now, as we head for the 21st century, the experts are beginning to think again.

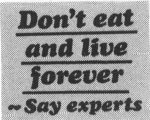

Don't eat and live forever
~ Say experts

Someone in a food shop. Will the shelves be empty in a few years time?

ABORIGINE

Over the years we have all heard reports of perfectly healthy people who eat nothing at all. There was the Australian Aborigine who hadn't eaten for fifty years who recently celebrated his 140th birthday. Or the story from Peru about the young child who fell down a well in 1938 and emerged forty years later a fully grown man, despite having had no food at all. And closer to home it is said that pop star Cliff Richard remains young by fasting all year round.

These are the rumours, but what about the facts? Scientists need concrete evidence before they begin to burn the cookery books. So we decided to carry out an experiment of our own to see exactly how long the human body can survive without food.

PERSON B

We used two people in our experiment — person A and person B. We allowed person A to eat as much food as they wanted, but allowed person B no food at all. After 2 hours we asked them both to complete a crossword puzzle. Then, 2 hours later, we invited them to play Space Invaders. Half-an-hour later we had to stop the experiment when person B remembered he had to go to the bank and left in a hurry.

MEALS

So we decided to take our experiment onto the streets to see what you, the public, thought. Pub landlord Bob Taylor told us that he ate three meals a day. When we asked him if he could do without food, he said "no".

PIE

Factory worker Eric Dunn had just purchased a steak & kidney pie at the bar. When we asked him whether he'd like to try and do without the pie he became abusive and we were forced to leave.

NEARBY

Later the manageress of a nearby bakers shop admitted that her business would suffer if people stopped buying food altogether. But she refused to comment any further, and then asked us to leave as there was a queue of customers building up behind us.

FOOD FOR THOUGHT

Nothing lasts forever — and food is certainly no exception. And whether we need it or not, we may find ourselves having to do without food sooner than we think.

BREAD

For experts predict that the Earth's food stocks will have started to run out by the turn of the century. And by the year 2200, supermarket shelves around the world will be completely empty. Food items that we all take for granted, like bread, sausage rolls and frozen pizzas will be like gold dust to our grandchildren.

HAM

And as supplies run out, so the prices will soar. A tin of cooked ham costing 60p today could fetch a staggering £100,000 if it were auctioned in 75 years time. And as food becomes more expensive, so people will have to find alternative things to eat.
Scientists have already started looking for the foods of the future. They believe that cork, rubber and cardboard boxes could soon be part of our regular diet. In contrast, researchers in the USSR believe that by the year 3000 humans will be leading a strange worm-like existance, living underground and eating soil.

WORMS

We tried eating soil to see what life would be like in the year 3000. But we got diarrhoea.

DOCTOR, I'M A SHADOW OF MY FORMER SELF.

ST. Shiloe

"With This Bling, I

LAST MONTH, readers of *OK* magazine drooled over page after page of photographs of the marriage between singer Peter Andre and breast model Jordan. Dubbed the Wedding of the Century, the ceremony at an expensive Berkshire Castle was the climax to a fairytale romanace that had started 18 months earlier in the jungles of Australia.

In the heavily copyright protected photographs, the D-list-celebrity-filled event appeared to go without a hitch. The pumpkin coach pulled by six white horses arrived on time, Jordan's 10 foot-wide dress sparkled with the the lights of 1000 Swarovski crystals, and the £50,000, 18-tier cake was cut without it toppling off its specially-made £100,000 gold-plated trestle table.

But according to one insider, away from the *OK* cameras the day went anything but smoothly.

Former SAS paratrooper **Frankie Liar** was employed by the couple to make sure than no gatecrashers or unwanted friends and family made it through the security cordon and into the star-studded bash.

According to Frankie, the day was a catalogue of cock-ups from start to finish. And in a hastily published book, he spills the beans on the Fairytale wedding that very nearly turned into a horror story. Here for the first time are exclusive extracts from his book that Peter and Jordan have not tried to ban.

> The day started badly. I was on a routine security sweep, and I was in Mr Andre's bedroom checking there were no *Hello* photographers in his wardrobe.

EXCLUSIVE!

By our Tacky Showbiz Correspondent
Duncan Smothley

When he got out of bed, I heard an almighty commotion. "They've gone! They've gone!" he kept shouting.

At first I thought that someone must have broken in and stolen the wedding rings, but then I realised he was talking about his abdominal muscles.

ABS

Where they had been there was just a small, flabby pot bellly. In all the excitement, Andre had forgotten to do his abs excercises the night before, and his trademark six-pack had turned to flab. But lucky for him, I was there to save the day. I taught him a secret SAS training technique to get his muscles back, involving doing half a million sit-ups in an hour. By breakfast, his stomach looked like a bag of six burger buns once more."

Everything went smoothly for a while. But about halfway through the morning, Frankie was doing a routine sweep of the castle when all Hell broke loose once more.

"We'd received a tip-off that a Sun photographer was hiding in the tea urn. That turned out to be a false alarm, but as I was leaving the kitchens I heard a commotion in the next room. It was Peter Andre yelling at the top of his voice. "Take them off! Take them off!" he kept shouting.

TCS

It turned out that former Big Brother contestant Bubble had turned up wearing a pair of shoes which had cost several thousand pounds more than Peter's. As the groom, he was adamant that his footwear should be the most expensive in the room.

He was furious, and because Bubble was refusing to take his shoes off, it looked like the wedding would have to be cancelled.

4WD

Luckily, one of the guests came up with the bright idea of increasing the value of Andre's shoes by stuffing them with fifty pound notes. It was a bit of a squeeze with nearly eight grand in each shoe, but he managed to get his feet in somehow. In fact, in the wedding photographs you can see Peter wincing as he hobbles down the aisle."

Once the bride and groom had made it to the wedding ceremony, you could be forgiven for thinking that nothing else could go wrong. But Jordan and Peter's troubles were only just beginning. Frankie takes up the story.

"During the service I was carrying out a routine check for hidden cameras under the vicar's cassock while the couple were standing at the

> *"...Peter was furious, and because Bubble was refusing to take his shoes off, it looked like the wedding would be cancelled."*

Don't they make a lovely pair: Tit model Jordan's wedding as it may have looked (above), and (below) a six pack similar to the one Andre lost. (Below right) Bling: Tit model Jordan's wedding ring as it may have looked.

altar. When they had exchanged their vows, Peter reached in his pocket and brought out a ring the size of a man's fist, covered in hundreds of diamonds. Talk about bling! It was the most showy thing I've ever seen.

SRS

But it didn't impress Jordan; she went absolutely ballistic, shouting and screaming about the ring, saying that it wasn't gaudy enough. She told Peter in no uncertain terms that the marriage was off unless he could find a more ostentatious wedding ring ... and quick!

The call went out, and within an hour a helicopter landed on the lawn with a very important cargo - a bucket of diamonds and a tube of superglue. Peter's best man, the winner of last year's *'Fame Academy'*, then

Thee Wed"

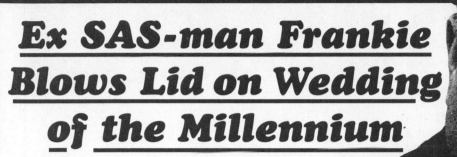

Ex SAS-man Frankie Blows Lid on Wedding of the Millennium

had to sit down and start sticking the jewels onto the ring. A couple of hours and about ten thousand diamonds later, Jordan decided it was tawdry enough, and the wedding ceremony went ahead."

But even as Mr and Mrs Andre left the church to get into their tacky pumpkin-shaped horse-drawn coach, things nearly went pear-shaped again.

"I was just checking up the horses' arses for concealed video equipment when Jordan and Peter came out of the church. Everyone was cheering and clapping and each guest had been given a box of very special confetti to throw over the newly-weds.

> **"...Peter handed a servant at the castle a paper shredder and £1million in notes and told him to get to work."**

Normal confetti is fine for a normal marriage, but the wedding of the century demanded something a little more tasteless. Earlier in the day, Peter had handed a servant at the castle a paper shredder and a million pounds in notes and told him to get to work. But, amazingly, as the guests showered the shredded cash over the happy couple, the emotional bride threw another wobbly.

AQUAFRESHES

Jordan had noticed that the confetti was made of twenties, and she started crying. She threatened to divorce Andre then and there, sobbing that if he really loved her it would have been a million pounds in fifties.

Peter kept apologising, but there was no time to shred a million pounds in fifties; the couple's farcical golden Cinderella coach couldn't be kept waiting any longer. It seemed like the marriage was going to be over before it had even begun.

EUTHYMOLS

Then Peter had a brainwave. He took out his wallet and wrote a cheque for a million pounds cash in fifties, tore it into little bits and sprinkled it over his wife's head. Everyone breathed a sigh of relief. The wedding was back on!

Copies of Frankie Liar's book *'Love at Any Price - Behind the Scenes with the SAS at the Wedding of the Century'* (Ex-SAS Man Publishing) are available from the author in the snug of the Winning Post, Clifton, every night from 6pm, price 1 pint.

FANS TURN SCREW ON BOULT

ANGRY MUSIC fans were last night calling for the sacking of conductor *Sir Adrian Boult* after the London Symphony Orchestra turned in another poor performance playing Schubert at the Festival Hall.

A crowd of 10,000 concertgoers watched the players give a lacklustre 1st half rendition of the 8th Symphony in D minor, and a stony-faced Boult was booed off the dais as he left the stage at the interval.

Insiders say the 58-year-old maestro gave the orchestra 'a proper roasting' in the dressing room. "He was absolutely furious with them," said one source. "I've been in the orchestra business for forty years and I've never seen a conductor lose it like that. He must of knew his job was on the line."

As the second half got underway, the orchestra began their rendition of Mozart's Requiem with more spirit than they had shown in the first half, and there was a feeling that the players had raised their game following Boult's hairdryer treatment at half time. But it wasn't long before the familiar silly mistakes began to creep back in. A fumble by the principal clarinettist during a slow movement led to the double basses missing a cue, and the angry cat-calls from the crowd began again.

Shortly afterwards, a trombone set-piece was well off the pace, and left Boult throwing his baton to the ground in despair. When the audience members were finally put out of their misery, the orchestra left the stage to a chorus of angry boos and whistles. Several members tore up their LSO season tickets in disgust and threw them at Sir Adrian.

Performance Leaves Conductor's Job on Line

BOULT: Is the maestro set for the nut-heap?

"This performance was unacceptable," fumed angry fan Quentin Cuthbert, who had travelled from Holland Park for the concert. *"Schubert's 8th should have been a walkover for players of this calibre. Instead, they played like a bloody school orchestra."*

As the crowd left the hall, there was no doubt who was being blamed for the poor performance.

"Boult played with four kettle drums at the back and only twelve principal violins. That left the bassoons wide open looking for their cues during the fast movements," raged Tarquin Bumboy. "What was he thinking of?"

And chairman of supporters' club *Friends of the LSO* Crispin Fortescue was equally disappointed. "I've followed the orchestra for years, and this is the worst I've ever seen them play," he said. "The violas were sloppy, the horn section was outclassed throughout, and the triangles simply weren't good enough. They simply weren't playing like an orchestra. It was a national embarrassment. Boult is the conductor and he has to take the rap."

The Trustees of the LSO were last night fending off an angry chorus of calls for Boult's dismissal. And although no official announcement has been made, it is thought it will not be long before they are forced to bow to public opinion and show Sir Adrian the door of the Royal Albert Hall.

LSO DEAR!
What Went Wrong on the Night

1st HALF — 14 MINUTES: *Crash!* Cymbal comes in a semitone late during Scherzo movement.

1st HALF — 37 MINUTES: *Bang!* Flute plays pianissimo during fortissimo overture.

2nd HALF — 88 MINUTES: *Wallop!* Harp glissando set-piece goes tits up.

SHOCKER: How the Sun reported the concert

Green Measures Hit UK Bagladies

BRITAIN'S BAGLADIES were last night up in arms after new figures showed that their supply of carriers has reached a record low.

Green measures brought in by high street shops and supermarket chains have seen numbers of disposable plastic bags fall dramatically in recent months. And while the widespread take-up of so-called 'bags for life' by shoppers has been good news for the environment, it has proved a major headache for the country's female vagrants.

carrier

"We bagladies have historically relied on a steady supply of carrier bags to carry our miscellaneous bits of crap around in," said Edna Speight, President of the British Institute of Bagladies and Ciderwomen. *"The turnover rate has been traditionally high. Because the handles tend to go after just a few days of being carted round the streets, they need replacing with new bags on a regular basis. Quite simply, thanks to the so-called green initiatives of our shops, old-fashioned carriers are getting harder and harder to source."*

EXCLUSIVE

Pausing to take a swig from a plastic bottle of White Lightning and shout incoherent abuse at a passer-by, Miss Speight continued: "If the availability of fresh carriers doesn't improve for my members soon, I can foresee a situation in the very near future when this country's precincts, parks and shop doorways will be empty of bagladies."

homing

But Leonard Tripper, spokesman for the Royal Society of Supermarket Managers, was last night adamant that his members would not be bringing back the old style carriers. He told us: "I'm very sympathetic to the plight of the bagladies, but at the end of the day the environment must come first. Our new bags for life are an important weapon in the fight against global warming. The fact that we can sell them for £1.99 and they last approximately two months before the handles snap and they have to go for landfill is neither here nor there."

But speaking from her piss-soaked bench in Gloucester Bus Station, Edna Speight dismissed Tripper's environmental claims. "All I've got to say about global warming is 'Bring it On!'," she said. "If, like me, you had to sleep in a cardboard box under the access ramp to the library every night, with six coats on and all newspapers stuck in your tights, you wouldn't mind the global warming being turned up a couple of degrees, I can tell you."

Walter

"Have you got ten pence for a cup of tea? Just ten pence, pet. I'm really struggling," she added.

Keep off the nuts this Christmas

If you're going to be driving this Christmas, stay off the nuts.

That's the shock warning being issued by convicted drink/driver Frank Parkington, after magistrates slapped a 2 year ban on him and fined him £200 for driving with three times the legal limit of alcohol in his blood.

RESERVATION

Parkington, 36, was stopped by police after driving the wrong way up a dual carriageway, crashing through the central reservation and driving 60 miles per hour in a pedestrian precinct. Despite being found guilty, Frank still maintains his innocence.

TEEPEE

"As I explained to the police, I'd deliberately stayed low that night 'cos I knew I was driving. I'd only had 6 or 7 pints, and a couple of shorts. I know my limit". But Frank claims that before he left the pub, he was offered some peanuts.

"I had a couple of packets, and the next thing I knew I was feeling drowsy. It was definitely the nuts that did it. People warn you about drink/driving, but they never warn you about the nuts.".

TOMAHAWK

Frank believes a rare allergic reaction to peanuts causes his blood to turn to alcohol. But magistrates were unimpressed by his claim and found him guilty — his third similar conviction in the last six years.

"It's bloody infuriating, but what can I do? In future if I go out for a pint, I'll just have to lay off the nuts altogether".

We rang crazy drink/driving comic Jim "Nick Nick' Davison for his comment, but he wasn't in.

Biscuit Barrel

● **WHEN OH WHEN** are the manufacturers of Ritz crackers going to put salt on the underside of the biscuit rather that the top? I for one am fed up of having to turn my Ritzes upside down before putting them on my tongue.

Barney Snitterby, Louth

● **I HAD TO LAUGH** the other day. My young son ate so many biscuits that he gave himself indigestion. What type of biscuits were they? Unfortunately, they were Rich Tea biscuits, but how ironic it would have been if they were digestives. It would certainly have been worth a letter in this column.

Margaret Toft, Spridlington

● **ARE NICE** biscuits pronounced Nice or Nice? I always pronounce it Nice, but my wife insists that it is Nice.

Barry Macbeth, Barnes

*I'm afraid your wife is right, Mr Macbeth. It is pronounced Nice.

● **I LOVE** biscuits. So much so that I wrote a poem about them.

If I did have some biscuits,
Oh, happy would I be,
From their tin I would take them,
And dunk them in my tea.

That's as far as I've got at the moment, but I'm working on the next verse. The second line ends in 'eat' and the fourth line ends in 'treat'. I'm just filling in the gaps.

Andrew Motion, London

● **WHY DON'T** Crawfords make their Cheddar biscuits two-inches thick? I am morbidly obese, and I love to shove a quarter of a packet in my mouth at once. My work colleagues say it makes me look uncouth and greedy, and I find this hurtful. They wouldn't be able to say this if I was eating a single, albeit two-inch thick, biscuit.

Kevin Bellend, Heartford

● **I LIKE** to eat biscuits whilst on the toilet. My favourite are Jaffa Cakes, as they are easy to get out of the packet using one hand - particularly important if you have already started wiping.

Morgan Cheeseborough, e-mail

TELL THEM I WANT IT DELIVERED BY 9.00 TOMORROW, OR I'LL GIVE THE CONTRACT TO AMALGAMATED IMPERIAL

THAT DOG'S DOING ITS BUSINESS IN OUR GARDEN AGAIN

A RIGHT ROYAL WASTE OF TAXPAYERS' CASH

This year Britain's Royal Family will spend an incredible £8 BILLION on Christmas decorations — more than every other household in the country put together.

And if you include the Royals' Christmas shopping – for things like presents, food, drink, tangerines, walnuts, party hats etc., the bill comes to a staggering £1000 billion, all of it paid for with taxpayers' money.

VAST

Buckingham Palace is a vast building, with many hundreds of rooms and miles of corridors. And all of it has to be decorated. That's why the Queen will need:

- 500 tins of spray on snow. Enough for every window.
- 600 miles of tinsel.
- 84 tons of glitter.
- Over 2000 Christmas trees, plus fairy lights.

As well as forking out for decorations, the Queen also has to pay people to put them up. The Royal household includes a permanent staff of 60 Christmas decorators. They spend six months of the year taking down old decorations, and the other six putting up the new ones.

BALLOONS

And with over 10 million balloons to blow up, 19,000 miles of paper chain to hang and 5,000 tons of Christmas tree decorations to put up, their task is not an easy one. "The Queen is very fussy about her decorations", one Palace insider told us. "Everything must be just right. All the trees must be perfectly shaped. If any of them are a bit thin, or bottom heavy, or if there's so much as a single needle out of place, there'll be trouble." Palace staff now employ a top London hairdresser to trim each tree individually before they are put on display.

DECORATIONS

The findings of an all-party Commons Committee on

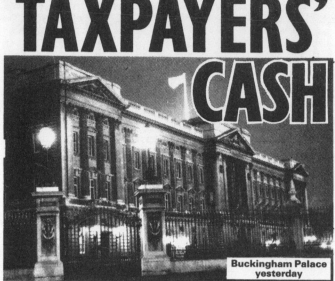

Buckingham Palace yesterday

the Queen's Christmas decorations are due to be published later this month. Among their recommendations are thought to be the proposal that the Royal Family make their own Christmas cards, plus a few Christmas tree decorations, using coloured paper, paint and glue etc. It is thought that this could produce an overall saving of around £160 million per year.

CRACKERS

Meanwhile, the Government has approved extra funding of £8 million to buy better quality Christmas crackers for the Royal family after one member complained about the poor design quality of the party hats.

Greedy Little Bastard

A copy of a letter written by Prince Harry and addressed to Santa Claus has come into our possession.

The letter, marked 'top secret' and written on Palace notepaper, was found lodged in a chimney at Buckingham Palace.

PRESENTS

In it Harry, youngest son of the Prince and Princess of Wales, lists the Christmas presents he expects to receive this year. It is a staggering list, containing over 700 separate items.

GREED

And the chimney sweep who found it was so incensed by the young Prince's greed he decided to hand the letter over to the press. Father of four Dick Poppins described the list as 'disgraceful'.
"I was shocked and appalled. There's not a single thing on that list that I could afford to buy for any of my six kids. And, of course it's tax payers like me who'll end up paying for it all".

The mammoth list, which runs to 8 pages, includes:

- A submarine, complete with crew, costing over £10 million.
- A half size exact replica gold Aston Martin car, with diamond spark plugs. Price £18 million.
- A large rural county in England, preferably Hampshire. Value £8 billion.
- A full set of Teenage Mutant Hero Turtles, price £90.
- A miniature rocket to take the young Prince on space adventures to Mars. Price £400 million.
- Fourteen tons of chocolates, fudge, toffees and liquorice allsorts.

Prince Harry. Or William.

"The most annoying thing is that the spoilt brat will get every single thing on that list", Dick fumed. "And after playing with the toys for half an hour he'll get bored and chuck them away. Meanwhile, my eight kids will get nothing, especially since I've lost my job".

LUST

Dick was sacked by Palace officials who accused him of having stolen a French Renaissance mantle clock worth £25,000. He was arrested by police after trying to sell it in a pub for £40.

"I never done it", he told us. "I had took the clock home to clean it, after I got soot on it. And I just nipped into the pub to show it to a mate of mine who likes clocks, that's all".

SLOTH

A spokesman for Buckingham Palace told us that details of Prince Harry's Christmas list were never made public. However a spokesman for Hampshire today confirmed that a mystery buyer had bought it.

Sex earthquake rocks quiet village

I made love to myself

~ while I watched

A 42 year old man has revealed how he made love to himself while his wife slept nearby.

And Reginald Thompson has stunned neighbours in the quiet village of Banwell with his saucy revelations. "He seemed so quiet", one neighbour told us last night.

FROLICS

Reg's sexy frolics with himself began one night in 1986 after he had spent a quiet evening at home with his wife, Carol. "Carol had gone to bed early and I was alone on the settee", he told us. "I'd had a few

Village torn apart by torrid sex tornado

The house in Banwell where Thompson's sordid sex sessions took place.

drinks and I was feeling very relaxed. The next thing I knew I felt my hand on my shoulder. Seconds later I was rolling around naked on the floor with myself. It seemed like the most natural thing in the world".

Reg began to have regular solo sex sessions whenever his wife was out of the house. "One minute I'd be in the garden mowing the lawn, and the next minute I'd be in the shed, fondling my own buttocks", he revealed.

Passion volcano erupts showering village with red hot love lava

On one occasion, Reg sat and watched while he made love to himself. "I'd just drunk a bottle of Vodka and was feeling uninhibited. I sat and watched myself in the mirror. It was a fantastic experience, although I had a headache afterwards". Soon Reg was having sex with himself up to three or four times a week. "Sometimes I would just sit in the cupboard under the stairs with my hand on my knee. At other times, I would roll around on the carpet in front of the fire for hours on end. On one occasion, I even knocked the coffee table over and spilt a glass of wine. I only just managed to clean it up before my wife got home".

SESSIONS

Locals at the Kings Head Hotel only 200 yards from Thompson's semi-detached house, were shocked whey they heard of Reg's steamy sex sessions. "He seemed so quiet", one customer told us.

TIPTON IS TOP FOR TOURISM

Ask any discerning holiday-maker where he's heading this Summer — and the answer won't be Benidorm, Bermuda or Barbados. It will be Tipton!

This is the astonishing claim made by Mr Hugo Guthrie, chairman of the Tipton Borough Council Committee on Tourism. And he believes that 'Terrific Tipton' will soon be top of the tourist tree.

"At Tipton we have a comprehensive range of amenities to suit holidaymakers of all ages. There's truly something for everyone at Tipton. We've got some smashing countryside with marvellous views only 20 miles the other side of Wolverhampton".

"And for the water sports enthusiast, we've got a very pleasant stretch of river, and plenty of canals. And did you know incidentally, that here in Tipton there are more miles of canal than there are in Stourbridge. Indeed, Tipton has been described as the Venice of the West Midlands".

Tip Top Tipton Tipped as Top Tourist Trap

And according to Mr Guthrie, Tipton certainly isn't lacking in nightlife. "Appearing for the whole Summer season at the Tipton Apollo Theatre, we hope to have the one and only Bernie Clifton and his Comedy Ostrich — as seen on TV".

BOOK EARLY

Holiday-makers intending to visit Tipton are advised to book early, as accommodation will be in short supply. And a final word from Mr Guthrie — he believes Tipton's attractive new tourist slogan says it all. "Tipton — it's terrific".

WE'RE SKINT, M'LUD!

Top lawyers plead poverty

Many top lawyers are having to take on part-time jobs in order to make ends meet. Highly trained barristers and top solicitors are resorting to bar jobs and part-time restaurant work in order to boost their paltry income.

GET BY

One solicitor we spoke to said he worked Saturday mornings collecting litter at MacDonald's restaurant in London's Piccadily, as well as delivering free newspapers during the evenings. "I still find it hard to get by", he told us. "A colleague has resorted to advertising his services in newsagent's windows as a part-time gardener and handyman", he added.

SURVIVE

Top lawyers feel that with rising mortgage rates and inflation, it's not possible for them to survive charging only £120 per hour for their services. "Unless there is a dramatic increase in our levels of pay, I can see ugly scenes reminiscent of the miners strike breaking out in the Court rooms and the offices of the legal profession", one told us this morning.

MAN 'HAD SEX WITH 2,500 WOMEN' ~claim

A Northampton man is claiming to have had sex with 2,500 women over the last five years.

Probably

"Come to think of it, it was probably more like 3,000", he told us yesterday.

ZOOO 2000!

Daddy's taking us to the zoo tomorrow, zoo tomorrow, zoo tomorrow. Daddy's taking us to the zoo tomorrow, and we can stay all day.

We're going to the zoo, zoo, zoo. How about you, you, you? You can come too, too, too. We're going to the zoo, zoo, zoo.

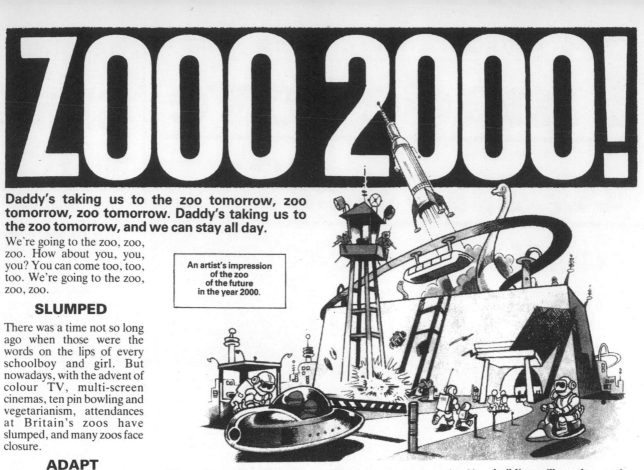

An artist's impression of the zoo of the future in the year 2000.

SLUMPED

There was a time not so long ago when those were the words on the lips of every schoolboy and girl. But nowadays, with the advent of colour TV, multi-screen cinemas, ten pin bowling and vegetarianism, attendances at Britain's zoos have slumped, and many zoos face closure.

ADAPT

So what is the future for Britain's zoos? How will they need to adapt to survive in the fiercely competitive entertainment world of the future? We asked top animal experts Johnny Morris and Richard Attenborough to predict for us the shape of zoos in the year 2000. This is their vision of the future.

PRISON

Visitors arriving at the zoos of the future may be mistaken for thinking they were entering a prison camp. For the zoo would be surrounded by 50 foot high walls and barbed wire, and armed soldiers will accompany zoo keepers on their rounds.

HUMILIATING

Visitors will undergo humiliating body searches as they enter the gates, and any cars left unattended in the car park will be blown up by bomb disposal experts.

The reason for this tight security is that by the year 2000 loony lefties calling themselves the Animal Liberation Army will have declared war on Britain's zoos, and they will be under constant threat of terrorist attack.

DEGRADING

Inside the zoo there will be some unfamiliar sights too. For as well as old fashioned animals like monkeys, lions and bears, there will be a new attraction – space animals.

MARS

Brought to Earth in giant space rockets from far away planets like Mars and the Sun, space creatures will come in all shapes and sizes, and a variety of brilliant colours never before seen on Earth. Some will have three eyes, others ten legs. And many will be able to sort of dematerialise into silver clouds – like on Star Trek.

The bars on space animal cages will be very close together. For no matter how big the space animals may be, they will be able to change shape at will, and so they will be able to get through the smallest of gaps.

TWIX

Old-fashioned Earth animals will be updated using modern technology. Visitors will be able to play safely with remote control lions, or watch while the zoo keeper changes the giraffe's batteries. And silicon chimps will no longer have messy tea parties. Instead they'll entertain us by playing chess, listening to Radio 4 and discussing last night's viewing on Channel 4.

KIT-KAT

In the Reptile House temperatures may appear cool – but the snakes and lizards will be hotter than ever. That's because the building will revolve – using the microwave principle – heating the animals more thoroughly, from the inside, outwards, in a few seconds as opposed to minutes.

AZTEC

And as well as the hiss of snakes, there'll be the roar of mighty dinosaurs. Examples of these extinct giants will have been captured using time machines and special tranquilizer ray guns, and brought back to the future.

INCA

Meanwhile, in the Aquarium there'll be a lack of water. That's because by the year 2000 fish will have evolved and developed lungs, and clambered out of the sea and onto the land.

MAYAN

New technology will mean a dream come true for Dr Doolittle, as visitors queue up to actually talk to the

Professor Piehead

OKAY JOE. ALL SET TO TEST MY NEW IMPACT RESISTANT FALSE TEETH?

SPRING LOADED HAMMER

CRACK!

CLICK

SPRING LOADED HAMMER

LETS HAVE ANOTHER LOOK AT THOSE MEASUREMENTS.

Zoo fan Johnny Austin (left) and some moon rabbits

animals. Using space technology, visitors will be able to communicate with the animals of their choice through a small decoder no bigger than a hearing aid. At the flick of a switch they will be able to talk in monkey, lion or any other animal language.

LIBERAL

And not only will we be talking to animals, in the zoo of the future we will also be having sex with them. Increasingly liberal attitudes towards sex will mean that by the year 2000 left wing councils will actually encourage beastiality in schools.

TORY

Schoolchildren will visit zoos regularly where they will learn about the important role all animals play in main-

taining our environment, and they will come to respect the animal neighbours with whom we share the planet.

LABOUR

And there'll also be a working abattoir where children can fire bolts into the heads of cattle and pigs. And a special audio visual display will show the process by which animals are chopped up and turned into pies and sausage rolls.

PLAID CYMRU

Last night, former Daktari star and zoo critic Virginia McKenna blasted plans for Britain's zoos of the future. "I'm appalled by the idea of space animals being captured and brought to Earth to be displayed in this humiliating and unnatural way. Space animals should be allowed to roam free in space, on planets like Mars and the Sun", she told us.

DAFT COW
Doris's window pane in the arse

A stupid woman who bought 200 double glazed windows has called for changes in the law governing door-to-door salesmen.

Doris Twatt and her husband Frank bought the windows despite the fact that their one bedroom flat is already double glazed.

PRESSURE

They claim they were the victims of high pressure selling techniques which included:

● **KNOCKING** at their door and ringing their bell.

● **ASKING** of the Twatts wanted to buy the windows.

● **THREATENING** to call back the next day if they couldn't decide.

Eventually Mrs Twatt gave in after the salesman told her the windows would be supplied at half the usual price if she bought 200. But that still left her owing the company £280,000. And now the windows are left blocking the drive of Mrs Twatt's home.

A spokesman for BBC TV's Watchdog programme told us they were investigating the case. "Be careful when buying toys for young children this Christmas", they added. "Be on the look-out for any small, sharp or dangerous pieces which could cause a child to choke".

Continued from page 10

SPOT the STAR

WIN TICKETS TO THE ZOO 2000

Here's a great competition – and a chance for you to win a free visit to one of Britain's zoos – if there's any left – in the year 2000.

IMAGINATION

All you have to do is draw what you think space animals of the future will look like. Use your imagination, and the more space age your animals look the better. Colour them in with silver paint or tin foil. Then send them to 'Viz Space Animals in the Zoo of the Future for the Year 2000 Drawings competition', P.O. Box 1PT, Newcastle upon Tyne, NE99 1PT, to arrive by no later than 31st December 1999.

Another GREAT competition

We'll ask Johnny Morris, if he's still alive, to judge the competition, and the winner will receive two tickets to the zoo of the future. Five lucky runners-up will each receive a goldfish in a plastic bag.

SHAKATAK

If you'd like your pictures returned don't bother sending them in the first place. This competition is not open to members of the Animal Liberation Front, their friends or relatives. The judges' decision is final.

Here's a fun-to-enter competition for all you star spotters out there. We've disguised some well known celebrities. Can you spot who they are?

Send your answers, on a post card, to 'Viz Crudely Disguised Celebrity Recognition Competition, P.O. Box 1PT, Newcastle upon Tyne, NE99 1PT. The first correct entry

drawn out of the hat on 1st February 1993 will receive our first prize – a guided tour of the homes of the stars, with full commentary and including a visit to the moon.

The competition is not open to former Blue Peter presenters, their friends, relatives or dogs. We reserve the right to forget about the competition completely and throw all entries in the bin.

JOHNNY BALL REVEALS ALL!

Johnny lifts the blue T-shirt over his head.

He strips for some grass-cutting action

JOHNNY BALL reveals all his charms as he strips off whilst mowing the lawn of his Buckinghamshire home.

The gorgeous telly babe slipped his blue top over his head to reveal a fine set of assets.

Bubbly 'Think of a Number' presenter Johnny, 61, showed that he has certainly got ONE figure worth thinking about.

One neighbour said: "All the men here go topless when doing their lawns, but Johnny really shone. He looked fantastic."

Johnny - taking a break after quitting T.V.'s 'Play School' in 1983 - later sat with wife Diane and had a nice cup of tea.

Pictures: ENRICO RATZORIZZO

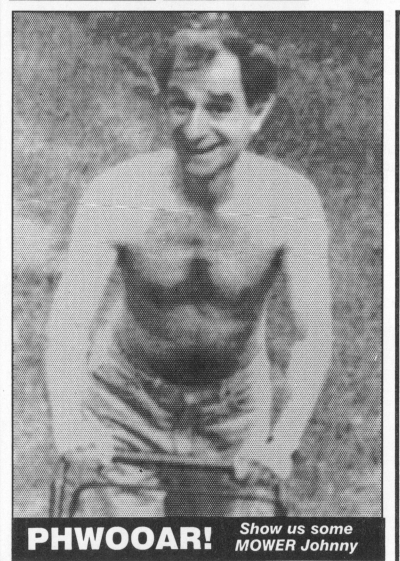

PHWOOAR! *Show us some MOWER Johnny*

PERV FALLS FOR BEAUTY

A PEEPING Tom fell 90ft to his death from a tree as he tried to spy on a topless beauty who was mowing the lawn.

Pervert

The filthy pervert had inched his way along a branch overlooking the garden, as he tried to snap pictures of the unsuspecting stunner.

Sicko

"We all go topless when mowing our lawns round here", said neighbour George Fisher," but you don't expect to be spied on by sickos."

Filthy

Another neighbour said: "I heard a scream from the tree, and saw a man desperately grabbing at a branch. Then he disappeared and the scream got fainter until I heard a thump. Serves him right."

OBITUARY

Enrico Ratzorizzo 1974 - 1999

VIZ SNAPPER Enrico Ratzorizzo - who has been killed in a tragic accident on an assignment in Buckinghamshire - had in his short but illustrious career earned himself a reputation for fearless professionalism and cold, ferret-like persistence, writes Picture Desk Editor, Ronnie Shit.

Loved

Over the past few years Enrico earned himself the title 'The People's Parasite' for his brutal disregard for the privacy or feelings of his victims.

Sensitive

Three-times winner of the prestigious Chuck Berry Award for Intrusive Photojournalism, Ratzorizzo was the lensman behind many front page scoops, including the first shots of Arthur Askey's legs in a hospital incinerator, and his sensational pictures of Christopher Reeve fighting for his life, taken from inside the air-conditioning system of the Intensive Care Unit.

Caring

But he will be best remembered for his sensitive coverage of Benny Hill's decaying corpse, photographed through the dead star's letterbox over the four day period he lay undiscovered.

Charity

He leaves a camera with an absolutely enormous lens, and a high-powered motorcycle with white Fiat Uno paint down the side.

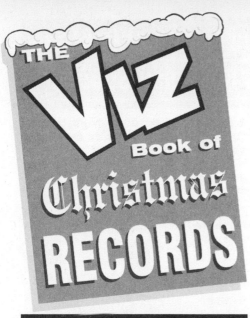

THE Viz Book of Christmas RECORDS

Least convincing Santa ● 12 year-old Dafydd Postgate fools almost no one as he takes up his position in the grotto at his Llandudno department store

Eating

Date discrepancy

The record for the biggest disparity between dates eaten on Christmas day versus the rest of the year is held by Phillip Runcorn (GB) of Derby. Between 8.30 am and 11.30 pm on 25th Dec 1998, he consumed a total of *23.456 Kg* (51lb 12½oz) of 'Eat Me' dates, despite not particularly liking them. During the previous 364 days he had eaten half a date weighing 7gm (0.247oz) whilst at his brother's wedding, an annual discrepancy of *23.449kg* or 335214.28%.

Presents

Longest gift-related huff

On November 29th 1968, Margaret Pierce of Gateshead (GB), told her husband Ron not to get her anything that year for Christmas. When on Christmas morning she found that he hadn't, she embarked upon a huff which as of Nov 14th 2001 has lasted for 11981 days. During this time man landed on the Moon, the Berlin wall has been torn down and a third world war has begun. Meanwhile, Mrs Pierce has sat with her arms folded, pretending to watch the telly and affecting not to hear her husband.

Most socks

On Christmas day 1996, Mr Brigham Osmond, a polygamous Mormon of Salt Lake City, Utah (US) received 847 pairs of socks from his ten wives, 84 children and 753 grandchildren. The following year, thanks to the addition of 2 more wives, six more children, 56 grandchildren and a great grandchild, he received 912 bottles of Pagan Man aftershave.

Children

Earliest wake-up

Henry and James Montgomery of Basildon (GB) were so excited about opening their presents from Santa on Christmas Day 1999, that they woke up at 2.33 am on October 25th 1998. Their bleary-eyed father Stephen told them to go back to bed, as there were still 14 months to go.

Least convincing Santa

In December 1998, in the Grotto at Claire's Department Store in Llandudno (GB), Dafydd Postgate managed to fool only three children, two of whom were blind, that he was the real Santa Claus. Postgate, the manager's 12-year-old son was ordered to put on a cotton wool beard after all staff had refused. His high-pitched Ho! Ho! Ho's did little to convince the 687 children who visited the grotto that year, many of whom left in tears. Postgate estimated that he had had his beard pulled off over 500 times by irate toddlers.

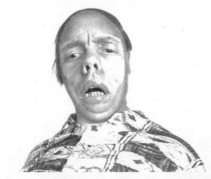

Most socks ● Ho! Ho! Hose! Brigham Osmond opens his last present on Christmas morning.

Trees

Most Christmas trees behind shed

On January 6th 1946, Ernest Sands of Northampton (GB) put his Christmas tree behind the garden shed, intending to take it to the municipal tip that weekend. At the time of writing, Mr. Sands has 56 similar trees behind the shed, which he has had to extend five times to accommodate them. He intends to definitely take them to the tip this year.

Parties

Arse photocopying

The record for the most photocopying of arses at an office Christmas party is held by the staff of Eversheds Solicitors in Newcastle upon Tyne (UK). On Dec 23rd 1987 a total of *6938* A3 buttock copies were produced on a standard Sharp SF2035 copier. During the 4-hour party, 14 reams of paper were used, the toner cartridges were changed 8 times and the service engineer was called out twice.

Christmas Day TV Choice
Your essential guide to what's on TV this Christmas...

9.00 BBC 2: I Love Scraping the Bottom of the Barrel Assorted B-list celebrities pretend to remember their childhood in precise detail whilst sitting in front of a coloured backdrop. **9.00 ITV: Branching Out** Ross Kemp stars as Robson Thaw, an unconventional tree surgeon who sometimes bends the rules but always gets results. **9.30 ITV: The Ross Kemp Story** Robson Green stars as Ross Kemp in the true life story of an unconventional actor who sometimes bends the rules but always gets results. **10.00 BBC 2: I Love Stuart Maconie's Opinions.** Assorted fame hungry K-list media tarts reminisce about some of Stuart Maconie's most memorable opinions on everything from Hula Hoops and Space Hoppers to Clackers and Chopper bikes. **11.30 ITV: Ross Kemp vehicle** Another fucking programme starring Ross Kemp. Details to be confirmed. **12.00 C4: Escape From Cardiff** Claudia Winkleman introduces an exciting adventure game in which two teams of whooping lycra-clad twats are given a week to attempt to get out of The Welsh capital and reach civilisation. **1.00 ITV: Robson Green, QC** Legal drama, starring Ross Kemp as John Thaw, an unconventional police astrologer who sometimes bends the rules but always gets results. **1.00 BBC 2: I Love Doing These Sorts of Programmes.** Desperate attention-seeking T-list celebrities reminisce about their favourite appearances on these sorts of programmes. **1.30 BBC 1: Christmas Pro-Celebrity Golf** Bruce Forsyth, Jimmy Tarbuck and Kenny Lynch play a round of golf with three prostitutes. Introduced by Peter Allis. **2.00 ITV: The weather** with Ross Kemp **3.00 ITV: The Queens Christmas message** The traditional Christmas Day broadcast to the commonwealth, starring Ross Kemp as Elizabeth Windsor, the unconventional monarch who sometimes bends the rules but always gets results. **3.30 BBC 2: I Love Not Decided at Time of Going to Press.** Tragic W-list nonentities and Jamie Theakston reminisce about something which had not been decided at time of going to press. **4.00 Carlton Food Network: Cooking for One** Anthea Turner cooks a Stilton cheese souffle with chick pea couscous whilst a single viewer watches. **6.00 ITV: Kemp it in the Family** Hilarious sitcom by Robson Green. Ross and Martin Kemp play Martin and Ross Kemp, two divorced brothers who marry each other's ex-wives, played by Amanda Burton and John Thaw. This week, Ross pretends he has forgotten his anniversary, but Martin has secretly planned a surprise party! **9.00 ITV: Search for a Kemp** It's down to the last two finalists, Martin and Ross. Who will impress the judges the most and win 8 badly-written drama series? On the panel are John Thaw, Amanda Burton, Robson Green, Sarah Lancashire and Ross Kemp. **10.00 ITV: What the Kemps Did for Us** Adam Hart-Davis looks at the contributions to civilization made by the Kemps, Ross and Martin. This week, bad acting in hoarse voices. **10.30 ITV: Morse** John Thaw plays Ken Morse, an unconventional rostrum cameraman who sometimes bends the rules but always gets results. **11.00 ITV: The Kemps - A Warning from History** Harrowing documentary outlining the circumstances that allowed two unversatile cockney baddy actors to seize power of an entire television network. Narrated by Ross Kemp. **12.30 ITV: Carry On Kemping** Late night saucy fun, starring Sid James as Ross Kemp and Jack Douglas as Martin Kemp.

WIG WATCH

Sender: *Graham Doyle, Liverpool*

Sender: *J. Harris, Welwyn Garden City*

syrups@viz.co.uk

HOW ABOUT THESE?

I'M NOT SURE. I'D BEST TRY THE SIZE HUNDRED AND FORTY TWO AND A HALFS.

Volca-NO to Howerd's controversial TV show

A woman who survived the Pompeii disaster yesterday branded comedian Frankie Howerd 'sick' after he announced plans to have made a comedy series about it in the seventies.

Octavia Johnson, 64, was only 3 years old when a massive volcanic eruption swept away her Roman villa and destroyed the ill-fated city where her entire family had lived. And now she's blown her top over plans for a crude television spoof of the infamous tragedy.

Speed

The volcano Vesuvius erupted with such speed that Octavia's father, a mechanic in the Roman army, was turned to stone as he worked underneath a chariot. Octavia only survived because her quick thinking mum managed to grab her and climb up an olive tree as the smegma approached.

Dope

That was many years ago, but Octavia, who now has two children and five grandchildren of her own, is haunted by memories of the disaster to this very day. And yesterday she described plans for a light hearted TV series entitled 'Up Pompeii' as "disgusting".

Whizz

Sources close to the BBC admit that the controversial series, which is due to be filmed in the seventies, will have contained jokes about volcanoes erupting and people being turned into stone.

Frankie Howerd is consoled by large breasted women as a storm brews over his proposed 70s TV show yesterday.

However Mrs Johnson certainly will not be having watched the show. "I know I was only 3 at the time and my memories are very vague, but I couldn't bear the thought of watching the TV and seeing Frankie Howerd make a joke about an actor who might of been portraying my father", she said last night as tears welled in her eyes.

Frankie Howerd, who was still alive when he made the TV series, has since died. But last night he was quick to defend the show against accusations of bad taste. "Ooooh, no missus... don't... stop it", he said, while putting his hands up his back.

A faded old Roman photograph (above) is all that Mrs Octavia Johnson (right) has to remind her of the brave father who she never knew. Yesterday.

Anne Nightingale stang in Berkeley Square

FORMER Radio One DJ Anne Nightingale was being comforted by friends late last night after being stang by a wasp in Berkeley Square.

Miss Nightingale, who presented a request show on Sunday evenings during the 1970s, had travelled from her home in Brighton to Berkeley Square for a cup of tea. Witnesses report hearing her say "Ouch" when the wasp stung her.

Bremner

Police had received several unconfirmed reports of a wasp in Berkeley Square prior to the attack. A Metropolitan Police spokesman said that he suspected Miss Nightingale, 70, may of ate a bun in Berkeley Square and that there might of been all jam on her mouth when the wasp struck.

Bunter

This is the second time that Miss Nightingale has been the victim of yellow and black stripey insects. For two years ago giant bees attacked her and kidnapped her legs during a visit to Peru.

Liar

Although they was never caught, an insect gang known as the 'Busy Bees' were believed to have been responsible. Under the leadership of the late Arthur Askey, bees the size of coconuts comb Peru looking for legs to kidnap and present to their leader at his Aztec temple hideaway. As each gift is received Askey grins through bottle bottom

Anne Nightingale - stunged yesterday in Berkeley Square

glasses and utters the ceremonial words "Ay thang-yow".

The Kid

A Foreign Office spokesman yesterday reassured British tourists that Peru remains a safe holiday destination, but recommended that intending travellers take what he called 'sensible' precautions.

"We would recommend that anyone visiting Peru visit their GP and have their legs inoculated against kidnap by coconut sized bees", he told us.

Modern Times

Both Miss Nightingale and Arthur Askey was last night unavailable for comment.

ANDY CRAPP   **by REG SHYTE**

Shyte.

DO YOU FANCY A DRINK, PET?

AYE! THE USUAL

OWT ELSE?

AYE! SOME CRISPS

CRISPS, EH?

AYE!

ARSES ON PEWS!

A controversial vicar is calling for dramatic changes in the way churches operate. For unless drastic steps are taken, he fears we could soon be witnessing the end of Christianity itself.

And vicar Dennis Randall believes that unless Holy men are prepared to move with the times, they will soon be left preaching to rows of empty pews.

PACKED

Christmas has traditionally meant big business for the churches, with standing room only in packed houses throughout the country. But all that is changing, and this year vicars are bracing themselves for record low attendances.

FALL

Over the last few years there has been a dramatic fall in the number of people going to church. And religious chiefs fear that unless action is taken to stop the rot, thousands of churches around Britain could soon go under.

STEEPLES

Rev. Randall believes several factors are responsible for the fall in attendances. "There's been a lack of investment" he told us. "Too much money has been spent on steeples, and not enough on the churches themselves. We're stuck with old, outdated buildings. Most of them lack even basic toilet facilities".

FORMAL

"Hymns are also outdated. Some of them are literally hundreds of years old. And I'm sure many young people are put off by the formal dress code. For instance, a church is probably the only place in Britain where you aren't allowed to wear a hat".

SHORTCOMING

Failure to compete in an increasingly competitive Sunday morning environment has been another major shortcoming, according to Rev. Randall. "DIY super-stores and Garden Centres are pulling in the punters in their thousands, he told us.

'That's what churches need' says controversial vicar

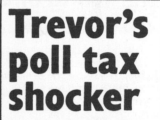

Could churches like this soon be closing their doors for the last time. (Inset) Rev. Randall yesterday.

"They offer shopping, refreshments, play areas for the kids and free car parking. And all we have to offer is a cold seat, a couple of hymns and a few stories about God you've probably heard a hundred times before. It's no wonder we're losing out".

OUTSKIRTS

Among many suggestions he has put forward is the construction of new, out of town 'super-churches'. "The whole idea of the little church on the corner is completely outdated. We should be building big, new churches on the outskirts of town, with late opening, seven nights a week, and free car parking".

SPACE

Steps should also be taken to attract people to church. "Prime land is wasted on cemeteries. We could use this space to have attractive garden displays, fun fairs for kids, and car washes. Everyone washes their car on a Sunday".

FINAL

Rev. Randall believes a huge commercial opportunity exists in the form of Sunday lunches. "If we served up good, basic, traditional nosh, at reasonable prices, we'd have the punters queuing up for it", he told us.

The Reverend also dreams of the day when churches will be granted drinks licenses. "It's ridiculous", he told us. "You can buy a drink in any pub in the country. But if you're in a church you can't. Britain must be the only country in the world that has such outdated licensing laws. God only knows what tourists make of it all".

FRONTIER

Rev. Randall believes the key to future success will be attracting young people back to church. "We must try to get families back. It's all well and good the old folks turning up – they're always welcome – but a lot of them are only interested in the free cup of tea afterwards. And they're not exactly the most generous people in the world when the collection plate comes round".

BISHOP

So far Rev. Randall's suggestions have met with a cautious response from the Arch Bishop of Canterbury. "He hasn't actually replied yet", Rev. Randall admitted, "but he's been very busy lately".

ROOK

Meanwhile, the Reverend tells us that he hopes to attract a bumper congregation to his church on Christmas Day, by lining up a troup of exotic dancers to top the bill. "There's nothing in the Bible to say thou shalt not have strippers on", he joked yesterday. "And besides, anything that puts arses on pews is good business in my book". To overcome the drinks ban Rev. Randall will be inviting parishioners to bring along their own bottle of wine.

Trevor's poll tax shocker

Unemployed gas fitter Trevor Tomlinson, from Prestatyn, North Wales, almost fell through the floor when he opened his Community Charge bill from the local council.

FAN

For as well as Trevor's Poll Tax, the bill contained an additional charge of £24,000 – for a Pifco plastic oscillating fan which Trevor keeps on the top of the fridge in the kitchen.

HOT

"It's ludicrous", he told us yesterday. "I only ever use the fan once or twice a year when it gets really hot. How I'm supposed to afford £24,000 I'll never know.

"I've already had to sell my car, most of my furniture and re-mortgage the house, but I haven't raised half the cash yet. I wish I'd never seen the bloody fan. I wouldn't care. It only cost me £12. And it doesn't even work, because all the stuff has all come out of the batteries".

WELSH

We asked for a comment from a spokeman for Prestatyn council, but it was in Welsh.

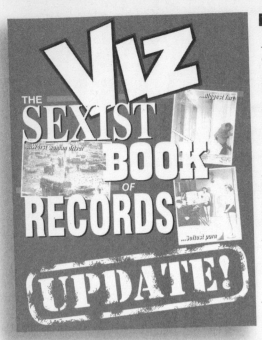

The World of Women

Night-Time Beauty Regime

The longest night-time beauty regime for a woman is one of 14hr 12min, practiced by Kylie Bradwell (GB) of Luton. The marathon exercise begins each night at 6.00pm when Ms Bradwell goes to bed, and the ensuing 7 hours sees her use 148 products to remove the day's make-up. The following 7 hours and 12 minutes is then taken up applying over 193 moisturisers, night creams, ungents and collagen-rich skin enhancers. She finally gets to bed at 8.12 the following morning.

• Cosmetic counter - A selection of some of the hundreds of beauty products that Kylie Bradwell applies to her skin on a nightly basis.

Longest diet

The longest diet ever embarked upon is one of 75yrs 6mths 2days, by Edith Bootree (GB) of Preston. At the age of 18 and weighing 98kg *14st 8lb*, Mrs Bootree bought a size 8 dress in a sale at her local Co-op. She then spent the next 75½ years attempting to slim into it, avoiding all fatty foods, but allowing herself the occasional treat. Over the course of her marathon weight-loss attempt, she tried over 286 faddy diets including the Atkin's, the Hay, the G-plan, the California and the Vanessa Feltz. She briefly lost 250g *8oz* in 1948 but quickly put it back on after celebrating with a chocolate eclair. Her diet eventually came to an end in March 2007 when she died weighing 98kg *14st 8lb*.

Make-up application

The longest make-up application regime for a woman is one of 9hr 30min undertaken by Kylie Bradwell (GB) of Luton. Rising at 8.30 each morning, Ms Bradwell begins to beautify herself with no less that 418 different foundations, blushers, highlighters, lowlighters, contour creams, eye shadows and lip glosses, stopping only to use the toilet.

The World of Men

Least Manly Sandwich

The most effeminate male-purchased sandwich was one ordered on 12th July 2006 by Grant Whistler (GB) in a delicatessen in Chelsea. It comprised of mixed salad leaves including rocket, radicchio and lambs' lettuce, sliced brie and fig relish in a balsamic vinegar dressing. Mr Whistler, a theatre choreographer, made things worse by having it on a wholemeal ciabatta finger roll.

Shiniest suit

A suit belonging to Alf Normans (GB) of Manchester is the recorded as being the shiniest in the world. The arse of the trousers has a reflectance value of 98.32%, which is equivalent to the reflective property of a silver coated mirror being struck by polarized light of a wavelength of 1000nm. Mr Normans bought the suit in 1939 for his wedding and has worn it for his daily visits to the betting shop ever since without once taking it to the dry cleaners.

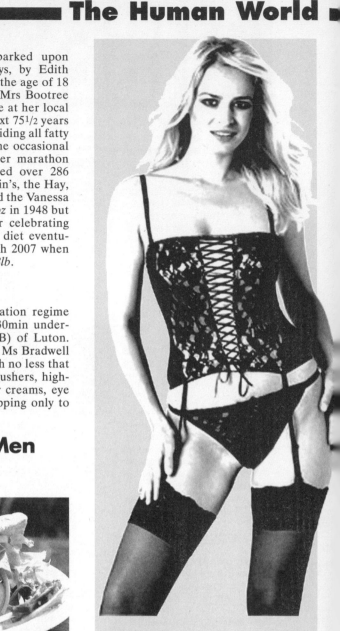

Lingerie Optimism

The most optimistic male purchase of ladies' underwear was made on the 18th December 2006 by Albert Plywood (GB). Aroused by a window display in the Derby city centre branch of La Senza, Plywood went in to buy something as a Christmas gift for his wife. The outfit, which consisted of a black, sheer-lace basque with balconette bra cups, matching heart-shaped thong and sheer Chantilly lace-topped stockings, cost him a total of £122.99. However, on Christmas day the gift did not go down well with his 58-year-old wife Maureen, who tutted and used them as dusters.

Mental Undressing

The record for the longest time that a man has resisted the temptation to mentally undress a nurse was set by Barry Haddock (GB) of Yorkshire. On 18th March 1993 at Leeds General Infirmary, Mr Haddock watched three nurses wheeling his wife Janice into the operating theatre for a routine operation. It took a full 8.6 seconds before, in his mind's eye, all three nurses were naked except for stockings, suspenders, high-heeled shoes and nurses' hats.

- Ooh, Matron - Barry Haddock of Leeds showed remarkable restraint when his wife was having an op.

Jazz film endurance

The world record for the longest time a man has trod water through a porn film is held by Manuel Espadrille (USA) who, on 16th August 2007, gently strummed through all 85 minutes of *Cock Hungry Nymphos* without fast-forwarding the DVD at any point, finally ejaculating as the credits came up. The British record is held by Lupin Sinstadt (GB) who, on 12th July 1999 on the BBC *Record Breakers* programme successfully watched 48 minutes of *Cum Stained Casting Couch* before going off.

Pornographic magazine

The most expensive pornographic magazine ever

Schoolboy Sexual Boasting Record Smashed

A NUNEATON schoolboy today smashed the record for the most outlandish sexual boast ever made.

In a playground conversation with several of his Year 10 friends, 14-year-old Ed Tiptree claimed to have bedded over 780 women, including 27 teachers at the school, most of the sixth form, several mothers of friends, the woodwork teacher's daughter and 50 girls he met at Butlins in the summer.

overjoyed

Staff at the Sir Les Patterson High School said they were overjoyed with the schoolboy's feat, which broke a national sexual boasting record made 28 years ago.

assembly

Tiptree's achievement was recognised by the headmaster at a special assembly where he received a certificate of commendation. He also received a cup, presented by the former record holder, 42-year-old Terry French who, in 1980, claimed that he "must of knobbed about 500 birds at least. Probably nearer 600."

purchased was a copy of the December 2002 issue of *Escort*, which cost its buyer a staggering £854.97. On November 14th 2002, Denis Architrave (GB) of Chester went into his local branch of Fourboys Newsagent with the intention of purchasing the Christmas edition of the erotic periodical, which had a cover price of £1.85. However, spotting a friend of his mother's talking to the shop assistant, Mr Architrave bought a packet of chewing gum and a tube of Pringles instead. He returned 10 minutes later but left with a box of Roses chocolates and a packet of pipecleaners after seeing his old primary school headmistress at the counter, paying her paper bill. Over the course of the following three weeks, Mr Architrave continued to visit the shop at approximately 10-minute intervals, but was thwarted at each attempt and forced to make a decoy purchase, eventually running up his record-breaking bill. He finally managed to purchase his copy of *Escort* on December 5th 2002 having amassed, during the preceding 21 days, a haul of unwanted goods including 706 packets of chewing gum, 36 bottles of Dandelion & Burdock, 112 Bic Biros, 355 spiral-bound reporters' notebooks, 45 combs and a pencil sharpener.

Sneaking a Peek

The longest that a man has managed to maintain eye contact with a woman during a conversation without surreptitiously glancing at her breasts is held by Barry Haddock (GB) of Yorkshire. On the 18th March 1993 at Leeds General Infirmary, whilst being informed of his wife Janice's death during a routine operation, Mr Haddock managed to look the nurse in the eye for 14.1 seconds before his eyes briefly checked out her knockers.

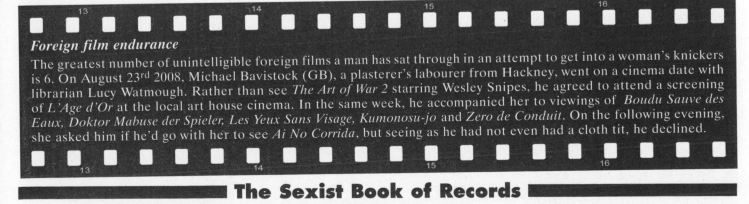

Foreign film endurance

The greatest number of unintelligible foreign films a man has sat through in an attempt to get into a woman's knickers is 6. On August 23rd 2008, Michael Bavistock (GB), a plasterer's labourer from Hackney, went on a cinema date with librarian Lucy Watmough. Rather than see *The Art of War 2* starring Wesley Snipes, he agreed to attend a screening of *L'Age d'Or* at the local art house cinema. In the same week, he accompanied her to viewings of *Boudu Sauve des Eaux, Doktor Mabuse der Spieler, Les Yeux Sans Visage, Kumonosu-jo* and *Zero de Conduit*. On the following evening, she asked him if he'd go with her to see *Ai No Corrida*, but seeing as he had not even had a cloth tit, he declined.

The Sexist Book of Records

SPENT! SPENT! SPENT!

'And now I'm skint' says pools win Trev

A Shrewsbury man claims he is flat broke — only months after celebrating a fairytale win on the pools.

Joy soon turned to sorrow for Trevor Singleton, and his wife Barbara, when their dream success on the pools suddenly became a nightmare.

"Winning the pools was the worst thing that ever happened to us", said Trevor, 46. "I wish I'd never set eyes on the money. It's brought us nothing but heartache".

JACKPOT

The Singleton's couldn't believe their luck when a pools official arrived on the doorstep of their two bedroomed terrace house with the news that they'd won an incredible £312. "I couldn't even remember filling in the coupon", Trevor told us. Immediately the celebrations – and the spending – began.

WINDFALL

"We went out and we spent, spent, spent. We'd never had money before, and we were throwing it about like confetti. In one week alone we bought a new vacuum cleaner, a new Thermos flask for work, and had our toaster repaired. We were spending like there was no tomorrow".

FORTUNE

Soon Trevor resigned from his job as Allotment Supervisor for the local council. Then, after news of their good fortune appeared in the Shrewsbury and Wellington Bugle, begging letters began to arrive. "Suddenly everyone needed money. Friends, relatives, even total strangers were asking for cash. And like a fool I gave them it", he recalls.

BONANZA

Meanwhile the lavish spending continued. Trevor's wife Barbara returned home one day to find a brand new cover on her ironing board, while in the garden Trevor splashed out on a new paraffin heater for his greenhouse, and some seed trays.

"It was only the best for me. No expense spared. I was living like a King, thinking it would last forever".

THE VIRGINIAN

Although he didn't realise it, the money was dwindling rapidly. Trevor continues, "I was going out at nights and buying drinks for my mates. On one occasion I even paid for a taxi home. I was loving every minute of it". But the real problems started when Trevor began to gamble.

Mr Singleton today – "I wish I'd never seen the money."

"I'd never gambled in my life. But one night I spotted a fruit machine in the pub, and that was it. Before I knew what has happened I'd stuffed £1.60 into it, and I had no change left at all. I had to go to the bar to get some more. By the end of the evening I must have lost over £2.50. When I awoke the next morning I was physically sick".

HIGH CHAPARAL

Only months after receiving his cheque, Trevor had squandered the lot. He insists that every penny has been spent, and all he has left to show for it is a pile of unpaid bills. "I'm up to my neck in debt now", he confessed, fighting back the tears. "I owe the newsagent for two weeks' papers, and I've had a final reminder for my phone bill which is £27. I'm flat broke. Apart from a small amount my wife and I have saved in a building society, we haven't got a penny to our names".

LANCER

Yet despite it all, Trevor admits that if he won the pools tomorrow he'd probably do the same again. "I must admit, I enjoyed it while it lasted", he told us.

GUNSMOKE

Now unemployed, and living on meagre social security hand-outs, Trevor's only real regret is not taking his wife on the dream holiday they had planned. "We'd always talked about going on a caravan holiday in Wales, as my wife has relatives in the area", he told us. "This was our once-in-a-lifetime chance to visit them. But now it looks like we never will".

Flashback to happier times – Mr Singleton celebrates his jackpot with wife Barbara and friends.

Garage gets Royal blessing

By Reg Dildo

A Dudley man's plans to build a 'lean to' garage in the back yard of his terraced home have received the Royal seal of approval, from Prince Charles.

GARAGE

Bob Chambers plans to build the garage to house his car, gardening tools and other equipment. But before going ahead he hit upon the idea of writing to Prince Charles seeking Royal approval for the venture. Bob sent the Prince a sketch he had prepared of the proposed garage.

GARAGE

"To my surprise the Prince wrote back thanking me for the plan, and wishing me the best of luck with my garage," a delighted Bob told us.

GARAGE

Bob hopes to have the garage, which will be built of wood and corrugated iron, complete in time for Christmas.

426

PEW! WHAT A SCORCHER

A man claiming to be an ex-vicar sacked by the Church for having sex with a parishoner, is also claiming he was unfairly dismissed.

"I've been made a scape-goat", claims 52-year-old Reg Potter. And now Reg is threatening to blow the lid off some of the seedy goings on he witnessed behind church doors during the six months he was employed as a vicar. "They sacked me for getting a bit fruity with the customers", says Reg, "but every vicar in the country is *at it*. Why they picked on me I just don't know".

EXCLUSIVE

Rev. Potter - Fruity

GRAVEYARD

Mr. Potter says he was arrested after churchgoers reported him to police having sex in a graveyard behind his church. "Since when was it illegal to have a bit of fun?" he said yesterday. "That's all I ever did, unlike some vicars I could name".

STEEPLE

According to Mr. Potter sex and adultery are commonplace in the Church today. "You wouldn't believe some of the things going on", he told us. "One vicar up the road from me was at it day and night. We used to call him Rev. Randy Bollocks he was that bad. Mind you, he had the biggest *steeple* I've ever seen. The birds used to love to *peel* off in his belfry, and he could *ring their bells* all night long. What a racket! Every now and then you'd hear bells ringing at odd times of the day. Everyone would look puzzled and check their watches, but not me. I knew it was just old Randy Bollocks giving some punter a *special service*.

TOWER

I must admit, I used to *play around a bit* in my belfry. One day I felt a bit kinky in the *bell tower* so I tied my knob to one of the bell ropes. Next thing I knew it was 12 o'clock and in walks the bishop to ring the bells. He caught me red handed, and I ended up getting *tolled off* in more ways than one!

ROTTEN

Another vicar mate of mine fancied his organist something rotten. So one Sunday, after the morning service, he asked her if she fancied a quick *session* on another *organ*. She jumped at the chance, I can tell you, *pulling out all the stops*. She had it pumped up and playing in a flash, the dirty old bag.

VICIOUS

Churches are ideal for sex, there's that many different places where you can *do it*. My favourite was always that big tub thing where they Christen all the babies and that. I gave quite a few birds a *bubble bath* in there, I can tell you.

SENSIBLE

And then there's the pulpit, the wooden bit where I say all the prayers. A bit *pokey*, but that's the name of the game! One day another vicar I know (if you're reading this, you know who you are!) was *up* in the pulpit, giving some bird a *proper seeing to*. Suddenly, in walks the Bishop.

Sex claim vicar has Bishop's knickers in a twist

My mate nearly shit a brick, I can tell you. He tried to keep as quiet as he could, but this bird started giggling, and gave the game away.

SCABIES

My mate thought he'd get the sack on the spot, but the Bish just smiled and told him to scarper. Next thing you know His Holiness was *on the job*, giving it *six nowt* with this bird. And needless to say, not a word was said afterwards. You see, there's an unwritten rule in the Church. You can do it *whenever you like*, with *whoever you like*, as long as you use a blob. That's what I reckon.

ECZEMA

"Mind, if you think vicars are bad, you should see what the left footers get up to. Our local priest was off sick one day (too much sex the night before, probably) so anyway, I had to nip down to his church and do some confessions for him. And boy, was I in for a surprise!

RINGWORM

"Waiting outside the confessions box was a queue of good looking birds a mile long, all waiting to *reveal all!* Most of them had nothing to confess when they arrived, but that soon changed once I got them inside my little box.

A church yesterday

You wouldn't believe what I got up to. *Mind you, if the Pope ever finds out, some of them birds will be saying "hail Mary" for the rest of their lives, I can tell you.*

DERMATITIS

To be honest I reckon *sex* is the only reason why people bother being vicars in the first place. The pay's crap, and saying prayers isn't exactly the most exciting job in the world. And then there's all the poncy gear you have to wear. I felt a right pratt in some of them dresses. Mind you, come to think of it they're ideal for wearing *on the job*. You can get your tackle out a lot quicker than if you were wearing pants".

DOHBI ITCH

Mr. Potter is claiming £500 compensation from the Church for unfair dismissal. "But I'd accept £100 cash", he told us. When we approached the Church for a comment, a vicar appeared and told us we were trespassing. "Go away or I'll call the police", he said.

I'M SICK OF ALL THESE TRAILERS ON THE TELEVISION

DOCTOR, I'VE GOT AN IRON DEFICIENCY

■ MY HUSBAND bought me a chair from Ikea, and when he got it home he was disappointed to see that he had to build it himself. It took him a long time and when he was finished he was so exhausted that he needed a little sit down.... in my new chair!

Doreen Erf, Tring

■ ON A RECENT shopping trip to London, my wife and I bought a couple of deckchairs. We decided to take them back to our hotel on the tube in the middle of the rush hour. The train was packed with commuters and there were no seats available, so we put up our deck chairs near the doors and sat down. I couldn't believe how rude some of the other passengers were, tutting as they pushed past us on and off the train. In the end we were made to feel so uncomfortable that we decided to get off and catch the bus. There, if anything, the other passengers' manners were even worse! That's the last time we go to London.

Hubert Greengauge, Harrogate

■ I BOUGHT a chair in 1942 and it's been sat on everyday since and it's still going strong. It certainly puts some of these modern chairs to shame!

Les Crayfish, Basingstoke

Sonny & Chairs

" Well, I woke up this mornin' and had me a whole mailbag full of your letters about chairs. So I took them out on the stoop and I sat me down in my big ol' rockin' chair to read them. Then I done remembered that I was blind, and that kinda gave me the blues. Anyhow, ain't no big thing 'cause my woman done read them for me. Here's some of the best I done received. "

with MISSISSIPPI BLUESMAN and chair enthusiast SONNY TERRY

■ I THINK the problem with television these days is that nobody sits in chairs any more. In my day there was Val Doonican in a rocking chair, Dave Allen on a high stool and Ronnie Corbett in a baggy armchair, and it was proper family entertainment. Now, whenever I turn my set on, it's just foul-mouthed smut from the likes of Ali G, Andrew 'Dice' Clay and Mike Strutter, all standing up and swearing with not a chair in sight.

Edith Scammell, Reakin

■ MY HUSBAND recently came home from the tip with an old, high-backed chair that he had found in one of the skips. I told him that we already had enough junk in the house, but he said he wanted to turn it into a throne. He took it into his shed where he sprayed it gold and spent hours gluing sequins and tinsel on it. I have to admit that, when he'd finished, it did look magnificent. He has put it in pride of place in front of the telly and he won't get out of it even to have his tea. However, I pointed out to him that it would never actually be a throne, as a throne is defined as the ceremonial seat of a king, queen or bishop. How foolish I felt when he reminded me that he was the current Archbishop of Canterbury.

Maureen Williams Lambeth Palace

■ WITH reference to my earlier letter in which I stated that a chair I bought in 1942 had been sat on every day since. This was slightly misleading, as my wife and I take a week's holiday in the Isle of Wight each year, during which time the chair is not sat on.

Les Crayfish, Basingstoke

■ I WAS very disturbed when I heard my husband, a lifelong campaigner against the death penalty, say, 'I wish they'd bring back the electric chair'. How relieved I was when he explained that the 'they' in question was a firm of local electricians who had taken his orthopaedic vibro-massage chair away to be mended.

Edna Bollocks, Cuntingford

■ SIX MONTHS ago, my husband died whilst sitting in his favourite chair by the fire. I've left it just the way it was the night he passed away on in it, as I've not had the heart to move it. Unfortunately, he died of a rectal prolapse following a particularly violent and prolonged attack of amoebic dysentery, and consequently the room stinks like a fucking sewer.

Mrs Doreen Mengele, Goole

■ WITH REFERENCE to my previous letters claiming to have sat on a chair everyday since 1942 except for a week's holiday in the Isle of

Three Chairs

THREE chairs to the kind young man who helped me pick up five wing-back armchairs when I dropped them on the pavement the other day whilst getting off the bus. He was so nice, making sure I was alright before lifting them back up onto my shoulders and he wouldn't even accept my offer of a reward.

Mrs G Capes Humberside

* Hip-hip...Chair! Hip-hip... Chair! Hip-hip...Chair!

Wight each year. My wife has just pointed out that in 1982 I spent five days in hospital after a penile cyst under my foreskin went septic, during which time she stayed with her sister in Chilcomb. So it wasn't used for those five days either. I would like to apologise to any readers that were misled by my earlier letters and hope that it doesn't lead them to think any less of my chair.

Les Crayfish, Basingstoke

■ 'I'M IN the chair,' announced my friend one night in our local pub. He meant, of course, that it was his turn to buy a round of drinks. The entire pub erupted in uncontrollable laughter, however, because he was literally 'sitting in a chair' when he said it! It still makes the entire pub erupt in uncontrollable laughter whenever we think about it even now, 38 years later.

Arnold Nottingham Arnold, Nottingham

■ I JUST remembered that whilst we were on holiday in the Isle of Wight in 1978 or 1979, someone broke into the house and stole our television set. It is possible that one or more of the burglars may have sat on my chair whilst they were in the house, although my wife doesn't think they went in the room where the chair was.

Les Crayfish, Basingstoke

■ FURTHER to Mr Crayfish's letter (*above*), I am the burglar who broke into his

YOU ask, WE answer... ABOUT CHAIRS!

Q. WHAT did circus lion tamers use to ward lions off before chairs were invented?

Martin Lacey, Nottingham

A. IN THE days before chairs, lion tamers are thought to have controlled lions using a small, flat piece of wood about the size of chair seat with four short posts attached to the corners. The lion tamer held onto it using a braced, wooden hoop fixed to the rear of the device.

house to steal his television set whilst he and his wife were on holiday in 1979. I took his TV, a set of cutlery and several irreplaceable items of sentimental value. I also laid a cable on the rug in front of the fireplace, but I can categorically state that I did not sit on any of his chairs whilst I was in his house. I hope this helps to clear up any confusion as to the admirability (or otherwise) of Mr Crayfish's chair.

T Johnson, Basingstoke

Do They Know it's Christmas?

It was the Supergroup to end all Supergroups. A glittering line up of Rock Royalty. Everyone who was anyone in the world of music, and Bananarama, turned up at Sarm Studios that day in 1985, and the rest is history.

The intervening years have been a rock and rollercoaster for many of the stars featured on the single. But after two decades and with the holiday season fast approaching, are the Band Aid stars still aware of what time of year it is? We called them up to ask… *Do They Know it's Christmas?*

Boneo, U2

"Yes, I know it's Christmas, but I don't celebrate it anymore because I've gone Jewish. Or possibly Buddhist. It's that one where you put red string round your wrist. Like Madonna and David Beckham have done. So I've had to give up Christmas because it's against the rules. But I'm still going to do Easter because I love chocolate eggs."

George Michael, Wham

"Of course I know it's Christmas. It's my favourite time of the year and the only time I get a break from my extensive recording and touring commitments. I usually have a very traditional Christmas surrounded by family and friends. We all get up early to open our presents, then everyone settles down for a traditional lunch of turkey and all the trimmings. We'll watch a film in the afternoon and then have turkey sandwiches for tea. Later on in the evening I might pop to a public lavatory to masturbate at a policeman."

Tony Hadley, Spandau Ballet

"Do I know it's Christmas? You betcha! It's the busiest time of the year for me. I spend all day playing Santa Claus in the grotto at our local garden centre. It may seem an unusual career move, but I think I had taken my music as far as I could and I wanted to explore new avenues. And at night I stack the freezers at Iceland in Smethwick."

Sting, The Police

"I hate the tacky commercialisation of Christmas which seems to start earlier every year. My wife Trudi and I long for a simple Christmas, one in tune with its humble beginnings and true meaning. So we've bought an island off the coast of Mustique and evicted all its inhabitants so as we can enjoy this special time undisturbed."

Rick Parfitt, Status Quo

"I know it's Christmas, but it doesn't mean a lot to me these days, to be honest. I had a heart transplant some years back and there was a bit of a mix up. To cut a long story short, I ended up with the heart out of Schnorbitz, Bernie Winters's dog. Now I have seven Christmases a year and the sparkle has gone."

Boy George, Culture Club

"No I don't know it's Christmas, and I don't even know what Christmas is. During my well-publicised period of excessive drug use, the bit of my brain that recognises religious festivals was burnt out by some bad acid. As a result, I not only don't know it's Christmas, I've no idea if it's Easter, Yom Kippur or Divali either. Coincidentally, another part of my brain was irreparably damaged at the same time, the bit that enables one to identify stupid hats."

Phil Collins, Genesis

"Christmas? Is it really? I had no idea. I live in Switzerland these days and Christmas is very different over here. There is no snow for a start, and the Swiss don't have a Christmas tree in the house, they have an enormous cuckoo clock made of cheese which they cover with tinsel and baubles. And you won't find a sixpence in your plum duff in this country- instead you'll find a bar of nazi gold."

Simon le Bon, Duran Duran

"Yes, I know it's Christmas. It's a very busy time of year for teen idols like myself. There are millions of Durannies out there who want to know if our Christmas single will beat that of our arch rivals, the Spands. And they want to know what nutty things me and the lads will get up to behind the scenes at the Top of the Pops Christmas special. I won't tell you what prank we've got up our sleeves, but look out Haircut 100 — you might find your dressing room filled with shaving foam!"

THAT's How Rich I Am!

~A series of profiles in which tawdry celebrities boast about the extent of their wealth.

No. 23
Noel Edmonds

"I've got an automatic tennis ball serving machine. However, I don't use it for tennis, I simply fill it full of tennis-ball sized diamonds and fire them into a lake, 24 hours a day. *That's* how rich I am."

Next week: Cilla Black~ *"I wipe my arse on Rembrandts."*

Dancer Blair lives at 58

Lionel Blair, one of Britain's best loved entertainers, was found alive and well at his London home early yesterday morning.

It is believed the 58 year old show business veteran went to bed yesterday evening and woke up after a peaceful sleep. Blair has been in good health for several years and family and friends were said to be delighted. Tributes from show business colleagues have been flooding in ever since the news of his continued well being broke yesterday.

"He is a wonderful man. A terrific dancer and a true star. And he does a lot of work for charity" said Bruce Forsyth, a close showbiz pal for over forty years. "I had lunch with him only two days ago and he was in good spirits, laughing and joking and talking about the future."

Blair, who began his career on stage and in later years carved out a successful career in television, stays with a wife and two daughters.

Blair - 'a true star'

IS ELVIS PRESLEY THI LOCH NESS MONSTER

Is it a floating log? A freak wave? An undiscovered life form or perhaps something from another planet? Everyone has their own theory about the Loch Ness Monster.

But despite over 50 years of investigations into possibly the world's most famous mystery, no-one can say for sure exactly what lurks beneath the dark and murky waters of Scotland's largests loch.

FIGMENT

Many sceptics claim that 'Nessie' is nothing more than a figment of the imagination. But over 4,000 unexplained sightings on the loch have convinced many experts that **something** is lurking beneath the waves.

U-BOAT

One of the most popular theories among local fishermen who spend many hours each day out fishing the loch is that a German 'U' boat became trapped in the loch during the second world war and, unaware that the war is over, still surfaces occasionally to search for enemy shipping. Many recent sightings have resembled a slightly bent submarine periscope.

TOILETRIES

Critics argued that food and supplies on the vessel would by now have run out. But there is a never ending supply

By Bob Twatt

of fish in the loch, and many isolated villages nearby where submariners could go ashore and purchase toiletries and other basic supplies. An increasing number of Germans dressed as tourists have been seen in the area in recent years.

ELEPHANT

However most scientists now go along with the theory that the loch has for many years been home to large form of animal. But according to top biologists, visitors to the loch have nothing to fear. For they believe that the Loch Ness Monster could in fact be nothing more than a harmless elephant which escaped from Edinburgh Zoo in 1929.

FLIPPER

Hugh McGrath, now 89, former head keeper at the Zoo believes that the elephant could have developed flipper shaped feet to enable it to swim effortlessly beneath the water. And without doubt some of the most well known pictures taken of 'Nessie' bear an uncanny resemblance to an elephants trunk.

Elvis — born again?

Hugh McGrath - Elephant theory

Using its trunk as a snorkel the elephant could stay under the water for long periods at a time, surfacing to feed on nearby trees and bushes during the night.

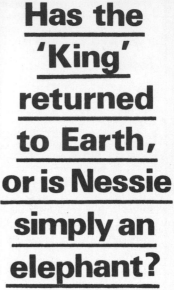

Has the 'King' returned to Earth, or is Nessie simply an elephant?

But perhaps the most fascinating explaination of all is that Elivs Presley, the late 'King' of Rock'n'Roll, has returned to Earth to haunt the loch in the form of a large, ungainly aquatic monster up to 45 feet in length.

FAN

This astonishing Elvis re-incarnation theory was first put forward by Archie Gubbins, a life long Elvis fan who first began to investigate the loch while on holiday in Scotland in 1983. On a subsequent visit, using advanced sound equipment, Archie was able to play

?

igh McGrath's vision of an underwater elephant. Scientists believe that
s could be the real 'Nessie'.

Win a fabulous camera!

Over the years thousands of photographers, professionals and amateurs alike have spent hours sitting on the shores of Loch Ness hoping to get the one picture that would prove the monster really exists.

Now we're offering all budding 'Nessie' photographers the chance to win this fabulous Helix Camera pencil box. All you have to do is send us your photos of the Loch Ness Monster.

Your pictures can be in colour or black and white, and you can send as many as you like. Please write your name and address on the back of each one, and send them to: Viz Comic, Loch Ness Monster Photograph Competition, 16 Lily Crescent, Newcastle upon Tyne NE2 2SP. We'll be printing all the best efforts in our August issue (No. 25) and there's a fiver paid for every picture we use.

The first prize will be awarded to the photographs which, in the view of our judges, proves beyond all reasonable doubt that the Loch Ness Monster exists.

Plus £100!

Additionally, we're offering a bumper cash prize of £100 to anyone who can capture the Loch Ness Monster alive. But remember, to win the prize the monster must be:

● Captured alive
● At least 45 feet in length
● A species of animal previously unknown to mankind

---

tapes of Presley's greatest hits including 'Blue Suede Shoes' and 'Are You Lonesome Tonight', through speakers lowered 200 feet below the surface of the loch.

EYEBROWS

The results of his investigations would have raised eyebrows among even his strongest critics had they not been lost when his boat sank, taking with it all his equipment. But despite this setback, Gubbins is quick to point out that existing evidence supports his theory.

ELVIS

"In his former life Elvis was quite a good swimmer, and I have it on good authority that he did on one occasion mention to a friend that he would one day like to visit Scotland."

ELEPHANT

Elephant or Elvis? Perhaps we'll never know. For the time being at least Nessie's secret is safe, deep down below the waters of Loch Ness.

How do the stars explain the Loch Ness monster?

We decided to ask a few famous faces how they would explain the mystery of the Loch Ness Monster. And here is what they said.

England manager **BOBBY ROBSON** took time off from a training session to tell us that Nessie may simply be a seal. "Either that or a giant eel of some sort", said the England Boss.

Pop singer **BOY GEORGE** who's single 'Everything I Own' is currently riding high in the charts, believes that Nessie may be a plesiosaur — a large, long necked aquatic dinosaur thought to have been extinct for over a million years. "It is possible that the species may have survived in Loch Ness undiscovered by man", says George.

LESLIE GRANTHAM, star of TV's Eastenders admitted to being completely baffled. "I haven't given it a great deal of thought", said Leslie, better known to millions of TV viewers as 'Dirty Den', landlord of the Queen Vic

Do you have a theory about the Loch Ness Monster? If so why not send it to: The person in charge of Loch Ness Monster theories, The British Museum, London.

BLIMEY! WHAT A FIND

The farmer who discovered Tommy Steele in a disused barn at his farm in Cumbria is today £20,000 richer after the vintage actor and singer was sold at auction yesterday.

Sam Armstrong uncovered the veteran star among piles of straw while clearing out derelict buildings at a farm near Penrith which had previously belonged to his father.

BARN

"I had no idea who he was at first until we pulled him out of the barn and dusted him off", Mr Armstrong told reporters yesterday. But a quick examination revealed that the veteran star was in remarkably good condition. "Someone has painted him pink, and quite a few bits and pieces were missing, but we got quite a surprise when we tried starting him up and he began singing and dancing first time", said Mr Armstrong.

SCHOOL

Enquiries revealed that Mr Armstrong's grandfather had bought Steele in the mid

Blimey! What a find he's got

What a great big find

Farmer Sam (left) with his dog and Tommy Steel (right) - found in barn

seventies from a fairground in Perthshire where he had been converted for use as a sword dancer. "After my grandfather died he must have simply been forgotten about and left in the barn to rust".

BACK

And while the remarkable find has meant a surprise financial windfall for Sam, there is also good news for fans of the singer who's best known hit was 'The Little White Bull'. For the star has been bought by members of the Tommy Steele Society and they plan to have him fully restored, singing and dancing live on stage, within five years. But the restoration work could prove to be expensive, as their secretary Bill Walsh explains.

SQUARE

"Unfortunately, Tommy's bowler hat and cheeky grin were missing when he was found. But we have already been in touch with one manufacturer in Poland who still makes that type of hat and we are confident of finding a suitable replacement grin, even if it means commissioning a new one from scratch. There is still a foundry in Shropshire equipped to cast a grin. The only problem is it costs a lot of money".

TRIANGLE

On top of the £20,000 already raised to purchase the star, Bill expects to spend a further £20,000 restoring him to his original condition. And he said the Tommy Steele Society would be grateful for any financial help that the public were willing to give.

ELDORADO

In 1982 the sixties singer Joe Brown was restored to full working order after being found in use as an advertising sign on a roundabout near Leeds. He is now a part of the National Collection and can regularly be seen on TV and making public appearances around the country.

Collins dug up

A farm labourer looking for a piece of a tractor may have uncovered the remains of seventies TV actor Lewis Collins.

Following the find in a field in Suffolk an actor, believed to be Collins, has been sent to the University of Warwickshire for scientific identification. If his identity is confirmed farm labourer Jim Marsden, who discovered the hard man actor, could be in line for a six figure reward.

Collins, star of TV's Professionals, has been preserved by dry clay which surrounded him. Mr Marsden was out searching for a missing bolt which had fallen from his tractor the previous day when he uncovered the actor using a metal detector.

If Collins is declared treasure trove then Mr Marsden will be free to sell the star and keep the proceeds. The last time a seventies TV detective came up for auction was in 1988 when Peter Wingard, star of the Jason King TV series, was sold to a private collector in Japan for £475,000 after being discovered rolled up in an attic in Kent.

20 THINGS YOU NEVER KNEW ABOUT ROOFS

They're on houses and huts, sheds and shopping centres. Angry people hit them, sales go through them. And arseholes jump off them with parachutes.

Yes, they're roofs. And whether you love them or hate them, there's simply no ignoring them. But what do you really know about these amazing structures that form or cover the top of a building? Here's twenty things you probably never new about roofs…

1 A 'flat roof' is not necessarily the roof of a flat. It is in fact a new, horizontal style roof which either leaks or has puddles on it, which recently replaced the old fashioned, cumbersome 'pitched roofs' which had previously been used, and had protected buildings from the weather without any problems for several hundred years.

2 A 'roofer' is a shifty, suspicious looking character who occasionally appears at your door with a broken slate in his hand and claims that, despite visual evidence to the contrary, it has fallen off your roof.

3 Next thing you know he'll charge you £25 to lean a ladder on your house, climb up it and make a banging noise for ten minutes, before pissing off to the pub.

4 If you call a 'Thatcher' to come and mend your roof, don't expect a bitter former lady Prime Minister, her drunken husband or her half-wit son to turn up with a ladder. A thatcher is in fact a traditional craftsman who will attempt to repair leaks with clumps of straw.

5 If someone tells you that they've just spent a night 'on the tiles', they haven't necessarily slept on the roof. Unless they are a prisoner in one of Britain's overcrowded jails, in which case they very probably have, and the remains of the roof are probably lying in the exercise yard.

6 If the vicar asks his choir to 'raise the roof', he isn't trying to get building work done on the cheap. He's simply asking them to sing loudly.

7 If the vicar asks you to contribute to his 'steeple restoration fund', he *is* trying to get building work done on the cheap, by getting *you* to pay for it.

8 'So long as you've got a roof over your head', someone once said. But orthodox Jews may well disagree. Their religion dictates that they cannot have a roof over their heads on a Saturday, or something like that.

9 And they don't eat pork. Not even sausages.

10 Houses are built from the bottom, upwards, finishing with the roof. This is to allow for any inaccuracy in the height of the building. For if the roof was built first, a couple of inches too high, then it would leave a small gap at the bottom of the walls. The only exception is Australia where they build the roof first, and have Christmas Day in the middle of June.

11 Roofs are no strangers to the UK pop charts. In 1970 The Supremes climbed 'Up The Ladder To The Roof', reaching No. 6 in the Top Ten.

12 And more recently Lionel Ritchie was 'Dancing On The Ceiling', which was sort of similar.

13 Indeed, in another remarkable pop/roof tie-in, The Beatles hit record 'Let It Be' was recorded on a roof.

14 However, 'Fiddler On The Roof' was not a

copy cat album by scruffy violinist Nigel Kennedy. It was in fact the hit musical in which that bloke with a beard called Toblerone or something sang that song that went "If I was a rich man, yaddle diddle diddle, yaddle yaddle diddle diddle dum. All day long I'd biddy biddy bum, if I was a wealthy man", etc.

15 People on 'That's Life' frequently claim that their dog can say "roof".

16 And "sausages".

17 Mountaineer Chris Bonnington has on more than one occasion climbed to the 'roof of the world' – the Himalayan mountains. And he occasionally comes back with the same number of people he set out with.

18 Singer David Bowie's roof maintenance bills are enormous. That's because Dave has 3 roofs. One on each of his houses in Australia, Mustique and Switzerland.

19 In fact he's got 4 roofs, if you count the roof of his mouth, with which he has sung such hits as 'Alvin Stardust' and 'Space Odessey'.

20 There is no cockney rhyming slang expression for roof. That's because 'roof', along with 'orange' and 'motorbike', are the only three words in the English language with which no other word rhymes. In fact, the Oxford English Dictionary have offered a £10,000 reward to anyone who can think of a word that rhymes with roof. If you think you know one, send it to The Oxford English Dictionary, Roof Rhyme £10,000 Reward Department, Oxford University Press, Horse Hoof Lane, Oxford. Please mark your envelope 'I've thought of a word that rhymes with roof'.

I'M SORRY. YOU'LL HAVE TO SPEAK UP. I'M ON MY MOBILE.

Food & Drunk

With JILLY GOOLDEN

This week, Jilly recommends her favourite hangover for under £15

3 bottles of Nigerian Cabernet Sauvignon. 1/2 bottle Woods Navy Rum. 4 tins of White Lightning. 1 bottle of cooking sherry. Morrissons £14.49

I AWOKE with this hangover with a distinct taste in my mouth. I was getting cupro-nickel, like sucking a handful of old two-pence pieces. The back of my front teeth were coated with sulphurous fur, like on a bee's back.

I tried to lift my head from the pillow, but I was getting rhythmic pulsating throbs, as if an all-in wrestler was trying to force sausage meat behind my eyes.

And there was a strong bouquet. I was getting Parmesan cheese and bad eggs, a sort of putrid, acrid smell, like a dairy farmer's slippers.

Then I realised my hair and ears were stuck to the pillow with congealed vomit. I swung my legs over the side of my bed and sat there waiting for my brain to catch up. I became aware of a strange feeling in my stomach. It was like Marlon Brando wearing a jumper soaked in sea water, trying to kick start a diesel Harley Davidson Fat Boy in two feet of porridge. I was getting hippopotamus's tongue licking canal water off my kidneys mixed with The Keystone Cops made out of omelette being chased out of my arse by a jelly tube train full of lead bricks. It was all in there.

And I was sweating like a Mother's Pride processed cheese sandwich wrapped in cling film and pressed into a driving instructor's arse stuck in a traffic jam on a hot bank holiday. When my brain caught up with my eyes, I was in a kaleidoscope. There was an increasing pressure in my head, culminating in an explosion of hot light behind one eye.

I was getting a sudden massive increase in heart rate accompanied by a terrifying spiral of anxiety, like a shark in a washing machine eating its own tail.

And for such a spicy hangover it had a very long finish. I was spewing Fairy Liquid till after tea time, and the feelings of depression and remorse lasted well into the next day.

Obviously for £15, it's not the most explosive hangover I've ever had, but it was cheeky and unpretentious and the ideal accompaniment to a few tentative sips from a cup of water. Very good value. ★★★

Lumley in New Leg Fear

THE WORLD of showbiz was rocked once again in the early hours of this morning as rumours that Joanna Lumley may have grown a third leg began to circulate.

It had earlier breathed a sigh of relief after speculation that she had possibly lost her left leg above the knee were scotched when she appeared at a hastily-convened press conference with two legs.

Baseless gossip about the 50-year-old Ab Fab star suggested that a freak third leg could have emerged from her lisk as a result of her perhaps accidentally drinking from a glass containing gamma radiation which might possibly have been left on a table at London's swanky Grosvenor House by an absent-minded scientist.

"If this proves to be correct then it will be a dark day for the world of entertainment," speculated her sombre co-star Jennifer Saunders.

"But Joanna's a real trouper. I'm sure her personal courage would enable her to easily overcome any obstacles that perhaps growing an extra leg might put in her path."

That's Shelves!!!

ALWAYS BRITAIN'S FIRST AND LIVELIEST SHELF FORUM

I've got a shelf that's two metres long. Can any of your readers beat that?

King Shelf
Newcastle

I put up a shelf the other day, and the wife said it didn't look straight. Imagine her surprise when I checked it with a spirit level and it was perfectly horizontal. It was an optical illusion! How we both laughed.

Arthur Twoshelves
Daventry

The things kids say! "Look, grandad, that mantlepiece has lost its fireplace," my 3-year-old grandson said the other day. He was pointing at a shelf! Do I win £5?

Ernie McShelf
Kinross

I won a really nice shelf in a raffle, but everything I put on it falls off. That's because I'm a lighthouse keeper or I live in a windmill, and the only way I could fasten it on the wall was to fit it vertically!

Eamonn O'Shelf
Edison Rock Lighthouse, Amsterdam

I haven't got any shelves in my house. That's because I work in a supermarket stacking shelves, and the last thing I want to look at when I get home is more shelves!

Rosemary Shelfson
St Asaph

That's nothing. I work in IKEA, stacking self-assembly shelves...onto shelves! If I came home and there was more shelves in my house, I'd probably flip, murder my wife and kids, and then turn the gun on myself!

Billy Bookcase, Gateshead

The things kids say! "Grandad, why has that shelf got four legs?" my 3-year-old grandson asked me in the kitchen the other day. He was pointing at the table! Now do I win £5?

Ernie McShelf
Kinross

I've been collecting shelves for forty years. "I've bought you something to help you store your collection," my wife said the other day. "I hope it's not another shelf," I replied. It was a box!

Sheldon Elf
Matlock Bath

All you shelf fans are saddos and wrinklies. They haven't even got doors. Get with it and get cupboarded up!

Kurt Cupboard
Cirencester

Miriam

SHELF HELP WITH MIRIAM STOPPARD

Dear Miriam...
This Christmas, my boyfriend bought me a pair of bookends to put on my shelf. He'd overlooked one fact; I don't have any books. However, I have got twelve videos which occasionally topple over, especially when I'm dusting. Do you think it would be okay to use these bookends as "video-ends" instead?

Sarah, Luton

LETTER OF THE DAY

Miriam writes...
My ex-husband playwright Tom Stoppard's old woodwork teacher always used to tell him to never use a tool for any purpose for which it wasn't intended. However, in this case I think you can safely ignore that advice. Your bookends will make marvellous "video-ends" to stop your tapes toppling over!

434

ADVICE ON ICE

Christmas should be a time of fun and laughter, of holly and ivy, of log fires, chestnuts and mistletoe.

But all too often the joy of Christmas turns to tears as dangerous conditions underfoot make the festive season a nightmare for pedestrians.

SLUSH

For every year icy pavements and slush covered pathways cause havoc for hundreds of would-be walkers. And these problems can be complicated by additional hazards, including drifted snow build ups against doorsteps and curbs, irregular mounds of treacherous frozen slush, fresh powdery snow concealing existing layers of ice, and clumsy pensioners sprawling around on the ground in front of you.

FIGHTING

It's every Christmas pedestrian's nightmare. That brief visit to the shops, or casual call on a neighbour, that ends in a fall. And all too often bruised elbows, grazed knees and twisted ankles are the result.

STEEPLE RESTORATION

Of course Christmas is a very busy time of year for the stars of showbusiness, with TV shows, public appearances and pantomime work leaving them more at risk than most when it comes to pedestrian accidents. So we decided to ask **them** what precautions they take, and how they go about reducing the risk of a fall during the inclement winter weather. They certainly have some useful things to say, so why not take a tip from the stars and stay on your feet this Christmas.

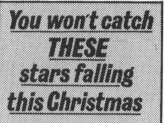

You won't catch **THESE** stars falling this Christmas

down on Christmas shopping. "Don't buy large or cumbersome presents. Stick to small things like pens and jewelry. Then you can pop them in your pockets and still have both hands free for balancing on your way home".

Comedy actor **ROY BARRACLOUGH**, alias The Street's no-nonsense barmaid Bett Gilroy has learnt his lessons from years spent in panto work across the country. "I've walked on ice, through snow, in slush, and in numerous treacherous combinations of the three", he told us. "I recommend your readers adapt their walking style to suit the conditions. Generally speaking, take shorter steps and raise your feet higher off the ground. A handy hint is to imagine you are wading through water. It works for me."

"Slow down on corners". That was the key advice given by high brow arts commentator **MELVYN BRAGG**. "The vast majority of accidents on ice and snow occur when people are changing direction. So think ahead. Begin to slow down early for a corner, and never hurry. It isn't worth it".

Showbiz tips to avoid the slips

Being a weatherman we thought **MICHAEL FISH** might have some useful advice on how to cope in dangerous walking conditions. And we were right. "Choose the right footwear", he told us. "Wellingtons for snow, something with a good grip for slush, and rubber soled shoes for ice. Also, try bending down, to lower your centre of gravity. Bend your knees, and tilt your head forwards. It may feel uncomfortable, but it works".

Former Holiday show host **CLIFF MICHELMORE** be-

lieves the secret of balance is all in the mind. "In icy conditions our muscles tend to tense up. Our leg movements become jolty and awkward. The answer is to be more relaxed. Think positive. Try skidding a few yards before you start walking. Take a run up and see how far you can slide. Once you've conquered your fear of falling you will then have the confidence to walk normally in these difficult conditions".

Award winning actor and playwright **COLIN WELLAND** admits that walking on slippery and hazardous pavements is one of his worst fears. Indeed, he has been known to stay indoors for months on end in order to avoid walking in nasty weather. "If I do have to go out and there's ice on the ground, I must admit, I chicken out", he confessed. "Rather than walk I just slide along on my bottom, propelling myself with my arms and legs".

Nutty crazy oddball Mr Bean actor comic **ROWAN ATKINSON** knows only too well what it's like to fall on an icy pavement. He's done it several times. His tip is to cut

This Christmas — Show them you care...

GIVE THEM FAGS

WHERE'S the TOILET?

Here's a chance for you to test your powers of observation.

Every issue we feature a photograph of a toilet taken somewhere in the UK. Do you recognise the urinals pictured below? Perhaps you've used them recently. Study the picture carefully and try to work out where it is. If you're stuck, the answer is printed below.

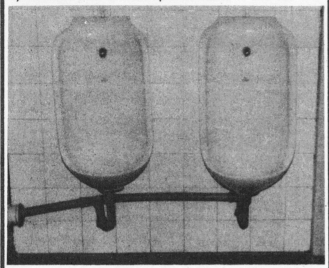

This issue's toilet is at Metro Radio, in Swalwell, Tyne and Wear.

FISH of the WEEK

What is your favourite fish? Perhaps it's haddock, herring or bass. Or maybe you're a fan of bream.

Write and tell us about the kind of fish you like. Every issue we'll be featuring a new fish as suggested by a reader. For our first 'Fish Of The Week' we have chosen carp.

No.1 Carp

Which fish would you like to see? Write to: Fish Of The Week, Viz Comic, Viz House, 16 Lily Crescent, Newcastle upon Tyne NE2 2SP. £1 paid for each letter we print.

KILLER WASP SEX VICAR IN GAY NAZI STORM

No Girls 'Danced Naked' In Moonlit Devil Ritual - claim

By our INVESTIGATIVE STAFF

A village vicar has denied taking part in moonlit ceremonies involving naked women.

And Rev. Stanley Compton has denied allegations that he is a leading member of a gay nazi movement in the quiet village of Todhamlet.

CHILDREN

And we were unable to find any evidence to support claims that 64 year old Compton was having sex with under age children, some of them boys. But in an outhouse adjoining his home we found firewood, two buckets and a rusty lawn mower.

HORROR

Local residents told us the Rev. Compton came to the village in 1954 and has lived there ever since. They described him as a quiet man, a non-drinker who didn't socialise much but often arranged jumble sales or garden fetes.

NAKED

Reg Dixon, owner of The Golden Lion Hotel in the village told us he had never heard rumours connecting the vicar with midnight sex ceremonies. And his daughter, a shapely 16 year old, told us she was unaware of any satanic activities in the village.

Her mother, Mabel Dixon, cook at the Golden Lion, confirmed that neither she nor her daughter had ever danced naked in the flickering flames of a pagan bonfire.

AXE

A regular at the Golden Lion, village vet Norman Taylor said he was "unaware" of any local girls dancing themselves into a hypnotic frenzy in the course of a gruesome moonlit satanic ritual.

VAMPIRES

When we spoke to Rev. Compton he strongly denied any involvement with the National Front. He was however later 'unavailable' to comment on the suggestion that he was breeding giant indestructable killer wasps in a garage adjoining his home.

Are You SEXY?

Fellas - Here's A Chance To Find Out!

What makes men sexy? What is that vital ingredient that turns women on? Is it their haircut? Or their eyes? Or perhaps the kind of trousers they wear. Even the experts are unsure.

Here's a chance for the fellas to find out just how sexy they really are. Simply answer the following questions A, B or C, tot up your final score and then see how your total compares with our experts' sex ratings.

1. Are your trousers:
   A. *Baggy round the waist*
   B. *Comfortable*
   C. *Tight and bulging*

2. Which of the following do you smoke?
   A. *No cigarettes*
   B. *Ordinary cigarettes*
   C. *Foreign cigarettes*

3. You are heading for the bus stop and a bus is about to pull away. Would you:
   A. *Run for it*
   B. *Walk and hope to catch it*
   C. *Stop and light up a cigarette*

4. If you nipped out of the house to buy a newspaper, how would you walk?
   A. *Hurriedly, tripping from time to time*
   B. *At a medium pace*
   C. *Slowly, swaying from side to side*

5. If you went out for a meal, how would you travel?
   A. *On a No. 14 bus*
   B. *In a taxi*
   C. *In a flashy sports car with a throbbing engine*

6. You are in a restaurant and the waiter offers you a starter. Which of the following would you choose?
   A. *A small bowl of pea soup*
   B. *A couple of poached eggs*
   C. *An enormous wedge of juicy melon*

7. You wake up one morning feeling sexy. What would you do?
   A. *Get up and have a cold bath*
   B. *Go back to sleep*
   C. *Go out and buy a pair of leather underpants*

8. You are at the cinema watching a movie with a friend. What would you do during the interval?
   A. *Go to the lavatory*
   B. *Buy an ice lolly·*
   C. *Put your sunglasses on*

9. Which of the following pets would you prefer to keep?
   A. *A parrot*
   B. *A small hampster*
   C. *Several large dogs*

10. If you went fishing, how much underwear would you put on?
    A. *Lots of warm, woolly underwear*
    B. *A small amount of underwear*
    C. *No underwear at all*

SCORING

A — 1 point

B — 2 points

C — 3 points

YOUR SEX RATINGS

Less than 10 — *Dull and unattractive.*   11 to 20 — *Rather ordinary.*
21 to 29 — *Pretty saucy.*   Maximum 30 — *Ooh la la!*

CONFESSIONS OF A MOVIE STAR SHAG-A-LIKE

A Rotherham window cleaner's resemblance to a top movie star is making him a fortune. For lookalike Burt Johnson is charging star-struck women up to £500 for sex.

Dead ringer Burt (right) as Robin, and the discreet ad (below) which he placed in a shop window.

And female fans desperate to fulfil their fantasies have literally been queuing up to pay for sex with Burt since he placed an ad in his local newsagents' window.

MOVIE STAR
SEX FOR SALE
ROBIN ASKWITH LOOK-A-LIKE
AVAILABLE FOR SEX.
TEL: ▮▮▮▮▮▮ (EVENINGS)

GIGOLO

The hunky 54 year old gigolo boosts his paltry £80 a week window cleaning income by charging desperate housewives hundreds of pounds to make their wildest sexual dreams come true. For Burt is a dead ringer for saucy seventies 'Confessions' star Robin Askwith, and for hundreds of sex starved housewives, Burt is as close to the real thing as they will ever get.

FIGARO

Our investigator rang the phone number which was given on the back of Johnson's post card and asked how much it would cost to book the Robin Askwith lookalike for one night. Johnson told her that the price would depend on what exactly she wanted.

ANGELO

"I could do you sex in the back of the car, with the windows steaming up, the bonnet springing open, the suspension bouncing up and down and steam coming out of the radiator, for £500. But doors falling off would be extra," he told her.

ROMEO

"If you don't want to spend that much I do straight sex in a lift for £250. That includes the lift going up and down very very fast and the lift indicator arrow frantically waving backwards and forwards. If you want a group of nuns to be waiting outside when the lift eventually stops that would be another £100."

Our investigator arranged to meet Johnson at a local hotel. He strolled in ten minutes late wearing bell bottom jeans, a large floppy cap and a cap sleeved T-shirt several sizes too small. He introduced himself as Robin, and looked down her cleavage before wolf whistling and wiping his brow.

FOXTROT

After being handed an advance payment of £200 cash Johnson agreed to accompany our reporter to a room on the second floor. At first he seemed nervous, but once inside he began to relax a little and said that he had been working as a lookalike escort for about a year. He said he worked for five or six nights a week, usually as Robin Askwith, but he did occasionally do Sid James or Kenneth Williams.

VICTOR

Our reporter then asked him if he would be prepared to do straight sex on the bed. He said he would, and that for £500 he would have sex with her until all the bed springs went 'boing', the legs collapsed, and the bed fell through the floor and into the room below where an old couple would be

The real McCoy - seventies sauce pot Robin Askwith.

watching TV. "If you want the old couple not to notice and keep watching telly as if nothing had happened that would be £25 extra," he said.

TANGO

When offered champagne Johnson suggested that he have sex with our reporter in order to make the cork spontaneously pop out of the bottle, causing champagne to spray all over her buttocks, but she declined.

In a taped conversation Johnson also boasted how in the past he had once had sex in a kitchen causing the kettle to boil, the whistle to blow furiously before popping out and flying across the room, and various cupboards to pop open and the contents, including flour, to scatter all over the place. "The food mixer started getting faster and faster, spraying us with chocolate sauce, and two bits of burnt toast popped out of the toaster at the same time" he said, gesturing with his hands.

CRESTA

He then described how on another occasion he had sex on a snooker table, causing all the balls to fly into the pockets.

PEPSI MAX

Our investigator suggested that she may be interested in something more kinky, and asked Johnson if there was anything 'special' he could recommend. His face cracked into a crude smile as he suggested they have sex backstage at a theatre among the props and costumes. "After a few minutes the backdrop will slowly rise to reveal the audience staring at us in disbelief. But I'll keep going, and after you come the audience will burst into a spontaneous standing ovation, at which point I'll grin and bow." He said that would cost £700, but he'd need a couple of days' notice to sort a theatre out. At this point our reporter made her excuses and left.

SOCCER SHOCKER!

New commitment rate hike kicks players where it hurts

PROFESSIONAL footballers were reeling last night after the Chancellor of the English Football Association raised the players' commitment rates for the third time this year. A rise of 50 percentage points means that all players must now give 250 per cent effort each time they take the field.

League

The decision was taken to bring the FA into line with the Bundesleague, which raised its own rate last week.

"We had little choice but to take this action" said David Davies, the only man left at the FA. "No one likes to raise commitment rates, but we must take these steps if we are to remain competitive in Europe."

Fathom

But many amateur clubs fear that this is beyond their players' means. "All our players hold down full time jobs." said Phil

Kegan: Thousand per cent.

Castiaux, secretary of Blyth Spartans. "They cannot possibly go to work during the week and then give 250 per cent on a Saturday. The level should be capped for part time players. They cannot be expected to give much more than 170."

By our sports staff, a fat red-faced drunk

Britain enjoyed a stable 100 per cent rate throughout the seventies. But in 1982 it was raised to 101 per cent by Trevor Francis during a controversial summing up on Match of the Day. The eighties saw the rate creep up to 110 per cent.

Chain

The highest ever commitment rates occured on 'Black Saturday', when comments by Kevin Keegan sent rates spiralling. The part-time England coach promised to give a thousand per cent in his new job, causing many clubs to panic and set their own commitment rates. By the end of play that afternoon, the rate had reached an unsustainable 10,000 per cent. Officials at Lancaster Gate finally stepped in and restored sanity by announcing a standard rate of 200 per cent.

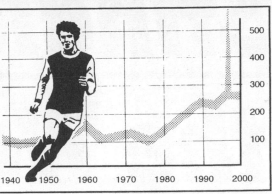

A graph yesterday.

We called Keegan at Bisham Abbey, to see how his 1000 per cent commitment to the England job was going, but we were told he was probably at Fulham F.C.

that day. "If he's not there, you might catch him at his racing stables in Hampshire or perhaps at home in Durham," the cleaning lady told us.

THE *WHEEL* SECRET BEHIND YOUR FELLAS LUNCHBOX

YOU can tell what a man packs in his lunchbox by watching how he holds his car steering wheel, researchers advised women yesterday.

BOTTOM of the lunchbox league is the anxious motorist who drives with one hand on the wheel and the other hovering over the horn. Verdict: "Dull and unimaginative packed lunch, limp cheese sandwiches, non-branded chocolate biscuit and a scotch egg."

STEER CLEAR of the man who grabs the wheel with both hands at exactly the same height. Verdict: "No appetite for lunch. A bag of crisps, a flask of tea and he's happy untill teatime."

BETTER is the guy who holds the top of the wheel with two hands close together. Verdict: "Adventurous sandwiches on unusual breads, fancy salads and little tomatoes, a Mr Kipling cake and a bag of Quavers."

Christie: Obligatory in lunchbox article.

BORING. Those who drive with both hands firmly clenching the bottom of the wheel. Verdict: "Same packed lunch every day. Ham, cheese and pickle sandwiches on Mother's Pride, raspberry yoghurt and an apple."

BEST EATERS drive with one hand at the 8 o'clock position and the other at 2 o'clock, says the Aston University study, which looked at the driving habits of 7 men, then asked their wives what they liked in their sandwiches. Verdict: "Doorstep sandwiches packed with filling, 2 sausage rolls, a can of pop, a Mars bar and a family bag of Cheesy Wotsits. And another sausage roll."

How DO you fit 6 HOURS of SEX into the busiest day of the year? Former Police frontman turned arse-hole STING reveals his festive tan-trics of the trade

Stingle All the Way

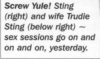

SEXMAS SEXCLUSIVE!

DECEMBER 25th is hectic at the best of times. What with making the dinner, opening your presents and watching the Queen's speech, it often feels like there aren't enough hours in the day. Imagine if you had to fit in a marathon *6-hours of sex* as well!

It sounds impossible, but that's just what eco-warrior-cum-Jaguar sales-man Sting manages to do each year. And if you want to live like a star this Christmas, here's his timetable to ensure a successfully stuffed turkey in the kitchen and a successfully stuffed wife in the bedroom.

8.00am

Christmas day starts early in the Sting household when the alarm goes off. There's a long busy day ahead of us, and we have to fit 6 hours of sex in somewhere! Me and my wife Trudie Sting squeeze in half an hour of heavy petting before popping downstairs for a breakfast of free-range sausages, fried organic eggs, granary toast and a cup of fairtrade tea.

Total nookie so far : 30 minutes

Fully breakfasted, there's just time to check that the turkey has defrosted and turn the Aga up before nipping back upstairs for another quarter of an hour of cunnilingus.

8.30am

Total nookie so far : 45 minutes

8.30am

Countdown to a Suc-sex-ful XXX-mas
-by Sting

9.00am

Back downstairs to get the veg ready while the kids open their presents. We only buy organic vegetables because we believe that modern agricultural pesticides can be very detrimental to health. I notch up another 45 minutes rubbing Trudie's breasts, 20 minutes on the left one, 25 on the right, while she's peeling the spuds, chopping the carrots and cutting them crosses in the sprouts.

Total nookie so far : 1 hour 30 minutes

The oven's reached 250 F, so it's in with the turkey. It's an opportunity to spend an hour or so opening my presents, but with another four and a half hours of sex to somehow fit in before bed, there's no time to waste. I get on with unwrapping my gifts while Trudie gets on with tickling my nuts.

9.45am

Total nookie so far : 2 hours 30 minutes

10.45am

Mid-morning, and just time to suck Trudie's nipples while she bastes the turkey. Only 5 minutes, but every little helps.

Total nookie so far : 2 hours 35 minutes

10.45am

With the dinner cooking nicely in the oven most people would take the opportunity to relax. But not in the Sting household. Trudie and I are back upstairs for 20 minutes of mutual masturbation and 5 minutes of her sitting on my face.

10.50am

Total nookie so far : 3 hours

We're half way there. With three hours under my belt, I allow myself a small sherry in bed while Trudie pops down to put the sprouts on. But then it's straight back to work, kneading her thighs sensuously for three quarters of an hour until the in-laws turn up.

11.15am

Total nookie so far : 3 hours 45 minutes

Dinner is served! Time to sit down and tuck into turkey with all the trimmings...for everyone except me. I've still got to fit in another 2 and a quarter hours of sex, so whilst I'm carving the turkey with my right hand I'm giving Trudie's bottom a kinky spank with my left. Then we cover

2.00pm

each other in Christmas dinner, and slowly lick it off like in 9 and a Half Weeks. We have to be careful with the pudding, though. It's easy to end up with singed pubes!

Total nookie so far : 4 hours 30 minutes

Everything stops for the Queen's speech. Except me and Trudie's marathon tantric bonk session! It's not that I don't respect the monarchy. Her majesty has her job to do, but I have mine as well. Whilst the Queen tells the nation about her travels and hopes for the coming year, Trudie and I tell each other our most intimate and outrageous sexual fantasies, using explicit language.

3.00pm

Total nookie so far : 4 hours 50 minutes

3.00pm
1ST

Time to get our breath back and snooze our way through the Bond film with a box of Matchmakers and a Terry's Chocolate Orange. We might try a bit of oral during a commercial break, but that's about it.

Total nookie so far : **4 hours 55 minutes**

3.30pm

It's back into the kitchen, and I sit on the worktop masturbating while Trudie makes the turkey sandwiches.

5.00pm

Total nookie so far : **5 hours 25 minutes**

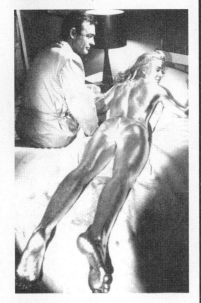

We eat our tea in the sitting room watching Christmas *You've Been Framed*. Trudie and me sit on the sofa and caress each other's genitals with our toes. That's another half hour out of the way. Only another 5 minutes left to go.

5.30pm

Total nookie so far : **5 hours 55 minutes**

Nearly there. The end's in sight, so it's upstairs for a 3 minute diddy-ride before the main event.

6.00pm

Total nookie so far : **5 hours 58 minutes**

At last. We have penetration. I scuttle Trudie for a good 2 minutes before shooting my bolt.

Total nookie so far : **6 hours**

6.03pm

Phew! It's all over and the evening is our own. But we've only got a few hours to get our strength back before starting our next 6 hour sex session... *at bedtime!*

6.05pm

"

10 things you never knew about...
Christmas Dinners

1 As a baby, Christ was brought sprouts by one of the Wise Men as a gift, but he threw up and cried, so the Wise Man gave him some frankincense instead. Since then, sprouts have been a great part of Christian tradition and frankincense has been largely forgotten – a bloody good thing too, since it stinks... *of sprouts!*

2 The eccentric Earl of Crackers *(1770-1825)* invented crackers during a card game. He didn't want to leave his place at the Quantoon table to get his hat, his magnifying glass and his favourite joke, so he famously asked his servant to bring him all three in an exploding toilet roll! The Earl also invented water biscuits and fireworks.

3 Bootylicious bint Beyoncé Knowles loves Xmas Dinner so much that it's the only thing she eats. The independent woman attributes her foxxy figure to a strict diet of THREE Xmas Dinners a day. And at 28,000 calories per meal, it's no wonder she looks so crazy in love!

4 And even Madonna is a Xmas Dinner fan! She first got a taste for them when she visited a traditional East End Xmas Dinner restaurant in 1999 with hubbie Shane Richie.

5 In America, Xmas is called Thanksgiving and starts in November. Americans eat buffalo instead of turkey and shout instead of sing. They listen to Bing Crosby instead of the Queen, and their crackers contain cowboy hats.

6 The now-traditional Xmas Dinner row was introduced by Prince Albert. In a ceremony that continues to this day, the King of Denmark visits the royal family on Xmas Day to start their row for them. Last year's festive punch-up was about the the difference between Cornish Puddings and Yorkshire Pasties.

7 In 1976, comedy dancers Morecambe & Wise got the highest ever TV viewing figures for their Xmas Dinner. Eric Morecambe had had two heart attacks that year, and his doctors advised against a big TV special, so the BBC broadcast the duo's Xmas Dinner instead. 26 million people watched them eating turkey and nuts with Robert Dougall, Andrew Preview and Angela Rippon. The next day, the papers were full of stories about Rippon's amazing pair of turkey drumsticks, which had previously been hidden behind the newsdesk. This record stood until 1986, when 68m people watched Delboy falling memorably through his Xmas Dinner.

8 For short-haired women, Xmas Dinner is just another day off. Lesbians have not celebrated Xmas Dinner since Queen Victoria declared that she didn't believe it existed.

9 Stuffing used to be made of stuff, but is now usually made of sage and onion. Stuff traditionally included paper, cloth, feathers, stones, owl pellets, twigs, sage and onion.

10 By the year 2000, experts predict that busy space travellers will be able to eat their Xmas Dinner in pill form, as they race towards the moon in time to collect space presents from Moon Santa. We will also have four toes.

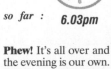

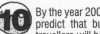

40 million dead... 50 million dead.

FLU, WHAT A SCORCHER!

MAKE the most of this Christmas, because it will almost certainly be your last. But amazingly, it's not a giant flaming meteorite, a terrorist dirty bomb or catastrophic global warming that's going to finish us all off, it's a few sneezing chickens. That's because experts reckon that 2006 will be the year that bird flu strikes the UK, and when it does we're all going to die.

Avian influenza, or to give it its correct scientific name *bird flu,* is spreading across Europe like wildfire. Fortunately, it's carried by chickens which can't fly. They can only walk slowly as they have to stop every few steps to peck at grit. But we can take little com-

By our We're All Going to Die Correspondent
CHICKEN LICKEN

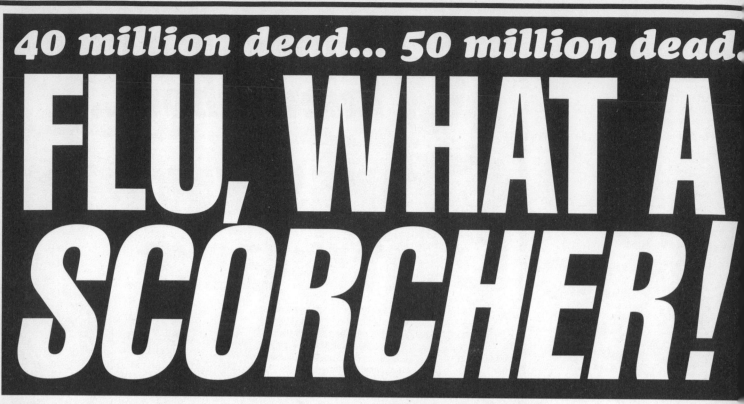

fort from this, according to government scientist Professor Jackie Pallo. He told us: "People shouldn't be alarmed about bird flu, but soon-

er or later a stray hen is going to walk through the Channel Tunnel and cough on someone. And when that happens it will be curtains for

the entire UK population within days."

But it's not all doom and gloom. A vaccine has been produced although the supply is very limited. Professor Pallo told us: "The Royal Family have all been inoculated to protect them from the bird flu. The remaining stocks of vaccine will be distributed strictly according to need. Politicians and celebrities from the worlds of entertainment, pop music and reality TV will all receive jabs before Christmas," he added. "Any stocks left over after that will be set aside for the Queen's dogs, horses and servants."

THE forthcoming pandemic may spell doom for everyone in the country, but it is good news for Britain's undertakers. With the prospect of 60 million burials to organise in the next twelve months, 2006 looks set to be a boom year for UK's funeral directors.

"We're very excited about bird flu," says British Embalmers' Society president Len Duxbury. "It's going to be a real shot in the arm for my members. Life expectancy has soared over the last few years, and the average person can now expect to live to a ripe old age. Sadly, as a result, our profits have taken a dive. An unstoppable fatal plague scything through

Swings and Roundabouts of Bird Flu

the population will be a great morale booster for the undertakers of Britain."

Also smiling is Ernie Shadrack, spokesman for the Federation of British Boiled Ham Producers. "Sixty million funeral teas means a lot of boiled ham," he laughs. "So my

members are looking forward to a bumper year. My advice to anyone expecting to succumb to avian flu this year is; Get your boiled ham in the larder with plenty of time to spare. People are going to be queuing round the block for it once the pandemic hits in earnest."

But the forthcoming plague will spell disaster for another traditional UK business. Birthday card manufacturers estimate that they will see sales plummet by up to more than 98% when bird flu reaches our shores. "The disease typically kills its victims within seven days," says British Birthday Card Trade Association head Billy Waterhouse. "That means that only one person in fifty-two will see their next birthday. With such a marked drop in demand for our products I can foresee widespread redundancies throughout the birthday card industry."

And tap shoe manufacturers are also expecting to find themselves out of pocket. "Bird flu is going to hit the tap shoe industry hard," says Tap Shoe Institute chief Lionel Blair. "When you're dead, the last thing you need is a new pair of taps. It's as simple as that."

60 million dead...

Chicken gravy: A hen in a cemetery yesterday

Blackpool Gears Up for Avian Flu

"Bring it on" ~ says Lord Mayor

One place that certainly isn't worrying about bird flu is Blackpool. The popular Lancashire resort town already has preparations well underway to deal with the worst that the forthcoming killer superbug can throw at it. And Lord Mayor Ivan Taylor has this message for the 100% fatal virus: "It'll take a bigger germ than H5N1 to stop the fun and games along the Golden Mile."

No matter what happens, councillor Taylor is determined that the north-west seaside town is going to keep smiling through. "The rest of the country might be feeling sick as a parrot," he told us, "but make no mistake, here in Blackpool it's going to be business as usual."

To ensure that the global pandemic affects the town's tourist trade as little as possible, the council has already prepared a 7-point action plan.

• *The Community Health Clinic on Tower Road will suspend its current appointments only system, replacing it with a more flexible 'drop-in' arrangement for people exhibiting symptoms of the fatal virus.*

• *Due to restrictions on the gathering of large crowds, the weekly Glamorous Granny competition at the Winter Gardens will be held behind closed doors, with a council official present to ensure fair play.*

• *Local boy scouts are presently going door to door at weekends and after school, collecting unwanted cotton sheets and shirts. These are going to be washed, cut up and hemmed by the members of Blackpool Women's Institute before being distributed to the population as emergency hankies. The ladies of the Blackpool, Lytham and Fleetwood Embroidery Circle have volunteered to apply personalised initials for a small charge, the proceeds of which will be presented to next year's BBC Children in Need appeal.*

• *Ice cream vans and parlours will be making and selling special Beecham's Hot Lemon ice lollies whilst the killer pandemic is at its height.*

• *The mayor's 11-year-old son has designed a special warning poster featuring a chicken with a thermometer sticking out of its beak, and the slogan 'Cluck Out! Bird Flu's About!' The poster is now on display in the foyer of the Town Hall on Corporation Street.*

• *The municipal dog catcher Mr Nicely will be issued with a smaller hoop, suitable for catching chickens.*

• *When the first case of bird flu is identified on the UK mainland the 15th, 16th, 17th and 18th holes of Reach Golf Course, between Normoss, Staining and Great Marton, will be closed to enable the excavation of a vast lime-lined pit capable of accommodating up to 250,000 corpses.*

"Blackpool holidaymakers are a hardy bunch," insisted councillor Taylor. "They're not going to let a worldwide fatal plague spoil their fortnight on the beach, that's for certain. Avian influenza may have a 100% mortality rate, but Blackpool's got a 110% fun rate, and that's not to be sneezed at," he added.

HOW YOU CAN PROTECT YOURSELF FROM BIRD FLU

Advice from the Government's Chief Medical Officer **Dr LIAM DOUBLEDAY**

1. Avoid large crowds of chickens. Places to steer clear of include chicken shows, battery farms, henhouses and pillow factories.

2. Wear a protective face-mask to stop you breathing in bird flu germs. If you can't find a facemask, a coffee filter secured round your ears with an elastic band is just as effective.

3. Set chicken traps to kill any wild hens that may come near your house. An ordinary rat trap baited with grit should do the trick.

4. Eat more chickens. It may sound like twisted logic, but think about it; the more chickens we eat, the less are left alive to sneeze their deadly germs at us.

5. If you have a bird table you can prevent your garden birds catching flu by mixing up a bit of Lem-Sip powder with their seed, and instead of hanging up a bag of peanuts for them to peck, hang up a bag of Tunes, Lockets or Mentholyptus.

6. If, despite all these precautions, you get bird flu symptoms such as a sore throat, tickly cough or runny nose this winter, write your name and national insurance number on your forehead with a marker pen, climb into a binbag and sit outside your local mortuary until you die.

Miriam

Dr MIRIAM STOPPARD, WHO QUALIFIED AS A DOCTOR SEVERAL DECADES AGO, ANSWERS YOUR QUESTIONS ABOUT BIRD FLU

Dear Miriam...

I HAVE heard a lot about this avian flu that is coming from Thailand. According to the paper, between 50,000 and 2 million britons will die. Should I panic?

Mrs Etherington, Rotherham

Miriam writes...

YES. Tests have shown that running around in small circles with your hair all stuck up whilst screaming at the top of your voice could reduce your susceptibility to H5N1 infection by up to 14%.

MY HEARTBEAT HELL!

"Seek thrills OR DIE!"

-Docs' stark warning to speed demon Damon

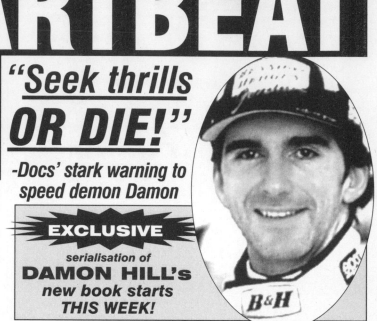

EX-FORMULA ONE champion Damon Hill today revealed his biggest fear since quitting the sport: *"Without thrills I'm a dead man."*

In 1997 Hill was at the top of the F1 tree. Every day brought him his fix of excitement behind the wheel of a 200mph racing car. But ironically, it's only since his retirement from the world's most dangerous sport that Hill has really begun to fear for his life.

In his new book, 'Thrillseeker - The Damon Hill Story' *(Cheese on Toast Books, £19.99)*, he tells how a high-octane F1 career has left him a helpless adrenalin junkie. And without the excitement his body craves, he fears his pulse rate may drop to zero and he will die.

I remember driving my last Grand Prix for Jordan in 1999, a pretty exciting race. Afterwards, I went back to my home in Dublin and sat down in the front room with my wife Georgie to do a crossword puzzle.

"I'd done about half the clues when everything went black. The next thing I remember, I was waking up in the intensive care unit of my local hospital.

pulse

"Apparently, the lack of excitement in just sitting at home had caused my pulse rate to plummet to the point where my heart stopped. The doctors had managed to get it started again, but they told me I was lucky to be alive. Next time I might not be so lucky. That's when I realised that without a constant rush of adrenalin, I was a dead man".

After his brush with death, Hill faced a choice. If he wanted to stay alive he had to drastically change his lifestyle.

"Even something as simple as drinking a cup of tea could be so mundane as to plunge me into a coma. Every time I fancy a cup now, my wife makes twenty and laces one of them with deadly poison. The thrill of knowing that I have a 1 in 20 chance of dying each time I have a cuppa is just enough to keep my heart from stopping.

"For most people, a trip to the garden centre is a relaxing way to spend a Sunday afternoon, but for me it could literally be a death sentence. When we go these days, Georgie takes the kids in the car, whilst I fire myself out of an enormous cannon and meet her there. It may sound extreme, but it's the only way I can shop for bedding plants and stay alive.

fruits

"My condition has made it very difficult for the whole family to share the fruits of my success. For instance, we have a swimming pool in the garden which the kids used to love to play in. But now they won't go anywhere near it. I've had to put ravenous crocodiles in it because simply splashing about on a lilo could prove fatally humdrum for an adrenalin

> **"Every time I fancy a cuppa, my wife makes twenty and laces one of them with deadly poison"**

junkie like me. The only way I can enjoy a relaxing swim is to be chased up and down with the imminent threat of being torn limb from limb by giant reptiles."

The hours of darkness offer no respite for thrill-dependent Hill. An uneventful forty winks could proove lethal, and so the former F1 champion has to go to extreme lengths to maintain a pulse until the morning.

"Like everyone I need sleep, but unlike everyone else, mine has to be action-packed. So these days I sleep in an enormous centrifuge. The G-forces recreate the sensations of driving round an F1 race track and that keeps my heart going. My wife Georgie is very understanding. She finds it difficult to get to sleep whilst being subjected to such huge centrifugal forces, but she knows that if I had anything less than 6G for the full 8 hours, I might not wake up at all."

sulphured apricots

In his book, Hill doesn't pretend that life is easy, and he openly admits that things have changed.

"I don't pretend that life is easy, and I openly admit that things have changed. Even collecting the post in the morning has turned into a dice with death. When I tread on the top step, a mechanism is triggered which releases a 10-ton stone ball from the attic which chases me down the stairs. I usually grab the post off the mat with a fraction of a second to spare before diving into the kitchen for my breakfast."

But despite his affliction, Hill is determined that he is going to live as normal a life as possible.

"It would be so easy for me to just throw in the towel and spend all day lying in my centrifuge. But that's not my style. I think it's important that I do all the things with my family that any normal person would do, although obviously I can't risk letting the excitement levels drop. For instance, I make a point of doing the weekly shop at Asda, but I have my own special shopping trolley. It's got an atom bomb underneath and it's wired up to explode if my speed drops below 20 mph. It may sound risky, but if it keeps my pulse rate high enough to get me to the checkout alive, then that's fine by me."

Damon Hill will be signing copies of his book whilst strapped to the wing of a pilotless 1922 Tiger Moth plummeting towards a tank of man-eating tiger sharks at Borders Books, Charing Cross Road on 15th September from 10.30am.

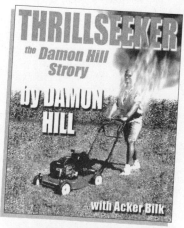

THRILLSEEKER the Damon Hill Story by DAMON HILL

with Acker Bilk

Extracts from 'Thrillseeker - the Damon Hill Story' © Damon Hill & Acker Bilk

Next Week: The Hit Men and Hill - *"The Ninja Assassins who attack me while I'm on the toilet... and I pick up the tab!"*

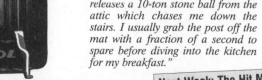

QUEEN HITS BACK!

WORLD EXCLUSIVE

'Anyone who says my job's easy is talking out their arse'

Over the years the Royal Family have been criticised for being overpaid, over priviliged and out of date. Yet no matter how vicious the attacks, the Royals have always remained silent, steadfastly refusing to answer their critics.

But now, for the first time ever the Queen has decided to break her silence and **HIT BACK** at her knockers. In an astonishing frank and forthright letter to this paper she has revealed exclusively **THE TRUTH** about life in the Royal Family. Here, in the first extract from her remarkable letter, the Queen puts the record straight about **MONEY** and the Royals.

PAYMENT

We would like to make it clear that the Queen has received no payment or fee from us, nor has she benefited in any other way financially from the publication of this letter.

BILLS

❝ People think I'm rolling in it, cos I'm the Queen. But once I've paid the bills an' that I've hardly got enough left to see me through the week. As often as not by Friday I'm on the scrounge.

PEANUTS

I get dead mad when people say "Aaah! Look at her. I bet she's coining it in". That's rubbish. You wouldn't believe the number of times I've had to go hundreds of miles to open a building or something, and I've been paid nowt. Even if they do decide to cough up a few bob, it's usually peanuts. At the end of the day I'm lucky if I come away with enough to cover me train fare.

FORTUNE

I had to open a bridge the other day. Of course the punters expect me to look the part — it's more than my jobs worth to be seen wearing the same hat twice. And posh frocks don't come cheap

The Queen at work yesterday

either. I spent a bleedin' fortune on a new outfit, and all I got for opening the bridge was a "Thank You Your Majesty". Try giving that to the bank manager.

NAPPIES

Don't get me wrong. I'm not just doing this for the money. There's easier ways of making a living than this, I can tell you. Opening things and waving at people is no picnic. I should know — I've been in this game over fifty years. I was opening buildings when Princess Di was still in nappies, but nowadays it's her and Fergie what get the headlines. I'm not kidding, them two probably get paid ten times as much as I do. Me, the Queen, and them just Princesses 'n all. I sometimes wonder why I bother.

FLACK

Whenever people slag off the Royals, it's always me what takes the flack. I don't care what they say — I do a rudy good job. For instance, if I'm booked to open something, I'm always there on time, looking the part. I don't mess around. I smile, I meet people

HM blasts the armchair critics

an' that, and when it comes to speeches I do a bloody good turn — at least twenty minutes, and no boring bits either. Not like the young 'uns. They turn up in their helicopters, smile for the telly then 'woosh!' They're off on holiday again. I can't remember the last time me and Phil had a decent holiday. We're lucky if we get a weekend off all year. Take it from me, anyone who says my job is easy is talking out their arse.

STAMPS

For instance, I never get a minute to myself during the day, and there's no chance of going out in the evening for a quiet drink or a meal in a posh restaurant. People see me all the time on money and stamps, and once they've recognised me they just won't leave me alone.

In the old days you got a bit of respect. We had a few bob back then 'n all. In them days if we had a banquet we used to put on a really good spread. Out would come the silver, there'd be seven or eight courses, with caviar, champagne, posh mints. The works. These days if someone like the King of France comes over he's lucky if he gets a bowl of soup before the main course. And more often than not there isn't even a choice of puddings.

POP STARS

I'm not one to grumble. After all, there's always thems what's worse off than yourself. But when you see the amount of money that pop stars and the like are making these days, it makes you wonder. You'd expect someone like the Queen would be taken care of. After all, being the Queen's not something I do for the good of me health you know.

In the next issue: How I hit the roof when I found out Fergie was pregnant. Plus them Royal Weddings — what a pain in the arse. ❞

Buckingan Patace, London.

Dear Sir
I am bloody furious, me. So I think its about time you and t exacktly about al what peo me and al

I don't have to pay (I'm the bloody Queen!)

The Editor
Viz
P.O. Box 1 PT
Newcasle.

TO PAY 24

HOW LONG HAVE YOU BEEN WORKING IN THIS FIELD PROFESSOR?

Celebrity scares!

Jimmy Nail is by no means the first showbusiness celebrity to reveal a secret phobia. *Celebrity phobias* – irrational fears suffered by the stars of show business – have been around as long as show business itself.

In the thirties Lancashire born **George Formby** dropped a bombshell when he revealed to stunned cinema audiences that he was afraid of shiny things. Movie makers went to great lengths to ensure that mirrors, polished surfaces, ball bearings and silverware were removed from film sets before Formby arrived.

Formby – leant on lampost

GEORDIE

Of today's celebrities **Geoff Capes** possibly boasts the most unusual phobia – a fear of budgies. Unlike many stars who try to hide their fears, former policeman Geoff has tackled his problem head on. Strong man Geoff decided to overcome his phobia by breeding budgies in the back yard of his home. And it worked. For Geoff, once a prisoner in his own home, now comes and goes freely, and isn't afraid of budgies anymore.

Blessed – shouty crackers

GOLDIE

Seventies pop idol **Gilbert O'Sullivan's** career was cut short – by a fear of anal germs. Obsessed with bottom hygiene, O'Sullivan would spend hours before and after gigs steeping his bum in a bath of Sterident. And the obsessive star would even book himself into a private clinic for expensive anal detoxification programmes every time he farted. Eventually his record sales began to fall as he spent more time in the bathroom and less in the recording studio.

Shouty crackers actor **Brian Blessed's** star phobia cost him his first major acting role. Brian was offered the part of TV's Captain Birdseye, but during rehearsals for a 30 second TV ad he began to sweat and suffered a panic attack. Top showbusiness doctors later diagnosed Brian as suffering from orpainophobia – an irrational fear of any frozen food coated in golden breadcrumbs. Blessed lost the job, and has since avoided all contact with fish fingers, Crispy Pancakes, scampi, Mini Kievs and Chicksticks.

FRANK GETS THE HUMP OVER CAMELS

Former soccer ace Frank Worthington is today puzzled about camels.

Worthington – camel headache

Worthington, the dashing goal hero who scored with his feet *and* with the ladies during the seventies, has swapped his football boots for books – about camels! But despite reading up on his favourite animal, Frank admits to being somewhat baffled by their humps.

HUMPED

In his soon to be published autobiography, the former England international admits he can't tell the difference between a dromedary and a bachtrian camel – the one and two humped varieties.

"For some reason throughout my career I've never been able to remember which has one hump and which has two. To this day I'm still unsure", admits Frank in his book, due for publication in the new year.

POKED

We rang Jersey zoo keeper and animal book writer Gerald Durrell and asked if he had any hints for Frank.

"There is a simple way to remember which is which", said Gerald. "The letter 'D' for dromedary, when laid on its back, resembles a single hump. The letter 'B' for bachtrian similarly resembles two humps. It's easy."

KNOBBED

When we asked Mr Durrell how many humps the camels in his zoo have got he didn't hear us, because he'd already hung up.

SAUCY CALENDAR IS A W 'I' OPENER!

LADIES from the Fulchester branch of the Woman's Institute this year came up with a cheeky idea to raise money for a Sunshine Coach for Orphans. - *by stripping off for a saucy calendar!*

And their fruity photos have proved so popular with the locals that the ladies have had to order a reprint.

Friends

"It was just a bit of fun, so we're amazed how popular it has become," said 68-year-old grandmother Doris Willis, who posed in the altogether with friends from her bowls club.

"We're none of us Spring chickens but we like to think we're game old birds."

Cheers

60-year-old Mavis Carlisle is chairwoman of the WI knitting circle, but didn't hesitate to pose without a stitch on when she heard about the project.

"Until the calendar came out, the only people

EXCLUSIVE!

Glamorous grannies raise eyebrows as well as cash!

who had seen me in my birthday suit were my husband and my doctor," said Mavis. "Now everybody in Fulchester has seen me in the pink! I'm on the February page with my legs wide open, masturbating two erect penises into my hair."

Seinfeld

And sales of the calendar look set to exceed the ladies' wildest dreams.

The Fulchester Women's Institute Calendar - a genteel cover, but the contents will raise a few eyebrows.

"We had 500 printed, and they sold out on the first day at the village fete," said Enid Marshall, who appears on the August picture performing fellatio on the chairman of the Parish Council and pulling her labia apart to expose her clitoris whilst her face and breasts are splashed with three men's ejaculate. "Now we're having another thousand printed, and it looks like they should sell out too."

Frasier

Local photographer Ernie Stewart is more used to taking pictures of local people's weddings,

so when he was asked to snap the naughty pics for the calendar he was more than a little surprised.

Bugner

"It was certainly an eye-opener," said Ernie, who gave his services for free. "I think I was more embarassed than they were at first. But we got the pictures done, and I think it's turned out really well. I hope it raises lots of money."

Jackson

Publican Brian Dougal has even hung up a copy of the cheeky calendar behind his bar at the Dog & Duck. "I'm particularly looking forward

to December, because that's the month my wife Noreen appears in a saucy snap taken on the steps of the cricket pavilion," he said.

Dellasandro

"She's urinating into the mouth of Mrs Preece from the flower shop, whilst Mr Williams the churchwarden felches his sperm from out of her anus."

Miss January
Edith Swain - spit roast

Miss April
Dolly Hill - Cleveland steamer

Miss October
Ida Simpson - cum bath

HEY, I'M GONNA CUT MYSELF WITH BROKEN GLASS AND WATCH MYSELF BLEED WHILE MASTURBATING TO MARYLIN MANSON, BACKWARDS.

MY ANKLES TWISTED

FAMOUS PEOPLE ON THE TOILET

NO. 142
ANNE ROBINSON

YOU ARE THE WEAKEST LINK. GOODBYE.

At last...it's OFFICIAL!

MORRISEY IS A TWAT

Cult pop singer Morrisey — hailed as hero by his fanatical fans — is a twat, according to experts.

And that will come as bad news to his many admirers who have worshipped the pop idol since he came to fame as lead singer of The Smiths.

VIDEO

Professor Ivan Sogorski of Barrow-in-Furness University's Department of Advanced Human Behavioural Studies came to his dramatic conclusion about the star after listening to many of his records and watching video footage of his TV appearances. And he summed up his professional opinion in a few short words.

TWAT

"The man is an absolute twat", he told us.

ARSEHOLE

Professor Sogorski cited examples of behaviour which had lead him to his controversial conclusion. "Take for example Mr Morrisey's appearance on Top Of The Pops in the early eighties when he wore oversized shirts, National Health glasses, a hearing aid, and

EXCLUSIVE

flailed about the stage with daffodils sticking out of his back pocket. Clearly, even the most casual analysis could only conclude this to be the behaviour of an arsehole', said the Professor.

CRAP

As a part of his painstaking research, Professor Sogorski consulted a colleague to obtain a second independent opinion. "I submitted manuscripts and recordings of many Morrissey songs to a leading Professor of Composition at the Royal College of Music, and he says they are crap".

BULLSHIT

The Professor quoted examples of Morrisey's song titles as further evidence to support his views. "Girl In A Coma. Big Mouth Strikes Again. Heaven Knows I'm Miserable Now. These are all bullshit", said Professor Sogorski.

During his career Morrisey has endeared himself to a huge cult following of pop fans, among them many students, and has also won artistic acclaim for his work.

WANKER

But Professor Sogorski's comments are bound to fuel speculation that whilst some

of his songs might be quite good, the man is, quite frankly, a bit of an arsehole. "I am convinced Morrisey is a twat, and anyone who says otherwise is a wanker", said the Professor yesterday. Professor Sogorski last hit the headlines in 1988 when he claimed that page three model Samantha Fox was a "boiler".

'Fuck' is OK

Britain's swearing chiefs are set to lift the ban on many rude words, among them 'fuck' and 'cunt'.

And the shock move will be a spanner in the works for many 'adult humour' magazines, 'alternative' comedians and Channel 4 programmes for whom rude words are vital ingredients. Indeed, the downgrading of words like 'fuck' from rude to slightly rude will leave many *blue* comics *red* faced.

BASTARD

In the past, rude works like 'shit' and 'bastard' have been downgraded, and are now in common usage, upsetting only Mary Whitehouse and few other old crocks. But the de-rudening of many remaining obscenities will leave genuine foul mouths with a limited arsenal of vulgarity from which to choose.

Swear bosses green light to blue comics

One English language expert believes that brand new expletives may have to be invented, or existing mild obscenities upgraded to replace ageing rude words. "there's a chance that words like 'kipper', 'snatch' or 'fanny' may soon be rude", he told us.

PISS FLAPS

When we asked him about 'beef curtains' he said he wasn't sure, and that he'd have to look it up.
